Lecture Notes in Mathematics 2258

More information about this series at http://www.springer.com/series/304

Huayi Chen • Atsushi Moriwaki

Arakelov Geometry over Adelic Curves

Springer

Huayi Chen
Institut de Mathématiques de Jussieu
— Paris Rive Gauche
Université de Paris
Paris, France

Atsushi Moriwaki
Department of Mathematics
Kyoto University
Kyoto, Japan

ISSN 0075-8434 ISSN 1617-9692 (electronic)
Lecture Notes in Mathematics
ISBN 978-981-15-1727-3 ISBN 978-981-15-1728-0 (eBook)
https://doi.org/10.1007/978-981-15-1728-0

Mathematics Subject Classification (2010): 14G40, 14G25, 11G50, 11G35, 37P30

This Springer imprint is published by the registered company Springer Nature Singapore Pte Ltd.
The registered company address is: 152 Beach Road, #21-01/04 Gateway East, Singapore 189721, Singapore

Contents

Introduction

The purpose of this book is to build the fundament of an Arakelov theory over adelic curves in order to provide a unified framework for the researches on arithmetic geometry in several directions.

Let us begin with a brief description of the main ideas of Arakelov geometry. In number theory, it is well known that number fields are similar to fields of rational functions over algebraic curves defined over a base field, which is often assumed to be finite. It is expected that the geometry of schemes of finite type over $\mathbb{Z}$ should be similar to the algebraic geometry of schemes of finite type over a regular projective curve. However, some nice properties, especially finiteness of cohomological groups (in the case where the base field is finite), fail to hold in the arithmetic setting, which prevents using geometrical methods to count the arithmetic objects. The core problem is that schemes over $\operatorname{Spec}\mathbb{Z}$, even projective, are not "compact", and in general it is not possible to "compactify" them in the category of schemes. The seminal works of Arakelov [4, 5] propose to "compactify" a scheme of finite type over $\operatorname{Spec}\mathbb{Z}$ by transcendental objects, such as the associated complex analytic variety, Hermitian metrics, Green functions, and differential forms etc. In the case of relative dimension zero, the idea of Arakelov corresponds to the classic approach in algebraic number theory to include the infinite places of a number field to obtain a product formula, and to introduce Hermitian norms on projective modules over an algebraic integer ring to study the geometry of numbers and counting problems. Most interestingly, the approach of Arakelov proposes an intersection theory for divisors on a projective arithmetic surface (relative dimension one case), which is similar to the intersection pairing of Cartier divisors on classic projective surfaces.

The works of Arakelov have opened a gate to a new geometric theory of arithmetic varieties (schemes of finite type over $\operatorname{Spec}\mathbb{Z}$). Inspired by the classic algebraic geometry, many results have been obtained and enriched Arakelov's geometry. Among the wide literature, we can mention for example the arithmetic Hodge index theorem by Faltings [58] and Hriljac [87], arithmetic intersection theory of higher dimensional arithmetic varieties and arithmetic Riemann-Roch theorem by Gillet and Soulé [68, 69], see also [67]. Arakelov geometry also provides an alternative approach (compared to the classic Weil height theory) to the height theory in arithmetic geometry, see [4,5,139,60,23], (see also the approach of Philippon [122,123], and [136] for the comparison between Philippon height and Arakelov height). The Arakelov height is often more precise than the Weil height machine since the choice of a Hermitian metric on a line bundle permits to construct an explicit height function associated with that line bundle (in the Weil heigh machine, the height function is defined only up to a bounded function).

These advancements have led to fruitful applications in number theory, such as the proof of Mordell's conjecture by Faltings [57,59] and the alternative proof by Vojta [146] (see also the proof of Bombieri [11] and the generalisation of Vojta's approach to the study of subvarieties in an Abelian variety [60]), equidistribution of algebraic points in an arithmetic variety and applications to Bogomolov's conjecture [140,144,157], algebraicity of formal leaves of algebraic foliation [18] etc.

Although the philosophy of Arakelov allows to inspire notions and results of algebraic geometry and has already led to a rich arithmetic theory, the realisation of Arakelov theory is rather different from that of the classic algebraic geometry and usually gets involved subtle tools in analysis. The transition of technics on the two sides is often obscure. For example, the *abc* conjecture, which can be easily established in the function field setting by algebraic geometry tools (see [103]), turns out to be very deep in the number field setting. Conversely, the Bogomolov's conjecture has been resolved in the number field setting, before the adaptation of its proof in the function field setting by using Berkovich analytic spaces (see [80,149,150]). It is therefore an interesting problem to provide a uniform fundament for Arakelov geometry, both in the function field and number field settings, and the adelic approach is a natural choice for this goal. We would like however to mention that Durov [54] has proposed an approach of different nature to algebrify the Arakelov geometry over number fields.

The theory of adèles in the study of global fields was firstly introduced by Chevalley [49, Chapitre III] for function fields and by Weil [147] for number fields. This theory allows to consider all places of a global filed in a unified way. It also leads to a uniform approach in the geometry of numbers in global fields, either via the adelic version of Minkowski's theorems and Siegel's lemma developed by McFeat [105], Bombieri-Vaaler [13], Thunder [142], Roy-Thunder [128], or via the study of adelic vector bundles developed by Gaudron [62], generalising the slope theory introduced by Bost [16,18].

Several works have been realised in the adelification of Arakelov theory. Besides the result of Gaudron on adelic vector bundles over global fields mentioned above, we can for example refer to [156] for adelic metrics on arithmetic line bundles and applications to the Bogomolov problem for cycles. Moreover, Moriwaki [113] has studied the birational geometry of adelic line bundles over arithmetic varieties. The key point is to consider an arithmetic variety as a scheme of finite type over a global field, together with a family of analytic varieties (possibly equipped with metrised vector bundles) associated with the scheme, which is parametrised by the set of all places of the global field. Classic objects in Arakelov geometry can be naturally considered in this setting. For example, given a Hermitian line bundle over a classic arithmetic variety (scheme of finite type over Spec $\mathbb{Z}$), the algebraic structure of the line bundle actually induces, for each finite place of $\mathbb{Q}$, a metric on the pull-back of the line bundle on the corresponding analytic space.

In this book, we introduce the notion of *adelic curves* and develop an Arakelov theory over them. By adelic curve we mean a field equipped with a family of absolute values parametrised by a measure space, such that the logarithmic absolute value of each non-zero element of the filed is an integrable function on the measure space, with 0 as its integral. This property is called a *product formula*. Note that this notion has been studied by Gubler [79] in the setting of height theory and is also considered by Ben Yaakov and Hrushovski [8, 88] in a recent work on model theory of global fields. Clearly the notion of adelic curve generalises the classic one of global field, where the measure space is given by the set of all places of the global field equipped with the discrete measure of local degrees. However, this is certainly not the only motivation for the general notion of adelic curves. Our choice is rather inspired by several bunches of researches which are apparently transversal to each other, which we will resume as follows (we will explain further the reason for the choice of terminology "adelic curve").

1. Finitely generated extensions of a number field. From a point of view of birational geometry, we expect that the field of rational functions of an algebraic variety determines the geometric properties of the variety. In Arakelov geometry, we consider integral schemes of finite type over $\operatorname{Spec}\mathbb{Q}$, whose function field is a finitely generated extension of $\mathbb{Q}$. Moriwaki [106] has developed an Arakelov height theory for varieties over a finitely generated extension of a number field and applied it to the study of Bogomolov problem over such a field (see [107], see also [108] for a panoramic view). Burgos, Philippon and Sombra [34] have expressed the height of cycles in a projective variety over a finitely generated extension of $\mathbb{Q}$ as an integral of local heights over the set of places of the field.

2. Trivially valued field. In number theory, we usually consider non-trivial absolute values on fields. Note that on any field there exists a trivial absolute value which takes value 1 on each non-zero element of the field. Note that a trivial product formula is satisfied in this setting. Although the trivially valued fields are very simple, the corresponding geometry of numbers is rather rich, which has wide interactions with the classic geometry of lattices or Hermitian vector bundles. In fact, given a finite-dimensional vector space over a trivially valued field, the ultrametric norms on it are canonically in bijection to the decreasing $\mathbb{R}$-filtrations on the vector space. The $\mathbb{R}$-filtration is a key method of the works [40, 37, 39], where the main idea consists in associating to each Hermitian vector bundle an $\mathbb{R}$-filtration on the generic fibre, which captures the arithmetic information such as successive minima or successive slopes.

3. Harder-Narasimhan theory for vector bundles on higher dimensional varieties. Harder and Narasimhan theory [84] is an important tool in the study of vector bundles on a projective curve. In the geometry of Euclidean lattices, the counterpart of Harder-Narasimhan theory has been proposed by Stuhler [138] and Grayson [71]. Later Bost [16] has generalised their works in the setting of Hermitian vector bundles on the spectrum of the ring of algebraic integers in a number field. Moreover, he has developed the slope inequalities in this framework and applied them to the study of algebraicity of formal schemes [18, 19, 20]. Note that the slope function and the notion of semistability can be naturally defined for torsion-free coherent sheaves on a polarised projective variety [141]. This allows Shatz [135] and Maruyama [102] to develop a Harder-Narasimhan theory for general torsion-free coherent sheaves. However, it seems that the analogue of their results in the arithmetic case is still missing.

4. Fields of algebraic numbers, Siegel fields. The geometry of numbers for algebraic (not necessarily finite) extensions plays an important role both in Diophantine problems and in Arakelov geometry. Recall that the Minkowski's theorem and Siegel's lemma in geometry of numbers admit an adelic version for number fields, see [13, 105]. They also have an absolute counterpart over $\overline{\mathbb{Q}}$, see [128, 129, 155]. In Arakelov geometry, a notion of Hermitian vector bundle over $\overline{\mathbb{Q}}$ has been proposed in the work [22] of Bost and Chen, on which the absolute Siegel's lemma applies and is useful in the study of tensorial semistability of classic Hermitian vector bundles. Similarly, in the approach of Gaudron and Rémond [64] to the tensorial semistability, the absolute Siegel's lemma is also a key argument. In [65], the notion of Siegel field has been proposed. A Siegel field is a subfield of $\overline{\mathbb{Q}}$ on which an analogue of Siegel's lemma is true. In order to formulate a geometry of numbers for a Siegel field, Gaudron and Rémond have introduced a topology on the space of all places of such a field and a Borel measure on it.

5. Algebraic extensions of function fields. In [51], Corvaja and Zannier have studied the arithmetic of algebraic extensions of function fields. They have characterised the infinite algebraic extensions of function fields of a curve which still satisfy a product formula. They have also discussed several examples of product formulas associated with algebraic surfaces in revealing the non-uniqueness of the extension of a product formula under finite field extensions.

The above results are obtained in various settings of arithmetic geometry. It turns out that these settings can be naturally included in the framework of adelic curves (see §3.2 for details) in order to treat the geometry of various fields analogously to that of vector bundles on projective curves. For example, on the field $\mathbb{Q}(T)$ of rational functions with coefficients in $\mathbb{Q}$, three types of absolute values are defined (see §3.2.5 for details): the valuation corresponding to closed points of $\mathbb{P}^1_{\mathbb{Q}}$, the natural extensions of p-adic absolute values, and the Archimedean absolute value corresponding to divers embeddings of $\mathbb{Q}(T)$ in $\mathbb{C}$. Note that Jensen's formula for Mahler measure shows that these absolute values, once suitably parametrised by a measure space, satisfy a product formula. Thus we can consider it as an adelic curve. This is actually a particular case of polarised arithmetic projective varieties, where the polarisation provides a structure of adelic curve on the field of rational functions on the projective variety. Moreover, algebraic coverings of an adelic curve can be naturally constructed (see Section 3.4), which provides a framework for the study of the arithmetic of algebraic extensions.

Note that Gubler [79] has introduced a similar notion of M-field and extended the Arakelov height theory to this setting. An M-field is a field equipped with a measure space and a family of functions parametrised by the measure space which are absolute values almost everywhere, and the height of an arithmetic variety is defined as the integration along the measure space of local heights. However, our main concern is to build up a suitable geometry of numbers while the purpose of [79] is to extend the Arakelov height theory in a sufficiently general setting in order to include the theory of Nevanlinna. In Diophantine geometry the geometry of numbers is as important as a the height theory, particularly in the geometrisation of the method of "auxiliary polynomials". We propose the notion of *adelic vector bundles* on adelic curves, which consist of a finite-dimensional vector space over the underlying field, equipped with a measurable family of norms parametrised by the measure space. Our choice facilites the study of algebraic constructions of adelic vector bundles. The height of arithmetic varieties is described in a global way by the asymptotic behaviour of graded linear series equipped with structures of adelic vector bundles, rather than the integral of local heights.

In the framework of model theory, Ben Yaakov and Hrushovski [8, 88] also consider the formalisme of a field equipped with a family of absolute value parametrised by a measure space, which satisfied a product formula (called *globally valued field* in their terminologies). Their work permits to considered classic Diophantine geometry objects (in particular heights) in the model theory setting.

In order to set up a theory of adelic vector bundles over adelic curves, we present in the first chapter various constructions and properties of seminormed vector spaces over a complete valued field. Although the constructions and results are basic, the subtleties in the interaction and the compatibility of divers algebraic constructions, such as restriction, quotient, dual, tensor product, exterior powers etc, have not been clarified in the literature in a systematic way. In particular, several classic results in the functional analysis over $\mathbb{C}$ are no longer true in the non-Archimedean setting. We choose carefully our approach of presentation to unify the treatment of non-Archimedean and Archimedean cases whenever possible, and specify the differences and highlight the subtleties in detail. A particular attention is paid to the two constructions of tensor product seminorms: the π-tensor product and the ε-tensor product. These notions have been firstly introduced by Grothendieck [73, 72] in the setting of functional analysis over $\mathbb{C}$. It turns out that similar constructions can be defined more generally over an arbitrary complete valued field, and they are useful for example in the study of seminorms on exterior powers.

The orthogonality is another theme discussed in the first chapter. Classically the orthogonality is a natural notion in the study of inner product spaces. We consider an equivalent form of this notion, which can be defined in the setting of finitely generated seminormed vector spaces over an arbitrary complete valued field. This reformulation has been used in [33] to study the arithmetic positivity on toric varieties. Here it will serve as a fundamental tool to study ultrametrically normed spaces, inner product spaces and the construction of orthogonal tensor products. In particular, an analogue of the Gram-Schmidt process holds for finite-dimensional ultrametrically seminormed spaces, which plays a key role in the compatibility of the determinant norm with respect to short exact sequences.

We also discuss extension of seminorms under a valued extension of scalars. We distinguish three extensions of seminorms, corresponding to the three types of tensor product. The compatibility of extension of scalars with respect to divers algebraic construction is also explained. These constructions are used in the pull-back of an adelic vector bundle by an algebraic covering of the adelic curve.

Note that in the classic Arakelov theory, usually we consider a vector space over a global field equipped with a family of norms. However, from the point of view of birational geometry, it is natural to consider metrics which admits singularity, that is, degenerates on a closed subscheme (which is usually the base locus of a linear series) to a family of seminorms. Moreover, in the study of algebraicity of formal leaves of an arithmetic foliation, the canonical "metrics" on the tangent bundle are often seminorms. We refer the readers to [21] for more details. Motivated by these observations, we choose to present a panoramic view on the tools about general seminormed vector spaces which could be useful in Arakelov geometry later.

The second chapter is devoted to a presentation of metrised line bundles on a projective scheme over a complete valued field. It could be considered as a higher dimensional version of the results presented in Chapter 1. We use Berkovich topology to define continuous metrics on a vector bundle. Note that in the case where the base field is $\mathbb{C}$, our definition coincides with the classic definition of continuous metric on a vector bundle over a complex analytic space (associated with a complex projective scheme).

The Fubini-Study metric is another important ingredient of Chapter 2. It is closely related to the positivity of metrics on line bundles. More precisely, a continuous metric on a line bundle over a projective scheme defined over a complete valued field is said to be *semipositive* if it can be written as a uniform limit of Fubini-Study metrics. In the case where the absolute value is Archimedean, this definition is equivalent to the semipositivity of the curvature current of the metric. In the case where the absolute value is non-Archimedean and non-trivial, it is equivalent to the semipositivity condition proposed in [36, Section 6.8] and [82, Section 6]. However, in the trivial valuation case, it seems that our formulation is crucial to study the positivity of the metrics.

In classic Hermitian geometry, the positivity is closely related to the extension of sections of an ample line bundle with a control on the supremum norms. We establish a non-Archimedean analogue of the extension property, generalising the main result of [48] to the non-necessarily reduced case.

The third chapter is devoted to the fundament of adelic curves. We first give the formal definition of this notion and illustrate by various examples. The algebraic coverings of adelic curves occupy an important part of the chapter. As mentioned above, an adelic curve is a field equipped with a family of absolute values parametrised by a measure space, which satisfies a product formula. Given an algebraic extension of the underlying field, there is a canonical family of absolute values parametrised by a measure space fibered on the initial measure space and equipped with a disintegration kernel. This construction is important in the height theory for algebraic points and in the study of Siegel and Northcott properties. Contrary to the approach of [65], we do not assume the structural measurable space of an adelic curve to be a topological space and do not adopt the topological construction of algebraic coverings. Although it is possible to reduce the construction to the case of finite extensions by an argument of passage to projective limit, even for the simplest case of finite separable extension of the underlying field, the problem is highly non-trivial. The main subtleties come from the measurability of the fibre integral, which neither follows from the classic disintegration theory, nor from the property of extension of absolute values in algebraic number theory. The difficulty is resolved by using symmetric polynomials and Vandermonde matrix.

The analogue in the adelic curve setting of the geometry of numbers occupies the main part of the fourth chapter. Given an adelic curve, for any finite-dimensional vector space over the underlying field, we consider families of norms indexed by the structural measure space of the adelic curve. Natural measurability and dominancy conditions are defined for such norm families. An adelic vector bundle is a finite-dimensional vector space over the underlying field of the adelic curve, equipped with a measurable and dominated norm family. In the case where the adelic curve arises from a global field, this notion corresponds essentially to the notion of adelic vector bundle in the work [62]. Note that in the classic global field case it is required that almost all norms in the structure of an adelic vector bundle come from a common integral model of the vector space. However, in our general setting of adelic curve, it is not adequate to discuss integral models since the integral ring in an adelic curves is not well defined. The condition of common integral model is replaced by the dominancy condition, which in the global field case can be considered as uniform limit of classic structure of adelic vector bundle.

The arithmetic invariants of adelic vector bundles are also discussed. For example, the Arakelov degree of an adelic vector bundle is defined as the integral of the logarithmic determinant norm of a non-zero maximal exterior power vector, similarly as in the classic case of Hermitian vector bundle over an arithmetic curve. Moreover, although the analogue of classic minima of lattices can not be reformulated in the adelic curve setting, due to the lack of integral models, the version of Roy and Thunder [128], which is based on the height function (or equivalently the Arakelov degree of the non-zero vectors), can be naturally generalised in our setting of adelic curves. However, it turns out that several fundamental results in geometry of numbers, such as Minkowski's theorems, are not true in the general setting, and the set of vectors in the adelic unit ball is not the good generalisation of lattice points of norm $\leqslant 1$. This phenomenon suggests that the slope method of Bost [16] might be more efficient in Diophantine geometry. In fact, inspired by the Harder-Narasimhan theory of vector bundles over curves, the notion of successive slopes has been proposed in [138, 71] for Euclidean lattices and generalised in [16, 18] with applications to the period and isogenies of abelian varieties, and algebraicity of formal schemes. In the setting of adelic vector bundles on adelic curves, we establish an analogue of Harder-Narasimhan theory and the slope method. In this sense, adelic vector bundles on adelic curves have very similar properties as those of vector bundles on a regular projective curve, or Hermitian vector bundles over an arithmetic curve. It is for this reason that we have chosen the terminology of adelic curve. However, although the successive minima and the successive slopes are close in the number field case (see [15,44]), they can differ much in the general adelic curve setting, even for the simple case of a field equipped with several copies of the trivial absolute value. Note that the semistability of adelic vector bundles over such adelic curves plays an important role in Diophantine geometry of projective spaces, as for example in the work of Faltings and Wüstholz [61] (although not written explicitly in the language of the slope theory). Our general setting of adelic vector bundles helps to understand the roles of different arithmetic invariants should play in a Diophantine argument.

The adelic curve consisting of the trivial absolute value is also closely related to the geometric invariant theory. In the fifth chapter of the book, we explain this link and apply it to the estimation of the minimal slope of the tensor product of two adelic vector bundles. In fact, an ultrametrically normed vector space over a trivially valued field can be considered as a decreasing $\mathbb{R}$-filtration of the vector space. In the geometric invariant theory, an action of the multiplicative group on a finite-dimensional vector space over a field corresponds to the decomposition of the vector space into the direct sum of eigensubspaces and thus determines an $\mathbb{R}$-filtration of the vector space by the eigenvalues. Therefore we can reformulate the Hilbert-Mumford criterion for general linear groups (or products of general linear groups) in terms of a slope inequality for adelic vector bundles on the adelic curve of one trivial absolute value.

Bogomolov (see [126]) has interpreted the semistability of a vector bundle over a projective curve as an inequality linking the $\mathbb{R}$-filtration and the Arakelov degree. This result can also be viewed as a link between the geometric invariant theory and the semistability in the theory of Harder-Narasimhan. Later Ramanan and Ramanathan [124] have given an algebraic proof of the semistability of the tensor product of two semistable vector bundles on a regular projective curve over a field of characteristic 0. In the number field case, Bost [17] has conjectured that the arithmetic analogue of the tensorial semistability is also true. This conjecture is equivalent to the statement that the tensor product of two Hermitian vector bundles has a minimal slope which is bounded from below by the minimal slopes of the two Hermitian vector bundles.

In the setting of adelic vector bundles over adelic curves, we can consider the natural generalisation of Bost's conjecture stating that, if the underlying base field of the adelic curve is perfect, then the tensor product of two semistable *Hermitian* adelic vector bundles is also semistable. Besides the function field case proved by Ramanan and Ramanathan, the generalised conjecture is also true in the case where the adelic curve is given by a perfect field equipped with a finite number of copies of the trivial absolute value (see [143]). We prove here a weaker version of this conjecture, showing that the minimal slope of the tensor product of two (non-necessarily Hermitian) adelic vector bundles is bounded from below by the sum of the minimal slopes of the two adelic vector bundles, minus three half of the logarithm of the rank of the tensor product bundle times the measure of Archimedean places. In particular, the conjecture is true if the base field is of characteristic zero and all absolute values in the adelic curve structure are non-Archimedean. This result is similar to the works [38, 3, 64, 22] in the case where the adelic curve comes from a number field. However, the strategy of proof is different. In fact, the common point of the works cited above is a geometric version of Siegel's lemma proved by Zhang [155], which could be considered as an absolute version of Minkowski's second theorem, which is false for general adelic curves. Our method relies on the geometric invariant theory of grassmannian (with Plücker coordinates) and combines the technics of [38] and [22].

The sixth chapter is devoted to the study of metrised line bundles on arithmetic varieties over adelic curves. In the classic setting of adelic metrics such as [156, 113], it was required that an adelic metric should coincides with an integral model metric for all but finitely many places. Again the integral model metric is not adequate in our setting of adelic curves, the suitable notions of dominancy and measurability occupy thus an important part of the chapter. An adelic line bundle on a projective variety is then defined to be an invertible sheaf equipped with a dominated and measurable family of metrics parametrised by the adelic curve. In the setting of global fields, our definition is slightly more general than the classic one, which includes the limits of classic adelic line bundles. The analogue of some classic geometric invariants, such as height function, essential minimum, and arithmetic volume function is also discussed. In particular, in the definition of the arithmetic volume function, we use the positive degree instead of the logarithmic cardinal of the small sections since the latter is no longer adequate in the general setting. Note that the failure of Minkowski's first theorem brings several technical difficulties, notably the filtration by minima and the filtration by slopes do not lead to the same arithmetic invariants, on the contrary of the case of number fields as in [44]. Our strategy consists in introducing a refinement of the method of arithmetic Newton-Okounkov bodies [26], which allows to treat the case of graded linear series equipped with filtrations which are not necessarily additive.

In the seventh and the last chapter, we relate the asymptotic minimal slope to the absolute minimum of the height function of an adelic line bundle, which could be considered as a generalisation of Nakai-Moishezon's criterion in the setting of Arakelov geometry over an adelic curve. In the case where the analogue of a strong version of Minkowski's first theorem holds for the adelic curve, we deduce from the criterion an analogue of Siegel's lemma for adelic vector bundles on the adelic curve. Our work clarifies the arguments of geometric nature from several fundamental result in the classic geometry of numbers.

Limited by the volume of the monograph, many aspects are not included in the current text. First of all, an arithmetic intersection theory should be developed in the setting of Arakelov geometry over an adelic curve, which allows to interpret the height of arithmetic varieties as the arithmetic intersection numbers. Secondly, by using the adelic curve of several copies of the trivial absolute value, we expect to incorporate the conditions and results of geometric invariant theory into the arithmetic setting. Thirdly, the geometry of adelic vector bundles should lead to a Diophantine approximation theory of adelic curves. Finally, the fundamental works achieved in the monograph could be applied to the study of Nevanlinna theory of M-field proposed by Gubler.

Acknowledgements. We are grateful to Jean-Benoît Bost, Carlo Gasbarri and Hugues Randriambololona for comments, and to Itaï Ben Yaacov and Ehud Hrushovski for having sent us their lecture notes and for letter communications. During the preparation of the monographie, Huayi Chen has benefitted from the visiting support of Beijing International Center for Mathematical Researches and the Department of Mathematics of Kyoto University. He is grateful to these institutions for hospitality. The second author was supported by JSPS KAKENHI Grant-in-Aid for Scientific Research(S) Grant Number JP16H06335. Finally we would like to express our thanks to the referees for their suggestions and comments.

Chapter 1
Metrized vector bundles: local theory

The purpose of this chapter is to explain the constructions and properties of normed vector spaces over a complete valued field. It will serve as the fundament for the global study of adelic vector bundles. Note that we need to consider both Archimedean and non-Archimedean cases. Hence we carefully choose the approach of presentation to unify the statements whenever possible, and to clarify the differences.

Throughout the chapter, let k be a field equipped with an absolute value $|\cdot|$. We assume that k is complete with respect to the topology induced by $|\cdot|$. We emphasise that $|\cdot|$ could be the trivial absolute value on k, namely $|a| = 1$ for any $a \in k \setminus \{0\}$. If the absolute value $|\cdot|$ is Archimedean, then k is either the field $\mathbb{R}$ of real numbers or the field $\mathbb{C}$ of complex numbers. For simplicity, we assume that $|\cdot|$ is the usual absolute value on $\mathbb{R}$ or $\mathbb{C}$ if it is Archimedean.

1.1 Norms and seminorms

Definition 1.1.1 Let V be a vector space over k. A map $\|\cdot\| : V \to \mathbb{R}_{\geqslant 0}$ is called a *seminorm* on V if the following conditions (a) and (b) are satisfied:

(a) for any $a \in k$ and any $x \in V$, one has $\|ax\| = |a| \cdot \|x\|$;
(b) the *triangle inequality*: for any $(x, y) \in V \times V$, one has $\|x + y\| \leqslant \|x\| + \|y\|$.

The couple $(V, \|\cdot\|)$ is called a *seminormed vector space* over k. If in addition the following *strong triangle inequality* is satisfied

$$\forall\, (x, y) \in V^2, \quad \|x + y\| \leqslant \max\{\|x\|, \|y\|\},$$

we say that the seminorm $\|\cdot\|$ is *ultrametric*. Note that the existence of a non-identically vanishing ultrametric seminorm on V implies that the absolute value $|\cdot|$ on k is non-Archimedean. Furthermore, if the following additional condition (c) is satisfied:

(c) for any $x \in V \setminus \{0\}$, one has $\|x\| > 0$,

H. Chen, A. Moriwaki, *Arakelov Geometry over Adelic Curves*, Lecture Notes in Mathematics 2258, https://doi.org/10.1007/978-981-15-1728-0_1

the seminorm $\|\cdot\|$ is called a *norm* on V, and the couple $(V, \|\cdot\|)$ is called a *normed vector space* over k.

If $(V, \|\cdot\|)$ is a seminormed vector space over k, then

$$N_{\|\cdot\|} := \{x \in V \ : \ \|x\| = 0\}$$

is a vector subspace of V, called the *null space* of $\|\cdot\|$. Moreover, if we denote by $\pi : V \to V/N_{\|\cdot\|}$ the linear map of projection, then there is a unique norm $\|\cdot\|^{\sim}$ on $V/N_{\|\cdot\|}$ such that $\|\cdot\| = \|\cdot\|^{\sim} \circ \pi$. The norm $\|\cdot\|^{\sim}$ is called the *norm associated with the seminorm* $\|\cdot\|$.

Definition 1.1.2 Let $f : W \to V$ be a linear map of vector spaces over k and $\|\cdot\|$ be a seminorm on V. We define $\|\cdot\|_f : W \to \mathbb{R}_{\geqslant 0}$ to be

$$\forall\, x \in W, \quad \|x\|_f := \|f(x)\|,$$

which is a seminorm on W, called the seminorm *induced* by f and $\|\cdot\|$. Clearly, if $\|\cdot\|$ is ultrametric, then also is $\|\cdot\|_f$. In the case where f is injective, $\|\cdot\|_f$ is often denoted by $\|\cdot\|_{W \hookrightarrow V}$ and is called the seminorm on W *induced* by $\|\cdot\|$, or the *restriction* of $\|\cdot\|$ to W.

Notation 1.1.3 Let $(V, \|\cdot\|)$ be a seminormed vector space over k. If ϵ is a non-negative real number, we denote by $(V, \|\cdot\|)_{\leqslant \epsilon}$ or simply by $V_{\leqslant \varepsilon}$ the closed ball $\{x \in V \ : \ \|x\| \leqslant \epsilon\}$ of radius ϵ centered at the origin. Similarly, we denote by $(V, \|\cdot\|)_{<\epsilon}$ or by $V_{<\epsilon}$ the open ball $\{x \in V \ : \ \|x\| < \epsilon\}$.

Proposition 1.1.4 *Assume that* $|\cdot|$ *is non-trivial. Let* $\lambda \in \,]0, 1[$ *such that*

$$\lambda < \sup\{|a| \ : \ a \in k^{\times},\ |a| < 1\}.$$

Let $(V, \|\cdot\|)$ *be a seminormed vector space over* k *and* x *be a vector in* V *such that* $\|x\| > 0$. *There exists* $b \in k^{\times}$ *such that* $\lambda \leqslant \|bx\| < 1$.

Proof Let a be an element in $k^{\times}$ such that $\lambda < |a| < 1$. We take $b = a^p$ with

$$p = \left\lfloor \frac{\ln(\lambda) - \ln \|x\|}{\ln |a|} \right\rfloor.$$

By definition one has $p \leqslant (\ln(\lambda) - \ln \|x\|)/\ln |a|$. Hence $|b| = |a|^p \geqslant \lambda/\|x\|$, which leads to $\|bx\| = |b| \cdot \|x\| \geqslant \lambda$. Moreover, since $\lambda < |a| < 1$ one has $\ln(\lambda) < \ln |a| < 0$. Hence $\ln(\lambda)/\ln |a| > 1$, which implies that $p > -\ln \|x\|/\ln |a|$. Hence $|b| = |a|^p < \|x\|^{-1}$, which leads to $\|bx\| < 1$. □

Proposition 1.1.5 *Let* $(V, \|\cdot\|)$ *be an ultrametrically seminormed vector space over* k.

(1) *If* $x_1, \ldots, x_n$ *are vectors of* V *such that the numbers* $\|x_1\|, \ldots, \|x_n\|$ *are distinct, then one has* $\|x_1 + \cdots + x_n\| = \max_{i \in \{1,\ldots,n\}} \|x_i\|$.

(2) *The cardinal of the image of the composed map*

$$V \setminus N_{\|\cdot\|} \xrightarrow{\|\cdot\|} \mathbb{R}_{>0} \longrightarrow \mathbb{R}_{>0}/|k^{\times}| \tag{1.1}$$

is not greater than the dimension of $V/N_{\|\cdot\|}$ *over* k*, where* $\mathbb{R}_{>0}$ *denotes the multiplicative group of positive real numbers, and* $|k^{\times}|$ *is the image of* $k^{\times}$ *by* $|\cdot|$*.*

Proof (1) The statement is trivial when $n = 1$. Moreover, by induction it suffices to treat the case where $n = 2$. Without loss of generality, we assume that $\|x_1\| < \|x_2\|$. Since $\|\cdot\|$ is ultrametric, one has $\|x_1 + x_2\| \leqslant \max\{\|x_1\|, \|x_2\|\} = \|x_2\|$. Moreover,

$$\|x_2\| = \|x_1 + x_2 + (-x_1)\| \leqslant \max\{\|x_1 + x_2\|, \|x_1\|\}.$$

Since $\|x_2\| > \|x_1\|$, one should have $\|x_2\| \leqslant \|x_1 + x_2\|$. Therefore

$$\|x_1 + x_2\| = \|x_2\| = \max\{\|x_1\|, \|x_2\|\}.$$

(2) By replacing V by $V/N_{\|\cdot\|}$ and $\|\cdot\|$ by the associated norm, we may assume that $\|\cdot\|$ is actually a norm. Denote by I the image of the composed map (1.1). For each element α in I, we pick a vector x_α in $V \setminus \{0\}$ such that the image of x_α by the composed map is α. We will show that the family $\{x_\alpha\}_{\alpha\in I}$ is linearly independent over k and hence the cardinal of I is not greater than the dimension of V over k. Assume that $\alpha_1, \ldots, \alpha_n$ are distinct elements of the set I and $\lambda_1, \ldots, \lambda_n$ are non-zero elements of k. Then the values $\|\lambda_1 x_{\alpha_1}\|, \ldots, \|\lambda_n x_{\alpha_n}\|$ are distinct. As the norm $\|\cdot\|$ is ultrametric, by (1) one has

$$\|\lambda_1 x_{\alpha_1} + \cdots + \lambda_n x_{\alpha_n}\| = \max_{i\in\{1,\ldots,n\}} \|\lambda_i x_i\| > 0.$$

Hence $\lambda_1 x_{\alpha_1} + \cdots + \lambda_n x_{\alpha_n}$ is non-zero. □

Corollary 1.1.6 *Let* $(V, \|\cdot\|)$ *be an ultrametrically seminormed vector space of finite dimension over* k*. Then we have the following:*

(1) *If* $|\cdot|$ *is a discrete valuation (namely* $|k^{\times}|$ *is a discrete subgroup of* $\mathbb{R}_{>0}$*), then the image of* $V \setminus N_{\|\cdot\|}$ *by* $\|\cdot\|$ *is a discrete subset of* $\mathbb{R}_{>0}$*.*
(2) *If* $|\cdot|$ *is the trivial absolute value, then the image of* V *by* $\|\cdot\|$ *is a finite set, whose cardinal does not exceed* $\dim_k(V/N_{\|\cdot\|}) + 1$*.*

1.1.1 Topology

Let $(V, \|\cdot\|)$ be a seminormed vector space over k. The seminorm $\|\cdot\|$ induces a pseudometric $\mathrm{dist}(\cdot,\cdot)$ on V such that $\mathrm{dist}(x, y) := \|x - y\|$ for any $(x, y) \in V^2$. We equip V with the most coarse topology which makes continuous the functions $(y \in V) \mapsto \|x - y\|$ for any $x \in V$. In other words, a subset U of V is open if and only if, for any $x \in U$, there is a positive number ϵ such that $\{y \in V \,:\, \|y - x\| < \epsilon\} \subseteq U$.

This topology is said to be *induced by* the seminorm $\|\cdot\|$. The set V equipped with this topology forms a topological vector space. For any vector subspace W of V, the closure of W is also a vector subspace of V. In particular, if W is a *hyperplane* in V (namely the kernel of a linear form), then either W is a closed vector subspace of V or W is dense in V. For any $x \in V$, the *pseudodistance* between W and x is defined as

$$\operatorname{dist}(x, W) := \inf\{\|x - y\| \ : \ y \in W\}.$$

Then $\operatorname{dist}(x, W) = 0$ if and only if x belongs to the closure of W. In particular, the null space of $(V, \|\cdot\|)$ is a closed subspace, which is the closure of the zero vector subspace $\{0\}$. Thus the topological vector space V is separated if and only if $\|\cdot\|$ is a norm.

Proposition 1.1.7 *Let $(V_1, \|\cdot\|_1)$ and $(V_2, \|\cdot\|_2)$ be seminormed vector spaces over k, and $f : V_1 \longrightarrow V_2$ be a k-linear map. Then we have the following:*

(1) *If the map f is continuous, then $f(N_{\|\cdot\|_1}) \subseteq N_{\|\cdot\|_2}$.*
(2) *If there is a non-negative constant C such that $\|f(x)\|_2 \leqslant C\|x\|_1$ for all $x \in V_1$, then the map f is continuous. The converse is true if either (i) the absolute value $|\cdot|$ is non-trivial or (ii) $\dim_k(V_2/N_{\|\cdot\|_2}) < \infty$.*

Proof (1) Since $N_{\|\cdot\|_2}$ is a closed subset of V_2, its inverse image by the continuous map f is a closed subset of V_1, which clearly contains $0 \in V_1$. Hence $f^{-1}(N_{\|\cdot\|_2})$ contains $N_{\|\cdot\|_1}$ since $N_{\|\cdot\|_1}$ is the closure of $\{0\}$ in V_1.

(2) Let $\{x_n\}_{n\in\mathbb{N}}$ be a sequence in V_1 which converges to a point $x \in V_1$. One has

$$\|f(x_n) - f(x)\|_2 = \|f(x_n - x)\|_2 \leqslant C\|x_n - x\|_1,$$

so that the sequence $\{f(x_n)\}_{n\in\mathbb{N}}$ converges to $f(x)$. Hence the map f is continuous.

Assume that f is continuous. First we consider the case where the absolute value $|\cdot|$ is not trivial. The set $f^{-1}((V_2, \|\cdot\|_2)_{<1})$ is an open subset of V_1 (see Notation 1.1.3). Hence there exists $\epsilon > 0$ such that $f^{-1}((V_2, \|\cdot\|_2)_{<1}) \supseteq (V_1, \|\cdot\|_1)_{<\epsilon}$. As the absolute value $|\cdot|$ is not trivial, there exists $a \in k$ such that $0 < |a| < 1$. Let us see that $\|f(x)\|_2 \leqslant (\epsilon|a|)^{-1}\|x\|_1$ for all $x \in V_1$. If $x \in N_{\|\cdot\|_1}$, then the assertion is obvious by (1), so that we may assume that $x \notin N_{\|\cdot\|_1}$. Then there exists a unique integer n such that

$$\|a^n x\|_1 < \epsilon \leqslant \|a^{n-1}x\|_1 = |a|^{n-1} \cdot \|x\|_1.$$

Thus $\|f(a^n x)\|_2 < 1$ and hence

$$\|f(x)\|_2 < |a|^{-n} \leqslant (\epsilon|a|)^{-1} \cdot \|x\|_1,$$

as desired.

Next we assume that the absolute value $|\cdot|$ is trivial and $\dim_k(V_2/N_{\|\cdot\|_2}) < \infty$. By (2) in Corollary 1.1.6 there exist positive numbers r and δ such that $\|y\| \leqslant r$ for any $y \in V_2$ and that $(V_2, \|\cdot\|_2)_{<\delta} = N_{\|\cdot\|_2}$. If f is continuous, then there exists $\epsilon > 0$ such that

$$f^{-1}(N_{\|\cdot\|_2}) = f^{-1}((V_2, \|\cdot\|_2)_{<\delta}) \supseteq (V_1, \|\cdot\|_1)_{<\epsilon}.$$

Therefore one has $\|f(x)\|_2 \leqslant (r/\epsilon)\|x\|_1$ for all $x \in V_1$. □

Remark 1.1.8 The hypothesis of non-triviality of the absolute value or

$$\dim_k(V_2/N_{\|\cdot\|_2}) < \infty$$

for the sufficiency part of the above proposition is essential. In fact, if V is an infinite-dimensional vector space over a trivially valued field k, equipped with the norm $\|\cdot\|$ such that $\|x\| = 1$ for any $x \in V \setminus \{0\}$, then the topology on V induced by the norm $\|\cdot\|$ is discrete. In particular, any k-linear map from V to a normed vector space over k is continuous. However, one can take a basis B of the vector space V (which is an infinite set) and define a new norm $\|\cdot\|'$ on V such that

$$\Big\| \sum_{x \in B} n_x x \Big\|' = \max_{x \in B,\, n_x \neq 0} \varphi(x),$$

where $\varphi : B \to]0, +\infty[$ is a map which is not bounded. If f is the identity map from $(V, \|\cdot\|)$ to $(V, \|\cdot\|')$, then one can not find a non-negative constant C such that $\|x\|' \leqslant C\|x\|$ for all $x \in V$.

1.1.2 Operator seminorm

Let $(V_1, \|\cdot\|_1)$ and $(V_2, \|\cdot\|_2)$ be seminormed vector spaces over k. Let $f : V_1 \longrightarrow V_2$ be a k-linear map. We say that the linear map f is *bounded* if there is a non-negative constant C such that $\|f(x)\|_2 \leqslant C\|x\|_1$ for all $x \in V_1$. Note that if f is bounded, then f is continuous and $f(N_{\|\cdot\|_1}) \subseteq N_{\|\cdot\|_2}$ by Proposition 1.1.7.

If $f(N_{\|\cdot\|_1}) \subseteq N_{\|\cdot\|_2}$, we denote by $\|f\|$ the element

$$\sup_{x \in V_1 \setminus N_{\|\cdot\|_1}} \frac{\|f(x)\|_2}{\|x\|_1} \in [0, +\infty].$$

If the relation $f(N_{\|\cdot\|_1}) \subseteq N_{\|\cdot\|_2}$ does not hold, then by convention $\|f\|$ is defined to be $+\infty$. With this notation, the linear map f is bounded if and only if $\|f\| < +\infty$.

We denote by $\mathscr{L}(V_1, V_2)$ the set of all bounded k-linear maps from V_1 to V_2, which forms a vector space over k since, for $(f, g) \in \mathscr{L}(V_1, V_2)^2$ and $x \in V_1 \setminus N_{\|\cdot\|_1}$,

$$\frac{\|(f+g)(x)\|_2}{\|x\|_1} \leqslant \begin{cases} \max\left\{\dfrac{\|f(x)\|_2}{\|x\|_1}, \dfrac{\|g(x)\|_2}{\|x\|_1}\right\} & \text{if } \|\cdot\|_2 \text{ is ultrametric,} \\[2ex] \dfrac{\|f(x)\|_2}{\|x\|_1} + \dfrac{\|g(x)\|_2}{\|x\|_1} & \text{otherwise.} \end{cases}$$

The map $\|\cdot\| : \mathscr{L}(V_1, V_2) \longrightarrow [0, +\infty[$ defined above is a seminorm, called the *operator seminorm*. Moreover, from the above formula, we observe that, if $\|\cdot\|_2$ is ultrametric, then the operator seminorm is also ultrametric. If $\|\cdot\|_2$ is a norm, then the operator seminorm is actually a norm, called *operator norm*.

In the case where either the absolute value $|\cdot|$ is non-trivial or $\dim_k(V_2/N_{\|\cdot\|_2}) < \infty$, the space $\mathscr{L}(V_1, V_2)$ identifies with the vector space of all continuous k-linear maps from V_1 to V_2 (see Proposition 1.1.7).

1.1.3 Quotient seminorm

Let $g : V \to Q$ be a surjective linear map of vector spaces over k and $\|\cdot\|$ be a seminorm on V. We define $\|\cdot\|_{V\twoheadrightarrow Q}$ to be

$$\forall\, y \in Q, \quad \|y\|_{V\twoheadrightarrow Q} := \inf\{\|x\| \ :\ x \in V,\ g(x) = y\}.$$

Then we have the following proposition:

Proposition 1.1.9 (1) *$\|\cdot\|_{V\twoheadrightarrow Q}$ is a seminorm on Q. Moreover, if $\|\cdot\|$ is ultrametric, then also is $\|\cdot\|_{V\twoheadrightarrow Q}$.*

(2) *Let $N_{\|\cdot\|_{V\twoheadrightarrow Q}}$ be the null space of $\|\cdot\|_{V\twoheadrightarrow Q}$. Then $g^{-1}(N_{\|\cdot\|_{V\twoheadrightarrow Q}})$ coincides with the closure of* Ker(g) *with respect to the topology induced by the seminorm $\|\cdot\|$. In particlar, if* Ker(g) *is closed, then $\|\cdot\|_{V\twoheadrightarrow Q}$ is a norm on Q.*

Proof (1) In order to see the condition (a) in Definition 1.1.1, we may assume that $a \neq 0$ since otherwise the assertion is obvious. Then

$$\begin{aligned}\|ay\|_{V\twoheadrightarrow Q} &= \inf\{\|x'\| \ :\ x' \in V,\ g(x') = ay\} = \inf\{\|ax\| \ :\ x \in V,\ g(x) = y\}\\ &= |a| \inf\{\|x\| \ :\ x \in V,\ g(x) = y\} = |a| \cdot \|y\|_{V\twoheadrightarrow Q}.\end{aligned}$$

Fix $(y, y') \in Q^2$. For any $\epsilon > 0$, we can find $(x, x') \in V^2$ such that $g(x) = y$, $g(x') = y'$, $\|x\| \leqslant \|y\|_{V\twoheadrightarrow Q} + \epsilon$ and $\|x'\| \leqslant \|y'\|_{V\twoheadrightarrow Q} + \epsilon$. Then $g(x + x') = y + y'$ and

$$\|y + y'\|_{V\twoheadrightarrow Q} \leqslant \|x + x'\| \leqslant \|x\| + \|x'\| \leqslant \|y\|_{V\twoheadrightarrow Q} + \|y'\|_{V\twoheadrightarrow Q} + 2\epsilon,$$

and hence (b) holds. If $\|\cdot\|$ is ultrametric, in a similar way we can see that $\|\cdot\|_{V\twoheadrightarrow Q}$ is also ultrametric.

(2) Let $x \in V$ and $y = g(x)$. It is easy to see $\|y\|_{V\twoheadrightarrow Q} = \mathrm{dist}(x, \mathrm{Ker}(g))$. Therefore

$$x \in \overline{\mathrm{Ker}(g)} \quad\Longleftrightarrow\quad \mathrm{dist}(x, \mathrm{Ker}(g)) = 0 \quad\Longleftrightarrow\quad \|y\|_{V\twoheadrightarrow Q} = 0,$$

as required. □

Given a vector subspace W of a seminormed vector space $(V, \|\cdot\|)$, the seminorm $\|\cdot\|_{V\twoheadrightarrow V/W}$ on V/W is called the *quotient seminorm* on V/W of the seminorm $\|\cdot\|$ on V. For simplicity, the seminorm $\|\cdot\|_{V\twoheadrightarrow V/W}$ is often denoted by $\|\cdot\|_{V/W}$. If the vector subspace W is closed, then the seminorm $\|\cdot\|_{V/W}$ is actually a norm, called the *quotient norm* of $\|\cdot\|$ by the quotient map $V \twoheadrightarrow V/W$. Note that the norm $\|\cdot\|^{\sim}$ identifies with the quotient norm of $\|\cdot\|$ by the quotient map $V \twoheadrightarrow V/N_{\|\cdot\|}$.

Proposition 1.1.10 *Let $(V, \|\cdot\|)$ be a seminormed vector space over k and W be a vector subspace of V. The topology on V/W defined by the quotient seminorm coincides with the quotient topology. In particular, the quotient map $V \to V/W$ is continuous if we equip V/W with the quotient seminorm.*

Proof Recall that the quotient topology is the finest topology on V/W which makes the quotient map $\pi : V \to V/W$ continuous. In other words, a subset U of V/W is open for the quotient topology if and only if $\pi^{-1}(U)$ is an open subset of V. If we equip V/W with the topology induced by the quotient seminorm, then the quotient map is continuous since $\|\pi\| \leqslant 1$ (see Proposition 1.1.7). Moreover, if U is a subset of V/W such that $\pi^{-1}(U)$ is an open subset of V, then, for any $u \in U$ and any $x_0 \in V$ such that $\pi(x_0) = u$, there exists $\epsilon > 0$ such that

$$\{x \in V \,:\, \|x - x_0\| < \epsilon\} \subseteq \pi^{-1}(U).$$

Hence for any $v \in V/W$ with $\|v - u\| < \epsilon$, there exists $x \in \pi^{-1}(U)$ such that $\pi(x) = v$. So U is an open subset of V/W for the topology defined by the quotient seminorm. The proposition is thus proved. □

1.1.4 Topology of normed vector spaces of finite dimension

If V is a finite-dimensional k-vector space, then all norms on V induce the same topology. More precisely, we have the following result (see [30] Chapter I, §2, no.3, Theorem 2 and the remark on the page I.15).

Proposition 1.1.11 *Assume that the vector space V is of finite dimension over k. If $\|\cdot\|$ and $\|\cdot\|'$ are norms on V, then there are positive constants C and C' such that $C\|\cdot\|' \leqslant \|\cdot\| \leqslant C'\|\cdot\|'$ on V. In particular, V is complete with respect to $\|\cdot\|$.* [1]

Proof Let $\{e_i\}_{i=1}^r$ be a basis of V and $f : k^r \to V$ be the isomorphism given by $f(a_1, \ldots, a_r) = a_1e_1 + \cdots + a_re_r$. Here we consider the product topology on k^r and the topology induced by any norm $\|\cdot\|$ on V. By Proposition 1.1.7, it is sufficient to show that f is a homeomorphism. Since

$$\|a_1e_1 + \cdots + a_re_r\| \leqslant \max\{|a_1|, \ldots, |a_r|\} \sum_{i=1}^{r} \|e_i\|,$$

[1] That is, for any sequence $\{x_n\}_{n\in\mathbb{N}}$ in V, if

$$\lim_{N\to+\infty} \sup_{\substack{(n,m)\in\mathbb{N}^2\\ n\geqslant N,\, m\geqslant N}} \|x_n - x_m\| = 0,$$

then there exists $x \in V$ such that $\lim_{n\to\infty} \|x_n - x\| = 0$.

f is continuous. It remains to show that f^{-1} is continuous.

We reason by induction on the dimension r of V. The case where $r = 0$ is trivial. In the case where $r = 1$, as $|a|/\|ae_1\| = 1/\|e_1\|$ for any $a \in k^\times$, f^{-1} is continuous by Proposition 1.1.7.

Assume that the proposition has been proved for vector spaces of dimension $< r$. Let W be the vector subspace of V generated by $e_1, \ldots, e_{r-1}$. By the induction hypothesis, the map $g : k^{r-1} \to W$ sending $(a_1, \ldots, a_{r-1}) \in k^{r-1}$ to $a_1e_1 + \cdots + a_{r-1}e_{r-1}$ is a homeomorphism. In particular, the topological vector space W is complete. As a consequence, W is a closed vector subspace of V. By the dimension 1 case of the proposition proved above, the map $\overline{f}$ from k to V/W sending $a \in k$ to $a[e_r]$ is a homeomorphism.

In the following, we show that, if U is an open neighbourhood of $(0, \ldots, 0) \in k^r$, then there exists $\epsilon > 0$ such that $f(U)$ contains all vectors $x \in V$ satisfying $\|x\| < \epsilon$. Without loss of generality, we may assume that U is the open multidisc B_δ^r, where $B_\delta = \{a \in k : |a| < \delta\}$ and $\delta > 0$. Since the map $g : k^{r-1} \to W$ is a homeomorphism, there exists $\epsilon_1 > 0$ such that

$$g(B_\delta^{r-1}) \supseteq \{y \in W : \|y\| < \epsilon_1\}.$$

Let $\delta' = \min\{\epsilon_1/(2\|e_r\|), \delta\}$. Since the map $\overline{f}$ is a homeomorphism, there exists $\epsilon_2 > 0$ such that

$$\overline{f}(B_{\delta'}) \supseteq \{u \in V/W : \|u\|_{V/W} < \epsilon_2\},$$

where we consider the quotient norm on V/W. We claim that

$$f(U) \supseteq \{x \in V : \|x\| < \epsilon\}$$

with $\epsilon = \frac{1}{2}\min\{\epsilon_1, \epsilon_2\}$. In fact, if x is an element of V such that $\|x\| < \epsilon$, then its class in V/W has norm $< \epsilon_2$. Hence there exists $a_r \in B_{\delta'}$ such that $[x] = a_r[e_r]$. Moreover, one has

$$\|x - a_re_r\| \leqslant \|x\| + |a_r| \cdot \|e_r\| < \frac{1}{2}\epsilon_1 + \delta'\|e_r\| \leqslant \epsilon_1.$$

Hence there exists $(a_1, \ldots, a_{r-1}) \in B_\delta^{r-1}$ such that $g(a_1, \ldots, a_{r-1}) = x - a_re_r$. Thus $(a_1, \ldots, a_r)$ is an element in B_δ^r such that $f(a_1, \ldots, a_r) = x$. The proposition is proved. □

Corollary 1.1.12 *Let $f : V_1 \to V_2$ be a linear map of vector spaces over k, and let $\|\cdot\|_1$ and $\|\cdot\|_2$ be seminorms on V_1 and V_2, respectively. We assume that $f(N_{\|\cdot\|_1}) \subseteq N_{\|\cdot\|_2}$ and $\dim_k(V_2/N_{\|\cdot\|_2}) < \infty$. Then the following conditions are equivalent:*

(a) *the map f is continuous;*
(b) *$f^{-1}(N_{\|\cdot\|_2})$ is a closed vector subspace of V;*
(c) *$\|f\|$ is finite.*

Proof "(a)$\Longrightarrow$(b)": Since $f^{-1}(N_{\|\cdot\|_2})$ is the inverse image by f of the closed subset $N_{\|\cdot\|_2}$ of V_2, if f is continuous, then it is a closed subset of V_1.

"(b)$\Longrightarrow$(c)": The assertion is trivial when $f(V_1) \subseteq N_{\|\cdot\|_2}$. In the following, we assume that $f(V_1) \not\subseteq N_{\|\cdot\|_2}$. We set

$$Q := f(V_1)/(f(V_1) \cap N_{\|\cdot\|_2}) \cong (f(V_1) + N_{\|\cdot\|_2})/N_{\|\cdot\|_2} \neq \{0\}.$$

Let $\|\cdot\|_Q$ be the quotient seminorm on Q induced by $V_1 \to f(V_1) \to Q$ and $\|\cdot\|_1$. By Proposition 1.1.9 and the condition (b), the seminorm $\|\cdot\|_Q$ is actually a norm. Moreover, we can consider Q as a vector subspace of $V_2/N_{\|\cdot\|_2}$. Let $\|\cdot\|'_Q$ be the restriction of $\|\cdot\|_2^{\sim}$ to Q, where $\|\cdot\|_2^{\sim}$ is the norm associated with the seminorm $\|\cdot\|_2$. By Proposition 1.1.11, there is a constant C with $\|\cdot\|'_Q \leqslant C\|\cdot\|_Q$. Thus, for any $x \in V_1 \setminus N_{\|\cdot\|_1}$, one has

$$\frac{\|f(x)\|_2}{\|x\|_1} = \frac{\|[f(x)]\|'_Q}{\|x\|_1} \leqslant \frac{C\|[f(x)]\|_Q}{\|x\|_1} \leqslant C,$$

which implies $\|f\| \leqslant C$.

"(c)$\Longrightarrow$(a)" follows from Proposition 1.1.7. □

Corollary 1.1.13 *Let $(V, \|\cdot\|)$ be a finite-dimensional seminormed vector space over k. Then we have the following:*

(1) *Every vector subspace of V containing $N_{\|\cdot\|}$ is closed.*
(2) *Let $(V', \|\cdot\|')$ be a seminormed vector space over k and $f : V \to V'$ be a linear map of vector spaces over k such that $f(N_{\|\cdot\|}) \subseteq N_{\|\cdot\|'}$. Then f is continuous and $\|f\| < +\infty$.*
(3) *A linear form on V is bounded if and only if its kernel contains $N_{\|\cdot\|}$.*

Proof (1) Let $\pi : V \to V/N_{\|\cdot\|}$ be the canonical projection map and $\|\cdot\|^{\sim}$ be the norm on $V/N_{\|\cdot\|}$ associated with $\|\cdot\|$. By Proposition 1.1.10, the linear map π is continuous. If W is a vector subspace of V containing $N_{\|\cdot\|}$, then one has $W = \pi^{-1}(\pi(W))$. By Proposition 1.1.11, $\pi(W)$ is complete with respect to the induced norm of $\|\cdot\|^{\sim}$ on $\pi(W)$, so that $\pi(W)$ is closed. Hence W is also closed since it is the inverse image of a closed subset of $V/N_{\|\cdot\|}$ by a continuous linear map.

(2) By replacing V' by $f(V)$, we may assume that $\dim_k(V') < \infty$. Thus the assertion follows from (1) and Corollary 1.1.12.

(3) Let $f : V \to k$ be a linear form. If f is bounded, by Corollary 1.1.12, the kernel of f is a closed vector subspace of V, hence it contains the closure of $\{0\}$, which is $N_{\|\cdot\|}$. Conversely, if $\mathrm{Ker}(f) \supseteq N_{\|\cdot\|}$, then by (2), the linear form f is bounded. □

Proposition 1.1.14 (1) *Let $V \xrightarrow{\alpha} W \xrightarrow{\beta} Q$ be a sequence of surjective linear maps of finite-dimensional vector spaces over k. For any seminorm $\|\cdot\|$ on V, one has $\|\cdot\|_{V\twoheadrightarrow W, W\twoheadrightarrow Q} = \|\cdot\|_{V\twoheadrightarrow Q}$.*

(2) *Let*

$$\begin{array}{ccc} V & \xrightarrow{\;f\;} & W \\ {\scriptstyle\alpha}\big\downarrow & & \big\downarrow{\scriptstyle\beta} \\ V' & \xrightarrow[\;g\;]{} & W' \end{array}$$

be a commutative diagram of linear maps of finite-dimensional vector spaces over k such that α and β are surjective. Then we have the following:

(2.a) *Let $\|\cdot\|_V$ and $\|\cdot\|_W$ be seminorms on V and W; let $\|\cdot\|_{V'}$ and $\|\cdot\|_{W'}$ be the quotient seminorms of $\|\cdot\|_V$ and $\|\cdot\|_W$ on V' and W', respectively. If $f(N_{\|\cdot\|_V}) \subseteq N_{\|\cdot\|_W}$, then $g(N_{\|\cdot\|_{V'}}) \subseteq N_{\|\cdot\|_{W'}}$ and $\|g\| \leqslant \|f\|$.*

(2.b) *We assume that f and g are injective. Let $\|\cdot\|_W$ be a seminorm on W. Then $\|\cdot\|_{W,V\hookrightarrow W,V\twoheadrightarrow V'} \geqslant \|\cdot\|_{W,W\twoheadrightarrow W',V'\hookrightarrow W'}$. Moreover, if $\mathrm{Ker}(\beta) \subseteq f(V)$, then the equality $\|\cdot\|_{W,V\hookrightarrow W,V\twoheadrightarrow V'} = \|\cdot\|_{W,W\twoheadrightarrow W',V'\hookrightarrow W'}$ holds.*

Proof (1) For $q \in Q$, one has

$$\begin{aligned}\|q\|_{V\twoheadrightarrow W,W\twoheadrightarrow Q} &= \inf_{y\in W,\,\beta(y)=q} \|y\|_{V\twoheadrightarrow W} = \inf_{\substack{y\in W\\ \beta(y)=q}} \inf_{\substack{x\in V\\ \alpha(x)=y}} \|x\|_V \\ &= \inf_{x\in V,\,\beta(\alpha(x))=q} \|x\|_V = \|q\|_{V\twoheadrightarrow Q},\end{aligned}$$

as desired.

(2.a) By Proposition 1.1.9, $\alpha^{-1}(N_{\|\cdot\|_{V'}})$ is the closure of $\mathrm{Ker}(\alpha)$ in V, hence is equal to $\mathrm{Ker}(\alpha) + N_{\|\cdot\|_V}$. Let y' be an element in $N_{\|\cdot\|_{V'}}$. There then exists $y \in N_{\|\cdot\|_V}$ such that $\alpha(y) = y'$. Therefore

$$g(y') = g(\alpha(y)) = \beta(f(y)) \in N_{\|\cdot\|_{W'}}$$

since $f(y) \in N_{\|\cdot\|_W}$.

It remains to prove that $\|g\| \leqslant \|f\|$. Let x' be an element of V'. For any $x \in V$ with $\alpha(x) = x'$, one has

$$\|g(x')\|_{W'} = \|g(\alpha(x))\|_{W'} = \|\beta(f(x))\|_{W'} \leqslant \|f(x)\|_W \leqslant \|f\| \cdot \|x\|_V,$$

which leads to

$$\|g(x')\|_{W'} \leqslant \|f\| \inf_{x\in V,\,\alpha(x)=x'} \|x\|_V = \|f\| \cdot \|x'\|_{V'}.$$

(2.b) Note that $f(\mathrm{Ker}(\alpha)) = f(V) \cap \mathrm{Ker}(\beta)$. Therefore, for $v \in V$,

$$\|\alpha(v)\|_{W,V\hookrightarrow W,V\twoheadrightarrow V'} = \inf\{\|x\|_W \;:\; x \in f(v) + (f(V) \cap \mathrm{Ker}(\beta))\}$$

and

$$\|\alpha(v)\|_{W,W\twoheadrightarrow W',V'\hookrightarrow W'} = \inf\{\|x\|_W \;:\; x \in f(v) + \mathrm{Ker}(\beta)\},$$

so that the first assertion follows. Moreover, if $\mathrm{Ker}(\beta) \subseteq f(V)$, then $f(V) \cap \mathrm{Ker}(\beta) = \mathrm{Ker}(\beta)$. Thus the second assertion holds. □

Proposition 1.1.15 (1) *Let $f : V \to W$ be a surjective linear map of vector spaces over k, $\|\cdot\|_V$ be a seminorm on V and $\|\cdot\|_W$ be the quotient seminorm of $\|\cdot\|_V$ on W. If the seminorm $\|\cdot\|_W$ does not vanish, then $\|f\| = 1$.*

(2) *Let*

$$\begin{array}{ccc} V & \xrightarrow{f_1} & W_1 \\ & \searrow^{f_2} & \downarrow g \\ & & W_2 \end{array}$$

be a commutative digram of linear maps of finite-dimensional vector spaces over k such that g is an isomorphism and $\dim_k(W_1) = \dim_k(W_2) = 1$. *Let $\|\cdot\|_V$, $\|\cdot\|_{W_1}$ and $\|\cdot\|_{W_2}$ be seminorms of V, W_1 and W_2, respectively. Then $\|f_2\| = \|f_1\| \cdot \|g\|$ provided that f_1, f_2 and g are continuous.*

Proof (1) Since $\|v\|_V \geqslant \|f(v)\|_W$ for any $v \in V$, one has $\|f\| \leqslant 1$. Let w be an element of W such that $\|w\|_W > 0$. Since

$$\|w\|_W = \inf_{v \in V,\, f(v)=w} \|v\|_V,$$

one has

$$1 = \inf_{v \in V,\, f(v)=w} \frac{\|v\|_V}{\|f(v)\|_W} \geqslant \|f\|^{-1},$$

which leads to $\|f\| \geqslant 1$.

(2) As g is an isomorphism and $\dim_k(W_1) = \dim_k(W_2) = 1$, for any $w_1 \in W_1$, $\|g\| \cdot \|w_1\|_{W_1} = \|g(w_1)\|_{W_2}$, Therefore,

$$\begin{aligned} \|f_2\| &= \sup_{v \in V \setminus N_{\|\cdot\|_V}} \frac{\|f_2(v)\|_{W_2}}{\|v\|_V} = \sup_{v \in V \setminus N_{\|\cdot\|_V}} \frac{\|g(f_1(v))\|_{W_2}}{\|v\|_V} \\ &= \sup_{v \in V \setminus N_{\|\cdot\|_V}} \|g\| \frac{\|f_1(v)\|_{W_1}}{\|v\|_V} = \|g\| \cdot \|f_1\|, \end{aligned}$$

as required. □

Proposition 1.1.16 *Let $(V, \|\cdot\|_V)$ be a finite-dimensional seminormed vector space over k, W be a vector subspace of V and Q be the quotient vector space V/W. We denote by $i : W \to V$ and $\pi : V \to Q$ the inclusion map and the projection map, respectively. Let $\|\cdot\|_W$ be the restriction of $\|\cdot\|_V$ to W and $\|\cdot\|_Q$ be the quotient seminorm of $\|\cdot\|_V$ on Q. Then one has $i(N_{\|\cdot\|_W}) \subseteq N_{\|\cdot\|_V}$ and $\pi(N_{\|\cdot\|_V}) \subseteq N_{\|\cdot\|_Q}$. Moreover, the linear maps i and π induce short exact sequences*

$$0 \longrightarrow N_{\|\cdot\|_W} \longrightarrow N_{\|\cdot\|_V} \longrightarrow N_{\|\cdot\|_Q} \longrightarrow 0 \tag{1.2}$$

and

$$0 \longrightarrow W/N_{\|\cdot\|_W} \longrightarrow V/N_{\|\cdot\|_V} \longrightarrow Q/N_{\|\cdot\|_Q} \longrightarrow 0, \tag{1.3}$$

and the induced norm (resp. quotient norm) of $\|\cdot\|_{\widetilde{V}}$ on $W/N_{\|\cdot\|_W}$ (resp. $Q/N_{\|\cdot\|_Q}$) identifies with $\|\cdot\|_{\widetilde{W}}$ (resp. $\|\cdot\|_{\widetilde{Q}}$).

Proof The relations $i(N_{\|\cdot\|_W}) \subseteq N_{\|\cdot\|_V}$ and $\pi(N_{\|\cdot\|_V}) \subseteq N_{\|\cdot\|_Q}$ follow directly from the definition of induced and quotient seminorms. Moreover, by definition one has $N_{\|\cdot\|_W} = N_{\|\cdot\|_V} \cap W$.

For any element $x \in V$, $\pi(x)$ lies in $N_{\|\cdot\|_Q}$ if and only if $x \in W + N_{\|\cdot\|_V}$ since $W + N_{\|\cdot\|_V}$ is the closure of W in V. Therefore one has

$$N_{\|\cdot\|_Q} \cong (W + N_{\|\cdot\|_V})/W \cong N_{\|\cdot\|_V}/(W \cap N_{\|\cdot\|_V}) = N_{\|\cdot\|_V}/N_{\|\cdot\|_W},$$

which proves that (1.2) is an exact sequence. The exactness of (1.2) implies that of (1.3). Moreover, if x is an element of W, then

$$\|[x]\|_{\widetilde{W}} = \|x\|_W = \|x\|_V = \|[x]\|_{\widetilde{V}}.$$

If u is an element in Q, then

$$\|[u]\|_{\widetilde{Q}} = \|u\|_Q = \inf_{y \in V,\, \pi(y)=u} \|y\|_V = \inf_{v \in V/N_{\|\cdot\|_V},\, \widetilde{\pi}(v)=[u]} \|v\|_{\widetilde{V}},$$

where $\widetilde{\pi} : V/N_{\|\cdot\|_V} \to Q/N_{\|\cdot\|_Q}$ is the linear map induced by π. Hence $\|\cdot\|_{\widetilde{Q}}$ identifies with the quotient norm of $\|\cdot\|_{\widetilde{V}}$. □

1.1.5 Dual norm

Let $(V, \|\cdot\|)$ be a seminormed vector space over k. We denote by V^* the vector space $\mathscr{L}(V, k)$ (where we consider $|\cdot|$ as a norm on k) of bounded k-linear forms on V (which necessarily vanish on $N_{\|\cdot\|}$), called the *dual* normed vector space of V. The operator norm on V^* is called the *dual norm* of $\|\cdot\|$, denoted by $\|\cdot\|_*$. Note that in general V^* is different from the (algebraic) dual vector space $V^\vee := \mathrm{Hom}_k(V, k)$. One has

$$V^* \subseteq (V/N_{\|\cdot\|})^\vee = \{\varphi \in V^\vee \;:\; \varphi|_{N_{\|\cdot\|}} = 0\}.$$

Proposition 1.1.17 *Let $(V, \|\cdot\|)$ be a finite-dimensional seminormed vector space over k. Then the map $(V/N_{\|\cdot\|})^\vee \to V^\vee$ sending $\varphi \in (V/N_{\|\cdot\|})^\vee$ to its composition with the projection map $V \to V/N_{\|\cdot\|}$ defines an isomorphism between $(V/N_{\|\cdot\|})^\vee$ and V^*. In particular, the equality $V^* = V^\vee$ holds when $\|\cdot\|$ is a norm.*

Proof By Corollary 1.1.13, a linear form on V is bounded if and only if its kernel contains $N_{\|\cdot\|}$. Therefore V^* is canonically isomorphic to $(V/N_{\|\cdot\|})^\vee$. □

If x is an element of V, for any $\alpha \in V^*$ one has

$$|\alpha(x)| \leqslant \|\alpha\|_* \cdot \|x\|. \tag{1.4}$$

Therefore the linear form on V^* sending $\alpha \in V^*$ to $\alpha(x) \in k$ is bounded. Hence one obtains a k-linear map from V to the double dual space V^{**} whose kernel contains $N_{\|\cdot\|}$. It is called the *canonical linear map* from V to V^{**}. The *double dual norm*

$\|\cdot\|_{**}$ on V^{**} induces by composition with the canonical k-linear map $V \to V^{**}$ a seminorm on V which we still denote by $\|\cdot\|_{**}$ by abuse of notation. Moreover, by (1.4) we obtain

$$\forall\, x \in V, \quad \|x\|_{**} \leqslant \|x\|. \tag{1.5}$$

We say that $(V, \|\cdot\|)$ is *reflexive* if the k-linear map $V \to V^{**}$ described above induces an isometric k-linear isomorphism between the normed vector spaces $(V/N_{\|\cdot\|}, \|\cdot\|^{\sim})$ and $(V^{**}, \|\cdot\|_{**})$.

The following proposition shows that, in the Archimedean case, the seminorm $\|\cdot\|_{**}$ on V identifies with $\|\cdot\|$. In particular, a finite-dimensional seminormed vector space over an Archimedean complete field is always reflexive. We will see further in Corollary 1.2.12 that any finite-dimensional ultrametrically seminormed vector space over k is also reflexive.

Proposition 1.1.18 *Assume that the absolute value $|\cdot|$ is Archimedean. Let $(V, \|\cdot\|)$ be a seminormed vector space over k. For any $x \in V$ one has $\|x\| = \|x\|_{**}$.*

Proof This is a direct consequence of Hahn-Banach theorem. In fact, if x is a vector in $V \setminus N_{\|\cdot\|}$, then by Hahn-Banach theorem there exists a k-linear form $\tilde{f} : V/N_{\|\cdot\|} \to k$ such that $\tilde{f}(\pi(x)) = \|\pi(x)\|^{\sim}$ and that $|\tilde{f}(\pi(y))| \leqslant \|\pi(y)\|^{\sim}$ for any $y \in V$, where $\pi : V \to V/N_{\|\cdot\|}$ is the canonical linear map. If we set $f = \tilde{f} \circ \pi$, then $f(x) = \|x\|$ and $|f(y)| \leqslant \|y\|$ for any $y \in V$. In particular, $\|f\|_* = 1$. Hence

$$\|x\|_{**} \geqslant \frac{|f(x)|}{\|f\|_*} = \|x\|.$$

Remark 1.1.19 The above proposition is not true when the absolute value $|\cdot|$ is non-Archimedean. Let $(V, \|\cdot\|)$ be a normed vector space over k. If the absolute value $|\cdot|$ is non-Archimedean, then the dual norm $\|\cdot\|_*$ is necessarily ultrametric (cf. Subsection 1.1.2). For the same reason, the double dual norm $\|\cdot\|_{**}$ is ultrametric, and hence cannot identify with $\|\cdot\|$ on V once the norm $\|\cdot\|$ is not ultrametric. In the next section, we will establish the analogue of the above proposition in the case where V is of finite dimension over k and $\|\cdot\|$ is ultrametric (see Corollary 1.2.12). We refer to [50] and [89] for more general results on non-Archimedean Hahn-Banach theorem.

Proposition 1.1.20 *Let $(V, \|\cdot\|_V)$ be a seminormed vector space over k, W be a vector subspace of V and $Q = V/W$ be the quotient space. Let $\|\cdot\|_Q$ be the quotient seminorm on Q induced by $\|\cdot\|_V$. Then the map $Q^* \to V^*$ sending $\varphi \in Q^*$ to the composition of φ with the projection map $V \to Q$ is an isometry from Q^* to its image (equipped with the induced norm), where we consider the dual norms $\|\cdot\|_{Q,*}$ and $\|\cdot\|_{V,*}$ on Q^* and V^*, respectively.*

Proof Note that $\|v\|_Q^{-1} = \sup_{x \in V,\, [x]=v} \|x\|_V^{-1}$ for $v \in Q \setminus N_{\|\cdot\|_Q}$. Thus, for $\varphi \in Q^*$,

$$\|\varphi\|_{Q,*} = \sup_{v \in Q \setminus N_{\|\cdot\|_Q}} \frac{|\varphi(v)|}{\|v\|_Q} = \sup_{x \in V \setminus N_{\|\cdot\|_V}} \frac{|\varphi([x])|}{\|x\|_V},$$

as required. □

Remark 1.1.21 The dual statement of the above proposition for the dual of a restricted seminorm is much more subtle. Let $(V, \|\cdot\|_V)$ be a seminormed vector space over k and W be a vector subspace of V. We denote by $\|\cdot\|_W$ the restriction of the seminorm $\|\cdot\|_V$ to W. Then the restriction to W of bounded linear forms on V defines a k-linear map π from V^* to W^*. We are interested in the nature of the dual norm $\|\cdot\|_{W,*}$. In the case where k is Archimedean, the k-linear map $\pi : V^* \to W^*$ is surjective and the norm $\|\cdot\|_{W,*}$ identifies with the quotient norm of $\|\cdot\|_{V,*}$. This is a direct consequence of Hahn-Banach theorem which asserts that any bounded linear form on W extends to V with the same operator norm (see Lemma 1.2.48 for more details). However, the non-Archimedean analogue of this result is not true, even in the case where V is finite-dimensional. In fact, assume that $(k, |\cdot|)$ is non-Archimedean and V is a finite-dimensional vector space of dimension $\geqslant 2$ over k, equipped with a norm $\|\cdot\|$ which is not ultrametric. Then the double dual norm $\|\cdot\|_{**}$ on V is bounded from above by $\|\cdot\|$ (see (1.5)), and there exists at least an element $x \in V$ such that $\|x\|_{**} < \|x\|$ since $\|\cdot\|_{**}$ is ultrametric but $\|\cdot\|$ is not. However, both norms $\|\cdot\|_{**}$ and $\|\cdot\|$ induce the same dual norm on $V^\vee$ (see Proposition 1.2.14). Therefore the quotient norm of $\|\cdot\|_*$ on $(kx)^\vee$ can not identify with the dual norm of the restriction of $\|\cdot\|$ to kx. We will show in Proposition 1.2.35 that the non-Archimedean analogue of the above statement is true when the norm on V is ultrametric.

1.1.6 Seminorm of the dual operator

Let $(V_1, \|\cdot\|_1)$ and $(V_2, \|\cdot\|_2)$ be seminormed vector spaces over k, and $f : V_1 \to V_2$ be a bounded linear map. Note that $f(N_{\|\cdot\|_1}) \subseteq N_{\|\cdot\|_2}$ by Proposition 1.1.7. For any $\alpha \in V_2^*$, we let $f^*(\alpha)$ be the linear form on V_1 which sends $x \in V_1$ to $\alpha(f(x))$. Note that for $x \in V_1$ one has

$$|f^*(\alpha)(x)| = |\alpha(f(x))| \leqslant \|\alpha\|_{2,*} \cdot \|f(x)\|_2 \leqslant \|\alpha\|_{2,*} \cdot \|f\| \cdot \|x\|_1. \tag{1.6}$$

Therefore $f^*(\alpha)$ is a bounded linear form on V_1. Thus f^* defines a linear map from V_2^* to V_1^*.

Proposition 1.1.22 *Let $(V_1, \|\cdot\|_1)$ and $(V_2, \|\cdot\|_2)$ be seminormed vector spaces over k and $f : V_1 \to V_2$ be a bounded linear map. Then one has $\|f^*\| \leqslant \|f\|$. The equality holds when $\|\cdot\|_2 = \|\cdot\|_{2,**}$ on V_2.*

Proof By (1.6) we obtain that, if α is an element of V_2^*, then one has

$$\|f^*(\alpha)\|_{1,*} \leqslant \|f\| \cdot \|\alpha\|_{2,*}.$$

Hence $\|f^*\| \leqslant \|f\|$. If we apply this inequality to f^*, we obtain $\|f^{**}\| \leqslant \|f^*\| \leqslant \|f\|$. Let $\iota_1 : V_1 \to V_1^{**}$ and $\iota_2 : V_2 \to V_2^{**}$ be the canonical linear maps. For any vector x in V_1, one has $f^{**}(\iota_1(x)) = \iota_2(f(x))$. Moreover, if $\|\cdot\|_2 = \|\cdot\|_{2,**}$ on V_2, then

$$\|f^{**}\| \geqslant \sup_{x\in V_1,\ \|x\|_{1,**}>0} \frac{\|f^{**}(\iota_1(x))\|_{2,**}}{\|x\|_{1,**}} \geqslant \sup_{x\in V_1,\ \|x\|_1>0} \frac{\|f(x)\|_2}{\|x\|_1} = \|f\|,$$

as required.

1.1.7 Lattices and norms

In this subsection, we assume that the absolute value $|\cdot|$ is non-Archimedean. Let $\mathfrak{o}_k := \{a \in k : |a| \leqslant 1\}$ be the closed unit ball of $(k, |\cdot|)$. It is a valuation ring, namely for any $a \in k \setminus \mathfrak{o}_k$ one has $a^{-1} \in \mathfrak{o}_k$ (see [27] Chapter IV, §1, no.2). It is a discrete valuation ring (namely a Noetherian valuation ring) if and only if the absolute value $|\cdot|$ is discrete, namely the image of $k^\times$ by $|\cdot|$ is a discrete subgroup of $(\mathbb{R}_{>0}, \times)$ (see [27] Chapter IV, §3, no.6). In this case, $\mathfrak{o}_k$ is a principal ideal domain. In particular, its maximal ideal $\{a \in k : |a| < 1\}$ is generated by one element ϖ, called a *uniformizing parameter* of k. Note that, if the absolute value $|\cdot|$ is not discrete, then $|k^\times|$ is a dense subgroup of $(\mathbb{R}_{>0}, \times)$. This results from the facts that a subgroup of $(\mathbb{R}, +)$ is either discrete or dense (cf. [32] Chapter V, §1, no.1 and §4, no.1) and that the exponential function defines an isomorphism between the topological groups $(\mathbb{R}, +)$ and $(\mathbb{R}_{>0}, \times)$.

Definition 1.1.23 Let V be a finite-dimensional vector space over k. A sub-$\mathfrak{o}_k$-module $\mathcal{V}$ of V is called a *lattice of* V if $\mathcal{V}$ generates V as a vector space over k (i.e. the natural linear map $\mathcal{V} \otimes_{\mathfrak{o}_k} k \to V$ is surjective) and $\mathcal{V}$ is bounded in V for a certain norm on V (or equivalently for any norm on V, see Proposition 1.1.11). In particular, if $\mathcal{V}$ is a sub-$\mathfrak{o}_k$-module of finite type of V, which generates V as a vector space over k, then it is a lattice in V. In this case, $\mathcal{V}$ is called a *finitely generated lattice of* V. If $\mathcal{V}$ is a lattice of V, we define a function $\|\cdot\|_{\mathcal{V}}$ on V as follows:[2]

$$\forall\, x \in V \setminus \{0\}, \quad \|x\|_{\mathcal{V}} := \inf\{|a| \,:\, a \in k^\times,\ a^{-1}x \in \mathcal{V}\}, \quad \text{and } \|0\|_{\mathcal{V}} := 0.$$

Clearly, if $\mathcal{V}$ and $\mathcal{V}'$ are lattices of V such that $\mathcal{V} \subseteq \mathcal{V}'$, then one has $\|\cdot\|_{\mathcal{V}} \geqslant \|\cdot\|_{\mathcal{V}'}$. Note that, if the absolute value $|\cdot|$ is trivial, then $\mathfrak{o}_k = k$ and the only lattice of V is V itself.

Proposition 1.1.24 *Let V be a finite-dimensional vector space over k, $\mathcal{V}$ be a lattice of V and $\|\cdot\|$ be a norm on V. Assume that $\mathcal{V}$ is contained in the unit ball of $(V, \|\cdot\|)$, then one has $\|\cdot\|_{\mathcal{V}} \geqslant \|\cdot\|$.*

Proof Let $x \in V \setminus \{0\}$ and a be an element of $k^\times$ such that $a^{-1}x \in \mathcal{V}$. One has

$$\|a^{-1}x\| = |a|^{-1} \cdot \|x\| \leqslant 1$$

[2] Note that, in the case where $|\cdot|$ is not the trivial absolute value, one has

$$\inf\{|a| \,:\, a \in k^\times,\ a^{-1}0 \in \mathcal{V}\} = 0.$$

However, this equality does not hold when $|\cdot|$ is trivial.

since $\mathcal{V}$ is contained in the unit ball of $(V, \|\cdot\|)$. Therefore $\|x\| \leqslant |a|$. Thus we deduce that

$$\|x\| \leqslant \inf\{|a| \,:\, a \in k^{\times},\, a^{-1}x \in \mathcal{V}\} = \|x\|_{\mathcal{V}}.$$

In the case where the absolute value $|\cdot|$ is non-trivial, the balls in an ultrametrically normed vector space are natural examples of lattices.

Proposition 1.1.25 *Assume that the absolute value $|\cdot|$ is non-trivial. Let V be a finite-dimensional vector space over k, equipped with an ultrametric norm $\|\cdot\|$. For any $\epsilon > 0$ the balls $V_{\leqslant\epsilon} = \{x \in V \,:\, \|x\| \leqslant \epsilon\}$ and $V_{<\epsilon} = \{x \in V \,:\, \|x\| < \epsilon\}$ are both lattices of V.*

Proof Since the norm $\|\cdot\|$ is ultrametric, both balls $V_{\leqslant\epsilon}$ and $V_{<\epsilon}$ are stable by addition. Clearly they are also stable by the multiplication by an element in $\mathfrak{o}_k$. Therefore they are sub-$\mathfrak{o}_k$-modules of V. Moreover, by definition they are bounded subsets of V. It remains to verify that they generate V as a vector space over k. It suffices to treat the open ball case. Let $\{e_i\}_{i=1}^r$ be a basis of V over k. Since the absolute value $|\cdot|$ is non-trivial, there exists a non-zero element $a \in k$ such that $|a| < 1$. For sufficiently large integer $n \in \mathbb{N}_{>0}$, one has $\|a^n e_i\| < \epsilon$ for any $i \in \{1, \ldots, r\}$. Hence $V_{<\epsilon}$ contains a basis of the vector space V. □

The following proposition shows that each lattice defines a norm on the underlying vector space.

Proposition 1.1.26 *Let V be a finite-dimensional vector space over k and $\mathcal{V}$ be a lattice of V. The map $\|\cdot\|_{\mathcal{V}}$ is an ultrametric norm on V. Moreover, $\mathcal{V}$ is contained in the unit ball of $(V, \|\cdot\|_{\mathcal{V}})$.*

Proof In the case where the absolute value $|\cdot|$ is trivial, one has $\mathcal{V} = V$ and the function $\|\cdot\|_{\mathcal{V}}$ takes value 1 on $V \setminus \{0\}$ and vanishes on $\{0\}$. The result is clearly true in this case. In the following, we assume that $|\cdot|$ is non-trivial. For any $x \in V$, let A_x be the set of all $a \in k^{\times}$ such that $a^{-1}x \in \mathcal{V}$. We claim that A_x is non-empty and hence $\|x\|_{\mathcal{V}}$ is finite. Let $\{e_i\}_{i=1}^r$ be a subset of $\mathcal{V}$ which forms a basis of V over k. We write x in the form $x = a_1e_1 + \cdots + a_re_r$ with $(a_1, \ldots, a_r) \in k^r$. Since k is the fraction field of $\mathfrak{o}_k$, there exists $b \in k^{\times}$ such that $ba_1, \ldots, ba_r$ are all in $\mathfrak{o}_k$. Thus $bx \in \mathcal{V}$ and hence $b^{-1} \in A_x$. Therefore $\|\cdot\|_{\mathcal{V}}$ is a map from V to $\mathbb{R}_{\geqslant 0}$.

Let x be an element of V and $a \in k^{\times}$. The map $b \mapsto ab$ defines a bijection between A_x and A_{ax}. Hence one has $\|ax\|_{\mathcal{V}} = |a| \cdot \|x\|_{\mathcal{V}}$.

Let x and y be elements of V, $a \in A_x$ and $b \in A_y$. One has $\{a^{-1}x, b^{-1}y\} \subseteq \mathcal{V}$. Note that

$$a^{-1}(x + y) = a^{-1}x + a^{-1}y = a^{-1}x + (a^{-1}b)(b^{-1}y),$$
$$b^{-1}(x + y) = b^{-1}x + b^{-1}y = (b^{-1}a)(a^{-1}x) + b^{-1}y.$$

Since $\mathfrak{o}_k$ is a valuation ring, either $b^{-1}a \in \mathfrak{o}_k$, or $a^{-1}b \in \mathfrak{o}_k$. Hence, either $a \in A_{x+y}$ or $b \in A_{x+y}$. Therefore $\|x + y\|_{\mathcal{V}} \leqslant \max\{\|x\|_{\mathcal{V}}, \|y\|_{\mathcal{V}}\}$.

It remains to verify that, if $\|x\|_{\mathcal{V}} = 0$ then $x = 0$. Assume that there exists a non-zero element $x \in V$ such that $\|x\|_{\mathcal{V}} = 0$. Then there exists a sequence $\{a_n\}_{n\in\mathbb{N}}$ in A_x such that $\lim_{n\to+\infty} |a_n| = 0$. However, one has $a_n^{-1}x \in \mathcal{V}$ for any $n \in \mathbb{N}$. This contradicts the assumption that $\mathcal{V}$ is bounded.

If x is an element in $\mathcal{V}$, then 1 belongs to A_x. Hence $\|x\|_{\mathcal{V}} \leqslant 1$. □

Definition 1.1.27 Let V be a finite-dimensional vector space over k and $\mathcal{V}$ be a lattice of V. We call $\|\cdot\|_{\mathcal{V}}$ the norm on V *induced by the lattice* $\mathcal{V}$.

Proposition 1.1.28 *Let V be a finite-dimensional vector space over k and r be its dimension over k. Let $\mathcal{V}$ be a lattice of V. Assume that $\mathcal{V}$ is an $\mathfrak{o}_k$-module of finite type. Then it is a free $\mathfrak{o}_k$-module of rank r.*

Proof Since $\mathcal{V}$ is a sub-$\mathfrak{o}_k$-module of V, it is torsion-free. By [27] Chapter VI, §4, no.6, Lemma 1, any torsion-free module of finite type over a valuation ring is free. Hence $\mathcal{V}$ is a free $\mathfrak{o}_k$-module. Finally, since $\mathcal{V}$ generates V as a vector space over k, any basis of $\mathcal{V}$ over $\mathfrak{o}_k$ is also a basis of V over k. Hence the rank of $\mathcal{V}$ over $\mathfrak{o}_k$ is r.□

Definition 1.1.29 Let $(V, \|\cdot\|)$ be a finite-dimensional seminormed vector space over k. We define the *default of purity* of $\|\cdot\|$ as

$$\mathrm{dpur}(\|\cdot\|) := \sup_{x\in V\setminus N_{\|\cdot\|}} \mathrm{dist}(\ln\|x\|, \ln|k^{\times}|),$$

with

$$\mathrm{dist}(\ln\|x\|, \ln|k^{\times}|) := \inf\{|\ln\|x\| - \ln|a|| \ : \ a \in k^{\times}\}.$$

We say that the seminorm $\|\cdot\|$ is *pure* if $\mathrm{dpur}(\|\cdot\|) = 0$, or equivalently, the image of $V \setminus N_{\|\cdot\|}$ by $\|\cdot\|$ is contained in the closure of $|k^{\times}|$ in $\mathbb{R}_{>0}$. By definition, if the absolute value $|\cdot|$ is not discrete, then any seminorm on V is pure; if $|\cdot|$ is discrete, then a seminorm $\|\cdot\|$ on V is pure if and only if its image is contained in that of $|\cdot|$. In the case where $|\cdot|$ is discrete, Moreover, for any lattice $\mathcal{V}$ of V, the norm $\|\cdot\|_{\mathcal{V}}$ is pure.

In the following, we study the correspondance between ultrametric norms and lattices of a finite-dimensional vector space over k. Note that the behaviour depends much on the discreteness of the absolute value $|\cdot|$.

Proposition 1.1.30 *Assume that the absolute value $|\cdot|$ is discrete.*

(1) *For any lattice $\mathcal{V}$ of V, one has $(V, \|\cdot\|_{\mathcal{V}})_{\leqslant 1} = \mathcal{V}$ (see Notation 1.1.3).*
(2) *Any lattice $\mathcal{V}$ of V is a free $\mathfrak{o}_k$-module of rank* $\dim_k(V)$.
(3) *Assume in addition that the absolute value $|\cdot|$ is non-trivial. Let $\|\cdot\|$ be an ultrametric norm on V and let $\mathcal{V} = (V, \|\cdot\|)_{\leqslant 1}$. Then one has $\|\cdot\| \leqslant \|\cdot\|_{\mathcal{V}} \leqslant |\varpi|^{-1}\|\cdot\|$, where ϖ is a uniformizing parameter of k. In particular, the default of purity of $\|\cdot\|$ is bounded from above by $-\ln|\varpi|$. Moreover, if the norm $\|\cdot\|$ is pure, then $\|\cdot\|_{\mathcal{V}} = \|\cdot\|$.*

Proof (1) By Proposition 1.1.26, one has $\mathcal{V} \subseteq (V, \|\cdot\|_{\mathcal{V}})_{\leqslant 1}$. Let x be an element of V such that $\|x\|_{\mathcal{V}} \leqslant 1$. In order to see that $x \in \mathcal{V}$, we may assume that $x \neq 0$. There is a sequence $\{\alpha_n\}_{n\in\mathbb{N}}$ in $k^\times$ such that $\alpha_n^{-1}x \in \mathcal{V}$ and $\lim_{n\to\infty} |\alpha_n| = \|x\|_{\mathcal{V}}$. As $|\cdot|$ is discrete, there is $n \in \mathbb{N}$ such that $|\alpha_n| = \|x\|_{\mathcal{V}}$, so that $\alpha_n \in \mathfrak{o}_k$ because $\|x\|_{\mathcal{V}} \leqslant 1$. Therefore, $x \in \alpha_n\mathcal{V} \subseteq \mathcal{V}$, and hence $(V, \|\cdot\|_{\mathcal{V}})_{\leqslant 1} \subseteq \mathcal{V}$.

(2) Let $\{e_i\}_{i=1}^r$ be a basis of V over k. We equip V with the norm $\|\cdot\|$ such that

$$\|\lambda_1 e_1 + \cdots + \lambda_r e_r\| = \max\{|\lambda_1|, \ldots, |\lambda_r|\}$$

for any $(\lambda_1, \ldots, \lambda_r) \in k^r$. For any $\epsilon > 0$, the ball

$$(V, \|\cdot\|)_{\leqslant\epsilon} = \{a \in k \,:\, |a| \leqslant \epsilon\}^r$$

is a free $\mathfrak{o}_k$-module of rank r since $\mathfrak{o}_k$ is a principal ideal domain. Let $\mathcal{V}$ be a lattice. Since it is bounded, it is contained in certain ball $(V, \|\cdot\|)_{\leqslant\epsilon}$. Thus $\mathcal{V}$ is an $\mathfrak{o}_k$-module of finite type, and hence a free $\mathfrak{o}_k$-module of rank r by Proposition 1.1.28.

(3) By the definition of the uniformizing element, one has $|k^\times| = \{|\varpi|^n \,:\, n \in \mathbb{Z}\}$. If x is a non-zero element in V and if A_x is the set of all $a \in k^\times$ such that

$$\|a^{-1}x\| = |a|^{-1} \cdot \|x\| \leqslant 1,$$

then one has

$$\{|a| \,:\, a \in A_x\} = \{|\varpi|^n \,:\, n \in \mathbb{Z},\ |\varpi|^n \geqslant \|x\|\}.$$

Since $\|x\|_{\mathcal{V}} = \inf\{|a| \,:\, a \in A_x\}$, one has $\|x\|_{\mathcal{V}} \geqslant \|x\| > |\varpi| \cdot \|x\|_{\mathcal{V}}$. Combined with the fact that the norm $\|\cdot\|_{\mathcal{V}}$ is pure, this implies the inequality $\mathrm{dpur}(\|\cdot\|) \leqslant -\ln|\varpi|$. If in addition the norm $\|\cdot\|$ is pure, $\|x\|_{\mathcal{V}}$ belongs to $\{|\varpi|^n \,:\, n \in \mathbb{Z}\}$. Hence $\|x\|_{\mathcal{V}} = \|x\|$. □

Remark 1.1.31 Let V be a finite-dimensional vector space over k. We denote by $\mathrm{Lat}(V)$ the set of all lattices of V, and by $\mathrm{Nor}(V)$ that of all ultrametric norms on V. The correspondance $(\mathcal{V} \in \mathrm{Lat}(V)) \mapsto \|\cdot\|_{\mathcal{V}}$ defines a map from $\mathrm{Lat}(V)$ to $\mathrm{Nor}(V)$. Proposition 1.1.30 shows that, if the absolute value $|\cdot|$ is discrete, then this map is injective, and its image is precisely the set of all pure ultrametric norms.

Proposition 1.1.32 *Assume that the absolute value $|\cdot|$ is not discrete.*

(1) *For any lattice $\mathcal{V}$ of V one has $(V, \|\cdot\|_{\mathcal{V}})_{<1} \subseteq \mathcal{V} \subseteq (V, \|\cdot\|_{\mathcal{V}})_{\leqslant 1}$. If in addition there exists an ultrametric norm $\|\cdot\|$ on V such that $\mathcal{V} = (V, \|\cdot\|)_{\leqslant 1}$, then one has $\mathcal{V} = (V, \|\cdot\|_{\mathcal{V}})_{\leqslant 1}$.*

(2) *Let $\|\cdot\|$ be an ultrametric norm on V and $\mathcal{V} = (V, \|\cdot\|)_{\leqslant 1}$. Then $\|\cdot\| = \|\cdot\|_{\mathcal{V}}$.*

Proof (1) If x is an element of $\mathcal{V}$, by the relation $1x = x \in \mathcal{V}$ we obtain that $\|x\|_{\mathcal{V}} \leqslant 1$. Hence $\mathcal{V} \subseteq (V, \|\cdot\|_{\mathcal{V}})_{\leqslant 1}$. In the following, we prove the inclusion relation $(V, \|\cdot\|_{\mathcal{V}})_{<1} \subseteq \mathcal{V}$. Let x be an element in V such that $\|x\|_{\mathcal{V}} < 1$. By definition there exists $a \in k^\times$, $|a| < 1$, such that $a^{-1}x \in \mathcal{V}$. Since $|a| < 1$ one has $a \in \mathfrak{o}_k$. Therefore $x = a(a^{-1}x) \in \mathcal{V}$.

The second assertion of (1) is a direct consequence of (2). In the following, we prove the statement (2). Let x be an element of V and

$$A_x = \{a \in k^\times \,:\, a^{-1}x \in \mathcal{V}\} = \{a \in k^\times \,:\, \|x\| \leqslant |a|\}.$$

Since the image of $|\cdot|$ is dense in $\mathbb{R}_+$, one has $\|x\|_{\mathcal{V}} = \inf\{|a| \,:\, a \in A_x\} = \|x\|$. Hence $\|\cdot\|_{\mathcal{V}} = \|\cdot\|$. □

Remark 1.1.33 Let V be a finite-dimensional vector space over k. Proposition 1.1.32 shows that, if the absolute value $|\cdot|$ is not discrete, the map $\mathrm{Lat}(V) \to \mathrm{Nor}(V)$, sending any lattice $\mathcal{V}$ of V to the norm $\|\cdot\|_{\mathcal{V}}$, is surjective (compare with Remark 1.1.31).

Proposition 1.1.34 *Let V be a finite-dimensional vector space over k and $\mathcal{V}$ be a lattice of V. Let $\mathcal{V}^\vee = \mathrm{Hom}_{\mathfrak{o}_k}(\mathcal{V}, \mathfrak{o}_k)$ be the dual $\mathfrak{o}_k$-module of $\mathcal{V}$. Then one has $\|\cdot\|_{\mathcal{V},*} = \|\cdot\|_{\mathcal{V}^\vee}$ on $V^\vee$.*

Proof Let f be a non-zero element of $V^\vee$. Assume that a is an element of $k^\times$ such that $a^{-1}f \in \mathcal{V}^\vee$. Then for any $x \in V$ and any $b \in k^\times$ such that $b^{-1}x \in \mathcal{V}$ one has $a^{-1}f(b^{-1}x) = (ab)^{-1}f(x) \in \mathfrak{o}_k$ and hence $|b| \geqslant |f(x)|/|a|$. Since b is arbitrary one has $\|x\|_{\mathcal{V}} \geqslant |f(x)|/|a|$ for any $x \in V$ and hence $|a| \geqslant \|f\|_{\mathcal{V},*}$. Since a is arbitrary we obtain $\|f\|_{\mathcal{V}^\vee} \geqslant \|f\|_{\mathcal{V},*}$.

Conversely, suppose that the operator norm of a non-zero linear form $f : V \to k$ is bounded from above by 1, where we consider the norm $\|\cdot\|_{\mathcal{V}}$ on V. Then for any $x \in \mathcal{V}$ one has $|f(x)| \leqslant \|x\|_{\mathcal{V}} \leqslant 1$ and hence $f(x) \in \mathfrak{o}_k$. This shows that $f \in \mathcal{V}^\vee$ and hence $\|f\|_{\mathcal{V}^\vee} \leqslant 1$. Therefore the unit ball of $\|\cdot\|_{\mathcal{V}^\vee}$ contains that of $\|\cdot\|_{\mathcal{V},*}$. Moreover, since the norm $\|\cdot\|_{\mathcal{V}}$ is pure, also is its dual norm $\|\cdot\|_{\mathcal{V},*}$. Therefore, the norm $\|\cdot\|_{\mathcal{V},*}$ coincides with the norm induced by its unit ball (see Propositions 1.1.30 and 1.1.32). Therefore, $\|\cdot\|_{\mathcal{V}^\vee} \leqslant \|\cdot\|_{\mathcal{V},*}$. The proposition is thus proved. □

1.1.8 Trivial valuation case

In this subsection, we study ultrametrically normed vector spaces over a trivially valued field. We fix a field k equipped with the trivial absolute value $|\cdot|$. If V is a vector space over k, we denote by $\Theta(V)$ the set of all non-zero vector subspaces of V. The set $\Theta(V)$ is equipped with the partial order of inclusion. If $\|\cdot\|$ is an ultrametric norm on V, we denote by $\Psi(V, \|\cdot\|)$ the set of closed balls of V (centered at the origin) which do not reduce to one point, namely (see Notation 1.1.3)

$$\Psi(V, \|\cdot\|) = \Big\{(V, \|\cdot\|)_{\leqslant r} \,:\, r > 0,\ (V, \|\cdot\|)_{\leqslant r} \neq \{0\}\Big\}.$$

Proposition 1.1.35 *Let V be a finite-dimensional vector space equipped with an ultrametric norm $\|\cdot\|$. The set $\Psi(V, \|\cdot\|)$ is a totally ordered subset of $\Theta(V)$, whose cardinal does not exceed the dimension of V over k.*

Proof By definition the set $\Psi(V, \|\cdot\|)$ is totally ordered with respect to the partial order of inclusion. In the following, we show that any element $W \in \Psi(V, \|\cdot\|)$ is a vector subspace of V and hence belongs to $\Theta(V)$. Assume that $W = (V, \|\cdot\|)_{\leqslant r}$ with $r > 0$. Since the absolute value on k is trivial, for any $x \in W$ and any $a \in k$

one has $\|ax\| \leqslant \|x\| \leqslant r$. Moreover, since the norm $\|\cdot\|$ is ultrametric, W is stable by addition. Hence $\Psi(V, \|\cdot\|)$ is a totally ordered subset of $\Theta(V)$. In particular, the function $\dim_k(\cdot) : \Psi(V, \|\cdot\|) \to \mathbb{N}_{\geqslant 1}$ is injective, which is bounded from above by $\dim_k(V)$. Therefore the cardinal of $\Psi(V, \|\cdot\|)$ does not exceed $\dim_k(V)$. □

The above proposition shows that the set $\Psi(V, \|\cdot\|)$ actually forms an increasing flag of non-zero vector subspaces of V. For any $W \in \Psi(V, \|\cdot\|)$, let

$$\varphi_{\|\cdot\|}(W) := \sup\{t \in \mathbb{R} : W \subseteq (V, \|\cdot\|)_{\leqslant e^{-t}}\}.$$

Then $\varphi_{\|\cdot\|}$ is a strictly decreasing function on $\Psi(V, \|\cdot\|)$ in the sense that, if W_1 and W_2 are two elements of $\Psi(V, \|\cdot\|)$ such that $W_1 \subsetneq W_2$, then one has $\varphi_{\|\cdot\|}(W_1) > \varphi_{\|\cdot\|}(W_2)$. The following proposition shows that the norm $\|\cdot\|$ is completely determined by the increasing flag $\Psi(V, \|\cdot\|)$ and the function $\varphi_{\|\cdot\|}$.

Proposition 1.1.36 *Let Ψ be a totally ordered subset of $\Theta(V)$ and $\varphi : (\Psi, \supseteq) \to (\mathbb{R}, \leqslant)$ be a function which preserves strictly the orders, that is, for any $(W_1, W_2) \in \Psi^2$ with $W_1 \subsetneq W_2$, one has $\varphi(W_1) > \varphi(W_2)$. Then there exists a unique ultrametric norm $\|\cdot\|$ on V such that $\Psi(V, \|\cdot\|) = \Psi$ and $\varphi_{\|\cdot\|} = \varphi$.*

Proof We write Ψ in the form of an increasing flag $V_1 \subsetneq \ldots \subsetneq V_n$. For $i \in \{1, \ldots, n\}$, let $a_i = \varphi(V_i)$. Since φ preserves strictly the orders, one has $a_1 > \ldots > a_n$. Let $\boldsymbol{e} = \{e_j\}_{j=1}^m$ be a basis of V which is compatible with the flag Ψ (namely $\operatorname{card}(\mathbf{e} \cap V_i) = \operatorname{rk}_k(V_i)$ for any $i \in \{1, \ldots, n\}$). For any $j \in \{1, \ldots, m\}$, there exists a unique $i \in \{1, \ldots, n\}$ such that $e_j \in V_i \setminus V_{i-1}$ (where $V_0 = \{0\}$ by convention) and we let $r_j = e^{-a_i}$. Let $\|\cdot\|$ be the ultrametric norm on V defined as

$$\forall (\lambda_1, \ldots, \lambda_m) \in k^m, \quad \|\lambda_1 e_1 + \cdots + \lambda_m e_m\| = \max_{\substack{j \in \{1, \ldots, m\} \\ \lambda_j \neq 0}} r_j.$$

Note that for $r \geqslant 0$ the ball $(V, \|\cdot\|)_{\leqslant r}$ identifies with the vector subspace generated by those e_j with $r_j \leqslant r$. Hence one has $\Psi(V, \|\cdot\|) = \Psi$. Moreover, for any $i \in \{1, \ldots, n\}$ one has

$$\varphi_{\|\cdot\|}(V_i) = \sup\{t \in \mathbb{R} : V_i \subseteq (V, \|\cdot\|)_{\leqslant e^{-t}}\} = a_i = \varphi(V_i).$$

Let $\|\cdot\|'$ be another ultrametric norm on V verifying the relations $\Psi(V, \|\cdot\|') = \Psi$ and $\varphi_{\|\cdot\|'} = \varphi$. For any $r \geqslant 0$, $(V, \|\cdot\|')_{\leqslant r} = V_i$ if and only if $r \in [e^{-a_i}, e^{-a_{i+1}}[$, with the convention $a_0 = +\infty$ and $a_{n+1} = -\infty$. Therefore one has $(V, \|\cdot\|)_{\leqslant r} = (V, \|\cdot\|')_{\leqslant r}$ for any $r \geqslant 0$, which leads to $\|\cdot\| = \|\cdot\|'$. □

Definition 1.1.37 Let V be a finite-dimensional vector space over k. A family $\mathcal{F} = \{\mathcal{F}^t(V)\}_{t \in \mathbb{R}}$ of vector subspaces of V parametrised by $\mathbb{R}$ is called an $\mathbb{R}$-*filtration* of V if it is separated ($\mathcal{F}^t(V) = \{0\}$ for sufficiently positive t), exhaustive ($\mathcal{F}^t(V) = V$ for sufficiently negative t) and left-continuous (the function $(t \in \mathbb{R}) \to \dim_k(\mathcal{F}^t(V))$ is left-continuous).

Definition 1.1.38 Let V be a finite-dimensional vector space over k and $\mathcal{F}$ be an $\mathbb{R}$-filtration on V. Let r be the dimension of V over k. We define a map $Z_{\mathcal{F}} : \{1, \ldots, r\} \to \mathbb{R}$ as follows:

$$\forall\, i \in \{1, \ldots, r\}, \quad Z_{\mathcal{F}}(i) := \sup\{t \in \mathbb{R} : \dim_k(\mathcal{F}^t(V)) \geqslant i\}.$$

By definition, for any $t \in \mathbb{R}$ and any $i \in \{1, \ldots, r\}$ one has

$$Z_{\mathcal{F}}(i) \geqslant t \Longleftrightarrow \dim_k(\mathcal{F}^t(V)) \geqslant i. \tag{1.7}$$

Proposition 1.1.39 *Let V be a finite-dimensional non-zero vector space over k and $\mathcal{F}$ and $\mathcal{G}$ be $\mathbb{R}$-filtrations on V. Let $a \in \mathbb{R}$ such that, for any $t \in \mathbb{R}$ one has $\mathcal{F}^t(V) \subseteq \mathcal{G}^{t-a}(V)$. Then one has $Z_{\mathcal{F}}(i) \leqslant Z_{\mathcal{G}}(i) + a$ for any $i \in \{1, \ldots, \dim_k(V)\}$.*

Proof By the relation (1.7), for any $i \in \{1, \ldots, \dim_k(V)\}$, if $Z_{\mathcal{F}}(i) \geqslant t$, then $\dim_k(\mathcal{F}^t(V)) \geqslant i$, which implies that $\dim_k(\mathcal{G}^{t-a}(V)) \geqslant i$ and hence (still by the relation (1.7)) $Z_{\mathcal{G}}(i) \geqslant t - a$. Therefore we obtain $Z_{\mathcal{F}}(i) - a \leqslant Z_{\mathcal{G}}(i)$. □

Remark 1.1.40 Let V be a finite-dimensional vector space over k. There are canonical bijections between the following three sets:

(A) the set of all pairs

$$(0 = V_0 \subsetneq V_1 \subsetneq \ldots \subsetneq V_n = V,\ \mu_1 > \ldots > \mu_n)$$

such that $0 = V_0 \subsetneq V_1 \subsetneq \ldots \subsetneq V_n = V$ is an increasing sequence of vector subspaces of V and $\mu_1 > \ldots > \mu_n$ is a decreasing sequence of real numbers.

(B) the set of all $\mathbb{R}$-filtrations $\mathcal{F}$ of V.

(C) the set of all ultrametric norms $\|\cdot\|$ of V over k.

In the following, we explain the construction of these canonical maps.

• (A)→(B): The associated $\mathbb{R}$-filtration $\mathcal{F}$ on V with the data $(V_0 \subsetneq \ldots \subsetneq V_n,\ \mu_1 > \ldots > \mu_n)$ is defined by $\mathcal{F}^t(V) := V_i$ if $t \in \,]\mu_{i+1}, \mu_i] \cap \mathbb{R}$, where $\mu_0 = +\infty$ and $\mu_{n+1} = -\infty$ by convention.

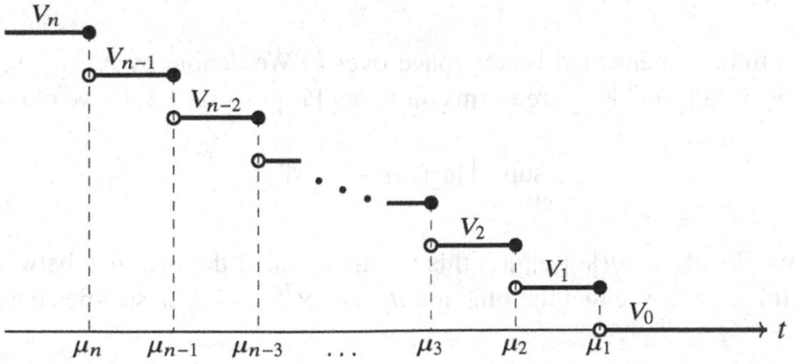

• (B)→(A): One has a sequence $0 = V_0 \subsetneq V_1 \subsetneq \ldots \subsetneq V_n = V$ such that $\{\mathcal{F}^t(V) : t \in \mathbb{R}\} = \{V_0, V_1, \ldots, V_n\}$. A sequence $\mu_1 > \ldots > \mu_n$ in $\mathbb{R}$ is given by $\mu_i = \sup\{t : \mathcal{F}^t(V) = V_i\}$ for $i \in \{1, \ldots, n\}$.

• (A)→(C): The corresponding norm $\|\cdot\|$ to the data $(V_0 \subsetneq \ldots \subsetneq V_n,\ \mu_1 > \ldots > \mu_n)$ is given by

$$\|x\| = \begin{cases} e^{-\mu_i} & \text{if } x \in V_i \setminus V_{i-1}, \\ 0 & \text{if } x = 0. \end{cases}$$

• (C)→(A): By Proposition 1.1.35, there is an increasing sequence

$$0 = V_0 \subsetneq V_1 \subsetneq \ldots \subsetneq V_n = V$$

of subspaces of V such that $\Psi(V, \|\cdot\|) = \{V_1, \ldots, V_n\}$. A decreasing sequence of real numbers is given by $\mu_i = \varphi_{\|\cdot\|}(V_i)$ for $i \in \{1, \ldots, n\}$.

• (B)→(C): We define a function $\lambda_{\mathcal{F}} : V \to \mathbb{R} \cup \{+\infty\}$ such that

$$\forall\, x \in V, \quad \lambda_{\mathcal{F}}(x) := \sup\{t \in \mathbb{R} \,:\, x \in \mathcal{F}^t(V)\}.$$

Then the ultrametric norm $\|\cdot\|$ on V corresponding to $\mathcal{F}$ is given by

$$\forall\, x \in V, \quad \|x\| = e^{-\lambda_{\mathcal{F}}(x)}.$$

• (C)→(B): The corresponding $\mathbb{R}$-filtration $\mathcal{F}$ to the norm $\|\cdot\|$ is given by $\mathcal{F}^t(V) = (V, \|\cdot\|)_{\leqslant e^{-t}}$.

Let $\mathcal{F}$ be an $\mathbb{R}$-filtration on V, which corresponds to an increasing flag $0 = V_0 \subsetneq V_1 \subsetneq \ldots \subsetneq V_n = V$ together with a decreasing sequence $\mu_1 > \ldots > \mu_n$ of real numbers. Note that the sets $\{\mu_1, \ldots, \mu_n\}$ and $\{Z_{\mathcal{F}}(1), \ldots, Z_{\mathcal{F}}(r)\}$ are actually equal, where r denotes the dimension of V over k. Moreover, the value μ_i appears exactly $\dim_k(V_i/V_{i-1})$ times in the sequence $Z_{\mathcal{F}}(1), \ldots, Z_{\mathcal{F}}(r)$.

1.1.9 Metric on the space of norms

Let V be a finite-dimensional vector space over k. We denote by $\mathcal{N}_V$ the set of all norms on V. If $\|\cdot\|_1$ and $\|\cdot\|_2$ are norms on V, by Proposition 1.1.11 we obtain that

$$\sup_{s \in V \setminus \{0\}} \Big| \ln \|s\|_1 - \ln \|s\|_2 \Big|$$

is finite. We denote by $d(\|\cdot\|_1, \|\cdot\|_2)$ this number, called the *distance* between $\|\cdot\|_1$ and $\|\cdot\|_2$. It is easy to see that the function $d : \mathcal{N}_V \times \mathcal{N}_V \to \mathbb{R}_{\geqslant 0}$ satisfies the axioms of metric.

Remark 1.1.41 Let V be a finite-dimensional vector space over k and $\|\cdot\|_0$ be a norm on V, which is the trivial norm if the absolute value $|\cdot|$ is trivial (namely $\|x\|_0 = 1$ for any $x \in V \setminus \{0\}$). Let λ be a real number in $]0, 1[$. If the absolute value $|\cdot|$ is non-trivial, we require in addition that $\lambda < \sup\{|a| \,:\, a \in k^{\times},\ |a| < 1\}$. We denote by C the annulus $\{x \in V \,:\, \lambda \leqslant \|x\|_0 \leqslant 1\}$. Note that one has $C = V \setminus \{0\}$ when $|\cdot|$ is trivial. For any norm $\|\cdot\|$ on V, the restriction of the function $\ln\|\cdot\|$ to C is bounded, and the norm $\|\cdot\|$ is uniquely determined by its restriction to C (this is a consequence of Proposition 1.1.4 when $|\cdot|$ is non-trivial). Thus we can identify $\mathcal{N}_V$ with a closed subset of $C_b(C)$, the space of bounded and continuous functions on C equipped with the sup norm. In particular, $\mathcal{N}_V$ is a complete metric space.

Proposition 1.1.42 *Let V be a finite-dimensional vector space over k, and $\|\cdot\|_1$ and $\|\cdot\|_2$ be norms on V.*

(1) *Let U be a vector subspace of V, $\|\cdot\|_{U,1}$ and $\|\cdot\|_{U,2}$ be the restrictions of $\|\cdot\|_1$ and $\|\cdot\|_2$ to U, respectively. Then one has $d(\|\cdot\|_{U,1}, \|\cdot\|_{U,2}) \leqslant d(\|\cdot\|_1, \|\cdot\|_2)$.*
(2) *Let W be a quotient vector space of V, $\|\cdot\|_{W,1}$ and $\|\cdot\|_{W,2}$ be quotient norms of $\|\cdot\|_1$ and $\|\cdot\|_2$ on W, respectively. Then one has $d(\|\cdot\|_{W,1}, \|\cdot\|_{W,2}) \leqslant d(\|\cdot\|_1, \|\cdot\|_2)$.*

Proof (1) follows directly from the definition of the distance function.

(2) It is sufficient to show that

$$\Big|\ln \|x\|_{W,1} - \ln \|x\|_{W,2}\Big| \leqslant d(\|\cdot\|_1, \|\cdot\|_2).$$

for $x \in W \setminus \{0\}$. Clearly we may assume that $\|x\|_{W,1} > \|x\|_{W,2}$. For $\epsilon > 0$, one can choose $s \in V$ such that $[s] = x$ and $\|s\|_2 \leqslant \mathrm{e}^{\epsilon}\|x\|_{2,W}$. Then

$$\begin{aligned} 0 < \ln \|x\|_{W,1} - \ln \|x\|_{W,2} &\leqslant \ln \|s\|_1 - \ln\left(\mathrm{e}^{-\epsilon}\|s\|_2\right) \\ &= (\ln \|s\|_1 - \ln \|s\|_2) + \epsilon \leqslant d(\|\cdot\|_1, \|\cdot\|_2) + \epsilon, \end{aligned}$$

as desired. □

Proposition 1.1.43 *Let V and W be finite-dimensional vector spaces over k, $\|\cdot\|_{V,1}$ and $\|\cdot\|_{V,2}$ be norms on V, and $\|\cdot\|$ be a norm on W. Let $\|\cdot\|_1$ and $\|\cdot\|_2$ be the operator norms on $\mathscr{L}(V, W)$, where we consider the norm $\|\cdot\|_W$ on W, and the norms $\|\cdot\|_{V,1}$ and $\|\cdot\|_{V,2}$ on V, respectively. Then one has*

$$d(\|\cdot\|_1, \|\cdot\|_2) \leqslant d(\|\cdot\|_{V,1}, \|\cdot\|_{V,2}). \tag{1.8}$$

In particular, one has

$$d(\|\cdot\|_{V,1,*}, \|\cdot\|_{V,2,*}) \leqslant d(\|\cdot\|_{V,1}.\|\cdot\|_{V,2}), \tag{1.9}$$

Moreover, the equality in (1.9) *holds when both norms $\|\cdot\|_{V,1}$ and $\|\cdot\|_{V,2}$ are reflexive.*

Proof For (1.8), it is sufficient to show

$$\Big|\ln \|f\|_1 - \ln \|f\|_2\Big| \leqslant d(\|\cdot\|_{V,1}, \|\cdot\|_{V,2})$$

for $f \in \mathscr{L}(V, W) \setminus \{0\}$. Clearly we may assume that $\|f\|_1 > \|f\|_2$. By definition one has

$$\|f\|_1 = \sup_{x \in V\setminus\{0\}} \frac{\|f(x)\|_W}{\|x\|_{V,1}} \quad \text{and} \quad \|f\|_2 = \sup_{x \in V\setminus\{0\}} \frac{\|f(x)\|_W}{\|x\|_{V,2}},$$

so that, for $\epsilon > 0$, one can find $x \in V \setminus \{0\}$ such that $\mathrm{e}^{-\epsilon}\|f\|_1 \leqslant \|f(x)\|_W/\|x\|_{V,1}$. Therefore,

$$\begin{aligned}0 < \ln\|f\|_1 - \ln\|f\|_2 &\leqslant \ln\left(\mathrm{e}^{\epsilon}\frac{\|f(x)\|_W}{\|x\|_{V,1}}\right) - \ln\left(\frac{\|f(x)\|_W}{\|x\|_{V,2}}\right)\\ &= (\ln\|x\|_{V,2} - \ln\|x\|_{V,1}) + \epsilon \leqslant d(\|\cdot\|_{V,1}, \|\cdot\|_{V,2}) + \epsilon,\end{aligned}$$

as desired.

In order to obtain (1.9), it suffices to apply (1.8) to the case where $(W, \|\cdot\|) = (k, |\cdot|)$. If in addition both norms $\|\cdot\|_{V,1}$ and $\|\cdot\|_{V,2}$ are reflexive, then one has

$$d(\|\cdot\|_{V,1,*}, \|\cdot\|_{V,2,*}) \geqslant d(\|\cdot\|_{V,1,**}, \|\cdot\|_{V,2,**}) = d(\|\cdot\|_{V,1}, \|\cdot\|_{V,2}).$$

Hence the equality holds. □

1.1.10 Direct sums

Let $\mathscr{S}$ be the set of all convex and continuous functions $\psi : [0,1] \to [0,1]$ such that $\max\{t, 1-t\} \leqslant \psi(t)$ for any $t \in [0,1]$.

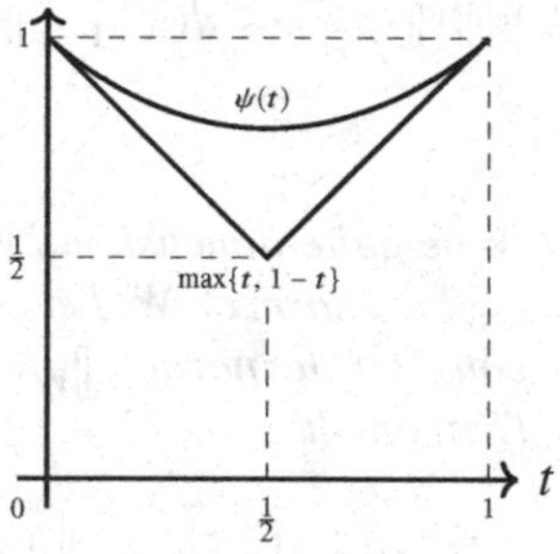

Let $\|\cdot\|$ be a norm on $\mathbb{R}^2$, where we consider the usual absolute value $|\cdot|_\infty$ on $\mathbb{R}$. We say that $\|\cdot\|$ is an *absolute normalised norm* if $\|(1,0)\| = \|(0,1)\| = 1$ and if

$$\forall\,(x,y) \in \mathbb{R}^2, \quad \|(x,y)\| = \|(|x|_\infty, |y|_\infty)\|.$$

By [14, §21, Lemma 3], the set of all absolute normalised norms on $\mathbb{R}^2$ can be parametrised by the functional space $\mathscr{S}$ (see [133] for the higher dimensional generalisation of this result). If $\|\cdot\|$ is the absolute normalised norm corresponding to $\psi \in \mathscr{S}$, one has

$$\|(x,y)\| = (|x| + |y|)\psi\Big(\frac{|x|}{|x|+|y|}\Big).$$

In particular, one always has

$$\|(x,y)\| \geqslant \max(|x|_\infty, |y|_\infty) \qquad (1.10)$$

Conversely, given an absolute normalised norm $\|\cdot\|$ on $\mathbb{R}^2$, the corresponding function in $\mathscr{S}$ is

$$(t \in [0,1]) \longmapsto \|(t, 1-t)\|. \tag{1.11}$$

For example, the function $\psi(t) = \max\{t, 1-t\}$, $t \in [0,1]$ corresponds to the norm $(x,y) \mapsto \max\{|x|_\infty, |y|_\infty\}$ on $\mathbb{R}^2$. If $p \geqslant 1$ is a real number, the function $\psi_p(t) = (t^p + (1-t)^p)^{1/p}$, $t \in [0,1]$ belongs to $\mathscr{S}$; it corresponds to the ℓ^p-norm $(x,y) \mapsto (|x|^p + |y|^p)^{1/p}$.

Given a function ψ in $\mathscr{S}$, or equivalently an absolute normalised norm on $\mathbb{R}^2$, for any couple of finite-dimensional seminormed vector spaces over k, one can naturally attach to the direct sum of the vector spaces a direct sum seminorm, which depends on the function ψ.

Lemma 1.1.44 *Let a, b, a' and b' be real numbers such that $0 \leqslant a \leqslant a'$ and $0 \leqslant b \leqslant b'$. We assume in addition that $a + b > 0$. If ψ is a function in $\mathscr{S}$, then*

$$(a+b)\psi\Big(\frac{a}{a+b}\Big) \leqslant (a'+b')\psi\Big(\frac{a'}{a'+b'}\Big). \tag{1.12}$$

Proof For any $t \in [0,1]$, the value $\psi(t)$ is bounded from below by t. Moreover, one has $\psi(1) = 1$. The function $t \mapsto \psi(t)/t$ on $]0,1]$ is non-increasing. In fact, for $0 < s \leqslant t$, by the convexity of the function ψ one has

$$\begin{aligned}\psi(t) &= \psi\Big(\frac{t-s}{1-s} + \frac{1-t}{1-s}s\Big) \leqslant \frac{t-s}{1-s}\psi(1) + \frac{1-t}{1-s}\psi(s)\\ &\leqslant \frac{t-s}{1-s}\cdot\frac{\psi(s)}{s} + \frac{1-t}{1-s}\psi(s) = \frac{t}{s}\psi(s).\end{aligned}$$

In particular, one has

$$(a+b)\psi\Big(\frac{a}{a+b}\Big) \leqslant (a+b')\psi\Big(\frac{a}{a+b'}\Big).$$

Moreover, the function from $[0,1]$ to itself sending $t \in [0,1]$ to $\psi(1-t)$ also belongs to $\mathscr{S}$. By the above argument, we obtain that the function $t \mapsto \psi(1-t)/t$ is also non-increasing. Therefore

$$(a+b')\psi\Big(\frac{a}{a+b'}\Big) = (a+b')\psi\Big(1 - \frac{b'}{a+b'}\Big) \leqslant (a'+b')\psi\Big(1 - \frac{b'}{a'+b'}\Big).$$

The inequality (1.12) is thus proved. □

Proposition 1.1.45 *Let $(V, \|\cdot\|_V)$ and $(W, \|\cdot\|_W)$ be finite-dimensional seminormed vector spaces over k. For any $\psi \in \mathscr{S}$, let $\|\cdot\|_\psi : V \oplus W \to \mathbb{R}_{\geqslant 0}$ be the map such that $\|(v,w)\|_\psi = 0$ for $(v,w) \in N_{\|\cdot\|_V} \oplus N_{\|\cdot\|_W}$ and that, for any $(x,y) \in (V \oplus W) \setminus (N_{\|\cdot\|_V} \oplus N_{\|\cdot\|_W})$,*

$$\|(x,y)\|_\psi := (\|x\| + \|y\|)\psi\Big(\frac{\|x\|}{\|x\| + \|y\|}\Big).$$

Then $\|\cdot\|_\psi$ *is a seminorm on* $V \oplus W$ *such that* $N_{\|\cdot\|_\psi} = N_{\|\cdot\|_V} \oplus N_{\|\cdot\|_W}$. *Moreover, for any* $(x, y) \in V \times W$ *one has*

$$\max\{\|x\|, \|y\|\} \leqslant \|(x, y)\|_\psi \leqslant \|x\| + \|y\|. \tag{1.13}$$

Proof By definition, for any $(x, y) \in V \oplus W$ and any $a \in k$, one has

$$\|(ax, ay)\|_\psi = |a| \cdot \|(x, y)\|_\psi.$$

Moreover, for $(x, y) \notin N_{\|\cdot\|_V} \oplus N_{\|\cdot\|_W}$, one has $\|(x, y)\|_\psi > 0$. Thus it remains to verify the triangle inequality.

Let (x_1, y_1) and (x_2, y_2) be two elements in $V \oplus W$ such that $(x_1 + x_2, y_1 + y_2)$ does not belong to $N_{\|\cdot\|_V} \oplus N_{\|\cdot\|_W}$. One has

$$\|(x_1 + x_2, y_1 + y_2)\|_\psi = (\|x_1 + x_2\| + \|y_1 + y_2\|)\psi(u),$$

where

$$u = \frac{\|x_1 + x_2\|}{\|x_1 + x_2\| + \|y_1 + y_2\|}.$$

Since $\|x_1 + x_2\| \leqslant \|x_1\| + \|x_2\|$ and $\|y_1 + y_2\| \leqslant \|y_1\| + \|y_2\|$, by Lemma 1.1.44 one obtains that $\|(x_1 + x_2, y_1 + y_2)\|_\psi$ is bounded from above by

$$(\|x_1\| + \|x_2\| + \|y_1\| + \|y_2\|)\psi(v),$$

with

$$v = \frac{\|x_1\| + \|x_2\|}{\|x_1\| + \|x_2\| + \|y_1\| + \|y_2\|} = \left(1 + \frac{\|y_1\| + \|y_2\|}{\|x_1\| + \|x_2\|}\right)^{-1}$$

if $\|x_1\| + \|x_2\| > 0$, and $v = 0$ otherwise. If $\|x_1\| > 0$, let

$$s = \frac{\|x_1\|}{\|x_1\| + \|y_1\|} = \left(1 + \frac{\|y_1\|}{\|x_1\|}\right)^{-1},$$

otherwise let $s = 0$. Similarly, if $\|x_2\| > 0$, let

$$t = \frac{\|x_2\|}{\|x_2\| + \|y_2\|} = \left(1 + \frac{\|y_2\|}{\|x_2\|}\right)^{-1},$$

otherwise let $t = 0$. In the case where $\|x_1\|$ and $\|x_2\|$ are both > 0, one has

$$\min\left\{\frac{\|y_1\|}{\|x_1\|}, \frac{\|y_2\|}{\|x_2\|}\right\} \leqslant \frac{\|y_1\| + \|y_2\|}{\|x_1\| + \|x_2\|} \leqslant \max\left\{\frac{\|y_1\|}{\|x_1\|}, \frac{\|y_2\|}{\|x_2\|}\right\},$$

and therefore $\min\{s, t\} \leqslant v \leqslant \max\{s, t\}$. By the convexity of the function ψ we obtain

$$\psi(v) \leqslant \frac{v - t}{s - t}\psi(s) + \frac{s - v}{s - t}\psi(t).$$

Note that

$$\frac{v-t}{s-t}=\frac{\|x_1\|+\|y_1\|}{\|x_1\|+\|x_2\|+\|y_1\|+\|y_2\|},\quad \frac{s-v}{s-t}=\frac{\|x_2\|+\|y_2\|}{\|x_1\|+\|x_2\|+\|y_1\|+\|y_2\|}.$$

Thus we obtain the triangle inequality $\|(x_1+x_2,y_1+y_2)\|_\psi \leqslant \|(x_1,y_1)\|_\psi+\|(x_2,y_2)\|_\psi$.

We now proceed with the proof of the inequalities (1.13). The second inequality comes from the fact that ψ takes values $\leqslant 1$. The first inequality is a consequence of Lemma 1.1.44. In fact, by (1.12), when $\|x\|>0$ one has

$$\|x\|=(\|x\|+0)\psi\Big(\frac{\|x\|}{\|x\|+0}\Big)\leqslant \|(x,y)\|_\psi.$$

Similarly, one has $\|y\|\leqslant\|(x,y)\|_\psi$. The proposition is thus proved. □

Definition 1.1.46 The seminorm $\|\cdot\|_\psi$ constructed in the above proposition is called the *ψ-direct sum* of the seminorms of V and W.

Proposition 1.1.47 *Let $\|\cdot\|$ be an absolute normalised norm on $\mathbb{R}^2$. Then the dual norm $\|\cdot\|_*$ is also an absolute normalised norm, where* $\mathrm{Hom}_{\mathbb{R}}(\mathbb{R}^2,\mathbb{R})$ *is identified with $\mathbb{R}^2$ by using the isomorphism $\iota:\mathbb{R}^2\to$* $\mathrm{Hom}_{\mathbb{R}}(\mathbb{R}^2,\mathbb{R})$ *given by $\iota(x,y)(a,b)=ax+by$.*

Proof Let (x,y) be an element of $\mathbb{R}^2$. One has (recall that $|\cdot|_\infty$ denotes the usual absolute value on $\mathbb{R}$)

$$\|(x,y)\|_*=\sup_{(0,0)\neq(a,b)\in\mathbb{R}^2}\frac{|ax+by|_\infty}{\|(a,b)\|}.$$

Since $\|\cdot\|$ is an absolute normalised norm on $\mathbb{R}^2$, from the above formula we deduce that $\|(x,y)\|_*=\|(|x|_\infty,|y|_\infty)\|_*$ for any $(x,y)\in\mathbb{R}^2$. Moreover, by (1.10) one has

$$\|(1,0)\|_*=\sup_{(0,0)\neq(a,b)\in\mathbb{R}}\frac{|a|_\infty}{\|(a,b)\|}=\sup_{0\neq a\in\mathbb{R}}\frac{|a|_\infty}{\|(a,0)\|}=1.$$

Similarly, $\|(0,1)\|_*=1$. Therefore, $\|\cdot\|_*$ is an absolute normalised norm on $\mathbb{R}^2$. □

Definition 1.1.48 Let ψ be an element of $\mathscr{S}$, which corresponds to an absolute normalised norm $\|\cdot\|$ on $\mathbb{R}^2$. The above proposition shows that the dual norm $\|\cdot\|_*$ is also an absolute normalised norm. We denote by ψ_* the element of $\mathscr{S}$ corresponding to this dual norm. Note that ψ_* is actually given by

$$\psi_*(t)=\sup_{\lambda\in]0,1[}\left\{\frac{\lambda t+(1-\lambda)(1-t)}{\psi(\lambda)}\right\}.$$

The following proposition studies the dual of a direct sum seminorm.

Proposition 1.1.49 *Let $(V,\|\cdot\|_V)$ and $(W,\|\cdot\|_W)$ be finite-dimensional seminormed vector spaces over k, ψ be an element in $\mathscr{S}$, and $\|\cdot\|_\psi$ be the ψ-direct sum of $\|\cdot\|_V$ and $\|\cdot\|_W$. Let $\psi_0\in\mathscr{S}$ such that $\psi_0(t)=\max\{t,1-t\}$ for any $t\in[0,1]$.*

(1) *Assume that the absolute value $|\cdot|$ is non-Archimedean. Then the dual norm $\|\cdot\|_{\psi,*}$ identifies with the ψ_0-direct sum of $\|\cdot\|_{V,*}$ and $\|\cdot\|_{W,*}$.*

(2) *Assume that the absolute value* $|\cdot|$ *is Archimedean. Then the dual norm* $\|\cdot\|_{\psi,*}$ *identifies with the* ψ_**-direct sum of* $\|\cdot\|_{V,*}$ *and* $\|\cdot\|_{W,*}$.

Proof Since the null space of the seminorm $\|\cdot\|$ is $N_{\|\cdot\|_V} \oplus N_{\|\cdot\|_W}$, we obtain that a linear form $(f,g) \in V^\vee \oplus W^\vee$ vanishes on $N_{\|\cdot\|_\psi}$ if and only if it belongs to $V^* \oplus W^*$. In other words, one has $(V \oplus W)^* = V^* \oplus W^*$.

(1) Let (f,g) be an element in $V^* \oplus W^*$, one has

$$\|(f,g)\|_{\psi,*} = \sup_{\substack{(s,t)\in V\oplus W\\ \max\{\|s\|_V,\|t\|_W\}>0}} \frac{|f(s)+g(t)|}{\|(s,t)\|_\psi}$$

$$\leqslant \sup_{\substack{(s,t)\in V\oplus W\\ \max\{\|s\|_V,\|t\|_W\}>0}} \frac{\max\{|f(s)|,|g(t)|\}}{\max\{\|s\|_V,\|t\|_W\}} \leqslant \max\{\|f\|_{V,*},\|g\|_{W,*}\},$$

where the first inequality comes from (1.13) and the fact that the absolute value $|\cdot|$ is non-Archimedean. Moreover, one has

$$\|(f,g)\|_{\psi,*} \geqslant \sup_{s\in V\setminus N_{\|\cdot\|_V}} \frac{|f(s)+g(0)|}{\|(s,0)\|_\psi} = \|f\|_{V,*}.$$

Similarly, one has

$$\|(f,g)\|_{\psi,*} \geqslant \sup_{t\in W\setminus N_{\|\cdot\|_W}} \frac{|f(0)+g(t)|}{\|(0,t)\|_\psi} = \|g\|_{W,*}.$$

Therefore $\|(f,g)\|_{\psi,*} = \max\{\|f\|_{V,*},\|g\|_{W,*}\}$.

(2) Let $\|\cdot\|$ be the absolute normalised norm on $\mathbb{R}^2$ corresponding to ψ and let $\|\cdot\|_*$ be its dual norm. For any $(s,t) \in V \oplus W$, one has $\|(s,t)\|_\psi = \|(\|s\|_V,\|t\|_W)\|$. Let (f,g) be an element in $V^* \oplus W^*$. One has

$$\|(f,g)\|_{\psi,*} = \sup_{\substack{(s,t)\in V\oplus W\\ \max\{\|s\|_V,\|t\|_W\}>0}} \frac{|f(s)+g(t)|}{\|(s,t)\|_\psi}$$

$$\leqslant \sup_{\substack{(s,t)\in V\oplus W\\ \max\{\|s\|_V,\|t\|_W\}>0}} \frac{\|f\|_{V,*}\cdot\|s\|_V + \|g\|_{W,*}\cdot\|t\|_W}{\|(\|s\|_V,\|t\|_W)\|} = \|(\|f\|_{V,*},\|g\|_{W,*})\|_*.$$

Moreover, since $k = \mathbb{R}$ or $\mathbb{C}$, by Hahn-Banach theorem, for any $a > 0$, there exists $s \in V$ such that $\|s\|_V = a$ and that $f(s) = \|f\|_{V,*}\cdot\|s\|_V$. Similarly, for any $b > 0$, there exists $t \in W$ such that $\|t\|_W = b$ and $g(t) = \|g\|_{W,*}\cdot\|t\|_W$. Therefore the inequality in the above formula is actually an equality. □

Proposition 1.1.50 *Let* $f : V \to V'$ *and* $g : W \to W'$ *be surjective linear maps of finite-dimensional vector spaces over k. Let* $\|\cdot\|_V$ *and* $\|\cdot\|_W$ *be seminorms on V and W, and let* $\|\cdot\|_{V'}$ *and* $\|\cdot\|_{W'}$ *be the quotient seminorms of* $\|\cdot\|_V$ *and* $\|\cdot\|_W$ *on* V' *and* W'*, respectively. Then the quotient seminorm* $\|\cdot\|_{V\oplus W,\psi,V\oplus W\twoheadrightarrow V'\oplus W'}$ *of* $\|\cdot\|_{V\oplus W,\psi}$ *on* $V'\oplus W'$ *coincides with* $\|\cdot\|_{V'\oplus W',\psi}$.

Proof It is sufficient to see that

$$\|(x', y')\|_{V'\oplus W',\psi} = \|(x', y')\|_{V\oplus W,\psi,V\oplus W\twoheadrightarrow V'\oplus W'}$$

for all $x' \in V'$ and $y' \in W'$ with $\|x'\|_{V'} + \|y'\|_{W'} > 0$. Let $x \in V$ and $y \in W$ with $f(x) = x'$ and $g(y) = y'$. Then, as $\|x\|_V \geqslant \|x'\|_{V'}$ and $\|y\|_W \geqslant \|y'\|_{W'}$, by Lemma 1.1.44, one has $\|(x', y')\|_{V'\oplus W',\psi} \leqslant \|(x, y)\|_{V\oplus W,\psi}$, so that

$$\|(x', y')\|_{V'\oplus W',\psi} \leqslant \|(x', y')\|_{V\oplus W,\psi,V\oplus W\twoheadrightarrow V'\oplus W'}.$$

Let us consider the converse inequality. We choose sequences $\{x_n\}_{n\in\mathbb{N}}$ and $\{y_n\}_{n\in\mathbb{N}}$ in V and W such that $f(x_n) = x'$, $g(y_n) = y'$, $\lim_{n\to\infty} \|x_n\|_V = \|x'\|_{V'}$ and $\lim_{n\to\infty} \|y_n\|_W = \|y'\|_{W'}$.

We assume that $\|x'\|_{V'} + \|y'\|_{W'} > 0$. Then as $\|x_n\|_V + \|y_n\|_W > 0$ for sufficiently large n and ψ is continuous, one has

$$\begin{aligned}
\|(x', y')\|_{V'\oplus W',\psi} &= (\|x'\|_{V'} + \|y'\|_{W'})\psi\left(\frac{\|x'\|_{V'}}{\|x'\|_{V'} + \|y'\|_{W'}}\right)\\
&= \lim_{n\to\infty} (\|x_n\|_V + \|y_n\|_W)\psi\left(\frac{\|x_n\|_V}{\|x_n\|_V + \|y_n\|_W}\right)\\
&= \lim_{n\to\infty} \|(x_n, y_n)\|_{V\oplus W,\psi} \geqslant \|(x', y')\|_{V\oplus W,\psi,V\oplus W\twoheadrightarrow V'\oplus W'},
\end{aligned}$$

as required. Otherwise, as

$$0 \leqslant \|(x', y')\|_{V\oplus W,\psi,V\oplus W\twoheadrightarrow V'\oplus W'} \leqslant \|(x_n, y_n)\|_{V\oplus W,\psi} \leqslant \|x_n\|_V + \|y_n\|_W$$

and $\lim_{n\to\infty} \|x_n\|_V + \|y_n\|_W = 0$, one has

$$\|(x', y')\|_{V\oplus W,\psi,V\oplus W\twoheadrightarrow V'\oplus W'} = 0,$$

as desired. □

Remark 1.1.51 Let ψ be an element of $\mathscr{S}$. Let $\{\psi_n\}_{n=1}^{\infty}$ be a sequence of functions given in the following ways:

$$\begin{cases}
\forall a \in \mathbb{R}_{\geqslant 0}, & \psi_1(a) = a,\\
\forall (a, b) \in \mathbb{R}^2_{\geqslant 0}, & \psi_2(a, b) = \begin{cases}(a+b)\psi\left(\dfrac{a}{a+b}\right) & \text{if } a + b > 0,\\ 0 & \text{if } a = b = 0,\end{cases}\\
\forall (a_1, \ldots, a_n) \in \mathbb{R}^n_{\geqslant 0}, & \psi_n(a_1, \ldots, a_n) = \psi_2(\psi_{n-1}(a_1, \ldots, a_{n-1}), a_n).
\end{cases}$$

Let $(V_1, \|\cdot\|_1), \ldots, (V_n, \|\cdot\|_n)$ be finite-dimensional normed vector spaces over k. If we define

$$\|(x_1, \ldots, x_n)\|_\psi := \psi_n(\|x_1\|_1, \ldots, \|x_n\|_n)$$

for $(x_1, \ldots, x_n) \in V_1 \oplus \cdots \oplus V_n$, then, by Proposition 1.1.45, it yields a norm on $V_1 \oplus \cdots \oplus V_n$.

We assume that

$$\forall\, a_1, a_2, a_3 \in \mathbb{R}_{\geqslant 0}, \quad \psi_2(a_1, \psi_2(a_2, a_3)) = \psi_2(\psi_2(a_1, a_2), a_3). \tag{1.14}$$

Then it is easy to see that $\psi_n(a_1, \ldots, a_n) = \psi_2(\psi_i(a_1, \ldots, a_i), \psi_{n-i}(a_{i+1}, \ldots, a_n))$ for $i \in \{1, \ldots, n-1\}$, so that the construction of the norm $\|\cdot\|_\psi$ is associative. If we assume $\psi_2(a, b) = \psi_2(b, a)$ for all $(a, b) \in \mathbb{R}^2_{\geqslant 0}$ in addition to (1.14), then $\psi_n(a_1, \ldots, a_n)$ is symmetric, that is, for any permutation σ, $\psi_n(a_{\sigma(1)}, \ldots, a_{\sigma(n)}) = \psi_n(a_1, \ldots, a_n)$, which means that its construction is order independent.

1.1.11 Tensor product seminorms

Let V and W be seminormed vector spaces of finite dimension over k. On the tensor product space $V \otimes_k W$ there are several natural ways to construct tensor product seminorms. We refer the readers to the original article [73] of Grothendieck for different constructions. In this subsection, we recall the π-tensor product and the ε-tensor product. We refer to the book [131] for a more detailed presentation in the Archimedean case.

Definition 1.1.52 Let $(V_1, \|\cdot\|_1), \ldots, (V_n, \|\cdot\|_n)$ be seminormed vector spaces over k. We define a map $\|\cdot\|_\pi : V_1 \otimes_k \cdots \otimes_k V_n \to [0, +\infty[$ such that, for any $\varphi \in V_1 \otimes_k \cdots \otimes_k V_n$,

$$\|\varphi\|_\pi := \inf\left\{ \sum_{i=1}^N \|x_1^{(i)}\|_1 \cdots \|x_n^{(i)}\|_n \;:\; \varphi = \sum_{i=1}^N x_1^{(i)} \otimes \cdots \otimes x_n^{(i)} \right\}. \tag{1.15}$$

Note that $\|\cdot\|_\pi$ is a seminorm on $V_1 \otimes_k \cdots \otimes_k V_n$. For example, the triangle inequality can be checked as follows: for $\varphi, \psi \in V_1 \otimes_k \cdots \otimes_k V_n$ and a positive number ϵ, we choose expressions

$$\varphi = \sum_{i=1}^N x_1^{(i)} \otimes \cdots \otimes x_n^{(i)} \quad \text{and} \quad \psi = \sum_{j=1}^M y_1^{(j)} \otimes \cdots \otimes y_n^{(j)}$$

such that

$$\sum_{i=1}^N \|x_1^{(i)}\|_1 \cdots \|x_n^{(i)}\|_n \leqslant \|\varphi\|_\pi + \epsilon \quad \text{and} \quad \sum_{j=1}^M \|y_1^{(j)}\|_1 \cdots \|y_n^{(j)}\|_n \leqslant \|\psi\|_\pi + \epsilon.$$

Then, as

$$\varphi + \psi = \sum_{i=1}^N x_1^{(i)} \otimes \cdots \otimes x_n^{(i)} + \sum_{j=1}^M y_1^{(j)} \otimes \cdots \otimes y_n^{(i)},$$

one has

$$\|\varphi+\psi\|_\pi \leqslant \sum_{i=1}^{N}\|x_1^{(i)}\|_1\cdots\|x_n^{(i)}\|_n+\sum_{j=1}^{M}\|y_1^{(j)}\|_1\cdots\|y_n^{(j)}\|_n \leqslant \|\varphi\|_\pi+\|\psi\|_\pi+2\epsilon,$$

as desired. We call $\|\cdot\|_\pi$ the *π-tensor product* of the seminorms $\|\cdot\|_1,\ldots,\|\cdot\|_n$.

Any element φ in the tensor product space $V_1\otimes_k\cdots\otimes_k V_n$ can be considered as a multilinear form on $V_1^*\times\cdots\times V_n^*$. In particular, if φ is of the form $x_1\otimes\cdots\otimes x_n$, the corresponding multilinear form sends $(f_1,\ldots,f_n)\in V_1^*\times\cdots\times V_n^*$ to $f_1(x_1)\cdots f_n(x_n)\in k$. For any $\varphi\in V_1\otimes_k\cdots\otimes_k V_n$, viewed as a k-multilinear form on $V_1^*\times\cdots\times V_n^*$, let

$$\|\varphi\|_\varepsilon := \sup_{\substack{(f_1,\ldots,f_n)\in V_1^*\times\cdots\times V_n^*\\ \forall\, i\in\{1,\ldots,n\},\ f_i\neq 0}} \frac{|\varphi(f_1,\ldots,f_n)|}{\|f_1\|_{1,*}\cdots\|f_n\|_{n,*}}.$$

Then $\|\cdot\|_\varepsilon$ is a seminorm on the tensor product space $V_1\otimes_k\cdots\otimes_k V_n$, called the *$\varepsilon$-tensor product* of seminorms $\|\cdot\|_1,\ldots,\|\cdot\|_n$. It is a norm once the seminorms $\|\cdot\|_1,\ldots,\|\cdot\|_n$ are norms. Similarly to the dual norm case, if the absolute value $|\cdot|$ is non-Archimedean, then the ε-tensor product $\|\cdot\|_\varepsilon$ is ultrametric. By Proposition 1.2.14 in the next section, we obtain that, in the case where all V_i are of finite type over k, the ε-tensor product of $\|\cdot\|_1,\ldots,\|\cdot\|_n$ identifies with that of $\|\cdot\|_{1,**},\ldots,\|\cdot\|_{n,**}$.

Remark 1.1.53 Let $(V_1,\|\cdot\|_1)$ and $(V_2,\|\cdot\|_2)$ be seminormed vector spaces over k. Let $\|\cdot\|$ be the operator seminorm on the vector space $\mathscr{L}(V_1^*,V_2)$, where we consider the dual norm $\|\cdot\|_{1,*}$ on V_1^* and the double dual seminorm $\|\cdot\|_{2,**}$ on V_2. One has a canonical k-linear map from $V_1\otimes_k V_2$ to $\mathscr{L}(V_1^*,V_2)$ sending $x\otimes y\in V_1\otimes_k V_2$ to the bounded linear map $(\alpha\in V_1^*)\mapsto\alpha(x)y$. We claim that the seminorm on $V_1\otimes_k V_2$ induced by $\|\cdot\|$ and the above canonical map identifies with the ε-tensor product $\|\cdot\|_\varepsilon$ of $\|\cdot\|_1$ and $\|\cdot\|_2$. In fact, for any $\varphi\in V_1\otimes_k V_2$ one has

$$\|\varphi\| = \sup_{f_1\in V_1^*\setminus\{0\}}\frac{\|\varphi(f_1)\|_{2,**}}{\|f_1\|_{1,*}} = \sup_{\substack{f_1\in V_1^*\setminus\{0\}\\ f_2\in V_2^*\setminus\{0\}}}\frac{|\varphi(f_1,f_2)|}{\|f_1\|_{1,*}\|f_2\|_{2,*}} = \|\varphi\|_\varepsilon.$$

In particular, if $\|\cdot\|_2=\|\cdot\|_{2,**}$ on V_2, then the ε-tensor product norm $\|\cdot\|_\varepsilon$ identifies with the operator seminorm if we consider tensors in $V_1\otimes_k V_2$ as k-linear operators from $(V_1^*,\|\cdot\|_{1,*})$ to $(V_2,\|\cdot\|_2)$.

Proposition 1.1.54 *We keep the notation of Definition 1.1.52. If $\|\cdot\|$ is a seminorm on $V_1\otimes_k\cdots\otimes_k V_n$ such that $\|x_1\otimes\cdots\otimes x_n\|\leqslant\|x_1\|_1\cdots\|x_n\|_n$ for any $(x_1,\ldots,x_n)\in V_1\times\cdots\times V_n$, then one has $\|\cdot\|\leqslant\|\cdot\|_\pi$. In particular, the seminorm $\|\cdot\|_\varepsilon$ is bounded from above by $\|\cdot\|_\pi$. Moreover, if $\|\cdot\|_1,\ldots,\|\cdot\|_n$ are norms, then $\|\cdot\|_\pi$ is also a norm.*

Proof Let φ be an element of $V_1\otimes_k\cdots\otimes_k V_n$. If φ is written in the form

$$\varphi=\sum_{i=1}^{N}x_1^{(i)}\otimes\cdots\otimes x_n^{(i)},$$

where $x_j^{(i)} \in V_j$ for any $j \in \{1, \ldots, n\}$, then one has

$$\|\varphi\| \leqslant \sum_{i=1}^{N} \|x_1^{(i)} \otimes \cdots \otimes x_n^{(i)}\| \leqslant \sum_{i=1}^{N} \|x_1^{(1)}\|_1 \cdots \|x_n^{(1)}\|_n.$$

Therefore we obtain $\|\cdot\| \leqslant \|\cdot\|_\pi$. Note that, for any $(x_1, \ldots, x_n) \in V_1 \times \cdots \times V_n$ one has

$$\begin{aligned}\|x_1 \otimes \cdots \otimes x_n\|_\varepsilon &= \sup_{\substack{(f_1,\ldots,f_n)\in V_1^*\times\cdots\times V_n^* \\ \forall i\in\{1,\ldots,n\},\, f_i\neq 0}} \frac{|f_1(x_1)|\cdots|f_n(x_n)|}{\|f_1\|_{1,*}\cdots\|f_n\|_{n,*}} \\ &= \|x_1\|_{1,**}\cdots\|x_n\|_{n,**} \leqslant \|x_1\|_1\cdots\|x_n\|_n.\end{aligned}$$

Therefore, one has $\|\cdot\|_\varepsilon \leqslant \|\cdot\|_\pi$. If the seminorms $\|\cdot\|_i$ ($i \in \{1, \ldots, n\}$) are norms, then $\|\cdot\|_\varepsilon$ is a norm and hence $\|\cdot\|_\pi$ is also a norm. □

Remark 1.1.55 From the definition we observe that the ε-tensor product and π-tensor product are commutative. Namely, if V_1 and V_2 are finite-dimensional normed vector spaces over k, then the canonical isomorphism $V_1 \otimes_k V_2 \to V_2 \otimes_k V_1$ is an isometry if we consider ε-tensor products or π-tensor product norms on both sides. The ε-tensor product and the π-tensor product are also associative. Namely, if V_1, V_2 and V_3 are finite-dimensional normed vector spaces over k, then the canonical isomorphisms $(V_1 \otimes_k V_2) \otimes_k V_3 \to V_1 \otimes_k V_2 \otimes_k V_3$ and $V_1 \otimes_k (V_2 \otimes_k V_3) \to V_1 \otimes_k V_2 \otimes_k V_3$ are both isometries.

Remark 1.1.56 Let $(V_1, \|\cdot\|_1), \ldots, (V_n, \|\cdot\|_n)$ be finite-dimensional seminormed vector spaces over k. From the definition, we observe that, if $(u_1, \ldots, u_n)$ is an element of $V_1 \times \cdots \times V_n$, then one has

$$\|u_1 \otimes \cdots \otimes u_n\|_\varepsilon = \|u_1\|_{1,**} \cdots \|u_n\|_{n,**}. \tag{1.16}$$

If the seminormed vector spaces $(V_1, \|\cdot\|_1), \ldots, (V_n, \|\cdot\|_n)$ are reflexive, by (1.16) and Proposition 1.1.54, we obtain that, for any $(u_1, \ldots, u_n) \in V_1 \times \cdots \times V_n$, one has

$$\prod_{i=1}^{n} \|u_i\|_i = \|u_1 \otimes \cdots \otimes u_n\|_\varepsilon \leqslant \|u_1 \otimes \cdots \otimes u_n\|_\pi.$$

Moreover, by definition one has $\|u_1 \otimes \cdots \otimes u_n\|_\pi \leqslant \|u_1\|_1 \cdots \|u_n\|_n$. Therefore

$$\|u_1 \otimes \cdots \otimes u_n\|_\varepsilon = \|u_1 \otimes \cdots \otimes u_n\|_\pi = \|u_1\|_1 \cdots \|u_n\|_n. \tag{1.17}$$

In particular, if $V_1, \ldots, V_n$ are seminormed vector spaces of dimension 1 over k (in this case they are necessarily reflexive), then their ε-tensor product and π-tensor product norms are the same. We simply call it the *tensor product* of the seminorms $\|\cdot\|_1, \ldots, \|\cdot\|_n$.

Proposition 1.1.57 *Let $(V_1, \|\cdot\|_1), \ldots, (V_n, \|\cdot\|_n)$ be finite-dimensional seminormed vector spaces over k. Let $\|\cdot\|_{*,\pi}$ and $\|\cdot\|_{*,\varepsilon}$ be respectively the π-tensor product and the ε-tensor product of the dual norms $\|\cdot\|_{1,*}, \ldots, \|\cdot\|_{n,*}$. The ε-tensor product of $\|\cdot\|_1, \ldots, \|\cdot\|_n$ identifies with the seminorm induced by the dual norm $\|\cdot\|_{*,\pi,*}$ on $(V_1^* \otimes_k \cdots \otimes_k V_n^*)^*$ by the natural linear map $V_1 \otimes_k \cdots \otimes_k V_n \to (V_1^* \otimes_k \cdots \otimes_k V_n^*)^*$. If the absolute value $|\cdot|$ is Archimedean, then the π-tensor product of $\|\cdot\|_1, \ldots, \|\cdot\|_n$ identifies with seminorm induced by the dual norm $\|\cdot\|_{*,\varepsilon,*}$ on $(V_1^* \otimes_k \cdots \otimes_k V_n^*)^*$ by the natural linear map $V_1 \otimes_k \cdots \otimes_k V_n \to (V_1^* \otimes_k \cdots \otimes_k V_n^*)^*$.*

Proof Let φ be an element in $V_1 \otimes_k \cdots \otimes_k V_n$, which can also be viewed as a k-multilinear form on $V_1^* \times \cdots \times V_n^*$ or a linear form on $V_1^* \otimes_k \cdots \otimes_k V_n^*$. Let α be an element in $V_1^* \otimes_k \cdots \otimes_k V_n^*$. If α is written in the form

$$\alpha = \sum_{i=1}^{N} f_1^{(i)} \otimes \cdots \otimes f_n^{(i)},$$

where $f_j^{(i)} \in V_j^*$, then one has $\varphi(\alpha) = \sum_{i=1}^N \varphi(f_1^{(i)}, \ldots, f_n^{(i)})$ and hence

$$|\varphi(\alpha)| \leqslant \sum_{i=1}^{N} |\varphi(f_1^{(i)}, \ldots, f_n^{(i)})|.$$

Thus we obtain

$$\frac{|\varphi(\alpha)|}{\sum_{i=1}^N \|f_1^{(i)}\|_{1,*} \cdots \|f_n^{(i)}\|_{n,*}} \leqslant \frac{\sum_{i=1}^N |\varphi(f_1^{(i)}, \ldots, f_n^{(i)})|}{\sum_{i=1}^N \|f_1^{(i)}\|_{1,*} \cdots \|f_n^{(i)}\|_{n,*}} \leqslant \|\varphi\|_\varepsilon.$$

Therefore φ is a bounded linear form on $(V_1^* \otimes_k \cdots \otimes_k V_n^*, \|\cdot\|_{*,\pi})$ and $\|\varphi\|_{*,\pi,*} \leqslant \|\varphi\|_\varepsilon$.

For any $(f_1, \ldots, f_n) \in (V_1^* \setminus \{0\}) \times \cdots \times (V_n^* \setminus \{0\})$ one has

$$\frac{|\varphi(f_1 \otimes \cdots \otimes f_n)|}{\|f_1 \otimes \cdots \otimes f_n\|_{*,\pi}} = \frac{|\varphi(f_1, \ldots, f_n)|}{\|f_1 \otimes \cdots \otimes f_n\|_{*,\pi}} \geqslant \frac{|\varphi(f_1, \ldots, f_n)|}{\|f_1\|_{1,*} \cdots \|f_n\|_{n,*}}.$$

Therefore one has $\|\varphi\|_\varepsilon \leqslant \|\varphi\|_{*,\pi,*}$. The first assertion is thus proved.

If $|\cdot|$ is Archimedean, any finite-dimensional normed vector space is reflexive. By the first assertion, the dual norm of the π-tensor product of $\|\cdot\|_1, \ldots, \|\cdot\|_n$ is the ε-tensor product of $\|\cdot\|_{1,*}, \ldots, \|\cdot\|_{n,*}$. By taking the double dual seminorm we obtain that the π-tensor product of $\|\cdot\|_1, \ldots, \|\cdot\|_n$ identifies with the seminorm induced by $\|\cdot\|_{*,\varepsilon,*}$. □

Proposition 1.1.58 *Let V and W be seminormed vector spaces over k, and Q be a quotient space of V, equipped with the quotient seminorm. Let V_0 be the kernel of the projection map $V \to Q$. Then the canonical isomorphism $(V \otimes_k W)/(V_0 \otimes_k W) \to Q \otimes_k W$ is an isometry, where we consider the π-tensor product seminorms on $V \otimes_k W$ and $Q \otimes_k W$, and the quotient seminorm on $(V \otimes_k W)/(V_0 \otimes_k W)$.*

Proof Let ψ be an element of $Q \otimes_k W$. One has

$$\|\psi\|_\pi = \inf\left\{\sum_{i=1}^N \|\alpha_i\|\cdot\|y_i\| \;:\; N\in\mathbb{N},\ \psi=\sum_{i=1}^N \alpha_i\otimes y_i\right\}$$

$$= \inf_{N\in\mathbb{N}} \inf_{\substack{(\alpha_i)_{i=1}^N\in Q^N\\ (y_i)_{i=1}^N\in W^N\\ \psi=\sum_{i=1}^N \alpha_i\otimes y_i}} \inf_{\substack{(x_i)_{i=1}^N\in V^N\\ [x_i]=\alpha_i}} \sum_{i=1}^N \|x_i\|\cdot\|y_i\| = \inf_{\substack{\varphi\in V\otimes W\\ [\varphi]=\psi}} \|\varphi\|_\pi.$$

Remark 1.1.59 We consider the ε-tensor product analogue of the above proposition. Let f be an element of $V\otimes_k W$, viewed as a k-bilinear form on $V^*\times W^*$. Then its image g in $Q\otimes_k W$ corresponds to the restriction of f to $Q^*\times W^*$. By Proposition 1.1.20, the dual norm on Q^* of the quotient seminorm identifies with the restriction to Q^* of the dual norm on V^*. Therefore, one has $\|g\|_\varepsilon \leqslant \|f\|_\varepsilon$. However, in the case where the absolute value $|\cdot|$ is Archimedean, in general the inequality

$$\|g\|_\varepsilon \leqslant \inf_{\substack{f\in V\otimes_k W\\ f|_{Q^*\times W^*}=g}} \|f\|_\varepsilon$$

is strict. In fact, this problem is closely related to the extension property of the normed vector space V^*, which consists of extending a linear operator defined on a vector subspace of V^* and valued in another seminormed vector space while keeping the operator seminorm. In the case where the linear operator is a linear form (namely valued in k), it is just a consequence of Hahn-Banach theorem. However, in general the extension property does not hold, except in the cases where $\dim_k(V)\leqslant 2$ or the norm on V comes from a symmetric semipositive bilinear form (see §1.2.1 for the notation). We refer the readers to [92, 132] for more details.

In the case where the absolute value $|\cdot|$ is non-Archimedean, any dual norm is ultrametric, we will give a proof for the ε-tensor product analogue of Proposition 1.1.58, by using the ultrametric Gram-Schmidt process (see Proposition 1.2.36).

Proposition 1.1.60 *Let $(V,\|\cdot\|_V)$ and $(W,\|\cdot\|_W)$ be seminormed vector spaces over k, V_0 be a vector subspace of V and $\|\cdot\|_{V_0}$ be the restriction of $\|\cdot\|_V$ to V_0.*

(1) *Let $\|\cdot\|_\pi$ be the π-tensor product of $\|\cdot\|_V$ and $\|\cdot\|_W$, $\|\cdot\|_{\pi,0}$ be the π-tensor product of $\|\cdot\|_{V_0}$ and $\|\cdot\|_W$. Then the seminorm $\|\cdot\|_{\pi,0}$ is bounded from below by the restriction of $\|\cdot\|_\pi$ to $V_0\otimes_k W$.*

(2) *Let $\|\cdot\|_\varepsilon$ be the ε-tensor product of $\|\cdot\|_V$ and $\|\cdot\|_W$, and $\|\cdot\|_{\varepsilon,0}$ be the ε-tensor product of $\|\cdot\|_{V_0}$ and $\|\cdot\|_W$. Then the seminorm $\|\cdot\|_{\varepsilon,0}$ is bounded from below by the restriction of $\|\cdot\|_\varepsilon$ to $V_0\otimes_k W$.*

Proof (1) Let φ be an element of $V_0\otimes_k W$. By definition, for any writing of φ as $\sum_{i=1}^N x_i\otimes y_i$ with $\{x_1,\ldots,x_N\}\subseteq V_0$ and $\{y_1,\ldots,y_N\}\subseteq W$, one has

$$\|\varphi\|_\pi \leqslant \sum_{i=1}^N \|x_i\|_{V_0}\cdot\|y_i\|_W.$$

Therefore $\|\varphi\|_\pi \leqslant \|\varphi\|_{\pi,0}$.

(2) We consider the canonical linear map $V^* \to V_0^*$ sending a bounded linear form on V to its restriction to V_0. Note that for any $f \in V^*$ one has $\|f_0\|_{V_0,*} \leqslant \|f\|_{V,*}$, where f_0 is the restriction of f to V_0. Therefore, for any element φ of $V_0 \otimes_k W$, viewed as a bilinear form on $V_0^* \times W^*$ or as a bilinear form on $V^* \times W^*$ via the inclusion $V_0 \otimes_k W \subseteq V \otimes_k W$, one has

$$\|\varphi\|_{\varepsilon,0} = \sup_{\substack{(f_0,g)\in V_0^*\times W^*\\ f_0\neq 0,\, g\neq 0}} \frac{|\varphi(f_0,g)|}{\|f\|_{V_0,*}\|g\|_{W,*}} \geqslant \sup_{\substack{(f,g)\in V^*\times W^*\\ f\neq 0,\, g\neq 0}} \frac{|\varphi(f,g)|}{\|f\|_{V,*}\|g\|_{W,*}} = \|\varphi\|_{\varepsilon}.$$

Proposition 1.1.61 *Let n be a positive integer and*

$$\{(V_j, \|\cdot\|_{V_j})\}_{j=1}^n \quad \text{and} \quad \{(W_j, \|\cdot\|_{W_j})\}_{j=1}^n$$

be finite-dimensional seminormed vector spaces over k. For any $j \in \{1, \ldots, n\}$, let $f_j : V_j \to W_j$ be a bounded k-linear map. Let $f : V_1 \otimes_k \cdots \otimes V_n \to W_1 \otimes_k \cdots \otimes_k W_n$ be the k-linear map sending $x_1 \otimes \cdots \otimes x_n$ to $f_1(x_1) \otimes \cdots \otimes f_n(x_n)$.

(1) We equip the vector spaces $V_1 \otimes_k \cdots \otimes_k V_n$ and $W_1 \otimes_k \cdots \otimes_k W_n$ with the π-tensor product seminorms of $\{\|\cdot\|_{V_j}\}_{j=1}^n$ and of $\{\|\cdot\|_{W_j}\}_{j=1}^n$, respectively. Then the operator seminorm of f is bounded from above by $\|f_1\| \cdots \|f_n\|$.

(2) We equip the vector spaces $V_1 \otimes_k \cdots \otimes_k V_n$ and $W_1 \otimes_k \cdots \otimes_k W_n$ with the ε-tensor product seminorms of $\{\|\cdot\|_{V_j}\}_{j=1}^n$ and of $\{\|\cdot\|_{W_j}\}_{j=1}^n$, respectively. Then the operator seminorm of f is bounded from above by $\|f_1^*\| \cdots \|f_n^*\|$.

Proof (1) Let φ be an element in $V_1 \otimes_k \cdots \otimes_k V_n$, which is written as

$$\varphi = \sum_{i=1}^N x_1^{(i)} \otimes \cdots \otimes x_n^{(i)}$$

where $x_j^{(i)} \in V_j$ for any $j \in \{1, \ldots, n\}$. By definition, one has

$$f(\varphi) = \sum_{i=1}^N f_1(x_1^{(i)}) \otimes \cdots \otimes f_n(x_n^{(i)}).$$

Therefore

$$\begin{aligned}\|f(\varphi)\|_\pi &\leqslant \sum_{i=1}^N \|f_1(x_1^{(i)})\|_{W_1} \cdots \|f_n(x_n^{(i)})\|_{W_n} \\ &\leqslant \Big(\prod_{i=1}^n \|f_i\|\Big) \sum_{i=1}^N \|x_1^{(i)}\|_{V_1} \cdots \|x_n^{(i)}\|_{V_n}.\end{aligned}$$

Thus $\|f(\varphi)\|_\pi \leqslant \|f_1\| \cdots \|f_n\| \cdot \|\varphi\|_\pi$.

(2) Let φ be an element in $V_1 \otimes_k \cdots \otimes_n V_n$, which can be viewed as a multilinear form on $V_1^* \times \cdots \times V_n^*$. Then the element $f(\varphi) \in W_1 \otimes_k \cdots \otimes_k W_n$, viewed as a multilinear

form on $W_1^* \times \cdots \times W_n^*$, sends $(\beta_1, \ldots, \beta_n) \in W_1^* \times \cdots \times W_n^*$ to $\varphi(f_1^*(\beta_1), \ldots, f_n^*(\beta_n))$. Thus for $(\beta_1, \ldots, \beta_n) \in (W_1^* \setminus \{0\}) \times \cdots \times (W_n^* \setminus \{0\})$, one has

$$\frac{|f(\varphi)(\beta_1, \ldots, \beta_n)|}{\|\beta_1\|_{W_1,*} \cdots \|\beta_n\|_{W_n,*}} \leqslant \frac{\|\varphi\|_\varepsilon \|f_1^*(\beta_1)\|_{V_1,*} \cdots \|f_n^*(\beta_n)\|_{V_n,*}}{\|\beta_1\|_{W_1,*} \cdots \|\beta_n\|_{W_n,*}}$$
$$\leqslant \|\varphi\|_\varepsilon \|f_1^*\| \cdots \|f_n^*\|,$$

so that $\|f(\varphi)\|_\varepsilon \leqslant \|\varphi\|_\varepsilon \|f_1^*\| \cdots \|f_n^*\|$, as required. □

1.1.12 Exterior power seminorm

Let V be a vector space over k and r be the dimension of V over k. For any $i \in \mathbb{N}$, we let $\Lambda^i V$ be the ith exterior power of the vector space V. It is a quotient vector space of $V^{\otimes i}$.

Definition 1.1.62 Let $\|\cdot\|$ be a seminorm on the vector space V and $\|\cdot\|_\pi$ be the π-tensor power of $\|\cdot\|$ on $V^{\otimes i}$. The ith *π-exterior power seminorm* of $\|\cdot\|$ on $\Lambda^i V$ is by definition the quotient seminorm on $\Lambda^i V$ of $\|\cdot\|_\pi$ induced by the canonical projection map $V^{\otimes i} \to \Lambda^i V$ sending $x_1 \otimes \cdots \otimes x_i$ to $x_1 \wedge \cdots \wedge x_i$, denoted by $\|\cdot\|_{\Lambda^i_\pi}$, or simply by $\|\cdot\|_{\Lambda^i}$. Similarly, the ε-tensor product seminorm $\|\cdot\|_\varepsilon$ on $V^{\otimes i}$ induces by quotient a seminorm on $\Lambda^i V$, called the ith *ε-exterior power* of $\|\cdot\|$, denoted by $\|\cdot\|_{\Lambda^i_\varepsilon}$.

Proposition 1.1.63 *Let $(V, \|\cdot\|)$ be a seminormed vector space over k and i be a natural number. For any $(x_1, \ldots, x_i) \in V^i$ one has*

$$\|x_1 \wedge \cdots \wedge x_i\|_{\Lambda^i_\varepsilon} \leqslant \|x_1 \wedge \cdots \wedge x_i\|_{\Lambda^i_\pi} \leqslant \|x_1\| \cdots \|x_i\|.$$

Proof The first inequality follows from Proposition 1.1.54.

Note that $x_1 \wedge \cdots \wedge x_i$ is the image of $x_1 \otimes \cdots \otimes x_i$ by the canonical projection map $V^{\otimes i} \to \Lambda^i V$. Therefore one has

$$\|x_1 \wedge \cdots \wedge x_i\|_{\Lambda^i_\pi} \leqslant \|x_1 \otimes \cdots \otimes x_i\|_\pi \leqslant \|x_1\| \cdots \|x_i\|,$$

where $\|\cdot\|_\pi$ denotes the π-tensor power of $\|\cdot\|$. □

Proposition 1.1.64 *Let V and W be seminormed vector spaces over k and $f : V \to W$ be a bounded k-linear map. Let i be a positive integer. The k-linear map f induces by passing to the ith exterior power a k-linear map $\Lambda^i f : \Lambda^i V \to \Lambda^i W$.*

(1) *If we equip $\Lambda^i V$ and $\Lambda^i W$ with the ith π-exterior power seminorms, then the operator seminorm of $\Lambda^i f$ is bounded from above by $\|f\|^i$.*
(2) *If we equip $\Lambda^i V$ and $\Lambda^i W$ with the ith ε-exterior power seminorms, then the operator seminorm of $\Lambda^i f$ is bounded from above by $\|f^*\|^i$.*

Proof Let us consider a commutative diagram:

$$\begin{array}{ccc} V^{\otimes i} & \xrightarrow{f^{\otimes i}} & W^{\otimes i} \\ \big\downarrow & & \big\downarrow \\ \Lambda^i V & \xrightarrow[\Lambda^i f]{} & \Lambda^i W \end{array}$$

By (2.a) in Proposition 1.1.14, we obtain $\|\Lambda^i f\| \leqslant \|f^{\otimes i}\|$. Thus the assertions follow from Proposition 1.1.61. □

1.1.13 Determinant seminorm

Let V be a finite-dimensional vector space over k. Recall that the *determinant* of V is defined as the maximal exterior power $\Lambda^r V$ of the vector space V, where r is the dimension of V over k. It is a quotient space of dimension 1 of $V^{\otimes r}$. We denote by $\det(V)$ the determinant of V.

Definition 1.1.65 Assume that the vector space V is equipped with a seminorm $\|\cdot\|$. We call the *determinant seminorm of* $\|\cdot\|$ on $\det(V)$ and we denote by $\|\cdot\|_{\det}$ the π-exterior power seminorm of $\|\cdot\|$, that is, quotient seminorm induced by the π-tensor power seminorm on $V^{\otimes r}$.

Proposition 1.1.66 (Hadamard's inequality) *Let $(V, \|\cdot\|)$ be a finite-dimensional seminormed vector space of dimension $r > 0$ over k. For any $\eta \in \det(V)$,*

$$\|\eta\|_{\det} = \inf\big\{\|x_1\|\cdots\|x_r\| \;:\; \eta = x_1 \wedge \cdots \wedge x_r\big\}.$$

In particular, the determinant seminorm is a norm if and only if $\|\cdot\|$ is a norm.

Proof If η is written in the form $\eta = x_1 \wedge \cdots \wedge x_r$, where $x_1, \ldots, x_r$ are elements in V, then it is the image of $x_1 \otimes \cdots \otimes x_r$ by the canonical projection $V^{\otimes r} \to \det(V)$. Therefore one has $\|\eta\|_{\det} \leqslant \|x_1\|\cdots\|x_r\|$. Thus we obtain

$$\|\eta\|_{\det} \leqslant \inf\big\{\|x_1\|\cdots\|x_r\| \;:\; \eta = x_1 \wedge \cdots \wedge x_r\big\}.$$

In the following, we prove the converse inequality. It suffices to treat the case where $\eta \neq 0$. By definition one has

$$\|\eta\|_{\det} = \inf\Big\{\sum_{i=1}^N \|x_1^{(i)}\|\cdots\|x_r^{(i)}\| \;:\; \eta = \sum_{i=1}^N x_1^{(i)} \wedge \cdots \wedge x_r^{(i)}\Big\}.$$

Let $\{x_j^{(i)}\}_{i\in\{1,\ldots,N\},\, j\in\{1,\ldots,r\}}$ be elements in V such that $\eta = \sum_{i=1}^N x_1^{(i)} \wedge\cdots\wedge x_r^{(i)}$. Let $\{e_j\}_{j=1}^r$ be a basis of V and $\eta_0 = e_1 \wedge \cdots \wedge e_r$. For any $i \in \{1, \ldots, N\}$, there exists $a_i \in k$ such that $x_1^{(i)} \wedge \cdots \wedge x_r^{(i)} = a_i \eta_0$. Without loss of generality, we may assume that all a_i are non-zero and that

$$\frac{\|x_1^{(1)}\|\cdots\|x_r^{(1)}\|}{|a_1|}=\min_{i\in\{1,\ldots,N\}}\frac{\|x_1^{(i)}\|\cdots\|x_r^{(i)}\|}{|a_i|}.$$

Note that one has

$$\eta=(a_1+\cdots+a_N)\eta_0=\Big(1+\frac{a_2}{a_1}+\cdots+\frac{a_N}{a_1}\Big)x_1^{(1)}\wedge\cdots\wedge x_r^{(1)},$$

and

$$\begin{aligned}&\Big|1+\frac{a_2}{a_1}+\cdots+\frac{a_N}{a_1}\Big|\cdot\|x_1^{(1)}\|\cdots\|x_r^{(1)}\|\\ &\leqslant\Big(1+\Big|\frac{a_2}{a_1}\Big|+\cdots+\Big|\frac{a_N}{a_1}\Big|\Big)\|x_1^{(1)}\|\cdots\|x_r^{(1)}\|\leqslant\sum_{i=1}^N\|x_1^{(i)}\|\cdots\|x_r^{(i)}\|.\end{aligned}$$

The proposition is thus proved. □

Remark 1.1.67 Let $(V,\|\cdot\|)$ be a non-zero finite-dimensional normed vector space over k. Let r be the dimension of V over k. Proposition 1.1.66 shows that

$$\inf\left\{\frac{\|x_1\|\cdots\|x_r\|}{\|x_1\wedge\cdots\wedge x_r\|_{\det}}\;:\;(x_1,\ldots,x_r)\in V^r,\;x_1\wedge\cdots\wedge x_r\neq 0\right\}=1.\tag{1.18}$$

If the infimum is attained by some $(e_1,\ldots,e_r)\in V^r$, then $\{e_i\}_{i=1}^r$ is called an *Hadamard basis* of $(V,\|\cdot\|)$. By convention, the empty subset of the zero normed vector space is considered as an Hadamard basis.

Corollary 1.1.68 *Let V be a finite-dimensional seminormed vector space over k and W be a vector subspace of V. The canonical isomorphism (see [28] Chapter III, §7, no.7)*

$$\det(W)\otimes\det(V/W)\longrightarrow\det(V)\tag{1.19}$$

has seminorm $\leqslant 1$, where we consider the determinant seminorm of the induced seminorm on $\det(W)$ *and that of the quotient seminorm on* $\det(V/W)$, *and the tensor product seminorm on* $\det(W)\otimes\det(V/W)$ *(see Remark 1.1.56).*

Proof Let $\{x_1,\ldots,x_n\}$ be a basis of W and $\{y_1,\ldots,y_m\}$ be elements in $V\setminus W$ whose image in V/W forms a basis of V/W. By Proposition 1.1.66 one has

$$\|x_1\wedge\cdots\wedge x_n\wedge y_1\wedge\cdots\wedge y_m\|_{\det}\leqslant\|x_1\|\cdots\|x_n\|\cdot\|y_1\|\cdots\|y_m\|.$$

Note that if we replace each y_i by an element y_i' in the same equivalent class, one has

$$x_1\wedge\cdots\wedge x_n\wedge y_1\wedge\cdots\wedge y_m=x_1\wedge\cdots\wedge x_n\wedge y_1'\wedge\cdots\wedge y_m'.$$

Hence we obtain

$$\|x_1\wedge\cdots\wedge x_n\wedge y_1\wedge\cdots\wedge y_m\|\leqslant\|x_1\|\cdots\|x_n\|\cdot\|[y_1]\|\cdots\|[y_m]\|.$$

Therefore, for any $\eta\in\det(W)$ and $\eta'\in\det(V/W)$ one has

$$\|\eta \wedge \eta'\|_{\det} \leqslant \Bigg(\inf_{\substack{(x_1,\ldots,x_n)\in W^n \\ x_1\wedge\cdots\wedge x_n=\eta}} \|x_1\| \cdots \|x_n\| \Bigg) \Bigg(\inf_{\substack{(y_1,\ldots,y_m)\in(V\setminus W)^m \\ [y_1]\wedge\cdots\wedge[y_m]=\eta'}} \|[y_1]\| \cdots \|[y_m]\| \Bigg),$$

which leads to, by Proposition 1.1.66, the inequality $\|\eta \wedge \eta'\|_{\det} \leqslant \|\eta\|_{\det} \cdot \|\eta'\|_{\det}$. □

Proposition 1.1.69 *Let* $(V, \|\cdot\|_V)$ *and* $(W, \|\cdot\|_W)$ *be finite-dimensional seminormed vector spaces over* k*, and* n *and* m *be respectively the dimensions of* V *and* W *over* k*. We equip* $V \otimes_k W$ *with the* π*-tensor product seminorm* $\|\cdot\|_\pi$*. Then the natural* k*-linear isomorphism* $\det(V \otimes_k W) \cong \det(V)^{\otimes m} \otimes_k \det(W)^{\otimes n}$ *is an isometry, where we consider the determinant seminorm of* $\|\cdot\|_\pi$ *on* $\det(V \otimes_k W)$ *and the tensor product of determinant seminorms on* $\det(V)^{\otimes m} \otimes_k \det(W)^{\otimes n}$.

Proof Let $\|\cdot\|'$ be the seminorm on $\det(V)^{\otimes m} \otimes \det(W)^{\otimes n}$ given by tensor product of determinant seminorms. By Proposition 1.1.58, the seminorm $\|\cdot\|'$ identifies with the quotient of the π-tensor power on $(V \otimes_k W)^{\otimes nm}$ of the seminorm $\|\cdot\|_\pi$ on $V \otimes_k W$. In other words, $\|\cdot\|'$ identifies with $\|\cdot\|_{\pi,\det}$. □

Proposition 1.1.70 *Let* $(V, \|\cdot\|)$ *be a finite-dimensional seminormed vector space over* k *and* r *be the dimension of* V *over* k*. Let* i *be a positive integer. Then the canonical* k*-linear isomorphism* $\det(\Lambda^i V) \to \det(V)^{\otimes\binom{r-1}{i-1}}$ *is an isometry, where we consider the* i^{th} π*-exterior power seminorm on* $\Lambda^i V$.

Proof Consider the following commutative diagram

$$\begin{array}{ccc} V^{\otimes i\binom{r}{i}} & \xrightarrow{\;p_1\;} & \det(V)^{\otimes\binom{r-1}{i-1}} \\ {\scriptstyle p_2}\big\downarrow & & \big\uparrow{\scriptstyle \cong} \\ (\Lambda^i V)^{\otimes\binom{r}{i}} & \xrightarrow[\;p_3\;]{} & \det(\Lambda^i V) \end{array}$$

By definition, if we equip $V^{\otimes i\binom{r}{i}}$ with the π-tensor product seminorm, then its quotient seminorm on $(\Lambda^i V)^{\otimes\binom{r}{i}}$ identifies with the π-tensor product of the π-exterior power seminorm. Moreover, by Proposition 1.1.58, the quotient seminorm on $\det(\Lambda^i V)$ (induced by p_3) of the tensor product of the π-exterior power seminorm identifies with the determinant seminorm of the latter. Still by the same proposition, the quotient seminorm on $\det(V)^{\otimes\binom{r-1}{i-1}}$ induced by p_1 identifies with the tensor power of the determinant seminorm. Therefore the natural isomorphism $\det(\Lambda^i V) \to \det(V)^{\otimes\binom{r-1}{i-1}}$ preserves actually the seminorms by using (1) in Proposition 1.1.14. □

1.1.14 Seminormed graded algebra

Let $R_\bullet = \bigoplus_{n\in\mathbb{N}} R_n$ be a graded k-algebra such that, for any $n \in \mathbb{N}$, R_n is of finite dimension over k. For any $n \in \mathbb{N}$, let $\|\cdot\|_n$ be a seminorm on R_n. We say that $\overline{R}_\bullet = \{(R_n, \|\cdot\|_n)\}_{n\in\mathbb{N}}$ is a *seminormed graded algebra over* k if the following submultiplicativity condition is satisfied: for any $(n, m) \in \mathbb{N}^2$ and any $(a, b) \in R_n \times R_m$, one has

$$\|a \cdot b\|_{n+m} \leqslant \|a\|_n \cdot \|b\|_m.$$

Furthermore, we say that $\overline{R}_\bullet$ is *of finite type* if the underlying graded k-algebra $R_\bullet$ is of finite type over k.

Let $M_\bullet = \bigoplus_{m\in\mathbb{Z}} M_m$ be a $\mathbb{Z}$-graded k-linear space and h be a positive integer. We say that $M_\bullet$ is an *h-graded $R_\bullet$-module* if $M_\bullet$ is equipped with a structure of $R_\bullet$-module such that

$$\forall\,(n, m) \in \mathbb{N}\times\mathbb{Z}, \quad \forall\,(a, x) \in R_n \times M_m, \quad ax \in M_{nh+m}.$$

Let $M_\bullet$ be an h-graded $R_\bullet$-module. Assume that each homogeneous component M_m is of finite dimension over k and is equipped with a seminorm $\|\cdot\|_{M_m}$. We say that $\overline{M}_\bullet = \{(M_m, \|\cdot\|_{M_m})\}_{m\in\mathbb{Z}}$ is a *seminormed h-graded $\overline{R}_\bullet$-module* if the following condition is satisfied: for any $(n, m) \in \mathbb{N}\times\mathbb{Z}$ and any $(a, x) \in R_n \times M_m$, one has

$$\|a \cdot x\|_{M_{nh+m}} \leqslant \|a\|_n \cdot \|x\|_{M_m}.$$

We say that an h-graded $\overline{R}_\bullet$-module $\overline{M}_\bullet$ is *of finite type* if the underlying h-graded $R_\bullet$-module $M_\bullet$ is of finite type.

Proposition 1.1.71 *Let $\overline{R}_\bullet = \{(R_n, \|\cdot\|_n)\}_{n\in\mathbb{N}}$ be a seminormed graded algebra over k. Let $I_\bullet$ be a homogenous ideal of $R_\bullet$ and $R'_\bullet := R_\bullet/I_\bullet$.*

(1) *Let $\|\cdot\|'_n$ be the quotient seminorm on R'_n induced by $\|\cdot\|_n$ and $R_n \to R'_n$. Then $\overline{R}'_\bullet = \{(R'_n, \|\cdot\|'_n)\}_{n\in\mathbb{N}}$ forms a seminormed graded algebra over k.*
(2) *Let $\overline{M}_\bullet = \{(M_m, \|\cdot\|_{M_m})\}_{m\in\mathbb{N}}$ be a normed h-graded $\overline{R}_\bullet$-module and $f_\bullet : M_\bullet \to N_\bullet$ be a homomorphism of h-graded modules over $R_\bullet$.*[3] *We assume that $I_\bullet \cdot N_\bullet = 0$ and $f_m : M_m \to N_m$ is surjective for all $m \in \mathbb{Z}$. Let $\|\cdot\|_{N_m}$ be the quotient seminorm on N_m induced by $\|\cdot\|_{M_m}$ and f_m. Then $\overline{N}_\bullet = \{(N_m, \|\cdot\|_{N_m})\}_{m\in\mathbb{N}}$ forms a seminormed h-graded $\overline{R}'_\bullet$-module.*

Proof First let us see the following:

$$\forall\,(n, m) \in \mathbb{N}\times\mathbb{Z},\ (a', y) \in R'_n \times N_m, \quad \|a' \cdot y\|_{N_{nh+m}} \leqslant \|a'\|'_n \cdot \|y\|_{N_m}. \tag{1.20}$$

Indeed, for a fixed positive number ϵ, one can find $a \in R_n$ and $x \in M_m$ such that

$$\begin{cases} [a] = a', & \|a\|_n \leqslant \mathrm{e}^{\epsilon}\|a'\|'_n, \\ f_m(x) = y, & \|x\|_{M_m} \leqslant \mathrm{e}^{\epsilon}\|y\|_{N_m}. \end{cases}$$

Then, as $f_m(a \cdot x) = a' \cdot y$,

$$\|a' \cdot y\|_{N_{nh+m}} \leqslant \|a \cdot x\|_{M_{nh+m}} \leqslant \|a\|_n \cdot \|x\|_{M_m} \leqslant \mathrm{e}^{2\epsilon}\|a'\|'_n \cdot \|y\|_{N_m},$$

[3] That is, for each $m \in \mathbb{Z}$, $f_m : M_m \to N_m$ is a k-linear map such that $f_{nh+m}(a \cdot x) = [a] \cdot f_m(x)$ for all $a \in R_n$ and $x \in M_m$.

which implies (1.20) because ϵ is an arbitrary positive number. Applying (1.20) to the case where $\overline{M}_{\bullet} = \overline{R}_{\bullet}$ and $\overline{N}_{\bullet} = \overline{R}'_{\bullet}$, one has

$$\forall\,(n,n') \in \mathbb{N}^2, \quad \forall\,(a',b') \in R'_n \times R'_{n'}, \quad \|a' \cdot b'\|'_{n+n'} \leqslant \|a'\|'_n \cdot \|b'\|'_{n'}. \tag{1.21}$$

Thus (1) is proved, so that (2) is also proved by (1.20). □

1.1.15 Norm of polynomial

Let $k[X]$ be the polynomial ring of one variable over k. For $f = a_nX^n + \cdots + a_1X + a_0 \in k[X]$, We define $\|f\|$ to be

$$\|f\| := \max\{|a_n|, \ldots, |a_1|, |a_0|\}.$$

It is easy to see that $\|\cdot\|$ yields a norm of $k[X]$ over k.

Proposition 1.1.72 *For $f, g \in k[X]$, one has the following:*

(1) *If the absolute value of k is Archimedean, then*

$$\|fg\| \leqslant \min\{\deg(f)+1, \deg(g)+1\}\|f\| \cdot \|g\|,$$

where the degree of the zero polynomial is defined to be -1 by convention.

(2) *If the absolute value of k is non-Archimedean, then $\|fg\| = \|f\| \cdot \|g\|$.*

Proof Clearly we may assume that $f \neq 0$, $g \neq 0$ and $\deg(f) \leqslant \deg(g)$. We set

$$\begin{cases} f = a_nX^n + \cdots + a_1X + a_0, \\ g = b_mX^m + \cdots + b_1X + b_0, \\ fg = c_{n+m}X^{n+m} + \cdots + c_1X + c_0, \end{cases}$$

where $n = \deg(f)$ and $m = \deg(g)$. Then

$$c_l = \sum_{(i,j)\in\Delta(l)} a_ib_j,$$

where

$$\Delta(l) = \Big\{(i,j) : i + j = l,\ i \in \{0,\ldots,n\},\ j \in \{0,\ldots,m\}\Big\},$$

so that, as $\mathrm{card}(\Delta(l)) \leqslant n+1$, one has

$$|c_l| \leqslant \begin{cases} \displaystyle\sum_{(i,j)\in\Delta(l)} |a_i| \cdot |b_j| \leqslant (n+1)\|f\| \cdot \|g\| & \text{(Archimedean case)}, \\ \displaystyle\max_{(i,j)\in\Delta(l)} \{|a_i| \cdot |b_j|\} \leqslant \|f\| \cdot \|g\| & \text{(non-Archimedean case)}. \end{cases}$$

Thus (1) and the inequality $\|fg\| \leqslant \|f\| \cdot \|g\|$ in the non-Archimedean case are obtained.

Finally let us consider the converse inequality in the non-Archimedean case. We set

$$\alpha = \min\{i : |a_i| = \|f\|\} \quad \text{and} \quad \beta = \min\{j : |b_j| = \|g\|\}.$$

Note that if $i + j = \alpha + \beta$ and $(i, j) \neq (\alpha, \beta)$, then $|a_i| \cdot |b_j| < \|f\| \cdot \|g\|$ because either $i < \alpha$ or $j < \beta$. Therefore, $|c_{\alpha+\beta}| = \|f\| \cdot \|g\|$ by Proposition 1.1.5 and hence $\|fg\| \geqslant \|f\| \cdot \|g\|$. □

1.2 Orthogonality

The orthogonality of bases plays an important role in the study of finite-dimensional normed vector spaces. In the classic functional analysis over $\mathbb{R}$ or $\mathbb{C}$, the orthogonality often refers to a property related to an inner product. This property actually has an equivalent form, which has an analogue in the non-Archimedean case. However, in a finite-dimensional normed vector space over a non-Archimedean valued field, there may not exist an orthogonal basis. One can remedy this problem by introducing an approximative variant of the orthogonality. This technic is useful in the study of determinant norms.

1.2.1 Inner product

In this subsection, we assume that the absolute value $|\cdot|$ is Archimedean. In this case the field k is either $\mathbb{R}$ or $\mathbb{C}$ and we assume that $|\cdot|$ is the usual absolute value.

Let V be a vector space over k. A map $\langle , \rangle : V \times V \to k$ is called a *semidefinite inner product* on V if the following conditions are satisfied:

(i) $\langle x, ay + bz\rangle = a\langle x, y\rangle + b\langle x, z\rangle$ for all $(x, y, z) \in V^3$ and $(a, b) \in k^2$.
(ii) $\langle x, y\rangle = \overline{\langle y, x\rangle}$ for any $(x, y) \in V^2$, where $\overline{\langle y, x\rangle}$ is the complex conjugation of $\langle y, x\rangle$.
(iii) $\langle x, x\rangle \in \mathbb{R}_{\geqslant 0}$ for any $x \in V$.

If $\langle x, x\rangle > 0$ for any $x \in V \setminus \{0\}$, we just say that $\langle , \rangle$ is an *inner product*. Namely, an inner product means either a scalar product or a Hermitian product according to $k = \mathbb{R}$ or $\mathbb{C}$. Note that the semidefinite inner product $\langle , \rangle$ induces a seminorm $\|\cdot\|$ on V such that $\|x\| = \langle x, x\rangle^{1/2}$ for any $x \in V$.

Proposition 1.2.1 *Let V be a vector space over k, $\langle , \rangle$ be a semidefinite inner product on V and $\|\cdot\|$ be the seminorm induced by $\langle , \rangle$.*

(1) *For any $x \in N_{\|\cdot\|}$ and any $y \in V$ one has $\langle x, y\rangle = \langle y, x\rangle = 0$.*
(2) *The semidefinite inner product $\langle , \rangle$ induces by passing to quotient an inner product $\langle , \rangle^{\sim}$ on $V/N_{\|\cdot\|}$ such that*

$$\forall (x,y) \in V^2, \quad \langle [x],[y]\rangle^{\sim} = \langle x,y\rangle,$$

where $[x]$ *and* $[y]$ *are the classes of* x *and* y *in* $V/N_{\|\cdot\|}$*, respectively. Moreover, one has* $(\|\alpha\|^{\sim})^2 = \langle \alpha,\alpha\rangle^{\sim}$ *for any* $\alpha \in V/N_{\|\cdot\|}$.

(3) *Assume that* V *is of finite dimension over* k*. For any bounded linear form* f *on* V *there exists an element* y *in* V *such that* $f(x) = \langle y,x\rangle$ *for any* $x \in V$*. Moreover, the element* y *is unique up to addition by an element in* $N_{\|\cdot\|}$.

Proof (1) By Cauchy-Schwarz inequality, one has $|\langle x,y\rangle|^2 \leqslant \|x\|^2 \cdot \|y\|^2 = 0$. Hence $\langle x,y\rangle = 0$. Similarly, $\langle y,x\rangle = 0$.

(2) By (1) and the properties (i) and (ii) of semidefinite inner product, we obtain that, if x, x', y and y' are vectors in V such that $x - x' \in N_{\|\cdot\|}$ and $y - y' \in N_{\|\cdot\|}$, then $\langle x,y\rangle = \langle x',y'\rangle$. Therefore the semidefinite inner product $\langle\,,\rangle$ induces by passing to quotient a function

$$\langle\,,\rangle^{\sim} : (V/N_{\|\cdot\|}) \times (V/N_{\|\cdot\|}) \longrightarrow k.$$

From the definition it is straightforward to check that $\langle\,,\rangle^{\sim}$ is a semidefinite inner product and $\langle \alpha,\alpha\rangle^{\sim} = (\|\alpha\|^{\sim})^2$ for any $\alpha \in V/N_{\|\cdot\|}$. It remains to verify that $\langle\,,\rangle^{\sim}$ is definite. Let x be an element in V such that $\langle [x],[x]\rangle^{\sim} = 0$. Then one has $\langle x,x\rangle = \|x\|^2 = 0$. Hence $\|x\| = 0$, namely $x \in N_{\|\cdot\|}$.

(3) Since f is a bounded linear form, it vanishes on $N_{\|\cdot\|}$. Hence there exists a unique linear form $\widetilde{f} : V/N_{\|\cdot\|} \to k$ such that $\widetilde{f} \circ \pi = f$, where $\pi : V \to V/N_{\|\cdot\|}$ is the projection map. Moreover, by Riesz's representation theorem for usual finite-dimensional inner product space, there exists a unique $\beta \in V/N_{\|\cdot\|}$ such that $\widetilde{f}(\alpha) = \langle \beta,\alpha\rangle^{\sim}$ for any $\alpha \in V/N_{\|\cdot\|}$. Hence we obtain that the equivalence class β equals the set of $y \in V$ such that $f(x) = \langle y,x\rangle$ for any $x \in V$. □

Let V be a finite-dimensional vector space over k equipped with a seminorm $\|\cdot\|$. We say that the seminorm $\|\cdot\|$ is *Euclidean* (resp. *Hermitian*) if $k = \mathbb{R}$ (resp. $k = \mathbb{C}$) and if the seminorm $\|\cdot\|$ is induced by a semidefinite inner product. Note that if a seminorm $\|\cdot\|$ on V is Euclidean (resp. Hermitian), then also is its dual norm on V^*. In fact, if $\langle\,,\rangle$ is a semidefinite inner product on V and $\|\cdot\|$ is the corresponding seminorm, then it induces (by Riesz's representation theorem) an $\mathbb{R}$-linear isometry $\iota : (V/N_{\|\cdot\|}, \|\cdot\|^{\sim}) \to (V^*, \|\cdot\|_*)$ such that

$$\forall (x,y) \in V^2, \quad \iota([x])(y) = \langle x,y\rangle.$$

Moreover, for $a \in k$ and $x \in V$ one has $\iota(ax) = \overline{a}\,\iota(x)$. Then the dual norm on V^* is induced by the following inner product $\langle\,,\rangle_*$:

$$\forall (\alpha,\beta) \in (V^*)^2, \quad \langle \alpha,\beta\rangle_* = \overline{\langle \iota^{-1}(\alpha), \iota^{-1}(\beta)\rangle^{\sim}}.$$

Remark 1.2.2 Let $\psi : [0,1] \to [0,1]$ be the function $t \mapsto (t^2 + (1-t)^2)^{1/2}$. If V and W are finite-dimensional vector spaces over k equipped with semidefinite inner products, then the direct sum seminorm $\|\cdot\|_{\psi}$ on $V \oplus W$ as constructed in §1.1.10 is induced by the semidefinite inner product on $V \oplus W$ defined as $\big\langle (x,y),(x',y')\big\rangle :=$

$\langle x, x'\rangle + \langle y, y'\rangle$. The seminorm $\|\cdot\|_\psi$ is called the *orthogonal direct sum* of the seminorms on V and W corresponding to their semidefinite inner products.

1.2.2 Orthogonal basis of an inner product

In this subsection, we assume that the absolute value $|\cdot|$ is Archimedean. Let V be a finite-dimensional vector space over k equipped with a semidefinite inner product $\langle , \rangle$. Let $\|\cdot\|$ be the seminorm induced by $\langle , \rangle$. We say that a basis $\{e_1, \ldots, e_r\}$ of V is *orthogonal* if $\langle e_i, e_j\rangle = 0$ for distinct indices i and j in $\{1, \ldots, r\}$. If in addition $\langle e_i, e_i\rangle = 1$ for any $i \in \{1, \ldots, r\}$ such that $e_i \in V \setminus N_{\|\cdot\|}$, we say that $\{e_1, \ldots, e_r\}$ is an *orthonormal basis*. Note that, if $\{e_1, \ldots, e_r\}$ is an orthogonal basis, then

$$\forall\, (\lambda_1, \ldots, \lambda_r) \in k^r, \quad \|\lambda_1 e_1 + \cdots + \lambda_r e_r\|^2 = \sum_{i=1}^{r} |\lambda_i|^2 \cdot \|e_i\|^2. \tag{1.22}$$

Moreover, by the Gram-Schmidt process, there always exists an orthonormal basis of V (cf. the proof of Proposition 1.2.30).

The following proposition provides an alternative form for the orthogonality condition of a basis in a finite-dimensional vector space equipped with a semidefinite inner product.

Proposition 1.2.3 *Let V be a finite-dimensional vector space over k, equipped with a semidefinite inner product $\langle , \rangle$. Let $\{e_i\}_{i=1}^r$ be a basis of V. Then it is an orthogonal basis if and only if the following condition is satisfied:*

$$\forall\, (\lambda_1, \ldots, \lambda_r) \in k^r, \quad \|\lambda_1 e_1 + \cdots + \lambda_r e_r\| \geqslant \max_{i \in \{1, \ldots, r\}} \|\lambda_i e_i\|. \tag{1.23}$$

Proof If $\{e_i\}_{i=1}^r$ is an orthogonal basis of V, then by (1.22) we obtain that the inequality (1.23) holds. Conversely, assume given a basis $\{e_i\}_{i=1}^r$ of V which verifies the condition (1.23). Then for any $(\lambda_1, \ldots, \lambda_{r-1}) \in k^{r-1}$, one has

$$\|\lambda_1 e_1 + \cdots + \lambda_{r-1} e_{r-1} + e_r\| \geqslant \|e_r\|,$$

which implies that e_r is orthogonal to the vector subspace generated by $e_1, \ldots, e_{r-1}$. Indeed, $\|(\pm\epsilon)e_i + e_r\| \geqslant \|e_r\|$ for $\epsilon > 0$, which implies that $\epsilon\|e_i\|^2 \pm 2\langle e_i, e_r\rangle \geqslant 0$, and hence $\pm\langle e_i, e_r\rangle \geqslant 0$ by taking the limit when $\epsilon \to 0$, as required. Therefore by induction we obtain that the basis $\{e_i\}_{i=1}^r$ is an orthogonal basis. □

1.2.3 Orthogonality in general cases

In this subsection, we consider a general valued field $(k, |\cdot|)$, which is not necessarily Archimedean. Let V be a finite-dimensional vector space over k and $\|\cdot\|$ be a seminorm on V. We say that a basis $\{e_i\}_{i=1}^r$ of V is *orthogonal* if for any $(a_1, \ldots, a_r) \in k^r$ one has

$$\|a_1 e_1 + \cdots + a_r e_r\| \geqslant \max_{i \in \{1,\ldots,r\}} \|a_i e_i\|.$$

If in addition $\|e_i\| = 1$ for any $i \in \{1, \ldots, r\}$ such that $e_i \in V \setminus N_{\|\cdot\|}$, we say that the basis $\{e_i\}_{i=1}^r$ is *orthonormal*. We have seen in Proposition 1.2.3 that this definition is equivalent to the definition in §1.2.2 when the absolute value $|\cdot|$ is Archimedean and the seminorm $\|\cdot\|$ is induced by a semidefinite inner product.

The existence of an orthogonal basis in the non-Archimedean case is not always true. We refer the readers to [120, Example 2.3.26] for a counter-example. Thus we need a refinement of the notion of orthogonality.

Definition 1.2.4 Let $(V, \|\cdot\|)$ be a finite-dimensional seminormed vector space over k, and $\alpha \in \,]0, 1]$. We say that a basis $\{e_1, \ldots, e_r\}$ of V is *α-orthogonal* if for any $(\lambda_1, \ldots, \lambda_r) \in k^r$ one has

$$\|\lambda_1 e_1 + \cdots + \lambda_r e_r\| \geqslant \alpha \max\{|\lambda_1| \cdot \|e_1\|, \ldots, |\lambda_r| \cdot \|e_r\|\}.$$

Note that the 1-orthogonality is just the orthogonality defined in the beginning of the subsection. We refer the readers to [120, §2.3] for more details about this notion.

Proposition 1.2.5 *Let $(V, \|\cdot\|)$ be a finite-dimensional seminormed vector space over k, α be an element in $]0, 1]$, and $\boldsymbol{e} = \{e_i\}_{i=1}^r$ be an α-orthogonal basis of $(V, \|\cdot\|)$. Then the intersection of $\boldsymbol{e}$ with $N_{\|\cdot\|}$ forms a basis of $N_{\|\cdot\|}$.*

Proof Without loss of generality, we assume that $\boldsymbol{e} \cap N_{\|\cdot\|} = \{e_1, \ldots, e_n\}$, where $n \in \mathbb{N}$, $n \leqslant r$. Suppose that $N_{\|\cdot\|}$ is not generated by $\boldsymbol{e} \cap N_{\|\cdot\|}$, then there exists an element $x = \lambda_1 e_1 + \cdots + \lambda_r e_r$ in $N_{\|\cdot\|}$ which does not belong to the vector subspace of V generated by $\boldsymbol{e} \cap N_{\|\cdot\|}$. Therefore there exists $i \in \{n+1, \ldots, r\}$ such that $\lambda_i \neq 0$. Since the basis $\boldsymbol{e}$ is α-orthogonal, one has

$$0 = \|x\| \geqslant \alpha|\lambda_i| \cdot \|e_i\| > 0,$$

which leads to a contradiction. □

Proposition 1.2.6 *Let $(V, \|\cdot\|)$ be a finite-dimensional seminormed vector space over k, $\alpha \in \,]0, 1]$ and $\boldsymbol{e}$ be an α-orthogonal basis of V. Let $\boldsymbol{e}'$ be a subset of $\boldsymbol{e}$ and W be the vector subspace of V generated by all vectors in $\boldsymbol{e}'$.*

(1) *The set $\boldsymbol{e}'$ is an α-orthogonal basis of W with respect to the restriction of $\|\cdot\|$ to W.*
(2) *The image of $\boldsymbol{e} \cap (V \setminus W)$ in V/W forms an α-orthogonal basis of V/W with respect to the quotient seminorm of $\|\cdot\|$. Moreover, for any $x \in \boldsymbol{e} \cap (V \setminus W)$, the quotient seminorm of the class of x is bounded from below by $\alpha\|x\|$. In particular, if $\alpha = 1$, namely $\boldsymbol{e}$ is an orthogonal basis, then for any element $x \in \boldsymbol{e} \cap (V \setminus W)$, the quotient seminorm of the class of x in V/W is equal to $\|x\|$.*

Proof (1) Assume that $\boldsymbol{e}' = \{e_1, \ldots, e_n\}$. Since W is generated by the vectors in $\boldsymbol{e}'$, $\{e_1, \ldots, e_n\}$ is a basis of W. Since $\boldsymbol{e}$ is an α-orthogonal basis of V, for any $(\lambda_1, \ldots, \lambda_n) \in k^n$ one has

$$\|\lambda_1 e_1 + \cdots + \lambda_n e_n\| \geqslant \alpha \max\{|\lambda_1| \cdot \|e_1\|, \ldots, |\lambda_n| \cdot \|e_n\|\}.$$

Therefore $\{e_1, \ldots, e_n\}$ is an α-orthogonal basis of W.

(2) Assume that $\boldsymbol{e}' = \{e_1, \ldots, e_n\}$ and $\boldsymbol{e} \cap (V \setminus W) = \{e_{n+1}, \ldots, e_r\}$. It is clear that the canonical image of $\boldsymbol{e} \cap (V \setminus W)$ in V/W forms a basis of V/W. It remains to show that it is an α-orthogonal basis. Let $\|\cdot\|'$ be the quotient seminorm of $\|\cdot\|$ on V/W. For any $i \in \{n+1, \ldots, r\}$, let y_i be the canonical image of e_i in V/W. Let $(\lambda_1, \ldots, \lambda_r)$ be an element in k^r. Since $\boldsymbol{e}$ is an α-orthogonal basis of $(V, \|\cdot\|)$, for any $(\lambda_1, \ldots, \lambda_r) \in k^r$ one has

$$\|\lambda_1 e_1 + \cdots + \lambda_r e_r\| \geqslant \alpha \max_{i \in \{1,\ldots,r\}} |\lambda_i| \cdot \|e_i\| \geqslant \alpha \max_{i \in \{n+1,\ldots,r\}} |\lambda_i| \cdot \|e_i\|.$$

Therefore, for any $(\lambda_{n+1}, \ldots, \lambda_r) \in k^{r-n}$ one has (note that $y_j = [e_j]$)

$$\|\lambda_{n+1} y_{n+1} + \cdots + \lambda_r y_r\|' \geqslant \alpha \max_{i \in \{n+1,\ldots,r\}} |\lambda_i| \cdot \|e_i\| \geqslant \alpha \max_{i \in \{n+1,\ldots,r\}} |\lambda_i| \cdot \|y_i\|'.$$

Hence $\{y_i\}_{i=n+1}^r$ is an α-orthogonal basis of V/W. The first inequality also implies that $\|y_i\|' \geqslant \alpha \|e_i\|$ for any $i \in \{n+1, \ldots, r\}$. If $\alpha = 1$, for any $i \in \{n+1, \ldots, r\}$ one has

$$\|y_i\|' \geqslant \|e_i\| \geqslant \|y_i\|',$$

which leads to the equality $\|y_i\|' = \|e_i\|$. □

The following proposition shows that, in the Archimedean case, any finite-dimensional normed vector space admits an orthogonal basis. In general case, for any $\alpha \in]0, 1[$, any finite-dimensional normed vector space admits an α-orthogonal basis.

Proposition 1.2.7 *Let $(V, \|\cdot\|)$ be a finite-dimensional normed vector space over k. Then we have the following:*

(1) *For any $\alpha \in]0, 1[$, there exists an α-orthogonal basis of V.*
(2) *Any Hadamard basis of V is orthogonal (see Remark 1.1.67).*
(3) *If the field k is locally compact and the absolute value $|\cdot|$ is not trivial, then V admits an Hadamard basis, which is also an orthogonal basis.*

Proof (1), (2) By Proposition 1.1.66, we can choose a basis $\boldsymbol{e} = \{e_i\}_{i=1}^r$ such that

$$\frac{\|e_1\| \cdots \|e_r\|}{\|e_1 \wedge \cdots \wedge e_r\|} \leqslant \alpha^{-1}.$$

We claim that $\boldsymbol{e} = \{e_i\}_{i=1}^r$ is α-orthogonal. Let $(\lambda_1, \ldots, \lambda_r)$ be an element in k^r and $x = \lambda_1 e_1 + \cdots + \lambda_r e_r$. For any $i \in \{1, \ldots, r\}$, by Proposition 1.1.66 one has

$$\frac{\|e_1\|\cdots\|e_r\|}{\|e_1\wedge\cdots\wedge e_r\|}\leqslant \alpha^{-1}\frac{\|e_1\|\cdots\|e_{i-1}\|\cdot\|x\|\cdot\|e_{i+1}\|\cdots\|e_r\|}{\|e_1\wedge\cdots e_{i-1}\wedge x\wedge e_{i+1}\wedge\cdots\wedge e_r\|}.$$

Therefore one has $\|x\| \geqslant \alpha|\lambda_i|\cdot\|e_i\|$. Since $i \in \{1,\ldots,r\}$ is arbitrary, we obtain that $\{e_i\}_{i=1}^r$ is an α-orthogonal basis. A similar argument also shows that an Hadamard basis is necessarily orthogonal.

(3) We assume that k is locally compact and $|\cdot|$ is not trivial. Then, as V is locally compact, there is $a_0 \in k^\times$ such that $(V, \|\cdot\|)_{\leqslant|a_0|}$ is compact. By Proposition 1.1.4, if we choose λ with $\lambda < \sup\{|a| : a \in k^\times, |a| < 1\}$, then, for any $x \in V \setminus \{0\}$, there is $b \in k^\times$ such that $\lambda \leqslant \|bx\| < 1$. Here we set

$$C = \{x \in V : \lambda|a_0| \leqslant \|x\| \leqslant |a_0|\},$$

which is a compact set in V. For $(x_1,\ldots,x_r) \in (V\setminus\{0\})^r$, there are $b_1,\ldots,b_r \in k^\times$ such that $\lambda \leqslant \|b_ix_i\| < 1$ for all i, so that $(a_0b_1x_1,\ldots,a_0b_rx_r) \in C^r$ and

$$\frac{\|(a_0b_1x_1)\wedge\cdots\wedge(a_0b_rx_r)\|}{\|a_0b_1x_1\|\cdots\|a_0b_rx_r\|} = \frac{\|x_1\wedge\cdots\wedge x_r\|}{\|x_1\|\cdots\|x_r\|}.$$

Hence the function

$$(x_1,\ldots,x_r)\longmapsto\frac{\|x_1\wedge\cdots\wedge x_r\|}{\|x_1\|\cdots\|x_r\|}$$

attains its maximal value on $(V\setminus\{0\})^r$, which is equal to 1. The proposition is thus proved. □

Remark 1.2.8 (1) In the case where $|\cdot|$ is trivial and $\|\cdot\|$ is ultrametric, there is an orthogonal basis $\boldsymbol{e}$ for $\|\cdot\|$ by Proposition 1.2.30. Thus, by Proposition 1.2.23, $\boldsymbol{e}$ is an Hadamard basis of $(V, \|\cdot\|)$.

(2) We assume that k is an infinite field and the absolute value $|\cdot|$ is trivial. Fix a map $\lambda : \mathbb{P}^1(k) \to [\frac{1}{2}, 1]$. Let $\pi : k^2 \setminus \{(0,0)\} \to \mathbb{P}^1(k)$ be the natural map. We set

$$\forall\, x \in k^2,\quad \|x\|_\lambda := \begin{cases}\lambda(\pi(x)) & \text{if } x \neq (0,0),\\ 0 & \text{if } x = (0,0).\end{cases}$$

It is easy to see that $\|\cdot\|_\lambda$ satisfies the axioms of norm: (1) $\|ax\|_\lambda = |a|\cdot\|x\|_\lambda$; (2) $\|x+y\|_\lambda \leqslant \|x\|_\lambda + \|y\|_\lambda$; (3) $\|x\|_\lambda = 0 \iff x = 0$. Choosing an infinite subset $S = \{\zeta_1, \zeta_2, \ldots, \zeta_n, \ldots\}$ of $\mathbb{P}^1(k)$, we consider λ given by

$$\lambda(\zeta) := \begin{cases}\frac{1}{2} + (\frac{1}{2})^n & \text{if } \zeta \in S \text{ and } \zeta = \zeta_n,\\ 1 & \text{otherwise.}\end{cases}$$

Then $\lambda(\mathbb{P}^1(k)) \subseteq]\frac{1}{2}, 1]$, and for any $\epsilon > 1/2$ there is $\zeta \in \mathbb{P}^1(k)$ with $\lambda(\zeta) < \epsilon$. Obviously $\{\|x\|_\lambda : x \in k^2\}$ is an infinite set, which means that (2) in Corollary 1.1.6 does not hold without the assumption that the norm is ultrametric. Moreover, let us see that there is no orthogonal basis for $\|\cdot\|_\lambda$. Indeed, we assume that $\{e_1, e_2\}$ is an orthogonal basis for $\|\cdot\|_\lambda$. By the property of λ, there is $x \in k^2 \setminus \{(0,0)\}$ such that

$\|x\|_\lambda < \min\{\|e_1\|_\lambda, \|e_2\|_\lambda\}$. If we set $x = ae_1 + be_2$ with $(a, b) \neq (0, 0)$, then

$$\|x\|_\lambda \geqslant \max\{|a| \cdot \|e_1\|_\lambda, |b| \cdot \|e_2\|_\lambda\} \geqslant \min\{\|e_1\|_\lambda, \|e_2\|_\lambda\},$$

which is a contradiction.

Corollary 1.2.9 *Let $(V, \|\cdot\|)$ be a finite-dimensional seminormed vector space over k. For any $\alpha \in \,]0, 1[$, there exists an α-orthogonal basis of $(V, \|\cdot\|)$. If the absolute value $|\cdot|$ is non-trivial and $(k, |\cdot|)$ is locally compact, then $(V, \|\cdot\|)$ admits an orthogonal basis.*

Proof If the absolute value $|\cdot|$ is non-trivial and $(k, |\cdot|)$ is locally compact, let α be an element in $]0, 1]$, otherwise let α be an element in $]0, 1[$. Let W be the quotient vector space $V/N_{\|\cdot\|}$, equipped with the quotient norm $\|\cdot\|^\sim$. By Proposition 1.2.7, the normed vector space $(W, \|\cdot\|^\sim)$ admits an α-orthogonal basis $\{x_i\}_{i=1}^n$. For any $i \in \{1, \ldots, n\}$, let e_i be an element in the class x_i. We also choose a basis $\{e_j\}_{j=n+1}^r$ of $N_{\|\cdot\|}$. Hence $\{e_i\}_{i=1}^r$ becomes a basis of V. For any $(\lambda_1, \ldots, \lambda_r) \in k^r$ one has

$$\begin{aligned}\|\lambda_1 e_1 + \cdots + \lambda_r e_r\| &= \|\lambda_1 x_1 + \cdots + \lambda_n x_n\|^\sim \geqslant \alpha \max_{i\in\{1,\ldots,n\}} |\lambda_i| \cdot \|x_i\|^\sim \\ &= \alpha \max_{i\in\{1,\ldots,n\}} |\lambda_i| \cdot \|e_i\| = \alpha \max_{i\in\{1,\ldots,r\}} |\lambda_i| \cdot \|e_i\|.\end{aligned}$$

Lemma 1.2.10 *Let $(V, \|\cdot\|)$ be a finite-dimensional seminormed vector space over k, and $\alpha \in \,]0, 1]$. If $\boldsymbol{e} = \{e_i\}_{i=1}^r$ is an α-orthogonal basis of V and if $\{e_i^\vee\}_{i=1}^r$ is its dual basis, then, for any $i \in \{1, \ldots, r\}$, $e_i \notin N_{\|\cdot\|}$ if and only if $e_i^\vee \in V^*$, and in this case one has*

$$1 \leqslant \|e_i^\vee\|_* \cdot \|e_i\| \leqslant \alpha^{-1}. \tag{1.24}$$

Proof The hypothesis that $e_i \notin N_{\|\cdot\|}$ actually implies that $e_i^\vee$ vanishes on $N_{\|\cdot\|}$ since $N_{\|\cdot\|}$ is generated by $\boldsymbol{e} \cap N_{\|\cdot\|}$ (see Proposition 1.2.5). By Corollary 1.1.13, $e_i^\vee$ belongs to V^*. Conversely, if e_i belongs to $N_{\|\cdot\|}$ then $e_i^\vee$ is not a bounded linear form on V since it takes non-zero value on $e_i \in N_{\|\cdot\|}$.

The first inequality of (1.24) comes from the formula (1.4) in §1.1.5. In the following, we prove the second inequality. For any $(\lambda_1, \ldots, \lambda_r) \in k^r$ one has

$$e_i^\vee(\lambda_1 e_1 + \cdots + \lambda_r e_r) = \lambda_i.$$

Hence

$$\|e_i^\vee\|_* = \sup_{\substack{(\lambda_1,\ldots,\lambda_r)\in k^r \\ \lambda_i \neq 0}} \frac{|\lambda_i|}{\|\lambda_1 e_1 + \cdots + \lambda_r e_r\|} \leqslant \alpha^{-1}\|e_i\|^{-1},$$

where the inequality comes from the hypothesis that the basis $\{e_i\}_{i=1}^r$ is α-orthogonal (so that $\|\lambda_1 e_1 + \cdots + \lambda_r e_r\| \geqslant \alpha|\lambda_i| \cdot \|e_i\|$). □

Proposition 1.2.11 *Let V be a finite-dimensional seminormed vector space over k and $\alpha \in \,]0, 1]$. If $\{e_i\}_{i=1}^r$ is an α-orthogonal basis of V and if $\{e_i^\vee\}_{i=1}^r$ is the dual* *basis of* $\{e_i\}_{i=1}^r$, *then* $\{e_i^\vee\}_{i=1}^r \cap V^*$ *is an* α*-orthogonal basis of* V^*. *Moreover,* $\{e_i\}_{i=1}^r$ *is an* α*-orthogonal basis of* $(V, \|\cdot\|_{**})$ *and one has*

$$\alpha\|e_i\| \leqslant \|e_i\|_{**} \leqslant \|e_i\|. \tag{1.25}$$

Proof By Proposition 1.2.5 and Lemma 1.2.10, the cardinal of $\{e_i^\vee\}_{i=1}^r \cap V^*$, which is equal to that of $\{e_i\}_{i=1}^r \cap (V \setminus N_{\|\cdot\|})$, is $\dim_k(V^*) = \dim_k(V/N_{\|\cdot\|})$. Therefore $\{e_i^\vee\}_{i=1}^r \cap V^*$ is a basis of V^*.

Consider $\xi = a_1 e_1^\vee + \cdots + a_r e_r^\vee$ in V^*. As $\xi(e_i) = a_i$ we get that

$$\|\xi\|_* \geqslant \frac{|a_i|}{\|e_i\|} \geqslant \alpha|a_i| \cdot \|e_i^\vee\|_*$$

for any $i \in \{1, \ldots, r\}$ such that $\|e_i\| \neq 0$, where the second inequality comes from Lemma 1.2.10. This implies that $\{e_i^\vee\}_{i=1}^r \cap V^*$ is an α-orthogonal basis of V^*.

Let $x = \lambda_1 e_1 + \cdots + \lambda_r e_r$ be an element in V. Without loss of generality, we assume that $\boldsymbol{e} \cap N_{\|\cdot\|} = \{e_{n+1}, \ldots, e_r\}$. By definition, for $i \in \{1, \ldots, n\}$ one has

$$\|x\|_{**} \geqslant \frac{|e_i^\vee(x)|}{\|e_i^\vee\|_*} = \frac{|\lambda_i|}{\|e_i^\vee\|_*}. \tag{1.26}$$

By Lemma 1.2.10, one has $1 \leqslant \|e_i^\vee\|_* \cdot \|e_i\| \leqslant \alpha^{-1}$ and hence

$$\|x\|_{**} \geqslant \alpha|\lambda_i| \cdot \|e_i\| \geqslant \alpha|\lambda_i| \cdot \|e_i\|_{**}, \tag{1.27}$$

where the last inequality comes from (1.5). For $i \in \{n+1, \ldots, r\}$ one has $\|e_i\|_{**} = \|e_i\| = 0$. Hence

$$\|x\|_{**} \geqslant \alpha \max_{i \in \{1,\ldots,r\}} |\lambda_i| \cdot \|e_i\|_{**},$$

which shows that $\{e_i\}_{i=1}^r$ is an α-orthogonal basis of $(V, \|\cdot\|_{**})$. Moreover, (1.27) also implies that $\|e_i\|_{**} \geqslant \alpha\|e_i\|$ for $i \in \{1, \ldots, n\}$, which, joint with the relation

$$\forall\, i \in \{n+1, \ldots, r\}, \quad \|e_i\| = \|e_i\|_{**} = 0,$$

leads to the first inequality of (1.25). The second inequality of (1.25) comes from (1.5). □

Corollary 1.2.12 *We suppose that the absolute value* $|\cdot|$ *is non-Archimedean. Let* $(V, \|\cdot\|)$ *be a finite-dimensional seminormed vector space over* k. *Then the double dual seminorm* $\|\cdot\|_{**}$ *on* V *is the largest ultrametric seminorm on* V *which is bounded from above by* $\|\cdot\|$, *and one has* $\|\cdot\| \leqslant \dim_k(V)\|\cdot\|_{**}$. *If the seminorm* $\|\cdot\|$ *is ultrametric, then one has* $\|\cdot\|_{**} = \|\cdot\|$.

Proof We have seen in Remark 1.1.19 that the double dual seminorm $\|\cdot\|_{**}$ is ultrametric, and in the formula (1.5) of §1.1.5 that it is bounded from above by $\|\cdot\|$. Let $\|\cdot\|'$ be an ultrametric seminorm on V such that $\|\cdot\|' \leqslant \|\cdot\|$. We will show that $\|\cdot\|' \leqslant \|\cdot\|_{**}$ and $\|\cdot\| \leqslant r\|\cdot\|_{**}$, where r is the dimension of V over k.

Let $\alpha \in\ ]0,1[$. By Proposition 1.2.7, there exists an α-orthogonal basis $\{e_i\}_{i=1}^r$ of $(V, \|\cdot\|)$. For any vector $x = \lambda_1 e_1 + \cdots + \lambda_r e_r$ in V one has

$$\alpha^2\|x\|' \leqslant \alpha^2 \max_{i\in\{1,\ldots,r\}} |\lambda_i|\cdot\|e_i\| \leqslant \alpha \max_{i\in\{1,\ldots,r\}} |\lambda_i|\cdot\|e_i\|_{**} \leqslant \|x\|_{**},$$

where the second inequality comes from (1.25) and the third inequality follows from the fact that $\{e_i\}_{i=1}^r$ is an α-orthogonal basis for $\|\cdot\|_{**}$ (see Proposition 1.2.11). Moreover, by the triangle inequality one has

$$\alpha^2\|x\| \leqslant \alpha^2\sum_{i=1}^r |\lambda_i|\cdot\|e_i\| \leqslant r\alpha^2 \max_{i\in\{1,\ldots,r\}} |\lambda_i|\cdot\|e_i\| \leqslant r\|x\|_{**}$$

Since $\alpha \in\ ]0,1[$ is arbitrary, we obtain $\|\cdot\|' \leqslant \|\cdot\|_{**}$ and $\|\cdot\| \leqslant r\|\cdot\|_{**}$. The first assertion of the proposition is thus proved.

If $\|\cdot\|$ is ultrametric, it is certainly the largest ultrametric norm bounded from above by $\|\cdot\|$. Hence one has $\|\cdot\| = \|\cdot\|_{**}$. □

Remark 1.2.13 In Corollary 1.2.12, the constant $\dim_k(V)$ in the inequality $\|\cdot\| \leqslant \dim_k(V)\|\cdot\|_{**}$ is optimal. We can consider for example the vector space $V = k^r$ equipped with the ℓ^1-norm

$$\forall\,(a_1,\ldots,a_r)\in k^r,\quad \|(a_1,\ldots,a_r)\|_{\ell^1} = |a_1|+\cdots+|a_r|.$$

Then its double dual norm is given by

$$\forall\,(a_1,\ldots,a_r)\in k^r,\quad \|(a_1,\ldots,a_r)\|_{\ell^1,**} = \max\{|a_1|,\ldots|a_r|\}.$$

In particular, one has $\|(1,\ldots,1)\|_{\ell^1} = r\|(1,\ldots,1)\|_{\ell^1,**}$.

Proposition 1.2.14 *Let $(V, \|\cdot\|)$ be a finite-dimensional seminormed vector space over the field k.*

(1) *The seminorm $\|\cdot\|$ and its double dual seminorm $\|\cdot\|_{**}$ induce the same dual norm on the vector space V^* of bounded linear forms.*
(2) *If W is a quotient space of dimension 1 of V, then the seminorms $\|\cdot\|$ and $\|\cdot\|_{**}$ induce the same quotient seminorm on W.*

Proof (1) The Archimedean case follows from Proposition 1.1.18. It suffices to treat the case where the absolute value $|\cdot|$ is non-Archimedean. Since $\|\cdot\|_*$ is ultrametric, by Corollary 1.2.12, one has $\|\cdot\|_* = \|\cdot\|_{*,**} = \|\cdot\|_{**,*}$.

(2) If the kernel of the quotient map $V \to W$ does not contain $N_{\|\cdot\|}$, then the quotient seminorm of $\|\cdot\|$ on W vanishes because $\dim_K W = 1$. The quotient seminorm of $\|\cdot\|_{**}$ on W also vanishes since we have observed in the proof of (1) that $N_{\|\cdot\|} = N_{\|\cdot\|_{**}}$. In the following we treat the case where the kernel of the projection map $V \to W$ contains $N_{\|\cdot\|}$, or equivalent, the quotient seminorms of $\|\cdot\|$ and $\|\cdot\|_{**}$ on W are actually norms. Since W is of dimension 1, any norm on W is uniquely determined by its dual norm on $W^\vee$. Let $\|\cdot\|_W$ be the quotient norm on W induced

by $\|\cdot\|$. By Proposition 1.1.20, the dual norm $\|\cdot\|_{W,*}$ identifies with the restriction of $\|\cdot\|_*$ to $W^\vee$ (viewed as a vector subspace of V^*). By (1), the norm $\|\cdot\|_*$ identifies with the dual norm of $\|\cdot\|_{**}$. As a consequence, $\|\cdot\|_W$ coincides with the quotient norm of $\|\cdot\|_{**}$. □

Proposition 1.2.15 *We assume that the absolute value $|\cdot|$ is non-Archimedean. Let $(V, \|\cdot\|)$ be a finite-dimensional seminormed vector space over k and let r be the dimension of V. Then the quotient seminorm of the ε-tensor product seminorm $\|\cdot\|_\varepsilon$ on $V^{\otimes r}$ by the canonical quotient map $V^{\otimes r} \to \det(V)$ identifies with the determinant seminorm on $\det(V)$ induced by $\|\cdot\|$. In particular, $\|\cdot\|$ and $\|\cdot\|_{**}$ induce the same determinant seminorm on $\det(V)$.*

Proof Denote by $\|\cdot\|_{\det_\varepsilon}$ the quotient seminorm on $\det(V)$ of the ε-tensor product seminorm on $V^{\otimes r}$. We have seen in Proposition 1.1.54 that the ε-tensor product seminorm is always bounded from above by the π-tensor product seminorm. Therefore, one has $\|\cdot\|_{\det_\varepsilon} \leqslant \|\cdot\|_{\det}$. Moreover, if $\|\cdot\|$ is not a norm, then the seminorm $\|\cdot\|_{\det}$ vanishes. Hence the seminorm $\|\cdot\|_{\det_\varepsilon}$ also vanishes. To prove the first assertion of the proposition, it remains to verify the inequality $\|\cdot\|_{\det} \leqslant \|\cdot\|_{\det_\varepsilon}$ in the case where $\|\cdot\|$ is a norm.

Consider a tensor vector φ in $V^{\otimes r}$, which is also viewed as a k-multilinear form on $(V^\vee)^r$. By definition, one has

$$\|\varphi\|_\varepsilon = \sup_{\substack{(f_1,\ldots,f_r)\in(V^\vee)^r \\ \forall\, i\in\{1,\ldots,r\},\ f_i\neq 0}} \frac{|\varphi(f_1,\ldots,f_r)|}{\|f_1\|_*\cdots\|f_r\|_*}.$$

Let $\alpha \in \,]0,1[$, $\{x_i\}_{i=1}^r$ be an α-orthogonal basis of V, and $\{x_i^\vee\}_{i=1}^r$ be its dual basis of $V^\vee$. Assume that φ is written in the form

$$\varphi = \sum_{I=(i_1,\ldots,i_r)\in\{1,\ldots,r\}^r} a_I(x_{i_1}\otimes\cdots\otimes x_{i_r}),$$

where $a_I \in k$. Then one has

$$\forall\,(i_1,\ldots,i_r)\in\{1,\ldots,r\}^r,\quad \varphi(x_{i_1}^\vee,\ldots,x_{i_r}^\vee) = a_{(i_1,\ldots,i_r)}.$$

In particular,

$$\forall\,(i_1,\ldots,i_r)\in\{1,\ldots,r\}^r,\quad \|\varphi\|_\varepsilon \geqslant \frac{|a_{(i_1,\ldots,i_r)}|}{\|x_{i_1}^\vee\|_*\cdots\|x_{i_r}^\vee\|_*}.$$

Note that the canonical image η of φ in $\det(V)$ is

$$\Big(\sum_{\sigma\in\mathfrak{S}_r} \operatorname{sgn}(\sigma)a_{(\sigma(1),\ldots,\sigma(r))}\Big)x_1\wedge\cdots\wedge x_r,$$

where $\mathfrak{S}_r$ is the symmetric group of order r, namely the group of all bijections from the set $\{1,\ldots,r\}$ to itself, and $\operatorname{sgn}(\cdot) : \mathfrak{S}_r \to \{\pm 1\}$ denotes the character of signature.

Hence

$$\begin{aligned}\|\eta\|_{\det} &= \Big|\sum_{\sigma\in\mathfrak{S}_r}\operatorname{sgn}(\sigma)a_{(\sigma(1),\ldots,\sigma(r))}\Big|\cdot\|x_1\wedge\cdots\wedge x_r\|_{\det},\\ &\leqslant \Big|\sum_{\sigma\in\mathfrak{S}_r}\operatorname{sgn}(\sigma)a_{(\sigma(1),\ldots,\sigma(r))}\Big|\cdot\|x_1\|\cdots\|x_r\|,\end{aligned} \tag{1.28}$$

where the inequality follows from (1.18). Since the absolute value $|\cdot|$ is non-Archimedean, one has

$$\|\eta\|_{\det}\leqslant\|\varphi\|_{\varepsilon}\cdot\|x_1\|\cdots\|x_r\|\cdot\|x_1^\vee\|_*\cdots\|x_r^\vee\|_*\leqslant\|\varphi\|_{\varepsilon}\alpha^{-r},$$

where the second inequality comes from Lemma 1.2.10. Since $\alpha\in\,]0,1[$ is arbitrary, the first assertion is proved.

We proceed with the proof of the second assertion. By Proposition 1.2.14, the seminorms $\|\cdot\|$ and $\|\cdot\|_{**}$ induce the same dual norm on V^*, and hence induce the same ε-tensor product seminorm on $V^{\otimes r}$. Therefore, by the first assertion of the proposition, we obtain that they induce the same determinant seminorm on $\det(V)$.□

Remark 1.2.16 In the above proposition, the non-Archimedean assumption on the absolute value is essential. In the Archimedean case, the inequality (1.28) only leads to a weaker estimate $\|\cdot\|_{\det}\leqslant r!\|\cdot\|_{\det_\varepsilon}$, where $\|\cdot\|_{\det_\varepsilon}$ is the quotient seminorm on $\det(V)$ induced by the ε-tensor product seminorm.

Definition 1.2.17 Let $(V,\|\cdot\|)$ be a finite-dimensional seminormed vector space over k and r be the dimension of V over k. We denote by $\|\cdot\|_{\det_\varepsilon}$ the quotient seminorm of the ε-tensor power of $\|\cdot\|$ on $V^{\otimes r}$ by the canonical projection map $V^{\otimes r}\rightarrow\det(V)$, called the *$\varepsilon$-determinant seminorm* of $\|\cdot\|$.

Proposition 1.2.18 *We assume that the absolute value $|\cdot|$ is Archimedean. Let $(V,\|\cdot\|)$ be a finite-dimensional seminormed vector space over k and let r be the dimension of $V/N_{\|\cdot\|}$ over k. Then the ε-determinant norm of the dual norm $\|\cdot\|_*$ on* $\det(V^*)$ *is bounded from below by $(r!)^{-1}\|\cdot\|^{\sim}_{\det,*}$, where $\|\cdot\|^{\sim}_{\det,*}$ is the dual norm of the determinant of the norm $\|\cdot\|^{\sim}$.*

Proof By Corollary 1.1.13, one has $V^*=(V/N_{\|\cdot\|})^\vee$. Moreover one has $\|\cdot\|_*=\|\cdot\|^{\sim}_*$. Hence, by replacing $(V,\|\cdot\|)$ by $(V/N_{\|\cdot\|},\|\cdot\|^{\sim})$, we may assume without loss of generality that $\|\cdot\|$ is a norm.

Let φ be an element in $V^{\vee\otimes r}$. Viewed as a k-multilinear form on V^r, one has

$$\|\varphi\|_{*,\varepsilon}=\sup_{(x_1,\ldots,x_r)\in(V\setminus\{0\})^r}\frac{|\varphi(x_1,\ldots,x_r)|}{\|x_1\|\cdots\|x_r\|}.$$

Let $\{e_i\}_{i=1}^r$ be a basis of V and $\{e_i^\vee\}_{i=1}^r$ be its dual basis of $V^\vee$. Assume that φ is written in the form

$$\varphi=\sum_{I=(i_1,\ldots,i_r)\in\{1,\ldots,r\}^r}a_I(e_{i_1}^\vee\otimes\cdots\otimes e_{i_r}^\vee),$$

where $a_I \in k$. Then for any $(i_1, \ldots, i_r)$ one has

$$\varphi(e_{i_1}, \ldots, e_{i_r}) = a_{(i_1,\ldots,i_r)}.$$

In particular,

$$\|\varphi\|_{*,\varepsilon} \geqslant \frac{|a_{(i_1,\ldots,i_r)}|}{\|e_{i_1}\| \cdots \|e_{i_r}\|}.$$

Note that the canonical image η of φ in $\det(V)$ is

$$\Big(\sum_{\sigma\in\mathfrak{S}_r} \operatorname{sgn}(\sigma) a_{(\sigma(1),\ldots,\sigma(r))}\Big) e_1^\vee \wedge \cdots \wedge e_r^\vee.$$

Therefore,

$$\begin{aligned}\|\eta\|_{\det,*} &= \Big|\sum_{\sigma\in\mathfrak{S}_r} \operatorname{sgn}(\sigma) a_{(\sigma(1),\ldots,\sigma(r))}\Big| \cdot \|e_1^\vee \wedge \cdots \wedge e_r^\vee\|_{\det,*} \\ &= \Big|\sum_{\sigma\in\mathfrak{S}_r} \operatorname{sgn}(\sigma) a_{(\sigma(1),\ldots,\sigma(r))}\Big| \cdot \frac{1}{\|e_1 \wedge \cdots \wedge e_r\|_{\det}} \\ &\leqslant r!\|\varphi\|_{*,\varepsilon} \frac{\|e_1\| \cdots \|e_r\|}{\|e_1 \wedge \cdots \wedge e_r\|_{\det}}.\end{aligned}$$

By (1.18), we obtain $\|\eta\|_{\det,*} \leqslant r!\|\varphi\|_{*,\varepsilon}$. The proposition is thus proved. □

Proposition 1.2.19 *Let $\{(V_j, \|\cdot\|_j)\}_{j=1}^d$ be a finite family of finite-dimensional seminormed k-vector spaces and let α be a real number in* $]0,1]$. *For any $j \in \{1, \ldots, d\}$, let $\{e_i^{(j)}\}_{i=1}^{n_j}$ be an α-orthogonal basis of V_j. Then*

$$e_{i_1}^{(1)} \otimes \cdots \otimes e_{i_d}^{(d)}, \quad (i_1, \ldots, i_d) \in \prod_{j=1}^d \{1, \ldots, n_j\}$$

form an α^d-orthogonal basis of $V_1 \otimes_k \cdots \otimes_k V_d$ for the ε-tensor product norm $\|\cdot\|_\varepsilon$ of $\{\|\cdot\|_j\}_{j=1}^d$. Moreover, if $\|\cdot\|$ is an ultrametric *norm on $V_1 \otimes_k \cdots \otimes_k V_d$ such that*

$$\|x_1 \otimes \cdots \otimes x_d\| \leqslant \|x_1\|_{1,**} \cdots \|x_d\|_{d,**}$$

for any $(x_1, \ldots, x_d) \in V_1 \times \cdots \times V_d$, then $\|\cdot\| \leqslant \|\cdot\|_\varepsilon$.

Proof Let

$$T = \sum_{(i_1,\ldots,i_d)\in\prod_{j=1}^d\{1,\ldots,n_j\}} a_{i_1,\ldots,i_d} e_{i_1}^{(1)} \otimes \cdots e_{i_d}^{(d)}$$

be a tensor in $V_1 \otimes_k \cdots \otimes_k V_d$, where $a_{i_1,\ldots,i_d} \in k$. We consider it as a k-multilinear form on $V_1^\vee \times \cdots \times V_d^\vee$. For any $j \in \{1, \ldots, d\}$, let $\{\varphi_i^{(j)}\}_{i=1}^{n_j}$ be the dual basis of $\{e_i^{(j)}\}_{i=1}^{n_j}$ and assume that

$$\{\varphi_i^{(j)}\}_{i=1}^{n_j} \cap V^* = \{\varphi_i^{(j)}\}_{i=1}^{n'_j}.$$

By Proposition 1.2.11, $\{\varphi_i^{(j)}\}_{i=1}^{n'_j}$ is an α-orthogonal basis of $V^{(j),*}$. For any $(i_1,\ldots,i_d) \in \prod_{j=1}^d \{1,\ldots,n'_j\}$ we have $T(\varphi_{i_1}^{(1)},\ldots,\varphi_{i_d}^{(d)}) = a_{i_1,\ldots,i_d}$, which leads to

$$\begin{aligned}\|T\|_\varepsilon &\geqslant \frac{|a_{i_1,\ldots,i_d}|}{\|\varphi_{i_1}^{(1)}\|_{1,*}\cdots\|\varphi_{i_d}^{(d)}\|_{d,*}} \geqslant \alpha^d |a_{i_1,\ldots,i_d}|\cdot\|e_{i_1}^{(1)}\|_{1,**}\cdots\|e_{i_d}^{(d)}\|_{d,**}\\ &= \alpha^d|a_{i_1,\ldots,i_d}|\cdot\|e_{i_1}^{(1)}\otimes\cdots\otimes e_{i_d}^{(d)}\|_\varepsilon,\end{aligned}$$

where the second inequality follows from (1.24) and the equality comes from Remark 1.1.56. This completes the proof of the proposition because

$$\|e_{i_1}^{(1)}\otimes\cdots\otimes e_{i_d}^{(d)}\|_\varepsilon = \|e_{i_1}^{(1)}\|_{1,**}\cdots\|e_{i_d}^{(d)}\|_{d,**}$$

vanishes once $(i_1,\ldots,i_d) \notin \prod_{j=1}^d\{1,\ldots,n'_j\}$ (see Lemma 1.2.10). Moreover, if $\|\cdot\|$ is an ultrametric norm on $V_1\otimes_k\cdots\otimes_k V_d$ such that

$$\|x_1\otimes\cdots\otimes x_d\| \leqslant \|x_1\|_{1,**}\cdots\|x_d\|_{d,**}$$

for any $(x_1,\ldots,x_d)\in V_1\times\cdots\times V_d$, then

$$\begin{aligned}\|T\| &\leqslant \max_{(i_1,\ldots,i_d)\in\prod_{j=1}^d\{1,\ldots,n_j\}} |a_{i_1,\ldots,i_d}|\cdot\|e_{i_1}^{(1)}\otimes\cdots\otimes e_{i_d}^{(d)}\|\\ &\leqslant \max_{(i_1,\ldots,i_d)\in\prod_{j=1}^d\{1,\ldots,n_j\}} |a_{i_1,\ldots,i_d}|\cdot\|e_{i_1}^{(1)}\|_{1,**}\cdot\|e_{i_d}\|_{d,**} \leqslant \alpha^{-d}\|T\|_\varepsilon\end{aligned}$$

By Proposition 1.2.7, for any $\alpha\in\,]0,1[$, there exist α-orthogonal bases of $V_1,\ldots,V_d$ respectively. Hence $\|\cdot\|\leqslant\|\cdot\|_\varepsilon$. □

Corollary 1.2.20 *Assume that the absolute value $|\cdot|$ is non-Archimedean. Let $\{(V_j,\|\cdot\|_j)\}_{j=1}^d$ be a finite family of finite-dimensional seminormed vector spaces over k, and $\|\cdot\|_\varepsilon$ be the ε-tensor product of the seminorms $\|\cdot\|_1,\ldots,\|\cdot\|_d$. Then the dual norm $\|\cdot\|_{\varepsilon,*}$ coincides with the ε-tensor product of the dual norms $\|\cdot\|_{1,*},\ldots,\|\cdot\|_{d,*}$.*

Proof Let α be an element of $]0,1[$. For $j\in\{1,\ldots,d\}$, let $\{e_i^{(j)}\}_{i=1}^{n_j}$ be an α-orthogonal basis of V_j over k (see Proposition 1.2.7) and $\{\varphi_i^{(j)}\}_{i=1}^{n_j}$ be the dual basis of $\{e_i^{(j)}\}_{i=1}^{n_j}$, and assume that $\{\varphi_i^{(j)}\}_{i=1}^{n_j}\cap V_j^* = \{\varphi_i^{(j)}\}_{i=1}^{n'_j}$. Note that for any $(i_1,\ldots,i_d)\in\prod_{j=1}^d\{1,\ldots,n_j\}$,

$$\|e_{i_1}^{(1)}\otimes\cdots\otimes e_{i_d}^{(d)}\|_\varepsilon = \|e_{i_1}\|_{1,**}\cdots\|e_{i_d}\|_{d,**}\neq 0$$

if and only if $(i_1,\ldots,i_d)\in\prod_{j=1}^d\{1,\ldots,n'_i\}$. Therefore,

$$\{\varphi_{i_1}^{(1)} \otimes \cdots \otimes \varphi_{i_d}^{(d)}\}_{(i_1,\ldots,i_d)\in\prod_{j=1}^d\{1,\ldots,n_j'\}}$$

forms a basis of the vector space of bounded linear forms on $(V_1 \otimes_k \cdots \otimes_k V_d, \|\cdot\|_\varepsilon)$, which shows that $(V_1 \otimes_k \cdots \otimes_k V_d)^* \cong V_1^* \otimes_k \cdots \otimes_k V_d^*$.

Let $\|\cdot\|'$ be the ε-tensor product of $\|\cdot\|_{1,*}, \ldots, \|\cdot\|_{d,*}$. By definition, for any T in $V_1^* \otimes_k \cdots \otimes_k V_d^*$, one has

$$\begin{aligned}\|T\|' &= \sup_{\substack{(s_1,\ldots,s_d)\in V_1\times\cdots\times V_d\\ \min\{\|s_1\|_1,\ldots,\|s_d\|_d\}>0}} \frac{|T(s_1,\ldots,s_d)|}{\|s_1\|_{1,**}\cdots\|s_d\|_{d,**}}\\ &= \sup_{\substack{(s_1,\ldots,s_d)\in V_1\times\cdots\times V_d\\ \min\{\|s_1\|_1,\ldots,\|s_d\|_d\}>0}} \frac{|T(s_1,\ldots,s_d)|}{\|s_1\otimes\cdots\otimes s_d\|_\varepsilon} \leqslant \|T\|_{\varepsilon,*},\end{aligned}$$

where the second equality comes from (1.16). Conversely, if T is of the form $\psi_1 \otimes \cdots \otimes \psi_d$, where $(\psi_1, \ldots, \psi_d) \in V_1^* \times \cdots \times V_d^*$, then

$$\|T\|_{\varepsilon,*} = \sup_{\substack{f\in V_1\otimes_k\cdots\otimes_k V_d\\ \|f\|_\varepsilon\neq 0}} \frac{|f(\psi_1,\ldots,\psi_d)|}{\|f\|_\varepsilon} \leqslant \|\psi_1\|_{1,*}\cdots\|\psi_d\|_{d,*}.$$

By Proposition 1.2.19, we obtain $\|\cdot\|_{\varepsilon,*} \leqslant \|\cdot\|'$. □

1.2.4 Orthogonality and lattice norms

In this subsection, we assume that the absolute value $|\cdot|$ is non-Archimedean and we denote by $\mathfrak{o}_k$ the valuation ring of $(k, |\cdot|)$. Let V be a finite-dimensional vector space over k and let r be the dimension of V.

Proposition 1.2.21 *Let $\mathcal{V}$ be a lattice of V which is an $\mathfrak{o}_k$-module of finite type (and hence a free $\mathfrak{o}_k$-module of rank r). Then any basis of $\mathcal{V}$ over $\mathfrak{o}_k$ is an orthonormal basis of $(V, \|\cdot\|_\mathcal{V})$.*

Proof Let $\{e_i\}_{i=1}^r$ be a basis of $\mathcal{V}$ over $\mathfrak{o}_k$. Note that an element $a_1e_1 + \cdots + a_re_r$ of V (with $(a_1, \ldots, a_r) \in k^r$) belongs to $\mathcal{V}$ if and only if all a_i are in $\mathfrak{o}_k$. Let $x = \lambda_1e_1 + \cdots + \lambda_re_r$ be an element of V. If a is an element of $k^\times$ such that $a^{-1}x$ belongs to $\mathcal{V}$, then one has $a^{-1}\lambda_i \in \mathfrak{o}_k$ and hence $|\lambda_i| \leqslant |a|$ for any $i \in \{1, \ldots, r\}$. Therefore $\max\{|\lambda_1|, \ldots, |\lambda_r|\} \leqslant \|x\|_\mathcal{V}$. Conversely, if $j \in \{1, \ldots, r\}$ is such that

$$|\lambda_j| = \max\{|\lambda_1|, \ldots, |\lambda_r|\} > 0,$$

then one has $\lambda_i\lambda_j^{-1} \in \mathfrak{o}_k$ for any $i \in \{1, \ldots, r\}$. Hence $\lambda_j^{-1}x \in \mathcal{V}$ and

$$\|x\|_\mathcal{V} \leqslant |\lambda_j| = \max\{|\lambda_1|, \ldots, |\lambda_r|\}.$$

Proposition 1.2.22 *Assume that the absolute value $|\cdot|$ is non-trivial. Let $\lambda \in]0,1[$ be a real number such that*

$$\lambda < \sup\{|a| \,:\, a \in k^{\times},\ |a| < 1\}. \tag{1.29}$$

Then, for any ultrametric norm $\|\cdot\|$ on V there exists a lattice of finite type $\mathcal{V}$ of V, such that $\|\cdot\| \leqslant \|\cdot\|_{\mathcal{V}} \leqslant \lambda^{-1}\|\cdot\|$.

Proof Let $\alpha \in]0,1[$ such that $\lambda/\alpha < \sup\{|a| \,:\, a \in k^{\times},\ |a| < 1\}$. Let $\{e_i\}_{i=1}^r$ be an α-orthogonal basis of V (the existence of which has been proved in Proposition 1.2.7). By Proposition 1.1.4, by dilating the vectors $e_i, i \in \{1,\ldots,r\}$, we may assume that $\lambda/\alpha \leqslant \|e_i\| < 1$ for any i. Let $\mathcal{V}$ be the free $\mathfrak{o}_k$-module generated by $\{e_i\}_{i=1}^r$. It is a lattice of V. Moreover, by Proposition 1.2.21, $\{e_i\}_{i=1}^r$ is an orthonormal basis of $(V, \|\cdot\|_{\mathcal{V}})$. In particular, for any vector $x = a_1e_1 + \cdots + a_re_r$ in V, one has

$$\|x\|_{\mathcal{V}} = \max_{i\in\{1,\ldots,r\}} |a_i| \geqslant \max_{i\in\{1,\ldots,r\}} |a_i|\cdot\|e_i\| \geqslant \|x\|.$$

Moreover, by the α-orthogonality of $\{e_i\}_{i=1}^r$ one has

$$\|x\| \geqslant \alpha \max_{i\in\{1,\ldots,r\}} |a_i|\cdot\|e_i\| \geqslant \lambda\|x\|_{\mathcal{V}}.$$

The proposition is thus proved. □

1.2.5 Orthogonality and Hadamard property

We have seen in Proposition 1.2.7 that an Hadamard basis of a finite-dimensional normed vector space is necessarily orthogonal. The converse of this assertion is also true when the absolute value $|\cdot|$ is non-Archimedean.

Proposition 1.2.23 *We assume that the absolute value $|\cdot|$ is non-Archimedean. Let $(V, \|\cdot\|)$ be a finite-dimensional seminormed vector space over k, and let r be the dimension of V over k. If α is an element in $]0,1]$ and if $\{x_i\}_{i=1}^r$ is an α-orthogonal basis of V, then one has*

$$\|x_1 \wedge \cdots \wedge x_r\| \geqslant \alpha^r \|x_1\| \cdots \|x_r\|. \tag{1.30}$$

In particular, if $\|\cdot\|$ is a norm, any orthogonal basis of V is an Hadamard basis.

Proof If $N_{\|\cdot\|}$ is non-zero, then the interserction of $\{x_i\}_{i=1}^r$ with $N_{\|\cdot\|}$ is not empty (see Proposition 1.2.5) and hence the inequality (1.30) holds. In the following, we assume that $\|\cdot\|$ is a norm. Note that the case where $V = \{0\}$ is trivial. Hence we may assume that $r > 0$. Let $\{x_i\}_{i=1}^r$ be an α-orthogonal basis of V. Let $\{y_i\}_{i=1}^r$ be an arbitrary basis of V and $A = (a_{ij})_{i\in\{1,\ldots,r\},\, j\in\{1,\ldots,r\}} \in k^{r\times r}$ be the transition matrix from $\{x_i\}_{i=1}^r$ to $\{y_i\}_{i=1}^r$, namely $y_i = \sum_{j=1}^r a_{ij}x_j$ for any $i \in \{1,\ldots,r\}$. By the α-orthogonality of the basis $\{x_i\}_{i=1}^r$ one has

$$|a_{ij}| \leqslant \alpha^{-1} \frac{\|y_i\|}{\|x_j\|}.$$

Thus, by the assumption that the absolute value $|\cdot|$ is non-Archimedean, one has

$$\|y_1 \wedge \cdots \wedge y_r\| = |\det(A)| \cdot \|x_1 \wedge \cdots \wedge x_r\| \leqslant \alpha^{-r} \frac{\|y_1\| \cdots \|y_r\|}{\|x_1\| \cdots \|x_r\|} \cdot \|x_1 \wedge \cdots \wedge x_r\|$$

and hence

$$\frac{\|x_1 \wedge \cdots \wedge x_r\|}{\|x_1\| \cdots \|x_r\|} \geqslant \alpha^r \frac{\|y_1 \wedge \cdots \wedge y_r\|}{\|y_1\| \cdots \|y_r\|}.$$

Since the basis $\{y_i\}_{i=1}^r$ is arbitrary, by Proposition 1.1.66 (see also Remark 1.1.67) we obtain that

$$\frac{\|x_1 \wedge \cdots \wedge x_r\|}{\|x_1\| \cdots \|x_r\|} \geqslant \alpha^r.$$

The proposition is thus proved. □

Remark 1.2.24 The Archimedean analogue of Proposition 1.2.23 is not true in general. One can consider for example the case where $V = \mathbb{R}^2$ equipped with the norm $\|\cdot\|$ such that $\|(a, b)\| = \max\{|a|, |b|\}$, where $|\cdot|$ is the usual absolute value on $\mathbb{R}$. Let $e_1 = (1, 0)$ and $e_2 = (0, 1)$. The basis $\{e_1, e_2\}$ is orthonormal. However,

$$\|e_1 \wedge e_2\| = \left\| \frac{1}{\sqrt{2}}(e_1 + e_2) \wedge \frac{1}{\sqrt{2}}(e_1 - e_2) \right\| = \frac{1}{2}.$$

Therefore $\{e_1, e_2\}$ is not an Hadamard basis.

The following proposition shows that the Archimedean analogue of Proposition 1.2.23 is true provided that the norm is induced by an inner product.

Proposition 1.2.25 *Let $(V, \|\cdot\|)$ be a finite-dimensional normed vector space over k. Assume that the absolute value $|\cdot|$ is Archimedean and that the norm $\|\cdot\|$ is induced by an inner product. Then any orthogonal basis of V is an Hadamard basis.*

Proof The field k is locally compact, therefore V admits an Hadamard basis $\boldsymbol{e} = \{e_i\}_{i=1}^r$, which is necessarily an orthogonal basis (see Proposition 1.2.7). Without loss of generality, we may assume that $\boldsymbol{e}$ is an orthonormal basis. Let $\boldsymbol{e}' = \{e_i'\}_{i=1}^r$ be another orthonormal basis. There exists a unitary matrix A such that $\boldsymbol{e}' = A\boldsymbol{e}$. One has $|\det(A)| = 1$ and hence

$$\|e_1' \wedge \cdots \wedge e_r'\|_{\det} = \|e_1 \wedge \cdots \wedge e_r\|_{\det} = 1 = \|e_1'\| \cdots \|e_r'\|.$$

Therefore the basis $\{e_i'\}_{i=1}^r$ is also an Hadamard basis. Thus we have proved that any orthonormal basis is an Hadamard basis. If $\{x_i\}_{i=1}^r$ is an orthogonal basis, and if $e_i = \|x_i\|^{-1} x_i$ for any $i \in \{1, \ldots, r\}$, then $\{e_i\}_{i=1}^r$ is an orthonormal basis of V, which is an Hadamard basis. We then deduce that $\{x_i\}_{i=1}^r$ is also an Hadamard basis. □

Proposition 1.2.26 *We assume that the absolute value $|\cdot|$ is trivial. Let $(V, \|\cdot\|)$ be an r-dimensional ($r \in \mathbb{N}_{>0}$), ultrametrically normed vector space over k, which corresponds to an increasing sequence*

$$0 = V_0 \subsetneq V_1 \subsetneq \ldots \subsetneq V_n = V$$

of vector subspaces of V and a decreasing sequence $\mu_1 > \ldots > \mu_n$ of real numbers as described in Remark 1.1.40.

(1) *A basis $\{x_j\}_{j=1}^r$ of V is orthogonal if and only if* $\mathrm{card}(\{x_j\}_{j=1}^r \cap V_i) = \dim_k(V_i)$ *for any $i \in \{1, \ldots, n\}$.*
(2) *Let α be an element of* $]0, 1]$ *such that*

$$\forall\, i \in \{1, \ldots, n\}, \quad \alpha > \mathrm{e}^{-(\mu_i - \mu_{i+1})/r},$$

where $\mu_{n+1} = -\infty$ by convention. Then any α-orthogonal basis of $(V, \|\cdot\|)$ is orthogonal.

Proof (1) Note that the restriction of $\|\cdot\|$ on each $V_i \setminus V_{i-1}$ is constant and takes $\mathrm{e}^{-\mu_i}$ as its value for $i \in \{1, \ldots, n\}$. Let

$$\lambda_1 \leqslant \ldots \leqslant \lambda_r$$

be the increasing sequence of positive real numbers such that $\mathrm{e}^{-\mu_i}$ appears exactly $\dim_k(V_i) - \dim_k(V_{i-1})$ times. Let $\{x_j\}_{j=1}^r$ be a basis of V such that $\|x_1\| \leqslant \ldots \leqslant \|x_r\|$. For each $i \in \{1, \ldots, n\}$, the cardinal of $\{x_j\}_{j=1}^r \cap V_i$ does not exceed $\dim_k(V_i)$, so that $\|x_j\| \geqslant \lambda_j$ for any $j \in \{1, \ldots, r\}$ and hence

$$\prod_{j=1}^r \|x_j\| \geqslant \prod_{j=1}^r \lambda_j.$$

Moreover, if the equality $\mathrm{card}(\{x_j\}_{j=1}^r \cap V_i) = \dim_k(V_i)$ holds for any $i \in \{1, \ldots, n\}$, then the basis $\{x_j\}_{j=1}^r$ is an Hadamard basis, and hence is an orthogonal basis (by Proposition 1.2.7 (2)). If there exists an index $i \in \{1, \ldots, n\}$ such that $\mathrm{card}(\{x_j\}_{j=1}^r \cap V_i) < \dim_k(V_i)$, then there exists an element

$$x = \lambda_1 x_1 + \cdots + \lambda_r x_r$$

of V_i and a $j \in \{1, \ldots, r\}$ such that $\lambda_j \neq 0$ and $x_j \in V \setminus V_i$. As $\|x\| \leqslant \mu_i < \|x_j\|$, the basis $\{x_j\}_{j=1}^r$ is not orthogonal.

(2) Let $\{e_j\}_{j=1}^r$ be an α-orthogonal basis of $(V, \|\cdot\|)$. Without loss of generality, we assume that $\|e_1\| \leqslant \ldots \leqslant \|e_r\|$. One has

$$\{\|e_1\|, \ldots, \|e_r\|\} \subseteq \{\lambda_1, \ldots, \lambda_r\} = \{\mathrm{e}^{-\mu_1}, \ldots, \mathrm{e}^{-\mu_n}\}. \tag{1.31}$$

Moreover, since

$$\mathrm{card}(\{e_j\}_{j=1}^r \cap V_i) \leqslant \dim_k(V_i)$$

for any $i \in \{1, \ldots, n\}$, one has $\|e_j\| \geqslant \lambda_j$ for any $j \in \{1, \ldots, r\}$. Therefore, if the strict inequality $\|e_j\| > \lambda_j$ holds for some $j \in \{1, \ldots, r\}$, then, by (1.31), one can find m such that $m > j$, $\|e_j\| \geqslant \lambda_m$ and $\lambda_m \lambda_j^{-1} = \mathrm{e}^{\mu_i - \mu_{i+1}}$ for some i, so that, by our assumption,

$$\|e_j\| \geqslant \lambda_j(\lambda_m \lambda_j^{-1}) = \lambda_j \mathrm{e}^{\mu_i - \mu_{i+1}} > \lambda_j \alpha^{-r}.$$

Therefore,

$$\prod_{\ell=1}^{r} \|e_\ell\| > \alpha^{-r} \prod_{\ell=1}^{r} \lambda_\ell = \alpha^{-r} \|e_1 \wedge \cdots \wedge e_r\|_{\det}.$$

This contradicts Proposition 1.2.23. Therefore one has $\|e_j\| = \lambda_j$ for any $j \in \{1, \ldots, r\}$ and hence, for each $i \in \{1, \ldots, n\}$, the cardinal of $\{e_j\}_{j=1}^r \cap V_i$ is equal to $\dim_k(V_i)$, which implies that $\{e_j\}_{j=1}^r$ is an orthogonal basis. □

1.2.6 Ultrametric Gram-Schimdt process

In this subsection, we consider a refinement of Proposition 1.2.7. First we recall the spherically completeness of a metric space.

Definition 1.2.27 We say that a metric space (X, d) is *spherically complete* if, for any decreasing sequence

$$B_1 \supseteq B_2 \supseteq \cdots \supseteq B_n \supseteq B_{n+1} \supseteq \cdots$$

of non-empty closed balls in X, one has $\bigcap_{n=1}^{\infty} B_n \neq \varnothing$. A normed vector space $(V, \|\cdot\|)$ over k is said to be *spherically complete* if $(V, \|\cdot\|)$ is spherically complete as a metric space. If $(k, |\cdot|)$, viewed as a normed vector space over k, is spherically complete, we say that the valued field $(k, |\cdot|)$ is spherically complete.

Remark 1.2.28 If $(k, |\cdot|)$ is a discrete valuation field, then $(k, |\cdot|)$ is spherically complete by [134, Proposition 20.2]. In particular, any locally compact non-Archimedean valued field is spherically complete.

Lemma 1.2.29 *Let $(V, \|\cdot\|)$ be an ultrametrically normed vector space of finite dimension over k. Then we have the following:*

(1) *Let W be a vector subspace of V over k. If W equipped with the restriction $\|\cdot\|_W$ of $\|\cdot\|$ to W is spherically complete, then, for $x \in V$, there is $w \in W$ such that* $\operatorname{dist}(x, W) = \|x - w\|$.
(2) *If $(V, \|\cdot\|)$ has an orthogonal basis $\{e_i\}_{i=1}^r$ and $(k, |\cdot|)$ is spherically complete, then $(V, \|\cdot\|)$ is also spherically complete.*

Proof For $a \in V$ and $\delta \in \mathbb{R}_{\geqslant 0}$, we set

$$B(a; \delta) := \{x \in V \,:\, \|x - a\| \leqslant \delta\}.$$

As $\|\cdot\|$ is ultrametric, we can easily see that

$$B(a;\delta) = B(a';\delta) \tag{1.32}$$

for all $\delta \in \mathbb{R}_{\geqslant 0}$ and $a, a' \in V$ with $\|a - a'\| \leqslant \delta$.

(1) We can choose a decreasing sequence $\{\delta_n\}_{n=1}^{\infty}$ of positive numbers and a sequence $\{w_n\}_{n=1}^{\infty}$ in W such that $\|x - w_n\| \leqslant \delta_n$ and $\lim_{n\to\infty} \delta_n = \operatorname{dist}(x, W)$. As $B(x;\delta_n) \cap W = B(w_n;\delta_n) \cap W$ by (1.32), $\{B(x;\delta_n) \cap W\}_{n=1}^{\infty}$ yields a decreasing sequence of non-empty closed balls in W. Thus, by our assumption, there is $w \in \bigcap_{n=1}^{\infty} B(x;\delta_n) \cap W$, so that $\|x - w\| \leqslant \delta_n$ for all n, that is, $\|x - w\| \leqslant \operatorname{dist}(x, W)$, as required.

(2) Note that $\|a_1e_1 + \cdots + a_re_r\| = \max\{|a_1| \cdot \|e_1\|, \ldots, |a_r| \cdot \|e_r\|\}$, so that

$$B(a;\delta) = B_k(a_1;\delta/\|e_1\|)e_1 + \cdots + B_k(a_r;\delta/\|e_r\|)e_r$$

for $a = a_1e_1 + \cdots + a_re_r \in V$ and $\delta \in \mathbb{R}_{\geqslant 0}$, where

$$B_k(\lambda;\delta') = \{t \in k \,:\, |t - \lambda| \leqslant \delta'\}$$

for $\lambda \in k$ and $\delta' \in \mathbb{R}_{\geqslant 0}$. Therefore the assertion follows. □

In the case of an ultrametrically normed finite-dimensional vector space, Proposition 1.2.7 has the following refined form. This could be considered as an ultrametric analogue of Gram-Schmidt orthogonalisation process.

Proposition 1.2.30 *Let* $(V, \|\cdot\|)$ *be an ultrametrically seminormed* k*-vector space of dimension* $r \geqslant 1$. *Let*

$$0 = V_0 \subsetneq V_1 \subsetneq V_2 \subsetneq \ldots \subsetneq V_r = V \tag{1.33}$$

be a complete flag of subspaces of V. *Fix a real number* α *such that*

$$\alpha \in \begin{cases}]0,1], & \text{if } (k,|\cdot|) \text{ is spherically complete,} \\ ]0,1[, & \text{otherwise.} \end{cases}$$

Then there exists an α*-orthogonal basis* $\boldsymbol{e}$ *of* V *such that, for any* $i \in \{1, \ldots, r\}$, $\operatorname{card}(V_i \cap \boldsymbol{e}) = i$.

Proof If a basis $\boldsymbol{e}$ of V is such that, for any $i \in \{1, \ldots, r\}$, $\operatorname{card}(V_i \cap \boldsymbol{e}) = i$, we say that the basis $\boldsymbol{e}$ is *compatible* with the flag (1.33).

We begin with the proof of the particular case where $\|\cdot\|$ is a norm by induction on r, the dimension of V over k. The case where $r = 1$ is trivial. Assume that the proposition holds for all vector spaces of dimension $< r$, where $r \geqslant 2$. Applying the induction hypothesis to V_{r-1} and the flag $0 = V_0 \subsetneq V_1 \subsetneq \ldots \subsetneq V_{r-1}$ we get a basis $\{e_1, \ldots, e_{r-1}\}$ of V_{r-1} compatible with the flag such that, for any $(\lambda_1, \ldots, \lambda_{r-1}) \in k^{r-1}$

$$\|\lambda_1e_1 + \cdots + \lambda_{r-1}e_{r-1}\| \geqslant \alpha^{1/2} \max_{i\in\{1,\ldots,r-1\}} |\lambda_i| \cdot \|e_i\|. \tag{1.34}$$

Let x be an element of $V \setminus V_{r-1}$. The distance between x and V_{r-1} is strictly positive since V_{r-1} is closed in V (see Proposition 1.1.11). Hence there exists $y \in V_{r-1}$ such that

$$\|x - y\| \leqslant \alpha^{-1/2}\mathrm{dist}(x, V_{r-1}). \tag{1.35}$$

In the case where $\alpha = 1$ and $(k, |\cdot|)$ is spherically complete, the existence of y follows from Lemma 1.2.29. We choose $e_r = x - y$. The basis $\{e_1, \ldots, e_r\}$ is compatible with the flag $0 = V_0 \subsetneq V_1 \subsetneq \ldots \subsetneq V_r = V$.

Let $(\lambda_1, \ldots, \lambda_r)$ be an element of k^r. We wish to find a lower bound for the norm of $z = \lambda_1 e_1 + \cdots + \lambda_r e_r$. By (1.35) we have that

$$\|z\| \geqslant |\lambda_r| \cdot \mathrm{dist}(x, V_{r-1}) \geqslant \alpha^{1/2}|\lambda_r| \cdot \|e_r\|.$$

This provides our lower bound when $\|\lambda_r e_r\| \geqslant \|\lambda_1 e_1 + \cdots + \lambda_{r-1}e_{r-1}\|$. If $\|\lambda_r e_r\| < \|\lambda_1 e_1 + \cdots + \lambda_{r-1}e_{r-1}\|$ then we have

$$\|z\| = \|\lambda_1 e_1 + \cdots + \lambda_{r-1}e_{r-1}\|$$

because the norm is ultrametric (see Proposition 1.1.5). By the induction hypothesis (1.34) we have that $\|z\| \geqslant \alpha|\lambda_i| \cdot \|e_i\|$ for any $i \in \{1, \ldots, r-1\}$. This completes the proof of the proposition in the case where $\|\cdot\|$ is a norm.

We now consider the general seminorm case. Let W be the quotient vector space $V/N_{\|\cdot\|}$. For each $i \in \{0, \ldots, r\}$, let W_i be $(V_i + N_{\|\cdot\|})/N_{\|\cdot\|}$. Applying the particular case of the proposition to $(W, \|\cdot\|^{\sim})$, we obtain the existence of an α-orthogonal basis $\widetilde{\boldsymbol{e}}$ of W such that $\mathrm{card}(\widetilde{\boldsymbol{e}} \cap W_i) = \dim_k(W_i)$. We set

$$I = \{i \in \{1, \ldots, r\} \,:\, W_{i-1} \subsetneq W_i\} \quad \text{and} \quad J = \{j \in \{1, \ldots, r\} \,:\, W_{j-1} = W_j\}.$$

If $i \in I$, then there is a unique element $u_i \in \widetilde{\boldsymbol{e}} \cap (W_i \setminus W_{i-1})$, so that we can choose $e_i \in V_i$ such that the class of e_i in $V/N_{\|\cdot\|}$ is u_i. If $j \in J$, then we can pick up $e_j \in (N_{\|\cdot\|} \cap V_j) \setminus V_{j-1}$. Indeed, as $V_j \setminus V_{j-1} \neq \varnothing$ and $V_j \subseteq V_{j-1} + N_{\|\cdot\|}$, we can find $x \in V_j \setminus V_{j-1}$, $y \in V_{j-1}$ and $e_j \in N_{\|\cdot\|}$ with $x = y + e_j$, and hence $e_j \in (N_{\|\cdot\|} \cap V_j) \setminus V_{j-1}$. By construction, $\boldsymbol{e} := \{e_i\}_{i=1}^r$ satisfies $\mathrm{card}(V_i \cap \boldsymbol{e}) = i$ for $i \in \{0, \ldots, r\}$. In particular, $\boldsymbol{e}$ forms a basis of V.

Let us see that $\boldsymbol{e}$ is α-orthogonal. For any $(\lambda_1, \ldots, \lambda_r) \in k^r$, if we let $x = \lambda_1 e_1 + \cdots + \lambda_r e_r$ and $u = \sum_{i \in I} \lambda_i u_i$, then one has

$$\|x\| = \|u\|^{\sim} \geqslant \alpha \max_{i \in I} |\lambda_i| \cdot \|u_i\|^{\sim} = \alpha \max_{i \in \{1,\ldots,r\}} |\lambda_i| \cdot \|e_i\|,$$

where the inequality comes from the α-orthogonality of $\widetilde{\boldsymbol{e}}$, as required. □

Corollary 1.2.31 *Let $(V, \|\cdot\|)$ be an ultrametrically seminormed vector space of finite dimension over k. If $(k, |\cdot|)$ is spherically complete, then $(V, \|\cdot\|)$ has an orthogonal basis. In particular, $(V, \|\cdot\|)$ is spherically complete.*

Remark 1.2.32 Assume that the absolute value $|\cdot|$ is Archimedean. Let V be a finite-dimensional vector space over k and $\|\cdot\|$ be a seminorm on V which is induced by an inner product. Given a complete flag

$$0 = V_0 \subsetneq V_1 \subsetneq V_2 \subsetneq \ldots \subsetneq V_r = V$$

of V, the Gram-Schmidt process permits to construct an orthogonal basis $\boldsymbol{e}$ of V such that $\operatorname{card}(\boldsymbol{e} \cap V_i) = i$ for any $i \in \{1, \ldots, r\}$, along the same line as in the proof of Proposition 1.2.30. The main point is that the field k is locally compact, and hence the distance in (1.35) is actually attained by some point in V_{r-1}. For a general seminorm, even though an orthogonal basis always exists, it is not always possible to find an orthogonal basis which is compatible with a given flag.

Proposition 1.2.30 and the usual Gram-Schmidt process lead to the following projection result.

Corollary 1.2.33 *Let V be a finite-dimensional vector space over k equipped with a seminorm $\|\cdot\|$ which is either ultrametric or induced by a semidefinite inner product. Let V_0 be a vector subspace of V. For any $\alpha \in \,]0, 1[$ there exists a k-linear projection $\pi : V \to V_0$ (namely π is a k-linear map and its restriction to V_0 is the identity map) such that $\|\pi\| \leqslant \alpha^{-1}$. If $(k, |\cdot|)$ is non-Archimedean and spherically complete, or if $\|\cdot\|$ is induced by a semidefinite inner product, we can choose the k-linear projection π such that $\|\pi\| \leqslant 1$.*

Proof We first consider the ultrametric case. By Proposition 1.2.30, there exists an α-orthogonal basis $\boldsymbol{e} = \{e_i\}_{i=1}^n$ of V such that $\operatorname{card}(\boldsymbol{e} \cap V_0) = \dim_k(V_0)$. Without loss of generality, we may assume that $\boldsymbol{e} \cap V_0 = \{e_i\}_{i=1}^m$, where $m = \dim_k(V_0)$. Let $\pi : V \to V_0$ be the k-linear map sending $\lambda_1 e_1 + \cdots + \lambda_n e_n \in V$ to $\lambda_1 e_1 + \cdots + \lambda_m e_m \in V_0$. Since the basis $\boldsymbol{e}$ is α-orthogonal, one has

$$\|\lambda_1 e_1 + \cdots + \lambda_n e_n\| \geqslant \alpha \max_{i \in \{1,\ldots,n\}} \|\lambda_i e_i\| \geqslant \alpha \|\lambda_1 e_1 + \cdots + \lambda_m e_m\|,$$

which implies that $\|\pi\| \leqslant \alpha^{-1}$.

If $(k, |\cdot|)$ is non-Archimedean and spherically complete, or if $\|\cdot\|$ is induced by an inner product, we use the existence of an orthogonal basis $\boldsymbol{e}$ such that $\operatorname{card}(\boldsymbol{e} \cap V_0) = \dim_k(V_0)$. By the same agrument as above, we obtain the existence of a linear projection $\pi : V \to V_0$ such that $\|\pi\| \leqslant 1$. □

Corollary 1.2.34 *Let V be a finite-dimensional vector space over k and $\|\cdot\|_1$ and $\|\cdot\|_2$ be two seminorms on V. We assume that $\|\cdot\|_1 \leqslant \|\cdot\|_2$ and that the seminorm $\|\cdot\|_2$ is either ultrametric or induced by a semidefinite inner product. If there exists a vector $x \in V$ such that $\|x\|_1 < \|x\|_2$, then one has $\|\cdot\|'_{1,\det} < \|\cdot\|^{\sim}_{2,\det}$ on $\det(V/N_{\|\cdot\|_2}) \setminus \{0\}$, where $\|\cdot\|'_1$ denotes the quotient seminorm of $\|\cdot\|_1$ on $V/N_{\|\cdot\|_2}$.*

Proof The condition $\|\cdot\|_1 \leqslant \|\cdot\|_2$ implies that $N_{\|\cdot\|_2} \subseteq N_{\|\cdot\|_1}$. In particular, for any $x \in V$, if we denote by $[x]$ the class of x in $V/N_{\|\cdot\|_2}$, then one has

$$\|[x]\|'_1 = \|x\|_1, \quad \|[x]\|^{\sim}_2 = \|x\|_2.$$

Therefore, by replacing V by $V/N_{\|\cdot\|_2}$, $\|\cdot\|_1$ by $\|\cdot\|_1'$, and $\|\cdot\|_2$ by $\|\cdot\|_2^{\sim}$, we may assume without loss of generality that $\|\cdot\|_2$ is a norm. Moreover, the case where $\|\cdot\|_1$ is not a norm is trivial since the seminorm $\|\cdot\|_{1,\det}$ vanishes. Hence it suffices to treat the case where both seminorms $\|\cdot\|_1$ and $\|\cdot\|_2$ are norms.

Let $\lambda = \|x\|_2/\|x\|_1 > 1$. We first consider the ultrametric case. By Proposition 1.2.30, for any $\alpha \in]0,1[$ there exists an α-orthogonal basis $\{e_i\}_{i=1}^r$ of $(V, \|\cdot\|_2)$ such that $e_1 = x$. Hence See Proposition 1.2.23 for the first inequality

$$\begin{aligned}\|e_1 \wedge \cdots \wedge e_r\|_{2,\det} &\geqslant \alpha^r \|e_1\|_2 \cdots \|e_r\|_2 \\ &\geqslant \lambda\alpha^r \|e_1\|_1 \cdots \|e_r\|_1 \geqslant \lambda\alpha^r \|e_1 \wedge \cdots \wedge e_r\|_{1,\det}.\end{aligned}$$

Since $\alpha \in]0,1[$ is arbitrary, one has $\|\cdot\|_{2,\det}/\|\cdot\|_{1,\det} \geqslant \lambda > 1$.

The Archimedean case is very similar. There exists an orthogonal basis $\{e_i\}_{i=1}^r$ of $(V, \|\cdot\|)$ such that $e_1 = x$. We then proceed as above in replacing α by 1. □

Proposition 1.2.35 *Let V be a finite-dimensional vector space over k, equipped with a seminorm $\|\cdot\|_V$, which is ultrametric or induced by a semidefinite inner product. Let $(W, \|\cdot\|_W)$ be a seminormed vector space over k. For any k-vector subspace V_0 of V, the k-linear map $\mathscr{L}(V,W) \to \mathscr{L}(V_0,W)$, sending $f \in \mathscr{L}(V,W)$ to its restriction to V_0, is surjective, and the operator seminorm on $\mathscr{L}(V_0,W)$ coincides with the quotient seminorm induced by the operator seminorm on $\mathscr{L}(V,W)$. In particular, the dual norm on V_0^* identifies with the quotient of the dual norm on V^* by the canonical quotient map $V^* \to V_0^*$.*

Proof For any $f \in \mathscr{L}(V,W)$, the operator seminorm of $f|_{V_0}$ does not exceed that of f. In the following, we show that, for any linear map $g \in \mathscr{L}(V_0,W)$ and any $\alpha \in]0,1[$, there exists a k-linear map $f : V \to W$ extending g such that $\alpha\|f\| \leqslant \|g\|$. By Corollary 1.2.33, there exists a k-linear projection $\pi : V \to V_0$ such that $\|\pi\| \leqslant \alpha^{-1}$. Let $f = g \circ \pi$. Then $\|f\| \leqslant \alpha^{-1}\|g\|$. The proposition is thus proved. □

Proposition 1.2.36 *Let V and W be finite-dimensional seminormed vector spaces over k, V_0 be a k-vector subspace of V, and Q be the quotient vector space V/V_0. We assume that, either the absolute value $|\cdot|$ is non-Archimedean, or the seminorm on V is induced by a semidefinite inner product. Then the canonique isomorphism $(V\otimes_k W)/(V_0\otimes_k W) \to Q\otimes_k W$ is an isometry, where we consider the ε-tensor product seminorms on $V\otimes_k W$ and $Q\otimes_k W$, and the quotient seminorm on $(V\otimes_k W)/(V_0\otimes_k W)$.*

Proof We have seen in Remark 1.1.59 that, for any $f \in V\otimes_k W$ viewed as a k-bilinear form on $V^* \times W^*$, its restriction to $Q^* \times W^*$ has an ε-tensor product norm not greater than that of f. In the following, we consider an element $g \in Q \otimes_k W$, viewed as a k-bilinear form on $Q^* \otimes_k W^*$. We will show that, for any $\alpha \in]0,1[$, there exists a k-bilinear form f on $V^*\times W^*$ extending g such that $\|f\|_\varepsilon \leqslant \alpha^{-1}\|g\|_\varepsilon$. By Proposition 1.1.20, the dual norm on Q^* identifies with the restriction of the dual norm on V^*. By Corollary 1.2.33, there exists a k-linear projection $\pi : V^* \to Q^*$ such that $\|\pi\| \leqslant \alpha^{-1}$ (in the non-Archimedean case, we use the fact that any dual norm is ultrametric). We let f be the k-bilinear form on $V^* \times W^*$ such that $f(\xi,\eta) = g(\pi(\xi),\eta)$. Then for $(\xi,\eta) \in (V^* \setminus \{0\}) \times (W^* \setminus \{0\})$ one has

$$\frac{|f(\xi,\eta)|}{\|\xi\|_*\|\eta\|_*}=\frac{|g(\pi(\xi),\eta)|}{\|\xi\|_*\|\eta\|_*}\leqslant\alpha^{-1}\frac{|g(\pi(\xi),\eta)|}{\|\pi(\xi)\|_*\|\eta\|_*}\leqslant\alpha^{-1}\|g\|_{\varepsilon}.$$

The proposition is thus proved. □

Corollary 1.2.37 *We assume that the absolute value $|\cdot|$ is non-Archimedean. Let $(V,\|\cdot\|)$ be a finite-dimensional seminormed vector space over k and r be the dimension of V over k. Let i be a positive integer. Then the canonical k-linear isomorphism* $\det(\Lambda^i V)\rightarrow\det(V)^{\otimes\binom{r-1}{i-1}}$ *is an isometry, where we consider the ith ε-exterior power seminorm on $\Lambda^i V$.*

Proof Consider the following commutative diagram

$$\begin{array}{ccc}
V^{\otimes i\binom{r}{i}} & \xrightarrow{\;p_1\;} & \det(V)^{\otimes\binom{r-1}{i-1}} \\
{\scriptstyle p_2}\downarrow & & \uparrow{\scriptstyle\simeq} \\
(\Lambda^i V)^{\otimes\binom{r}{i}} & \xrightarrow[\;p_3\;]{} & \det(\Lambda^i V)
\end{array}$$

By Proposition 1.2.36, if we equip $V^{\otimes i\binom{r}{i}}$ with the ε-tensor product seminorm, then its quotient seminorm on $(\Lambda^i V)^{\otimes\binom{r}{i}}$ identifies with the ε-tensor product of the ε-exterior power seminorm. Moreover, by Proposition 1.2.15, the quotient seminorm on $\det(\Lambda^i V)$ (induced by p_3) of the tensor product of the ε-exterior power seminorm identifies with the determinant seminorm of the latter. Still by the same proposition, the quotient seminorm on $\det(V)^{\otimes\binom{r-1}{i-1}}$ induced by p_1 identifies with the tensor power of the determinant seminorm. Therefore the natural isomorphism $\det(\Lambda^i V)\rightarrow\det(V)^{\otimes\binom{r-1}{i-1}}$ is actually an isometry by using (1) in Proposition 1.1.14. □

Corollary 1.2.38 *We assume that the absolute value $|\cdot|$ is non-Archimedean. Let $(V,\|\cdot\|)$ be a finite-dimensional normed vector space over k and i be a positive integer. Then the ith ε-exterior power norm on $\Lambda^i V$ is the double dual norm of the ith π-exterior power norm.*

Proof By definition, the ith ε-exterior power norm $\|\cdot\|_{\Lambda^i_\varepsilon}$ on $\Lambda^i V$ is ultrametric and is bounded from above by the ith π-exterior power norm. By Corollary 1.2.12, $\|\cdot\|_{\Lambda^i_\pi,**}$ is the largest ultrametric norm bounded from above by the ith π-exterior power norm $\|\cdot\|_{\Lambda^i_\pi}$. In particular, one has $\|\cdot\|_{\Lambda^i_\pi,**}\geqslant\|\cdot\|_{\Lambda^i_\varepsilon}$ and hence

$$\|\cdot\|_{\Lambda^i_\varepsilon,\det}\leqslant\|\cdot\|_{\Lambda^i_\pi,**,\det}\leqslant\|\cdot\|_{\Lambda^i_\pi,\det}.$$

By Corollary 1.2.37, one has $\|\cdot\|_{\Lambda^i_\varepsilon,\det}=\|\cdot\|_{\Lambda^i_\pi,\det}$ and hence the above inequalities are actually equalities. By Corollary 1.2.34, we obtain $\|\cdot\|_{\Lambda^i_\varepsilon}=\|\cdot\|_{\Lambda^i_\pi,**}$. □

Proposition 1.2.39 *Assume that $|\cdot|$ is non-Archimedean. Let V and W be finite-dimensional seminormed vector spaces over k and n and m be respectively the dimensions of V and W over k. We equip $V\otimes_k W$ with the ε-tensor product seminorm*

$\|\cdot\|_\varepsilon$. *Then the natural k-linear isomorphism* $\det(V\otimes_k W)\cong\det(V)^{\otimes m}\otimes_k\det(W)^{\otimes n}$ *is an isometry, where we consider the determinant seminorm of* $\|\cdot\|_\varepsilon$ *on* $\det(V\otimes_k W)$ *and the tensor product of determinant seminorms on* $\det(V)^{\otimes m}\otimes_k\det(W)^{\otimes n}$.

Proof Let $\|\cdot\|'$ be the seminorm on $\det(V)^{\otimes m}\otimes\det(W)^{\otimes n}$ induced by tensor product of determinant seminorms. By Propositions 1.2.36 and 1.2.15, the seminorm $\|\cdot\|'$ identifies with the quotient of the ε-tensor power seminorm on $(V\otimes_k W)^{\otimes nm}$ of $\|\cdot\|_\varepsilon$. Therefore, by Proposition 1.2.15 the seminorm $\|\cdot\|'$ identifies with $\|\cdot\|_{\varepsilon,\det}$. □

Proposition 1.2.40 *Let V be a finite-dimensional vector space over k and* $\|\cdot\|$ *be a seminorm on V. We assume that the seminorm* $\|\cdot\|$ *is either ultrametric or induced by a semidefinite inner product. For any vector subspace W of V, the canonical isomorphism*

$$\det(W)\otimes_k\det(V/W)\longrightarrow\det(V)$$

is an isometry, where we consider the determinant seminorm of the induced seminorm on $\det(W)$ *and that of the quotient seminorm on* $\det(V/W)$, *and the tensor product seminorm on the tensor product space* $\det(W)\otimes_k\det(V/W)$ *(see Remark 1.1.56).*

Proof By Proposition 1.1.16, if the seminorm $\|\cdot\|$ is not a norm, then either its restriction to W is not a norm, or its quotient seminorm on V/W is not a norm. In both cases, the seminorms on $\det(W)\otimes_k\det(V/W)$ and on $\det(V)$ vanish. Therefore we may assume without loss of generality that $\|\cdot\|$ is a norm.

Let $f:\det(W)\otimes_k\det(V/W)\to\det(V)$ be the canonical isomorphism. We have seen in Corollary 1.1.68 that the operator norm of f is $\leqslant 1$. Since f is an isomorphism between vector spaces of dimension 1 over k, to prove that f is an isometry, it suffices to verify that $\|f\|\geqslant 1$.

We first treat the case where the norm $\|\cdot\|$ is ultrametric. By Proposition 1.2.30, for any $\alpha\in\,]0,1[$, there exists an α-orthogonal basis $\boldsymbol{e}=\{e_i\}_{i=1}^r$ of V such that $\mathrm{card}(\boldsymbol{e}\cap W)=\dim_k(W)$. Without loss of generality, we assume that $\{e_1,\ldots,e_n\}$ forms a basis of W, and $e_{n+1},\ldots,e_r$ are vectors in $V\setminus W$. For any $i\in\{n+1,\ldots,r\}$, let $\overline{e}_i$ be the image of e_i in V/W. By Proposition 1.2.23, one has

$$\begin{aligned}\|e_1\wedge\cdots\wedge e_r\|_{\det}&\geqslant\alpha^r\cdot\|e_1\|\cdots\|e_r\|\geqslant\alpha^r\|e_1\|\cdots\|e_n\|\cdot\|\overline{e}_{n+1}\|\cdots\|\overline{e}_r\|\\&\geqslant\alpha^r\|e_1\wedge\cdots\wedge e_n\|_{\det}\cdot\|\overline{e}_{n+1}\wedge\cdots\wedge\overline{e}_r\|_{\det},\end{aligned}$$

where the last equality comes from Corollary 1.1.68. Therefore the operator norm of f is bounded from below by α^r. Since $\alpha\in\,]0,1[$ is arbitrary, one has $\|f\|\geqslant 1$.

For the Archimedean case where the norm $\|\cdot\|$ is induced by an inner product, by the classic Gram-Schmidt process we can construct an orthonormal basis $\boldsymbol{e}$ of V such that $\mathrm{card}(\boldsymbol{e}\cap W)=\dim_k(W)$. By Proposition 1.2.23, $\boldsymbol{e}$ is an Hadamard basis. We then proceed as above in replacing α by 1 to conclude. □

In Proposition 1.2.40, the assumption on the seminorm is crucial. In order to study the behaviour of the determinant seminorms of an exact sequence of general seminormed vector spaces, we introduce the following invariant.

Definition 1.2.41 Let $(V, \|\cdot\|)$ be a finite-dimensional seminormed vector space over k. Let $\mathcal{H}(V, \|\cdot\|)$ be the set of all normes $\|\cdot\|_h$ on $V/N_{\|\cdot\|}$ which are either ultrametric or induced by an inner product, and such that $\|\cdot\|_h \geqslant \|\cdot\|^{\sim}$. We define $\Delta(V, \|\cdot\|)$ to be the number (if $\|\cdot\|$ vanishe, by convention $\Delta(V, \|\cdot\|)$ is defined to be 1)

$$\inf\left\{\frac{\|\cdot\|_{h,\det}}{\|\cdot\|^{\sim}_{\det}} : \|\cdot\|_h \in \mathcal{H}(V, \|\cdot\|)\right\} \in [1, +\infty[,$$

where $\|\cdot\|_{h,\det}$ and $\|\cdot\|^{\sim}_{\det}$ are respectively determinant norms on $\det(V/N_{\|\cdot\|})$ induced by the norms $\|\cdot\|_h$ and $\|\cdot\|^{\sim}$. By definition, one has $\Delta(V, \|\cdot\|) = \Delta(V/N_{\|\cdot\|}, \|\cdot\|^{\sim})$. Moreover, if $\|\cdot\|$ is ultrametric or induced by a semidefinite inner product, then $\Delta(V, \|\cdot\|) = 1$.

Proposition 1.2.42 *Assume that $|\cdot|$ is non-Archimedean. Let $(V, \|\cdot\|)$ be a finite-dimensional seminormed vector space over k. One has*

$$\ln \Delta(V, \|\cdot\|) \leqslant \dim_k(V/N_{\|\cdot\|}) \sup_{x \in V \setminus N_{\|\cdot\|}} \big(\ln\|x\| - \ln\|x\|_{**}\big). \tag{1.36}$$

In particular, $\ln \Delta(V, \|\cdot\|) \leqslant \dim_k(V/N_{\|\cdot\|}) \ln(\dim_k(V/N_{\|\cdot\|}))$.

Proof By replacing $(V, \|\cdot\|)$ by $(V/N_{\|\cdot\|}, \|\cdot\|^{\sim})$, we may assume without loss of generality that $\|\cdot\|$ is a norm. Let

$$\lambda = \sup_{x \in V \setminus \{0\}} \big(\ln\|x\| - \ln\|x\|_{**}\big).$$

By definition one has $\|\cdot\| \leqslant e^{\lambda}\|\cdot\|_{**}$. Note that the norm $e^{\lambda}\|\cdot\|_{**}$ is ultrametric. Therefore

$$\Delta(V, \|\cdot\|) \leqslant \frac{e^{r\lambda}\|\cdot\|_{**,\det}}{\|\cdot\|_{\det}} = e^{r\lambda},$$

where r is the dimension of V over k, and the equality comes from Proposition 1.2.15. The inequality (1.36) is thus proved. The last inequality results from (1.36) and Corollary 1.2.12. □

Proposition 1.2.43 *Let $(V, \|\cdot\|)$ be a finite-dimensional normed vector space over k. For any vector subspace W of V, the norm of the canonical isomorphism*

$$f : \det(W) \otimes \det(V/W) \longrightarrow \det(V)$$

is bounded from below by

$$\frac{\Delta(W, \|\cdot\|_W)\Delta(V/W, \|\cdot\|_{V/W})}{\Delta(V, \|\cdot\|)} \geqslant \Delta(V, \|\cdot\|)^{-1},$$

where $\|\cdot\|_W$ is the restriction of the norm $\|\cdot\|$ to the vector subspace W and $\|\cdot\|_{V/W}$ is the quotient norm of $\|\cdot\|$ on the quotient space V/W.

Proof Let $\|\cdot\|_h$ be a norm in $\mathcal{H}(V, \|\cdot\|)$. Let $\|\cdot\|_{h,W}$ and $\|\cdot\|_{h,V/W}$ be respectively the restriction of $\|\cdot\|_h$ to W and the quotient norm of $\|\cdot\|_h$ on V/W. By Proposition 1.2.40, the canonical isomorphism

$$\det(W, \|\cdot\|_{h,W}) \otimes \det(V/W, \|\cdot\|_{h,V/W}) \longrightarrow \det(V, \|\cdot\|_{V,h})$$

is an isometry. Hence

$$\frac{\|\cdot\|_{h,\det}}{\|\cdot\|_{\det}} = \frac{\|\cdot\|_{h,W,\det}\|\cdot\|_{h,V/W,\det}}{\|f\| \cdot \|\cdot\|_{W,\det}\|\cdot\|_{V/W,\det}} \geqslant \frac{1}{\|f\|} \cdot \Delta(W, \|\cdot\|_W)\Delta(V/W, \|\cdot\|_{V/W}).$$

Since $\|\cdot\|_h \in \mathcal{H}(V, \|\cdot\|)$ is arbitrary, we obtain the lower bound announced in the proposition. □

Corollary 1.2.44 *Let $(V, \|\cdot\|_V)$ be a finite-dimensional seminormed vector space over k, W be a vector subspace of V, $\|\cdot\|_W$ be the restriction of $\|\cdot\|_V$ to W, and $\|\cdot\|_{V/W}$ be the quotient of $\|\cdot\|_V$ on V/W. One has*

$$\Delta(W, \|\cdot\|_W)\Delta(V/W, \|\cdot\|_{V/W}) \leqslant \Delta(V, \|\cdot\|). \tag{1.37}$$

In particular, $\Delta(W, \|\cdot\|_W) \leqslant \Delta(V, \|\cdot\|)$ and $\Delta(V/W, \|\cdot\|_{V/W}) \leqslant \Delta(V, \|\cdot\|)$.

Proof By Proposition 1.1.16, we can assume without loss of generality that $\|\cdot\|_V$ is a norm. By Corollary 1.1.68, if we denote by $f : \det(W) \otimes \det(V/W) \to \det(V)$ the canonical isomorphism, then $\|f\| \leqslant 1$. The inequality (1.37) thus follows from Proposition 1.2.43. Finally, by definition one has $\Delta(W, \|\cdot\|_W) \geqslant 1$ and $\Delta(V/W, \|\cdot\|_{V/W}) \geqslant 1$, thus we deduce from (1.37) the last two inequalities stated in the corollary. □

Remark 1.2.45 Let $(V, \|\cdot\|)$ be a finite-dimensional normed vector space over k. We assume that the dimension r of $V/N_{\|\cdot\|}$ is positive. In the case where the absolute value $|\cdot|$ is non-Archimedean, Corollary 1.2.12 provides the upper bound $\Delta(V, \|\cdot\|) \leqslant r^r$. This result is also true in the Archimedean case (which follows from the existence of an orthogonal basis, see the beginning of §1.2.8 for details). However, as we will see in the next subsection (cf. Theorem 1.2.54), in the Archimedean case one has a better upper bound $\Delta(V, \|\cdot\|) \leqslant r^{r/2}$.

1.2.7 Dual determinant norm

Let $(V, \|\cdot\|)$ be a finite-dimensional seminormed vector space over k. We denote by $\|\cdot\|_{\det}^{\sim}$ the determinant norm on $\det(V/N_{\|\cdot\|})$ induced by $\|\cdot\|^{\sim}$, and denote by $\|\cdot\|_{\det,*}^{\sim}$ the dual norm of $\|\cdot\|_{\det}^{\sim}$. Let $\|\cdot\|_{*,\det}$ be the determinant norm on $\det(V^*) \cong \det(V/N_{\|\cdot\|})^*$ of the dual norm $\|\cdot\|_*$ on V^*. The purpose of this subsection is to compare these two norms. We denote by $\delta(V, \|\cdot\|)$ the ratio

$$\delta(V, \|\cdot\|) := \frac{\|\cdot\|^{\sim}_{\det,*}}{\|\cdot\|_{*,\det}}.$$

In the case where there is no ambiguity on the seminorm $\|\cdot\|$ on V, we also use the abbreviate notation $\delta(V)$ to denote $\delta(V, \|\cdot\|)$. By definition, if η is a non-zero element in $\det(V/N_{\|\cdot\|})$ and if $\eta^\vee$ is its dual element in $\det(V^*)$, then one has

$$\delta(V, \|\cdot\|)^{-1} = \|\eta\|^{\sim}_{\det} \cdot \|\eta^\vee\|_{*,\det}. \tag{1.38}$$

In particular, one has (see Proposition 1.2.15)

$$\delta(V, \|\cdot\|) = \delta(V/N_{\|\cdot\|}, \|\cdot\|^{\sim}) = \delta(V^*, \|\cdot\|_*). \tag{1.39}$$

Proposition 1.2.46 *Let $(V, \|\cdot\|)$ be a finite-dimensional seminormed vector space over k. One has $\delta(V, \|\cdot\|) \geqslant 1$.*

Proof By (1.39) we may assume without loss of generality that $\|\cdot\|$ is a norm.

Let $\{e_i\}_{i=1}^r$ be a basis of V, and $\{e_i^\vee\}_{i=1}^r$ be its dual basis. One has

$$\|e_1^\vee \wedge \cdots \wedge e_r^\vee\|_{\det,*} = \|e_1 \wedge \cdots \wedge e_r\|_{\det}^{-1}.$$

Therefore

$$\begin{aligned}\delta(V, \|\cdot\|)^{-1} &= \|e_1 \wedge \cdots \wedge e_r\|_{\det} \cdot \|e_1^\vee \wedge \cdots \wedge e_r^\vee\|_{*,\det} \\ &\leqslant \|e_1\| \cdots \|e_r\| \cdot \|e_1^\vee\|_* \cdots \|e_r^\vee\|_*,\end{aligned}$$

where the inequality comes from Proposition 1.1.66. If the basis $\{e_i\}_{i=1}^r$ is α-orthogonal, where $\alpha \in \,]0, 1[$, by Lemma 1.2.10 one has $\delta(V, \|\cdot\|)^{-1} \leqslant \alpha^{-r}$. Since for any $\alpha \in \,]0, 1[$ there exists an α-orthogonal basis (see Proposition 1.2.7), one has $\delta(V, \|\cdot\|) \geqslant 1$. □

Proposition 1.2.47 *Let $(V, \|\cdot\|)$ be a finite-dimensional seminormed vector space over k. Assume that the absolute value $|\cdot|$ is non-Archimedean, or the seminorm $\|\cdot\|$ is induced by a semidefinite inner product. Then one has $\delta(V, \|\cdot\|) = 1$.*

Proof By (1.39) we may assume without loss of generality that $\|\cdot\|$ is a norm.

We first treat the case where the absolute value $|\cdot|$ is non-Archimedean. Let $\alpha \in \,]0, 1[$ and $\{e_i\}_{i=1}^r$ be an α-orthogonal basis of $(V, \|\cdot\|)$ (see Proposition 1.2.7 for the existence of an α-orthogonal basis). Then the dual basis $\{e_i^\vee\}_{i=1}^r$ is α-orthogonal with respect to the dual norm $\|\cdot\|_*$ (see Proposition 1.2.11). In particular, by Proposition 1.2.23 one has

$$\frac{\|e_1 \wedge \cdots \wedge e_r\|_{\det}}{\|e_1\| \cdots \|e_r\|} \geqslant \alpha^r, \quad \frac{\|e_1^\vee \wedge \cdots \wedge e_r^\vee\|_{*,\det}}{\|e_1^\vee\|_* \cdots \|e_r^\vee\|_*} \geqslant \alpha^r.$$

Therefore

$$\delta(V, \|\cdot\|) = \frac{\|e_1 \wedge \cdots \wedge e_r\|_{\det}^{-1}}{\|e_1^\vee \wedge \cdots \wedge e_r^\vee\|_{*,\det}} \leqslant \alpha^{-2r} \frac{1}{\|e_1\| \cdots \|e_r\| \cdot \|e_1^\vee\|_* \cdots \|e_r^\vee\|_*} \leqslant \alpha^{-2r},$$

where the last inequality comes from Lemma 1.2.10. Since α is arbitrary, one has $\delta(V, \|\cdot\|) \leqslant 1$.

The proof of the Archimedean case is quite similar, where we use the existence of an orthogonal basis, which is also an Hadamard basis (see Proposition 1.2.25). We omit the details. □

The following Lemma is the Archimedean counterpart of Proposition 1.2.35 (see also the comparison in Remark 1.1.21).

Lemma 1.2.48 *Assume that the absolute value $|\cdot|$ is Archimedean. Let $(V, \|\cdot\|_V)$ be a finite-dimensional seminormed vector space over k, W be a vector subspace of V, and $\|\cdot\|_W$ be the restriction of the seminorm $\|\cdot\|_V$ to W. Then the map $F : V^* \to W^*$, which sends $\varphi \in V^*$ to its restriction to W, is surjective. Moreover, the quotient norm on W^* induced by the dual norm $\|\cdot\|_{V,*}$ coincides with the norm $\|\cdot\|_{W,*}$.*

Proof Let ψ be an element in W^*. If φ is an element in V^* which extends ψ, then clearly one has $\|\varphi\|_{V,*} \geqslant \|\psi\|_{W,*}$. Moreover, by Hahn-Banach theorem, there exists $\varphi_0 \in V^*$ which extends ψ and such that $\|\varphi_0\|_{V,*} = \|\psi\|_{W,*}$. Therefore, the map F is surjective and the quotient norm on $W^\vee$ induced by $\|\cdot\|_{V,*}$ coincides with $\|\cdot\|_{W,*}$. □

Proposition 1.2.49 *Let $(V, \|\cdot\|_V)$ and $(W, \|\cdot\|_W)$ be finite-dimensional seminormed vector spaces over k, V_0 be a k-vector subspace of V and $\|\cdot\|_{V_0}$ be the restriction of $\|\cdot\|_V$ on V_0. Denote by $\|\cdot\|_\varepsilon$ and $\|\cdot\|_\pi$ the ε-tensor product and the π-tensor product of the seminorms $\|\cdot\|_V$ and $\|\cdot\|_W$, respectively.*

(1) Assume that, either the absolute value $|\cdot|$ is Archimedean, or the seminorm $\|\cdot\|_V$ is ultrametric. Then the ε-tensor product $\|\cdot\|_{\varepsilon,0}$ of $\|\cdot\|_{V_0}$ and $\|\cdot\|_W$ identifies with the restriction of $\|\cdot\|_\varepsilon$ to $V_0 \otimes_k W$.
(2) Assume that the seminorm $\|\cdot\|_V$ is either ultrametric or induced by a semidefinite inner product. Then the π-tensor product $\|\cdot\|_{\pi,0}$ of $\|\cdot\|_{V_0}$ and $\|\cdot\|_W$ coincides with the restriction of $\|\cdot\|_\pi$ to $V_0 \otimes_k W$.

Proof (1) Let φ be a tensor in $V_0 \otimes_k W$, viewed as a bilinear form on $V_0^* \times W^*$. By definition, one has

$$\|\varphi\|_{\varepsilon,0} = \sup_{\substack{(f_0,g)\in V_0^*\times W^*\\ f_0\neq 0,\ g\neq 0}} \frac{|\varphi(f_0, g)|}{\|f_0\|_{V_0,*} \cdot \|g\|_{W,*}}.$$

Since the absolute value $|\cdot|$ is Archimedean or the norm $\|\cdot\|_V$ is ultrametric, by Proposition 1.2.35 (for the ultrametric case) and Lemma 1.2.48 (for the Archimedean case), the norm $\|\cdot\|_{V_0,*}$ identifies with the quotient of $\|\cdot\|_{V,*}$ by the canonical surjective map $V^* \to V_0^*$. Therefore, one has

$$\sup_{\substack{(f_0,g)\in V_0^*\times W^*\\ f_0\neq 0,\ g\neq 0}} \frac{|\varphi(f_0, g)|}{\|f_0\|_{V_0,*} \cdot \|g\|_{W,*}} = \sup_{\substack{(f,g)\in V^*\times W^*\\ f\neq 0,\ g\neq 0}} \frac{|\varphi(f, g)|}{\|f\|_{V,*} \cdot \|g\|_{W,*}},$$

which shows $\|\varphi\|_{\varepsilon,0} = \|\varphi\|_\varepsilon$.

(2) We have already seen in Proposition 1.1.60 (1) that $\|\cdot\|_{\pi,0}$ is bounded from below by the restriction of $\|\cdot\|_\pi$ to $V_0 \otimes_k W$. Let T be an element of $V_0 \otimes_k W$, which is written, as an element of $V \otimes_k W$, in the form $T = \sum_{i=1}^N x_i \otimes y_i$, where $\{x_1, \ldots, x_N\} \subseteq V$ and $\{y_1, \ldots, y_N\} \subseteq W$. By Corollary 1.2.33, for any $\alpha \in]0, 1[$, there exists a linear projection $\pi_\alpha : V \to V_0$ such that $\|\pi_\alpha\| \leqslant \alpha^{-1}$. Since T belongs to $V_0 \otimes_k W$ one has $T = \sum_{i=1}^N \pi_\alpha(x_i) \otimes y_i$. Moreover,

$$\|T\|_{0,\pi} \leqslant \sum_{i=1}^N \|\pi_\alpha(x_i)\|_{V_0} \cdot \|y_i\|_W \leqslant \alpha^{-1} \sum_{i=1}^N \|x_i\|_V \cdot \|y_i\|_W.$$

Since α and the writing $T = \sum_{i=1}^N x_i \otimes y_i$ are arbitrary, we obtain $\|T\|_{0,\pi} \leqslant \|T\|_\pi$. □

Proposition 1.2.50 *Let $(V, \|\cdot\|_V)$ and $(W, \|\cdot\|_W)$ be seminormed vector spaces over k, and $\|\cdot\|_\pi$ be the π-tensor product norm of $\|\cdot\|_V$. We assume that $\|\cdot\|_W$ is ultrametric. For any $(x, y) \in V \times W$, one has $\|x \otimes y\|_\pi = \|x\|_V \cdot \|y\|_W$.*

Proof By definition on has $\|x \otimes y\|_\pi \leqslant \|x\|_V \cdot \|y\|_W$. It then suffices to show that, for any writing of $x \otimes y$ as

$$\sum_{i=1}^N x_i \otimes y_i,$$

with $(x_1, \ldots, x_n) \in V^n$ and $(y_1, \ldots, y_n) \in W^n$, one has

$$\|x\|_V \cdot \|y\|_W \leqslant \sum_{i=1}^N \|x_i\|_V \cdot \|y_i\|_W.$$

Therefore we may assume without loss of generality that V and W are finite-dimensional vector spaces over k. Consider the k-linear map ℓ from W^* to V sending $\varphi \in W^*$ to

$$\varphi(y)x = \sum_{i=1}^N \varphi(y_i)x_i.$$

We equip W^* with the dual norm $\|\cdot\|_{W,*}$ and consider the operator norm of ℓ. On one hand, one has

$$\|\ell\| = \sup_{\varphi \in W^* \setminus \{0\}} \frac{\|\varphi(y)x\|_V}{\|\varphi\|_{W,*}} = \sup_{\varphi \in W^* \setminus \{0\}} \frac{|\varphi(y)| \cdot \|x\|_V}{\|\varphi\|_{W,*}}$$
$$= \|y\|_{W,**} \cdot \|x\|_V = \|y\|_W \cdot \|x\|_V,$$

where the last equality comes from Corollary 1.2.12 and the hypothesis that $\|\cdot\|_W$ is ultrametric. On the other hand, one has

$$\|\ell\| = \sup_{\varphi\in W^*\setminus\{0\}} \frac{\|\varphi(y_1)x_1+\cdots+\varphi(y_N)x_N\|_V}{\|\varphi\|_{W,*}}$$
$$\leqslant \sup_{\varphi\in W^*\setminus\{0\}} \sum_{i=1}^N \frac{|\varphi(y_i)|\cdot\|x_i\|_V}{\|\varphi\|_{W,*}} \leqslant \sum_{i=1}^N \|x_i\|_V\cdot\|y_i\|_{W,**} = \sum_{i=1}^N \|x_i\|_V\cdot\|y_i\|_W,$$

where the last equality follows from Corollary 1.2.12 and the hypothesis that $\|\cdot\|_W$ is ultrametric again. The proposition is thus proved. □

The following result provides a variant of Proposition 1.2.43. Note that it generalises (by using Proposition 1.2.47) Proposition 1.2.40.

Proposition 1.2.51 *Let $(V,\|\cdot\|)$ be a finite-dimensional normed vector space over k. Assume that the absolute value $|\cdot|$ is Archimedean or the norm $\|\cdot\|$ is ultrametric. For any vector subspace W of V, the norm of the canonical isomorphism*

$$f : \det(W)\otimes\det(V/W) \longrightarrow \det(V)$$

is bounded from below by

$$\frac{\delta(W,\|\cdot\|_W)\delta(V/W,\|\cdot\|_{V/W})}{\delta(V,\|\cdot\|)} \geqslant \delta(V,\|\cdot\|)^{-1},$$

where we consider the restriction $\|\cdot\|_W$ of the norm $\|\cdot\|$ to the vector subspace W and the quotient norm $\|\cdot\|_{V/W}$ of $\|\cdot\|$ on the quotient space V/W. In particular, one has

$$\max\left\{\delta(W,\|\cdot\|_W),\delta(V/W,\|\cdot\|_{V/W})\right\} \leqslant \delta(V,\|\cdot\|).$$

Proof Let $\|\cdot\|_{V/W}$ be the quotient norm on V/W induced by $\|\cdot\|_V$. By Proposition 1.1.20, the dual norm $\|\cdot\|_{V/W,*}$ coincides with the restriction of the norm $\|\cdot\|_*$ to $(V/W)^\vee$. Moreover, by Lemma 1.2.48 (for the Archimedean case) and Proposition 1.2.35 (for the non-Archimedean case), the quotient norm on $W^\vee$ induced by $\|\cdot\|_*$ identifies with the dual norm $\|\cdot\|_{W,*}$. Let α and β be respectively non-zero elements in $\det(W)$ and $\det(V/W)$. Let $\alpha^\vee\in\det(W^\vee)$ and $\beta^\vee\in\det((V/W)^\vee)$ be their dual elements, η be the image of $\alpha\otimes\beta$ by the canonical isomorphism $\det(W)\otimes\det(V/W)\to\det(V)$, and $\eta^\vee$ be the image of $\alpha^\vee\otimes\beta^\vee$ by the canonical isomorphism $\det(W^\vee)\otimes\det((V/W)^\vee)\to\det(V^\vee)$. Then $\eta^\vee$ is the dual element of η.

By Proposition 1.1.68, one has

$$\|\eta^\vee\|_{*,\det} \leqslant \|\alpha^\vee\|_{W,*,\det}\cdot\|\beta^\vee\|_{V/W,*,\det}.$$

Hence by (1.38) one has

$$\frac{\delta(W,\|\cdot\|_W)\delta(V/W,\|\cdot\|_{V/W})}{\delta(V,\|\cdot\|)} = \frac{\|\eta^\vee\|_{*,\det}\cdot\|\eta\|_{\det}}{\|\alpha^\vee\|_{W,*,\det}\|\alpha\|_{W,\det}\cdot\|\beta\|_{V/W,\det}\|\beta^\vee\|_{V/W,*,\det}}$$
$$\leqslant \frac{\|\eta\|_{\det}}{\|\alpha\|_{W,\det}\cdot\|\beta\|_{V/W,\det}} = \|f\|.$$

Finally, by Corollary 1.1.68, we obtain

$$\delta(W, \|\cdot\|_W)\delta(V/W, \|\cdot\|_{V/W}) \leqslant \delta(V, \|\cdot\|).$$

Since $\delta(W, \|\cdot\|_W)$ and $\delta(V/W, \|\cdot\|_{V/W})$ are $\geqslant 1$ (see Proposition 1.2.46), we obtain the last inequality. □

Corollary 1.2.52 *Let V be a finite-dimensional vector space over k and $\|\cdot\|$ be a norm on V. We assume that, either the norm $\|\cdot\|$ is ultrametric or the absolute value $|\cdot|$ is Archimedean. If W_1 and W_2 are two k-vector subspaces of V, then the canonical isomorphism*

$$\det(W_1) \otimes \det(W_2) \longrightarrow \det(W_1 \cap W_2) \otimes \det(W_1 + W_2) \tag{1.40}$$

induced by the short exact sequence

$$0 \longrightarrow W_1 \cap W_2 \longrightarrow W_1 \oplus W_2 \longrightarrow W_1 + W_2 \longrightarrow 0$$

has operator norm $\leqslant \min\{\delta(W_1), \delta(W_2)\}/\delta(W_1 \cap W_2)$, where in the above formulae we consider the restricted norms on the vector subspaces of V. In particular, if $\|\cdot\|$ is an ultrametric norm, then the linear map (1.40) *has norm $\leqslant 1$.*

Proof Consider the short exact sequence

$$0 \longrightarrow W_1 \cap W_2 \longrightarrow W_1 \longrightarrow W_1/(W_1 \cap W_2) \longrightarrow 0\,.$$

By Proposition 1.2.51, the canonical element η in

$$\det(W_1)^\vee \otimes \det(W_1 \cap W_2) \otimes \det(G)$$

has norm $\leqslant \delta(W_1)/\delta(W_1 \cap W_2)\delta(G)$, where G denotes the vector space $W_1/(W_1 \cap W_2)$ equipped with the quotient norm $\|\cdot\|_G$.

Similarly, consider the short exact sequence

$$0 \longrightarrow W_2 \longrightarrow W_1 + W_2 \longrightarrow (W_1 + W_2)/W_2 \longrightarrow 0\,.$$

By Corollary 1.1.68, the canonical element η' in

$$\det(W_2)^\vee \otimes \det(G')^\vee \otimes \det(W_1 + W_2)$$

has norm $\leqslant 1$, where G' denotes the vector space $(W_1 + W_2)/W_2$ equipped with the quotient norm $\|\cdot\|_{G'}$. Therefore we obtain

$$\|\eta \otimes \eta'\| \leqslant \delta(W_1)/\delta(W_1 \cap W_2).$$

Let $f : G \to G'$ be the canonical isomorphism. One has $\|f(x)\|_{G'} \leqslant \|x\|_G$ for any $x \in G$. In particular, the canonical element of $\det(G) \otimes \det(G')^\vee$ has norme $\geqslant 1$. We deduce that the canonical element of

$$\det(W_1)^\vee \otimes \det(W_2)^\vee \otimes \det(W_1 \cap W_2) \otimes \det(W_1 + W_2)$$

has norm $\leqslant \delta(W_1)/\delta(W_1 \cap W_2)$. By the symmetry between W_1 and W_2, we obtain the announced inequality. □

Remark 1.2.53 We assume that the absolute value $|\cdot|$ is non-Archimedean. The result of Corollary 1.2.52 is not true in general if the norm $\|\cdot\|$ is not ultrametic. However, we can combine the proof of Corollary 1.2.52 and Proposition 1.2.43 to show that the canonical isomorphism (1.40) in Corollary 1.2.52 has an operator norm bounded from above by

$$\frac{\min\{\Delta(W_1), \Delta(W_2)\}}{\Delta(W_1 \cap W_2)}.$$

The same argument also works in the Archimedean case.

1.2.8 Ellipsoid of John and Löwner

We assume that the absolute value $|\cdot|$ is Archimedean. Let V be a finite-dimensional vector space over k, equipped with a norm $\|\cdot\|$. In this subsection, we discuss the approximation of the norm $\|\cdot\|$ by Euclidean or Hermitian norms. Note that Proposition 1.2.7 provides a result in this direction. Let $\{e_i\}_{i=1}^r$ be an orthonormal basis of V. Let $\langle , \rangle$ be an inner product on V such that $\{e_i\}_{i=1}^r$ is orthogonal with respect to the inner product, and that $\langle e_i, e_i \rangle = r$ for any $i \in \{1, \ldots, r\}$. If $\|\cdot\|_h$ denotes the norm on V induced by the inner product $\langle , \rangle$, then for any $x = \lambda_1 e_1 + \cdots + \lambda_r e_r \in V$ one has

$$\frac{1}{r}\|x\|_h = \left(\frac{|\lambda_1|^2 + \cdots + |\lambda_r|^2}{r}\right)^{1/2} \leqslant \max\{|\lambda_1|, \ldots, |\lambda_r|\} \leqslant \|x\|$$

and $\|x\| \leqslant |\lambda_1| + \cdots + |\lambda_r| \leqslant r^{1/2}(|\lambda_1|^2 + \cdots + |\lambda_r|^2)^{1/2} = \|x\|_h$.

The works of John [91] and Löwner provide a stronger result on the comparison of inner product norms and general norms. We refer to the expository article of Henk [86] for the history of this theory. For the convenience of the readers, we include the statement and the proof of this result.

Theorem 1.2.54 (John-Löwner) *Let V be a non-zero finite-dimensional vector space over k, equipped with a norm $\|\cdot\|$. There exists a unique Euclidean or Hermitian norm $\|\cdot\|_J$ bounded from above by $\|\cdot\|$ such that, for any Euclidean or Hermitian norm $\|\cdot\|_h$ satisfying $\|\cdot\|_h \leqslant \|\cdot\|$, one has $\|\cdot\|_{h,\det} \leqslant \|\cdot\|_{J,\det}$. Moreover, for any $x \in V$, one has $\|x\|_h \leqslant \|x\| \leqslant r^{1/2}\|x\|_h$, where r is the dimension of V over k.*

Proof We fix an arbitrary inner product $\langle , \rangle'$ on V and denote by Θ the vector space (over $\mathbb{R}$) of all endomorphisms of V which are self-adjoint with respect to the inner product $\langle , \rangle'$. Recall that a k-linear map $u : V \to V$ is said to be *self-adjoint* with respect to $\langle , \rangle'$ if and only if

$$\forall (x, y) \in V^2, \quad \langle u(x), y\rangle' = \langle x, u(y)\rangle'.$$

Let Θ^+ be the set of all positive definite self-adjoint operators. Since any pair of self-adjoint operator can be simultaneously diagonalised by a basis of V, we obtain that Θ^+ is a convex open subset of Θ and that the function $\log\det(\cdot)$ is strictly concave on Θ^+.

Let $B = \{x \in V : \|x\| \leqslant 1\}$ be the unit ball of the norm $\|\cdot\|$. For any $u \in \Theta^+$, let $B_u = \{x \in V : \langle x, u(x)\rangle' \leqslant 1\}$, which is the unit ball of the Euclidean or Hermitian norm $\|\cdot\|_u$ on V defined as

$$\forall x \in V, \quad \|x\|_u^2 = \langle x, u(x)\rangle'.$$

Let Θ_0 be the set of all $u \in \Theta^+$ such that $B_u \supseteq B$. Then for any $u_0 \in \Theta_0$, the set

$$\Theta(u_0) := \{u \in \Theta_0 \ :\ \det(u) \geqslant \det(u_0)\}$$

is a convex and compact subset of Θ. In fact, from the concavity and the continuity of the function $\log\det(\cdot)$ we obtain that the set $\Theta(u_0)$ is convex and closed. Moreover, the condition $B_u \supseteq B$ for $u \in \Theta_0$ implies that the set $\Theta(u_0)$ is bounded in Θ. Therefore the restriction of the function $\det(\cdot)$ to Θ_0 attains its maximal value at a unique point $u_1 \in \Theta_0$.

Let $\langle\,,\rangle$ be the inner product on V such that

$$\forall (x, y) \in V \times V, \quad \langle x, y\rangle = \langle x, u_1(y)\rangle'.$$

We call it the *John inner product* associated with the norm $\|\cdot\|$. The corresponding Euclidean or Hermitian norm $\|\cdot\|_J$ is called the *John norm* associated with $\|\cdot\|$.

In the following, we prove the relation

$$\forall x \in V, \quad \|x\|_J \leqslant \|x\| \leqslant r^{1/2}\|x\|_J$$

under the supplementary assumption that the unit ball B is the convex hull of finitely many orbits of the action of $\{a \in k : |a| = 1\}$ on V.

Without loss of generality, we assume that $\langle\,,\rangle' = \langle\,,\rangle$. For any $x \in V$ such that $\|x\| \leqslant 1$, let $\varphi_x : \Theta \to \mathbb{R}$ be the linear functional which sends $u \in \Theta$ to $\langle x, u(x)\rangle$. If $u : V \to V$ is a self-adjoint linear operator such that $\varphi_x(u) \leqslant 0$ for any $x \in B$ such that $\langle x, x\rangle = 1$, then one has $\mathrm{Tr}(u) \leqslant 0$. In fact, the condition

$$\forall x \in B, \quad \langle x, x\rangle = 1 \Longrightarrow \varphi_x(u) \leqslant 0$$

implies that $\mathrm{Id} + \varepsilon u \in \Theta_0$ for sufficiently small $\varepsilon > 0$ (here we use the supplementary assumption that the convex body B is spanned by a finite number of orbits). Therefore one has $\det(\mathrm{Id} + \varepsilon u) \leqslant \det(\mathrm{Id}) = 1$, which leads to $\mathrm{Tr}(u) \leqslant 0$. Therefore, the linear form $\mathrm{Tr}(\cdot)$ lies in the closure of the positive cone of $\Theta^\vee$ generated by $\varphi_x(\cdot)$ ($x \in B$, $\langle x, x\rangle = 1$), namely there exist a sequence of elements $\{x_n\}_{n\in\mathbb{N}}$ in

$$B \cap \{x \in V \ :\ \langle x, x\rangle = 1\}$$

and a sequence $\{\lambda_n\}_{n\geqslant 0}$ of real numbers such that

$$\mathrm{Tr}(u) = \sum_{n\in\mathbb{N}} \lambda_n \langle x_n, u(x_n)\rangle \tag{1.41}$$

for any $u \in \Theta$. If we apply the identity to $u = \mathrm{Id}$, we obtain

$$r = \sum_{n\in\mathbb{N}} \lambda_n \langle x_n, x_n\rangle = \sum_{n\in\mathbb{N}} \lambda_n. \tag{1.42}$$

Let y be an element in V such that $\langle y, y\rangle = 1$. We apply the identity (1.41) to the linear map $u(x) = \langle y, x\rangle y$, and obtain

$$1 = \sum_{n\in\mathbb{N}} \lambda_n |\langle x_n, y\rangle|_\infty^2.$$

Thus there should exist $n \in \mathbb{N}$ such that $|\langle x_n, y\rangle|_\infty \geqslant r^{-1/2}$ since otherwise we have

$$1 < \sum_{n\in\mathbb{N}} \frac{1}{r}\lambda_n = \frac{1}{r}\cdot r = 1,$$

where the first equality comes from (1.42), which leads to a contradiction. Since the unit ball $B = \{x \in V : \|x\| = 1\}$ is invariant by the multiplication by any $\lambda \in k$ with $|\lambda| = 1$, we obtain that, for any $y \in V$ such that $\langle y, y\rangle = 1$, there exists $x \in B$ such that $\mathrm{Re}\langle y, x\rangle \geqslant r^{-1/2}$.

We claim that the unit ball $B = \{x \in V : \|x\| \leqslant 1\}$ contains the set of all $x \in V$ such that $\langle x, x\rangle \leqslant 1/r$. In fact, if $x_0 \in V$ is a point such that $\langle x_0, x_0\rangle \leqslant 1/r$ and that $\|x\| > 1$, we can choose an $\mathbb{R}$-affine function $f : V \to \mathbb{R}$ such that $f(x_0) = 0$ and that $f(x) < 0$ for any $x \in B$. Note that $\mathrm{Re}\langle , \rangle$ defines an inner product on V, where V is viewed as a vector space over $\mathbb{R}$ if $k = \mathbb{C}$. By Riesz's theorem there exists $y \in V$ such that

$$\forall\, x \in V, \quad f(x) = \mathrm{Re}\langle y, x\rangle + f(0).$$

Without loss of generality, we may assume that $\langle y, y\rangle = 1$. One has

$$0 = f(x_0) = \mathrm{Re}\langle y, x_0\rangle + f(0) \leqslant \langle y, y\rangle^{1/2}\langle x_0, x_0\rangle^{1/2} + f(0) = \frac{1}{\sqrt{r}} + f(0).$$

Hence $f(0) \geqslant -r^{-1/2}$. However, the above argument shows that there exists $x \in B$ such that $\mathrm{Re}\langle y, x\rangle \geqslant r^{-1/2}$. Hence one has

$$0 > f(x) = \mathrm{Re}\langle y, x\rangle + f(0) \geqslant 0,$$

which leads a contradiction.

Since $B \subseteq \{x \in V : \langle x, x\rangle \leqslant 1\}$, one has $\|x\|_J \leqslant \|x\|$ for any $x \in V$. Moreover, the relation

$$\{x \in V : \langle x, x\rangle \leqslant 1/r\} \subseteq B = \{x \in V : \|x\| \leqslant 1\}$$

implies that $\|x\| \leqslant r^{1/2}\|x\|_J$. The theorem is thus proved under the supplementary hypothesis.

For the general case, we can construct a decreasing sequence of norms $\{\|\cdot\|_n\}_{n\in\mathbb{N}}$ such that each unit ball $\{x \in V : \|x\|_n \leqslant 1\}$ verifies the supplementary hypothesis mentioned above and that the sequence

$$\sup_{0\neq x\in V} \frac{\|x\|_n}{\|x\|}$$

converges to 1 when $n \to +\infty$. For each $n \in \mathbb{N}$, let $\|\cdot\|_{n,J}$ be the John norm associated to the norm $\|\cdot\|_n$. If we identify the set of Euclidean or Hermitian norms on V with Θ^+, we obtain that these John norms actually lies in a bounded subset of Θ. Therefore there exists a subsequence of $\{\|\cdot\|_{n,J}\}_{n\in\mathbb{N}}$ which converges in Θ, whose limite should be the John norm associated with $\|\cdot\|$ by the uniqueness of the John norm. Without loss of generality we may assume that $\{\|\cdot\|_{n,J}\}_{n\in\mathbb{N}}$ converges in Θ. By what we have established above, for any $n \in \mathbb{N}$ one has

$$\forall\, x \in V, \quad \|x\|_{n,J} \leqslant \|x\|_n \leqslant r^{1/2}\|x\|_{n,J}.$$

By taking the limit when $n \to +\infty$, we obtain the result announced in the theorem. □

Remark 1.2.55 Let $(V, \|\cdot\|)$ be a finite-dimensional normed vector space over $\mathbb{R}$ or $\mathbb{C}$ (equipped with the usual absolute value). Since $(V, \|\cdot\|)$ is reflexive (see Proposition 1.1.18), we deduce that the dual norm $\|\cdot\|_{J,*}$ is the unique norm on $V^\vee$ which is bounded from below by $\|\cdot\|_*$ and such that the corresponding determinant norm $\|\cdot\|_{J,*,\det}$ is minimal. In particular, one has

$$\Delta(V^\vee, \|\cdot\|_*) = \frac{\|\cdot\|_{J,*,\det}}{\|\cdot\|_{*,\det}}.$$

Similarly, one has

$$\Delta(V, \|\cdot\|) = \frac{\|\cdot\|_{L,\det}}{\|\cdot\|_{\det}}, \tag{1.43}$$

where $\|\cdot\|_L$ is the unique Euclidean or Hermitian norm on V which is bounded from below by $\|\cdot\|$ and such that $\|\cdot\|_{L,\det}$ is minimal (called the *Löwner norm* of $\|\cdot\|$), which is also equal to $\|\cdot\|_{*,J,*}$. Theorem 1.2.54 then leads to

$$\max\{\Delta(V, \|\cdot\|), \Delta(V^\vee, \|\cdot\|_*)\} \leqslant \dim_k(V)^{\dim_k(V)/2} \tag{1.44}$$

We denote by $\lambda(V, \|\cdot\|)$ the constant $\|\eta\|_{L,\det} \cdot \|\eta^\vee\|_{J,*,\det}$, where η is an arbitrary non-zero element in $\det(V)$, and $\eta^\vee$ is its dual element in $\det(V^\vee)$. With this notation, by (1.38) in §1.2.7 one has

$$\Delta(V^\vee, \|\cdot\|_*)\Delta(V, \|\cdot\|) = \lambda(V, \|\cdot\|)\delta(V, \|\cdot\|). \tag{1.45}$$

Note that one has $\|\cdot\|_J \leqslant \|\cdot\|$ by definition. Hence we obtain

$$\lambda(V, \|\cdot\|) = \frac{\|\cdot\|_{L,\det}}{\|\cdot\|_{J,\det}} \geqslant \frac{\|\cdot\|_{L,\det}}{\|\cdot\|_{\det}} = \Delta(V, \|\cdot\|), \tag{1.46}$$

where the first equality comes from Proposition 1.2.47. Therefore the relation (1.45) leads to $\Delta(V^\vee, \|\cdot\|_*) \geqslant \delta(V, \|\cdot\|)$. Since $\delta(V, \|\cdot\|)$ and $\lambda(V, \|\cdot\|)$ are both invariant by duality, one obtains

$$\begin{aligned}\delta(V, \|\cdot\|) &\leqslant \min\{\Delta(V, \|\cdot\|), \Delta(V^\vee, \|\cdot\|_*)\} \\ &\leqslant \max\{\Delta(V, \|\cdot\|), \Delta(V^\vee, \|\cdot\|_*)\} \leqslant \lambda(V, \|\cdot\|).\end{aligned} \tag{1.47}$$

Remark 1.2.56 We can deduce from Theorem 1.2.54 a similar result for seminorms. Let $(V, \|\cdot\|)$ be a finite-dimensional seminormed vector space over k. Let $\|\cdot\|_J^\sim$ be the John norm associated with $\|\cdot\|^\sim$. It is induced by an inner product on $V/N_{\|\cdot\|}$. Let $\|\cdot\|_J$ be the seminorm on V given by the composition of $\|\cdot\|_J^\sim$ with the canonical projection $V \to V/\|\cdot\|$. It is a seminorm induced by a semidefinite inner product. Moreover, the following inequalities hold

$$\|\cdot\|_J \leqslant \|\cdot\| \leqslant \dim_k(V/N_{\|\cdot\|})^{1/2}\|\cdot\|_J.$$

1.2.9 Hilbert-Schmidt tensor norm

In this subsection, we assume that the absolute value $|\cdot|$ is Archimedean.

Let V and W be finite-dimensional vector spaces over k, equipped with semidefinite inner products. For $f \in \mathrm{Hom}_k(V^*, W)$, the adjoint operator $f^* : W \to V^*$ of f is defined by $\langle f(\alpha), y\rangle = \langle \alpha, f^*(y)\rangle_*$ for all $\alpha \in V^*$ and $y \in W$. Note that the adjoint operator f^* exists for any f because the product $\langle\,,\rangle_*$ on V^* is positive definite. We can equip $\mathrm{Hom}_k(V^*, W)$ with the following semidefinite inner product $\langle\,,\rangle_{\mathrm{HS}}$:

$$\forall\, f, g \in \mathrm{Hom}_k(V^*, W), \quad \langle f, g\rangle_{\mathrm{HS}} := \mathrm{Tr}(f^* \circ g).$$

This semidefinite inner product defines a seminorm on $\mathrm{Hom}_k(V^*, W)$, which induces by the canonical linear map $V \otimes_k W \to \mathrm{Hom}_k(V^*, W)$ a seminorm $\|\cdot\|_{\mathrm{HS}}$ on $V \otimes_K W$, called the *orthogonal tensor product* of the seminorms of V and W, or *Hilbert-Schmidt seminorm*. Note that if $\{x_i\}_{i=1}^n$ and $\{y_j\}_{j=1}^m$ are respectively orthogonal basis of V and W, then $\{x_i \otimes y_j\}_{i\in\{1,\dots,n\},\, j\in\{1,\dots,m\}}$ is an orthogonal basis of $V \otimes_k W$ with respect to $\langle\,,\rangle_{\mathrm{HS}}$. Moreover, for $x \in V$ and $y \in W$ one has

$$\|x \otimes y\|_{\mathrm{HS}} = \|x\| \cdot \|y\|. \tag{1.48}$$

In particular, if V and W are both of dimension 1 over k, then the orthogonal tensor product seminorm on $V \otimes_k W$ coincides with the ε-tensor product and the π-tensor product seminorms. In this case we just call it the *tensor product seminorm*.

The dual norm on $V^* \otimes_k W^*$ of the Hilbert-Schmidt seminorm on $V \otimes_k W$ coincides with the orthogonal tensor product of the dual norms on V^* and W^*. Moreover, the orthogonal tensor product is commutative, namely the isomorphism from $V \otimes_k W$ to $W \otimes_k V$ given by the transposition is actually an isometry under orthogonal tensor product seminorms. Similarly, the orthogonal tensor product is associative. More precisely, given three finite-dimensional vector spaces U, V and W over k, equipped with semidefinite inner products, the natural isomorphism from $(U \otimes_k V) \otimes_k W$ to $U \otimes_k (V \otimes_k W)$ is an isometry for orthogonal tensor product seminorms.

The following assertion, which is similar to Proposition 1.2.36, studies the quotient norm of the orthogonal tensor product.

Proposition 1.2.57 *Let V and W be finite-dimensional seminormed vector spaces over k, V_0 be a k-vector subspace of V, and Q be the quotient vector space V/V_0 equipped with the quotient seminorm. We assume that the seminorms of V and W are induced by semidefinite inner products. Then the canonical isomorphism $(V \otimes_k W)/(V_0 \otimes_k W) \to Q \otimes_k W$ is an isometry, where we consider the orthogonal tensor product seminorms on $V \otimes_k W$ and $Q \otimes_k W$, and the quotient seminorm on $(V \otimes_k W)/(V_0 \otimes_k W)$.*

Proof By the Gram-Schmidt process we can identify the quotient space Q with the orthogonal supplementary of V_0 in V. Let $\boldsymbol{e} = \{e_i\}_{i=1}^n$ be an orthogonal basis of V such that $\mathrm{card}(\boldsymbol{e} \cap V_0) = \dim_k(V_0)$. Then the projection $V \to Q$ defines an isometry between Q and the vector subspace V_1 of V generated by $\boldsymbol{e} \setminus V_0$. Let $\boldsymbol{f} = \{f_j\}_{j=1}^m$ be an orthogonal basis of W. Then the basis $\boldsymbol{e} \otimes \boldsymbol{f} = \{e_i \otimes f_j\}_{(i,j)\in\{1,\ldots,n\}\times\{1,\ldots,m\}}$ of $V \otimes_k W$ is orthogonal. Moreover, one has

$$\mathrm{card}((\boldsymbol{e} \otimes \boldsymbol{f}) \cap (V_0 \otimes_k W)) = \dim_k(V_0 \otimes_k W).$$

Thus $(\boldsymbol{e} \setminus V_0) \otimes \boldsymbol{f}$ forms an orthogonal basis of $Q \otimes_k W$ equipped with the quotient seminorm (where we identify Q with $V_0^{\perp}$). Hence the quotient seminorm on $Q \otimes_k W$ identifies with the orthogonal tensor product seminorm. □

Proposition 1.2.58 *Let $(V, \|\cdot\|_V)$ and $(W, \|\cdot\|_W)$ be finite-dimensional seminormed vector space over k, V_0 be a k-vector subspace of V and $\|\cdot\|_{V_0}$ be the restriction of $\|\cdot\|_V$ on V_0. We assume that the absolute value $|\cdot|$ is Archimedean and that the seminorms $\|\cdot\|_V$ and $\|\cdot\|_W$ are induced by semidefinite inner products. Let $\|\cdot\|$ be the orthogonal tensor product of $\|\cdot\|_V$ and $\|\cdot\|_W$, and $\|\cdot\|_0$ be the orthogonal tensor product of $\|\cdot\|_{V_0}$ and $\|\cdot\|_W$. Then $\|\cdot\|_0$ identifies with the restriction of $\|\cdot\|$ to $V_0 \otimes_k W$.*

Proof Note that $\|\cdot\|_*$ identifies with the orthogonal tensor product of $\|\cdot\|_{V,*}$ and $\|\cdot\|_{W,*}$, and $\|\cdot\|_{0,*}$ identifies with the orthogonal tensor product of $\|\cdot\|_{V_0,*}$ and $\|\cdot\|_{W,*}$. Moreover, by Lemma 1.2.48, $\|\cdot\|_{V_0,*}$ identifies with the quotient norm of $\|\cdot\|_*$ by the canonical surjective map $V^* \to V_0^*$. By Proposition 1.2.57, we obtain that $\|\cdot\|_{0,*}$ identifies with the quotient norm of $\|\cdot\|_*$ by the canonical surjective map $V^* \otimes_k W^* \to V_0^* \otimes_k W^*$. Therefore, by Proposition 1.1.20, $\|\cdot\|_0$ is the restriction of $\|\cdot\|$ to $V_0 \otimes_k W$. □

The following proposition compares ε-tensor product to orthogonal tensor product.

Proposition 1.2.59 *Let V and W be finite-dimensional vector spaces over k, equipped with semidefinite inner products. Let* $\|\cdot\|_\varepsilon$ *and* $\|\cdot\|_{\mathrm{HS}}$ *be respectively the ε-tensor product seminorm and the orthogonal tensor product seminorm on* $V \otimes_k W$*. Then* $\|\cdot\|_\varepsilon \leqslant \|\cdot\|_{\mathrm{HS}} \leqslant \min\{\dim_k(V^*), \dim_k(W^*)\}^{1/2}\|\cdot\|_\varepsilon$.

Proof Without loss of generality, we may assume that $\dim_k(V^*) \leqslant \dim_k(W^*)$. Let φ be an element of $V \otimes_k W$, viewed as a k-linear map from V^* to W. Let $\lambda_1 \geqslant \ldots \geqslant \lambda_r$ be the eigenvalues of the positive semidefinite operator $\varphi^* \circ \varphi$. By definition, the Hilbert-Schmidt seminorm of φ is $\|\varphi\|_{\mathrm{HS}} = (\lambda_1 + \ldots + \lambda_r)^{1/2}$. Moreover, the operator seminorm of φ is $\lambda_1^{1/2}$. In fact, if $\alpha_1, \ldots, \alpha_r$ are eigenvectors of $\varphi^* \circ \varphi$ of eigenvalues $\lambda_1, \ldots, \lambda_r$, respectively, then for any $(a_1, \ldots, a_r) \in k^r$ one has

$$\begin{aligned}&\langle \varphi(a_1\alpha_1 + \ldots + a_r\alpha_r), \varphi(a_1\alpha_1 + \ldots + a_r\alpha_r)\rangle \\ &= \langle \varphi^*(\varphi(a_1\alpha_1 + \ldots + a_r\alpha_r)), a_1\alpha_1 + \ldots + a_r\alpha_r\rangle = \sum_{i=1}^r |a_i|^2\lambda_i.\end{aligned}$$

Therefore one has $\|\varphi\|_\varepsilon \leqslant \|\varphi\|_{\mathrm{HS}} \leqslant \sqrt{r}\|\varphi\|_\varepsilon$. □

By using the duality between the ε-tensor product and π-tensor product (see Proposition 1.1.57), we deduce from the previous proposition the following corollary.

Corollary 1.2.60 *Let V and W be finite-dimensional vector spaces over k, equipped with semidefinite inner products. Let* $\|\cdot\|_\pi$ *and* $\|\cdot\|_{\mathrm{HS}}$ *be respectively the π-tensor product and the orthogonal tensor product norms on* $V \otimes_k W$*. Then one has*

$$\|\cdot\|_\pi \geqslant \|\cdot\|_{\mathrm{HS}} \geqslant \min\{\dim_k(V^*), \dim_k(W^*)\}^{-1/2}\|\cdot\|_\pi.$$

The following proposition expresses the Hilbert-Schmidt norm of endomorphisms in terms of the operator norm.

Proposition 1.2.61 *Let V be a vector space of finite dimension r over k, equipped with an inner product* $\langle\,,\rangle$*. Let* $f : V \to V$ *be an endomorphism of V. Then one has*

$$\langle f, f\rangle_{\mathrm{HS}} = \sum_{i=1}^r \inf_{\substack{g \in \mathrm{End}_k(V) \\ \mathrm{rk}(g) \leqslant i-1}} \|f - g\|^2,$$

where $\|\cdot\|$ *denotes the operator norm on* $\mathrm{End}_k(V)$.

Proof Let $\{e_i\}_{i=1}^r$ be an orthonormal basis of V consisting of the eigenvectors of the self-adjoint operator $f^* \circ f$. For any $i \in \{1, \ldots, r\}$, let λ_i be the eigenvalue of $f^* \circ f$ corresponding to the eigenvector e_i. Without loss of generality, we may assume that $\lambda_1 \geqslant \ldots \geqslant \lambda_r$. Since the self-adjoint operator $f^* \circ f$ is positive semidefinite, one has $\lambda_r \geqslant 0$. By definition, one has $\langle f, f\rangle_{\mathrm{HS}} = \sum_{i=1}^r \lambda_i$. In the following, we prove that, for any $i \in \{1, \ldots, r\}$, one has

$$\inf_{\substack{g \in \mathrm{End}_k(V) \\ \mathrm{rk}(g) \leqslant i-1}} \|f - g\|^2 = \lambda_i.$$

Let π be the orthogonal projection of V to the vector subspace generated by $\{e_1, \ldots, e_{i-1}\}$. Then the endomorphism $f \circ \pi$ has rank $\leqslant i-1$. Moreover, since any orthogonal projection is self-adjoint, one has

$$(f - f\pi)^*(f - f\pi) = (f^* - \pi f^*)(f - f\pi) = f^* f + \pi f^* f\pi - \pi f^* f - f^* f\pi.$$

In particular, the linear endomorphism $(f - f\pi)^*(f - f\pi)$ sends an element $a_1 e_1 + \cdots + a_r e_r$ in V to $a_i\lambda_i e_i + \cdots + a_r\lambda_r e_r$. Hence the operator norm of $(f - f\pi)^*(f - f\pi)$, which is equal to the square of the operator norm of $f - f\pi$, is λ_i.

It remains to prove that, for any k-linear endomorphism $g \in \mathrm{End}_k(V)$ of rank $\leqslant i-1$, one has $\|f - g\|^2 \geqslant \lambda_i$. Let W be the vector subspace of V generated by $\{e_1, \ldots, e_i\}$. Since g has rank $\leqslant i-1$, one has $\mathrm{Ker}(g) \cap W \neq \{0\}$. Let x be a non-zero vector in $\mathrm{Ker}(g) \cap W$. One has

$$\|(f - g)(x)\|^2 = \|f(x)\|^2 = \langle f(x), f(x)\rangle = \langle f^*(f(x)), x\rangle.$$

Since $x \in W$, one obtains $\|(f - g)(x)\|^2 \geqslant \lambda_i\|x\|^2$. Therefore $\|f - g\|^2 \geqslant \lambda_i$. The proposition is thus proved. □

Proposition 1.2.62 *Let V be a finite-dimensional vector space over k, equipped with a semidefinite inner product $\langle\,,\rangle$, r be the dimension of V, and $\|\cdot\|_{\mathrm{det}'}$ be the quotient seminorm of the orthogonal tensor product seminorm on $V^{\otimes r}$ by the canonical quotient map $V^{\otimes r} \to \det(V)$. Then one has $\|\cdot\|_{\det} = (r!)^{1/2}\|\cdot\|_{\mathrm{det}'}$.*

Proof If the seminorm associated with the semidefinite inner product on V is not a norm, then both seminorms $\|\cdot\|_{\det}$ and $\|\cdot\|_{\mathrm{det}'}$ vanish. It then suffices to treat the case where $\langle\,,\rangle$ is an inner product.

Let φ be an element in $V^{\otimes r}$. Let $\{e_i\}_{i=1}^r$ be an orthonormal basis of V. We write φ into the form

$$\varphi = \sum_{I=(i_1,\ldots,i_r)\in\{1,\ldots,r\}^r} a_I(e_{i_1} \otimes \cdots e_{i_r}).$$

Then the canonical image η of φ in $\det(V)$ is

$$\Big(\sum_{\sigma\in\mathfrak{S}_r} \mathrm{sgn}(\sigma)a_{(\sigma(1),\ldots,\sigma(r))}\Big)e_1 \wedge \cdots \wedge e_r,$$

where $\mathfrak{S}_r$ is the symmetric group of order r. Hence the Cauchy-Schwarz inequality leads to

$$\|\eta\|_{\det} = \Big|\sum_{\sigma\in\mathfrak{S}_r} \mathrm{sgn}(\sigma)a_{(\sigma(1),\ldots,\sigma(r))}\Big| \leqslant (r!)^{1/2}\|\varphi\|_{\mathrm{HS}},$$

where $\|\cdot\|_{\mathrm{HS}}$ denotes the orthogonal tensor product norm on $V^{\otimes r}$. The equality is attained when φ is of the form $\sum_{\sigma\in\mathfrak{S}_r} \mathrm{sgn}(\sigma)e_{\sigma(1)} \otimes \cdots \otimes e_{\sigma(r)}$. The proposition is thus proved. □

Proposition 1.2.63 *Let V and W be finite-dimensional seminormed vector spaces. We assume that the seminorms of V and W are induced by semidefinite inner products.*

Let n and m be respectively the dimensions of V and W over k. We equip the tensor product $V \otimes_k W$ with the orthogonal tensor product seminorm $\|\cdot\|_{\mathrm{HS}}$. Then the canonical isomorphism $\det(V \otimes_k W) \to \det(V)^{\otimes m} \otimes \det(W)^{\otimes n}$ *is an isometry, where we consider the determinant of the Hilbert-Schmidt seminorm on* $\det(V \otimes_k W)$*, and the tensor product $\|\cdot\|'$ of determinant seminorms on* $\det(V)^{\otimes m} \otimes \det(W)^{\otimes n}$.

Proof The assertion is trivial when at least one of the seminorms of V and W is not a norm since in this case both seminorms $\|\cdot\|_{\mathrm{HS,det}}$ and $\|\cdot\|'$ vanish.

In the following, we assume that V and W are equipped with inner products. Let $\{e_i\}_{i=1}^n$ and $\{f_j\}_{j=1}^m$ be respectively orthonormal bases of V and W, which are also Hadamard bases (by Proposition 1.2.25). Then $\{e_i \otimes f_j\}_{(i,j)\in\{1,\ldots,n\}\times\{1,\ldots,m\}}$ is an orthonormal basis of $V \otimes_k W$. By Proposition 1.2.25, it is also an Hadamard basis. Hence one has

$$\Big\|\bigwedge_{i=1}^{n}\bigwedge_{j=1}^{m}(e_i \otimes f_j)\Big\|_{\mathrm{HS,det}} = 1 = \Big\|(e_1 \wedge \cdots \wedge e_n)^{\otimes m} \otimes (f_1 \wedge \cdots \wedge f_m)^{\otimes n}\Big\|'.$$

The proposition is thus proved. □

1.3 Extension of scalars

In this section, we suppose given a field extension K of k equipped with a complete absolute value which extends $|\cdot|$ on k. By abuse of notation, we still use the notation $|\cdot|$ to denote the extended absolute value on K. We can thus consider K as a normed vector space over k, which is ultrametric if and only if the absolute value $|\cdot|$ on k is non-Archimedean.

Let $(V, \|\cdot\|)$ be a finite-dimensional seminormed vector space over k. We consider the natural K-linear map from $V_K = V \otimes_k K$ to $\mathscr{L}(V^*, K)$ which sends $x \otimes a \in V \otimes_k K$ (with $x \in V$ and $a \in K$) to the k-linear map $(f \in V^*) \mapsto af(x)$. We equip V^* with the dual norm and $\mathscr{L}(V^*, K)$ with the operator norm, which induces by this natural K-linear map a seminorm on $V \otimes_k K$ denoted by $\|\cdot\|_{K,\varepsilon}$ and called the seminorm *induced by $\|\cdot\|$ by ε-extension of scalars*. Note that the seminorm $\|\cdot\|_{K,\varepsilon}$ is necessarily ultrametric if k is non-Archimedean. Moreover, if $(K, |\cdot|)$ is reflexive as normed vector space over k (this condition is satisfied notably when K/k is a finite extension), then the seminorm $\|\cdot\|_{K,\varepsilon}$ is the ε-tensor product of $\|\cdot\|$ and the absolute value on K (viewed as a norm on the k-vector space K), see Remark 1.1.53.

We denote by $\|\cdot\|_{K,\pi}$ the π-tensor product seminorm on $V \otimes_k K$ of the seminorm $\|\cdot\|$ on V and the absolute value $|\cdot|$ on K, called the seminorm *induced by $\|\cdot\|$ by π-extension of scalars*. If $|\cdot|$ is Archimedean and if the seminorm $\|\cdot\|$ is induced by a semidefinite inner product, we denote by $\|\cdot\|_{K,\mathrm{HS}}$ the orthogonal tensor product of the seminorm $\|\cdot\|$ on V and the absolute value $|\cdot|$ on K (in the Archimedean case the extension K/k is always finite), called the seminorm *induced by $\|\cdot\|$ by orthogonal extension of scalars*.

In what follows, an element $x \in V$ is often considered as an element of $V_K = V \otimes_k K$ by the inclusion map $V \to V \otimes_k K$ sending x to $x \otimes 1$.

1.3.1 Basic properties

In this subsection, we discuss some basic behaviour of norms induced by extension of scalars.

Proposition 1.3.1 *Let $(V, \|\cdot\|)$ be a finite-dimensional seminormed vector space over k.*

(1) *For any $x \in V$ one has $\|x\|_{K,\varepsilon} = \|x\|_{**}$, where $\|\cdot\|_{**}$ denotes the double dual seminorm of $\|\cdot\|$. In particular, if either $(k, |\cdot|)$ is Archimedean or $(V, \|\cdot\|)$ is ultrametric, then one has $\|x\|_{K,\varepsilon} = \|x\|$ for any $x \in V$.*
(2) *For any $x \in V$ one has $\|x\|_{K,\pi} = \|x\|$. If $|\cdot|$ is Archimedean and $\|\cdot\|$ is induced by a semidefinite inner product, for any $x \in V$ one has $\|x\|_{K,\mathrm{HS}} = \|x\|$.*
(3) *For any $y \in V_K$ one has $\|y\|_{K,\varepsilon} \leqslant \|y\|_{K,\pi}$. If $(k, |\cdot|)$ is $\mathbb{R}$ equipped with the usual absolute value, $K = \mathbb{C}$, and $\|\cdot\|$ is induced by a semidefinite inner product, then for any $y \in V_{\mathbb{C}}$ one has*

$$\|y\|_{\mathbb{C},\varepsilon} \leqslant \|y\|_{\mathbb{C},\mathrm{HS}} \leqslant \|y\|_{\mathbb{C},\pi}, \tag{1.49}$$

$$\min\{\dim_{\mathbb{R}}(V^*), 2\}^{-1/2}\|y\|_{\mathbb{C},\pi} \leqslant \|y\|_{\mathbb{C},\mathrm{HS}} \leqslant \min\{\dim_{\mathbb{R}}(V^*), 2\}^{1/2}\|y\|_{\mathbb{C},\varepsilon}. \tag{1.50}$$

Proof (1) Let $\ell_x : V^* \to k$ be the k-linear map sending any bounded linear form $f \in V^*$ to $f(x)$. Let $\widetilde{\ell_x} : V^* \to K$ be the composition ℓ_x with the inclusion map $k \to K$. The operator norms of ℓ_x and $\widetilde{\ell_x}$ are the same. Therefore one has $\|x\|_K = \|x\|_{**}$. The last assertion comes from Proposition 1.1.18 and Corollary 1.2.12.

The first assertion of (2) follow from Remark 1.1.56 in the Archimedean case and from Proposition 1.2.50 in the non-Archimedean case (note that the absolute value on K, viewed as a norm when we consider K as a vector space over k, is ultrametric once $|\cdot|$ is non-Archimedean). The second assertion follows from (1.48) in §1.2.9.

(3) The first assertion follows from (1) and Proposition 1.1.54.

In the case where $(k, |\cdot|)$ is $\mathbb{R}$ equipped with the usual absolute value, $K = \mathbb{C}$, and $\|\cdot\|$ is induced by an inner product, the inequalities follow from Proposition 1.2.59 and Corollary 1.2.60. □

Remark 1.3.2

(1) Note that $\|\cdot\|$ and its double dual seminorm $\|\cdot\|_{**}$ induce the same dual norm on V^* (see Proposition 1.2.14). Hence they induce the same seminorm on V_K by ε-extension of scalars. Moreover, if $K = k$, then $\|\cdot\|_{K,\varepsilon}$ identifies with the double dual seminorm of $\|\cdot\|$ on V.

(2) Assume that $k = \mathbb{R}$, $K = \mathbb{C}$ and $|\cdot|$ is the usual absolute value on $\mathbb{R}$. Suppose that the norm $\|\cdot\|$ is induced by a semidefinite inner product $\langle\,,\rangle$. Note that $\langle\,,\rangle$ induces a semidefinite inner product $\langle\,,\rangle_{\mathbb{C}}$, given by

$$\forall\, x, y, x', y' \in V, \quad \langle x+iy, x'+iy'\rangle_{\mathbb{C}} = \langle x, x'\rangle + \langle y, y'\rangle + i\big(\langle x, y'\rangle - \langle y, x'\rangle\big).$$

Note that the seminorm corresponding to $\langle\,,\rangle_{\mathbb{C}}$ identifies with the orthogonal tensor product $\|\cdot\|_{\mathbb{C},\mathrm{HS}}$ of $\|\cdot\|$ and $|\cdot|$. Moreover, an orthogonal basis of $(V, \langle\,,\rangle)$ remains to be an orthogonal basis of $(V_{\mathbb{C}}, \langle\,,\rangle_{\mathbb{C}})$, which implies that $\|\cdot\|_{\mathbb{C},\mathrm{HS}}$ is the unique seminorm on $V_{\mathbb{C}}$ extending $\|\cdot\|$ which is induced by a semidefinite inner product. In particular, one has $\langle\,,\rangle_{*,\mathbb{C}} = \langle\,,\rangle_{\mathbb{C},*}$, where $\langle\,,\rangle_*$ denotes the dual inner product of $\langle\,,\rangle$ (see §1.2.1), and hence $\|\cdot\|_{\mathbb{C},\mathrm{HS},*} = \|\cdot\|_{*,\mathbb{C},\mathrm{HS}}$.

(3) Let V be a seminormed vector space of dimension 1 on k. Then the norms $\|\cdot\|_{K,\varepsilon}$ and $\|\cdot\|_{K,\pi}$ are the same since they take the same value on a non-zero vector of V (by Proposition 1.3.1). Similarly, if $|\cdot|$ is Archimedean then one has $\|\cdot\|_{K,\varepsilon} = \|\cdot\|_{K,\mathrm{HS}} = \|\cdot\|_{K,\pi}$. We just call this seminorm the seminorm *induced by $\|\cdot\|$ by extension of scalars* and denote it by $\|\cdot\|_K$.

Proposition 1.3.3 *Let $(V, \|\cdot\|)$ be a finite-dimensional seminormed vector space over k and $N = N_{\|\cdot\|}$ be the null space of $\|\cdot\|$.*

(1) *The null spaces of the seminorms $\|\cdot\|_{K,\varepsilon}$ and $\|\cdot\|_{K,\pi}$ are both equal to N_K.*
(2) *A linear form on the K-vector space V_K is bounded with respect to the seminorm $\|\cdot\|_{K,\varepsilon}$ if and only if it is bounded with respect to $\|\cdot\|_{K,\pi}$. Moreover, the underlying vector spaces of $(V_K, \|\cdot\|_{K,\varepsilon})^*$ and $(V_K, \|\cdot\|_{K,\pi})^*$ are both canonically isomorphic to $(V_K/N_K)^\vee$.*
(3) *The quotient norm on V_K/N_K induced by $\|\cdot\|_{K,\varepsilon}$ (resp. $\|\cdot\|_{K,\pi}$) identifies with the ε-extension of scalars $\|\cdot\|^{\sim}_{K,\varepsilon}$ (resp. the π-extension of scalars $\|\cdot\|^{\sim}_{K,\pi}$) of the norm $\|\cdot\|^{\sim}$.*

Proof (1) Note that the relation $\|\cdot\|_{K,\varepsilon} \leqslant \|\cdot\|_{K,\pi}$ holds (see Proposition 1.3.1 (3)), so that it is sufficient to see that (i) $\|x\|_{K,\pi} = 0$ for $x \in N_K$ and (ii) $\|x\|_{K,\varepsilon} > 0$ for $x \in V_K \setminus N_K$. Let $\{e_i\}_{i=1}^n$ be a basis of V such that $\{e_i\}_{i=1}^r$ forms a basis of N.

(i) We write x in the form $x = \lambda_1 e_1 + \cdots + \lambda_r e_r$ with $(\lambda_1, \ldots, \lambda_r) \in K^r$. One has

$$0 \leqslant \|x\|_{K,\pi} \leqslant \sum_{i=1}^r |\lambda_i| \cdot \|e_i\|_{K,\pi} = \sum_{i=1}^r |\lambda_i| \cdot \|e_i\| = 0,$$

where the first equality comes from Proposition 1.3.1 (2).

(ii) We set $x = \lambda_1 e_1 + \cdots + \lambda_n e_n$ $(\lambda_1, \ldots, \lambda_n \in K)$. If x does not belong to N_K, then there exists $j \in \{r+1, \ldots, n\}$ such that $\lambda_j \neq 0$. Note that $e_j^\vee$ belongs to V^*. Hence

$$\|x\|_{K,\varepsilon} \geqslant \frac{|\lambda_j|}{\|e_j^\vee\|_*} > 0.$$

(2) By Corollary 1.1.13 (3), a linear form on a finite-dimensional seminormed vector space is bounded if and only if it vanishes on the null space of the seminorm.

By (1) we obtain that both seminorms $\|\cdot\|_{K,\varepsilon}$ and $\|\cdot\|_{K,\pi}$ admit N_K as the null space. Hence we obtain the required result.

(3) We identify V^* with $(V/N)^\vee$ and then the norm $\|\cdot\|_*$ identifies with the dual norm of $\|\cdot\|^\sim$. Therefore by definition for any $x \in V_K$ one has $\|x\|_{K,\varepsilon} = \|[x]\|^\sim_{K,\varepsilon}$, where $[x]$ denotes the class of x in V_K/N_K. The case of π-extension of scalars comes from Proposition 1.1.58. □

The following proposition proves a universal property of the π-extension of scalars.

Proposition 1.3.4 *Let $(V, \|\cdot\|)$ be a finite-dimensional seminormed vector space over k. If $\|\cdot\|'$ is a seminorm on V_K whose restriction to V is bounded from above by $\|\cdot\|$, then the seminorm $\|\cdot\|'$ is bounded from above by $\|\cdot\|_{K,\pi}$. In particular, $\|\cdot\|_{K,\pi}$ is the largest seminorm on $V_K = V \otimes_k K$ extending $\|\cdot\|$.*

Proof For any $x \in V$ and $a \in K$ one has

$$\|x \otimes a\|' = |a| \cdot \|x \otimes 1\|' \leqslant |a| \cdot \|x\|.$$

By Proposition 1.1.54, we obtain $\|\cdot\|' \leqslant \|\cdot\|_{K,\pi}$. □

Proposition 1.3.5 *Let $(V_1, \|\cdot\|_1)$ and $(V_2, \|\cdot\|_2)$ be finite-dimensional seminormed vector spaces over k, and $\|\cdot\|$ be the π-tensor product seminorm of $\|\cdot\|_1$ and $\|\cdot\|_2$. Then the norm $\|\cdot\|_{K,\pi}$ identifies with the π-tensor product of $\|\cdot\|_{1,K,\pi}$ and $\|\cdot\|_{2,K,\pi}$.*

Proof Let $\|\cdot\|'$ be the π-tensor product of $\|\cdot\|_{1,K,\pi}$ and $\|\cdot\|_{2,K,\pi}$. If s is an element of $V_1 \otimes_k V_2$, which is written as $s = x_1 \otimes y_1 + \cdots + x_n \otimes y_n$, with $(x_1, \ldots, x_n) \in V_1^n$ and $(y_1, \ldots, y_2) \in V_2^n$. Then one has

$$\|s\|' \leqslant \sum_{i=1}^n \|x_i\|_{1,K,\pi} \cdot \|y_i\|_{2,K,\pi} = \sum_{i=1}^n \|x_i\|_1 \cdot \|y_i\|_2.$$

Therefore one has $\|s\|' \leqslant \|s\|$. By Proposition 1.3.4, the norm $\|\cdot\|'$ is bounded from above by $\|\cdot\|_{K,\pi}$.

To prove the converse inequality, by Proposition 1.1.54, it suffices to show that, for any split tensor $u \otimes v$ in $V_{1,K} \otimes_K V_{2,K}$ one has $\|u \otimes v\|_{K,\pi} \leqslant \|u\|_{1,K,\pi} \cdot \|v\|_{2,K,\pi}$. Assume that u and v are written as $u = \lambda_1 x_1 + \cdots + \lambda_n x_n$ and $v = \mu_1 y_1 + \cdots + \mu_m y_m$ with $(\lambda_1, \ldots, \lambda_n) \in K^n$, $(x_1, \ldots, x_n) \in V_1^n$, $(\mu_1, \ldots, \mu_m) \in K^m$ and $(y_1, \ldots, y_m) \in V_2^m$. Then one has

$$\begin{aligned}\|u \otimes v\|_{K,\pi} &\leqslant \sum_{i=1}^n \sum_{j=1}^m |\lambda_i \mu_j| \cdot \|x_i \otimes y_j\|_{K,\pi} = \sum_{i=1}^n \sum_{j=1}^m |\lambda_i \mu_j| \cdot \|x_i \otimes y_j\| \\ &= \sum_{i=1}^n \sum_{j=1}^m |\lambda_i \mu_j| \cdot \|x_i\|_1 \cdot \|y_j\|_2 \\ &= \Big(\sum_{i=1}^n |\lambda_i| \cdot \|x_i\|_1\Big)\Big(\sum_{j=1}^m |\mu_j| \cdot \|y_j\|_2\Big).\end{aligned}$$

Since the decompositions $u = \lambda_1 x_1 + \cdots + \lambda_n x_n$ and $v = \mu_1 y_1 + \cdots + \mu_m y_m$ are arbitrary, we obtain

$$\|u \otimes v\|_{K,\pi} \leqslant \|u\|_{1,K,\pi} \cdot \|v\|_{2,K,\pi}.$$

Proposition 1.3.6 *Assume that $(k, |\cdot|)$ is the field $\mathbb{R}$ equipped with the usual absolute value and $K = \mathbb{C}$. Let $(V_1, \langle,\rangle_1)$ and $(V_2, \langle,\rangle_2)$ be finite-dimensional vector spaces over $\mathbb{R}$ equipped with semidefinite inner products, $\|\cdot\|_1$ and $\|\cdot\|_2$ be seminorms corresponding to $\langle,\rangle_1$ and $\langle,\rangle_2$, respectively, and $\|\cdot\|$ be the orthogonal tensor product of $\|\cdot\|_1$ and $\|\cdot\|_2$. Then the seminorm $\|\cdot\|_{\mathbb{C},\mathrm{HS}}$ identifies with the orthogonal tensor product of $\|\cdot\|_{1,\mathbb{C},\mathrm{HS}}$ and $\|\cdot\|_{2,\mathbb{C},\mathrm{HS}}$.*

Proof Let $\{x_i\}_{i=1}^n$ and $\{y_j\}_{j=1}^m$ be orthonormal bases of $(V_1, \|\cdot\|_1)$ and $(V_2, \|\cdot\|_2)$, respectively. Then $\{x_i \otimes y_j\}_{(i,j)\in\{1,\ldots,n\}\times\{1,\ldots,m\}}$ is an orthonormal basis of $(V_1 \otimes_{\mathbb{R}} V_2, \|\cdot\|)$ and hence is an orthonormal basis of $(V_{1,\mathbb{C}} \otimes_{\mathbb{C}} V_{2,\mathbb{C}}, \|\cdot\|_{\mathbb{C},\mathrm{HS}})$. Moreover, $\{x_i\}_{i=1}^n$ and $\{y_j\}_{j=1}^m$ are also orthonormal bases of $(V_{1,\mathbb{C}}, \|\cdot\|_{1,\mathbb{C},\mathrm{HS}})$ and $(V_{2,\mathbb{C}}, \|\cdot\|_{2,\mathbb{C},\mathrm{HS}})$, respectively. Hence $\{x_i \otimes y_j\}_{(i,j)\in\{1,\ldots,n\}\times\{1,\ldots,m\}}$ is an orthonormal basis of $V_{1,\mathbb{C}} \otimes_{\mathbb{C}} V_{2,\mathbb{C}}$ with respect to the orthogonal tensor product of $\|\cdot\|_{1,\mathbb{C},\mathrm{HS}}$ and $\|\cdot\|_{2,\mathbb{C},\mathrm{HS}}$. The proposition is thus proved. □

Proposition 1.3.7 *Let V be a finite-dimensional vector space over k, and $\|\cdot\|_1$ and $\|\cdot\|_2$ be two norms on V.*

(1) *One has*

$$d(\|\cdot\|_{1,K,\varepsilon}, \|\cdot\|_{2,K,\varepsilon}) = d(\|\cdot\|_{1,**}, \|\cdot\|_{2,**}) \leqslant d(\|\cdot\|_1, \|\cdot\|_2).$$

In particular, if both norms $\|\cdot\|_1$ and $\|\cdot\|_2$ are reflexive, then

$$d(\|\cdot\|_{1,K,\varepsilon}, \|\cdot\|_{2,K,\varepsilon}) = d(\|\cdot\|_1, \|\cdot\|_2).$$

(2) *One has $d(\|\cdot\|_{1,K,\pi}, \|\cdot\|_{2,K,\pi}) = d(\|\cdot\|_1, \|\cdot\|_2)$.*

(3) *Assume that $|\cdot|$ is Archimedean and that $\|\cdot\|_1$ and $\|\cdot\|_2$ are induced by inner products. Then $d(\|\cdot\|_{1,K,\mathrm{HS}}, \|\cdot\|_{2,K,\mathrm{HS}}) = d(\|\cdot\|_1, \|\cdot\|_2)$.*

Proof (1) By Proposition 1.1.43, one has $d(\|\cdot\|_{1,*}, \|\cdot\|_{2,*}) \leqslant d(\|\cdot\|_1, \|\cdot\|_2)$. By the same argument as that of the proof of Proposition 1.1.43, we can show that $d(\|\cdot\|_{1,K,\varepsilon}, \|\cdot\|_{2,K,\varepsilon}) \leqslant d(\|\cdot\|_{1,*}, \|\cdot\|_{2,*})$. Hence we obtain the inequality

$$d(\|\cdot\|_{1,K,\varepsilon}, \|\cdot\|_{2,K,\varepsilon}) \leqslant d(\|\cdot\|_1, \|\cdot\|_2).$$

By Proposition 1.2.14, for $i \in \{1,2\}$, $\|\cdot\|_i$ and $\|\cdot\|_{i,**}$ induce the same dual norm on V^*, and hence $\|\cdot\|_{i,K,\varepsilon} = \|\cdot\|_{i,**,K,\varepsilon}$. Therefore the above argument actually leads to $d(\|\cdot\|_{1,K,\varepsilon}, \|\cdot\|_{2,K,\varepsilon}) \leqslant d(\|\cdot\|_{1,**}, \|\cdot\|_{2,**})$. Conversely, by Proposition 1.3.1 (1) we obtain that $\|\cdot\|_{1,K,\varepsilon}$ and $\|\cdot\|_{2,K,\varepsilon}$ extend $\|\cdot\|_{1,**}$ and $\|\cdot\|_{2,**}$, respectively. Hence one has $d(\|\cdot\|_{1,K,\varepsilon}, \|\cdot\|_{2,K,\varepsilon}) \geqslant d(\|\cdot\|_{1,**}, \|\cdot\|_{2,**})$.

The inequality $d(\|\cdot\|_{1,**}, \|\cdot\|_{2,**}) \leqslant d(\|\cdot\|_1, \|\cdot\|_2)$ comes from Proposition 1.1.43. The equality holds when both norms $\|\cdot\|_1$ and $\|\cdot\|_2$ are reflexive.

(2) By Proposition 1.3.1 (2), $\|\cdot\|_{1,K,\pi}$ and $\|\cdot\|_{2,K,\pi}$ extend $\|\cdot\|_1$ and $\|\cdot\|_2$, respectively, and hence $d(\|\cdot\|_{1,K,\pi}, \|\cdot\|_{2,K,\pi}) \geqslant d(\|\cdot\|_1, \|\cdot\|_2)$. In the following, we prove the converse inequality. If we set $\delta = d(\|\cdot\|_1, \|\cdot\|_2)$, then $\mathrm{e}^{-\delta} \leqslant \|s\|_1/\|s\|_2 \leqslant \mathrm{e}^{\delta}$ for $s \in V \setminus \{0\}$, that is, $\|\cdot\|_1 \leqslant \mathrm{e}^{\delta}\|\cdot\|_2$ and $\|\cdot\|_2 \leqslant \mathrm{e}^{\delta}\|\cdot\|_1$. By Proposition 1.3.4, one has $\|\cdot\|_{1,K,\pi} \leqslant \mathrm{e}^{\delta}\|\cdot\|_{2,K,\pi}$. By the same reason, $\|\cdot\|_{2,K,\pi} \leqslant \mathrm{e}^{\delta}\|\cdot\|_{1,K,\pi}$. Hence the inequality $d(\|\cdot\|_{1,K,\pi}, \|\cdot\|_{2,K,\pi}) \leqslant \delta = d(\|\cdot\|_1, \|\cdot\|_2)$ holds.

(3) It suffices to treat the case where $k = \mathbb{R}$ and $K = \mathbb{C}$. By Proposition 1.3.1 (2), $\|\cdot\|_{1,K,\mathrm{HS}}$ and $\|\cdot\|_{2,K,\mathrm{HS}}$ extend $\|\cdot\|_1$ and $\|\cdot\|_2$, respectively, and hence $d(\|\cdot\|_{1,K,\mathrm{HS}}, \|\cdot\|_{2,K,\mathrm{HS}}) \geqslant d(\|\cdot\|_1, \|\cdot\|_2)$. As in (2), if we set $\delta = d(\|\cdot\|_1, \|\cdot\|_2)$, then $\|\cdot\|_1 \leqslant \mathrm{e}^{\delta}\|\cdot\|_2$ and $\|\cdot\|_2 \leqslant \mathrm{e}^{\delta}\|\cdot\|_1$. Let z be an element of $V_{\mathbb{C}}$, which is written as $z = x + iy$, where x and y are vectors in V. Then one has

$$\|z\|_{1,\mathbb{C},\mathrm{HS}}^2 = \|x\|_1^2 + \|y\|_1^2 \leqslant \mathrm{e}^{2\delta}(\|x\|_2^2 + \|y\|_2^2) = \mathrm{e}^{2\delta}\|z\|_{2,\mathbb{C},\mathrm{HS}}^2.$$

Therefore $\|\cdot\|_{1,\mathrm{HS},\mathbb{C}} \leqslant \mathrm{e}^{\delta}\|\cdot\|_{2,\mathrm{HS},\mathbb{C}}$. Similarly, $\|\cdot\|_{2,\mathbb{C},\mathrm{HS}} \leqslant \mathrm{e}^{\delta}\|\cdot\|_{1,\mathbb{C},\mathrm{HS}}$, so that the inequality $d(\|\cdot\|_{1,\mathbb{C},\mathrm{HS}}, \|\cdot\|_{2,\mathbb{C},\mathrm{HS}}) \leqslant \delta = d(\|\cdot\|_1, \|\cdot\|_2)$ holds. □

Proposition 1.3.8 *Let $(V_1, \|\cdot\|_1)$ and $(V_2, \|\cdot\|_2)$ be finite-dimensional seminormed vector spaces over k, and $f : V_1 \to V_2$ be a bounded k-linear map. Let $f_K : V_{1,K} \to V_{2,K}$ be the K-linear map induced by f.*

(1) *If we consider the seminorms $\|\cdot\|_{1,K,\varepsilon}$ and $\|\cdot\|_{2,K,\varepsilon}$ on $V_{1,K}$ and $V_{2,K}$, respectively, then the operator seminorm of f_K is bounded from above by that of f^* (which is bounded from above by $\|f\|$, see Proposition 1.1.22). The equality $\|f_K\| = \|f\|$ holds when $(V_2, \|\cdot\|_2)$ is reflexive.*
(2) *If we consider the seminorms $\|\cdot\|_{1,K,\pi}$ and $\|\cdot\|_{2,K,\pi}$ on $V_{1,K}$ and $V_{2,K}$, respectively, then the operator seminorms of f_K and f are the same.*
(3) *Assume that $(k, |\cdot|)$ is $\mathbb{R}$ equipped with the usual absolute value, $K = \mathbb{C}$ and that $\|\cdot\|_1$ and $\|\cdot\|_2$ are induced by semidefinite inner products. If we consider the norms $\|\cdot\|_{1,K,\mathrm{HS}}$ and $\|\cdot\|_{2,K,\mathrm{HS}}$ on $V_{1,K}$ and $V_{2,K}$, respectively, then the operator seminorms of f_K and f are the same.*

Proof (1) Let φ be an element of $V_{1,K}$, viewed as a k-linear map from V_1^* to K. Then the element $f_K(\varphi) \in V_{2,K}$, viewed as a k-linear form from V_2^* to K, sends $\beta \in V_2^*$ to $\varphi(f^*(\beta)) \in K$. One has

$$|\varphi(f^*(\beta))| \leqslant \|\varphi\|_{1,K,\varepsilon} \cdot \|f^*\| \cdot \|\beta\|_{2,*}.$$

Therefore $\|f_K(\varphi)\|_{2,K,\varepsilon} \leqslant \|f^*\| \cdot \|\varphi\|_{1,K,\varepsilon}$. Since φ is arbitrary, one has $\|f_K\| \leqslant \|f^*\|$. The first assertion is thus proved.

Assume that $(V_2, \|\cdot\|_2)$ is reflexive. For any element $x \in V_1$ one has

$$\begin{cases} \|x\|_{1,K} = \|x\|_{1,**} \leqslant \|x\|_1, \\ \|f_K(x)\|_{2,K,\varepsilon} = \|f(x)\|_{2,K,\varepsilon} = \|f(x)\|_{2,**} = \|f(x)\|_2 \end{cases}$$

since $(V_2, \|\cdot\|_2)$ is reflexive. Therefore one has

$$\|f_K\| \geqslant \sup_{x \in V_1 \setminus N_{\|\cdot\|_1}} \frac{\|f_K(x)\|_{2,K}}{\|x\|_{1,K}} \geqslant \sup_{x \in V_1 \setminus N_{\|\cdot\|_1}} \frac{\|f(x)\|_2}{\|x\|_1} = \|f\|.$$

(2) Since the norms $\|\cdot\|_{1,K,\pi}$ and $\|\cdot\|_{2,K,\pi}$ extend $\|\cdot\|_1$ and $\|\cdot\|_2$, respectively (see Proposition 1.3.1), the operator seminorm $\|f\|$ is bounded from above by $\|f_K\|$. It suffices to prove the converse inequality. Let y be an element in $V_{1,K}$, which is written as $y = x_1 \otimes a_1 + \cdots + x_n \otimes a_n$, where $(x_1, \ldots, x_n) \in V_1^n$ and $(a_1, \ldots, a_n) \in K^n$. Then one has $f_K(y) = f(x_1) \otimes a_1 + \cdots + f(x_n) \otimes a_n$. Hence

$$\|f_K(y)\|_{2,K,\pi} \leqslant \sum_{i=1}^n |a_i| \cdot \|f(x_i)\|_2 \leqslant \|f\| \sum_{i=1}^n |a_i| \cdot \|x_i\|_1.$$

As the decomposition $y = x_1 \otimes a_1 + \cdots + x_n \otimes a_n$ is arbitrary, we obtain

$$\|f(y)\|_{2,K,\pi} \leqslant \|f\| \cdot \|y\|_{1,K,\pi}.$$

(3) Since the seminorms $\|\cdot\|_{1,\mathbb{C},\mathrm{HS}}$ and $\|\cdot\|_{2,\mathbb{C},\mathrm{HS}}$ extend $\|\cdot\|_1$ and $\|\cdot\|_2$, respectively (see Proposition 1.3.1), the operator seminorm $\|f\|$ is bounded from above by $\|f_{\mathbb{C}}\|$. Let z be an element of $V_{1,\mathbb{C}}$, written as $u + iv$, where u and v are vectors in V_1. Then one has $f_{\mathbb{C}}(z) = f(u) + if(v)$. Therefore

$$\|f_{\mathbb{C}}(z)\|^2 = \|f(u)\|_2^2 + \|f(v)\|_2^2 \leqslant \|f\|^2(\|u\|_1^2 + \|u\|_2^2) = \|f\|^2 \cdot \|z\|_{1,\mathbb{C},\mathrm{FS}}^2.$$

Hence $\|f_{\mathbb{C}}\|^2 = \|f\|^2$. □

1.3.2 Direct sums

In this subsection, we discuss the behaviour of direct sums under scalar extension. We fix a continuous and convex function $\psi : [0,1] \to [0,1]$ such that $\max\{t, 1-t\} \leqslant \psi(t)$ for any $t \in [0,1]$ (cf. §1.1.10).

Proposition 1.3.9 *Let $(V, \|\cdot\|_V)$ and $(W, \|\cdot\|_W)$ be finite-dimensional seminormed vector spaces over k. Let $\|\cdot\|_\psi$ be the ψ-direct sum of $\|\cdot\|_V$ and $\|\cdot\|_W$. Then for $(f, g) \in V_K \oplus W_K$ one has*

$$\max\{\|f\|_{V,K,\varepsilon}, \|g\|_{W,K,\varepsilon}\} \leqslant \|(f,g)\|_{\psi,K,\varepsilon}. \quad (1.51)$$

The equality holds if either $(k, |\cdot|)$ is non-Archimedean or $\psi(t) = \max\{t, 1-t\}$ for any $t \in [0,1]$. Moreover, for any $(f, g) \in V_K \oplus W_K$ one has

$$\|(f,0)\|_{\psi,K,\varepsilon} = \|f\|_{V,K,\varepsilon}, \quad \|(0,g)\|_{\psi,K,\varepsilon} = \|g\|_{W,K,\varepsilon}. \quad (1.52)$$

Proof By Proposition 1.1.49, the dual norm $\|\cdot\|_{\psi,*}$ is a certain direct sum of $\|\cdot\|_{V,*}$ and $\|\cdot\|_{W,*}$. Hence one has

$$\|\alpha\|_{V,*} + \|\beta\|_{W,*} \geqslant \|(\alpha,\beta)\|_{\psi,*} \geqslant \max\{\|\alpha\|_{V,*}, \|\beta\|_{V,*}\}. \tag{1.53}$$

Therefore, for any $(f,g) \in V_K \oplus W_K$ one has

$$\|(f,g)\|_{\psi,K,\varepsilon} \geqslant \sup_{\substack{(\alpha,\beta)\in V^*\oplus W^* \\ (\alpha,\beta)\neq(0,0)}} \frac{|f(\alpha)+g(\beta)|}{\|\alpha\|_{V,*} + \|\beta\|_{W,*}} = \max\{\|f\|_{V,K,\varepsilon}, \|g\|_{V,K,\varepsilon}\}$$

which proves (1.51). Moreover, for any $f \in V_K$ one has

$$\|(f,0)\|_{\psi,K,\varepsilon} \leqslant \sup_{\substack{(\alpha,\beta)\in V^*\oplus W^* \\ (\alpha,\beta)\neq(0,0)}} \frac{|f(\alpha)|}{\max\{\|\alpha\|_{V,*}, \|\beta\|_{W,*}\}} = \|f\|_{V,K,\varepsilon}.$$

Therefore, by (1.51) we obtain the equality $\|(f,0)\|_{\psi,K,\varepsilon} = \|f\|_{V,K}$. Similarly, for any $g \in W_K$ one has $\|(0,g)\|_{\psi,K,\varepsilon} = \|g\|_{W,K,\varepsilon}$.

Finally, we proceed with the proof of the equality part of (1.51). If $(k,|\cdot|)$ is non-Archimedean, then the seminorm $\|\cdot\|_{\psi,K,\varepsilon}$ is ultrametric and hence by (1.52) one has

$$\forall\,(f,g) \in V_K \oplus W_K, \quad \|(f,g)\|_{\psi,K,\varepsilon} \leqslant \max\{\|f\|_{V,K,\varepsilon}, \|g\|_{W,K,\varepsilon}\},$$

which leads to (by (1.51)) the equality

$$\forall\,(f,g) \in V_K \oplus W_K, \quad \|(f,g)\|_{\psi,K,\varepsilon} = \max\{\|f\|_{V,K,\varepsilon}, \|g\|_{W,K,\varepsilon}\}.$$

In the case where k is Archimedean and $\psi(t) = \max\{t, 1-t\}$ for any $t \in [0,1]$, one has $\|(\alpha,\beta)\|_{\psi,*} = \|\alpha\|_{V,*} + \|\beta\|_{W,*}$ for any $(\alpha,\beta) \in V^\vee \oplus W^\vee$. Therefore

$$\|(f,g)\|_{\psi,K,\varepsilon} = \sup_{\substack{(\alpha,\beta)\in V^*\oplus W^* \\ (\alpha,\beta)\neq(0,0)}} \frac{|f(\alpha)+g(\beta)|}{\|\alpha\|_{V,*} + \|\beta\|_{W,*}} = \max\{\|f\|_{V,K,\varepsilon}, \|g\|_{V,K,\varepsilon}\}.$$

Remark 1.3.10 Let ψ be an element of $\mathscr{S}$ (see §1.1.10), which corresponds to an absolute normalised norm $\|\cdot\|$ on $\mathbb{R}^2$. Let ψ_* be the element of $\mathscr{S}$ corresponding to the dual norm $\|\cdot\|_*$ (see Definition 1.1.48). Suppose given finite-dimensional seminormed vector spaces $(V, \|\cdot\|_V)$ and $(W, \|\cdot\|_W)$ over $\mathbb{R}$ (equipped with the usual absolute value). By Proposition 1.1.49 (2), the dual norm of $\|\cdot\|_\psi$ (the ψ-direct sum of $\|\cdot\|_V$ and $\|\cdot\|_W$) identifies with the ψ_*-direct sum of $\|\cdot\|_{V,*}$ and $\|\cdot\|_{W,*}$. Therefore, for any $(f,g) \in V_{\mathbb{C}} \oplus W_{\mathbb{C}}$, one has

$$\begin{aligned}\|(f,g)\|_{\psi,\mathbb{C},\varepsilon} &= \sup_{\substack{(\alpha,\beta)\in V^*\oplus W^* \\ (\alpha,\beta)\neq(0,0)}} \frac{|f(\alpha)+g(\beta)|}{\|(\|\alpha\|_{V,*}, \|\beta\|_{W,*})\|_*} \\ &\leqslant \sup_{\substack{(\alpha,\beta)\in V^*\oplus W^* \\ (\alpha,\beta)\neq(0,0)}} \frac{\|f\|_{V,\mathbb{C},\varepsilon}\cdot\|\alpha\|_{V,*} + \|g\|_{W,\mathbb{C},\varepsilon}\cdot\|\beta\|_{W,*}}{\|(\|\alpha\|_{V,*}, \|\beta\|_{W,*})\|_*}\end{aligned}$$

$$= \|(\|f\|_{V,\mathbb{C},\varepsilon}, \|g\|_{W,\mathbb{C},\varepsilon})\|.$$

In other words, the seminorm $\|\cdot\|_{\psi,\mathbb{C},\varepsilon}$ is bounded from above by the ψ-direct sum of $\|\cdot\|_{V,\mathbb{C},\varepsilon}$ and $\|\cdot\|_{W,\mathbb{C},\varepsilon}$.

Proposition 1.3.11 *Let $(V, \|\cdot\|_V)$ and $(W, \|\cdot\|_W)$ be finite-dimensional seminormed vector spaces over k. Let $\|\cdot\|_\psi$ be the ψ-direct sum of $\|\cdot\|_V$ and $\|\cdot\|_W$, and $\|\cdot\|_{K,\pi,\psi}$ be the ψ-direct sum of $\|\cdot\|_{V,K,\pi}$ and $\|\cdot\|_{W,K,\pi}$. Then $\|\cdot\|_{K,\pi,\psi} \leqslant \|\cdot\|_{\psi,K,\pi}$.*

Proof Let (x, y) be an element of $V \oplus W$. One has

$$\begin{aligned}\|(x,y)\|_{K,\pi,\psi} &= (\|x\|_{V,K,\pi} + \|y\|_{W,K,\pi})\psi\Big(\frac{\|x\|_{V,K,\pi}}{\|x\|_{V,K,\pi} + \|y\|_{W,K,\pi}}\Big)\\ &= (\|x\|_V + \|y\|_W)\psi\Big(\frac{\|x\|_V}{\|x\|_V + \|y\|_W}\Big) = \|(x,y)\|_\psi,\end{aligned}$$

where the second equality comes from Proposition 1.3.1 (2). Therefore the seminorm $\|\cdot\|_{K,\pi,\psi}$ extends $\|\cdot\|_\psi$. By Proposition 1.3.4, it is bounded from above by $\|\cdot\|_{\psi,K,\pi}$. □

Proposition 1.3.12 *Assume that $(k, |\cdot|)$ is the real field $\mathbb{R}$ equipped with the usual absolute value. Let $(V, \langle\,,\rangle_V)$ and $(W, \langle\,,\rangle_W)$ be finite-dimensional vector spaces over $\mathbb{R}$, equipped with semidefinite inner products, $\|\cdot\|_V$ and $\|\cdot\|_W$ be seminorms associated with $\langle\,,\rangle_V$ and $\langle\,,\rangle_W$, respectively, and $\|\cdot\|$ be the orthogonal direct sum of $\|\cdot\|_V$ and $\|\cdot\|_W$. Then $\|\cdot\|_{\mathbb{C},\mathrm{HS}}$ is the orthogonal direct sum of $\|\cdot\|_{V,\mathbb{C},\mathrm{HS}}$ and $\|\cdot\|_{W,\mathbb{C},\mathrm{HS}}$.*

Proof Let $\|\cdot\|'$ be the orthogonal direct sum of $\|\cdot\|_{V,\mathbb{C},\mathrm{HS}}$ and $\|\cdot\|_{W,\mathbb{C},\mathrm{HS}}$. It is a seminorm on $V_\mathbb{C} \oplus W_\mathbb{C}$ which is induced by a semidefinite inner product. Moreover, for any $(x, y) \in V \oplus W$ one has

$$\|(x,y)\|' = (\|x\|^2_{V,\mathbb{C},\mathrm{HS}} + \|y\|^2_{V,\mathbb{C},\mathrm{HS}})^{1/2} = (\|x\|^2_V + \|y\|^2_V)^{1/2} = \|(x,y)\|,$$

where the second equality comes from Proposition 1.3.1 (2). Therefore, $\|\cdot\|'$ is a seminorm extending $\|\cdot\|$ which is induced by a semidefinite inner product and hence one has $\|\cdot\|' = \|\cdot\|_{\mathbb{C},\mathrm{HS}}$ (see Remark 1.3.2). □

1.3.3 Orthogonality

In this subsection, we discuss the preservation of the orthogonality under extension of scalars, and its consequences. We have seen in Remark 1.3.2 (2) that the orthonormality is preserved by the orthogonal extension of scalars.

Proposition 1.3.13 *Let $(V, \|\cdot\|)$ be a finite-dimensional seminormed vector space over k, and α be a real number in $]0, 1]$. If $\boldsymbol{e} = \{e_i\}_{i=1}^r$ is an α-orthogonal basis of V with respect to the norm $\|\cdot\|$, then it is also an α-orthogonal basis of V_K with respect to the norms $\|\cdot\|_{K,\varepsilon}$ and $\|\cdot\|_{K,\pi}$.*

Proof Let $\boldsymbol{e}^\vee = \{e_i^\vee\}_{i=1}^r$ be the dual basis of $\boldsymbol{e}$. By Proposition 1.2.11, the intersection $\boldsymbol{e}^\vee \cap V^*$ is an α-orthogonal bases of V^*, and one has $\|e_i^\vee\|_* \leqslant \alpha^{-1}\|e_i\|^{-1}$ for any $e_i^\vee \in \boldsymbol{e}^\vee \cap V^*$ (see Lemma 1.2.10). If $x = a_1e_1 + \cdots + a_re_r$ is an element of V_K, where $(a_1, \ldots, a_r) \in K^r$, and if $\ell_x : V^* \to K$ is the k-linear map sending $\varphi \in V^*$ to $a_1\varphi(e_1)+\cdots+a_r\varphi(e_r)$, then for any $i \in \{1, \ldots, r\}$ such that $e_i^\vee \in V^*$ (or equivalently, $e_i \notin N_{\|\cdot\|}$) one has

$$\|\ell_x\|_{K,\varepsilon} \geqslant \frac{|\ell_x(e_i^\vee)|}{\|e_i^\vee\|_*} = \frac{|a_i|}{\|e_i^\vee\|_*} \geqslant \alpha|a_i| \cdot \|e_i\| \geqslant \alpha|a_i| \cdot \|e_i\|_{K,\varepsilon},$$

where the last inequality comes from Proposition 1.3.1 and the relation (1.5). Therefore $\boldsymbol{e}$ is also an α-orthogonal basis for $\|\cdot\|_{K,\varepsilon}$.

By Proposition 1.3.1 (3), one has $\|\cdot\|_{K,\varepsilon} \leqslant \|\cdot\|_{K,\pi}$. Therefore

$$\|x\|_{K,\pi} \geqslant \|x\|_{K,\varepsilon} \geqslant \alpha \max_{i\in\{1,\ldots,r\}} |a_i| \cdot \|e_i\| = \alpha \max_{i\in\{1,\ldots,r\}} |a_i| \cdot \|e_i\|_{K,\pi},$$

where the last equality comes from Proposition 1.3.1 (2). □

By using the preservation of orthogonality of bases, we prove an universal property of ε-extension of scalars, which is an ultrametric analogue of Proposition 1.3.4.

Proposition 1.3.14 *Assume that the absolute value $|\cdot|$ is non-Archimedean. Let V be a finite-dimensional vector space over k, equipped with a seminorm $\|\cdot\|$. Let $\|\cdot\|_K'$ be an ultrametric seminorm on V_K whose restriction to V is bounded from above by $\|\cdot\|_{**}$. Then one has $\|\cdot\|_K' \leqslant \|\cdot\|_{K,\varepsilon}$. In particular, $\|\cdot\|_{K,\varepsilon}$ is the largest ultrametric seminorm on V_K which extends $\|\cdot\|_{**}$.*

Proof By Proposition 1.3.13, if α is an element of $]0, 1[$ and if $\{e_i\}_{i=1}^r$ is an α-orthogonal basis of $(V, \|\cdot\|)$, then $\{e_i\}_{i=1}^r$ is also an α-orthogonal basis of $(V_K, \|\cdot\|_{K,\varepsilon})$. In particular, for any $(\lambda_1, \ldots, \lambda_r) \in K^r$ one has

$$\|\lambda_1e_1 + \cdots + \lambda_re_r\|_K' \leqslant \max_{i\in\{1,\ldots,r\}} |\lambda_i| \cdot \|e_i\|_{**} \leqslant \alpha^{-1}\|\lambda_1e_1 + \cdots + \lambda_re_r\|_{K,\varepsilon}.$$

Since $(V, \|\cdot\|)$ admits an α-orthogonal basis for any $\alpha \in]0, 1[$, we obtain $\|\cdot\|_K' \leqslant \|\cdot\|_{K,\varepsilon}$ for any ultrametric seminorm $\|\cdot\|_K'$ with $\|\cdot\|_K' \leqslant \|\cdot\|_{**}$ on V. □

Corollary 1.3.15 *Let K' be an extension of K equipped with a complete absolute value extending that on K. Let $(V, \|\cdot\|)$ be a finite-dimensional seminormed vector space over k. One has $\|\cdot\|_{K,\natural,K',\natural} = \|\cdot\|_{K',\natural}$ on $V_{K'}$, where $\natural = \varepsilon$ or π.*

Proof The assertion is trivial when the absolute value $|\cdot|$ is Archimedean since in this case $k = \mathbb{R}$ or $\mathbb{C}$ and hence either $k = K$ or $K = K'$. In the following, we assume that the absolue value $|\cdot|$ is non-Archimedean.

By Proposition 1.3.14, $\|\cdot\|_{K',\varepsilon}$ is the largest ultrametric seminorm on $V_{K'}$ extending the seminorm $\|\cdot\|_{**}$ on V. Moreover, by Proposition 1.3.1, $\|\cdot\|_{K,\varepsilon}$ is an ultrametric seminorm on V_K extending $\|\cdot\|_{**}$, and the seminorm $\|\cdot\|_{K,\varepsilon,K',\varepsilon}$ extends $\|\cdot\|_{K,\varepsilon}$. Therefore one has $\|\cdot\|_{K,\varepsilon,K',\varepsilon} \leqslant \|\cdot\|_{K',\varepsilon}$. By the same reason, as the norm

$\|\cdot\|_{K',\varepsilon}$ extends $\|\cdot\|_{**}$, its restriction to V_K is bounded from above by $\|\cdot\|_{K,\varepsilon}$ and hence the restriction of $\|\cdot\|_{K',\varepsilon}$ to V_K coincides with $\|\cdot\|_{K,\varepsilon}$ (since we have already shown that $\|\cdot\|_{K,\varepsilon,K',\varepsilon} \leqslant \|\cdot\|_{K',\varepsilon}$). Therefore one has $\|\cdot\|_{K,\varepsilon,K',\varepsilon} \geqslant \|\cdot\|_{K',\varepsilon}$, still by the maximality property (for $\|\cdot\|_{K,\varepsilon,K',\varepsilon}$) proved in Proposition 1.3.14.

The case of π-extension of scalars is quite similar. By Proposition 1.3.1 (2), the seminorm $\|\cdot\|_{K,\pi,K',\pi}$ extends $\|\cdot\|_{K,\pi}$ on V_K and hence extends $\|\cdot\|$ on V. By the maximality property proved in Proposition 1.3.4, we obtain that $\|\cdot\|_{K,\pi,K',\pi} \leqslant \|\cdot\|_{K',\pi}$. In particular, the restriction of $\|\cdot\|_{K',\pi}$ to V_K is bounded from below by $\|\cdot\|_{K,\pi}$. Moreover, this restricted seminorm extends $\|\cdot\|$. Still by the maximality property proved in Proposition 1.3.4, we obtain that the restriction of $\|\cdot\|_{K',\pi}$ to V_K is bounded from above by $\|\cdot\|_{K,\pi}$. Therefore the restriction of $\|\cdot\|_{K',\pi}$ to V_K coincides with $\|\cdot\|_{K,\pi}$. By Proposition 1.3.4, the norm $\|\cdot\|_{K',\pi}$ is bounded from above by $\|\cdot\|_{K,\pi,K',\pi}$. The proposition is thus proved. □

Proposition 1.3.16 *Let $(V, \|\cdot\|_V)$ be a finite-dimensional seminormed vector space over k, Q be a quotient vector space of V, and $\|\cdot\|_Q$ be the quotient seminorm of $\|\cdot\|_V$ on Q.*

(1) *The seminorm $\|\cdot\|_{Q,K,\pi}$ identifies with the quotient of $\|\cdot\|_{V,K,\pi}$ on Q_K.*
(2) *The seminorm $\|\cdot\|_{Q,K,\varepsilon}$ is bounded from above by the quotient seminorm of $\|\cdot\|_{V,K,\varepsilon}$ on Q_K. The equality holds if one of the following conditions is satisfied:* (i) *$|\cdot|$ is non-Archimedean;* (ii) *$|\cdot|$ is Archimedean and $\|\cdot\|_V$ is induced by a semidefinite inner product;* (iii) *Q is of dimension* 1 *over k.*
(3) *Assume that $|\cdot|$ is Archimedean and $\|\cdot\|_V$ is induced by a semidefinite inner product. Then $\|\cdot\|_{Q,K,\mathrm{HS}}$ identifies with the quotient of $\|\cdot\|_{V,K,\mathrm{HS}}$ on Q_K.*

Proof (1) follows directly from Proposition 1.1.58.

(2) Let $\|\cdot\|'_{Q,K}$ be the quotient of the seminorm $\|\cdot\|_{V,K,\varepsilon}$ on Q_K. Let $p : V \to Q$ be the canonical linear map. Note that $Q^* \subseteq V^*$ via $\psi \mapsto \psi \circ p$. Moreover, by Proposition 1.1.20, $\|\psi \circ p\|_{V,*} = \|\psi\|_{Q,*}$ for $\psi \in Q^*$. Thus, for $s \in Q_K$,

$$\begin{aligned}\|s\|'_{Q,K} &= \inf_{\substack{x\in V_K\\ p_K(x)=s}} \sup_{\varphi\in V^*\setminus\{0\}} \frac{|\varphi_K(x)|}{\|\varphi\|_{V,*}} \geqslant \inf_{\substack{x\in V_K\\ p_K(x)=s}} \sup_{\psi\in Q^*\setminus\{0\}} \frac{|\psi_K \circ p_K(x)|}{\|\psi \circ p\|_{V,*}}\\ &= \sup_{\psi\in Q^*\setminus\{0\}} \frac{|\psi_K(s)|}{\|\psi\|_{Q,*}} = \|s\|_{Q,K,\varepsilon},\end{aligned}$$

and hence the first assertion holds.

In the following, we prove the equality $\|\cdot\|'_{Q,K} = \|\cdot\|_{Q,K,\varepsilon}$ under each of the three conditions (i), (ii) and (iii). We first assume that the condition (i) or (ii) is satisfied. By Proposition 1.1.20, the dual norm $\|\cdot\|_{Q,*}$ identifies with the restriction of $\|\cdot\|_{V,*}$ to Q^*. By Proposition 1.2.35, we obtain that the seminorm $\|\cdot\|_{Q,K,\varepsilon}$ identifies with the quotient seminorm of $\|\cdot\|_{V,K,\varepsilon}$ on Q_K.

Assume that the condition (iii) is satisfied and that the absolute value $|\cdot|$ is Archimedean (the non-Archimedean case has already been proved above). Let f be a continuous k-linear operator from Q^* to K. Since Q is assumed to be of dimension 1 over k, the image of f is contained in a k-linear subspace of dimension 1 in K.

Therefore by Hahn-Banach theorem we obtain that there exists a continuous k-linear map $g : V^* \to K$ extending f such that f and g have the same operator seminorm. Hence the seminorm $\|\cdot\|_{Q,K,\varepsilon}$ identifies with the quotient seminorm of $\|\cdot\|_{V,K,\varepsilon}$ on Q_K.

(3) follows directly from Proposition 1.2.57. □

Proposition 1.3.17 *Let $(V, \|\cdot\|_V)$ be a finite-dimensional seminormed vector space over k and W be a vector subspaces of V. Let $\|\cdot\|_W$ be the restriction of $\|\cdot\|_V$ to W.*

(1) *The restriction of $\|\cdot\|_{V,K,\varepsilon}$ to W_K is bounded from above by $\|\cdot\|_{W,K,\varepsilon}$. If $|\cdot|$ is Archimedean or $\|\cdot\|_V$ is ultrametric, then the restriction of $\|\cdot\|_{V,K,\varepsilon}$ to W_K coincides with $\|\cdot\|_{W,K,\varepsilon}$.*
(2) *The restriction of $\|\cdot\|_{V,K,\pi}$ to W_K is bounded from above by $\|\cdot\|_{W,K,\pi}$. It identifies with $\|\cdot\|_{W,K,\pi}$ if $\|\cdot\|_V$ is ultrametric or induced by a semidefinite inner product.*
(3) *Assume that $|\cdot|$ is Archimedean and that $\|\cdot\|_V$ is induced by a semidefinite inner product. Then the restriction of $\|\cdot\|_{V,K,\mathrm{HS}}$ to W_K identifies with $\|\cdot\|_{W,K,\mathrm{HS}}$.*

Proof (1) Assume that $|\cdot|$ is non-Archimedean. By Proposition 1.3.1 (1), the seminorm $\|\cdot\|_{V,K,\varepsilon}$ extends $\|\cdot\|_{V,**}$. The restriction of $\|\cdot\|_{V,K,\varepsilon}$ to V is then bounded from above by $\|\cdot\|_V$, which implies that the restriction of $\|\cdot\|_{V,K,\varepsilon}$ to W is bounded from above by $\|\cdot\|_W$. Since $\|\cdot\|_{W,**}$ is the largest ultrametric seminorm on W which is bounded from above by $\|\cdot\|_W$ (see Corollary 1.2.12), we deduce from Proposition 1.3.14 that the restriction of $\|\cdot\|_{V,K,\varepsilon}$ to W is bounded from above by $\|\cdot\|_{W,**}$. By Proposition 1.3.14, we obtain that the restriction of $\|\cdot\|_{V,K,\varepsilon}$ to W_K is bounded from above by $\|\cdot\|_{W,K,\varepsilon}$.

If $\|\cdot\|_V$ is ultrametric or $|\cdot|$ is Archimedean, the dual norm $\|\cdot\|_{W,*}$ coincides with the quotient norm of $\|\cdot\|_{V,*}$ induced by the canonical quotient map $V^* \to W^*$ (see Proposition 1.2.35 for the ultrametric case, and Remark 1.1.21 for the Archimedean case). Therefore, any $f \in W_K$, viewed as a k-linear operator from W^* to K or as a k-linear operator from V^* to K, has the same operator norm. In other words, the restriction of $\|\cdot\|_{V,K}$ to W_K coincides with $\|\cdot\|_{W,K}$.

(2) follows directly from Proposition 1.2.49 (2).

(3) follows directly from Proposition 1.2.58. □

Proposition 1.3.18 *Let $(V, \|\cdot\|)$ be a finite-dimensional normed vector space over k. We assume that either $|\cdot|$ is non-Archimedean or the norm $\|\cdot\|$ is induced by an inner product. If $\{e_i\}_{i=1}^r$ is an Hadamard basis of V, then it is also an Hadamard basis of V_K with respect to the norm $\|\cdot\|_{K,\varepsilon}$.*

Proof By Proposition 1.2.7, $\{e_i\}_{i=1}^r$ is an orthogonal basis with respect to $\|\cdot\|$. By Proposition 1.3.13, it is also an orthogonal basis with respect to $\|\cdot\|_{K,\varepsilon}$. Hence it is an Hadamard basis with respect to $\|\cdot\|_{K,\varepsilon}$ (see Propositions 1.2.23 and 1.2.25). □

Proposition 1.3.19 *Let $(V, \|\cdot\|)$ be a finite-dimensional seminormed vector space over k. Let $\|\cdot\|_{\det,K}$ be the seminorm induced by the determinant seminorm $\|\cdot\|_{\det}$ of $\|\cdot\|$ by extension of scalars.*

(1) *If either* $|\cdot|$ *is non-Archimedean or the seminorm* $\|\cdot\|$ *is induced by a semidefinite inner product, then the determinant seminorm* $\|\cdot\|_{K,\varepsilon,\det}$ *of* $\|\cdot\|_{K,\varepsilon}$ *on* $\det(V_K)$ *coincides with* $\|\cdot\|_{\det,K}$.
(2) *The determinant seminorm* $\|\cdot\|_{K,\pi,\det}$ *of* $\|\cdot\|_{K,\pi}$ *coincides with* $\|\cdot\|_{\det,K}$.
(3) *Assume that* $(k,|\cdot|)$ *is* $\mathbb{R}$ *equipped with the usual absolute value and* $\|\cdot\|$ *is a seminorm associated with a semidefinite inner product* $\langle\,,\rangle$. *Then the determinant seminorm* $\|\cdot\|_{\mathbb{C},\mathrm{HS},\det}$ *of* $\|\cdot\|_{\mathbb{C},\mathrm{HS}}$ *coincides with* $\|\cdot\|_{\det,\mathbb{C}}$.

Proof (1) If $\|\cdot\|$ is not a norm, then $\|\cdot\|_{K,\varepsilon}$ is not a norm either. In this case both seminorms $\|\cdot\|_{K,\varepsilon,\det}$ and $\|\cdot\|_{\det,K}$ vanish. Hence we may assume without loss of generality that $\|\cdot\|$ is a norm.

We first assume that $(V,\|\cdot\|)$ admits an Hadamard basis $\{e_i\}_{i=1}^r$. By Proposition 1.3.18, it is also an Hadamard basis of $(V_K,\|\cdot\|_{K,\varepsilon})$. Moreover, by Propositions 1.2.11 and 1.3.1, for any $i\in\{1,\ldots,r\}$, one has $\|e_i\|=\|e_i\|_{K,\varepsilon}$. In particular, the vector $e_1\wedge\cdots\wedge e_r$ has the same length under the determinant norms induced by $\|\cdot\|$ and $\|\cdot\|_{K,\varepsilon}$. This establishes the proposition in the particular case where $(V,\|\cdot\|)$ admits an Hadamard basis (and hence in the case where $\|\cdot\|$ is induced by an inner product, see Proposition 1.2.7).

In the following, we assume that the absolute value $|\cdot|$ is non-Archimedean. Let α be an element in $]0,1[$ and $\{e_i\}_{i=1}^r$ be an α-orthogonal basis of $(V,\|\cdot\|)$. By Proposition 1.3.13, it is also an α-orthogonal basis of $(V,\|\cdot\|_{K,\varepsilon})$. By Proposition 1.2.23, one has

$$\begin{aligned}\|e_1\wedge\cdots\wedge e_r\|_{K,\varepsilon,\det}&\geqslant\alpha^r\|e_1\|_{K,\varepsilon}\cdots\|e_r\|_{K,\varepsilon}\\&\geqslant\alpha^{2r}\|e_1\|\cdots\|e_r\|\geqslant\alpha^{2r}\|e_1\wedge\cdots\wedge e_r\|_{\det},\end{aligned}$$

where the second inequality comes from Propositions 1.3.1 and 1.2.11. Conversely, one has

$$\|e_1\wedge\cdots e_r\|_{K,\varepsilon\det}\leqslant\|e_1\|_{K,\varepsilon}\cdots\|e_r\|_{K,\varepsilon}\leqslant\|e_1\|\cdots\|e_r\|\leqslant\alpha^{-r}\|e_1\wedge\cdots\wedge e_r\|_{\det},$$

where the second inequality comes from Proposition 1.3.1 and the formula (1.5) in §1.1.5, and the last inequality results from Proposition 1.2.23 . Thus one has

$$\alpha^{-r}\|\cdot\|_{\det}\geqslant\|\cdot\|_{K,\det}\geqslant\alpha^{2r}\|\cdot\|_{\det}.$$

Since $\alpha\in\,]0,1[$ is arbitrary, we obtain $\|\cdot\|_{\det,K}=\|\cdot\|_{K,\varepsilon,\det}$.

(2) Let r be the dimension of V over k. Note that the r-th π-tensor power of the norm $\|\cdot\|_{K,\pi}$ on $V_K^{\otimes_K r}\cong(V^{\otimes_k r})\otimes_k K$ coincides with the π-tensor product of r copies of $\|\cdot\|$ and the absolute value $|\cdot|$ on K (see Proposition 1.3.5). Hence by Proposition 1.1.58 its quotient norm on $\det(V_K)$ coincides with $\|\cdot\|_{\det,K}$.

(3) Let r be the dimension of V over k. Note that the r-th orthogonal tensor power of the norm $\|\cdot\|_{\mathbb{C},\pi}$ on $V_{\mathbb{C}}^{\otimes_{\mathbb{C}} r}\cong(V^{\otimes_{\mathbb{R}} r})\otimes_{\mathbb{R}}\mathbb{C}$ coincides with the orthogonal tensor product of r copies of $\|\cdot\|$ and the usual absolute value $|\cdot|$ on $\mathbb{C}$ (see Proposition 1.3.6). Hence by Proposition 1.2.62 its quotient seminorm on $\det(V_{\mathbb{C}})$ coincides with $\|\cdot\|_{\det,\mathbb{C}}$. □

Proposition 1.3.20 *Let $(V, \|\cdot\|)$ be a finite-dimensional seminormed vector space over k.*

(1) *Let $\|\cdot\|_{K,\varepsilon,*}$ be the dual norm of $\|\cdot\|_{K,\varepsilon}$ and $\|\cdot\|_{*,K,\varepsilon}$ be the norm induced by $\|\cdot\|_*$ by the ε-extension of scalars. Then we have $\|\cdot\|_{K,\varepsilon,*} \geqslant \|\cdot\|_{*,K,\varepsilon}$, and the restrictions to V^* of these two norms are both equal to the dual norm $\|\cdot\|_*$. Moreover, the equality $\|\cdot\|_{K,\varepsilon,*} = \|\cdot\|_{*,K,\varepsilon}$ holds if $|\cdot|$ is non-Archimedean or if V is of dimension 1 over k.*

(2) *The dual norm $\|\cdot\|_{K,\pi,*}$ of $\|\cdot\|_{K,\pi}$ is equal to $\|\cdot\|_{*,K,\varepsilon}$ on V_K.*

Proof (1) Let φ be an element in V_K^*. By definition one has

$$\|\varphi\|_{K,\varepsilon,*} = \sup_{x\in V_K\setminus N_{\|\cdot\|_{K,\varepsilon}}} \frac{|\varphi(x)|}{\|x\|_K} \geqslant \sup_{x\in V\setminus N_{\|\cdot\|_{**}}} \frac{|\varphi(x)|}{\|x\|_{**}} = \|\varphi\|_{*,K,\varepsilon}.$$

Note that for any $x \in V_K$ one has

$$\|x\|_{K,\varepsilon} = \sup_{\alpha\in V^*\setminus\{0\}} \frac{|\alpha(x)|}{\|\alpha\|_*}.$$

Therefore, if $\varphi \in V^* \setminus \{0\}$ then one has $\|x\|_{K,\varepsilon} \geqslant |\varphi(x)|/\|\varphi\|_*$, which leads to

$$\|\varphi\|_{K,\varepsilon,*} \leqslant \|\varphi\|_* = \|\varphi\|_{*,K,\varepsilon},$$

where the equality comes from Proposition 1.3.1 (in the non-Archimedean case we use the fact that the norm $\|\cdot\|_*$ is ultrametric).

In the following we prove the equality $\|\cdot\|_{K,\varepsilon,*} = \|\cdot\|_{*,K,\varepsilon}$ under the assumption that $|\cdot|$ is non-Archimedean or $\dim_k(V) = 1$. We treat firstly the case where $\dim_k(V) = 1$. In this case, either the seminorm $\|\cdot\|$ vanishes and V_K^* is the trivial vector space, which has only one norm, or the seminorm $\|\cdot\|$ is a norm and for any non-zero element η in V one has

$$\|\eta^\vee\|_{K,\varepsilon,*} = \|\eta\|_{K,\varepsilon}^{-1} = \|\eta\|^{-1} = \|\eta^\vee\|_* = \|\eta^\vee\|_{*,K,\varepsilon},$$

where $\eta^\vee$ denotes the dual element of η in $V^* = V^\vee$. Hence the equality $\|\cdot\|_{K,\varepsilon,*} = \|\cdot\|_{*,K,\varepsilon}$ always holds.

We now treat the case where the absolute value $|\cdot|$ is non-Archimedean. Note that $\|\cdot\|_{K,\varepsilon,*}$ is an ultrametric norm on $V^* \otimes_k K$ extending $\|\cdot\|_*$. Hence by Proposition 1.3.14 one has $\|\cdot\|_{K,\varepsilon,*} \leqslant \|\cdot\|_{*,K,\varepsilon}$. Therefore the equality $\|\cdot\|_{*,K,\varepsilon} = \|\cdot\|_{K,\varepsilon,*}$ holds.

(2) If $(k, |\cdot|)$ is $\mathbb{R}$ equipped with the usual absolute value and if $K = \mathbb{C}$, then by Proposition 1.1.57, the norm $\|\cdot\|_{*,\mathbb{C},\varepsilon,*}$ identifies with the π-tensor product of $\|\cdot\|_*$ and $|\cdot|$ (here we consider the absolute value $|\cdot|$ on $\mathbb{C}$ as a norm on a vector space over $\mathbb{R}$). Hence it is equal to the norm $\|\cdot\|_{\mathbb{C},\pi,**}$ on $V_{\mathbb{C}}^{**}$, which implies the equality $\|\cdot\|_{*,\mathbb{C},\varepsilon} = \|\cdot\|_{\mathbb{C},\pi,*}$ since any finite-dimensional normed vector space over $\mathbb{R}$ is reflexive.

Assume that $|\cdot|$ is non-Archimedean. By (1) and the fact that $\|\cdot\|_{K,\varepsilon} \leqslant \|\cdot\|_{K,\pi}$ (which results from Proposition 1.3.4 and Proposition 1.3.1 (1)), one has

$$\|\cdot\|_{*,K,\varepsilon} = \|\cdot\|_{K,\varepsilon,*} \geqslant \|\cdot\|_{K,\pi,*},$$

which leads to

$$\|\cdot\|_{K,\pi,**} \geqslant \|\cdot\|_{K,\varepsilon,**} = \|\cdot\|_{K,\varepsilon},$$

where the equality comes from the fact that the norm $\|\cdot\|_{K,\varepsilon}$ is ultrametric. Note the the restriction of $\|\cdot\|_{K,\pi,**}$ to V is bounded from above by $\|\cdot\|$ since $\|\cdot\|$ identifies with the restriction of $\|\cdot\|_{K,\pi}$ to V (see Proposition 1.3.1 (2)). As $\|\cdot\|_{K,\pi,**}$ is ultrametric, by Proposition 1.3.14 we obtain $\|\cdot\|_{K,\pi,**} \leqslant \|\cdot\|_{K,\varepsilon}$, which leads to the equality $\|\cdot\|_{K,\pi,**} = \|\cdot\|_{K,\varepsilon}$. By passing to the dual norms, using Proposition 1.2.14 (1) we obtain $\|\cdot\|_{K,\pi,*} = \|\cdot\|_{K,\varepsilon,*} = \|\cdot\|_{*,K,\varepsilon}$. □

The following proposition is an ε-tensor analogue of Propositions 1.3.5 and 1.3.6.

Proposition 1.3.21 *We assume that the absolute value $|\cdot|$ on k is non-Archimedean. Let $(V_1, \|\cdot\|_1)$ and $(V_2, \|\cdot\|_2)$ be finite-dimensional ultrametrically seminormed vector space over k, and $\|\cdot\|$ be the ε-tensor product norm of $\|\cdot\|_1$ and $\|\cdot\|_2$. Then $\|\cdot\|_{K,\varepsilon}$ identifies with the ε-tensor product of $\|\cdot\|_{1,K,\varepsilon}$ and $\|\cdot\|_{2,K,\varepsilon}$.*

Proof Let $\|\cdot\|'_\varepsilon$ be the ε-tensor product of the norms $\|\cdot\|_{1,K,\varepsilon}$ and $\|\cdot\|_{2,K,\varepsilon}$. By Remark 1.1.53, it identifies with the seminorm induced by the operator seminorm on the K-vector space $\mathrm{Hom}_K(V_{1,K}^*, V_{2,K})$ by the canonical K-linear map

$$V_{1,K} \otimes_K V_{2,K} \longrightarrow \mathrm{Hom}_K(V_{1,K}^*, V_{2,K}),$$

where we consider the dual norm of $\|\cdot\|_{1,K,\varepsilon}$ on $V_{1,K}^*$, which identifies with the norm $\|\cdot\|_{1,*,K,\varepsilon}$ induced by $\|\cdot\|_{1,*}$ by ε-extension of scalars (see (1) in Proposition 1.3.20). By Proposition 1.3.8, for any $f \in \mathrm{Hom}_k(V_1^*, V_2)$, the seminorm of f_K identifies with that of f. Therefore $\|\cdot\|'_\varepsilon$ is an ultrametric norm on $V_{1,K} \otimes_K V_{2,K}$ which extends the ε-tensor product $\|\cdot\|_\varepsilon$ of $\|\cdot\|_1$ and $\|\cdot\|_2$. By Proposition 1.3.14, one has $\|\cdot\|'_\varepsilon \leqslant \|\cdot\|_{K,\varepsilon}$.

In the following, we prove the converse inequality $\|\cdot\|_{K,\varepsilon} \leqslant \|\cdot\|'_\varepsilon$. Let $\alpha \in \left]0,1\right[$ and $\{e_i\}_{i=1}^n$ and $\{f_j\}_{j=1}^m$ be respectively α-orthogonal bases of $(V_1, \|\cdot\|_1)$ and $(V_2, \|\cdot\|_2)$. By Proposition 1.2.19, they are also α-orthogonal bases of $(V_{1,K}, \|\cdot\|_{1,K,\varepsilon})$ and $(V_{2,K}, \|\cdot\|_{2,K,\varepsilon})$, respectively. By Propsition 1.2.19, the basis $\{e_i \otimes f_j\}_{i\in\{1,\ldots,n\},\, j\in\{1,\ldots,m\}}$ of $V_{1,K} \otimes_K V_{2,K}$ is α^2-orthogonal with respect to the seminorm $\|\cdot\|'_\varepsilon$. Hence for $(a_{ij})_{i\in\{1,\ldots,n\},\, j\in\{1,\ldots,m\}} \in K^{n\times m}$ and $T = \sum_{i,j} a_{ij} e_i \otimes f_j \in V_{1,K} \otimes_K V_{2,K}$, one has

$$\|T\|'_\varepsilon \geqslant \alpha^2 \max_{\substack{i\in\{1,\ldots,n\}\\ j\in\{1,\ldots,m\}}} |a_{ij}| \cdot \|e_i \otimes f_j\|_\varepsilon \geqslant \alpha^2 \|T\|_{K,\varepsilon}.$$

Since $\alpha \in \left]0,1\right[$ is arbitrary, we obtain the inequality $\|\cdot\|'_\varepsilon \geqslant \|\cdot\|_{K,\varepsilon}$. □

Proposition 1.3.22 *We assume that the absolute value $|\cdot|$ of k is trivial. Let $(V, \|\cdot\|)$ be an ultrametrically seminormed vector space of finite dimension over k. Let $(K, |\cdot|_K)$ be an extension of $(k, |\cdot|)$ such that $|\cdot|_K$ is non-trivial and complete. Let $\mathfrak{o}_K$ be the valuation ring of $(K, |\cdot|_K)$ and $\mathfrak{m}_K$ be the maximal ideal of $\mathfrak{o}_K$. Suppose the following assumptions* (1) *and* (2)*:*

(1) *the natural map* $k \to \mathfrak{o}_K$ *induces an isomorphism* $k \xrightarrow{\sim} \mathfrak{o}_K/\mathfrak{m}_K$,
(2) $\{\|v'\|/\|v\| \,:\, v, v' \in V \setminus N_{\|\cdot\|}\} \cap |K^\times|_K \subseteq \{1\}$.

Let $\|\cdot\|_{K,\varepsilon}$ *be the seminorm of* V_K *induced by* $\|\cdot\|$ *by* ε*-extension of scalars. Then* $\|\cdot\|_{K,\varepsilon}$ *is the only ultrametric seminorm on* V_K *extending* $\|\cdot\|$.

Proof We prove the assertion by induction on the dimension n of V over k. The case where $n = 1$ is trivial. In the following, we suppose that the assertion has been proved for seminormed vector spaces of dimension $< n$ over k. Since $\|\cdot\|$ is ultrametric, one has $\|\cdot\| = \|\cdot\|_{**}$ (see Corollary 1.2.12). Let $\|\cdot\|'$ be another ultrametric seminorm on V_K extending $\|\cdot\|$. By Proposition 1.3.14, one has $\|\cdot\|' \leqslant \|\cdot\|_{K,\varepsilon}$.

Let r be the dimension of $V/N_{\|\cdot\|}$ and $\{e_i\}_{i=1}^n$ be an orthogonal basis of V such that $\{e_i\}_{i=r+1}^n$ forms a basis of $N_{\|\cdot\|}$ (see Proposition 1.2.5). If the equality $\|\cdot\|' = \|\cdot\|_{K,\varepsilon}$ does not hold, then there exists a vector $x \in V_K$ such that $\|x\|' < \|x\|_{K,\varepsilon}$. We write x in the form $x = a_1e_1 + \cdots + a_ne_n$ with $(a_1, \ldots, a_n) \in K^n$. Note that

$$\|a_{r+1}e_{r+1} + \cdots + a_ne_n\|' \leqslant \max_{i \in \{r+1,\ldots,n\}} |a_i| \cdot \|e_i\| = 0.$$

For the same reason, $\|a_{r+1}e_{r+1} + \cdots + a_ne_n\|_{K,\varepsilon} = 0$. Therefore one has

$$\|a_1e_1 + \cdots + a_re_r\|' = \|x\|' < \|x\|_{K,\varepsilon} = \|a_1e_1 + \cdots + a_re_r\|_{K,\varepsilon}.$$

By replacing x by $a_1e_1 + \cdots + a_re_r$ we many assume without loss of generality that $a_{r+1} = \cdots = a_n = 0$.

We will prove that $|a_i|_K \cdot \|e_i\|$ are the same for $i \in \{1, \ldots, r\}$ by contradiction. Without loss of generality, we assume on the contrary that

$$|a_1|_K \cdot \|e_1\| \leqslant \cdots \leqslant |a_j|_K \cdot \|e_j\| < |a_{j+1}|_K \cdot \|e_{j+1}\| = \cdots = |a_r|_K \cdot \|e_r\|$$

with $j \in \{1, \ldots, r-1\}$. Note that

$$\|x\|' < \|x\|_{K,\varepsilon} = \max_{i \in \{1,\ldots,r\}} |a_i|_K \cdot \|e_i\| = |a_r|_K \cdot \|e_r\|.$$

Moreover, by the induction hypothesis, the norms $\|\cdot\|'$ and $\|\cdot\|_{K,\varepsilon}$ coincide on $Ke_{j+1} + \cdots + Ke_r$. In particular, one has $\|a_{j+1}e_{j+1} + \cdots + a_re_r\|' = |a_r|_K \cdot \|e_r\|$. Therefore, if we let $y = a_1e_1 + \cdots + a_je_j$, then one has

$$\begin{aligned}\|y\|' = \|x - (a_{j+1}e_{j+1} + \cdots + a_re_r)\|' &= |a_r|_K \cdot \|e_r\| \\ &> \max_{i \in \{1,\ldots,j\}} |a_i|_K \cdot \|e_i\| = \|y\|_{K,\varepsilon},\end{aligned}$$

which leads to a contradiction since $\|\cdot\|' \leqslant \|\cdot\|_{K,\varepsilon}$. Hence we should have

$$|a_1|_K \cdot \|e_1\| = \cdots = |a_r|_K \cdot \|e_r\|.$$

By the condition (2), we have $\|e_1\| = \cdots = \|e_r\|$ (namely the function $\|\cdot\|$ is constant on $V \setminus N_{\|\cdot\|}$) and hence $|a_1|_K = \cdots = |a_r|_K > 0$. As $|a_i/a_r|_K = 1$

for any $i \in \{1, \ldots, r\}$, by the assumption (1), there exists a $b_i \in k^\times$ such that $|a_i/a_r - b_i|_K < 1$, that is, $|a_i - b_i a_r|_K < |a_r|_K$. Thus, by Proposition 1.1.5,

$$\|x\|' = \left\|a_r \sum_{i=1}^r b_i e_i + \sum_{i=1}^r (a_i - b_i a_r) e_i\right\|' = |a_r|_K \cdot \|e_r\| = \|x\|_{K,\varepsilon}$$

because

$$\left\|a_r \sum_{i=1}^r b_i e_i\right\|' = |a_r|_K \left\|\sum_{i=1}^r b_i e_i\right\|' = |a_r|_K \|e_r\|$$

and

$$\left\|\sum_{i=1}^r (a_i - b_i a_r) e_i\right\|' < |a_r|_K \|e_r\|.$$

This leads to a contradiction. The proposition is thus proved. □

Proposition 1.3.23 *We assume that $k = \mathbb{R}$ and that $|\cdot|$ is the usual absolute value. Let $\{(V_i, \|\cdot\|_i)\}_{i=1}^n$ be finite-dimensional seminormed vector spaces over k. We assume that the seminorms $\|\cdot\|_i$ are induced by semidefinite inner products $\langle\,,\,\rangle_i$ and we let $\|\cdot\|_{\mathrm{HS}}$ be their orthogonal tensor product. For $i \in \{1, \ldots, n\}$, let $\pi_i : V_{i,\mathbb{C}} \to W_i$ be a quotient spaces of dimension 1 of $V_{i,\mathbb{C}}$, and $\|\cdot\|_{W_i}$ be the quotient seminorm on W_i induced by $\|\cdot\|_{i,\mathbb{C}}$. Let $\|\cdot\|_W$ be the quotient seminorm on $W = \bigotimes_{i=1}^n W_i$ induced by $\|\cdot\|_{\mathrm{HS},\mathbb{C}}$ and let $\|\cdot\|$ be the tensor product of $\|\cdot\|_{W_i}$. Then one has*

$$\frac{1}{\sqrt{2}}\|\cdot\| \leqslant \|\cdot\|_W \leqslant (\sqrt{2})^n \|\cdot\|.$$

Proof For any $i \in \{1, \ldots, n\}$, let $\|\cdot\|_i'$ be the seminorm on $V_{i,\mathbb{C}}$ induced by the semidefinite inner product $\langle\,,\,\rangle_{i,\mathbb{C}}$. One has $\|\cdot\|_{i,\mathbb{C}} \leqslant \|\cdot\|_i' \leqslant \sqrt{2}\|\cdot\|_{i,\mathbb{C}}$ (see Remark 1.3.2 (2)). Let $\langle\,,\,\rangle_{\mathrm{HS}}$ be the semidefinite inner product corresponding to the orthogonal tensor product seminorm $\|\cdot\|_{\mathrm{HS}}$ and $\|\cdot\|'$ be the seminorm on $\bigotimes_{i=1}^n V_{i,\mathbb{C}}$ induced by $\langle\,,\,\rangle_{\mathrm{HS},\mathbb{C}}$. Still by Remark 1.3.2 (2) one has $\|\cdot\|_{\mathrm{HS},\mathbb{C}} \leqslant \|\cdot\|' \leqslant \sqrt{2}\|\cdot\|_{\mathrm{HS},\mathbb{C}}$. Moreover, $\|\cdot\|'$ coincides with the orthogonal tensor product of the seminorms $\|\cdot\|_i'$.

For $i \in \{1, \ldots, n\}$, let $\|\cdot\|_{W_i}'$ be the quotient seminorms on W_i induced by $\|\cdot\|_i'$. Let $\|\cdot\|_W'$ be the quotient seminorm on W induced by $\|\cdot\|'$. By Proposition 1.2.57, $\|\cdot\|_W'$ coincides with the tensor product of the seminorms $\|\cdot\|_{W_i}'$. Moreover, by the relations $\|\cdot\|_{i,\mathbb{C}} \leqslant \|\cdot\|_i' \leqslant \sqrt{2}\|\cdot\|_{i,\mathbb{C}}$ we obtain $\|\cdot\|_{W_i} \leqslant \|\cdot\|_{W_i}' \leqslant \sqrt{2}\|\cdot\|_{W_i}$, which implies

$$\|\cdot\| \leqslant \|\cdot\|_W' \leqslant (\sqrt{2})^n \|\cdot\|; \tag{1.54}$$

by the relation $\|\cdot\|_{\mathrm{HS},\mathbb{C}} \leqslant \|\cdot\|' \leqslant \sqrt{2}\|\cdot\|_{\mathrm{HS},\mathbb{C}}$ we obtain

$$\|\cdot\|_W \leqslant \|\cdot\|_W' \leqslant \sqrt{2}\|\cdot\|_W. \tag{1.55}$$

Combining (1.54) and (1.55), we obtain $\frac{1}{\sqrt{2}}\|\cdot\| \leqslant \|\cdot\|_W \leqslant (\sqrt{2})^n\|\cdot\|$. The proposition is thus proved. □

Proposition 1.3.24 *Let $(V, \|\cdot\|)$ be a finite-dimensional seminormed vector space over k and W be a quotient vector space of dimension 1 of V. Let $\|\cdot\|_W$ be the quotient seminorm on W induced by $\|\cdot\|$.*

(1) *The seminorm $\|\cdot\|_{W,K}$ coincides with the quotient seminorm on W_K induced by $\|\cdot\|_{K,\natural}$, where $\natural = \varepsilon$ or π.*
(2) *Assume that $(k, |\cdot|)$ is $\mathbb{R}$ equipped with the usual absolute value, $K = \mathbb{C}$, and $\|\cdot\|$ is induced by a semidefinite inner product. Then the seminorm $\|\cdot\|_{W,\mathbb{C}}$ coincides with the quotient seminorm on $W_{\mathbb{C}}$ induced by $\|\cdot\|_{\mathbb{C},\mathrm{HS}}$.*

Proof (1) The case where $\natural = \pi$ follows directly from Proposition 1.1.58. In the following, we consider the case where $\natural = \varepsilon$.

Let $\|\cdot\|_{W_K}$ be the quotient seminorm on W_K induced by the seminorm $\|\cdot\|_{K,\varepsilon}$. If the kernel of the quotient map $V \to W$ does not contain $N_{\|\cdot\|}$, then the quotient seminorm $\|\cdot\|_W$ vanishes since W is of dimension 1 over k. In this case the quotient seminorm $\|\cdot\|_{W_K}$ also vanishes since the kernel of the quotient map $V_K \to W_K$ does not contain $N_{\|\cdot\|_{K,\varepsilon}} = N_{\|\cdot\|} \otimes_k K$ (see Proposition 1.3.3).

In the following, we assume that the seminorm $\|\cdot\|_W$ is a norm. In this case $\|\cdot\|_{W_K}$ is also a norm since the kernel of the quotient map $V_K \to W_K$ contains $N_{\|\cdot\|_{K,\varepsilon}} = N_{\|\cdot\|} \otimes_k K$. We will show that the dual norms $\|\cdot\|_{W_K,*}$ and $\|\cdot\|_{W,K,\varepsilon,*}$ on $W_K^\vee$ are equal. Since W is a vector space of dimension 1, it suffices to show that the restrictions of these norms to $W^\vee$ are the same. We identify $W_K^\vee$ with a vector subspace of dimension 1 of $V_K^\vee$. By Proposition 1.1.20, the norm $\|\cdot\|_{W_K,*}$ coincides with the restriction of $\|\cdot\|_{K,\varepsilon,*}$ to $W_K^\vee$, where $\|\cdot\|_{K,\varepsilon,*}$ denotes the dual seminorm of $\|\cdot\|_{K,\varepsilon}$. By (1) in Proposition 1.3.20, the restriction of $\|\cdot\|_{K,\varepsilon,*}$ to $V^\vee$ coincides with $\|\cdot\|_*$. Therefore, the restriction of $\|\cdot\|_{K,\varepsilon,*}$ to $W^\vee$ coincides with the restriction of $\|\cdot\|_*$ to $W^\vee$, which identifies with the dual norm of $\|\cdot\|_W$ (by Proposition 1.1.20). By (1) in Proposition 1.3.20, one has $\|\cdot\|_{W,K,*} = \|\cdot\|_{W,*,K}$. Finally, since W is of dimension 1, if k is non-Archimedean, then any norm on $W^\vee$ is ultrametric. Hence by Proposition 1.3.1 (for both Archimedean and non-Archimedean cases), the restriction of $\|\cdot\|_{W,K,*} = \|\cdot\|_{W,*,K}$ to $W^\vee$ identifies with $\|\cdot\|_{W,*}$. The assertion is thus proved.

(2) follows directly from Proposition 1.2.57. □

Proposition 1.3.25 *Let $(V, \|\cdot\|)$ be a finite-dimensional seminormed vector space over k and W be a quotient space of dimension 1 of $V_K = V \otimes_k K$. We equip V_K with the seminorm $\|\cdot\|_{K,\pi}$ induced by $\|\cdot\|$ by π-extension of scalars and W with the quotient seminorm $\|\cdot\|_W$ of $\|\cdot\|_{K,\pi}$. Then for any $\ell \in W$ one has*

$$\|\ell\|_W = \inf_{\substack{s \in V,\ \lambda \in K^\times \\ [s]=\lambda\ell}} |\lambda|^{-1}\|s\|.$$

Proof By definition one has

$$\|\ell\|_W = \inf_{\substack{s\in V_K,\,\lambda\in K^\times\\ [s]=\lambda\ell}} |\lambda|^{-1}\cdot\|s\|_{K,\pi} \leqslant \inf_{\substack{s\in V,\,\lambda\in K^\times\\ [s]=\lambda\ell}} |\lambda|^{-1}\cdot\|s\|_{K,\pi}$$
$$= \inf_{\substack{s\in V,\,\lambda\in K^\times\\ [s]=\lambda\ell}} |\lambda|^{-1}\cdot\|s\|,$$

where the last equality comes from Proposition 1.3.1.

Without loss of generality, we may assume that $\ell \neq 0$. Let s be an element in V_K, which is written as $s = a_1x_1 + \cdots + a_nx_n$, where $(a_1,\ldots,a_n) \in K^n$ and $(x_1,\ldots,x_n) \in V$. For any $i \in \{1,\ldots,n\}$, let λ_i be the element of K such that $[x_i] = \lambda_i\ell$. Then $[s] = \lambda\ell$ with $\lambda = a_1\lambda_1 + \cdots + a_r\lambda_r$. Let

$$h = \inf_{\substack{t\in V,\,\lambda\in K^\times\\ [t]=\lambda\ell}} |\lambda|^{-1}\cdot\|t\|.$$

For any $i \in \{1,\ldots,n\}$ one has $\|x_i\| \geqslant |\lambda_i|h$. Hence

$$|\lambda|^{-1}\sum_{i=1}^n |a_i|\cdot\|x_i\| \geqslant |\lambda|^{-1}\sum_{i=1}^n |a_i|\cdot|\lambda_i|h \geqslant h.$$

The proposition is thus proved. □

Proposition 1.3.26 *Let $(V,\|\cdot\|)$ be a finite-dimensional seminormed vector space over k. We assume one of the following conditions:*

(i) *$(k,|\cdot|)$ is non-Archimedean;*
(ii) *$k = \mathbb{C}$ equipped with the usual absolute value.*

Let W be a quotient space of dimension 1 of $V\otimes_k K$. Let $\|\cdot\|_W$ be the quotient seminorm on W induced by $\|\cdot\|_{K,\varepsilon}$ (the seminorm on $V\otimes_k K$ induced by $\|\cdot\|$ by ε-extension of scalars). Then, for any $\ell\in W$ one has

$$\|\ell\|_W = \inf_{\substack{s\in V,\,\lambda\in K^\times\\ [s]=\lambda\ell}} |\lambda|^{-1}\|s\|.$$

Proof The case where $k = \mathbb{C}$ equipped with the usual absolute value is trivial since $K = k$. In the following, we assume that $(k,|\cdot|)$ is non-Archimedean.

By definition one has

$$\|\ell\|_W = \inf_{\substack{s\in V_K,\,\lambda\in K^\times\\ [s]=\lambda\ell}} |\lambda|^{-1}\cdot\|s\|_{K,\varepsilon} \leqslant \inf_{\substack{s\in V,\,\lambda\in K^\times\\ [s]=\lambda\ell}} |\lambda|^{-1}\cdot\|s\|_{K,\varepsilon}$$
$$= \inf_{\substack{s\in V,\,\lambda\in K^\times\\ [s]=\lambda\ell}} |\lambda|^{-1}\cdot\|s\| \leqslant \inf_{\substack{s\in V,\,\lambda\in K^\times\\ [s]=\lambda\ell}} |\lambda|^{-1}\cdot\|s\|,$$

where the last equality comes from Proposition 1.3.1.

We then prove the converse inequality. Let α be a real number in $]0,1[$. By Proposition 1.2.7 , there exists an α-orthogonal basis $\{s_i\}_{i=1}^r$ of $(V,\|\cdot\|)$. By Proposition

1.3.13, $\{s_i\}_{i=1}^r$ is also an α-orthogonal basis of $(V_K, \|\cdot\|_K)$. For each $i \in \{1, \ldots, r\}$, let $\lambda_i \in K$ such that $[s_i] = \lambda_i \ell$. Let $s = a_1 s_1 + \cdots + a_r s_r$ be an element in $V \otimes_k K$, where $(a_1, \ldots, a_r) \in K^r$. Assume that $[s]$ is of the form $\lambda \ell$, where $\lambda \in K^\times$. Then one has $\lambda = a_1\lambda_1 + \cdots + a_r\lambda_r$, which leads to $|\lambda| \leqslant \max_{i\in\{1,\ldots,r\}} |a_i| \cdot |\lambda_i|$ since the absolute value is non-Archimedean. By the α-orthogonality of the basis $\{s_i\}_{i=1}^r$, we obtain

$$\begin{aligned}|\lambda|^{-1} \cdot \|s\|_{K,\varepsilon} &\geqslant \frac{\alpha}{|\lambda|} \max_{i\in\{1,\ldots,r\}} |a_i| \cdot \|s_i\|_{**} \\ &\geqslant \alpha \min_{i\in\{1,\ldots,r\}} |\lambda_i|^{-1}\|s_i\|_{**} \geqslant \alpha^2 \min_{i\in\{1,\ldots,r\}} |\lambda_i|^{-1}\|s_i\|,\end{aligned}$$

where the last inequality comes from Proposition 1.2.11. The proposition is thus proved. □

Corollary 1.3.27 *We keep the notation and hypotheses of Proposition 1.3.26. Let V' be a quotient k-vector space of V, equipped with the quotient seminorm $\|\cdot\|'$ induced by $\|\cdot\|$. We assume that the projection map $\pi : V_K \to W$ factorises through V'_K. Then the quotient seminorm on W induced by $\|\cdot\|'_{K,\varepsilon}$ coincides with $\|\cdot\|_W$.*

Proof Let $\|\cdot\|'_W$ be the quotient seminorm on W induced by $\|\cdot\|'_{K,\varepsilon}$. We apply Proposition 1.3.26 to $(V', \|\cdot\|')$ and W to obtain that, for any $\ell \in W$, one has

$$\|\ell\|'_W = \inf_{\substack{t\in V',\,\lambda\in K^\times\\ [t]=\lambda\ell}} |\lambda|^{-1}\|t\|' = \inf_{\substack{s\in V,\,\lambda\in K^\times\\ [s]=\lambda\ell}} |\lambda|^{-1}\|s\|.$$

Still by Proposition 1.3.26, we obtain $\|\ell\|'_W = \|\ell\|_W$. □

1.3.4 Extension of scalars in the real case

In this subsection, we assume that $(k, |\cdot|)$ is the field $\mathbb{R}$ of real numbers equipped with the usual absolute value.

Definition 1.3.28 Let V be a vector space over $\mathbb{R}$. We say that a seminorm $\|\cdot\|$ on $V_{\mathbb{C}} := V \otimes_{\mathbb{R}} \mathbb{C}$ is *invariant under the complex conjugation* if the equality $\|x + iy\| = \|x - iy\|$ holds for any $(x, y) \in V^2$.

Proposition 1.3.29 *Let $(V, \|\cdot\|)$ be a finite-dimensional seminormed vector space over $\mathbb{R}$. The seminorms $\|\cdot\|_{\mathbb{C},\varepsilon}$ and $\|\cdot\|_{\mathbb{C},\pi}$ are invariant under the complex conjugation. If $\|\cdot\|$ is induced by a semidefinite inner product, then $\|\cdot\|_{\mathbb{C},\mathrm{HS}}$ is invariant under the complex conjugation.*

Proof These statements follow directly from the definition of different tensor product seminorms and the fact that the absolute value on $\mathbb{C}$ is invariant under the complex conjugation (namely $|a + ib| = |a - ib|$ for any $(a, b) \in \mathbb{R}^2$). □

Proposition 1.3.30 *Let* $(V, \|\cdot\|)$ *be a finite-dimensional vector space over* $\mathbb{R}$ *(equipped with the usual absolute value) and* $\|\cdot\|'$ *be a seminorm on* $V_{\mathbb{C}}$ *extending* $\|\cdot\|$. *Assume that* $\|\cdot\|'$ *is invariant under the complex conjugation. Then for any* $(x, y) \in V^2$ *one has* $\max\{\|x\|, \|y\|\} \leqslant \|x + iy\|' \leqslant \|x\| + \|y\|$.

Proof One has

$$
\begin{aligned}
2\|x\| &= \|2x\| = \|2x\|' \leqslant \|x + iy\|' + \|x - iy\|' = 2\|x + iy\|', \\
2\|y\| &= \|2y\| = \|2iy\|' \leqslant \|x + iy\|' + \|iy - x\|' = 2\|x + iy\|'.
\end{aligned}
$$

Therefore $\|x + iy\|' \geqslant \max\{\|x\|, \|y\|\}$. The relation $\|x + iy\|' \leqslant \|x\| + \|y\|$ comes from the triangle inequality. □

Proposition 1.3.31 *Let* $(V, \|\cdot\|)$ *be a seminormed vector space over* $\mathbb{R}$. *For any* $(x, y) \in V^2$ *one has*

$$
\max\{\|x\|, \|y\|\} \leqslant \|x + iy\|_{\mathbb{C},\varepsilon} \leqslant (\|x\|^2 + \|y\|^2)^{1/2}. \tag{1.56}
$$

Moreover, for any seminorm $\|\cdot\|'$ *on* $V_{\mathbb{C}}$ *extending* $\|\cdot\|$ *which is invariant under the complex conjugation, one has*

$$
\frac{1}{2}\|\cdot\|' \leqslant \|\cdot\|_{\mathbb{C},\varepsilon} \leqslant \sqrt{2}\|\cdot\|', \tag{1.57}
$$

$$
\|\cdot\|' \leqslant \|\cdot\|_{\mathbb{C},\pi} \leqslant 2\|\cdot\|'. \tag{1.58}
$$

Proof The first inequality of (1.56) comes from Propositions 1.3.29 and 1.3.30. Moreover, one has

$$
\begin{aligned}
\|x + iy\|_{\mathbb{C},\varepsilon} &= \sup_{\varphi \in V^* \setminus \{0\}} \frac{\sqrt{\varphi(x)^2 + \varphi(y)^2}}{\|\varphi\|_*} \\
&\leqslant \sup_{(\varphi_1,\varphi_2) \in (V^* \setminus \{0\})^2} \left(\frac{\varphi_1(x)^2}{\|\varphi_1\|_*^2} + \frac{\varphi_2(y)^2}{\|\varphi_2\|_*^2} \right)^{1/2} = (\|x\|^2 + \|y\|^2)^{1/2},
\end{aligned}
$$

which proves the second inequality of (1.56).

By Proposition 1.3.30 , for any $(x, y) \in V^2$, one has

$$
\frac{1}{2}\|x + iy\|' \leqslant \frac{1}{2}(\|x\| + \|y\|) \leqslant \max\{\|x\|, \|y\|\} \leqslant \|x + iy\|_{\mathbb{C},\varepsilon},
$$

where the last inequality comes from (1.56). Moreover, still by (1.56) one has

$$
\|x + iy\|_{\mathbb{C},\varepsilon} \leqslant (\|x\|^2 + \|y\|^2)^{1/2} \leqslant \sqrt{2}\max\{\|x\|, \|y\|\} \leqslant \|x + iy\|',
$$

where the last inequality comes from Proposition 1.3.30. Hence (1.57) is proved.

Since the seminorm $\|\cdot\|'$ extends $\|\cdot\|$, by Proposition 1.3.4 one has $\|\cdot\|' \leqslant \|\cdot\|_{\mathbb{C},\pi}$. Moreover, for any $(x, y) \in V^2$ one has

$$
\|x + iy\|_{\mathbb{C},\pi} \leqslant \|x\| + \|y\| \leqslant 2\max\{\|x\|, \|y\|\} \leqslant 2\|x + iy\|',
$$

where the last inequality comes from Proposition 1.3.30. Hence (1.58) is proved. □

Proposition 1.3.32 *Let $(V, \|\cdot\|_V)$ be a finite-dimensional seminormed vector space over $\mathbb{R}$, Q be a quotient vector space of V and $\|\cdot\|_Q$ be the quotient seminorm of $\|\cdot\|_V$ on Q. Let $\|\cdot\|$ be a seminorm on $V_{\mathbb{C}}$ extending $\|\cdot\|_V$, which is invariant under the complex conjugation. Then the quotient seminorm of $\|\cdot\|$ on $Q_{\mathbb{C}}$ extends $\|\cdot\|_Q$. It is moreover invariant under the complex conjugation.*

Proof Denote by $\|\cdot\|'$ the quotient seminorm of $\|\cdot\|$ on $Q_{\mathbb{C}}$. For $q \in Q$ one has

$$\|q\|' = \inf_{\substack{(x,y)\in V^2,\\ [x]=q,\,[y]=0}} \|x+iy\| \leqslant \inf_{x\in V,\,[x]=q} \|x\|.$$

Since $\|\cdot\|$ is invariant under the complex conjugation, for any $(x, y) \in V^2$ one has $\|x+iy\| \geqslant \|x\|$. Hence $\|q\|' \geqslant \inf_{x\in V,\,[x]=q} \|x\|$, so that $\|q\|' = \inf_{x\in V,\,[x]=q} \|x\|$. Therefore,

$$\|q\|' = \inf_{x\in V,\,[x]=q} \|x\| = \inf_{x\in V,\,[x]=q} \|x\|_V = \|q\|_Q.$$

Finally, for any $(p, q) \in Q^2$ one has

$$\|p+iq\|' = \inf_{\substack{(x,y)\in V^2\\ ([x],[y])=(p,q)}} \|x+iy\| = \inf_{\substack{(x,y)\in V^2\\ ([x],[y])=(p,q)}} \|x-iy\| = \|p-iq\|'.$$

Remark 1.3.33 Let $(V, \|\cdot\|)$ be finite-dimensional seminormed vector space over $\mathbb{R}$, W be a quotient vector space of dimension one of $V_{\mathbb{C}} := V \otimes_{\mathbb{R}} \mathbb{C}$. Let $\|\cdot\|_W$ be the quotient seminorm on W induced by $\|\cdot\|_{\mathbb{C},\varepsilon}$. If ℓ is a vector of W, then clearly one has

$$\|\ell\|_W \leqslant \inf_{\substack{s\in V,\,\lambda\in\mathbb{C}^\times\\ [s]=\lambda\ell}} |\lambda|^{-1}\|s\|.$$

The equality is in general not satisfied (see the counter-example in Remark 1.3.37). However, we can show that

$$2\|\ell\|_W \geqslant \inf_{\substack{s\in V,\,\lambda\in\mathbb{C}^\times\\ [s]=\lambda\ell}} |\lambda|^{-1}\|s\|. \tag{1.59}$$

In fact, by definition one has

$$\|\ell\|_W = \inf_{\substack{s\in V_{\mathbb{C}},\,\lambda\in\mathbb{C}^\times\\ [s]=\lambda\ell}} |\lambda|^{-1}\|s\|_{\mathbb{C},\varepsilon}.$$

Let s be an element in $V_{\mathbb{C}}$, which is written as $s = s_1 + is_2$, where s_1 and s_2 are vectors in V. Assume that λ_1 and λ_2 are complex numbers such that $[s_1] = \lambda_1\ell$ and $[s_2] = \lambda_2\ell$. Then one has $[s] = (\lambda_1 + i\lambda_2)\ell$. By Proposition 1.3.31,

$$\|s\|_{\mathbb{C},\varepsilon} \geqslant \max\{\|s_1\|, \|s_2\|\} \geqslant \frac{1}{2}(\|s_1\| + \|s_2\|),$$

and $|\lambda_1 + i\lambda_2| \leqslant |\lambda_1| + |\lambda_2|$. Hence

$$\frac{\|s\|_{\mathbb{C},\varepsilon}}{|\lambda_1 + i\lambda_2|} \geqslant \frac{1}{2} \cdot \frac{\|s_1\| + \|s_2\|}{|\lambda_1| + |\lambda_2|} \geqslant \frac{1}{2} \inf_{\substack{s\in V,\, \lambda\in\mathbb{C}^\times\\ [s]=\lambda\ell}} |\lambda|^{-1}\|s\|.$$

Thus we obtain (1.59).

In particular, if V' is a quotient vector space of V such that the projection map $V_{\mathbb{C}} \to W$ factorises through $V'_{\mathbb{C}}$, $\|\cdot\|'$ is the quotient seminorm on V' induced by $\|\cdot\|$, and $\|\cdot\|'_W$ is the quotient seminorm on W induced by $\|\cdot\|'_{\mathbb{C},\varepsilon}$, then one has

$$\|\cdot\|'_W \leqslant \|\cdot\|_W \leqslant 2\|\cdot\|'_W.$$

In fact, by the above argument, for any non-zero element $\ell \in W$ one has

$$\|\ell\|'_W \leqslant \inf_{\substack{t\in V',\, \lambda\in\mathbb{C}^\times\\ [t]=\lambda\ell}} |\lambda|^{-1}\|t\| = \inf_{\substack{s\in V,\, \lambda\in\mathbb{C}^\times\\ [s]=\lambda\ell}} |\lambda|^{-1}\|s\| \leqslant 2\|\ell\|'_W.$$

The following proposition should be compared with (2) and (3) in Propositions 1.3.19.

Proposition 1.3.34 *Let $(V, \|\cdot\|)$ be a finite-dimensional seminormed vector space over $\mathbb{R}$. Denote by $\|\cdot\|_{\det}$ and $\|\cdot\|_{\mathbb{C},\varepsilon,\det}$ the determinant seminorms induced by $\|\cdot\|$ and $\|\cdot\|_{\mathbb{C},\varepsilon}$, respectively. Then one has*

$$\|\cdot\|_{\mathbb{C},\varepsilon,\det} \leqslant \|\cdot\|_{\det,\mathbb{C}} \leqslant \frac{\delta(V_{\mathbb{C}}, \|\cdot\|_{\mathbb{C},\varepsilon})}{\delta(V, \|\cdot\|)} \|\cdot\|_{\mathbb{C},\varepsilon,\det}, \tag{1.60}$$

where r is the dimension of V and $\|\cdot\|_{\det,\mathbb{C}}$ is the seminorm on $\det(V) \otimes_{\mathbb{R}} \mathbb{C}$ induced by $\|\cdot\|_{\det}$ by extension of scalars.

Proof In the case where $\|\cdot\|$ is not a norm, both seminorms $\|\cdot\|_{\mathbb{C},\varepsilon,\det}$ and $\|\cdot\|_{\det,\mathbb{C}}$ vanish. In the following, we treat the case where $\|\cdot\|$ is a norm.

Let $\{e_i\}_{i=1}^r$ be an Hadamard basis of $(V, \|\cdot\|)$. One has

$$\|e_1 \wedge \cdots \wedge e_r\|_{\mathbb{C},\varepsilon,\det} \leqslant \|e_1\|_{\mathbb{C},\varepsilon} \cdots \|e_r\|_{\mathbb{C},\varepsilon} = \|e_1\| \cdots \|e_r\| = \|e_1 \wedge \cdots \wedge e_r\|_{\det},$$

where the first equality comes from Propositions 1.3.1 and 1.1.18. Hence we obtain

$$\|\cdot\|_{\mathbb{C},\varepsilon,\det} \leqslant \|\cdot\|_{\det,\mathbb{C}}.$$

Similarly, if $\{\alpha_i\}_{i=1}^r$ is an Hadamard basis of $(V^\vee, \|\cdot\|_*)$, one has

$$\|\alpha_1 \wedge \cdots \wedge \alpha_r\|_{*,\det,\mathbb{C}} = \|\alpha_1 \wedge \cdots \wedge \alpha_r\|_{*,\det} = \|\alpha_1\|_* \cdots \|\alpha_r\|_*,$$

where $\|\cdot\|_{*,\det}$ denotes the determinant norm of $\|\cdot\|_*$. Since $\alpha_1, \ldots, \alpha_r$ are elements in $V^\vee$, by (1) in Proposition 1.3.20 one has $\|\alpha_i\|_* = \|\alpha_i\|_{\mathbb{C},\varepsilon,*}$ for any $i \in \{1, \ldots, r\}$, where $\|\cdot\|_{\mathbb{C},\varepsilon,*}$ is the dual norm of $\|\cdot\|_{\mathbb{C},\varepsilon}$. Hence we obtain

$$\|\alpha_1 \wedge \cdots \wedge \alpha_r\|_{*,\det,\mathbb{C}} = \|\alpha_1\|_{\mathbb{C},*} \cdots \|\alpha_r\|_{\mathbb{C},\varepsilon,*} \geqslant \|\alpha_1 \wedge \cdots \wedge \alpha_r\|_{\mathbb{C},\varepsilon,*,\det}, \tag{1.61}$$

where $\|\cdot\|_{\mathbb{C},\varepsilon,*,\det}$ denotes the determinant norm of $\|\cdot\|_{\mathbb{C},\varepsilon,*}$.

Let η be a non-zero element of $\det(V)$, and $\eta^\vee$ be its dual element in $\det(V^\vee)$. By definition (see §1.2.7) one has

$$\|\eta\|_{\det} = \delta(V, \|\cdot\|)^{-1}\|\eta^\vee\|_{*,\det}^{-1}, \tag{1.62}$$

where $\|\cdot\|_{*,\det}$ is the determinant norm of the dual norm $\|\cdot\|_*$ on $V^\vee$. Since $\eta^\vee$ belongs to $V^\vee$, by (1.61) we obtain $\|\eta^\vee\|_{*,\det} = \|\eta^\vee\|_{*,\det,\mathbb{C}} \geqslant \|\eta^\vee\|_{\mathbb{C},\varepsilon,*,\det}$. Hence we obtain

$$\|\eta\|_{\det} \leqslant \delta(V, \|\cdot\|)^{-1}\|\eta^\vee\|_{\mathbb{C},\varepsilon,*,\det}^{-1} = \frac{\delta(V_{\mathbb{C}}, \|\cdot\|_{\mathbb{C}})}{\delta(V, \|\cdot\|)}\|\eta\|_{\mathbb{C},\varepsilon,\det}.$$

The proposition is thus proved. □

The following proposition shows that, in the Archimedean case, the norm obtained by extension of scalars is "almost the largest" norm extending the initial one.

Proposition 1.3.35 *Let $(V, \|\cdot\|)$ be a finite-dimensional seminormed vector space over $\mathbb{R}$. Let $\|\cdot\|'$ be a seminorm on $V_{\mathbb{C}}$ which extends $\|\cdot\|$. Then one has $\|\cdot\|' \leqslant 2\|\cdot\|_{\mathbb{C},\varepsilon}$.*

Proof Let $s+it$ be an element of $V_{\mathbb{C}}$, where $(s,t) \in V^2$. One has $\|s+it\|' \leqslant \|s\| + \|t\|$. By Proposition 1.3.31, $\max\{\|s\|, \|t\|\} \leqslant \|s+it\|_{\mathbb{C},\varepsilon}$. Hence we obtain $\|s+it\|' \leqslant 2\|s+it\|_{\mathbb{C},\varepsilon}$. □

Proposition 1.3.36 *Let $(V, \|\cdot\|)$ be a finite-dimensional seminormed vector space over $\mathbb{R}$. Let $\|\cdot\|_{\mathbb{C},\varepsilon,*}$ be the dual norm of $\|\cdot\|_{\mathbb{C},\varepsilon}$ and $\|\cdot\|_{*,\mathbb{C},\varepsilon}$ be the norm on $E_{\mathbb{C}}^*$ induced by $\|\cdot\|_*$ by ε-extension of scalars. One has $\|\cdot\|_{\mathbb{C},\varepsilon,*} \leqslant 2\|\cdot\|_{*,\mathbb{C},\varepsilon}$.*

Proof By (1) in Proposition 1.3.20, the restriction of $\|\cdot\|_{\mathbb{C},\varepsilon,*}$ to V^* coincides with $\|\cdot\|_*$. Hence Proposition 1.3.35 leads to the inequality $\|\cdot\|_{\mathbb{C},\varepsilon,*} \leqslant 2\|\cdot\|_{*,\mathbb{C},\varepsilon}$. □

Remark 1.3.37 The results of Proposition 1.3.19 is not necessarily true for a general seminormed vector space over an Archimedean valued field. Consider the vector space $V = \mathbb{R}^2$ equipped with the norm $\|\cdot\|$ such that

$$\forall\,(a,b) \in \mathbb{R}^2, \quad \|(a,b)\| = (\max\{a,b,0\}^2 + \min\{a,b,0\}^2)^{1/2}.$$

In other words, if a and b have the same sign, one has $\|(a,b)\| = \max\{|a|,|b|\}$; otherwise $\|(a,b)\| = (a^2+b^2)^{1/2}$. The unit disc of this norm is represented by Figure 1.1. Let $\{e_1, e_2\}$ be the canonical basis of $\mathbb{R}^2$, where $e_1 = (1,0)$ and $e_1 = (0,1)$. One has

$$\|e_1 \wedge e_2\|_{\det} = \inf_{ad-bc\neq 0} \frac{\|(a,b)\| \cdot \|(c,d)\|}{|ad-bc|},$$

where $\|\cdot\|_{\det}$ is the determinant norm induced by $\|\cdot\|$. Note that if a,b,c,d are four real numbers such that $\max\{|a|,|b|,|c|,|d|\} \leqslant 1$ and that $abcd \geqslant 0$, then one has $|ad-bc| \leqslant \max\{|ad|,|bc|\} \leqslant 1$ since ad and bc have the same sign. Hence

Fig. 1.1 Unit ball of the norm $\|\cdot\|$

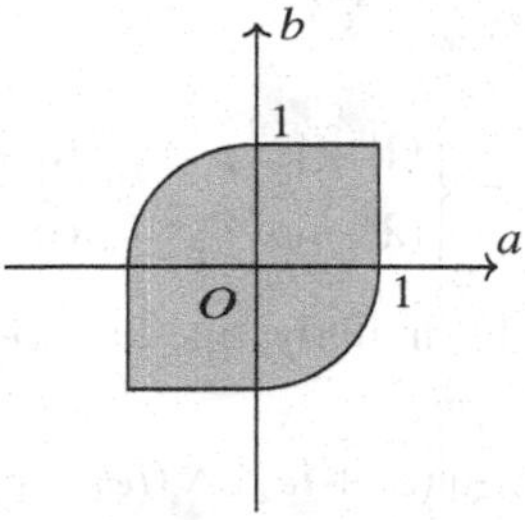

$$\|e_1 \wedge e_2\|_{\det} = \inf_{\substack{ad-bc\neq 0\\ abcd<0}} \frac{\|(a,b)\|\cdot\|(c,d)\|}{|ad-bc|} = \frac{1}{\sqrt{2}}.$$

Moreover, $(e_1 + e_2, e_1 - e_2)$ forms an Hadamard basis of $(V, \|\cdot\|)$.

The dual norm of $\|\cdot\|$ is given by the following formula

$$\forall\,(\lambda,\mu)\in\mathbb{R}^2,\quad \|\lambda e_1^\vee + \mu e_2^\vee\|_* = \begin{cases} |\lambda| + |\mu|, & \lambda\mu < 0,\\ (\lambda^2+\mu^2)^{1/2}, & \lambda\mu \geqslant 0.\end{cases}$$

The unit disc of the dual norm is represented by Figure 1.2.

Fig. 1.2 Unit ball of the norm $\|\cdot\|_*$

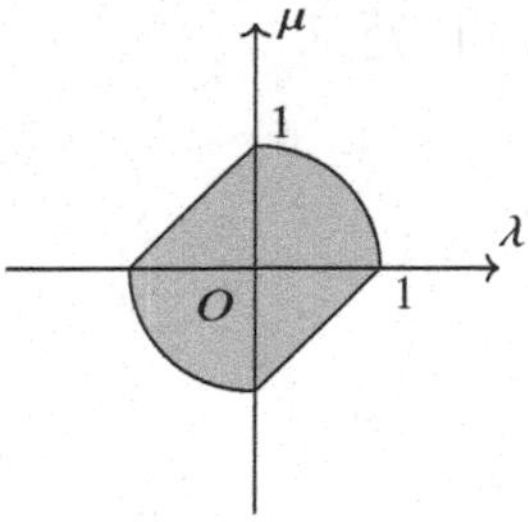

Consider now a vector $x + iy \in V \otimes_{\mathbb{R}} \mathbb{C}$, where x and y are vectors in V, and i is the imaginary unit. One has

$$\|x+iy\|_{\mathbb{C},\varepsilon} = \sup_{\varphi\in V^\vee\setminus\{0\}} \frac{|\varphi(x)+i\varphi(y)|}{\|\varphi\|_*} = \sup_{\varphi\in V^\vee\setminus\{0\}} \frac{(\varphi(x)^2+\varphi(y)^2)^{1/2}}{\|\varphi\|_*}.$$

In particular, one has

$$\|e_1 + ie_2\|_{\mathbb{C},\varepsilon} = \sup_{(\lambda,\mu)\neq(0,0)} \frac{(\lambda^2+\mu^2)^{1/2}}{f(\lambda,\mu)},$$

where

$$f(\lambda,\mu) = \begin{cases} |\lambda| + |\mu|, & \lambda\mu < 0, \\ (\lambda^2+\mu^2)^{1/2}, & \lambda\mu \geqslant 0. \end{cases}$$

Hence one has $\|e_1 + ie_2\|_{\mathbb{C},\varepsilon} = 1$. Similarly, one has $\|ie_1 + e_2\|_{\mathbb{C},\varepsilon} = 1$. Therefore

$$\|e_1 \wedge e_2\|_{\mathbb{C},\varepsilon,\det} = \frac{1}{2}\|(e_1 + ie_2) \wedge (ie_1 + e_2)\|_{\mathbb{C},\varepsilon,\det} \leqslant \frac{1}{2},$$

where $\|\cdot\|_{\mathbb{C},\varepsilon,\det}$ is the determinant norm associated with $\|\cdot\|_{\mathbb{C},\varepsilon}$. In particular, $(e_1 + e_2, e_1 - e_2)$ is no longer an Hadamard basis of $(V \otimes_{\mathbb{R}} \mathbb{C}, \|\cdot\|_{\mathbb{C},\varepsilon})$.

The above construction also provides a counter-example to the statement of Proposition 1.3.26 in the case where $(k, |\cdot|)$ is $\mathbb{R}$ equipped with the usual absolute value and $K = \mathbb{C}$. Consider the surjective $\mathbb{C}$-linear map π from $\mathbb{C}^2$ to $\mathbb{C}$ which sends $(z_1, z_2) \in \mathbb{C}^2$ to $z_1 - iz_2$. Let $\|\cdot\|'$ be the quotient norm on $\mathbb{C}$ induced by $\|\cdot\|_{\mathbb{C},\varepsilon}$. Since $\pi(1, i) = 2$ we obtain that

$$\|1\|' \leqslant \frac{1}{2}\|(1, i)\|_{\mathbb{C},\varepsilon} = \frac{1}{2}.$$

However, for any non-zero element $(\lambda, \mu) \in \mathbb{R}^2$ one has

$$\frac{\|(\lambda,\mu)\|}{|\pi(\lambda,\mu)|} = \frac{\|(\lambda,\mu)\|}{\sqrt{\lambda^2+\mu^2}} = \begin{cases} \max(|\lambda|,|\mu|)/\sqrt{\lambda^2+\mu^2}, & \lambda\mu \geqslant 0, \\ 1, & \lambda\mu < 0, \end{cases}$$

which is bounded from below by $1/\sqrt{2}$.

Chapter 2
Local metrics

Throughout the chapter, let k be a field equipped with an absolute value $|\cdot|$. We assume that k is complete with respect to this absolute value. If $|\cdot|$ is Archimedean, we assume that it is the usual absolute value on $\mathbb{R}$ or $\mathbb{C}$.

2.1 Metrised vector bundles

2.1.1 Berkovich space associated with a scheme

Definition 2.1.1 Let X be a k-scheme. The *Berkovich space* (as a set) associated with X is defined as the set of pairs $x = (p, |\cdot|_x)$, where p is a scheme point of X and $|\cdot|_x$ is an absolute value on the residue field $\kappa(p)$, which extends the absolue value $|\cdot|$ on k. If $x = (p, |\cdot|_x)$ is an element of X^{an}, we denote by $\kappa(x)$ the residue field $\kappa(p)$ and by $\widehat{\kappa}(x)$ the completion $\kappa(p)$ with respect to the absolute value $|\cdot|_x$. Note that the absolute value $|\cdot|_x$ extends naturally on $\widehat{\kappa}(x)$. Denote by $j : X^{\mathrm{an}} \to X$ the map sending $(p, |\cdot|_x) \in X^{\mathrm{an}}$ to $p \in X$, called the *specification map*.

On the Berkovich space X^{an}, one can naturally define the Zariski topology, which is the most coarse topology making the specification map $j : X^{\mathrm{an}} \to X$ continuous. Moreover, according to Berkovich [9], the construction of X^{an} allows to define a finer topology, which we describe as follows. Let U be an open subscheme of X. Each regular function $f \in \mathcal{O}_X(U)$ determines a real-valued function $|f|$ on $j^{-1}(U)$ which sends any $x \in j^{-1}(U) = U^{\mathrm{an}}$ to $|f(x)|_x$, where $f(x)$ denotes the residue class of f in $\kappa(x)$.

Definition 2.1.2 Let X be a scheme over $\operatorname{Spec} k$. The *Berkovich topology* on X^{an} is defined as the most coarse topology on X^{an} which makes the specification map $j : X^{\mathrm{an}} \to X$ and all functions of the form $|f|$ continuous, where f runs over the set of all regular functions on Zariski open subsets of the scheme X. We refer the readers to [9, §3.4] for more details.

H. Chen, A. Moriwaki, *Arakelov Geometry over Adelic Curves*, Lecture Notes in Mathematics 2258, https://doi.org/10.1007/978-981-15-1728-0_2

The construction of Berkovich topological spaces associated with k-schemes is functorial. Let X and Y be k-schemes and $\varphi : X \longrightarrow Y$ be a k-morphism. It induces a map $\varphi^{\mathrm{an}} : X^{\mathrm{an}} \longrightarrow Y^{\mathrm{an}}$, which sends $(p, |\cdot|_x) \in X^{\mathrm{an}}$ to the couple consisting of $f(p)$ and the restriction of $|\cdot|_x$ on $\kappa(f(p))$. This map is called the *map associated with* the morphism of k-schemes $X \longrightarrow Y$.

Proposition 2.1.3 *Let $\varphi : X \longrightarrow Y$ be a morphism of k-schemes. Then the map φ^{an} between Berkovich spaces is continuous with respect to the Berkovich topologies.*

Proof Clearly the map φ^{an} is continuous with respect to the Zariski topologies. It suffices to prove that, for any regular function f on a Zariski open subset U of Y, the function $|f| \circ \varphi^{\mathrm{an}}$ is continuous. Let g be the image of f by the morphism of sheaves $\mathcal{O}_Y \longrightarrow \varphi_*(\mathcal{O}_X)$ in the structure of the morphism of schemes φ. It is a regular function on $\varphi^{-1}(U)$. For any $x \in \varphi^{-1}(U)^{\mathrm{an}}$, the residue field $\kappa(x)$ is a valued extension of $\kappa(y)$ with $y = \varphi^{\mathrm{an}}(x)$. Moreover, $g(x)$ is the canonical image of $f(y)$ in $\kappa(x)$. Therefore, one has $|f| \circ \varphi^{\mathrm{an}} = |g|$, which is a continuous function. □

Remark 2.1.4 Assume that k is an Archimedean valued field, that is, $k = \mathbb{R}$ or $\mathbb{C}$. If X is a k-scheme, then the Berkovich space X^{an} identifies (as a set) with the set $X(\mathbb{C})$ of all complex points of X modulo the action of the Galois group $\mathrm{Gal}(\mathbb{C}/k)$. In particular, if X is the affine line $\mathbb{A}^1_k$, then the Berkovich space associated with X is $\mathbb{C}$ when $k = \mathbb{C}$, and is $\mathbb{C}/\tau$ when $k = \mathbb{R}$, where τ denotes the complex conjugation. In this case the Berkovich topology on X^{an} is generated by functions of the form $|P(z)|$, where P is a polynomial in $k[z]$. Therefore, it coincides with the usual topology on $\mathbb{C}$ or on $\mathbb{C}/\tau$. In fact, in the case where $k = \mathbb{C}$, the usual topology on $\mathbb{C}$ is generated by the functions $(z \in \mathbb{C}) \mapsto |z - a|$ (where $a \in \mathbb{C}$). In the case where $k = \mathbb{R}$, the usual topology on $\mathbb{C}/\tau$ is generated by the functions

$$z \longmapsto |z - a| \cdot |z - \overline{a}| = |z^2 - 2\mathrm{Re}(a)z + a\overline{a}|, \quad \text{where } a \in \mathbb{C}.$$

For a general k-scheme X, any regular function f on X determines a function f^{an} on X^{an} valued in $\mathbb{C}$ (in the case where $k = \mathbb{R}$, we identify $(\mathbb{A}^1_k)^{\mathrm{an}}$ with the upper half-plane in $\mathbb{C}$). By Proposition 2.1.3, the map f^{an} is a continuous complex function on X^{an}.

Proposition 2.1.5 *Let X be a k-scheme, X_{red} be the reduced scheme associated with X, and $i : X_{\mathrm{red}} \longrightarrow X$ be the canonical morphism. Then the associated continuous map of Berkovich spaces $i^{\mathrm{an}} : X^{\mathrm{an}}_{\mathrm{red}} \longrightarrow X^{\mathrm{an}}$ is a homeomorphism.*

Proof Note that the restrictions of the functors F_X and $F_{X_{\mathrm{red}}}$ to $\mathbf{E}_k$ are the same. Therefore i^{an} is a bijection of sets. Moreover, it is an homeomorphism if we equip $X^{\mathrm{an}}_{\mathrm{red}}$ and X^{an} with the Zariski topologies. Let U be a Zariski open subset of X. By definition $\mathcal{O}_{X_{\mathrm{red}}}(U)$ is the reduced ring associated with $\mathcal{O}_X(U)$. For any nilpotent element s in $\mathcal{O}_X(U)$ one has $s(x) = 0$ for any $x \in X$. As a consequence, if f is a regular function of X on U and if $\overline{f}$ is its canonical image in $\mathcal{O}_{X_{\mathrm{red}}}(U)$, then one has $|\overline{f}| = |f|$ on U^{an}. Thus the Berkovich topologies on X^{an} and $X^{\mathrm{an}}_{\mathrm{red}}$ are the same. □

Remark 2.1.6 We assume that the absolute value $|\cdot|$ is non-Archimedean and non-trivial. Let $\mathfrak{o}_k$ be the valuation ring of $(k, |\cdot|)$. Let $\mathscr{A}$ be a finitely generated $\mathfrak{o}_k$-algebra, which contains $\mathfrak{o}_k$ as a subring. Let $A = \mathscr{A} \otimes_{\mathfrak{o}_k} k$, which identifies with the localisation of $\mathscr{A}$ with respect to $\mathfrak{o}_k \setminus \{0\}$. Note that the Berkovich space $(\operatorname{Spec} A)^{\mathrm{an}}$ identifies with the set of all multiplicative seminorms on A extending the absolute value $|\cdot|$ on k. If x is a point of $(\operatorname{Spec} A)^{\mathrm{an}}$, we denote by $\widehat{\kappa}(x)$ the completion of the residue field $\kappa(x)$ with respect to the absolute value $|\cdot|_x$, and we let p_x be the k-morphism from $\operatorname{Spec} \widehat{\kappa}(x)$ to $\operatorname{Spec} A$ corresponding to the point $j(x) \in \operatorname{Spec} A$ (see Definition 2.1.1 for the specification map j). Then p_x extends to an $\mathfrak{o}_k$-morphism $\mathscr{P}_x$ from $\operatorname{Spec} \mathfrak{o}_x$ to $\operatorname{Spec} \mathscr{A}$ if and only if $|a|_x \leqslant 1$ for any $a \in \mathscr{A}$. In this case the image of the maximal ideal $\mathfrak{m}_x$ of $\mathfrak{o}_x$ by the morphism $\mathscr{P}_x : \operatorname{Spec} \mathfrak{o}_x \to \operatorname{Spec} \mathscr{A}$ identifies with the prime ideal

$$(\mathscr{A}, |\cdot|_x)_{<1} := \{a \in \mathscr{A} : |a|_x < 1\} \text{ of } \mathscr{A},$$

which lies in the fibre $(\operatorname{Spec} \mathscr{A})_\circ$ of $\operatorname{Spec} \mathscr{A}$ over the maximal ideal of $\mathfrak{o}_k$. We denote by $(\operatorname{Spec} A)^{\mathrm{an}}_{\mathscr{A}}$ the subset of $(\operatorname{Spec} A)^{\mathrm{an}}$ of points x such that

$$\sup_{a \in \mathscr{A}} |a|_x \leqslant 1$$

and by $r_{\mathscr{A}} : (\operatorname{Spec} A)^{\mathrm{an}}_{\mathscr{A}} \to (\operatorname{Spec} \mathscr{A})_\circ$ the map sending $x \in (\operatorname{Spec} A)^{\mathrm{an}}_{\mathscr{A}}$ to $(\mathscr{A}, |\cdot|_x)_{<1}$, called the *reduction map*. Note that the reduction map is always surjective (cf. [9, Proposition 2.4.4] or [81, 4.13 and Proposition 4.14]).

Proposition 2.1.7 *We assume that the absolute value $|\cdot|$ is non-trivial and non-Archimedean and let $\mathfrak{o}_k$ be the valuation ring of $(k, |\cdot|)$. Let $\mathscr{A}$ be a finitely generated $\mathfrak{o}_k$-algebra and A be the localisation of $\mathscr{A}$ with respect to $\mathfrak{o}_k \setminus \{0\}$. Then the integral closure of $\mathscr{A}$ in A (that is, the integral closure of the image of $\mathscr{A} \to A$) identifies with*

$$\bigcap_{x \in (\operatorname{Spec} A)^{\mathrm{an}}_{\mathscr{A}}} (A, |\cdot|_x)_{\leqslant 1},$$

where

$$(A, |\cdot|_x)_{\leqslant 1} = \{a \in A : |a|_x \leqslant 1\}.$$

In particular, if $(k, |\cdot|)$ is discrete, $\mathscr{A}$ is flat over $\mathfrak{o}_k$ and $\mathscr{A}/\varpi\mathscr{A}$ is reduced, then

$$\mathscr{A} = \bigcap_{x \in (\operatorname{Spec} A)^{\mathrm{an}}_{\mathscr{A}}} (A, |\cdot|_x)_{\leqslant 1},$$

where ϖ is a uniformizing parameter of $(k, |\cdot|)$.

Proof Let $\mathscr{B}$ be the integral closure of $\mathscr{A}$ in A. We first show that $\mathscr{B}$ is contained in $(A, |\cdot|_x)_{\leqslant 1}$ for any $x \in (\operatorname{Spec} A)^{\mathrm{an}}_{\mathscr{A}}$. If $a \in \mathscr{B}$, then there are $a_1, \ldots, a_n \in \mathscr{A}$ such that $a^n + a_1 a^{n-1} + \cdots + a_n = 0$. Therefore

$$|a|_x^n = |a^n|_x = |a_1 a^{n-1} + \cdots + a_n|_x \leqslant \max_{i \in \{1,\ldots,n\}} |a_i|_x \cdot |a|_x^{n-i} \leqslant \max_{i \in \{1,\ldots,n\}} |a|_x^{n-i},$$

which implies that $|a|_x \leqslant 1$.

Let $a \in A$ such that a is not integral over $\mathscr{A}$. Since A is a k-algebra of finite type, it is a Noetherian ring which is non-zero (since $a \in A$). In particular, it admits only finitely many minimal prime ideals $S^{-1}\mathfrak{p}_1, \ldots, S^{-1}\mathfrak{p}_n$, where $\mathfrak{p}_1, \ldots, \mathfrak{p}_n$ are prime ideals of $\mathscr{A}$ which do not intersect $S := \mathfrak{o}_k \setminus \{0\}$. We show that there exists $j \in \{1, \ldots, n\}$ such that the canonical image of a in $A/S^{-1}\mathfrak{p}_j$ is not integral over $\mathscr{A}/\mathfrak{p}_j$. Assume that, for any $i \in \{1, \ldots, n\}$, f_i is a monic polynomial in $(\mathscr{A}/\mathfrak{p}_i)[T]$ such that $f_i(\lambda_i) = 0$, where λ_i is the class of a in $A/S^{-1}\mathfrak{p}_i$. Let F_i be a monic polynomial in $\mathscr{A}[T]$ whose reduction modulo $\mathfrak{p}_i[T]$ coincides with f_i. One has $F_i(a) \in S^{-1}\mathfrak{p}_i$ for any $i \in \{1, \ldots, n\}$. Let F be the product of the polynomials $F_1, \ldots, F_n$. Then $F(a)$ belongs to the intersection $\bigcap_{i=1}^n S^{-1}\mathfrak{p}_i$, hence is nilpotent, which implies that a is integral over $\mathscr{A}$. To show that there exists $x \in (\operatorname{Spec} A)^{\mathrm{an}}_{\mathscr{A}}$ such that $|a|_x > 1$ we may replace $\mathscr{A}$ (resp. A) by $\mathscr{A}/\mathfrak{p}_j$ (resp. $A/S^{-1}\mathfrak{p}_j$) and hence assume that $\mathscr{A}$ is an integral domain without loss of generality.

Let $b = a^{-1}$ in the fraction field of A. We assert that

$$b\mathscr{A}[b] \cap \mathfrak{o}_k \neq \{0\} \quad \text{and} \quad 1 \notin b\mathscr{A}[b].$$

We set $a = a'/s$ for some $a' \in \mathscr{A}$ and $s \in S$. Then $s = ba' \in b\mathscr{A}[b] \cap \mathfrak{o}_k$, so that $b\mathscr{A}[b] \cap \mathfrak{o}_k \neq \{0\}$. Next we assume that $1 \in b\mathscr{A}[b]$. Then there exist $m \in \mathbb{N}_{\geqslant 1}$ and $(a'_1, \ldots, a'_m) \in \mathscr{A}^m$ such that

$$1 = a'_1 b + a'_2 b^2 + \cdots + a'_m b^m,$$

or equivalently $a^m = a'_1 a^{m-1} + \cdots + a'_m$, which is a contradiction.

Let $\mathfrak{p}$ be a maximal ideal of $\mathscr{A}[b]$ such that $b\mathscr{A}[b] \subseteq \mathfrak{p}$. As $\mathfrak{p} \cap \mathfrak{o}_k \neq \{0\}$ and $\mathfrak{p} \cap \mathfrak{o}_k \subseteq \mathfrak{m}_k$ (where $\mathfrak{m}_k$ is the maximal ideal of $\mathfrak{o}_k$), we have $\mathfrak{p} \cap \mathfrak{o}_k = \mathfrak{m}_k$ (since the Krull dimension of $\mathfrak{o}_k$ is 1[1]), and hence $\mathfrak{p}$ lies in the fibre $(\operatorname{Spec} \mathscr{A}[b])_\circ$ of $\operatorname{Spec} \mathscr{A}[b]$ over $\mathfrak{m}_k$. Note that $\mathscr{A}[b]$ is finitely generated over $\mathfrak{o}_k$ and $\mathscr{A}[b] \otimes_{\mathfrak{o}_k} k = A[b]$. Thus, since the reduction map

$$r_{\mathscr{A}[b]} : (\operatorname{Spec} A[b])^{\mathrm{an}}_{\mathscr{A}[b]} \longrightarrow (\operatorname{Spec} \mathscr{A}[b])_\circ$$

is surjective, there is $x \in (\operatorname{Spec} A[b])^{\mathrm{an}}_{\mathscr{A}[b]}$ such that $r_{\mathscr{A}[b]}(x) = \mathfrak{p}$. Clearly $x \in (\operatorname{Spec} A)^{\mathrm{an}}_{\mathscr{A}}$. As $b \in \mathfrak{p}$, we have $|b|_x < 1$, so that $|a|_x > 1$ because $ab = 1$. Therefore,

$$a \notin \bigcap_{x \in (\operatorname{Spec} A)^{\mathrm{an}}_{\mathscr{A}}} (A, |\cdot|_x)_{\leqslant 1},$$

as required.

Finally we consider the last assertion. We assume that there is

$$a \in \bigcap_{x \in (\operatorname{Spec} A)^{\mathrm{an}}_{\mathscr{A}}} (A, |\cdot|_x)_{\leqslant 1} \setminus \mathscr{A}.$$

---

[1] It suffices to see $\mathfrak{m}_k \subseteq \mathfrak{p}$ for a non-zero prime ideal $\mathfrak{p}$ of $\mathfrak{o}_k$. Fix $e \in \mathfrak{p} \setminus \{0\}$. If $x \in \mathfrak{m}_k$, then $x^n e^{-1} \in \mathfrak{o}_k$ for some positive integer n because $|x| < 1$, so that $x^n \in \mathfrak{o}_k e \subseteq \mathfrak{p}$, and hence $x \in \mathfrak{p}$.

By the previous result, there are $a_1, \ldots, a_n \in \mathscr{A}$ such that

$$a^n + a_1 a^{n-1} + \cdots + a_{n-1}a + a_n = 0.$$

One can choose a positive integer e such that $\varpi^e a \in \mathscr{A}$ and $\varpi^{e-1}a \notin \mathscr{A}$. As

$$(\varpi^e a)^n + \varpi^e a_1(\varpi^e a)^{n-1} + \cdots + \varpi^{e(n-1)}a_{n-1}(\varpi^e a) + \varpi^{en}a_n = 0,$$

$(\varpi^e a)^n = 0$ in $\mathscr{A}/\varpi\mathscr{A}$, so that $\varpi^e a = 0$ in $\mathscr{A}/\varpi\mathscr{A}$ because $\mathscr{A}/\varpi\mathscr{A}$ is reduced. Therefore there is $a' \in \mathscr{A}$ such that $\varpi^e a = \varpi a'$, and hence $\varpi^{e-1}a = a' \in \mathscr{A}$ because $\mathscr{A}$ is flat over $\mathfrak{o}_k$. This is a contradiction.

2.1.2 Metric on a vector bundle

Let X be a scheme over $\operatorname{Spec} k$. We denote by $\mathcal{F}_{X^{\mathrm{an}}}$ the sheaf of all real-valued functions on the Berkovich topological space X^{an}. Let $C^0_{X^{\mathrm{an}}}$ be the subsheaf of $\mathcal{F}_{X^{\mathrm{an}}}$ of continuous functions.

Definition 2.1.8 Let E be a locally free $\mathcal{O}_X$-module of finite rank. We call *metric* on E any family $\varphi = \{|\cdot|_\varphi(x)\}_{x \in X^{\mathrm{an}}}$, where each $|\cdot|_\varphi(x)$ is a norm on $E(x) := E \otimes \widehat{\kappa}(x)$, $\widehat{\kappa}(x)$ being the completion of $\kappa(x)$ with respect to the absolute value $|\cdot|_x$. We use the symbol $|\cdot|_\varphi$ instead of the usual double bar symbol in order to distinguish a local norm from a global seminorm (cf. Definition 2.1.15).

Note that the family φ actually defines a morphism of sheaves (of sets) from E to $j_*(\mathcal{F}_{X^{\mathrm{an}}})$, which sends each section s of E over a Zariski open subset U of X to the function $|s|_\varphi : U^{\mathrm{an}} \to \mathbb{R}_{\geqslant 0}$ sending $x \in U^{\mathrm{an}}$ to

$$|s|_\varphi(x) := |s(x)|_\varphi(x),$$

where $s(x)$ denotes the reduction of s in $E(x)$. If this morphism of sheaves takes values in $j_*(C^0_{X^{\mathrm{an}}})$ (namely, for any section s of E on a Zariski open subset of X, the function $|s|_\varphi$ is continuous with respect to the Berkovich topology), we say that the metric φ is *continuous*.

Remark 2.1.9 Let E be a locally free $\mathcal{O}_X$-module of finite rank, equipped with a continuous metric φ. Let F be a locally free sub-$\mathcal{O}_X$-module of E. For any $x \in X^{\mathrm{an}}$, the restriction of the norm $|\cdot|_\varphi(x)$ to $F(x)$ defines a norm on $F(x)$. These norms actually define a continuous metric on F. However, we don't know if, for any quotient vector bundle of E, the quotient norms of $|\cdot|_\varphi(x)$ $(x \in X^{\mathrm{an}})$ define a continuous metric on the quotient bundle.

The following lemma is used in the proof of Proposition 2.1.12.

Lemma 2.1.10 *Let M be a topological space and f be a non-negative function on M. Suppose that, for any $\alpha \in \,]0,1[$, there exists a continuous function f_α on M such that $\alpha f_\alpha \leqslant f \leqslant f_\alpha$. Then the function f is continuous.*

Proof Let x_0 be a point of M. From the inequalities $\alpha f_\alpha \leqslant f \leqslant f_\alpha$, we deduce

$$\liminf_{x\to x_0} \alpha f_\alpha(x) \leqslant \liminf_{x\to x_0} f(x) \leqslant \limsup_{x\to x_0} f(x) \leqslant \limsup_{x\to x_0} f_\alpha(x).$$

Since the function f_α is continuous, we obtain

$$\alpha f_\alpha(x_0) \leqslant \liminf_{x\to x_0} f(x) \leqslant \limsup_{x\to x_0} f(x) \leqslant f_\alpha(x_0).$$

Moreover, one has $\alpha f_\alpha(x_0) \leqslant f(x_0) \leqslant f_\alpha(x_0)$. Hence

$$\liminf_{x\to x_0} f(x) \leqslant \limsup_{x\to x_0} f(x) \leqslant \alpha^{-1} f(x_0) \leqslant \alpha^{-1} f_\alpha(x_0) \leqslant \alpha^{-2} \liminf_{x\to x_0} f(x).$$

Since $\alpha \in \,]0,1[$ is arbitrary and $\liminf_{x\to x_0} f(x)$ is finite, we obtain

$$\liminf_{x\to x_0} f(x) = \limsup_{x\to x_0} f(x) = f(x_0).$$

The proposition is thus proved. □

Definition 2.1.11 Let $\pi : X \to \mathrm{Spec}(k)$ be a k-scheme and $\overline{V} = (V, \|\cdot\|)$ be a finite-dimensional normed vector space over k. For any $x \in X^{\mathrm{an}}$, let $|\cdot|_{\overline{V},\varepsilon}(x)$ be the norm on $V \otimes_k \widehat{\kappa}(x)$ induced by $\|\cdot\|$ by ε-extension of scalars, and $|\cdot|_{\overline{V},\pi}$ be the norm on $V \otimes_k \widehat{\kappa}(x)$ induced by $\|\cdot\|$ by π-extension of scalars (see §1.3). If $|\cdot|$ is Archimedean and if the norm $\|\cdot\|$ is induced by an inner product, for any $x \in X^{\mathrm{an}}$, we denote by $|\cdot|_{\overline{V},\mathrm{HS}}(x)$ the norm on $V \otimes_k \widehat{\kappa}(x)$ induced by $\|\cdot\|$ by orthogonal extension of scalars. For simplicity, the norms $|\cdot|_{\overline{V},\varepsilon}(x)$, $|\cdot|_{\overline{V},\pi}(x)$ and $|\cdot|_{\overline{V},\mathrm{HS}}(x)$ are often denoted by $|\cdot|_\varepsilon(x)$, $|\cdot|_\pi(x)$ and $|\cdot|_{\mathrm{HS}}(x)$, respectively.

Proposition 2.1.12 *The norms $|\cdot|_{\overline{V},\varepsilon}(x)$, $x \in X^{\mathrm{an}}$ define a continuous metric on the locally free $\mathcal{O}_X$-module $\pi^*(V)$.*

Proof Let U be a Zariski open subset of X and s be a section of $\pi^*(V)$ over U. It suffices to prove that the function $(x \in U^{\mathrm{an}}) \mapsto |s|_{\overline{V},\varepsilon}(x)$ is continuous on U^{an}.

We first treat the non-Archimedean case. By Proposition 1.2.7, for any $\alpha \in \,]0,1[$, there exists an α-orthogonal basis $\{e_i\}_{i=1}^n$ of V. By Proposition 1.3.13, for any $x \in X^{\mathrm{an}}$, $\{e_i\}_{i=1}^n$ is also an α-orthogonal basis of $(V \otimes_k \widehat{\kappa}(x), |\cdot|_{\overline{V},\varepsilon}(x))$. We can write s into the form $s = f_1 e_1 + \cdots + f_n e_n$, where $f_1, \ldots, f_n$ are regular functions on U. Since $\{e_i\}_{i=1}^n$ is an α-orthogonal basis and the norm $|\cdot|_{\overline{V},\varepsilon}(x)$ is ultrametric for any $x \in U^{\mathrm{an}}$, one has

$$\forall\, x \in U^{\mathrm{an}}, \quad \alpha \max_{i\in\{1,\ldots,n\}} |f_i|_x \cdot |e_i|_{\overline{V},\varepsilon}(x) \leqslant |s|_{\overline{V},\varepsilon}(x) \leqslant \max_{i\in\{1,\ldots,n\}} |f_i|_x \cdot |e_i|_{\overline{V},\varepsilon}(x).$$

By Proposition 1.3.1, one has $|e_i|_{\overline{V},\varepsilon}(x) = \|e_i\|_{**}$ for any $x \in X^{\mathrm{an}}$. Hence

$$\forall\, x \in U^{\mathrm{an}}, \quad \alpha \max_{i\in\{1,\ldots,n\}} |f_i|_x \cdot \|e_i\|_{**} \leqslant |s|_{\overline{V},\varepsilon}(x) \leqslant \max_{i\in\{1,\ldots,n\}} |f_i|_x \cdot \|e_i\|_{**}.$$

Note that the function $(x \in U^{\mathrm{an}}) \mapsto |f_i|_x$ is continuous for any i. Hence the function

$$(x \in U^{\mathrm{an}}) \longmapsto \max_{i \in \{1,\ldots,n\}} |f_i|_x \cdot \|e_i\|_{**}$$

is also continuous. Since $\alpha \in]0,1[$ is arbitrary, by Lemma 2.1.10 we obtain that the function $(x \in U^{\mathrm{an}}) \mapsto |s|_{\overline{V},\varepsilon}(x)$ is continuous.

We now consider the Archimedean case. Let $\{e_i\}_{i=1}^n$ be a basis of V. We write the section s in the form $f_1e_1 + \cdots + f_ne_n$, where $f_1, \ldots, f_n$ are regular functions on U. Note that $f_1^{\mathrm{an}}, \ldots, f_n^{\mathrm{an}}$ are continuous complex functions on U^{an}. Since the norm $\|\cdot\|_{\mathbb{C},\varepsilon}$ is a continuous function on $V_{\mathbb{C}}$, we obtain that the map (see Remark 2.1.4)

$$(x \in X^{\mathrm{an}}) \longmapsto |s|_{\overline{V},\varepsilon}(x) = \|f_1^{\mathrm{an}}(x)e_1 + \cdots + f_n^{\mathrm{an}}(x)e_n\|_{\mathbb{C},\varepsilon}$$

is a continuous function on U^{an}. The proposition is thus proved. □

Proposition 2.1.13 *We assume that the absolute value $|\cdot|$ is Archimedean. Let $\overline{V} = (V, \|\cdot\|)$ be a finite-dimensional normed vector space over k.*

(1) *The norms $|\cdot|_{\overline{V},\pi}(x)$, $x \in X^{\mathrm{an}}$ define a continuous metric on the locally free $\mathcal{O}_X$-module $\pi^*(V)$.*
(2) *If the norm $\|\cdot\|$ is induced by an inner product, then the norms $|\cdot|_{\overline{V},\mathrm{HS}}(x)$, $x \in X^{\mathrm{an}}$ define a continuous metric on the locally free $\mathcal{O}_X$-module $\pi^*(V)$.*

Proof The proof is quite similar to the second part of the proof of Proposition 2.1.12, where we use the continuity of the norms $\|\cdot\|_{\mathbb{C},\pi}$ and $\|\cdot\|_{\mathbb{C},\mathrm{HS}}$ (in the case where the norm $\|\cdot\|$ is induced by an inner product) on the topological space $V_{\mathbb{C}}$. □

Proposition 2.1.14 *We assume that the field k is Archimedean. Let $\pi : X \to \mathrm{Spec}(k)$ be a k-scheme and $\overline{V} = (V, \|\cdot\|)$ be a finite-dimensional normed vector space over $\mathbb{R}$. For any $x \in X^{\mathrm{an}}$, let $|\cdot|_{\overline{V}}(x)$ be the norm on $V \otimes_k \widehat{\kappa}(x)$ induced by $\|\cdot\|$ by $\natural$-extension of scalars, where $\natural = \varepsilon$, π or HS (in the case where $\|\cdot\|$ is induced by an inner product) and let $|\cdot|_{\overline{V}}(x)_*$ be the dual norm on $V^\vee \otimes_k \widehat{\kappa}(x)$ of $|\cdot|_{\overline{V}}(x)$. Then the norms $|\cdot|_{\overline{V}}(x)_*$, $x \in X^{\mathrm{an}}$ define a continuous metric on the locally free $\mathcal{O}_X$-module $\pi^*(V^\vee)$.*

Proof Let $\{\alpha_i\}_{i=1}^n$ be a basis of $V^\vee$. Locally on a Zariski open subset U of X, any element $s \in H^0(U, \pi^*(V^\vee))$ can be written in the form $s = f_1\alpha_1 + \cdots + f_n\alpha_n$, where $f_1, \ldots, f_n$ are regular functions on U. Let $\|\cdot\|_{\mathbb{C},\natural,*}$ be the dual norm of $\|\cdot\|_{\mathbb{C},\natural}$ (the norm on $V_{\mathbb{C}}$ induced by $\|\cdot\|$ by $\natural$-extension of scalars). We claim that

$$|s|_{\overline{V}}(x)_* = \|f_1^{\mathrm{an}}(x)\alpha_1 + \cdots + f_n^{\mathrm{an}}(x)\alpha_n\|_{\mathbb{C},\natural,*}. \tag{2.1}$$

The equality follows from the definition of $|\cdot|_{\overline{V}}(x)_*$ when $\kappa(x) = \mathbb{C}$. In the case where $\kappa(x) = \mathbb{R}$, the norm $|\cdot|_{\overline{V}}(x)_*$ is the dual norm of $\|\cdot\|$. Hence it coincides with the restriction of $\|\cdot\|_{\mathbb{C},\natural,*}$ to $V^\vee$ (see Proposition 1.3.20 (1), (2) and Remark 1.3.2 (2) for the cases $\natural = \varepsilon$, π and HS, respectively). Thus the equality (2.1) also holds in this case. Since the norm $\|\cdot\|_{\mathbb{C},\natural,*}$ is a continuous function on $V_{\mathbb{C}}^\vee$, we obtain that the function $(x \in U^{\mathrm{an}}) \mapsto |s|_{\overline{V}}(x)_*$ is continuous. □

Definition 2.1.15 If the k-scheme X is proper, then the associated Berkovich space X^{an} is compact (see [9] Proposition 3.4.8). In particular, if E is a locally free $\mathcal{O}_X$-module equipped with a continuous metric φ, for any global section $s \in H^0(X, E)$, the number

$$\|s\|_\varphi := \sup_{x \in X^{\mathrm{an}}} |s|_\varphi(x)$$

is finite. Thus we obtain a map $\|\cdot\|_\varphi : H^0(X, E) \to \mathbb{R}_+$, which is actually a seminorm on the k-vector space $H^0(X, E)$.

Let X_{red} be the reduced scheme associated with X and $E_{\mathrm{red}} := E \otimes_{\mathcal{O}_X} \mathcal{O}_{X_{\mathrm{red}}}$. Note that the natural morphism $X_{\mathrm{red}}^{\mathrm{an}} \to X^{\mathrm{an}}$ is a homeomorphism (see Proposition 2.1.5), so that to give a continuous metric of E on X^{an} is equivalent to give a continuous metric of E_{red} on $X_{\mathrm{red}}^{\mathrm{an}}$. The corresponding metric of E_{red} is denoted by φ_{red}. Moreover, if X is proper and we denote the natural homomorphism $H^0(X, E) \to H^0(X_{\mathrm{red}}, E_{\mathrm{red}})$ by γ_E, then it is easy to see that $\|s\|_\varphi = \|\gamma_E(s)\|_{\varphi_{\mathrm{red}}}$ for any $s \in H^0(X, E)$. By (1) in the following proposition, the null space of $\|\cdot\|_\varphi$ coincides with the kernel of γ_E, which is denoted by $\mathcal{N}(X, E)$. The induced norm on $H^0(X, E)/\mathcal{N}(X, E)$ is denoted by $\|\cdot\|_{\widetilde{\varphi}}$.

Proposition 2.1.16 (1) *If X is reduced, then $\|\cdot\|_\varphi$ is actually a norm.*

(2) *For any $x \in X^{\mathrm{an}}$, the image of $\mathcal{N}(X, E) \otimes_k \widehat{\kappa}(x)$ by the natural homomorphism $H^0(X, E) \otimes_k \widehat{\kappa}(x) \to E \otimes_{\mathcal{O}_X} \widehat{\kappa}(x)$ is zero, so that one has the induced homomorphism $(H^0(X, E)/\mathcal{N}(X, E)) \otimes_k \widehat{\kappa}(x) \to E \otimes_{\mathcal{O}_X} \widehat{\kappa}(x)$. Moreover, if E is generated by global sections, then $(H^0(X, E)/\mathcal{N}(X, E)) \otimes_k \widehat{\kappa}(x) \to E \otimes_{\mathcal{O}_X} \widehat{\kappa}(x)$ is surjective for all $x \in X^{\mathrm{an}}$.*

(3) *If E is invertible and $s \in \mathcal{N}(X, E)$, then there is a positive integer n_0 such that $s^{\otimes n} = 0$ for any integer n such that $n \geqslant n_0$.*

Proof (1) It is sufficient to see that if $\|s\|_\varphi = 0$, then $s = 0$. Let $\eta_1, \ldots, \eta_r$ be the generic points of the irreducible components of X. Let $\tilde{\eta}_i$ be a point of X^{an} such that $j(\tilde{\eta}_i) = \eta_i$. By our assumption, $|s|_\varphi(\tilde{\eta}_i) = 0$, so that $s(\eta_i) = 0$ for all i. Therefore, one has the assertion because X is reduced.

(2) We denote the natural homeomorphism $X^{\mathrm{an}} \to X_{\mathrm{red}}^{\mathrm{an}}$ by p. Then we have the following commutative diagram:

$$\begin{array}{ccccc}
H^0(X, E) \otimes_k \widehat{\kappa}(x) & \xrightarrow{\sim} & H^0(X, E) \otimes_k \widehat{\kappa}(p(x)) & \xrightarrow{\gamma_E \otimes \mathrm{id}} & H^0(X_{\mathrm{red}}, E_{\mathrm{red}}) \otimes_k \widehat{\kappa}(p(x)) \\
\downarrow & & \downarrow & \swarrow & \\
E \otimes_{\mathcal{O}_X} \widehat{\kappa}(x) & \xrightarrow{\sim} & E_{\mathrm{red}} \otimes_{\mathcal{O}_{X_{\mathrm{red}}}} \widehat{\kappa}(p(x)) & &
\end{array}$$

because $E_{\mathrm{red}} = E \otimes_{\mathcal{O}_X} \mathcal{O}_{X_{\mathrm{red}}}$. Therefore one has (2).

(3) Let $X = \bigcup_{i=1}^N \mathrm{Spec}(A_i)$ be an affine open covering of X such that, for each i, there is a local basis ω_i of E over $\mathrm{Spec}(A_i)$. For any $i \in \{1, \ldots, N\}$, let $a_i \in A_i$ such that $s = a_i \omega_i$ on $\mathrm{Spec}\, A_i$. As $s|_{X_{\mathrm{red}}} = 0$, a_i is a nilpotent element of A_i, so that we can find a positive integer n_0 such that $s^{\otimes n_0} = 0$. Therefore $s^{\otimes n} = 0$ for $n \geqslant n_0$ because $s^{\otimes n} = s^{\otimes n_0} \otimes s^{\otimes n - n_0}$. □

2.1.3 Base change

Let k'/k be a field extension equipped with an absolute value $|\cdot|'$ which extends $|\cdot|$ on k. We assume that k' is complete with respect to this absolute value.

Let X be a scheme over $\operatorname{Spec} k$, X' be the fibre product $X \times_{\operatorname{Spec} k} \operatorname{Spec} k'$ and $p : X' \to X$ be the morphism of projection. If $(K, |\cdot|_K)$ is a valued extension of k' and $f : \operatorname{Spec} K \to X'$ is a k'-point of X' valued in K, then the composition morphism $\pi \circ f$ is a k-point of X valued in K. This construction is functorial and thus determines by passing to colimit a surjective map between Berkovich spaces from X'^{an} to X^{an} which we denote by $p^\natural$. We emphasise that X'^{an} is constructed from the projective k'-scheme X'. Thus $p^\natural$ differs from the map between Berkovich spaces associated with p (considered as a k-morphism of schemes) as in Definition 2.1.2.

Proposition 2.1.17 *The map $p^\natural : X'^{\mathrm{an}} \to X^{\mathrm{an}}$ defined above is continuous with respect to the Berkovich topology.*

Proof Let U be a Zariski open subset of X and g be a regular function on U. We denote by g' the pull-back of g by p, which is a regular function on $p^{-1}(U)$. For any $y \in X'^{\mathrm{an}}$ one has $|g'|(y) = |g|(p^\natural(y))$. Hence $|g| \circ p^\natural$ is a continuous function on $p^{-1}(U)^{\mathrm{an}}$. Therefore the map $p^\natural$ is continuous. □

Definition 2.1.18 Let E be a locally free $\mathcal{O}_X$-module of finite rank, equipped with a metric φ and let $E_{k'}$ be the pull-back of E by the projection morphism $p : X' \to X$. If y is a point of X'^{an} and if $x = p^\natural(y)$, then the norm $|\cdot|_\varphi(x)$ on $E(x) = E \otimes \widehat{\kappa}(x)$ induces by ε-extension (resp. π-extension) of scalars a norm on $E_{k'}(y) \cong E(x) \otimes_{\widehat{\kappa}(x)} \widehat{\kappa}(y)$, denoted by $|\cdot|_{\varphi_{k',\varepsilon}}(y)$ (resp. $|\cdot|_{\varphi_{k',\pi}}(y)$). These norms define a metric on $E_{k'}$, denoted by $\varphi_{k',\varepsilon}$ (resp. $\varphi_{k',\pi}$), called the metric *induced by φ by ε-extension (resp. π-extension)* of scalars.

Assume that the norm $|\cdot|_\varphi(x)$ on $E(x)$ is induced by an inner product. For any point $y \in X'^{\mathrm{an}}$ such that $x = p^\natural(y)$, we let $|\cdot|_{\varphi_{k',\mathrm{HS}}}(y)$ be the norm on $E_{k'}(y)$ induced by $|\cdot|_\varphi(x)$ by orthogonal extension of scalars. These norms define a metric on $E_{k'}$, denoted by $\varphi_{k',\mathrm{HS}}$ and called the metric *induced by φ by orthogonal extension of scalars.*

In the case where E is an invertible $\mathcal{O}_X$-module, the three metrics $\varphi_{k',\varepsilon}$, $\varphi_{k',\pi}$ and $\varphi_{k,\mathrm{HS}}$ are the same (see Remark 1.3.2 (3)), and are just denoted by $\varphi_{k'}$.

Proposition 2.1.19 *Let L be an invertible $\mathcal{O}_X$-module equipped with a continuous metric φ. Then one has the following:*

(1) *The metric $\varphi_{k'}$ is continuous.*
(2) *For all $s \in H^0(X, L)$, $\|s\|_\varphi = \|p^*(s)\|_{\varphi_{k'}}$.*

Proof Let U be a Zariski open subset of X and $s \in H^0(U, L)$. Then one has

$$|p^*(s)|_{\varphi_{k'}} = |s|_\varphi \circ p^\natural|_{p^{-1}(U)^{\mathrm{an}}}, \tag{2.2}$$

so that the assertion (2) follows. If we assume that L is trivialised by s over U, then $p^*(s)$ is a section in $H^0(p^{-1}(U), L_{k'})$ which trivialises $L_{k'}$ on $p^{-1}(U)$. Moreover, by the above (2.2) together with Proposition 2.1.17, we obtain that $|p^*(s)|_{\varphi_{k'}}$ is a continuous function on $p^{-1}(U)^{\mathrm{an}}$. Therefore one has the assertion (1). □

2.2 Metrics on invertible sheaves

Let X be a scheme over $\operatorname{Spec} k$. In this section, we discuss constructions and properties of metrics on invertible $\mathcal{O}_X$-modules.

2.2.1 Dual metric and tensor product metric

Let L be an invertible $\mathcal{O}_X$-module and φ be a metric on L. Note that for any $x \in X^{\mathrm{an}}$, the norm $|\cdot|_\varphi(x)$ is determined by its value on any non-zero element of $L \otimes \widehat{\kappa}(x)$. In particular, to verify that the metric φ is continuous, it suffices to prove that there exists a covering $\{U_i\}_{i\in I}$ of X by affine open subsets and for each $i \in I$ there exists a section $s_i \in H^0(U_i, L)$ which trivialises the invertible sheaf L on U_i such that the function $|s_i|_\varphi$ is continuous on the topological space U_i^{an}.

Definition 2.2.1 Let L be an invertible $\mathcal{O}_X$-module. If φ is a metric on L, then the dual $\mathcal{O}_X$-module $L^\vee$ is naturally equipped with a metric $\varphi^\vee$ such that, for sections α and s of $L^\vee$ and L over a Zariski open subset U of X respectively, one has

$$\forall\, x \in U^{\mathrm{an}}, \quad |\alpha(s)|(x) = |s|_\varphi(x) \cdot |\alpha|_{\varphi^\vee}(x).$$

We call $\varphi^\vee$ the *dual metric* of φ and we also use the expression $-\varphi$ to denote the metric $\varphi^\vee$.

Proposition 2.2.2 *Let X be a k-scheme, L be an invertible $\mathcal{O}_X$-module and φ be a metric on L. If φ is a continuous metric, then $\varphi^\vee$ is also continuous.*

Proof Let U be a Zariski open subset of X on which the invertible sheaves L and $L^\vee$ are trivialised by sections $s \in \Gamma(U, L)$ and $\alpha \in \Gamma(U, L^\vee)$ respectively. Then $\alpha(s)$ is a regular function, and

$$|\alpha|_{\varphi^\vee} = \frac{|\alpha(s)|}{|s|_\varphi}$$

on U^{an}. Since the functions $|\alpha(s)|$ and $|s|_\varphi$ are all continuous, also is $|\alpha|_{\varphi^\vee}$. Since U is arbitrary, we obtain that $\varphi^\vee$ is a continuous metric. □

Definition 2.2.3 Let L be an invertible $\mathcal{O}_X$-module and n be a positive integer. Suppose given a metric φ on $L^{\otimes n}$. Then the maps

$$(s \in H^0(U, L)) \longmapsto |s^n|_\varphi^{1/n},$$

with U running over the set of all Zariski open subsets of X, define a metric on L, denoted by $\frac{1}{n}\varphi$. If the metric φ is continuous, then also is $\frac{1}{n}\varphi$.

Definition 2.2.4 Suppose given two invertible $\mathcal{O}_X$-modules L_1 and L_2, equipped with metrics φ_1 and φ_2 respectively. We denote by $\varphi_1 + \varphi_2$ the metric on $L_1 \otimes L_2$ such that, for any Zariski open subset U of X and all sections $s_1 \in H^0(U, L_1)$, $s_2 \in H^0(U, L_2)$, one has

$$\forall\, x \in U^{\mathrm{an}}, \quad |s_1 \cdot s_2|_{\varphi_1+\varphi_2}(x) = |s_1|_{\varphi_1}(x) \cdot |s_2|_{\varphi_2}(x).$$

The metric $\varphi_1 + \varphi_2$ is called *tensor product* of φ_1 and φ_2. Note that, if the metrics φ_1 and φ_2 are continuous, then also is $\varphi_1 + \varphi_2$. We also use the expression $\varphi_1 - \varphi_2$ to denote the metric $\varphi_1 + \varphi_2^\vee$ on $L_1 \otimes L_2^\vee$. If L is an invertible $\mathcal{O}_X$-module equipped with a metric φ, for any integer $n \in \mathbb{N}_{\geqslant 1}$, we use the expression $n\varphi$ to denote the metric $\varphi + \cdots + \varphi$ (n copies) on $L^{\otimes n}$.

Proposition 2.2.5 *Let X be a scheme over $\operatorname{Spec} k$, L_1 and L_2 be invertible $\mathcal{O}_X$-modules, and φ_1 and φ_2 be continuous metrics on L_1 and L_2, respectively. Then the canonical k-linear homomorphism $H^0(X, L_1) \otimes_k H^0(X, L_2) \to H^0(X, L_1 \otimes L_2)$, sending $s_1 \otimes s_2 \in H^0(X, L_1) \otimes_k H^0(X, L_2)$ to $s_1 \cdot s_2$, has operator norm $\leqslant 1$, where we consider the π-tensor product of $\|\cdot\|_{\varphi_1}$ and $\|\cdot\|_{\varphi_2}$ on the tensor product space, and the norm $\|\cdot\|_{\varphi_1+\varphi_2}$ on $H^0(X, L_1 \otimes L_2)$. In particular, if s_1 and s_2 are elements in $H^0(X, L_1)$ and $H^0(X, L_2)$, respectively, then the following inequality holds*

$$\|s_1 \cdot s_2\|_{\varphi_1+\varphi_2} \leqslant \|s_1\|_{\varphi_1} \cdot \|s_2\|_{\varphi_2}. \tag{2.3}$$

Proof Let η be an element of $H^0(X, L_1) \otimes_k H^0(X, L_2)$, which is written as

$$\eta = \sum_{i=1}^{N} s_1^{(i)} \otimes s_2^{(i)},$$

where $s_1^{(1)}, \ldots, s_1^{(N)}$ are elements in $H^0(X, L_1)$, $s_2^{(1)}, \ldots, s_2^{(N)}$ are elements in $H^0(X, L_2)$. Let s be the element

$$\sum_{i=1}^{N} s_1^{(i)} \cdot s_2^{(i)}$$

in $H^0(X, L_1 \otimes L_2)$, which is the image of η by the canonical homomorphism

$$H^0(X, L_1) \otimes_k H^0(X, L_2) \longrightarrow H^0(X, L_1 \otimes L_2).$$

For any $x \in X^{\mathrm{an}}$ one has

$$|s|_{\varphi_1+\varphi_2}(x) = \Big|\sum_{i=1}^N s_1^{(i)} \cdot s_2^{(i)}\Big|_{\varphi_1+\varphi_2}(x) \leqslant \sum_{i=1}^N |s_1^{(i)} \cdot s_2^{(i)}|_{\varphi_1+\varphi_2}(x)$$
$$= \sum_{i=1}^N |s_1^{(i)}|_{\varphi_1}(x) \cdot |s_2^{(i)}|_{\varphi_2}(x) \leqslant \sum_{i=1}^N \|s_1^{(i)}\|_{\varphi_1} \cdot \|s_2^{(i)}\|_{\varphi_2}.$$

Since $x \in X^{\mathrm{an}}$ is arbitrary, we obtain

$$\|s\|_{\varphi_1+\varphi_2} \leqslant \sum_{i=1}^N \|s_1^{(i)}\|_{\varphi_1} \cdot \|s_2^{(i)}\|_{\varphi_2}.$$

Therefore $\|s\|_{\varphi_1+\varphi_2} \leqslant \|\eta\|_\pi$, where $\|\cdot\|_\pi$ denotes the π-tensor product of $\|\cdot\|_{\varphi_1}$ and $\|\cdot\|_{\varphi_2}$. The first assertion is thus proved.

If s_1 and s_2 are elements in $H^0(X, L_1)$ and $H^0(X, L_2)$ respectively, then one has

$$\|s_1 \cdot s_2\|_{\varphi_1+\varphi_2} = \sup_{x\in X^{\mathrm{an}}} |s_1 \cdot s_2|_{\varphi_1+\varphi_2}(x)$$
$$= \sup_{x\in X^{\mathrm{an}}} |s_1|_{\varphi_1}(x) \cdot |s_2|_{\varphi_2}(x) \leqslant \|s_1\|_{\varphi_1} \cdot \|s_2\|_{\varphi_2},$$

as required. □

Remark 2.2.6 Assume that the absolute value $|\cdot|$ is non-Archimedean. The statement of Proposition 2.2.5 remains true if we consider the ε-tensor product $\|\cdot\|_\varepsilon$ of $\|\cdot\|_{\varphi_1}$ and $\|\cdot\|_{\varphi_2}$ on the tensor product space $H^0(X, L_1) \otimes_k H^0(X, L_2)$. In fact, by Proposition 1.2.19, if $\{e_i\}_{i=1}^n$ and $\{f_j\}_{j=1}^m$ are α-orthogonal basis of $(H^0(X, L_1), \|\cdot\|_{\varphi_1})$ and $(H^0(X, L_2), \|\cdot\|_{\varphi_2})$ respectively, then

$$\{e_i \otimes f_j\}_{(i,j)\in\{1,\ldots,n\}\times\{1,\ldots,m\}}$$

is an α^2-orthogonal basis with respect to $\|\cdot\|_\varepsilon$, where $\alpha \in \,]0, 1[$. For any

$$\eta = \sum_{i=1}^n \sum_{j=1}^m a_{ij} e_i \otimes e_j \in H^0(X, L_1) \otimes_k H^0(X, L_2),$$

one has

$$\|\eta\|_\varepsilon \geqslant \alpha^2 \max_{(i,j)\in\{1,\ldots,n\}\times\{1,\ldots,m\}} |a_{ij}| \cdot \|e_i\| \cdot \|f_j\|,$$

which is bounded from below by α^2 times the norm of the canonical image of η in $H^0(X, L_1 \otimes L_2)$ since the norm $\|\cdot\|_{\varphi_1+\varphi_2}$ is ultrametric.

2.2.2 Distance between metrics

Let φ be a metric on $\mathcal{O}_X$. Then $-\ln|\mathbf{1}|_\varphi$ is a function on X^{an}, where $\mathbf{1}$ denotes the section of unity of $\mathcal{O}_X$. If φ is a continuous metric, then $-\ln|\mathbf{1}|_\varphi$ is a continuous function. Conversely, any real-valued function g on X^{an} determines a metric on $\mathcal{O}_X$ such that the norm at $x \in X^{\mathrm{an}}$ of the section of unity of $\mathcal{O}_X$ is $\mathrm{e}^{-g(x)}$. The metric is continuous if and only if the function g is continuous. Therefore the set of all metrics on $\mathcal{O}_X$ is canonically in bijection with the set of all real-valued function on X^{an}. This correspondance also maps bijectively the set of all continuous metrics on $\mathcal{O}_X$ to the set $C^0(X^{\mathrm{an}})$ of all continuous real-valued functions on X^{an}.

Definition 2.2.7 Let L be an invertible $\mathcal{O}_X$-module. If φ and φ' are two metrics on L, then $\varphi' - \varphi$ is a metric on $L \otimes L^\vee \cong \mathcal{O}_X$, hence corresponds to a real valued function on X^{an}. By abuse of notation, we use the expression $\varphi' - \varphi$ to denote this function. We say that the metric φ' is *larger* than φ if $\varphi' - \varphi$ is a non-negative function and we use the expressions $\varphi' \geqslant \varphi$ or $\varphi \leqslant \varphi'$ to denote the relation "*φ' is larger than φ*". If φ and φ' are metrics on L, we denote by $d(\varphi, \varphi')$ the element

$$\sup_{x \in X^{\mathrm{an}}} |\varphi' - \varphi|(x) \in \mathbb{R}_{\geqslant 0} \cup \{+\infty\},$$

called the *distance* between φ and φ'. Note that one has

$$d(\varphi, \varphi') = \sup_{x \in X^{\mathrm{an}}} \big| \ln|\cdot|_\varphi(x) - \ln|\cdot|_{\varphi'}(x) \big|. \tag{2.4}$$

Proposition 2.2.8 *If the k-scheme X is proper (so that the sup seminorms are defined), then*

$$d(\|\cdot\|_{\widetilde{\varphi}}, \|\cdot\|_{\widetilde{\varphi'}}) \leqslant d(\varphi, \varphi') \tag{2.5}$$

(see §1.1.9 for the notion of distance between two norms and §1.1 for the notion of norm associated with a seminorm).

Proof Fix $s \in H^0(X, L) \setminus \mathcal{N}(X, L)$. For $\epsilon > 0$, one can choose $x \in X^{\mathrm{an}}$ such that $\mathrm{e}^{-\epsilon}\|s\|_{\varphi_1} \leqslant |s|_{\varphi_1}(x)$. Then

$$\ln\|s\|_{\varphi_1} - \ln\|s\|_{\varphi_2} \leqslant \ln|s|_{\varphi_1}(x) - \ln|s|_{\varphi_2}(x) + \epsilon \leqslant d(\varphi_1, \varphi_2) + \epsilon,$$

so that one has $\ln\|s\|_{\varphi_1} - \ln\|s\|_{\varphi_2} \leqslant d(\varphi_1, \varphi_2)$ because ϵ is an arbitrary positive number. In the same way, $\ln\|s\|_{\varphi_2} - \ln\|s\|_{\varphi_1} \leqslant d(\varphi_1, \varphi_2)$, and hence one obtains

$$\big| \ln\|s\|_{\varphi_1} - \ln\|s\|_{\varphi_2} \big| \leqslant d(\varphi_1, \varphi_2),$$

which implies the assertion of the proposition. □

For any integer $n \in \mathbb{Z}$ one has $n\varphi' - n\varphi = n(\varphi' - \varphi)$ and hence

$$d(n\varphi', n\varphi) = |n| d(\varphi', \varphi). \tag{2.6}$$

The distance function verifies the triangle inequality: if φ_1, φ_2 and φ_3 are three continuous metrics on L, then one has

$$d(\varphi_1, \varphi_3) \leqslant d(\varphi_1, \varphi_2) + d(\varphi_2, \varphi_3) \tag{2.7}$$

because

$$\big|\ln|\cdot|_{\varphi_1}(x) - \ln|\cdot|_{\varphi_3}(x)\big| \leqslant \big|\ln|\cdot|_{\varphi_1}(x) - \ln|\cdot|_{\varphi_2}(x)\big| + \big|\ln|\cdot|_{\varphi_2}(x) - \ln|\cdot|_{\varphi_3}(x)\big|$$

for any $x \in X^{\mathrm{an}}$.

Definition 2.2.9 Let Y and X be two schemes over $\operatorname{Spec} k$, and $f : Y \to X$ be a k-morphism. Suppose given an invertible $\mathcal{O}_X$-module L, equipped with a metric φ. Then the metric φ induces by pull-back a metric $f^*(\varphi)$ on Y such that, for any $y \in Y^{\mathrm{an}}$, the norm $|\cdot|_{f^*(\varphi)}(y)$ is induced by $|\cdot|_{\varphi}(f(y))$ by extension of scalars. The metric $f^*(\varphi)$ is called the *pull-back* of φ by f. For any section s of L on a Zariski open subset U of X, one has

$$|f^*(s)|_{f^*(\varphi)} = |s|_{\varphi} \circ f^{\mathrm{an}}|_{f^{-1}(U)^{\mathrm{an}}}. \tag{2.8}$$

In particular, if the metric φ is continuous, then also is $f^*(\varphi)$.

Proposition 2.2.10 *Let Y and X be two schemes over* $\operatorname{Spec} k$, *$f : Y \to X$ be a k-morphism, L be an invertible $\mathcal{O}_X$-module, and φ and φ' be two metrics on L. Then one has*

$$d(f^*(\varphi), f^*(\varphi')) \leqslant d(\varphi, \varphi').$$

Moreover, the equality holds if $f : Y \to X$ is surjective.

Proof By (2.8), one has $f^*(\varphi) - f^*(\varphi') = (\varphi - \varphi') \circ f^{\mathrm{an}}$. Hence

$$d(f^*(\varphi), f^*(\varphi')) = \sup_{y \in Y^{\mathrm{an}}} |f^*(\varphi) - f^*(\varphi')|(y) \leqslant \sup_{y \in Y^{\mathrm{an}}} |\varphi - \varphi'|(y) = d(\varphi, \varphi').$$

If $f : Y \to X$ is surjective, then $f^{\mathrm{an}} : Y^{\mathrm{an}} \to X^{\mathrm{an}}$ is also surjective, so that the last assertion follows. □

2.2.3 Fubini-Study metric

Let V be a finite-dimensional vector space over k. We denote by $\pi : \mathbb{P}(V) \to \operatorname{Spec}(k)$ the projective space of V. Note that the functor $F_{\mathbb{P}(V)}$ from the category $\mathbf{A}_k$ of k-algebras to the category of sets corresponding to $\mathbb{P}(V)$ (see §2.1) sends any k-algebra A to the set of all projective quotient A-modules of $V \otimes_k A$ which are of rank 1. By gluing morphisms of schemes, we obtain that, for any k-scheme $f : X \to \operatorname{Spec}(k)$, the set of all k-morphisms from X to $\mathbb{P}(V)$ is in functorial bijection with the set of all invertible quotient $\mathcal{O}_X$-module of $f^*(V)$. In the case where X is the projective space $\mathbb{P}(V)$, the invertible quotient $\mathcal{O}_X$-module of $\pi^*(V)$ corresponding to the identity map

$\mathbb{P}(V) \to \mathbb{P}(V)$ is called the *universal invertible sheaf*, denoted by $\mathcal{O}_V(1)$. It verifies the following universal property: for any k-scheme $f : X \to \operatorname{Spec} k$, a k-morphism $g : X \to \mathbb{P}(V)$ corresponds to the invertible quotient

$$g^*(p) : g^*(\pi^*(V)) \cong f^*(V) \longrightarrow g^*(\mathcal{O}_V(1)),$$

where $p : \pi^*(V) \to \mathcal{O}_V(1)$ is the quotient homomorphism defining the universal invertible sheaf.

Let $\overline{V} = (V, \|\cdot\|)$ be a normed vector space of finite dimension over k. For any point x in the Berkovich space $\mathbb{P}(V)^{\mathrm{an}}$, if the absolute value $|\cdot|$ is non-Archimedean, we denote by $|\cdot|_{\overline{V}}(x)$ the norm on $V \otimes_k \widehat{\kappa}(x)$ induced by $\|\cdot\|$ by ε-extension of scalars; if the absolute value $|\cdot|$ is Archimedean, we denote by $|\cdot|_{\overline{V}}(x)$ the norm on $V \otimes_k \widehat{\kappa}(x)$ induced by $\|\cdot\|$ by π-extension of scalars. We emphasise that, in the case where $\widehat{\kappa}(x) = k$ (namely x corresponds to a rational point of $\mathbb{P}(V)$), the vector space $V \otimes_k \widehat{\kappa}(x)$ is canonically isomorphic to V and the norm $|\cdot|_{\overline{V}}(x)$ identifies with the double dual norm of $\|\cdot\|$. We denote by $|\cdot|_{\overline{V},\mathrm{FS}}(x)$ the quotient norm on $\mathcal{O}_V(1)(x) = \mathcal{O}_V(1) \otimes_{\mathcal{O}_{\mathbb{P}(V)}} \widehat{\kappa}(x)$ induced by the norm $|\cdot|_{\overline{V}}(x)$ on $V \otimes_k \widehat{\kappa}(x)$, called the *Fubini-Study norm* on $\mathcal{O}_V(1)(x)$ induced by $\|\cdot\|$. For simplicity, the norm $|\cdot|_{\overline{V},\mathrm{FS}}(x)$ is often denoted by $|\cdot|_{\mathrm{FS}}(x)$.

Remark 2.2.11 It is a natural question to determine if the Fubini-Study metric can be defined in a uniform way (for non-Archimedean and Archimedean cases). Let $\overline{V} = (V, \|\cdot\|)$ be a finite-dimensional vector space over k. For any point $x \in X^{\mathrm{an}}$, we let $|\cdot|_{\overline{V},\varepsilon}(x)$ and $|\cdot|_{\overline{V},\pi}(x)$ be the norms on $V \otimes_k \widehat{\kappa}(x)$ induced by $\|\cdot\|$ by ε-extension and π-extension of scalars respectively. If the absolute value $|\cdot|$ is non-Archimedean, then both norms $|\cdot|_{\overline{V},\varepsilon}(x)$ and $|\cdot|_{\overline{V},\pi}(x)$ induce the same quotient norm on $\mathcal{O}_V(1)(x)$. In fact, by Proposition 1.3.20 (1), (2), the dual norms of both $|\cdot|_{\overline{V},\varepsilon}$ and $|\cdot|_{\overline{V},\pi}$ identify with $\|\cdot\|_{*,\widehat{\kappa}(x),\varepsilon}$, and hence induce the same restricted norm on $\mathcal{O}_V(1)(x)^\vee$. By Proposition 1.1.20, the dual norms of the quotient norms on $\mathcal{O}_V(1)(x)$ of $|\cdot|_{\overline{V},\varepsilon}$ and $|\cdot|_{\overline{V},\pi}$ are the same. Since $\mathcal{O}_V(1)(x)$ is a vector space of dimension 1 over $\widehat{\kappa}(x)$, we obtain that these quotient norms are the same. In other words, in both the Archimedean and non-Archimedean cases, we may use the π-extension of scalars to define the Fubini-Study metric. However, for the reason of applications in the study of adelic vector bundles, it is more convenient to consider the ε-extension of scalars for the non-Archimedean case. We emphasis however that, in the Archimedean case, if we apply the ε-extension of scalars instead of the π-extension of scalars, in general we obtain a different metric from the Fubini-Study metric.

Proposition 2.2.12 *Let $\overline{V} = (V, \|\cdot\|)$ be a finite-dimensional normed vector space over k. Then the norms $|\cdot|_{\overline{V},\mathrm{FS}}(x)$, $x \in \mathbb{P}(V)^{\mathrm{an}}$ described above define a continuous metric on the universal invertible sheaf $\mathcal{O}_V(1)$.*

Proof By Proposition 1.2.14 (see also Remark 1.3.2), for any $x \in \mathbb{P}(V)^{\mathrm{an}}$, the norms $\|\cdot\|$ and $\|\cdot\|_{**}$ induce the same Fubini-Study norm on $\mathcal{O}_V(1)(x)$. Hence we may assume without loss of generality that the norm $\|\cdot\|$ is ultrametric when $(k, |\cdot|)$ is non-Archimedean.

For any $x \in \mathbb{P}(V)^{\mathrm{an}}$, let $|\cdot|_{\overline{V}}(x)$ be the norm on $V \otimes_k \widehat{\kappa}(x)$ induced by $\|\cdot\|$ by π-extension of scalars, and let $|\cdot|_{\overline{V}}(x)_*$ be the dual norm of $|\cdot|_{\overline{V}}(x)$. The norms $|\cdot|_{\overline{V}}(x)_*$ define a metric φ on $\pi^*(V^\vee)$. By Proposition 1.3.20 (1), (2), the norm $|\cdot|_{\overline{V}}(x)_*$ coincides with the norm induced by $\|\cdot\|_*$ by ε-extension of scalars. Therefore, by Proposition 2.1.12, we obtain that the metric φ is continuous.

The dual norm of the Fubini-Study norm $|\cdot|_{\overline{V},\mathrm{FS}}(x)$ then coincides with the restriction of $|\cdot|_{\overline{V}}(x)_*$ to $\mathcal{O}_V(1)^\vee \otimes \widehat{\kappa}(x)$ by Proposition 1.1.20. Hence these dual norms (for $x \in \mathbb{P}(V)^{\mathrm{an}}$) form a continuous metric on $\mathcal{O}_V(1)^\vee$. Therefore the Fubini-Study norms $|\cdot|_{\overline{V},\mathrm{FS}}(x)$, $x \in \mathbb{P}(V)^{\mathrm{an}}$ define a continuous metric on $\mathcal{O}_V(1)$ (see Proposition 2.2.2). □

Definition 2.2.13 Let $\overline{V} = (V, \|\cdot\|)$ be a finite-dimensional normed vector space over k. The continuous metric on $\mathcal{O}_V(1)$ formed by the Fubini-Study norms $|\cdot|_{\overline{V},\mathrm{FS}}(x)$ with $x \in \mathbb{P}(V)^{\mathrm{an}}$ is called the *Fubini-Study metric* on $\mathcal{O}_V(1)$ associated with the norm $\|\cdot\|$ on V.

Remark 2.2.14 Let $\overline{V} = (V, \|\cdot\|)$ be a normed vector space over k and s be an element of V. For any $x \in \mathbb{P}(V)^{\mathrm{an}}$ such that $s(x) \neq 0$, by definition one has

$$|s|_{\overline{V},\mathrm{FS}}(x) = \inf_{\substack{t \in V \otimes_k \widehat{\kappa}(x) \\ t(x)=s(x)}} \|t\|_{\widehat{\kappa}(x),\natural}$$

with $\natural = \varepsilon$ if $|\cdot|$ is non-Archimedean, and $\natural = \pi$ if $|\cdot|$ is Archimedean. In particular, one has (see Proposition 1.3.1)

$$|s|_{\overline{V},\mathrm{FS}}(x) \leqslant \|s\|_{\widehat{\kappa}(x),\natural} = \|s\|_{**}. \tag{2.9}$$

Moreover, any rational point $y \in \mathbb{P}(V)(k)$ corresponds to a non-zero element $\beta_y : V \to k$ in the dual vector space $V^\vee$. The dual norm of β_y identifies with the inverse of the quotient norm of $1 \in k$. Therefore one has

$$|s|_{\overline{V},\mathrm{FS}}(y) = \frac{|\beta_y(s)|}{\|\beta_y\|_*}$$

and hence

$$\sup_{y \in \mathbb{P}(V)(k)} |s(y)|_{\overline{V},\mathrm{FS}}(y) = \sup_{y \in \mathbb{P}(V)(k)} \frac{|\beta_y(s)|}{\|\beta_y\|_*} = \|s\|_{**}.$$

Combing with (2.9), we obtain $\|\cdot\|_{\overline{V},\mathrm{FS}} = \|\cdot\|_{**}$.

Let $(k', |\cdot|)$ be a valued extension of $(k, |\cdot|)$. Note that the fibre product $\mathbb{P}(V) \times_{\operatorname{Spec} k} \operatorname{Spec} k'$ identifies with the projective space of $V' := V \otimes_k k'$. The Fubini-Study metric on $\mathcal{O}_V(1)$ induces by base change a continuous metric on $\mathcal{O}_{V'}(1)$ which we denote by $\{|\cdot|_{\overline{V},\mathrm{FS},k'}(x)\}_{x \in \mathbb{P}(V')^{\mathrm{an}}}$. By Corollary 1.3.15 and Proposition 1.3.24, we obtain that this metric coincides with the Fubini-Study metric associated with the norm $\|\cdot\|_{k',\natural}$ on V', where $\natural = \varepsilon$ if $|\cdot|$ is non-Archimedean and $\natural = \pi$ if $|\cdot|$ is Archimedean. In particular, one has

$$\|\cdot\|_{\overline{V},\mathrm{FS},k'} = \|\cdot\|_{\overline{V}',\mathrm{FS}}. \tag{2.10}$$

Definition 2.2.15 Let $f : X \to \operatorname{Spec} k$ be a k-scheme and L be an invertible $\mathcal{O}_X$-module. Suppose given a finite-dimensional vector space V over k and a surjective $\mathcal{O}_X$-homomorphism $\beta : f^*(V) \to L$. Then the homomorphism β corresponds to a k-morphism of schemes $g : X \to \mathbb{P}(V)$ such that $g^*(\mathcal{O}_V(1))$ is canonically isomorphic to L. If V is equipped with a norm $\|\cdot\|$, then the Fubini-Study metric on $\mathcal{O}_V(1)$ induces by pull-back a continuous metric on L, called the *quotient metric* induced by the normed vector space $(V, \|\cdot\|)$ and the surjective homomorphism β.

Definition 2.2.16 Let ℓ be a section of L over a Zariski open set U, which trivialises the invertible sheaf L. The section ℓ yields the isomorphism $\iota : \mathcal{O}_U \to L|_U$ given by $a \mapsto a\ell$. We define $\boldsymbol{\ell} : f^*(V)|_U \to \mathcal{O}_U$ by $\iota^{-1} \circ \beta_U$, that is, the following diagram is commutative:

$$\begin{array}{ccc} f^*(V)|_U & \xrightarrow{\ \boldsymbol{\ell}\ } & \mathcal{O}_U \\ & {\scriptstyle \beta_U} \searrow & \downarrow {\scriptstyle \iota} \\ & & L|_U \end{array}$$

If $\{e_i\}_{i=1}^r$ is a basis of V, $\{e_i^\vee\}_{i=1}^r$ is the dual basis of $\{e_i\}_{i=1}^r$ and $\beta_U(e_i) = a_i\ell$ for $i \in \{1, \ldots, n\}$, then $\boldsymbol{\ell}$ is given by

$$\boldsymbol{\ell} = a_1 e_1^\vee + \cdots + a_r e_r^\vee.$$

For each $x \in U^{\mathrm{an}}$, the evaluation of $\boldsymbol{\ell}$ at x is denoted by $\boldsymbol{\ell}_x$, that is,

$$\boldsymbol{\ell}_x = a_1(x)e_1^\vee + \cdots + a_r(x)e_r^\vee \in \operatorname{Hom}_{\widehat{\kappa}(x)}(V_{\widehat{\kappa}(x)}, \widehat{\kappa}(x)).$$

Proposition 2.2.17 *Let $f : X \to \operatorname{Spec} k$ be a k-scheme and L be an invertible $\mathcal{O}_X$-module. Suppose given a finite-dimensional normed vector space $(V, \|\cdot\|)$ over k and a surjective $\mathcal{O}_X$-homomorphism $\beta : f^*(V) \to L$. Let φ be the quotient metric induced by $(V, \|\cdot\|)$ and β. For any section ℓ of L on a Zariski open subset U of X which trivialises L on U, one has*

$$\forall\, x \in U^{\mathrm{an}}, \quad |\ell|_\varphi(x) = \|\boldsymbol{\ell}_x\|^{-1}_{\widehat{\kappa}(x),\natural,*} = \|\boldsymbol{\ell}_x\|^{-1}_{*,\widehat{\kappa}(x),\varepsilon}, \tag{2.11}$$

where $\natural = \varepsilon$ if $|\cdot|$ is non-Archimedean and $\natural = \pi$ if $|\cdot|$ is Archimedean.

Proof For each $x \in U^{\mathrm{an}}$, one has the following commutative diagram:

$$\begin{array}{ccc} V_{\widehat{\kappa}(x)} & \xrightarrow{\ \boldsymbol{\ell}_x\ } & \widehat{\kappa}(x) \\ & {\scriptstyle \beta_x} \searrow & \downarrow {\scriptstyle \iota_x} \\ & & L \otimes_{\mathcal{O}_U} \widehat{\kappa}(x) \end{array}$$

By (1) in Lemma 1.1.15, the operator norm of β_x is 1. Moreover, the operator norm of $\boldsymbol{\ell}_x$ is $\|\boldsymbol{\ell}_x\|_{\widehat{\kappa}(x),\natural,*}$ and that of ι_x is $|\ell|_\varphi(x)$. As the operator norm of β_x is the product of the operator norms of $\boldsymbol{\ell}_x$ and ι_x by (2) in Proposition 1.1.15, we obtain $|\ell|_\varphi(x) = \|\boldsymbol{\ell}_x\|_{\widehat{\kappa}(x),\natural,*}$. The equality $\|\boldsymbol{\ell}_x\|^{-1}_{\widehat{\kappa}(x),\natural,*} = \|\boldsymbol{\ell}_x\|_{*,\widehat{\kappa}(x),\varepsilon}$ follows from Proposition 1.3.20 (1), (2) for the non-Archimedean and Archimedean cases, respectively. □

Remark 2.2.18 Let X be a quasi-projective k-scheme. The above construction shows that any ample invertible $\mathcal{O}_X$-module admits a continuous metric. By Proposition 2.2.2, we deduce that, more generally, any invertible $\mathcal{O}_X$-module admits a continuous metric.

Remark 2.2.19 Let $f : X \to \operatorname{Spec} k$ be a k-scheme, L be an invertible $\mathcal{O}_X$-module, V be a finite-dimensional vector space over k, and $\beta : f^*(V) \to L$ be a surjective $\mathcal{O}_X$-homomorphism. Let n be an integer, $n \geqslant 1$. Then β induces a surjective $\mathcal{O}_X$-homomorphism $\beta^{\otimes n} : f^*(V^{\otimes n}) \to L^{\otimes n}$. Let $\|\cdot\|$ be a norm on V and φ be the quotient metric induced by $(V, \|\cdot\|)$ and β. We claim that $n\varphi$ is the quotient metric induced by $(V^{\otimes n}, \|\cdot\|_\natural)$ and $\beta^{\otimes n}$, where $\|\cdot\|_\natural$ denotes the $\natural$-tensor power of the norm $\|\cdot\|$, $\natural = \varepsilon$ if $|\cdot|$ is non-Archimedean and $\natural = \pi$ if $|\cdot|$ is Archimedean. In fact, if x is an point in X^{an} and ℓ is a non-zero element of $L \otimes \widehat{\kappa}(x)$, then $\ell^{\otimes n}$ is a non-zero element of $L^{\otimes n} \otimes \widehat{\kappa}(x)$. By Propositions 1.3.21 and 1.3.5, the norm $\|\cdot\|_{\natural,\widehat{\kappa}(x),\natural}$ on $V^{\otimes n} \otimes \widehat{\kappa}(x)$ coincides with the $\natural$-tensor power of $\|\cdot\|_{\widehat{\kappa}(x),\natural}$. Consider the dual homomorphism

$$\beta_x^{\vee\otimes n} : (L^\vee \otimes \widehat{\kappa}(x))^{\otimes n} \longrightarrow (V^\vee \otimes \widehat{\kappa}(x))^{\otimes n}.$$

By Proposition 1.1.57 and Corollary 1.2.20, the dual norm $\|\cdot\|_{\natural,\widehat{\kappa}(x),\natural,*}$ coincides with the ε-tensor power of $\|\cdot\|_{\widehat{\kappa}(x),\natural,*} = \|\cdot\|_{*,\widehat{\kappa}(x),\varepsilon}$ (see Proposition 1.3.20 (1),(2)). In both cases one has

$$\|\beta_x^{\vee\otimes n}(\ell^{\vee\otimes n})\|_{\natural,\widehat{\kappa}(x),\natural,*} = \|\beta_x^\vee(\ell^\vee)\|^n_{\widehat{\kappa}(x),\natural,*} = |\ell^{\otimes n}|_{n\varphi}(x)^{-1},$$

where the first equality comes from Remark 1.1.56, and the second comes from Proposition 1.1.20.

Proposition 2.2.20 *Let $f : X \to \operatorname{Spec} k$ be a scheme over $\operatorname{Spec} k$ and L be an invertible $\mathcal{O}_X$-module. Let V be a finite-dimensional vector space and $\beta : f^*(V) \to L$ be a surjective homomorphism. If $\|\cdot\|$ and $\|\cdot\|'$ are two norms on V and if φ and φ' are quotient metrics on L induced by $\overline{V} = (V, \|\cdot\|)$ and $\overline{V}' = (V, \|\cdot\|')$ (and the surjective homomorphism β) respectively, then one has $d(\varphi, \varphi') \leqslant d(\|\cdot\|, \|\cdot\|')$.*

Proof Let x be a point of X^{an}, $|\cdot|_{\overline{V}}(x)$ and $|\cdot|_{\overline{V}'}(x)$ be the norms on $V \otimes \widehat{\kappa}(x)$ induced by $\|\cdot\|$ and $\|\cdot\|'$ by extension of scalars. Proposition 1.3.7 leads to

$$d(|\cdot|_{\overline{V}}(x), |\cdot|_{\overline{V}'}(x)) \leqslant d(\|\cdot\|, \|\cdot\|').$$

Since $|\cdot|_\varphi(x)$ and $|\cdot|_{\varphi'}(x)$ are respectively the quotient norms of $|\cdot|_{\overline{V}}(x)$ and $|\cdot|_{\overline{V}'}(x)$, by Proposition 1.1.42 one has $d(|\cdot|_\varphi(x), |\cdot|_{\varphi'}(x)) \leqslant d(|\cdot|_{\overline{V}}(x), |\cdot|_{\overline{V}'}(x))$. Therefore

$$d(\varphi, \varphi') = \sup_{x \in X^{\mathrm{an}}} d(|\cdot|_\varphi(x), |\cdot|_{\varphi'}(x)) \leqslant \sup_{x \in X^{\mathrm{an}}} d(|\cdot|_{\overline{V}}(x), |\cdot|_{\overline{V}'}(x)) \leqslant d(\|\cdot\|, \|\cdot\|'),$$

as required. □

Definition 2.2.21 Let $\pi : X \to \mathrm{Spec}(k)$ be a projective k-scheme, L be an invertible $\mathcal{O}_X$-module, which is generated by global sections, and φ be a continuous metric on L. By Proposition 2.1.16, for each $x \in X^{\mathrm{an}}$, the homomorphism $(H^0(X, L)/\mathcal{N}(X, L)) \otimes \widehat{\kappa}(x) \to L \otimes \widehat{\kappa}(x)$ induced by $H^0(X, L) \otimes \widehat{\kappa}(x) \to L \otimes \widehat{\kappa}(x)$ is surjective, so that one has a quotient norm $|\cdot|'(x)$ on $L \otimes \widehat{\kappa}(x)$ induced by $\|\cdot\|_{\varphi}^{\sim}$. The family $\{|\cdot|'(x)\}_{x \in X^{\mathrm{an}}}$ of metrics is denoted by φ_{FS}, that is, $|\cdot|_{\varphi_{\mathrm{FS}}}(x) := |\cdot|'(x)$ for $x \in X^{\mathrm{an}}$, called the *Fubini-Study metric associated with* φ. Let X_{red} be the reduced scheme associated with X and $L_{\mathrm{red}} := L \otimes_{\mathcal{O}_X} \mathcal{O}_{X_{\mathrm{red}}}$. Let V be the image of $H^0(X, L) \to H^0(X_{\mathrm{red}}, L_{\mathrm{red}})$. Then $\|\cdot\|_{\varphi}^{\sim}$ is a norm of V and φ_{FS} is the quotient metric induced by the surjection $V \otimes \mathcal{O}_{X_{\mathrm{red}}} \to L_{\mathrm{red}}$ and $\|\cdot\|_{\varphi}^{\sim}$. In particular, φ_{FS} is continuous. For an integer $n \geqslant 1$, we set $\varphi_n = (n\varphi)_{\mathrm{FS}}$.

By Propositions 1.3.26 and 1.3.25, for any point $x \in X^{\mathrm{an}}$ and any non-zero element $\ell \in L \otimes_{\mathcal{O}_X} \widehat{\kappa}(x)$, one has

$$|\ell|_{\varphi_{\mathrm{FS}}}(x) = \inf_{\substack{s \in H^0(X,L),\ \lambda \in \widehat{\kappa}(x)^{\times} \\ s(x) = \lambda \ell}} |\lambda|_x^{-1} \cdot \|s\|_{\varphi}. \tag{2.12}$$

This equality is fundamental in the study of quotient metrics.

Proposition 2.2.22 *Let $\pi : X \to \mathrm{Spec}\, k$ be a projective k-scheme and L be an invertible $\mathcal{O}_X$-module generated by global sections, equipped with a continuous metric φ. Then the following assertions hold.*

(1) *For any integer $n \in \mathbb{N}_{\geqslant 1}$, one has $\varphi_n \geqslant n\varphi$, where φ_n denotes the Fubini-Study metric associated with $n\varphi$.*
(2) *The sup seminorm $\|\cdot\|_{\varphi_n}$ on $V_n := H^0(X, L^{\otimes n})$ induced by φ_n coincides with $\|\cdot\|_{n\varphi}$.*
(3) *Let M be another invertible $\mathcal{O}_X$-module generated by global sections, equipped with a continuous metric ψ. Then $(\varphi + \psi)_{\mathrm{FS}} \leqslant \varphi_{\mathrm{FS}} + \psi_{\mathrm{FS}}$. In particular, for any pair (m, n) of positive integers one has $\varphi_{n+m} \leqslant \varphi_n + \varphi_m$.*
(4) *For any integer $n \geqslant 1$, one has $d(\varphi_n, n\varphi) \leqslant nd(\varphi_1, \varphi)$. In particular, if $\varphi_1 = \varphi$ then $\varphi_n = n\varphi$ for any $n \in \mathbb{N}_{\geqslant 1}$.*
(5) *Let φ' be another continuous metric on L. Then one has $d(\varphi_n, \varphi_n') \leqslant nd(\varphi, \varphi')$ for any $n \in \mathbb{N}_{\geqslant 1}$.*

Proof For any $n \in \mathbb{N}_{\geqslant 1}$, we denote by V_n the vector space $H^0(X, L^{\otimes n})$ over k.

(1) Let x be a point of X^{an} and ℓ be an element of $L^{\otimes n} \otimes \widehat{\kappa}(x)$. Note that, for any $s \in V_n$, one has $|s|_{n\varphi}(x) \leqslant \|s\|_{n\varphi}$, so that, by (2.12), one obtains

$$|\ell|_{\varphi_n}(x) = \inf_{\substack{s \in V_n,\ \lambda \in \widehat{\kappa}(x)^{\times}, \\ s(x) = \lambda \ell}} |\lambda|_x^{-1} \cdot \|s\|_{n\varphi} \geqslant \inf_{\substack{s \in V_n,\ \lambda \in \widehat{\kappa}(x)^{\times}, \\ s(x) = \lambda \ell}} |\lambda|_x^{-1} \cdot |s|_{n\varphi}(x) = |\ell|_{n\varphi}(x),$$

as desired.

(2) By (1), one has $\|\cdot\|_{\varphi_n} \geqslant \|\cdot\|_{n\varphi}$. In the following, we prove the converse inequality. If s is a global section of $L^{\otimes n}$, for any $x \in X^{\mathrm{an}}$, by (2.12), one has

$$|s|_{\varphi_n}(x) = \inf_{\substack{t\in V_n,\ \lambda\in\widehat{\kappa}(x)^\times \\ t(x)=\lambda s(x)}} |\lambda|_x^{-1}\cdot\|t\|_{n\varphi} \leqslant \|s\|_{n\varphi}.$$

Hence $\|s\|_{\varphi_n} = \sup_{x\in X^{\mathrm{an}}} |s|_{\varphi_n}(x) \leqslant \|s\|_{n\varphi}$.

(3) Let x be a point of X^{an}, ℓ and ℓ' be elements of $L\otimes\widehat{\kappa}(x)$ and $M\otimes\widehat{\kappa}(x)$ respectively. By (2.12) together with (2.3) one has

$$\begin{aligned}
|\ell\cdot\ell'|_{(\varphi+\psi)_{\mathrm{FS}}}(x) &= \inf_{\substack{s\in H^0(X,L\otimes M),\ \lambda\in\widehat{\kappa}(x)^\times \\ s(x)=\lambda\ell\cdot\ell'}} |\lambda|_x^{-1}\cdot\|s\|_{\varphi+\psi} \\
&\leqslant \inf_{\substack{(t,t')\in H^0(X,L)\times H^0(X,M) \\ (\mu,\eta)\in(\widehat{\kappa}(x)^\times)^2 \\ t(x)=\mu\ell,\ t'(x)=\eta\ell'}} |\mu\eta|_x^{-1}\cdot\|t\cdot t'\|_{\varphi+\psi} \\
&\leqslant \inf_{\substack{(t,t')\in H^0(X,L)\times H^0(X,M) \\ (\mu,\eta)\in(\widehat{\kappa}(x)^\times)^2 \\ t(x)=\mu\ell,\ t'(x)=\eta\ell'}} \Big(|\mu|_x^{-1}\cdot\|t\|_{\varphi}\Big)\Big(|\eta|_x^{-1}\cdot\|t'\|_{\psi}\Big) \\
&= |\ell|_{\varphi_{\mathrm{FS}}}(x)\cdot|\ell'|_{\psi_{\mathrm{FS}}}(x),
\end{aligned}$$

For the last assertion, note that

$$\varphi_{n+m} = (n\varphi+m\varphi)_{\mathrm{FS}} \leqslant (n\varphi)_{\mathrm{FS}} + (m\varphi)_{\mathrm{FS}} = \varphi_n + \varphi_m.$$

(4) By (3), one has $\varphi_n \leqslant n\varphi_1$. Moreover, by (1), one has $\varphi_n \geqslant n\varphi$ and $\varphi_1 \geqslant \varphi$. Hence $0 \leqslant \varphi_n - n\varphi \leqslant n\varphi_1 - n\varphi = n(\varphi_1-\varphi)$, which implies

$$d(\varphi_n, n\varphi) = \sup_{x\in X^{\mathrm{an}}} (\varphi_n - n\varphi)(x) \leqslant n \sup_{x\in X^{\mathrm{an}}} (\varphi_1-\varphi)(x) = nd(\varphi_1,\varphi).$$

(5) By Proposition 2.2.20 together with (2.5), one has

$$d(\varphi_n,\varphi_n') \leqslant d(\|\cdot\|_{n\varphi}^{\sim}, \|\cdot\|_{n\varphi'}^{\sim}) \leqslant d(n\varphi, n\varphi') = nd(\varphi,\varphi'),$$

where the equality comes from (2.6). □

Proposition 2.2.23 *Let $\pi : X \to \operatorname{Spec} k$ be a projective k-scheme and L be an invertible $\mathcal{O}_X$-module. Suppose given a normed vector space $(V,\|\cdot\|)$ and a surjective homomorphism $\beta : \pi^*(V) \to L$. Let φ be the quotient metric on L induced by $(V,\|\cdot\|)$ and β. Then, one has the following:*

(1) *Let $f : V \to H^0(X,L)$ be the adjoint homomorphism of $\beta : \pi^*(V) \to L$. Then, $\|f(v)\|_\varphi \leqslant \|v\|$ for any $v\in V$.*
(2) *For any integer $n \geqslant 1$, $\varphi_n = n\varphi$.*

Proof (1) By Propositions 1.3.26 and 1.3.25, for $x\in X^{\mathrm{an}}$,

$$|f(v)|_\varphi(x) = \inf_{\substack{t\in V,\ \lambda\in\widehat{\kappa}(x)^\times \\ f(t)(x)=\lambda f(v)(x)}} |\lambda|_x^{-1}\cdot\|t\| \leqslant \|v\|,$$

so that one has (1).

(2) By Proposition 2.2.22 (4), it suffices to verify that $\varphi_1 = \varphi$. Note that $\varphi_1 \geqslant \varphi$ (by Proposition 2.2.22 (1)). In the following, we prove the converse inequality. Let x be a point of X^{an} and ℓ be an element of $L \otimes \widehat{\kappa}(x)$. By (1) and (2.12) together with Propositions 1.3.26 and 1.3.25, one has

$$\begin{aligned}|\ell|_{\varphi_1}(x) &= \inf_{\substack{s \in H^0(X,L)\\ \lambda \in \widehat{\kappa}(x)^\times\\ s(x)=\lambda\ell}} |\lambda|_x^{-1} \cdot \|s\|_\varphi \leqslant \inf_{\substack{s' \in V,\, \lambda \in \widehat{\kappa}(x)^\times\\ f(s')(x)=\lambda\ell}} |\lambda|_x^{-1} \cdot \|f(s')\|_\varphi \\ &\leqslant \inf_{\substack{s' \in V,\, \lambda \in \widehat{\kappa}(x)^\times\\ f(s')(x)=\lambda\ell}} |\lambda|_x^{-1} \cdot \|s'\| = |\ell|_\varphi(x).\end{aligned}$$

Therefore one has $\varphi_1 = \varphi$. □

2.3 Semi-positive metrics

Let $(k, |\cdot|)$ be a complete valued field and $\pi : X \to \operatorname{Spec} k$ be a projective k-scheme. In this section, we discuss positivity conditions of continuous metrics on invertible $\mathcal{O}_X$-modules.

2.3.1 Definition and basic properties

Let L be an invertible $\mathcal{O}_X$-module equipped with a continuous metric φ. We assume that L is generated by global sections. We have constructed in Definition 2.2.21 a sequence of quotient metrics $\{\varphi_n\}_{n \in \mathbb{N}_{\geqslant 1}}$. By Proposition 2.2.22 (1) and (4), we obtain that $\{d(\varphi_n, n\varphi)\}_{n \in \mathbb{N}_{\geqslant 1}}$ is a sub-additive non-negative sequence and the normalised sequence $\{d(\varphi_n, n\varphi)/n\}_{n \in \mathbb{N}_{\geqslant 1}}$ is bounded from above. Hence the sequence $\{d(\varphi_n, n\varphi)/n\}_{n \in \mathbb{N}_{\geqslant 1}}$ converges in $\mathbb{R}_+$. We denote by $\mathrm{dp}(\varphi)$ the limit

$$\mathrm{dp}(\varphi) := \lim_{n \to +\infty} \frac{1}{n} d(\varphi_n, n\varphi), \tag{2.13}$$

called the *default of positivity* of the metric φ. By definition, for any integer $m \in \mathbb{N}_{\geqslant 1}$, one has

$$\mathrm{dp}(m\varphi) = m\,\mathrm{dp}(\varphi). \tag{2.14}$$

We say that the metric φ is *semipositive* if one has $\mathrm{dp}(\varphi) = 0$. Clearly, if the metric φ is semipositive, then for any integer $m \geqslant 1$, the metric $m\varphi$ on $L^{\otimes m}$ is also semipositive. Conversely, if there is an integer $m \geqslant 1$ such that $m\varphi$ is semipositive, then the metric φ is also semipositive.

More generally, we assume that L is semiample (namely a positive tensor power of L is generated by global sections). Let n be a positive integer such that $L^{\otimes n}$ is generated by global sections. The quantity $\mathrm{dp}(n\varphi)/n$ does not depend on the choice of n by (2.14), so that we define $\mathrm{dp}(\varphi)$ to be $\mathrm{dp}(n\varphi)/n$. It is easy to see that (2.14) still holds under the assumption that L is semiample. We say that φ is *semipositive* if $\mathrm{dp}(\varphi) = 0$.

Remark 2.3.1 Let $(V, \|\cdot\|)$ be a normed vector space of finite dimension over k and $\beta : \pi^*(V) \to L$ be a surjective homomorphism. Let φ be the quotient metric on L induced by $(V, \|\cdot\|)$ and β. Then, by Proposition 2.2.23, φ is semipositive.

Proposition 2.3.2 *Let L be a semiample invertible $\mathcal{O}_X$-module, equipped with a continuous metric φ. If φ is semipositive, then $n\varphi$ is semipositive for any $n \in \mathbb{N}_{\geqslant 1}$. Conversely, if there exists an integer $n \in \mathbb{N}_{\geqslant 1}$ such that $n\varphi$ is semipositive, then the metric φ is also semipositive.*

Proof This follows from (2.14), which also holds for a continuous metric on a semiample invertible $\mathcal{O}_X$-module. □

The following proposition shows that semipositive metrics form a closed subset in the topological space of continuous metrics.

Proposition 2.3.3 *Let L be a semiample invertible $\mathcal{O}_X$-module, equipped with a continuous metric φ. Suppose that there is a sequence of semipositive metrics $\{\varphi^{(m)}\}_{n\in\mathbb{N}}$ on L such that*

$$\lim_{m\to+\infty} d(\varphi^{(m)}, \varphi) = 0.$$

Then the metric φ is also semipositive.

Proof For any integer $p \geqslant 1$, one has

$$d(p\varphi^{(m)}, p\varphi) = p d(\varphi^{(m)}, \varphi).$$

Therefore, by replacing L by a certain tensor power $L^{\otimes p}$ and φ by $p\varphi$, we may assume without loss of generality that L is generated by global sections. Thus the metrics $\varphi_n^{(m)}$ and φ_n are well defined for any $m \in \mathbb{N}$ and any $n \in \mathbb{N}_{\geqslant 1}$. Moreover, by Proposition 2.2.22 (5) we obtain that

$$d(\varphi_n^{(m)}, \varphi_n) \leqslant n d(\varphi^{(m)}, \varphi).$$

Note that for $m \in \mathbb{N}$ and $n \in \mathbb{N}_{\geqslant 1}$ one has

$$\begin{aligned} d(\varphi_n, n\varphi) &\leqslant d(\varphi_n, \varphi_n^{(m)}) + d(\varphi_n^{(m)}, n\varphi^{(m)}) + d(n\varphi^{(m)}, n\varphi) \\ &\leqslant d(\varphi_n^{(m)}, n\varphi^{(m)}) + 2d(\varphi^{(m)}, \varphi). \end{aligned}$$

By taking the limit when n tends to the infinity, we obtain

$$\mathrm{dp}(\varphi) \leqslant 2d(\varphi^{(m)}, \varphi) + \mathrm{dp}(\varphi^{(m)}) = 2d(\varphi^{(m)}, \varphi),$$

where the equality comes from the hypothesis that the metrics $\varphi^{(m)}$ are semipositive. By taking the limit when m tends to the infinity, we obtain the semipositivity of the metric φ. □

Remark 2.3.4 Proposition 2.2.23 shows that quotient metrics on an invertible $\mathcal{O}_X$-module are semipositive. Let L be a semiample invertible $\mathcal{O}_X$-module and φ be a continuous metric on L. For any $n \in \mathbb{N}_{\geqslant 1}$ such that $L^{\otimes n}$ is generated by global sections, let $\varphi^{(n)}$ be a continuous metric on L such that $n\varphi^{(n)}$ is a quotient metric. If $\lim_{n \to +\infty} d(\varphi^{(n)}, \varphi) = 0$, then the metric φ is semipositive. This is a consequence of Propositions 2.3.2 and 2.3.3.

Proposition 2.3.5 *Let X be a projective k-scheme, L and L' be semiample invertible $\mathcal{O}_X$-modules, equipped with continuous metrics φ and φ', respectively. One has* $\mathrm{dp}(\varphi+\varphi') \leqslant \mathrm{dp}(\varphi)+\mathrm{dp}(\varphi')$. *In particular, if both metrics φ and φ' are semipositive, then the metric $\varphi + \varphi'$ on the tensor product $L \otimes L'$ is also semipositive.*

Proof By (2.14), we may assume that L and L' are generated by global sections. For any integer $n \geqslant 1$, one has a natural k-linear homomorphism

$$H^0(X, L^{\otimes n}) \otimes H^0(X, L'^{\otimes n}) \longrightarrow H^0(X, (L \otimes L')^{\otimes n})$$

given by the tensor product. Moreover, by Proposition 2.2.5, for $s \in H^0(X, L^{\otimes n})$ and $s' \in H^0(X, L'^{\otimes n})$ one has

$$\|ss'\|_{n(\varphi+\varphi')} \leqslant \|s\|_{n\varphi} \cdot \|s'\|_{n\varphi'}.$$

By Proposition 2.2.22 (3), we obtain $(\varphi + \varphi')_n \leqslant \varphi_n + \varphi'_n$ and hence

$$d((\varphi + \varphi')_n, n(\varphi + \varphi')) \leqslant d(\varphi_n, n\varphi) + d(\varphi'_n, n\varphi').$$

Dividing the two sides of the inequality by n, by passing to limit when n tends to the infinity we obtain $\mathrm{dp}(\varphi + \varphi') \leqslant \mathrm{dp}(\varphi) + \mathrm{dp}(\varphi')$. □

Proposition 2.3.6 *Let L be a semiample invertible $\mathcal{O}_X$-module, equipped with a continuous metric φ. Then the following are equivalent:*

(1) *The metric φ is semipositive.*
(2) *For any $\epsilon > 0$, there is a positive integer n such that, for all $x \in X^{\mathrm{an}}$, we can find $s \in H^0(X, L^{\otimes n})_{\widehat{\kappa}(x)} \setminus \{0\}$ with $\|s\|_{n\varphi,\widehat{\kappa}(x)} \leqslant \mathrm{e}^{n\epsilon}|s|_{n\varphi}(x)$.*

Proof (1) $\Longrightarrow$ (2): By our assumption, there is a positive integer n such that

$$|\cdot|_{n\varphi}(x) \leqslant |\cdot|_{\varphi_n}(x) \leqslant \mathrm{e}^{n\epsilon/2}|\cdot|_{n\varphi}(x)$$

for all $x \in X^{\mathrm{an}}$. Moreover, there is an $s \in H^0(X, L^{\otimes n})_{\widehat{\kappa}(x)} \setminus \{0\}$ such that $\|s\|_{n\varphi,\widehat{\kappa}(x)} \leqslant \mathrm{e}^{n\epsilon/2}|s|_{\varphi_n}(x)$. Therefore,

$$\|s\|_{n\varphi,\widehat{\kappa}(x)} \leqslant \mathrm{e}^{n\epsilon/2}|s|_{\varphi_n}(x) \leqslant \mathrm{e}^{n\epsilon}|s|_{n\varphi}(x).$$

(2) $\Longrightarrow$ (1): For a positive integer m, there is a positive integer a_m such that, for any $x \in X^{\mathrm{an}}$, we can find $s \in H^0(X, L^{\otimes a_m})_{\widehat{\kappa}(x)} \setminus \{0\}$ with $\|s\|_{a_m\varphi,\widehat{\kappa}(x)} \leqslant \mathrm{e}^{a_m/m}|s|_{a_m\varphi}(x)$. Note that

$$|s|_{a_m\varphi}(x) \leqslant |s|_{\varphi_{a_m}}(x) \leqslant e^{a_m/m}|s|_{a_m\varphi}(x),$$

which implies that

$$0 \leqslant \frac{1}{a_m}\left(\ln|\cdot|_{\varphi_{a_m}}(x) - \ln|\cdot|_{a_m\varphi}(x)\right) \leqslant \frac{1}{m}$$

for all $x \in X^{\mathrm{an}}$, so that $\mathrm{dp}(\varphi) = 0$. □

Theorem 2.3.7 *Let X be an irreducible and reduced projective scheme over* $\operatorname{Spec}\mathbb{C}$, *$L$ be a semiample invertible $\mathcal{O}_X$-module and φ be a continuous metric of L. Then the following are equivalent:*

(1) *The first Chern current $c_1(L, \varphi)$ is positive.*
(2) *For any positive number $\epsilon > 0$, there is a positive integer n such that, for all $x \in X$, we can find $s \in H^0(X, L^{\otimes n}) \setminus \{0\}$ with $\|s\|_{n\varphi} \leqslant e^{n\epsilon}|s|_{n\varphi}(x)$.*
(3) *The metric φ is semipositive.*

Proof The proof of "(1) $\Longrightarrow$ (2)" is very technical. For the proof, we refer to the papers [155] and [112, Theorem 0.2]. "(2) $\Longrightarrow$ (3)" is nothing more than Proposition 2.3.6. Here let us consider the following claim:

Claim 2.3.8 *Let M be an invertible $\mathcal{O}_X$-module, $\overline{V} = (V, \|\cdot\|)$ be a finite-dimensional normed vector space over $\mathbb{C}$ and $V \otimes_{\mathbb{C}} \mathcal{O}_X \to M$ be a surjective homomorphism. We assume that there is a basis $\{e_i\}_{i=1}^r$ of V such that*

$$\forall\, (a_1, \ldots, a_r) \in \mathbb{C}^r, \quad \|a_1e_1 + \cdots + a_re_r\| = \max\{|a_1|, \ldots, |a_r|\}.$$

Let ψ be the quotient metric of M induced by $\overline{V}$ and $V \otimes_{\mathbb{C}} \mathcal{O}_X \to M$. Then the first Chern current $c_1(M, \psi)$ is semipositive. □

Proof Let $\{e_i^\vee\}_{i=1}^r$ be the dual basis of V. Then it is easy to see that the dual norm $\|\cdot\|_*$ of $\|\cdot\|$ is given by

$$\forall a_1, \ldots, a_r \in \mathbb{C}, \quad \|a_1e_1^\vee + \cdots + a_re_r^\vee\|_* = |a_1| + \cdots + |a_r|.$$

For $v \in V$, the global section of M over X corresponding to v is denoted by $\tilde{v}$. Let s be a local basis of M over a Zariski open set U. We set $\tilde{e}_i = a_is$ for some holomorphic function a_i on U. Then, by Proposition 2.2.17, the function

$$x \longmapsto -\ln|s|_\psi(x) = \ln(|a_1|(x) + \cdots + |a_r|(x))$$

is plurisubharmonic on U^{an} because $\ln(|a_1|(\cdot) + \cdots + |a_r|(\cdot))$ is plurisubharmonic on U^{an}. □

Let us see that (3) $\Longrightarrow$ (1). Clearly we may assume that L is generated by global sections. For each $n \geqslant 1$, let $r_n := \dim_{\mathbb{C}} H^0(X, L^{\otimes n})$ and $\{e_{n,i}\}_{i=1}^{r_n}$ be an orthonormal basis of $H^0(X, L^{\otimes n})$ with respect to $\|\cdot\|_{n\varphi}$. If we set

$$\|a_1e_{n,1} + \cdots + a_{r_n}e_{n,r}\|_n' := \max\{|a_1|, \ldots, |a_{r_n}|\}$$

for $a_1, \ldots, a_{r_n} \in \mathbb{C}$, then $\|\cdot\|_n' \leqslant \|\cdot\|_{n\varphi} \leqslant r_n \|\cdot\|_n'$. Let ψ_n be the quotient metric of $L^{\otimes n}$ by $\|\cdot\|_n'$. Then $d(\varphi_n, \psi_n) \leqslant \ln(r_n)$ because $\psi_n \leqslant \varphi_n \leqslant r_n \psi_n$. Therefore, as

$$d(\tfrac{1}{n}\psi_n, \varphi) \leqslant d(\tfrac{1}{n}\psi_n, \tfrac{1}{n}\varphi_n) + d(\tfrac{1}{n}\varphi_n, \varphi) \leqslant \tfrac{1}{n}\ln(r_n) + d(\tfrac{1}{n}\varphi_n, \varphi),$$

one has $\lim_{n\to\infty} d(\frac{1}{n}\psi_n, \varphi) = 0$ by our assumption. This means that, for a local basis s of L over an open set U, the sequence $\{-\frac{1}{n}\ln|s^{\otimes n}|_{\psi_n}\}_{n=1}^{\infty}$ converges to $-\ln|s|_\varphi$ uniformly on any compact set in U. As $-\frac{1}{n}\ln|s^{\otimes n}|_{\psi_n}$ is plurisubharmonic by the above claim, $-\ln|s|_\varphi$ is also plurisubharmonic, as required. □

Corollary 2.3.9 *Let T be a reduced complex analytic space and $\|\cdot\|$ be a norm of $\mathbb{C}^n$. If $f_1, \ldots, f_n$ are holomorphic functions on T, then* $\log\|(f_1, \ldots, f_n)\|$ *is plurisubharmonic on T.*

Proof First of all, recall the following fact (cf. [99, Corollary 2.9.5]):

> *If u is a plurisubharmonic function on $\mathbb{C}^n$ and $f_1, \ldots, f_n$ are holomorphic functions on T, then $u(f_1, \ldots, f_n)$ is plurisubharmonic on T.*

Thus it is sufficient to see that $f(z_1, \ldots, z_n) := \log\|(z_1, \ldots, z_n)\|$ is plurisubharmonic on $\mathbb{C}^n \setminus \{(0, \ldots, 0)\}$. Let $\mathbb{C}^n \otimes_{\mathbb{C}} \mathcal{O}_{\mathbb{P}^{n-1}_{\mathbb{C}}} \to \mathcal{O}_{\mathbb{P}^{n-1}_{\mathbb{C}}}(1)$ be the surjective homomorphism given by $e_i \mapsto X_i$, where $\{e_i\}_{i=1}^n$ is the standard basis of $\mathbb{C}^n$ and $(X_1 : \cdots : X_n)$ is a homogeneous coordinate of $\mathbb{P}^{n-1}_{\mathbb{C}}$. Let us consider the dual norm $\|\cdot\|_*$ of $\|\cdot\|$ on $\mathbb{C}^n$, that is, we identify the dual space $(\mathbb{C}^n)^\vee$ with $\mathbb{C}^n$ in the natural way. Let φ be the quotient metric of $\mathcal{O}_{\mathbb{P}^{n-1}_{\mathbb{C}}}(1)$ induced by $\|\cdot\|_*$ and $\mathbb{C}^n \otimes_{\mathbb{C}} \mathcal{O}_{\mathbb{P}^{n-1}_{\mathbb{C}}} \to \mathcal{O}_{\mathbb{P}^{n-1}_{\mathbb{C}}}(1)$. Note that X_i gives a local basis of $\mathcal{O}_{\mathbb{P}^{n-1}_{\mathbb{C}}}(1)$ over $\{X_i \neq 0\}$ and $\mathbf{X}_i = \sum_{j=1}^n (X_j/X_i)e_j$ (see Definition 2.2.16), so that by Proposition 2.2.17 together with the fact $\|\cdot\|_{**} = \|\cdot\|$, one has $-\log|X_i|_\varphi = \log\|\mathbf{X}_i\|$ on $\{X_i \neq 0\}$. Therefore, by Theorem 2.3.7 together with the previous fact, the function

$$(z_1, \ldots, z_n) \longmapsto \log\|(\tfrac{z_1}{z_i}, \ldots, \tfrac{z_{i-1}}{z_i}, 1, \tfrac{z_{i+1}}{z_i}, \ldots, \tfrac{z_n}{z_i})\|$$

is plurisubharmonic on $\mathbb{C}^n \setminus \{z_i = 0\}$. Note that

$$f(z_1, \ldots, z_n) = \log|z_i| + \log\|(\tfrac{z_1}{z_i}, \ldots, \tfrac{z_{i-1}}{z_i}, 1, \tfrac{z_{i+1}}{z_i}, \ldots, \tfrac{z_n}{z_i})\|,$$

so that f is plurisubharmonic on $\mathbb{C}^n \setminus \{z_i = 0\}$ for all i, and hence f is plurisubharmonic on $\mathbb{C}^n \setminus \{(0, \ldots, 0)\}$. □

2.3.2 Model metrics

In this subsection, we assume that the absolute value $|\cdot|$ is non-Archimedean and *non-trivial*. We denote by $\mathfrak{o}_k$ the valuation ring of $(k, |\cdot|)$. Let $X \to \mathrm{Spec}(k)$ be a projective k-scheme and L be an invertible $\mathcal{O}_X$-module. By *model* of (X, L), we refer to a projective $\mathfrak{o}_k$-scheme $\mathscr{X}$ equipped with an invertible $\mathcal{O}_{\mathscr{X}}$-module $\mathscr{L}$ such that the generic fibre of $\mathscr{X}$ coincides with X and that the restriction of $\mathscr{L}$ to X coincides with L. If $\mathscr{X} \to \mathrm{Spec}(\mathfrak{o}_k)$ is of finite presentation (resp. flat), then the model $(\mathscr{X}, \mathscr{L})$ is called a *coherent model* (resp. *flat model*) of (X, L). If $(k, |\cdot|)$ is discrete, then any model is a coherent model because $\mathfrak{o}_k$ is noetherian. Note that if $(\mathscr{X}, \mathscr{L})$ is a coherent model, then $O_{\mathscr{X}}$ is coherent (cf. Remark 2.3.10, (1)). Let $(\mathscr{X}, \mathscr{L})$ be a model of (X, L) and $(O_{\mathscr{X}})_{\mathrm{tor}}$ be the torsion part of $O_{\mathscr{X}}$ as an $\mathfrak{o}_k$-module. Note that $(O_{\mathscr{X}})_{\mathrm{tor}}$ is an ideal of $O_{\mathscr{X}}$. If we set $\mathscr{X}' = \mathrm{Spec}(O_{\mathscr{X}}/(O_{\mathscr{X}})_{\mathrm{tor}})$ and $\mathscr{L}' = \mathscr{L}|_{\mathscr{X}'}$, then $(\mathscr{X}', \mathscr{L}')$ is a flat model of (X, L) (cf. Remark 2.3.10, (1)).

As in Definition 2.1.1, we denote by $j : X^{\mathrm{an}} \to X$ the specification map. Let x be a point in X^{an} and let $p_x : \mathrm{Spec}\,\widehat{\kappa}(x) \to X$ be the k-morphism of schemes defined by x, where $\widehat{\kappa}(x)$ is the completion of the residue field $\kappa(x)$ of $j(x)$ with respect to the absolute value $|\cdot|_x$. Let $(\mathscr{X}, \mathscr{L})$ be a model of (X, L). The composition of p_x with the inclusion morphism $X \to \mathscr{X}$ then defines a $\mathfrak{o}_k$-morphism from $\mathrm{Spec}\,\widehat{\kappa}(x)$ to $\mathscr{X}$. By definition $L \otimes \widehat{\kappa}(x)$ is the pull-back sheaf $p_x^*(L)$. By the valuative criterion of properness (see [74] Chapter II, Theorem 7.3.8), there exists a unique $\mathfrak{o}_k$-morphism $\mathscr{P}_x$ from $\mathrm{Spec}(\mathfrak{o}_x)$ to $\mathscr{X}$ which identifies with p_x on the generic fibre, where $\mathfrak{o}_x$ is the valuation ring of $\widehat{\kappa}(x)$. The image of the maximal ideal of $\mathfrak{o}_x$ by $\mathscr{P}_x$, denoted by $r_{\mathscr{X}}(x)$, is called the *reduction point* of x. Note that $r_{\mathscr{X}}(x)$ belongs to the special fibre of $\mathscr{X} \to \mathrm{Spec}(\mathfrak{o}_k)$. Furthermore $\mathscr{P}_x^*(\mathscr{L})$ is a lattice in $L \otimes \widehat{\kappa}(x)$ (see §1.1.7). We denote by $|\cdot|_{\mathscr{L}}(x)$ the norm on $L \otimes \widehat{\kappa}(x)$ defined by this lattice, namely

$$\forall\, \ell \in L \otimes \widehat{\kappa}(x), \quad |\ell|_{\mathscr{L}}(x) := \inf\{|a|_x \,:\, a \in \widehat{\kappa}(x)^{\times},\ a^{-1}\ell \in \mathscr{P}_x^*(\mathscr{L})\}.$$

The family of norms $\{|\cdot|_{\mathscr{L}}(x)\}_{x \in X^{\mathrm{an}}}$ forms a metric on L which we denote by $\varphi_{\mathscr{L}}$, called the metric *induced* by the model $(\mathscr{X}, \mathscr{L})$. If we consider $(\mathscr{X}', \mathscr{L}')$ as before, then one can easily see that $\varphi_{\mathscr{L}} = \varphi_{\mathscr{L}'}$

Let $(\mathscr{X}, \mathscr{L})$ be a model of (X, L). Let $H^0(\mathscr{X}, \mathscr{L})_{\mathrm{tor}}$ be the torsion part of $H^0(\mathscr{X}, \mathscr{L})$. We denote $H^0(\mathscr{X}, \mathscr{L})/H^0(\mathscr{X}, \mathscr{L})_{\mathrm{tor}}$ by $\widetilde{H}^0(\mathscr{X}, \mathscr{L})$. Note that $\widetilde{H}^0(\mathscr{X}, \mathscr{L})$ is an $\mathfrak{o}_k$-submodule of $H^0(X, L)$ such that $H^0(X, L) = \widetilde{H}^0(\mathscr{X}, \mathscr{L}) \otimes_{\mathfrak{o}_k} k$ (cf. [78, Chapter I, Proposition 9.3.2]).

Remark 2.3.10 (1) Recall that a valuation ring is a Prüfer domain, which is a generalisation of Dedekind domain (non-necessarily Noetherian). In particular, an $\mathfrak{o}_k$-module is flat if and only if it is torsion free (see [27] Chapter VII, §2, Exercices 12 and 14). Further, if an $\mathfrak{o}_k$-module is flat and finitely generated, then it is a free $\mathfrak{o}_k$-module of finite rank. Moreover, by [70, Theorem 7.3.3], $\mathfrak{o}_k$ is stably coherent, that is, any polynomial ring with finite variables is a coherent ring, so that any $\mathfrak{o}_k$-algebra of finite presentation is coherent (cf. [1, Proposition 1.4.2]).

(2) Let $\mathscr{A}$ be an $\mathfrak{o}_k$-algebra such that $\mathscr{A}$ is of finite presentation over $\mathfrak{o}_k$. Let $S = \mathfrak{o}_k \setminus \{0\}$ and $A = S^{-1}\mathscr{A}$. Note that A is a finitely generated k-algebra. Let I be an ideal of A. We choose $f_1, \ldots, f_r \in \mathscr{A}$ such that I is generated by $f_1, \ldots, f_r$ in A. Let $B = A/I$ and $\mathscr{B} = \mathscr{A}/(\mathscr{A} f_1 + \cdots + \mathscr{A} f_r)$. Then $B = S^{-1}\mathscr{B}$ and $\mathscr{B}$ is of finite presentation over $\mathfrak{o}_k$. In particular, $\mathscr{B}$ is a coherent ring. The scheme $\mathrm{Spec}(\mathscr{B})$ is called a *coherent extension* of $\mathrm{Spec}(B)$ over $\mathfrak{o}_k$. If we set $\mathscr{I} = I \cap \mathscr{A}$, then the quotient ring $\mathscr{B}' = \mathscr{A}/\mathscr{I}$ is a torsion-free $\mathfrak{o}_k$-module because one has

an injective ring homomorphism $\mathscr{B} \hookrightarrow B$, and hence it is flat. The scheme $\mathrm{Spec}(\mathscr{B}')$ is called the *Zariski closure of* $\mathrm{Spec}(B)$ *in* $\mathrm{Spec}(\mathscr{A})$.

(3) Let $(\mathscr{X}, \mathscr{L})$ be a coherent model of (X, L). Then, by [1, Corollaire 1.4.8], $H^0(\mathscr{X}, \mathscr{L})$ is a finitely generated $\mathfrak{o}_k$-module, so that $\widetilde{H}^0(\mathscr{X}, \mathscr{L})$ is a free $\mathfrak{o}_k$-module of finite rank (cf. (1)). Moreover, as $H^0(X, L) = \widetilde{H}^0(\mathscr{X}, \mathscr{L}) \otimes_{\mathfrak{o}_k} k$, $\widetilde{H}^0(\mathscr{X}, \mathscr{L})$ is a finitely generated lattice of $H^0(X, L)$ (see Definition 1.1.23).

Remark 2.3.11 Let $\mathscr{X}$ and $\mathscr{Y}$ be projective $\mathfrak{o}_k$-schemes and $f : \mathscr{Y} \to \mathscr{X}$ an $\mathfrak{o}_k$-morphisme. Let X and Y be the generic fibres of $\mathscr{X}$ and $\mathscr{Y}$ respectively, and $f_k : X \to Y$ be the morphism induced by f. Let $\mathscr{L}$ be an invertible sheaf on $\mathscr{X}$ and L be the restriction of $\mathscr{L}$ to the generic fibre X. Then the model $(\mathscr{X}, \mathscr{L})$ induces a metric $|\cdot|_{\mathscr{L}}$ on the invertible sheaf L. The couple $(\mathscr{Y}, f^*(\mathscr{L}))$ forms a model of $(Y, f_k^*(L))$. Note that the model metric $|\cdot|_{f^*(\mathscr{L})}$ on $f_k^*(L)$ coincides with the pull-back of the metric $|\cdot|_{\mathscr{L}}$ by f_k (see Definition 2.2.9).

Proposition 2.3.12 *Let L be an invertible O_X-module which is generated by global sections. Let $\varphi_{\mathscr{L}}$ be the metric induced by a model $(\mathscr{X}, \mathscr{L})$ of (X, L). Let φ be a continuous metric of L and $\mathscr{H} := \{s \in \widetilde{H}^0(\mathscr{X}, \mathscr{L}) : \|s\|_\varphi \leqslant 1\}$. Moreover, let $\mathcal{E}$ be a lattice of $H^0(X, L)$ such that $\mathcal{E} \subseteq \widetilde{H}^0(\mathscr{X}, \mathscr{L})$. Then one has the following:*

(1) *If $\mathscr{H} \otimes_{\mathfrak{o}_k} O_{\mathscr{X}} \to \mathscr{L}$ is surjective, then $\varphi \leqslant \varphi_{\mathscr{L}}$.*

(2) *If φ is the quotient metric on L induced by $\|\cdot\|_{\mathcal{E}}$ (see Definition 1.1.27 for the norm induced by a lattice), then $\varphi \geqslant \varphi_{\mathscr{L}}$.*

(3) *If φ is the quotient metric on L induced by $\|\cdot\|_{\mathcal{E}}$, and the natural homomorphism $\mathcal{E} \otimes_{\mathfrak{o}_k} O_{\mathscr{X}} \to \mathscr{L}$ is surjective, then $\varphi = \varphi_{\mathscr{L}}$.*

Proof For $x \in X^{\mathrm{an}}$, let $p_x : \mathrm{Spec}\,\widehat{\kappa}(x) \to X$ be the k-morphism of schemes defined by x, and $\mathscr{P}_x : \mathrm{Spec}\,\mathfrak{o}_x \to \mathscr{X}$ be the $\mathfrak{o}_k$-morphism extending p_x. Moreover, let $\pi_x : H^0(X, L) \otimes_k \widehat{\kappa}(x) \to p_x^*(L)$ be the natural homomorphism.

(1) By our assumption, $\mathscr{H} \otimes_{\mathfrak{o}_k} \mathfrak{o}_x \to \mathscr{P}_x^*(\mathscr{L})$ is surjective, so that, if ℓ lies in $\mathscr{P}_x^*(\mathscr{L})$, then there exist $s_1, \ldots, s_n$ in $\mathscr{H}$ and $a_1, \ldots, a_n$ in $\mathfrak{o}_x$ such that $\pi_x(a_1 s_1 + \cdots + a_n s_n) = \ell$. For any $i \in \{1, \ldots, n\}$, let $\ell_i = \pi_x(s_i)$. Then one has $\ell = a_1\ell_1 + \cdots + a_n\ell_n$. As $s_i \in \mathscr{H}$, one has $|\ell_i|_\varphi(x) \leqslant 1$ for any i, which leads to $|\ell|_\varphi(x) \leqslant 1$. By Proposition 1.1.24, one obtains $|\cdot|_\varphi(x) \leqslant |\cdot|_{\varphi_{\mathscr{L}}}(x)$. The assertion (1) is thus proved.

(2) Note that $p_x^*(L)$ is a quotient vector space of dimension 1 of $H^0(X, L) \otimes_k \widehat{\kappa}(x)$ and $|\cdot|_\varphi(x)$ is the quotient norm on $p_x^*(L)$ induced by $\|\cdot\|_{\mathscr{E}, \widehat{\kappa}(x)}$. By Proposition 1.3.26, for $\ell \in p_x^*(L) \setminus \{0\}$ one has

$$|\ell|_\varphi(x) = \inf_{\substack{s \in H^0(X,L),\ \lambda \in \widehat{\kappa}(x)^\times \\ \pi_x(s) = \lambda\ell}} |\lambda|^{-1} \|s\|_{\mathscr{E}}. \tag{2.15}$$

Let $s \in H^0(X, L)$ and $\lambda \in \widehat{\kappa}(x)^\times$ such that $\pi_x(s) = \lambda\ell$. By definition one has

$$\|s\|_{\mathscr{E}} = \inf\{|a| \ : \ a \in k^\times,\ a^{-1}s \in \mathscr{E}\}.$$

If a is an element in $k^\times$ such that $a^{-1}s \in \mathscr{E} \subseteq \widetilde{H}^0(\mathscr{X}, \mathscr{L})$, then $a^{-1}\lambda\ell \in \mathscr{P}_x^*(\mathscr{L})$ because $\mathscr{P}_x^*(\mathscr{L})$ contains the image of $H^0(\mathscr{X}, \mathscr{L}) \otimes_{\mathfrak{o}_k} \mathfrak{o}_x$ in $p_x^*(L)$ by π_x. Hence

$$|a^{-1}\lambda\ell|_{\varphi_{\mathscr{L}}}(x) = |a|^{-1}|\lambda| \cdot |\ell|_{\varphi_{\mathscr{L}}}(x) \leqslant 1,$$

which implies that $|\ell|_{\varphi_{\mathscr{L}}}(x) \leqslant |\lambda|^{-1}|a|$. Since a is arbitrary with $a^{-1}s \in \mathscr{E}$, one obtains $|\ell|_{\varphi_{\mathscr{L}}}(x) \leqslant |\lambda|^{-1}\|s\|_{\mathscr{E}}$, which leads to $|\ell|_{\varphi_{\mathscr{L}}}(x) \leqslant |\ell|_{\varphi}(x)$.

(3) By (2), it is sufficient to see $\varphi \leqslant \varphi_{\mathscr{L}}$. Note that for $s \in \mathcal{E}$, one has $\|s\|_{\varphi} \leqslant 1$, so that $\mathcal{E} \subseteq \mathscr{H}$. Thus, by (1), one obtains $\varphi \leqslant \varphi_{\mathscr{L}}$ because $\mathscr{H} \otimes_{\mathfrak{o}_k} \mathcal{O}_{\mathscr{X}} \to \mathscr{L}$ is surjective. □

Corollary 2.3.13 *Let E be a finite-dimensional vector space over k, $\mathcal{E}$ be a finitely generated lattice in E and $\|\cdot\|_{\mathcal{E}}$ be the norm on E induced by the lattice $\mathcal{E}$ (see Definition 1.1.27). Then the Fubini-Study metric (see Definition 2.2.13) on the invertible $\mathcal{O}_{\mathbb{P}(E)}$-module $\mathcal{O}_E(1)$ induced by $\|\cdot\|_{\mathcal{E}}$ coincides with the metric induced by the model $(\mathbb{P}(\mathcal{E}), \mathcal{O}_{\mathcal{E}}(1))$ of $(\mathbb{P}(E), \mathcal{O}_E(1))$.*

Remark 2.3.14 We assume that L is very ample. Let $\mathcal{E}$ be a finitely generated lattice in $H^0(X, L)$. Fix a free basis $e_0, \ldots, e_N$ of $\mathcal{E}$ as an $\mathfrak{o}_k$-module. Then $\mathbb{P}(H^0(X, L))$ and $\mathbb{P}(\mathcal{E})$ can be identified with $\mathbb{P}^N_k$ and $\mathbb{P}^N_{\mathfrak{o}_k}$, respectively. Let I be the defining homogeneous ideal of X in $\mathbb{P}^N_k$ and $f_1, \ldots, f_r$ be homogeneous polynomials in $\mathfrak{o}_k[X_0, \ldots, X_N]$ such that I is generated by $f_1, \ldots, f_r$ in $k[X_0, \ldots, X_N]$, where $X_0, \ldots, X_N$ are the indeterminates corresponding to the basis $e_0, \ldots, e_N$. Let

$$\mathscr{X} = \mathrm{Proj}(\mathfrak{o}_k[X_0, \ldots, X_N]/(f_1, \ldots, f_r)) \quad \text{and} \quad \mathscr{L} = \mathcal{O}_{\mathbb{P}^N_{\mathfrak{o}_k}}(1)\Big|_{\mathscr{X}}.$$

Then $\mathscr{X}$ is called a *coherent extension* of X and $(\mathscr{X}, \mathscr{L})$ is a coherent model of (X, L). Moreover, Proposition 2.3.12 shows that the quotient metric on L induced by the norm $\|\cdot\|_{\mathcal{E}}$ coincides with the metric induced by the model $(\mathscr{X}, \mathscr{L})$.

Proposition 2.3.15 *Let L and M are two invertible $\mathcal{O}_X$-modules. Suppose that $\mathscr{X}$ is a projective $\mathfrak{o}_k$-scheme such that $\mathscr{X}_k = X$. If $\mathscr{L}$ and $\mathscr{M}$ are invertible $\mathcal{O}_{\mathscr{X}}$-modules such that $(\mathscr{X}, \mathscr{L})$ and $(\mathscr{X}, \mathscr{M})$ are models of (X, L) and (X, M) respectively, then one has $\varphi_{\mathscr{L}\otimes\mathscr{M}} = \varphi_{\mathscr{L}} + \varphi_{\mathscr{M}}$.*

Proof Let x be a point of X^{an}, s and t be element in $\mathscr{L}_{r_{\mathscr{X}}(x)}$ and $\mathscr{M}_{r_{\mathscr{X}}(x)}$ which trivialise the invertible sheaves $\mathscr{L}$ and $\mathscr{M}$ around $r_{\mathscr{X}}(x)$ respectively, and $\ell = \mathscr{P}^*_x(s)$, $m = \mathscr{P}^*_x(t)$, where $\mathscr{P}_x : \mathrm{Spec}(\mathfrak{o}_x) \to \mathscr{X}$ is the unique $\mathfrak{o}_k$-morphism extending the k-morphism $\mathrm{Spec}\,\widehat{\kappa}(x) \to X$ corresponding to the point x. Since ℓ and m are generators of the free $\mathfrak{o}_x$-modules (of rank 1) $\mathscr{P}^*_x(\mathscr{L})$ and $\mathscr{P}^*_x(\mathscr{M})$, one has $|\ell|_{\mathscr{L}}(x) = |m|_{\mathscr{M}}(x) = 1$. Moreover, $\mathscr{P}^*_x(s \otimes t)$ is a generator of the free $\mathfrak{o}_x$-module $\mathscr{P}^*_x(\mathscr{L} \otimes \mathscr{M})$, we obtain that $|\ell \otimes m|_{\mathscr{L}\otimes\mathscr{M}}(x) = 1$. Hence one has $\varphi_{\mathscr{L}\otimes\mathscr{M}} = \varphi_{\mathscr{L}} + \varphi_{\mathscr{M}}$. □

Proposition 2.3.16 *Let X be a projective k-scheme and L be an invertible $\mathcal{O}_X$-module. Let $(\mathscr{X}, \mathscr{L})$ be a model of (X, L).*

(1) *The metric $\varphi_{\mathscr{L}}$ on L is continuous.*
(2) *If $(\mathscr{X}, \mathscr{L})$ is coherent, then the $\mathfrak{o}_k$-module $\widetilde{H}^0(\mathscr{X}, \mathscr{L})$ is contained in the unit ball of $H^0(X, L)$ with respect to the seminorm $\|\cdot\|_{\varphi_{\mathscr{L}}}$.*

(3) *If $(\mathscr{X}, \mathscr{L})$ is flat, the absolute value $|\cdot|$ is discrete, X is reduced and the central fibre of $\mathscr{X} \to \mathrm{Spec}(\mathfrak{o}_k)$ is reduced, then $H^0(\mathscr{X}, \mathscr{L})$ coincides with the unit ball of $H^0(X, L)$ with respect to $\|\cdot\|_{\varphi_\mathscr{L}}$ and $\|\cdot\|_{\varphi_\mathscr{L}} = \|\cdot\|_{H^0(\mathscr{X},\mathscr{L})}$.*

Proof (1) We choose an ample invertible $O_\mathscr{X}$-module $\mathscr{A}$ such that $\mathscr{L} \otimes \mathscr{A}$ and $\mathscr{A}$ are generated by global sections (cf. [74, Corollarie 4.5.11]). Let $\mathscr{E}$ and $\mathscr{F}$ be finitely generated lattices of $H^0(X, L \otimes A)$ and $H^0(X, A)$, respectively, such that $\mathscr{E} \subseteq H^0(\mathscr{X}, \mathscr{L} \otimes \mathscr{A})$, $\mathscr{F} \subseteq H^0(\mathscr{X}, \mathscr{A})$ and $\mathscr{E} \otimes O_\mathscr{X} \to \mathscr{L} \otimes \mathscr{A}$ and $\mathscr{F} \otimes O_\mathscr{X} \to \mathscr{A}$ are surjective, where $A = \mathscr{A}|_X$. Let φ_1 and φ_2 be the quotient metrics of $L \otimes A$ and A induced by $\|\cdot\|_\mathscr{E}$ and $\|\cdot\|_\mathscr{F}$, respectively. Then φ_1 and φ_2 are continuous by Proposition 2.2.12 and Definition 2.2.15. Moreover, by Proposition 2.3.12, $\varphi_{\mathscr{L}\otimes\mathscr{A}} = \varphi_1$ and $\varphi_\mathscr{A} = \varphi_2$. Therefore one has the assertion because $\varphi_\mathscr{L} = \varphi_{\mathscr{L}\otimes\mathscr{A}} - \varphi_\mathscr{A} = \varphi_1 - \varphi_2$ by Proposition 2.3.15.

(2) Let s be a section in $H^0(\mathscr{X}, \mathscr{L})$, viewed as an element in $H^0(X, L)$, by definition one has

$$\forall\, x \in X^{\mathrm{an}}, \quad \|s\|_{\varphi_\mathscr{L}}(x) \leqslant 1.$$

Hence $H^0(\mathscr{X}, \mathscr{L})$ is contained in the closed unit ball of $(H^0(X, L), \|\cdot\|_{\varphi_\mathscr{L}})$.

(3) Let $\mathscr{E}$ be the unit ball of $H^0(X, L)$ with respect to $\|\cdot\|_{\varphi_\mathscr{L}}$. First let us see that

$$H^0(\mathscr{X}, \mathscr{L}) = \mathscr{E}. \tag{2.16}$$

By (2), $H^0(\mathscr{X}, \mathscr{L}) \subseteq \mathscr{E}$. Let $\mathscr{X} = \bigcup_{i=1}^N \mathrm{Spec}(\mathscr{A}_i)$ be an affine open covering of $\mathscr{X}$ such that $\mathscr{A}_i$ is of finite type over $\mathfrak{o}_k$ and $\mathscr{L}$ has a local basis ℓ_i over $\mathrm{Spec}(\mathscr{A}_i)$. For $s \in \mathscr{E}$, we set $s = f_i \ell_i$ and $f_i \in A_i := S^{-1}\mathscr{A}_i$, where $S = \mathfrak{o}_k \setminus \{0\}$. Then, for $x \in (\mathrm{Spec}\, A_i)^{\mathrm{an}}_{\mathscr{A}_i}$ (cf. Remark 2.1.6), $|s|_{\varphi_\mathscr{L}}(x) = |f_i|_x \leqslant 1$. Therefore, as the central fibre $\mathscr{X}_\circ$ of $\mathscr{X} \to \mathrm{Spec}(\mathfrak{o}_k)$ is reduced, by the last assertion of Proposition 2.1.7, one has $f_i \in \mathscr{A}_i$, and hence $s \in H^0(\mathscr{X}, \mathscr{L})$.

Next we need to see that

$$\|\cdot\|_{\varphi_\mathscr{L}} = \|\cdot\|_\mathscr{E}. \tag{2.17}$$

Let ϖ be a uniformising parameter of $\mathfrak{o}_k$, $\mathscr{X}_\circ$ be the fibre of $\mathscr{X}$ over the maximal ideal of $\mathfrak{o}_k$ and $\mathscr{L}_\circ$ be the restriction of $\mathscr{L}$ to $\mathscr{X}_\circ$, that is, $\mathscr{L}_\circ = \mathscr{L}/\varpi\mathscr{L}$. The short exact sequence

$$0 \longrightarrow \mathscr{L} \xrightarrow{\varpi\cdot} \mathscr{L} \longrightarrow \mathscr{L}_\circ \longrightarrow 0$$

gives rise to an exact sequence:

$$0 \longrightarrow H^0(\mathscr{X}, \mathscr{L}) \xrightarrow{\varpi\cdot} H^0(\mathscr{X}, \mathscr{L}) \longrightarrow H^0(\mathscr{X}_\circ, \mathscr{L}_\circ),$$

that is, the natural homomorphism

$$H^0(\mathscr{X}, \mathscr{L})/\varpi H^0(\mathscr{X}, \mathscr{L}) \longrightarrow H^0(\mathscr{X}_\circ, \mathscr{L}_\circ) \tag{2.18}$$

is injective. Moreover, by Proposition 1.1.30, one has

$$\|\cdot\|_{\varphi_\mathscr{L}} \leqslant \|\cdot\|_\mathscr{E}. \tag{2.19}$$

Here we claim that if $\|s\|_{\mathscr{E}} = 1$ for $s \in H^0(X, L)$, then $\|s\|_{\varphi_{\mathscr{L}}} = 1$. Obviously $\|s\|_{\varphi_{\mathscr{L}}} \leqslant 1$ by (2.19). As $H^0(\mathscr{X}, \mathscr{L}) = \mathscr{E}$ by (2.16), one has $s \in H^0(\mathscr{X}, \mathscr{L})$ and s is not zero in $H^0(\mathscr{X}, \mathscr{L})/\varpi H^0(\mathscr{X}, \mathscr{L})$, so that by the injectivity of (2.18), s is not zero in $H^0(\mathscr{X}_\circ, \mathscr{L}_\circ)$. Let ξ be a closed point $\mathscr{X}_\circ$ with $s(\xi) \neq 0$. Let ℓ_ξ be a local basis of $\mathscr{L}$ around ξ. Then $s = f\ell_\xi$ and $f \in \mathcal{O}_{\mathscr{X},\xi}^\times$. On the other hand, since the reduction map $r : X^{\mathrm{an}} \to \mathscr{X}_\circ$ is surjective, one can find $x \in X^{\mathrm{an}}$ with $r(x) = \xi$. Then $|s|_{\varphi_{\mathscr{L}}}(x) = |f|_x = 1$, so that $\|s\|_{\varphi_{\mathscr{L}}} = 1$, as desired.

In general, for $s \in H^0(X, L) \setminus \{0\}$, there is an integer e such that $\|\varpi^e s\|_{\mathscr{E}} = 1$, so that $\|\varpi^e s\|_{\varphi_{\mathscr{L}}} = 1$, and hence $\|s\|_{\mathscr{E}} = \|s\|_{\varphi_{\mathscr{L}}} = |\varpi|^{-e}$. □

Proposition 2.3.17 *Let X be a projective k-scheme, L be an ample invertible $\mathcal{O}_X$-module and $(\mathscr{X}, \mathscr{L})$ be a model of (X, L). Assume that there exists an invertible $\mathcal{O}_{\mathscr{X}}$-module $\mathscr{M}$ such that $\mathscr{L}^{\otimes n} \otimes \mathscr{M}$ is ample for any integer $n \geqslant 1$. Then the metric $\varphi_{\mathscr{L}}$ is semipositive.*

Proof Let $\pi : \mathscr{X} \to \operatorname{Spec} \mathfrak{o}_k$ be the structural morphism. First we assume that $\mathscr{L}$ is ample. We choose a positive integer n such that $\mathscr{L}^{\otimes n}$ is very ample. Then we have a closed embedding $\iota : \mathscr{X} \to \mathbb{P}(\mathcal{E}_n)$ with $\mathcal{E}_n := \widetilde{H}^0(\mathscr{X}, \mathscr{L}^{\otimes n})$, which is induced by the canonical (surjective) homomorphism $\pi^*(\pi_*(\mathscr{L}^{\otimes n})) = \pi^*(\mathcal{E}_n) \to \mathscr{L}^{\otimes n}$. Note that one has $\mathscr{L}^{\otimes n} = \iota^*(\mathcal{O}_{\mathcal{E}_n}(1))$. Moreover, $\mathcal{E}_n$ is a lattice in $E_n := H^0(X, L^{\otimes n})$. By Proposition 2.3.12, the metric $\varphi_{\mathscr{L}^{\otimes n}} = n\varphi_{\mathscr{L}}$ (see Proposition 2.3.15 for this equality) coincides with the quotient metric on $L^{\otimes n}$ induced by the norm $\|\cdot\|_{\mathcal{E}_n}$, hence is semipositive (see Proposition 2.2.23). By Proposition 2.3.2, we obtain that the metric $\varphi_{\mathscr{L}}$ is also semipositive.

Let us see a general case. Let M be the restriction of $\mathscr{M}$ to X. Since L is ample, for a sufficiently positive integer $n_0 \geqslant 1$, the invertible $\mathcal{O}_X$-module $L^{\otimes n_0} \otimes M^\vee$ is ample. Thus for any integer $n > n_0$ one has (by Proposition 2.3.5)

$$n\,\mathrm{dp}(\varphi_{\mathscr{L}}) = \mathrm{dp}(\varphi_{\mathscr{L}^{\otimes n}}) \leqslant \mathrm{dp}(\varphi_{\mathscr{L}^{\otimes n_0} \otimes \mathscr{M}^\vee}) + \mathrm{dp}(\varphi_{\mathscr{L}^{\otimes (n-n_0)} \otimes \mathscr{M}}) = \mathrm{dp}(\varphi_{\mathscr{L}^{\otimes n_0} \otimes \mathscr{M}^\vee}),$$

where the first equality comes from (2.14), and the second equality comes from the previous observation (i.e. the semi-positivity of the metric $\varphi_{\mathscr{L}^{\otimes (n-n_0)} \otimes \mathscr{M}}$). Since $n \geqslant n_0$ is arbitrary, we obtain that $\mathrm{dp}(\varphi_{\mathscr{L}}) = 0$, namely $\varphi_{\mathscr{L}}$ is a semipositive metric. □

Remark 2.3.18 Note that if one can find an invertible $\mathcal{O}_{\mathscr{X}}$-module $\mathscr{M}$ such that $\mathscr{L}^{\otimes n} \otimes \mathscr{M}$ is ample for any integer $n \geqslant 1$, then it is easy to see that $\mathscr{L}$ is nef. Under the assumption that $\pi : \mathscr{X} \to \operatorname{Spec} \mathfrak{o}_k$ is of finite presentation (for example, the absolute value $|\cdot|$ is discrete), the converse holds by [76, Corollaire 9.6.4].

Proposition 2.3.19 *Let $\pi : X \to$ Spec k be a projective scheme over* Spec k *and L be an ample invertible $\mathcal{O}_X$-module, equipped with a continuous metric φ which is semipositive. Then, for a sufficiently positive integer n_0, there exist sequences $\{(\mathscr{X}_n, \mathscr{L}_n)\}_{n \in \mathbb{N},\, n \geqslant n_0}$ and $\{\mathcal{E}_n\}_{n \in \mathbb{N},\, n \geqslant n_0}$ starting from n_0 such that, for each $n \geqslant n_0$, $L^{\otimes n}$ is very ample, $\mathcal{E}_n$ is a finitely generated lattice of $H^0(X, L^{\otimes n})$, $(\mathscr{X}_n, \mathscr{L}_n)$ is a coherent model of $(X, L^{\otimes n})$ as constructed in Remark 2.3.14 (note that $\mathscr{L}_n$ is very ample), and that*

$$\lim_{n \to +\infty} \frac{1}{n} d(\varphi_{\mathscr{L}_n}, n\varphi) = 0. \tag{2.20}$$

Proof Let $\lambda \in]0, 1[$ be a number such that

$$\lambda < \sup\{|a| \;:\; a \in k^{\times},\ |a| < 1\}.$$

For any $n \in \mathbb{N}_{\geqslant 1}$, let $V_n = H^0(X, L^{\otimes n})$. Since L is ample, one can find a sufficiently positive integer n_0 such that, for $n \geqslant n_0$, the canonical homomorphism $\pi^*(V_n) \to L^{\otimes n}$ is surjective and the corresponding k-morphism $X \to \mathbb{P}(V_n)$ is a closed embedding. By Proposition 1.2.22, there exists a lattice of finite type $\mathcal{E}_n$ of V_n such that

$$d(\|\cdot\|_{\mathcal{E}_n}, \|\cdot\|_{n\varphi}) \leqslant \ln(\lambda^{-1}).$$

Let $\mathscr{X}_n$ be a coherent extension of X in $\mathbb{P}(\mathcal{E}_n)$ (see Remark 2.3.14) and $\mathscr{L}_n$ be the restriction of $\mathcal{O}_{\mathbb{P}(\mathcal{E}_n)}(1)$ to $\mathscr{X}_n$. Moreover, the metric on $L^{\otimes n}$ induced by $\mathscr{L}_n$ coincides with the quotient metric on $L^{\otimes n}$ induced by $(V_n, \|\cdot\|_{\mathcal{E}_n})$ and the canonical quotient homomorphism $\pi^*(V_n) \to L^{\otimes n}$ (see Remark 2.3.14). Therefore by Proposition 2.2.20 one has

$$d(\varphi_{\mathscr{L}_n}, n\varphi) \leqslant d(\|\cdot\|_{\mathcal{E}_n}, \|\cdot\|_{n\varphi}) \leqslant \ln(\lambda^{-1})$$

as required. □

2.3.3 Purity

Let $X \to \operatorname{Spec} k$ be a projective k-scheme, L be an invertible $\mathcal{O}_X$-module and φ be a continuous metric on X. If the norm $\|\cdot\|_{\varphi}$ on $H^0(X, L)$ is pure, we say that the metric φ is *pure*. If $n\varphi$ is pure for all $n \in \mathbb{N}_{\geqslant 1}$, we say that φ is *stably pure*. Note that, if the absolute value $|\cdot|$ is not discrete, then any continuous metric on L is stably pure (cf. Proposition 1.1.32).

Proposition 2.3.20 *We assume that the absolute value $|\cdot|$ is discrete. Let X be a projective and reduced k-scheme and L be an invertible $\mathcal{O}_X$-module. If $(\mathscr{X}, \mathscr{L})$ is a model of (X, L) such that the central fibre of $\mathscr{X} \to \operatorname{Spec} \mathfrak{o}_k$ is reduced, then the metric $\varphi_{\mathscr{L}}$ is stably pure.*

Proof Note that for any $n \in \mathbb{N}_{\geqslant 1}$ one has $n\varphi_{\mathscr{L}} = \varphi_{\mathscr{L}^{\otimes n}}$ and $(\mathscr{X}, \mathscr{L}^{\otimes n})$ is a model of $(X, L^{\otimes n})$. Therefore it suffices to show that $\varphi_{\mathscr{L}}$ is a pure metric. By Proposition 2.3.16, the norm $\|\cdot\|_{\varphi_{\mathscr{L}}}$ is induced by the lattice $H^0(\mathscr{X}, \mathscr{L})$, hence it is pure. □

2.3.4 Extension property

In this subsection, we introduce the extension property of an ample invertible module with a semipositive continuous metric, that is, an extension of a section with a control on the norm.

Throughout this subsection, let $\pi : X \to \mathrm{Spec}(k)$ be a projective k-scheme and L be an invertible $\mathcal{O}_X$-module, equipped with a continuous metric φ. Let us begin with the following lemma.

Lemma 2.3.21 *Let Y be a closed subscheme of X. For $n \in \mathbb{N}$, $n \geqslant 1$, let $\gamma_n : H^0(X, L^{\otimes n}) \to H^0(Y, L^{\otimes n}|_Y)$ be the restriction map. For any element ℓ of $H^0(Y, L|_Y) \setminus \mathcal{N}(Y, L|_Y)$, we define $a_{\varphi,n}(\ell) \in [0, \infty]$ to be*

$$a_{\varphi,n}(\ell) := \begin{cases} \infty & \text{if } \gamma_n^{-1}(\{\ell^{\otimes n}\}) = \varnothing, \\ \inf\limits_{\substack{s \in H^0(X, L^{\otimes n}) \\ s|_Y = \ell^{\otimes n}}} \left(\ln \|s\|_{n\varphi} - \ln \|\ell\|^n_{\varphi|_Y} \right) & \text{otherwise,} \end{cases}$$

where $\varphi|_Y$ denotes the restriction *of φ to $L|_Y$, defined as the pull-back of φ by the inclusion morphisme $Y \to X$ (see Definition 2.2.9). Then we have the following:*

(1) *The sequence $\{a_{\varphi,n}(\ell)\}_{n \in \mathbb{N}}$ is subadditive, that is, $a_{\varphi,n+n'}(\ell) \leqslant a_{\varphi,n}(\ell) + a_{\varphi,n'}(\ell)$ for $n, n' \in \mathbb{N}$.*
(2) *Let φ' be another continuous metric of L. If $\gamma_n^{-1}(\{\ell^{\otimes n}\}) \neq \varnothing$, then*

$$|a_{\varphi,n}(\ell) - a_{\varphi',n}(\ell)| \leqslant 2n\, d(\varphi, \varphi').$$

Proof (1) Clearly we may assume that $\gamma_n^{-1}(\{\ell^{\otimes n}\}) \neq \varnothing$ and $\gamma_{n'}^{-1}(\{\ell^{\otimes n'}\}) \neq \varnothing$. Then $\gamma_{n+n'}^{-1}(\{\ell^{\otimes n+n'}\}) \neq \varnothing$, so that

$$\begin{aligned} a_{\varphi,n+n'}(\ell) &= \inf_{\substack{s'' \in H^0(L^{\otimes n+n'}) \\ s''|_Y = \ell^{\otimes n+n'}}} \left(\ln \|s''\|_{(n+n')\varphi} - \ln \|\ell\|^{n+n'}_{\varphi|_Y} \right) \\ &\leqslant \inf_{\substack{(s,s') \in H^0(L^{\otimes n}) \times H^0(L^{\otimes n'}) \\ s|_Y = \ell^{\otimes n}, s'|_Y = \ell^{\otimes n'}}} \left(\ln \|s \otimes s'\|_{(n+n')\varphi} - \ln \|\ell\|^{n+n'}_{\varphi|_Y} \right) \\ &\leqslant \inf_{\substack{(s,s') \in H^0(L^{\otimes n}) \times H^0(L^{\otimes n'}) \\ s|_Y = \ell^{\otimes n}, s'|_Y = \ell^{\otimes n'}}} \left(\ln \|s\|_{n\varphi} + \ln \|s'\|_{n'\varphi} - \ln \|\ell\|^{n+n'}_{\varphi|_Y} \right) \\ &= a_{\varphi,n}(\ell) + a_{\varphi,n'}(\ell), \end{aligned}$$

as required.

(2) Clearly we may assume that $a_{\varphi,n}(\ell) \geqslant a_{\varphi',n}(\ell)$. For any $\epsilon > 0$, choose $s \in H^0(X, L^{\otimes n})$ such that $s|_Y = \ell^{\otimes n}$ and

$$\ln \|s\|_{n\varphi'} - \ln \|\ell\|^n_{\varphi'|_Y} \leqslant a_{\varphi',n}(\ell) + \epsilon.$$

Then, by using (2.5) and (2.6),

$$\begin{aligned} a_{\varphi,n}(\ell) - a_{\varphi',n}(\ell) &\leqslant (\ln \|s\|_{n\varphi} - \ln \|\ell\|^n_{\varphi|_Y}) - (\ln \|s\|_{n\varphi'} - \ln \|\ell\|^n_{\varphi'|_Y}) + \epsilon \\ &\leqslant \left| \ln \|s\|_{n\varphi} - \ln \|s\|_{n\varphi'} \right| + \left| \ln \|\ell\|^n_{\varphi|_Y} - \ln \|\ell\|^n_{\varphi'|_Y} \right| + \epsilon \end{aligned}$$

$$\leqslant d(\|\cdot\|_{n\varphi}, \|\cdot\|_{n\varphi'}) + nd(\|\cdot\|_{\varphi|_Y}, \|\cdot\|_{\varphi'|_Y}) + \epsilon$$
$$\leqslant d(n\varphi, n\varphi') + nd(\varphi|_Y, \varphi'|_Y) + \epsilon \leqslant 2n\, d(\varphi, \varphi') + \epsilon,$$

so that the assertion follows because ϵ is an arbitrary positive number. □

Definition 2.3.22 Let Y be a closed subscheme of X. For $\ell \in H^0(Y, L|_Y)$, we say that ℓ has the *extension property* for the metric φ if, for any $\epsilon > 0$, there exists $n_0 \in \mathbb{N}$, $n_0 \geqslant 1$, such that for any integer n, $n \geqslant n_0$, there exists a section $s \in H^0(X, L^{\otimes n})$ satisfying

$$s|_Y = \ell^{\otimes n} \quad \text{and} \quad \|s\|_{n\varphi} \leqslant \mathrm{e}^{\epsilon n} \|\ell\|^n_{\varphi|_Y}. \tag{2.21}$$

If $\ell \in \mathcal{N}(Y, L|_Y)$, then ℓ has the extension property for the metric φ. Indeed, by Proposition 2.1.16 (3), there is a positive integer n_0 such that $\ell^{\otimes n} = 0$ for all integer $n \geqslant n_0$, so that if we choose $s = 0 \in H^0(X, L^{\otimes n})$, then the above properties (2.21) hold. In this sense, in order to check the extension property, we may assume that $\ell \notin \mathcal{N}(Y, L|_Y)$.

For any non-zero element ℓ of $H^0(Y, L|_Y) \setminus \mathcal{N}(Y, L|_Y)$, we let

$$\lambda_\varphi(\ell) = \limsup_{n \to +\infty} \frac{a_{\varphi,n}(\ell)}{n} \in [0, +\infty]. \tag{2.22}$$

We call $\lambda_\varphi(\ell)$ the *extension obstruction index* of ℓ.

Definition 2.3.23 We assume that $H^0(X, L^{\otimes n}) \to H^0(Y, L|_Y^{\otimes n})$ is surjective for all $n \geqslant 1$. Let $\|\cdot\|_{n\varphi,\mathrm{quot}}$ be the quotient seminorm of $H^0(Y, L|_Y^{\otimes n})$ induced by $\|\cdot\|_{n\varphi}$ and the surjective homomorphism $H^0(X, L^{\otimes n}) \to H^0(Y, L|_Y^{\otimes n})$. For $\ell \in H^0(Y, L|_Y^{\otimes n})$, we define $\|\ell\|^{(n)}_{\varphi,\mathrm{quot}}$ to be

$$\|\ell\|^{(n)}_{\varphi,\mathrm{quot}} := \left(\|\ell^{\otimes n}\|_{n\varphi,\mathrm{quot}}\right)^{1/n}.$$

It is easy to see that

$$\|\ell\|^{(\infty)}_{\varphi,\mathrm{quot}} := \lim_{n\to\infty} \|\ell\|^{(n)}_{\varphi,\mathrm{quot}} = \inf_{n>0} \|\ell\|^{(n)}_{\varphi,\mathrm{quot}} \in \mathbb{R}_{\geqslant 0} \tag{2.23}$$

and

$$\|\ell\|_{\varphi|_Y} \leqslant \|\ell\|^{(\infty)}_{\varphi,\mathrm{quot}} \tag{2.24}$$

because

$$\|\ell_n \otimes \ell_{n'}\|_{(n+n')\varphi,\mathrm{quot}} \leqslant \|\ell_n\|_{n\varphi,\mathrm{quot}} \|\ell_{n'}\|_{n'\varphi,\mathrm{quot}} \text{ and } \|\ell_n\|_{n\varphi|_Y} \leqslant \|\ell_n\|_{n\varphi,\mathrm{quot}}$$

for all $\ell_n \in H^0(Y, L|_Y^{\otimes n})$ and $\ell_{n'} \in H^0(Y, L|_Y^{\otimes n'})$.

Proposition 2.3.24 *We assume that $\ell \notin \mathcal{N}(Y, L|_Y)$ and there exists a positive integer n_1 such that, for all $n \geqslant n_1$, $\ell^{\otimes n}$ lies in the image of the restriction map $H^0(X, L^{\otimes n}) \to H^0(Y, L|_Y^{\otimes n})$. Then one has the following:*

(1) $\displaystyle \lambda_\varphi(\ell) = \lim_{n\to+\infty} \frac{a_{\varphi,n}(\ell)}{n} = \inf_{n\geqslant 1} \frac{a_{\varphi,n}(\ell)}{n}$.

(2) *The following are equivalent:*

(2.a) *ℓ has the extension property for φ.*
(2.b) *For any $\epsilon > 0$, there are a positive integer n and a section $s \in H^0(X, L^{\otimes n})$ such that $s|_Y = \ell^{\otimes n}$ and $\|s\|_{n\varphi} \leqslant e^{\epsilon n}\|\ell\|^n_{\varphi|_Y}$.*
(2.c) *$\lambda_\varphi(\ell) = 0$.*

(3) *We assume that $H^0(X, L^{\otimes n}) \to H^0(Y, L|_Y^{\otimes n})$ is surjective for all $n \geqslant 1$. Then, the above equivalent properties are also equivalent to $\|\ell\|_{\varphi|_Y} = \|\ell\|^{(\infty)}_{\varphi,\mathrm{quot}}$.*

Proof (1) is a consequence of Fekete's lemma because the sequence $\{a_{\varphi,n}\}_{n\in\mathbb{N}}$ is subadditive by Lemma 2.3.21.

"(2.a) $\Longrightarrow$ (2.b)" is obvious.

"(2.b) $\Longrightarrow$ (2.c)": For any $\epsilon > 0$, there is a positive integer n such that $a_{\varphi,n}(\ell) \leqslant n\epsilon$, so that, by (1),

$$0 \leqslant \lambda_\varphi(\ell) = \inf_{n\geqslant 1} \frac{a_{\varphi,n}(\ell)}{n} \leqslant \epsilon,$$

and hence one has (2.c).

"(2.c) $\Longrightarrow$ (2.a)": Since $\lambda_\varphi(\ell) = \lim\limits_{n\to+\infty} \dfrac{a_{\varphi,n}(\ell)}{n}$ by (1), we can see (2.a).

(3) Note that $a_{\varphi,n}(\ell)/n = \ln\|\ell\|^{(n)}_{\varphi,\mathrm{qout}} - \ln\|\ell\|_{\varphi|_Y}$, so that $\lambda_\varphi(\ell) = \ln\|\ell\|^{(\infty)}_{\varphi,\mathrm{qout}} - \ln\|\ell\|_{\varphi|_Y}$. Thus the assertion follows. □

Remark 2.3.25 (1) Let φ' be another metric on L. By Lemma 2.3.21 one has

$$|a_{\varphi,n}(\ell) - a_{\varphi',n}(\ell)| \leqslant 2n\, d(\varphi, \varphi'),$$

provided that $a_{\varphi,n}(\ell)$ or $a_{\varphi',n}(\ell)$ is finite. We deduce from this inequality that, $\lambda_\varphi(\ell)$ is finite if and only if $\lambda_{\varphi'}(\ell)$ is finite. Moreover, when these numbers are finite, one has

$$|\lambda_\varphi(\ell) - \lambda_{\varphi'}(\ell)| \leqslant 2d(\varphi, \varphi'). \tag{2.25}$$

(2) We assume that $\lambda_\varphi(\ell) < \infty$. Then one has

$$\lambda_{n\varphi}(\ell^{\otimes n}) = n\lambda_\varphi(\ell) \tag{2.26}$$

for all $n > 0$. Indeed,

$$\begin{aligned}\lambda_{n\varphi}(\ell^{\otimes n}) = \lim_{m\to\infty} \frac{a_{n\varphi,m}(\ell^{\otimes n})}{m} &= \lim_{m\to\infty} \frac{a_{\varphi,nm}(\ell)}{m}\\ &= n\lim_{m\to\infty} \frac{a_{\varphi,nm}(\ell)}{nm} = n\lambda_\varphi(\ell).\end{aligned}$$

(3) Let X' be a closed subscheme of X such that $Y \subseteq X'$. We assume that there is a positive integer n_0 such that, for all $n \geqslant n_0$, $H^0(X, L^{\otimes n}) \to H^0(X', L^{\otimes n}|_{X'})$ is surjective. Then

$$\lambda_{\varphi|_{X'}}(\ell) \leqslant \lambda_{\varphi}(\ell). \tag{2.27}$$

Indeed, as $\|s|_{X'}\|_{\varphi|_{X'}} \leqslant \|s\|_{\varphi}$ for all $s \in H^0(X, L^{\otimes n})$ and $H^0(X, L^{\otimes n}) \to H^0(X', L^{\otimes n}|_{X'})$ is surjective for all $n \geqslant n_0$, one has $a_{\varphi|_{X'},n}(\ell) \leqslant a_{\varphi,n}(\ell)$ for all $n \geqslant n_0$, so that the assertion follows.

2.3.4.1 A generalisation of a result in [155] and [111]

Let X be a d-dimensional integral smooth scheme over $\mathbb{C}$. Let Y be a closed and reduced subscheme of X defined by an ideal sheaf I on X, that is, $I = \sqrt{I}$ and $Y = \operatorname{Spec}(\mathcal{O}_X/I)$. Let $\mu_I : X_I \to X$ be the blowing-up along I, that is, $X_I = \operatorname{Proj}\left(\bigoplus_{m=0}^{\infty} I^m\right)$. Let $\tilde{\mu} : \widetilde{X}_I \to X_I$ be the normalisation of X_I. Furthermore, let $\mu' : X' \to \widetilde{X}_I$ be a desingularisation of $\widetilde{X}_I$ such that μ' yields an isomorphism

$$X' \setminus \mu'^{-1}(\operatorname{Sing}(\widetilde{X}_I)) \xrightarrow{\sim} \widetilde{X}_I \setminus \operatorname{Sing}(\widetilde{X}_I).$$

We denote the compositions

$$X' \xrightarrow{\mu'} \widetilde{X}_I \xrightarrow{\tilde{\mu}} X_I \xrightarrow{\mu_I} X.$$

by μ, that is, $\mu := \mu_I \circ \tilde{\mu} \circ \mu'$. Note that $X' \setminus \mu^{-1}(Y) \xrightarrow{\sim} X \setminus Y$ via μ.

Lemma 2.3.26 *There are positive integers m_0 and c such that $\mu_*(I^m \mathcal{O}_{X'}) \subseteq I^{m-c}$ for all $m \geqslant m_0 + c$.*

Proof Let us consider the following claim:

Claim 2.3.27 (a) *$\mu'_*(I^m \mathcal{O}_{X'}) = I^m \mathcal{O}_{\widetilde{X}_I}$ for all integer $m \geqslant 0$.*

(b) *There is a positive integer c such that $\tilde{\mu}_*(I^m \mathcal{O}_{\widetilde{X}_I}) \cap \mathcal{O}_{X_I} \subseteq I^{m-c}\mathcal{O}_{X_I}$ for all integer $m \geqslant c$.*

(c) *There is a positive integer m_0 such that $\mu_{I,*}(I^m \mathcal{O}_{X_I}) = I^m$ for all integer $m \geqslant m_0$.* □

Proof (a) Note that $I^m \mathcal{O}_{\widetilde{X}_I}$ is invertible and $I^m \mathcal{O}_{X'} = \mu'^*(I^m \mathcal{O}_{\widetilde{X}_I})$. Moreover as $\widetilde{X}_I$ is normal, $\mu'_*(\mathcal{O}_{X'}) = \mathcal{O}_{\widetilde{X}_I}$, so that the assertion follows from the projection formula.

(b) We choose an affine open covering $X_I = \bigcup_{i=1}^N \operatorname{Spec}(A_i)$. Let $\widetilde{A}_i$ be the normalisation of A_i. Then $\widetilde{X}_I = \bigcup_{i=1}^N \operatorname{Spec}(\widetilde{A}_i)$ is an affine open covering. Note that $\widetilde{A}_i$ is a finitely generated A_i-module, so that, by Artin-Lees lemma (cf. [7, Corollary 10.10]), there is a positive constant c_i such that $I^m \widetilde{A}_i \cap A_i = I^{m-c_i}(I^c \widetilde{A}_i \cap A_i)$ for all $m \geqslant c_i$, which implies $I^m \widetilde{A}_i \cap A_i \subseteq I^{m-c_i} A_i$. Therefore, if we set $c = \max\{c_1, \ldots, c_N\}$, then one has the assertion.

(c) This is essentially proved in [85, Chapter II, Theorem 5.19]. Indeed, at the final line in the proof of the above reference, it says that "$S'_n = S_n$ for all sufficiently large n", which is nothing more than the assertion of (c) because $\mathcal{O}_{X_I}(m) = I^m \mathcal{O}_{X_I}$. □

Let us go back to the proof of the lemma. This is a local question, so that we may assume that $X = \mathrm{Spec}(A)$ for some finitely generated regular $\mathbb{C}$-algebra A. Note that $\mu_*(I^m O_{X'}) \subseteq O_X$. Therefore, it is sufficient to see that, if $f \in I^m O_{X'}$ for $f \in A$, then $f \in I^{m-c}$. First of all, by (a), $f \in I^m O_{\widetilde{X}_I}$, so that $f \in \tilde{\mu}_*(I^m O_{\widetilde{X}_I}) \cap O_{X_I}$, and hence, by (b), $f \in I^{m-c} O_{X_I}$. Note that $m - c \geqslant m_0$. Therefore, one has $f \in I^{m-c}$, as required. □

We assume that X is projective. Let L be an ample invertible O_X-module and φ be a C^∞-metric of L such that $c_1(L, \varphi)$ is positive. Let U be an open set (in the sense of the analytic topology) of X such that $Y \subseteq U$. The proof of [111, Theorem 7.6] works well even if we change the exponent d of ρ by a positive number δ except (3) in Claim 2, which should be

"If $\delta \geqslant d$, then $\rho^{-\delta}$ is not integrable on any neighborhood of Y".

At page 231, line 6 from the bottom, one constructs a C^∞-section l' of $L^{\otimes n}$ over X, which is holomorphic on U' and satisfies the integrability condition

$$\int_X |l'|^2 \rho^{-\delta} \Phi < \infty. \tag{2.28}$$

Let $E_1, \ldots, E_r$ be irreducible components of $\mu^{-1}(Y)$. We set

$$IO_{X'} = -(a_1 E_1 + \cdots + a_r E_r) \quad \text{and} \quad K_{X'} = \mu^*(K_X) + b_1 E_1 + \cdots + b_r E_r.$$

Note that $a_i > 0$ and $b_i > 0$ for all i.

Lemma 2.3.28 *If e_i is the multiplicity of l' along E_i, then $e_i > a_i\delta - b_i - 1$ for $i = 1, \ldots, r$.*

Proof Let η be a closed point of $E_i \setminus \mathrm{Sing}(E_1 + \cdots + E_r)$ and $\xi = \mu(\eta)$. Let $\{y_1, \ldots, y_d\}$ be a local coordinate of X' on an open neighbourhood W of η such that E_i is defined by $y_1 = 0$. Let $x_1, \ldots, x_d$ be a local coordinate of X' on an open neighborhood V of ξ. In the following, if it is necessary, we will shrink V and W freely. First of all, we may assume that

$$\mu(W) \subseteq V. \tag{2.29}$$

Moreover, we can find a positive constant C such that

$$\Phi \geqslant C(\sqrt{-1})^d (dx_1 \wedge d\bar{x}_1) \wedge \cdots \wedge (dx_d \wedge d\bar{x}_d) \tag{2.30}$$

on V. Let ω be a local basis of L on V. Note that ρ can be writen by

$$\rho = (|f_1|^2 + \cdots + |f_N|^2)|\omega^m|$$

on V, where $f_1, \ldots, f_N$ are generators of I on V, so that there is a positive constant C' such that

$$\mu^*(\rho) \leqslant C' |y_1|^{2a_i} \tag{2.31}$$

on W. Further one has

$$\mu^*((\sqrt{-1})^d(dz_1 \wedge d\bar{z}_1) \wedge \cdots \wedge (dz_d \wedge d\bar{z}_d)) = |y_1|^{2b_i}|u|^2(\sqrt{-1})^d(dy_1 \wedge d\bar{y}_1) \wedge \cdots \wedge (dy_d \wedge d\bar{y}_d)$$

on W, where u is a nowhere vanishing holomorphic function on W, so that

$$\mu^*((\sqrt{-1})^d(dz_1 \wedge d\bar{z}_1) \wedge \cdots \wedge (dz_d \wedge d\bar{z}_d)) \geqslant C''|y_1|^{2b_i}(\sqrt{-1})^d(dy_1 \wedge d\bar{y}_1) \wedge \cdots \wedge (dy_d \wedge d\bar{y}_d) \quad (2.32)$$

holds on W for some positive constant C''. If we set $l' = f'\omega^n$, then $f' = y_1^{e_i} g$ on W such that g is not identically zero on $E_i|_W$, so that one can find $(0, \alpha_2, \ldots, \alpha_n) \in E_i|_W$ and a positive number r such that $g \neq 0$ on

$$W_r = \{(y_1, \ldots, y_r) \in W \,:\, |y_j - \alpha_j| \leqslant r \text{ for all } j = 1, \ldots, d\},$$

where $\alpha_1 = 0$. Therefore one can find a positive constant C''' such that

$$\mu^*(|l'|^2) \geqslant C'''|y_1|^{2e_i} \quad (2.33)$$

on W_r. Thus, if we set $y_j - \alpha_j = r_j \exp(\sqrt{-1}\theta_j)$ for $j = 1, \ldots, d$, then, by (2.28), (2.29), (2.30), (2.31), (2.32) and (2.33),

$$\begin{aligned}
\infty &> \int_X |l'|^2\rho^{-\delta}\Phi \geqslant \int_V |l'|^2\rho^{-\delta}\Phi \\
&\geqslant \int_V C|l'|^2\rho^{-\delta}(\sqrt{-1})^d(dx_1 \wedge d\bar{x}_1) \wedge \cdots \wedge (dx_d \wedge d\bar{x}_d) \\
&\geqslant C\int_{W_r} \mu^*(|l'|^2\rho^{-\delta}(\sqrt{-1})^d(dx_1 \wedge d\bar{x}_1) \wedge \cdots \wedge (dx_d \wedge d\bar{x}_d)) \\
&\geqslant CC'^{-\delta}C''C''' \int_{W_r} |y_1|^{2e_i+2b_i-2a_i\delta}(\sqrt{-1})^d(dy_1 \wedge d\bar{y}_1) \wedge \cdots \wedge (dy_d \wedge d\bar{y}_d) \\
&\geqslant CC'^{-\delta}C''C''' \int_{[0,r]^d \times [0,2\pi]^d} r_1^{2e_i+2b_i-2a_i\delta+1} r_2 \cdots r_d dr_1 \cdots dr_d d\theta_1 \cdots d\theta_d \\
&\geqslant CC'^{-\delta}C''C'''2\pi^d r^{2(d-1)} \int_{[0,r]} r_1^{2e_i+2b_i-2a_i\delta+1} dr_1,
\end{aligned}$$

so that $2e_i + 2b_i - 2a_i\delta + 1 > -1$, as required. □

By virtue of Lemma 2.3.28 together with Lemma 2.3.26, one has the following generalisation of [155, Lemma 2.6] and [111, Theorem 7.6].

Theorem 2.3.29 *Let Y' be a closed subscheme of X defined by an ideal sheaf J on X, that is, $Y' = \operatorname{Spec}(\mathcal{O}_X/J)$. Let U be a Zariski open subset of X containing Y'. Then there are a positive integer n_0 and a positive constant C such that, for any integer $n \geqslant n_0$ and $l_U \in H^0(U, L^{\otimes n})$ with $\|l_U\|_{n\varphi_U} < \infty$, there is $l \in H^0(X, L^{\otimes n})$ such that*

$l|_{Y'} = l_U|_{Y'}$ *and* $\|l\|_{n\varphi} \leqslant Cn^{2d}\|l_U\|_{n\varphi_U}$, *where* $\|l_U\|_{n\varphi_U} := \sup\{|l_U|_{n\varphi}(x) : x \in U^{\mathrm{an}}\}$.

Proof Let $I := \sqrt{J}$ and $Y := \mathrm{Spec}(O_X/I)$. One can find $a \in \mathbb{Z}_{\geqslant 1}$ such that $I^a \subseteq J$. We fix a positive number δ with

$$\delta \geqslant \max_{i=1,\ldots,r}\left\{\frac{b_i+1}{a_i}\right\} + c + m_0 + a,$$

where c and m_0 are the positive integers in Lemma 2.3.26. The proof of [111, Theorem 7.6] is carried out by using the exponent δ instead of d. The point is to show that $l|_{Y'} = l_U|_{Y'}$. By Lemma 2.3.28,

$$e_i > a_i\delta - b_i - 1 \geqslant a_i(c + m_1 + a),$$

so that, there is a Zariski closed set Z of $\mu^{-1}(Y)$ such that $\dim Z \leqslant d-2$ and

$$\mu^*(l') \in I^{c+m_1+a}\mu^*(L^{\otimes n})\big|_{\mu^{-1}(U')\setminus Z}.$$

As $I^{c+m_1+a}\mu^*(L^{\otimes n})$ is invertible, one can see that $\mu^*(l') \in I^{c+m_1+a}\mu^*(L^{\otimes n})\big|_{\mu^{-1}(U')}$, and hence, by Lemma 2.3.26, $l' \in I^{m_1+a}L^{\otimes n} \subseteq I^a L^{\otimes n}$. Therefore the class of l' in $L^{\otimes n}/JL^{\otimes n}$ is zero over Y', and hence $l|_{Y'} = l_U|_{Y'}$. The remaining estimates are same as the proof of [111, Theorem 7.6]. □

Corollary 2.3.30 *Let X, L and Y' be the same as in Theorem 2.3.29. Let φ a continuous metric of L such that the first Chern current $c_1(L,\varphi)$ is positive. Then, for $\ell \in H^0(Y', L|_{Y'})$, ℓ has the extension property for φ.*

Proof Clearly we may assume that $\ell \notin \mathcal{N}(Y', L|_{Y'})$. As L is ample, there is a C^∞-metric ψ on L such that $c_1(L,\psi)$ is a positive form.

Claim 2.3.31 *If the corollary holds for any C^∞-metric of L with the semipositive Chern form, then the corollary holds in general.* □

Proof Let ϕ be a continuous function such that $\psi - \varphi = \phi$. It is well known that there is a sequence $\{\phi_n\}_{n=1}^\infty$ of C^∞-functions on X^{an} such that $\varphi_n := \psi - \phi_n$ is a C^∞-metric of L with the semipositive Chern form and $\{\phi_n\}_{n=1}^\infty$ converges uniformly to ϕ (for example, see [10, Theorem 1] or [109, Lemma 4.2]). Thus $\lim_{n\to\infty} d(\varphi,\varphi_n) = 0$. By our assumption, $\lambda_{\varphi_n}(\ell) = 0$, so that $\lambda_\varphi(\ell) \leqslant 2d(\varphi,\varphi_n)$, and hence the assertion follows. □

We fix a positive number ϵ. By the above claim, we may assume that φ is C^∞, so that if we set $f = \psi - \varphi$, then f is a C^∞ function. Note that for $\lambda \in \,]0,1[$, $\varphi + \lambda f$ gives rise to a positive Chern form because $\varphi + \lambda f = (1-\lambda)\varphi + \lambda\psi$. We choose $\lambda_0 \in \,]0,1[$ such that

$$\lambda_0 \sup\{|f(x)| : x \in X^{\mathrm{an}}\} \leqslant \epsilon.$$

We set $\varphi' = \varphi + \lambda_0 f$. Then

$$e^{-\epsilon}|\cdot|_{\varphi}(x) \leqslant |\cdot|_{\varphi'}(x) \leqslant e^{\epsilon}|\cdot|_{\varphi}(x) \tag{2.34}$$

for all $x \in X^{\mathrm{an}}$. We choose a positive integer a such that

$$H^0(X, L^{\otimes a}) \to H^0(Y', L^{\otimes a}|_{Y'})$$

is surjective, so that one can find t such that $t \in H^0(X, L^{\otimes a})$ and $t|_{Y'} = \ell^{\otimes a}$. We also choose an open set U of X such that $Y' \subseteq U$ and

$$\|t\|_{a\varphi'_U} \leqslant e^{a\epsilon}\|l^{\otimes a}\|_{a\varphi'|_{Y'}}. \tag{2.35}$$

By the above theorem, there are a positive integer n_1 and a positive constant C such that, for any $n \geqslant n_1$, one can find $s \in H^0(X, L^{\otimes an})$ such that $s|_{Y'} = \ell^{\otimes an}$ and

$$\|s\|_{na\varphi'} \leqslant Cn^{2d}\|t^{\otimes n}\|_{na\varphi'_U}. \tag{2.36}$$

Let n_2 be a positive integer such that $n_2 \geqslant n_1$ and

$$Cn^{2d} \leqslant e^{\epsilon an} \tag{2.37}$$

for $n \geqslant n_2$. Therefore, using (2.34), (2.35), (2.36) and (2.37), one has

$$\begin{aligned}\|s\|_{na\varphi} &\leqslant e^{na\epsilon}\|s\|_{na\varphi'} \leqslant e^{na\epsilon}\left(Cn^{2d}\|t\|^n_{a\varphi'_U}\right)\\ &\leqslant e^{2na\epsilon}\|t\|^n_{a\varphi'_U} \leqslant e^{3na\epsilon}\|l\|^{an}_{\varphi'|_{Y'}} \leqslant e^{4na\epsilon}\|l\|^{an}_{\varphi|_{Y'}},\end{aligned}$$

which means that $\lambda_{a\varphi}(l^{\otimes a}) \leqslant 4a\epsilon$, so that $\lambda_{\varphi}(l) \leqslant 4\epsilon$. Therefore one has $\lambda_{\varphi}(l) = 0$ because ϵ is an arbitrary positive number. □

2.3.4.2 Extension property over an Archimedean field

We assume that k is either $\mathbb{R}$ or $\mathbb{C}$ and the absolute value of k is the standard absolute value.

Theorem 2.3.32 *Let X be a projective scheme over k, L be an ample invertible $\mathcal{O}_X$-module and φ be a semipositive continuous metric metric of L. For any closed subscheme Y of X and any $\ell \in H^0(Y, L|_Y)$, ℓ has the extension property for φ.*

Proof Clearly we may assume that $\ell \notin \mathcal{N}(Y, L|_Y)$. Let us see the following claim:

Claim 2.3.33 (1) *We assume that $k = \mathbb{C}$, $X = \mathbb{P}^n$, $L = \mathcal{O}(1)$ and φ is the Fubini-Study metric arising from a norm $\|\cdot\|$ on $H^0(\mathbb{P}^n, \mathcal{O}(1))$. Then the assertion of the theorem holds.*

(2) *We assume that $k = \mathbb{R}$, $X = \mathbb{P}^n$, $L = \mathcal{O}(1)$ and φ is the Fubini-Study metric arising from a norm $\|\cdot\|$ on $H^0(\mathbb{P}^n, \mathcal{O}(1))$. Then the assertion of the theorem holds.* □

Proof (1) By Theorem 2.3.7, the first Chern current $c_1(L, \varphi)$ is positive, so that (1) is a consequence of Corollary 2.3.30.

(2) We consider $X_{\mathbb{C}}$, $L_{\mathbb{C}}$ and $\varphi_{\mathbb{C}}$. Then $\varphi_{\mathbb{C}}$ is the Fubini-Study metric induced by the norm $\|\cdot\|_{\mathbb{C}}$ on $H^0(X, L) \otimes_{\mathbb{R}} \mathbb{C}$ by Proposition 1.3.24. Thus, by using the case (1), for any $\epsilon > 0$, there is a positive integer n_0 such that, for any $n \geqslant n_0$, we can find $s \in H^0(X_{\mathbb{C}}, L_{\mathbb{C}}^{\otimes n})$ with $s|_{Y_{\mathbb{C}}} = \ell^{\otimes n}$ and $\|s\|_{n\varphi_{\mathbb{C}}} \leqslant e^{n\epsilon}\|\ell\|^n_{\varphi_{\mathbb{C}}|_Y}$. First of all, note that $\|\ell\|_{\varphi_{\mathbb{C}}|_Y} = \|\ell\|_{\varphi|_Y}$. If we set $s = \sigma + i\tau$ ($\sigma, \tau \in H^0(X, L^{\otimes n})$), then $(\sigma|_Y) + i(\tau|_Y) = \ell^{\otimes n}$, so that $\tau|_Y = 0$, and hence $\sigma|_Y = \ell^{\otimes n}$. Moreover, for any $x \in X_{\mathbb{C}}^{\mathrm{an}}$ one has $\|\sigma + i\tau\|_{n\varphi_{\mathbb{C}}}(x) = \|\sigma - i\tau\|_{n\varphi_{\mathbb{C}}}(\overline{x})$, so that $\|\sigma + i\tau\|_{n\varphi_{\mathbb{C}}} = \|\sigma - i\tau\|_{n\varphi_{\mathbb{C}}}$. We then deduce that

$$2\|\sigma\|_{n\varphi} = 2\|\sigma_{n\varphi_{\mathbb{C}}}\| \leqslant \|\sigma + i\tau\|_{n\varphi_{\mathbb{C}}} + \|\sigma - i\tau\|_{n\varphi_{\mathbb{C}}} = 2\|s\|_{n\varphi_{\mathbb{C}}}.$$

Thus one has the assertion in this case. □

We choose n_1 such that, for all $n \geqslant n_1$, $L^{\otimes n}$ is very ample. Then φ_n is the restriction of the Fubini-Study metric $\varphi_{\|\cdot\|_{n\varphi}}$ of $\mathcal{O}(1)$ to $\mathbb{P}(H^0(X, L^{\otimes n}))$ induced by the norm $\|\cdot\|_{n\varphi}$. Thus, by the above claim together with (2.27),

$$0 \leqslant \lambda_{\varphi_n}(\ell^{\otimes n}) \leqslant \lambda_{\varphi_{\|\cdot\|_{n\varphi}}}(\ell^{\otimes n}) = 0,$$

and hence $\lambda_{n\varphi}(\ell^{\otimes n}) \leqslant 2d(n\varphi, \varphi_n)$ by (2.25). Since $\lambda_{n\varphi}(\ell^{\otimes n}) = n\lambda_{\varphi}(\ell)$ by (2.26), one has

$$0 \leqslant \lambda_{\varphi}(\ell) \leqslant 2d(\varphi, \tfrac{1}{n}\varphi_n).$$

Therefore, the assertion follows. □

2.3.4.3 Extension property over a non-Archimedean field

In this subsection, we fix a field k equipped with a *non-Archimedean* absolute value $|\cdot|$, under which the field k is complete.

Proposition 2.3.34 *We assume that $|\cdot|$ is non-trivial. Let X be a projective k-scheme, L be an invertible $\mathcal{O}_X$-module and $(\mathscr{X}, \mathscr{L})$ be a model of (X, L). Let s be a global section of L such that $\|s\|_{\varphi_{\mathscr{L}}} \leqslant 1$. Then there exists an element $a \in \mathfrak{o}_k \setminus \{0\}$ such that as^n belongs to $H^0(\mathscr{X}, \mathscr{L}^{\otimes n})$ for all integers $n \geqslant 1$.*

Proof Let $(\mathscr{U}_i)_{i=1}^N$ be a covering of $\mathscr{X}$ by affine open subsets, such that $\mathscr{U}_i = \mathrm{Spec}(\mathscr{A}_i)$ and the invertible sheaf $\mathscr{L}$ trivialises on each $\mathscr{U}_i$, that is, $\mathscr{L}|_{\mathscr{U}_i} = \mathscr{A}_i s_i$ for some $s_i \in \mathscr{L}(\mathscr{U}_i)$. Then the restriction of s to $U_i := \mathscr{U}_i \cap X$ can be written in the form $\lambda_i s_i$, where $\lambda_i \in A_i = S^{-1}\mathscr{A}_i$ and $S = \mathfrak{o}_k \setminus \{0\}$. Note that for $x \in (U_i)^{\mathrm{an}}_{\mathscr{A}_i}$, the reduction point of x is in $\mathscr{U}_i$ (cf. Remark 2.1.6), so that since $\|s\|_{\varphi_{\mathscr{L}}} \leqslant 1$, we obtain that

$$|\lambda_i|_x = |\lambda_i|_x \cdot |s_i|_{\varphi_{\mathscr{L}}}(x) = |s|_{\varphi_{\mathscr{L}}}(x) \leqslant \|s\|_{\varphi_{\mathscr{L}}} \leqslant 1$$

for any $x \in (U_i)^{\mathrm{an}}_{\mathscr{A}_i}$. By Proposition 2.1.7, λ_i is integral over the ring $\mathscr{A}_i$, namely $\mathscr{A}_i[\lambda_i]$ is an $\mathscr{A}_i$-module of finite type. In particular, there exists an integer $d_i \geqslant 1$

such that, for any integer $n \geqslant 1$,

$$\lambda_i^n \in \mathscr{A}_i + \mathscr{A}_i\lambda_i + \cdots + \mathscr{A}_i\lambda_i^{d_i}.$$

Moreover, there exists $a_i \in \mathfrak{o}_k \setminus \{0\}$ such that

$$\{a_i\lambda_i, \ldots, a_i\lambda_i^{d_i}\} \subset \mathscr{A}_i.$$

We then obtain that $a_i\lambda_i^n \in \mathscr{A}_i$ for any integer $n \geqslant 1$. Finally, let $a = \prod_{i=1}^N a_i \in \mathfrak{o}_k \setminus \{0\}$. For any integer $n \geqslant 1$ and any $i \in \{1, \ldots, N\}$, one has

$$(as^n)|_{U_i} = (a\lambda_i^n)s_i^n \in H^0(\mathscr{U}_i, \mathscr{L}^{\otimes n}).$$

Since $\mathscr{X}$ is flat over $\mathrm{Spec}(\mathfrak{o}_k)$, these sections glue together to be a global section of $\mathscr{L}^{\otimes n}$. The proposition is thus proved. □

Proposition 2.3.35 *We assume that $|\cdot|$ is non-trivial. Let X be a projective k-scheme, L be an ample invertible O_X-module and Y be a closed subscheme of X. Let $u \geqslant 1$ be an integer such that $L^{\otimes u}$ is very ample, and $(\mathscr{X}, \mathscr{L})$ be a coherent model of $(X, L^{\otimes u})$ as constructed in Remark 2.3.14. Assume that $\varphi = \frac{1}{u}\varphi_{\mathscr{L}}$. Let φ_Y be the restriction of the metric φ to $L|_Y$. For any positive number ϵ and any $\ell \in H^0(Y, L|_Y)$, there exists an integer $n \geqslant 1$ and a section $s \in H^0(X, L^{\otimes n})$ such that $s|_Y = \ell^n$ and*

$$\|s\|_{n\varphi} \leqslant \mathrm{e}^{n\epsilon}\|\ell\|_{\varphi_Y}^n.$$

In other words, one has $\lambda_\varphi(\ell) = 0$ if $\ell \notin N(Y, L|_Y)$.

Proof Clearly we may assume that $\ell \notin \mathcal{N}(Y, L|_Y)$. We choose a positive integer m such that

$$\mathrm{e}^{-m\epsilon/2} < \sup\{|a| \ : \ a \in k^\times, \ |a| < 1\}.$$

By Proposition 1.1.4, on $H^0(Y, L|_Y^{\otimes m})/\mathcal{N}(Y, L|_Y^{\otimes m})$, there is $\alpha \in k^\times$ such that

$$\mathrm{e}^{-m\epsilon/2} \leqslant \|\alpha\ell^m\|_{m\varphi_Y} \leqslant 1. \tag{2.38}$$

One can find a coherent extension $\mathscr{Y}$ of Y such that $\mathscr{Y}$ is a closed subscheme of $\mathscr{X}$ (see Remark 2.3.14). By Proposition 2.3.34, there exists an element $\beta \in \mathfrak{o}_k \setminus \{0\}$ such that

$$\beta(\alpha\ell^m)^{pu} \in H^0(\mathscr{Y}, \mathscr{L}^{\otimes mp}|_{\mathscr{Y}})$$

for any integer $p \geqslant 1$. Moreover, since the invertible sheaf $\mathscr{L}$ is ample, for sufficiently positive integer p, by [1, Corollaire 1.4.10], the restriction map

$$H^0(\mathscr{X}, \mathscr{L}^{\otimes mp}) \longrightarrow H^0(\mathscr{Y}, \mathscr{L}^{\otimes mp}|_{\mathscr{Y}})$$

is surjective because the defining ideal of $\mathscr{Y}$ is coherent. Hence we can choose $p \in \mathbb{N}_{\geqslant 1}$ such that

$$|\beta|^{-1} \leqslant \mathrm{e}^{mpu\epsilon/2} \tag{2.39}$$

and that there exists $t \in H^0(\mathscr{X}, \mathscr{L}^{\otimes mp})$ verifying $t|_{\mathscr{Y}} = \beta(\alpha\ell^m)^{pu}$. We then take $n = mpu$ and $s = \beta^{-1}\alpha^{-pu}t \in H^0(X, L^{\otimes n})$. One has $s|_{\mathscr{Y}} = \ell^n$ and

$$\begin{aligned}\|s\|_{n\varphi} &= |\beta|^{-1} \cdot |\alpha|^{-pu} \cdot \|t\|_{n\varphi} \leqslant |\beta|^{-1} \cdot |\alpha|^{-pu} \\ &\leqslant \mathrm{e}^{mpu\epsilon/2} \cdot \left(\mathrm{e}^{m\epsilon/2}\|\ell^m\|_{m\varphi_Y}\right)^{pu} = \mathrm{e}^{n\varepsilon} \cdot \|\ell^m\|_{m\varphi_Y}^{pu} \leqslant \mathrm{e}^{n\epsilon}\|\ell\|_{\varphi_Y}^n,\end{aligned}$$

where the first inequality comes from Proposition 2.3.16 (2), the second one from (2.39) and (2.38), and the last one from (2.3). The first part of the proposition is thus proved, so that the last assertion follows from Proposition 2.3.24. □

Theorem 2.3.36 *Let X be a projective k-scheme and L be an ample invertible O_X-module, equipped with a semipositive continuous metric φ. Let Y be a closed subscheme of X and ℓ be an element in $H^0(Y, L|_Y)$. Then ℓ has the extension property for φ.*

Proof Clearly we may assume that $\ell \notin \mathcal{N}(Y, L|_Y)$. First we assume that the absolute value $|\cdot|$ on k is non-trivial. By Proposition 2.3.19, for sufficiently positive integer n, there exists a sequence $(\mathscr{X}_n, \mathscr{L}_n)$ of coherent models of $(X, L^{\otimes n})$ as constructed in Remark 2.3.14, and that

$$\lim_{n\to+\infty} \frac{1}{n} d(\varphi_{\mathscr{L}_n}, n\varphi) = 0. \tag{2.40}$$

For any $n \in \mathbb{N}_{\geqslant 1}$, let $\varphi^{(n)} = \frac{1}{n}\varphi_{\mathscr{L}_n}$. By Proposition 2.3.35 (see also Remark 2.3.25), one has $\lambda_{\varphi^{(n)}}(\ell) = 0$. Therefore, by the relations (2.25) and (2.40), we obtain that $\lambda_\varphi(\ell) = 0$.

In the following, we treat the trivial valuation case. The main idea is to introduce the field of formal Laurent series over k in order to reduce the problem to the non-trivial valuation case. We assume that the absolute value $|\cdot|$ is trivial. We denote by k' the field $k(\!(T)\!)$ of formal Laurent series over k, namely k' is the fraction field of the ring $k[\![T]\!]$ of formal series over k. Note that $k[\![T]\!]$ is a discrete valuation ring.

Claim 2.3.37 *The field extension $k \subseteq k'$ is separable.* □

Proof We may assume that the characteristic p of k is positive. First let us see the following claim:

SubClaim 2.3.38 *Let E be a finite extension of k and $\{\omega_i\}_{i=1}^e$ be a basis of E over k. Then we have the following:*

(i) *Let $(g_1, \ldots, g_e) \in (k')^e$. If $\omega_1 g_1 + \cdots + \omega_e g_e = 0$ in $E(\!(T)\!)$, then $g_1 = \cdots = g_e = 0$.*
(ii) *Let $\{f_i\}_{i=1}^s$ be a family of elements in k' which is linearly independent over k and $(c'_1, \ldots, c'_s)$ be an element of E^s. If $c'_1 f_1 + \cdots + c'_s f_s = 0$ in $E(\!(T)\!)$, then $c'_1 = \cdots = c'_s = 0$.* □

Proof (i) is trivial if we consider the coefficients of $g_1, \ldots, g_e$.

(ii) We set $c'_i = \sum_{j=1}^e c_{ij}\omega_j$ for some $c_{ij} \in k$. Then

$$\sum_{i=1}^{s} c_i' f_i = \sum_{j=1}^{e} \left(\sum_{i=1}^{s} c_{ij} f_i \right) \omega_j = 0,$$

so that, by (i), $\sum_{i=1}^{s} c_{ij} f_i = 0$ for all j. Therefore $c_{ij} = 0$ for all i, j, as desired. □

By [29, Théorème 2 in Chapter V, §25, n°4], it is sufficient to see that if $f_1, \ldots, f_s \in k(\!(T)\!)$ are linearly independent over k, then $f_1^p, \ldots, f_s^p$ are linearly independent over k. We assume that $c_1 f_1^p + \cdots + c_s f_s^p = 0$ for some $c_1, \ldots, c_s \in k$. Let E be a finite extension field of k such that we can find $c_i' \in E$ with $c_i = (c_i')^p$ for all i. Then $\sum_{i=1}^{s} c_i' f_i = 0$ because

$$0 = \sum_{i=1}^{s} c_i f_i^p = \left(\sum_{i=1}^{s} c_i' f_i \right)^p.$$

Thus, by (ii), one has $c_i' = 0$, as requested. □

Let us consider a subset Σ of $\mathbb{R}$ given by

$$\Sigma = \bigcup_{n=0}^{\infty} \bigcup_{(v,v') \in (H^0(X,L^{\otimes n}) \setminus \mathcal{N}(X,L^{\otimes n}))^2} \mathbb{Q}(\ln \|v\|_{n\varphi} - \ln \|v'\|_{n\varphi}).$$

Since

$$\{\|v\|_{n\varphi} : v \in H^0(X, L^{\otimes n})\}$$

is a finite set by Corollary 1.1.6, one has $\#(\Sigma) \leqslant \aleph_0$, so that one can choose $\alpha \in \mathbb{R}_{>0} \setminus \Sigma$. We denote by $v_T(\cdot)$ the corresponding valuation on k', and by $|\cdot|'$ the absolute value on k' defined as

$$\forall\, a \in k', \ |a|' = \mathrm{e}^{-\alpha v_T(a)}.$$

Note that this absolute value extends the trivial absolute value on k. We denote by $X_{k'}$ and $Y_{k'}$ the fibre products $X \times_{\operatorname{Spec} k} \operatorname{Spec} k'$ and $Y \times_{\operatorname{Spec} k} \operatorname{Spec} k'$, respectively, and by $p : X_{k'} \to X$ and $p_Y : Y_{k'} \to Y$ the morphism of projections.

As explained in §2.1.3, the morphism p corresponds to a map $p^\natural : X_{k'}^{\mathrm{an}} \to X^{\mathrm{an}}$. This map is actually surjective. In fact, if K is a field extension of k, equipped with an absolute value extending the trivial absolute value on k, then we can equip the field $K(T)$ of rational functions of one variable with the absolute value such that

$$\forall\, F = a_0 + a_1 T + \cdots + a_n T^n \in K[T], \quad |F| = \max_{i \in \{0,\ldots,n\}} |a_i| \cdot \mathrm{e}^{-\alpha i}.$$

This absolue value extends the restriction of $|\cdot|'$ to $k(T)$. Hence the completion $\widehat{K(T)}$ of $K(T)$ with this absoute value is a valued extension of k'. If $f : \operatorname{Spec} K \to X$ is a k-morphism defining a point x in X^{an}, then it gives rise to a k'-morphism from $\operatorname{Spec} \widehat{K(T)}$ to $X_{k'}$, which defines a point y in X'^{an} such that $p^\natural(y) = x$.

The surjectivity of $p^\natural$ implies that the restriction of the seminorm $\|\cdot\|_{n\varphi_{k'}}$ to $H^0(X, L)$ coincides with $\|\cdot\|_{n\varphi}$. In fact, if s is a section in $H^0(X, L^{\otimes n})$, then one has $\|p^*(s)\|_{n\varphi_{k'}} = \|s\|_{n\varphi} \circ p^\natural$. Therefore

$$\|p^*(s)\|_{n\varphi_{k'}} = \sup_{y\in X'^{\mathrm{an}}} \|p^*(s)\|_{n\varphi_{k'}}(y) = \sup_{y\in X'^{\mathrm{an}}} \|s\|_{n\varphi}(p^\natural(y)) = \|s\|_{n\varphi},$$

where the last equality comes from the surjectivity of the map $p^\natural$. For $(v,v') \in (H^0(X,L^{\otimes n}) \setminus \mathcal{N}(X,L^{\otimes n}))^2$, if $\|v\|_{n\varphi}/\|v'\|_{n\varphi} \in |k'^\times|$, then

$$\ln\|v\|_{n\varphi} - \ln\|v'\|_{n\varphi} = -\alpha\, v_T(a(T))$$

for some $a(T) \in k'^\times$. As

$$\alpha \notin \bigcup_{(v,v')\in(H^0(X,L^{\otimes n})\setminus\mathcal{N}(X,L^{\otimes n}))^2} \mathbb{Q}(\ln\|v\|_{n\varphi} - \ln\|v'\|_{n\varphi}),$$

we obtain $v_T(a(T)) = 0$, so that $\|v\|_{n\varphi} = \|v'\|_{n\varphi}$. By Proposition 1.3.22, the seminorm $\|\cdot\|_{n\varphi_{k'}}$ identifies with $\|\cdot\|_{n\varphi,k',\varepsilon}$, the ε-extension of scalars of $\|\cdot\|_{n\varphi}$.

Let X_{red} and Y_{red} be the reduced schemes associated with X and Y, respectively. By Claim 2.3.37, $X_{\mathrm{red},k'} := X_{\mathrm{red}} \times_{\mathrm{Spec}\,k} \mathrm{Spec}\,k'$ and $Y_{\mathrm{red},k'} := Y_{\mathrm{red}} \times_{\mathrm{Spec}\,k} \mathrm{Spec}\,k'$ are reduced (see [75, Proposition IV.(4.6.1)]), so that

$$\mathcal{N}\big(X_{k'}, L_{k'}^{\otimes n}\big) = \mathcal{N}(X,L^{\otimes n}) \otimes_k k' \quad\text{and}\quad \mathcal{N}\big(Y_{k'}, L_{k'}|_{Y_{k'}}^{\otimes n}\big) = \mathcal{N}(Y, L|_Y) \otimes_k k',$$

where $L_{k'} = L \otimes_k k'$.

By (2.25), without loss of generality, we may assume that L is very ample and that φ is the quotient metric on L induced by a ultrametric norm $\|\cdot\|$ on $V = H^0(X,L)/\mathcal{N}(X,L)$ and the natural surjection $\beta : V \otimes \mathcal{O}_{X_{\mathrm{red}}} \to L|_{X_{\mathrm{red}}}$. We may also assume that the restriction map $H^0(X,L^{\otimes n}) \to H^0(Y, L|_Y^{\otimes n})$ is surjective for all $n \geqslant 1$.

For $n \geqslant 1$, let

$$V_n := H^0(X,L^{\otimes n})/\mathcal{N}(X,L^{\otimes n}) \quad\text{and}\quad V_{Y,n} = H^0(Y, L|_Y^{\otimes n})/\mathcal{N}(Y, L|_Y^{\otimes n}).$$

Note that $V_1 = V$, and V_n and $V_{Y,n}$ are isomorphic to the images of

$$H^0(X,L^{\otimes n}) \to H^0(X_{\mathrm{red}}, L|_{X_{\mathrm{red}}}^{\otimes n}) \quad\text{and}\quad H^0(Y, L|_Y^{\otimes n}) \to H^0(Y_{\mathrm{red}}, L|_{Y_{\mathrm{red}}}^{\otimes n}),$$

respectively. Let $V_{k'}$ be the vector space $V \otimes_k k'$ and let $\|\cdot\|_{k',\varepsilon}$ be the norm on $V_{k'}$ induced by V by ε-extension of scalars. Then the surjective homomorphism $\beta : V \otimes \mathcal{O}_{X_{\mathrm{red}}} \to L|_{X_{\mathrm{red}}}$ induces a surjective homomorphism $\beta_{k'} : V_{k'} \otimes \mathcal{O}_{X_{\mathrm{red},k'}} \to L_{k'}|_{X_{\mathrm{red},k'}}$. By Proposition 1.3.24, the metric $\varphi_{k'}$ of $L_{k'}$ obtained by φ by extension of scalars coincides with the quotient metric on $L_{k'}$ induced by $(V_{k'}, \|\cdot\|_{k',\varepsilon})$ and $\beta_{k'}$. Therefore, by Theorem 2.3.36, for any $\epsilon > 0$, there exist an integer $n \geqslant 1$ and a section $s' \in H^0(X_{k'}, L_{k'}^{\otimes n})$ such that $s'|_{Y_{k'}} = p_Y^*(\ell)^n$ and

$$\|s'\|_{n\varphi_{k'}} \leqslant e^{n\epsilon} \|p_Y^*(\ell)\|_{\varphi_{Y,k'}}^n$$

where $\varphi_{Y,k'}$ is the metric on $(L|_Y) \otimes_k k' \cong L_{k'}|_{Y_{k'}}$ induced by φ_Y by extension of scalars, which equals the restriction of $\varphi_{k'}$ to $Y_{k'}$.

Let $\{e_1, \ldots, e_{\alpha_n}, f_1, \ldots, f_{\beta_n}\}$ be an orthogonal basis of $H^0(X, L^{\otimes n})$ with respect to $\|\cdot\|_{n\varphi}$ such that $\{f_1, \ldots, f_{\beta_n}\}$ form a basis of the kernel of the restriction map $H^0(X, L^{\otimes n}) \to H^0(Y, L|_Y^{\otimes n})$ (see Proposition 1.2.30 for the existence of such an orthogonal basis). We set

$$s' = \sum_{i=1}^{\alpha_n} a_i(T) e_i + \sum_{j=1}^{\beta_n} b_j(T) f_j,$$

where $(a_1(T), \ldots, a_{\alpha_n}(T), b_1(T), \ldots, b_{\beta_n}(T)) \in (k')^{\alpha_n+\beta_n}$. Note that

$$s'|_{Y_{k'}} = a_1(T)\, e_1|_Y + \cdots + a_{\alpha_n}(T)\, e_{\alpha_n}\big|_Y$$

and $\{e_1|_Y, \ldots, e_{\alpha_n}|_Y\}$ forms a basis of $H^0(Y, L|_Y^{\otimes n})$. Since the restriction of s' to $Y_{k'}$ can be written as the pull-back of an element in $H^0(Y, L|_Y^{\otimes n})$, one can see that $a_1(T), \ldots, a_{\alpha_n}(T) \in k$, so that $a_1(T), \ldots, a_{\alpha_n}(T)$ are denoted by $a_1, \ldots, a_{\alpha_n}$. Therefore if we set $s = a_1 e_1 + \cdots + a_{\alpha_n} e_{\alpha_n}$, then $s \in H^0(X, L^{\otimes n})$, $s|_Y = \ell^{\otimes n}$, and

$$\begin{aligned}
\|s\|_{n\varphi} &\leqslant \max_{i \in \{1,\ldots,\alpha_n\}} \{|a_i| \cdot \|e_i\|_{n\varphi}\} \\
&\leqslant \max\left\{ \max_{i\in\{1,\ldots,\alpha_n\}} |a_i| \cdot \|e_i\|_{n\varphi},\ \max_{j\in\{1,\ldots,\beta_n\}} |b_j(T)|' \cdot \|f_j\|_{n\varphi} \right\} \\
&= \|s'\|_{n\varphi_{k'}} \leqslant e^{\epsilon n} \|p_Y^*(\ell)\|^n_{\varphi_{Y,k'}} = e^{\epsilon n} \|\ell\|^n_{\varphi_Y},
\end{aligned}$$

where the first equality comes from the fact that $\{e_1, \ldots, e_{\alpha_n}, f_1, \ldots, f_{\beta_n}\}$ forms an orthogonal basis of $H^0(X_{k'}, L_{k'}^{\otimes n})$ with respect to $\|\cdot\|_{n\varphi_{k'}}$ (see Proposition 1.3.13). The theorem is thus proved. □

Remark 2.3.39 Let X be a scheme of finite type over k and $\mathcal{F}$ be a coherent $\mathcal{O}_X$-module. Let X_{red} be the reduced structure of X. The *reduced i-th cohomology group of* $\mathcal{F}$, denoted by $H^i_{\mathrm{red}}(X, \mathcal{F})$, is defined to be the image of the homomorphism

$$H^i(X, \mathcal{F}) \to H^i\left(X_{\mathrm{red}}, F|_{X_{\mathrm{red}}}\right).$$

Using the reduced cohomology group, one has the following variant of Theorem 2.3.36:

We assume that X is projective. Let L be an ample invertible $\mathcal{O}_X$-module, equipped with a semipositive continuous metric φ. Let Y be a closed subscheme of X and ℓ be an element of $H^0_{\mathrm{red}}(Y, L|_Y)$. Then, for any $\epsilon > 0$, there exist a positive integer n and $s \in H^0_{\mathrm{red}}(X, L^{\otimes n})$ such that $s|_Y = l^{\otimes n}$ and $\|s\|'_{n\varphi} \leqslant e^{n\epsilon}(\|l\|'_{\varphi_Y})^n$, where $s|_Y$ is the image of s by the homomorphism $H^0_{\mathrm{red}}(X, L^{\otimes n}) \to H^0_{\mathrm{red}}(Y, L|_Y^{\otimes n})$.

Since $H^0_{\mathrm{red}}(X, L^{\otimes n})$ and $H^0_{\mathrm{red}}(Y, L|_Y^{\otimes n})$ are subgroups of

$$H^0(X_{\mathrm{red}}, L|_{X_{\mathrm{red}}}^{\otimes n}) \quad \text{and} \quad H^0(Y_{\mathrm{red}}, L|_{Y_{\mathrm{red}}}^{\otimes n}),$$

respectively, the proof of the above result can be done in the similar way as Theorem 2.3.36.

2.4 Cartier divisors

In this section, we recall Cartier divisors and linear systems. Further we introduce $\mathbb{Q}$-Cartier and $\mathbb{R}$-Cartier divisors on an integral scheme and study their basic properties.

2.4.1 Reminder on Cartier divisors

Let X be a scheme. For any open subset U of X, we denote by $S_X(U)$ the set of all elements $a \in O_X(U)$ such that the homomorphisme of O_U-modules $O_U \to O_U$ defined as the homothety by a is injective. This is a multiplicative subset of $O_X(U)$. Moreover, S_X is a subsheaf of sets of O_X. We denote by $\mathscr{M}_X$ the sheaf of rings associated with the presheaf

$$U \longmapsto O_X(U)[S_X(U)^{-1}],$$

called the sheaf of *meromorphic functions*[2] on X. The canonical homomorphisms $O_X(U) \to O_X(U)[S_X(U)^{-1}]$ of rings induce a homomorphism of sheaves of rings $O_X \to \mathscr{M}_X$. The sections of $\mathscr{M}_X$ on an open subset U of X are called *meromorphic fonctions* on U.

Remark 2.4.1 Let U be an open subset of X. Any element a in $S_X(U)$ is a regular element (namely the homothety $O_X(U) \to O_X(U)$ defined by a is injective). of $O_X(U)$. It is not true in general that $S_X(U)$ contains all regular elements of $O_X(U)$. However, if U is an affine open subset of X, $S_X(U)$ identifies with the set of all regular elements in $O_X(U)$. In fact, an element $a \in O_X(U)$ is in $S_X(U)$ if and only if its image in the local ring $O_{X,x}$ is a regular element for any $x \in U$. The announced property thus results from the faithful flatness of

$$O_X(U) \longrightarrow \bigoplus_{x \in U} O_{X,x},$$

provided that U is an affine open subset. In particular, if X is an integral scheme, then $\mathscr{M}_X$ is the constant sheaf associated to the local ring of the generic point of X (which is a field). We use the notation $R(X)$ to denote the field of all meromorphic functions on X. If X is defined over a field K, $R(X)$ is often denoted by $K(X)$.

Denote by $O_X^{\times}$ the sheaf of abelian groupes described as follows. For any open subset U of X, $O_X^{\times}(U)$ is the set of elements $a \in O_X(U)$ such that the homothety $O_U \to O_U$ defined by a is an isomorphism. Similarly, denote by $\mathscr{M}_X^{\times}$ the sheaf of abelian groupes on X whose section space over any open subset $U \subseteq X$ is the set of all meromorphic functions φ on U such that the homothety $\mathscr{M}_U \to \mathscr{M}_U$ defined by φ is an isomorphism. This is a subsheaf of multiplicative monoids of $\mathscr{M}_X$.

[2] The definition in [77, IV.20] is not adequate, see [97] for details.

Definition 2.4.2 We call *Cartier divisor* on X any global section of the quotient sheaf $\mathscr{M}_X^\times/\mathcal{O}_X^\times$, that is, a data of a Zariski open covering $X = \bigcup_\alpha U_\alpha$ of X and a section $s_\alpha \in \mathscr{M}_X^\times(U_\alpha)$ over U_α for each α, which is called a *local equation* over U_α, such that $s_\alpha s_\beta^{-1} \in \mathcal{O}_X^\times(U_\alpha \cap U_\beta)$ for all α and β. The group of all Cartier divisors on X is denoted by $\mathrm{Div}(X)$, where the group law is written additively. We say that a Cartier divisor D is *effective* if it is a section of $(\mathscr{M}_X^\times \cap \mathcal{O}_X)/\mathcal{O}_X^\times$. We use the expression $D \geqslant 0$ to denote the condition "D is effective". Moreover, for $D_1, D_2 \in \mathrm{Div}(X)$, an expression $D_1 \geqslant D_2$ is defined by $D_1 - D_2 \geqslant 0$.

The Cartier divisors are closely related to invertible sheaves. Let D be a Cartier divisor on X. Denote by $\mathcal{O}_X(D)$ the sub-$\mathcal{O}_X$-module of $\mathscr{M}_X$ generated by $-D$. Namely, if $X = \bigcup_\alpha U_\alpha$ is an open covering of X and s_α is a local equation of D over U_α, then $\mathcal{O}_X(D)|_{U_\alpha} = \mathcal{O}_{U_\alpha} s_\alpha^{-1}$. Note that $\mathcal{O}_X(D)$ is an invertible $\mathcal{O}_X$-module since it is locally generated by a regular element. We say that the Cartier divisor D is *ample* (resp. *very ample*) if the invertible sheaf $\mathcal{O}_X(D)$ is ample (resp. very ample). By definition, one has $\mathcal{O}_X(-D) \cong \mathcal{O}_X(D)^\vee$. Moreover, if D_1 and D_2 are two Cartier divisors, then $\mathcal{O}_X(D_1+D_2) \cong \mathcal{O}_X(D_1)\otimes\mathcal{O}_X(D_2)$. Thus the map sending a divisor D to the isomorphism class of $\mathcal{O}_X(D)$ defines a homomorphism from $\mathrm{Div}(X)$ to the Picard group $\mathrm{Pic}(X)$ (namely the group of isomorphism classes of invertible $\mathcal{O}_X$-modules). This homomorphism is surjective notably when X is a reduced scheme with locally finite irreducible components, or a quasi-projective scheme over a Noetherian ring (cf. [77, IV.21.3.4-5]). We recall a simple proof of this result for the particular case where X is an integral scheme.

Proposition 2.4.3 *Let X be an integral scheme. Then the homomorphism* $\mathrm{Div}(X) \to \mathrm{Pic}(X)$ *constructed above is surjective.*

Proof Let η be the generic point of X and $R(X)$ be the field of all meromorphic functions on X. Let L be an invertible sheaf and s be a non-zero element in L_η. Then the maps $H^0(U, L) \to R(X), t \mapsto t_\eta/s$ (where U denotes an open subset of X) define a $\mathcal{O}_X$-linear homomorphisme from L to $\mathscr{M}_X$. The images of local trivialisations of L by this homomorphism define a global section of $\mathscr{M}_X^\times/\mathcal{O}_X^\times$, whose opposite D is a Cartier divisor such that $\mathcal{O}_X(D) \cong L$. □

Remark 2.4.4 Let X be an integral scheme and L be an invertible $\mathcal{O}_X$-module. Let η be the generic point of X. We call *rational section* of L any element in L_η. Note that for any non-empty open subset U of X, the restriction map $H^0(U, L) \to L_\eta$ is injective. By abuse of language, we also call a section of L on a non-empty open subset of X a rational section of L. The proof of the above proposition shows that any non-zero rational section s of L defines a Cartier divisor of X, which we denote by $\mathrm{div}(s)$. One can verify that, if L and L' are two invertible $\mathcal{O}_X$-modules and if s and s' are respectively non-zero rational sections of L and L', then one has $\mathrm{div}(ss') = \mathrm{div}(s) + \mathrm{div}(s')$.

The exact sequence of abelian sheaves

$$1 \longrightarrow \mathcal{O}_X^\times \longrightarrow \mathscr{M}_X^\times \longrightarrow \mathscr{M}_X^\times/\mathcal{O}_X^\times \longrightarrow 0$$

induces a long exact sequence of cohomology groups

$$1 \longrightarrow H^0(X, \mathcal{O}_X^\times) \longrightarrow H^0(X, \mathscr{M}_X^\times) \xrightarrow{\text{div}} \text{Div}(X) \xrightarrow{\theta} H^1(X, \mathcal{O}_X^\times) . \quad (2.41)$$

Note that the cohomology group $H^1(X, \mathcal{O}_X^\times)$ identifies with the Picard group $\text{Pic}(X)$ of X (cf. [78, 0.5.6.3]), and θ is just the group homomorphism sending any Cartier divisor D to the isomorphism class of the invertible $\mathcal{O}_X$-module $\mathcal{O}_X(D)$. The image of the group homomorphism $\text{div}(\cdot)$ is denoted by $\text{PDiv}(X)$. The divisors in $\text{PDiv}(X)$ are called *principal divisors*. The quotient group $\text{Div}(X)/\text{PDiv}(X)$ is called the *divisor class group* of X, denoted by $\text{Cl}(X)$. The exactness of the sequence (2.41) shows that the homomorphism from $\text{Cl}(X)$ to $\text{Pic}(X)$, sending the equivalent class of a Cartier divisor D to the isomorphism class of the invertible sheaf $\mathcal{O}_X(D)$ is injective. It is an isomorphism once X is a reduced scheme with locally finite irreducible components, or a quasi-projective scheme over a Noetherian ring. We write this result as a corollary of Proposition 2.4.3 in the particular case where X is an integral scheme.

Corollary 2.4.5 *Let X be an integral scheme. The homomorphism from* $\text{Cl}(X)$ *to* $\text{Pic}(X)$ *sending the equivalence class of a divisor class D to the isomorphism class of* $\mathcal{O}_X(D)$ *is an isomorphism.*

If two Cartier divisors D and D' of X differ by a principal divisor, namely lie in the same class in $\text{Cl}(X)$, we say that they are *linearly equivalent*.

Proposition 2.4.6 *We assume that X is locally Noetherian and normal. Let D be a Cartier divisor on X and* $D = \sum_{\Gamma \in X^{(1)}} a_\Gamma \Gamma$ *be the expansion as a Weil divisor, where* $X^{(1)}$ *is the set of all codimension one points of X. Then* $D \geqslant 0$ *if and only if* $a_\Gamma \geqslant 0$ *for all* $\Gamma \in X^{(1)}$.

Proof It is suuffuent to show that if $a_\Gamma \geqslant 0$ for all $\Gamma \in X^{(1)}$, then $D \geqslant 0$. Let f_x be a local equation of D at $x \in X$. By our assumption, $f_x \in \mathcal{O}_{X,\Gamma}$ for all $\Gamma \in X^{(1)}$ and $x \in \overline{\{\Gamma\}}$, so that, by virtue of [104, THEOREM 38],

$$f_x \in \bigcap_{x \in \overline{\{\Gamma\}},\, \Gamma \in X^{(1)}} \mathcal{O}_{X,\Gamma} = \mathcal{O}_{X,x},$$

and hence the assertion follows. □

2.4.2 Linear system of a divisor

In this subsection, we fix an integral scheme X, and denote by $R(X)$ the field of all rational functions on X.

Definition 2.4.7 Let D be a Cartier divisor of X. We define

$$H^0(D) := \{f \in R(X)^\times : \operatorname{div}(f) + D \geqslant 0\} \cup \{0\},$$

called the *complete linear system* of the divisor D. It forms a subgroup of $R(X)$ with respect to the additive composition law and is invariant by the multiplication by a scalar in K. Hence it is a K-vector subspace of $R(X)$.

We obtain from the definition that, if D and D' are two Cartier divisors which are linearly equivalent, and g is a non-zero rational function such that $D' = D + \operatorname{div}(g)$. Then the map $f \mapsto fg$ defines a bijection from $H^0(D)$ to $H^0(D')$.

Let D be a Cartier divisor of X. Being a sub-$\mathcal{O}_X$-module of $\mathscr{M}_X$, the invertible sheaf $\mathcal{O}_X(D)$ shares the same generic fibre with $\mathscr{M}_X$, which is also canonically isomorphic to the field $R(X)$. Therefore the unit element in $R(X)$ defines a rational section of D which we denote by s_D. One can verify that $\operatorname{div}(s_D) = D$ and $s_{D+D'} = s_D s_{D'}$ for any couple (D, D') of Cartier divisors of X.

Lemma 2.4.8 *Let D be a Cartier divisor of X and s_D be the meromorphic section of $\mathcal{O}_X(D)$ constructed above. Then D is an effective divisor if and only if s_D extends to a global section of $\mathcal{O}_X(D)$.*

Proof Assume that D is an effective divisor. Then the invertible $\mathcal{O}_X$-module $\mathcal{O}_X(-D)$ is actually an invertible ideal sheaf of $\mathcal{O}_X$ since it is generated by D. Let $s : \mathcal{O}_X \rightarrow \mathcal{O}_X(D)$ be the homomorphism of $\mathcal{O}_X$-modules which is dual to the inclusion map $\mathcal{O}_X(-D) \rightarrow \mathcal{O}_X$. It defines a global section of $\mathcal{O}_X(D)$ whose value at the generic point coincides with s_D.

Conversely, if L is an invertible sheaf on X and if s is a non-zero global section of L, then $\operatorname{div}(s)$ is an effective Cartier divisor. In particular, if s_D extends to a global section of $\mathcal{O}_X(D)$, then $D = \operatorname{div}(s_D)$ is an effective divisor. □

Proposition 2.4.9 *Let D be a Cartier divisor of X. A rational function f lies in $H^0(D)$ if and only if fs_D extends to a global section of $\mathcal{O}_X(D)$.*

Proof By definition, for any non-zero meromorphic function $f \in K$, the relation $fs_D = s_{\operatorname{div}(f)+D}$ holds. The Lemma 2.4.8 shows that fs_D extends to a global section of $\mathcal{O}_X \otimes \mathcal{O}_X(D) \cong \mathcal{O}_X(D)$ if and only if $\operatorname{div}(f) + D$ is an effective divisor. The proposition is thus proved. □

Remark 2.4.10 (1) Let L be an invertible $\mathcal{O}_X$-module. The Proposition 2.4.9 shows that, if s is a non-zero meromorphic section of L, then the relation $t \mapsto t/s$ defines an isomorphism between the groups $H^0(X, L)$ and $H^0(\operatorname{div}(s))$. In particular, if D is a Cartier divisor, then $H^0(D)$ is canonically isomorphic to $H^0(X, \mathcal{O}_X(D))$.

(2) Assume that the scheme X is defined over a ground field k, then the field of rational functions $R(X)$ is an extension of k. Moreover, for any Cartier divisor D of X, $H^0(D)$ is a k-vector subspace of $R(X)$.

2.4.3 $\mathbb{Q}$-Cartier and $\mathbb{R}$-Cartier divisors

As in the previous subsection, X denotes an integral scheme. Let $\mathbb{K}$ be either $\mathbb{Z}$, $\mathbb{Q}$ or $\mathbb{R}$. An element of $\mathrm{Div}_{\mathbb{K}}(X) := \mathrm{Div}(X) \otimes_{\mathbb{Z}} \mathbb{K}$ is called a $\mathbb{K}$-*Cartier divisor* on X. Note that a $\mathbb{Z}$-Cartier divisor is a usual Cartier divisor. A $\mathbb{K}$-Cartier divisor can be regarded as an element of

$$H^0(X, (\mathscr{M}_X^\times/\mathcal{O}_X^\times) \otimes_{\mathbb{Z}} \mathbb{K}) = H^0(X, (\mathscr{M}_X^\times \otimes_{\mathbb{Z}} \mathbb{K})/(\mathcal{O}_X^\times \otimes_{\mathbb{Z}} \mathbb{K})),$$

so that, for any point $x \in X$, there are an open neighborhood U of x and $f \in (\mathscr{M}_X^\times \otimes_{\mathbb{Z}} \mathbb{K})|_U$ such that D is defined by f over U. Note that if $f' \in (\mathscr{M}_X^\times \otimes_{\mathbb{Z}} \mathbb{K})|_U$ also defines D over U, then $f/f' \in (\mathcal{O}_X^\times \otimes_{\mathbb{Z}} \mathbb{K})|_U$. The element f is called a *local equation* of D. Moreover, the morphism of groups $K(X)^\times \to \mathrm{Div}(X)$ induces by extension of scalars a $\mathbb{K}$-linear map $K(X)^\times \otimes_{\mathbb{Z}} \mathbb{K} \to \mathrm{Div}_{\mathbb{K}}(X)$ which we denote by $\mathrm{div}_{\mathbb{K}}(\cdot)$.

Let D be a $\mathbb{K}$-Cartier divisor on X. Let f_x be a local equation of D around x. Note that the condition $f_x \in \mathcal{O}_X^\times \otimes_{\mathbb{Z}} \mathbb{K}$ does not depend on the choice of the local equation of D around x, so that we define $\mathrm{Supp}_{\mathbb{K}}(D)$ to be

$$\mathrm{Supp}_{\mathbb{K}}(D) = \{x \in X \,:\, f_x \notin \mathcal{O}_X^\times \otimes_{\mathbb{Z}} \mathbb{K}\}.$$

Proposition 2.4.11 (1) $\mathrm{Supp}_{\mathbb{K}}(D)$ *is a closed subset of* X.

(2) *If D is a Cartier divisor, then* $\mathrm{Supp}_{\mathbb{Q}}(D) = \bigcap_{n=1}^{\infty} \mathrm{Supp}_{\mathbb{Z}}(nD)$. *In particular,* $\mathrm{Supp}_{\mathbb{Q}}(D) \subseteq \mathrm{Supp}_{\mathbb{Z}}(D)$. *Moreover, if X is normal, then* $\mathrm{Supp}_{\mathbb{Q}}(D) = \mathrm{Supp}_{\mathbb{Z}}(D)$.

(3) *If D is a $\mathbb{Q}$-Cartier divisor, then* $\mathrm{Supp}_{\mathbb{Q}}(D) = \mathrm{Supp}_{\mathbb{R}}(D)$.

Proof The proof can be found in [113, Section 1.2]. □

Definition 2.4.12 Let D be a $\mathbb{K}$-Cartier divisor on X. We say that D is $\mathbb{K}$-effective, denoted by $D \geqslant_{\mathbb{K}} 0$, if, for every $x \in X$, a local equation of D can be expressed by $f_1^{a_1} \cdots f_r^{a_r}$, where $f_1, \ldots, f_r \in \mathcal{O}_{X,x} \setminus \{0\}$ and $a_1, \ldots, a_r \in \mathbb{R}_{>0}$. Similarly as Definition 2.4.7, we define $H^0_{\mathbb{K}}(D)$ to be

$$H^0_{\mathbb{K}}(X, D) := \{\varphi \in R(X)^\times \,:\, \mathrm{div}(\varphi) + D \geqslant_{\mathbb{K}} 0\} \cup \{0\}.$$

Note that in the case where $\mathbb{K} = \mathbb{Z}$, $H^0_{\mathbb{Z}}(X, D)$ coincides with $H^0(D)$ in Definition 2.4.7.

Proposition 2.4.13 *Let D be a $\mathbb{K}$-Cartier divisor on X. Then we have the following:*

(1) *We assume that $\mathbb{K} = \mathbb{Q}$. Then $D \geqslant_{\mathbb{Q}} 0$ if and only if $D \geqslant_{\mathbb{R}} 0$.*
(2) *We assume that $\mathbb{K} = \mathbb{Q}$. Then the natural map $H^0_{\mathbb{Q}}(X, D) \to H^0_{\mathbb{R}}(X, D)$ is bijective.*
(3) *We assume that $\mathbb{K} = \mathbb{Z}$ and X is locally Noetherian and normal. Then $D \geqslant_{\mathbb{Z}} 0$ if and only if $D \geqslant_{\mathbb{Q}} 0$.*
(4) *We assume that $\mathbb{K} = \mathbb{Z}$ and X is locally Noetherian and normal. Then the natural map $H^0_{\mathbb{Z}}(X, D) \to H^0_{\mathbb{Q}}(X, D)$ is bijective.*
(5) *If $a \in H^0(X, \mathcal{O}_X)$ and $\varphi \in H^0_{\mathbb{K}}(X, D)$, then $a\varphi \in H^0_{\mathbb{K}}(X, D)$.*

(6) *If X is locally Noetherian and normal, then $H^0_{\mathbb{K}}(X, D)$ forms a $H^0(X, \mathcal{O}_X)$-submodule of $R(X)$.*

Proof (1) Obviously $D \geqslant_{\mathbb{Q}} 0$ implies $D \geqslant_{\mathbb{R}} 0$. Conversely we assume that $D \geqslant_{\mathbb{R}} 0$. Let f_x be a local equation of D at $x \in X$. Then there are $f_1, \ldots, f_r \in \mathcal{O}_{X,x} \setminus \{0\}$ and $a_1, \ldots, a_r \in \mathbb{R}_{>0}$ such that $f_x = f_1^{a_1} \cdots f_r^{a_r}$. Note that $f_x \in R(X)^\times \otimes_{\mathbb{Z}} \mathbb{Q}$, so that, by Lemma 2.4.14 as below, there are $a'_1, \ldots, a'_r \in \mathbb{Q}_{>0}$ such that $f_x = f_1^{a'_1} \cdots f_r^{a'_r}$. Therefore, D is $\mathbb{Q}$-effective.

(2) Let $\varphi \in H^0_{\mathbb{R}}(X, D) \setminus \{0\}$. Then $D + \operatorname{div}(\varphi) \geqslant_{\mathbb{R}} 0$. Note that $D + \operatorname{div}(\varphi)$ is a $\mathbb{Q}$-Cartier divisor, so that, by (1), $D + \operatorname{div}(\varphi) \geqslant_{\mathbb{Q}} 0$, which means $\varphi \in H^0_{\mathbb{Q}}(X, D) \setminus \{0\}$.

(3) We assume that $D \geqslant_{\mathbb{Q}} 0$. Let $D = \sum_\Gamma a_\Gamma \Gamma$ be the expansion as a Weil divisor. Then $a_\Gamma \geqslant 0$ for all Γ. Thus $D \geqslant 0$ by Proposition 2.4.6.

(4) is a consequence of (3).

(5) is obvious.

(6) By (5), it is sufficient to show that if $\varphi, \psi \in H^0_{\mathbb{K}}(X, D)$, then $\varphi + \psi \in H^0_{\mathbb{K}}(X, D)$. If we set $D = \sum_\Gamma \alpha_\Gamma \Gamma$ ($\alpha_\Gamma \in \mathbb{K}$) and $\operatorname{div}(\varphi) = \sum_\Gamma \operatorname{ord}_\Gamma(\varphi)\Gamma$ as a Weil divisor for $\varphi \in R(X)^\times$, then

$$\operatorname{div}(\varphi) + D \geqslant_{\mathbb{K}} 0 \iff \forall\, \Gamma,\ \operatorname{ord}_\Gamma(\varphi) + \alpha_\Gamma \geqslant 0$$

by [113, Lemma 1.2.4] together with (1). Moreover, for $\varphi, \psi \in R(X)$,

$$\operatorname{ord}_\Gamma(\varphi + \psi) \geqslant \min\{\operatorname{ord}_\Gamma(\varphi), \operatorname{ord}_\Gamma(\psi)\}.$$

Therefore (6) follows. □

Lemma 2.4.14 *Let V be a vector space over $\mathbb{Q}$. Then we have the following:*

(1) *$W_{\mathbb{R}} \cap V = W$ for any vector subspace W of V.*

(2) *Let $x, x_1, \ldots, x_r \in V$ such that $x = a_1x_1 + \cdots + a_rx_r$ for some $a_1, \ldots, a_r \in \mathbb{R}$. Then, for any $\epsilon > 0$, there are $a'_1, \ldots, a'_r \in \mathbb{Q}$ such that $x = a'_1x_1 + \cdots + a'_rx_r$ and $|a'_i - a_i| \leqslant \epsilon$ for all i.*

Proof (1) is obvious because $V/W \to (V/W)_{\mathbb{R}}$ is injective and $(V/W)_{\mathbb{R}} = V_{\mathbb{R}}/W_{\mathbb{R}}$.

(2) We consider the homomorphism $\psi : \mathbb{Q}^r \to V$ sending $(t_1, \ldots, t_r) \in \mathbb{Q}^r$ to $t_1x_1 + \cdots + t_rx_r$. Denote by W the image of ψ. By (1), the point x belongs to W. We pick an element b in $\psi^{-1}(\{x\})$. Let $\psi_{\mathbb{R}} : \mathbb{R}^r \to V_{\mathbb{R}}$ be the scalar extension of ψ, that is, $\psi_{\mathbb{R}}(\alpha_1, \ldots, \alpha_r) = \alpha_1x_1 + \cdots + \alpha_rx_r$ for any $(\alpha_1, \ldots, \alpha_r) \in \mathbb{R}^r$, whose image is $W_{\mathbb{R}}$. As $\operatorname{Ker}(\psi_{\mathbb{R}}) = \operatorname{Ker}(\psi)_{\mathbb{R}}$, $\operatorname{Ker}(\psi)$ is dense in $\operatorname{Ker}(\psi_{\mathbb{R}})$. Therefore, $\psi^{-1}(\{x\}) = b + \operatorname{Ker}(\psi)$ is dense in $\psi_{\mathbb{R}}^{-1}(\{x\}) = b + \operatorname{Ker}(\psi_{\mathbb{R}})$, which implies the assertion of (2). □

Example 2.4.15 The study of effective $\mathbb{Q}$-Cartier or $\mathbb{R}$-Cartier divisors on non-normal schemes is more subtle than that in the normal case. This phenomenon can be shown by the following examples, which have been discussed in [47]. Let $X := \operatorname{Proj}(k[T_0, T_1, T_2]/(T_0T_2^2 - T_1^3))$ over a field k, $U_i := \{T_i \neq 0\} \cap X$ ($i = 0, 1, 2$) and

$x := T_1/T_0, y := T_2/T_0$ on U_0. Then $U_0 = X \setminus \{(0:0:1)\}$ and $U_2 = X \setminus \{(1:0:0)\}$, so that $X = U_0 \cup U_2$. Note that y/x is not regular at $(1:0:0)$ and $y/x \in \mathcal{O}_{X,\zeta}^{\times}$ for all $\zeta \in U_0 \cap U_2$. Let D be a Cartier divisor on X given by

$$D = \begin{cases} (y/x) & \text{on } U_0, \\ (1) & \text{on } U_2. \end{cases}$$

(1) As y/x is not regular at $(1:0:0)$, D is not effective as a Cartier divisor. On the other hand, since

$$2D = \begin{cases} (x) & \text{on } U_0, \\ (1) & \text{on } U_2, \end{cases}$$

D is effective as a $\mathbb{Q}$-Cartier divisor. As a consequence, $1 \notin H^0(X, D)$ and $1 \in H^0_{\mathbb{Q}}(X, D)$, that is, $H^0(X, D) \to H^0_{\mathbb{Q}}(X, D)$ is not surjective.

(2) We assume that $\operatorname{char}(k) = 0$. We set $\varphi := x/y$. As $\varphi = T_1/T_2$ is regular on U_2, $\varphi \in H^0(X, D)$. Here let us see $1 + \varphi \notin H^0_{\mathbb{Q}}(X, D)$. We assume the contrary, that is, $1 + \varphi \in H^0_{\mathbb{Q}}(X, D)$. Then

$$(1 + \varphi)(y/x) = 1 + y/x$$

is $\mathbb{Q}$-effective on U_0, so that there is a positive integer N such that $(1 + y/x)^N$ is regular on U_0. Here we claim that $(y/x)^i$ is regular over U_0 for an integer $i \geqslant 2$. Indeed, we set $i = 2j + \epsilon$, where $j \geqslant 1$ and $\epsilon \in \{0, 1\}$. Then as

$$(y/x)^i = (y/x)^{2j+\epsilon} = (y^2)^j y^{\epsilon} x^{-2j-\epsilon} = x^{j-\epsilon} y^{\epsilon},$$

the assertion follows. Note that

$$y/x = (1/N)\left((1 + y/x)^N - 1 - \sum_{i=2}^{N} \binom{N}{i} (y/x)^i\right),$$

so that y/x is regular on U_0. This is a contradiction because y/x is not regular on U_0.

(3) Next we assume that $\operatorname{char}(k) = 2$. We set $U_0' := U_0 \setminus \{(1:1:1)\}$. Note that $X = U_0' \cup U_2$ and $1 + y/x \in \mathcal{O}_{X,\zeta}^{\times}$ for all $\zeta \in U_0' \cap U_2$, so that we set

$$D' := \begin{cases} (1 + y/x) & \text{on } U_0', \\ (1) & \text{on } U_2. \end{cases}$$

Since y/x is not regular at $(1:0:0)$, we have $D' \neq 0$. Moreover, as $(1 + y/x)^2 = 1 + x$, we have

$$2D' = \begin{cases} (1 + x) & \text{on } U_0', \\ (1) & \text{on } U_2, \end{cases}$$

and hence $2D' = 0$ because $1 + x \in O_{X,\zeta}^{\times}$ for all $\zeta \in U_0'$. Therefore, the natural homomorphism $\mathrm{Div}(X) \to \mathrm{Div}_{\mathbb{K}}(X)$ is not injective. Furthermore $\mathrm{Supp}_{\mathbb{K}}(D') = \emptyset$, but $\mathrm{Supp}_{\mathbb{Z}}(D') = \{(1 : 0 : 0)\}$.

Proposition 2.4.16 *We assume that X is locally Noetherian and normal. Let D be an $\mathbb{R}$-effective $\mathbb{R}$-Cartier divisor on X. Then there are effective Cartier divisors $D_1, \ldots, D_n$ and positive real numbers $a_1, \ldots, a_n$ such that $D = a_1D_1 + \cdots + a_nD_n$.*

Proof If $D = 0$, then the assertion is obvious, so that we may assume that $D \neq 0$. We choose prime divisors $\Gamma_1, \ldots, \Gamma_n$ and $a_1, \ldots, a_n \in \mathbb{R}_{>0}$ such that $D = a_1\Gamma_1 + \cdots + a_n\Gamma_n$ as a Weil divisor. We set

$$\begin{cases} V = \{E = c_1\Gamma_1 + \cdots + c_n\Gamma_n \ : \ (c_1, \ldots, c_n) \in \mathbb{Q}^n \text{ and } E \text{ is a } \mathbb{Q}\text{-Cartier divisor}\}, \\ V_{\mathbb{R}} := V \otimes_{\mathbb{Q}} \mathbb{R}, \quad P = V_{\mathbb{R}} \cap (\mathbb{R}_{>0}\Gamma_1 + \cdots + \mathbb{R}_{>0}\Gamma_n). \end{cases}$$

Then P is an open cone in $V_{\mathbb{R}}$ and $D \in P$. Thus the assertion follows. □

We assume that X is projective over a field k. An $\mathbb{K}$-Cartier divisor D on X is said to be *ample* if there are ample Cartier divisors $D_1, \ldots, D_n$ and $(a_1, \ldots, a_n) \in \mathbb{K}_{>0}^n$ such that $D = a_1D_1 + \cdots + a_nD_n$.

Proposition 2.4.17 *Let A be an ample $\mathbb{R}$-Cartier divisor on X and $D_1, \ldots, D_m$ be Cartier divisors on X. Then there is a positive number δ such that $A + \sum_{j=1}^m \delta_j D_j$ is ample for all $\delta_1, \ldots, \delta_m \in \mathbb{R}$ with $|\delta_1| + \cdots + |\delta_m| < \delta$. In particular, the ampleness of $\mathbb{R}$-Cartier divisors is an open condition.*

Proof We choose ample Cartier divisors $A_1, \ldots, A_n$ and $(a_1, \ldots, a_n) \in \mathbb{R}_{>0}^n$ such that $A = a_1A_1 + \cdots + a_nA_n$. Let l be a positive rational number such that $lA_1 \pm D_j$ is ample for any $j \in \{1, \ldots, m\}$. Note that

$$A + \sum_{j=1}^m \delta_j D_j = \sum_{j=1}^m |\delta_j| \left(lA_1 + \mathrm{sign}(\delta_j)D_j\right) + (a_1 - l(|\delta_1| + \cdots + |\delta_m|))\, A_1 + \sum_{i=2}^m a_iA_i,$$

where

$$\mathrm{sign}(a) = \begin{cases} 1 & \text{if } a \geqslant 0, \\ -1 & \text{if } a < 0. \end{cases}$$

Therefore, if we choose $\delta = a_1/l$, then $A + \sum_{j=1}^m \delta_j D_j$ is ample. □

Proposition 2.4.18 *We assume that X is locally Noetherian and normal. Let D be an $\mathbb{R}$-effective $\mathbb{R}$-Cartier divisor on X. Let $s_1, \ldots, s_n \in \mathrm{Rat}(X)^{\times} \otimes_{\mathbb{Z}} \mathbb{Q}$ and $(a_1, \ldots, a_n) \in \mathbb{R}^n$ such that $a_1, \ldots, a_n$ are linearly independent over $\mathbb{Q}$ and $D + (s_1^{a_1} \cdots s_n^{a_n})$ is $\mathbb{R}$-effective. Then, for any $\epsilon > 0$, there is a positive number δ such that if $|a_1' - a_1| + \cdots + |a_n' - a_n| \leqslant \delta$, then $(1 + \epsilon)D + (s_1^{a_1'} \cdots s_n^{a_n'})$ is $\mathbb{R}$-effective.*

Proof We set $\phi = s_1^{a_1} \cdots s_n^{a_n}$. Let us see that $\mathrm{Supp}((s_i)) \subseteq \mathrm{Supp}((\phi))$ for all i. Otherwise there is a prime divisor Γ such that $\mathrm{ord}_\Gamma(s_i) \neq 0$ and $\mathrm{ord}_\Gamma(\phi) = 0$, so that $\sum_{j=1}^n a_j \,\mathrm{ord}_\Gamma(s_j) = 0$, which contradicts to the linear independency of $a_1, \ldots, a_n$ over $\mathbb{Q}$. If $\mathrm{Supp}((\phi)) = \emptyset$, then $\mathrm{Supp}((s_i)) = \emptyset$ for all i, and hence the assertion is obvious, so that we may assume that $\mathrm{Supp}((\phi)) \neq \emptyset$. Let $\Gamma_1, \ldots, \Gamma_m$ be distinct prime divisors such that $\mathrm{Supp}((\phi)) = \Gamma_1 \cup \cdots \cup \Gamma_m$. Then we can set $(s_i) = \sum_{l=1}^m h_{il}\Gamma_l$ for some $h_{il} \in \mathbb{Q}$. If we set $\gamma_l = \sum_{i=1}^n a_i h_{il}$, then $((\phi)) = \sum_{l=1}^m \gamma_l \Gamma_l$. As $\mathrm{Supp}((\phi)) = \Gamma_1 \cup \cdots \cup \Gamma_m$, one has $\gamma_l \neq 0$ for all l. We set

$$L_+ = \{l \in \{1, \ldots, m\} \,:\, \gamma_l > 0\} \quad \text{and} \quad L_- = \{l \in \{1, \ldots, m\} \,:\, \gamma_l < 0\}.$$

As $\mathrm{ord}_{\Gamma_l}(D) + \gamma_l \geqslant 0$, one has $\mathrm{ord}_{\Gamma_l}(D) > 0$ for all $l \in L_-$. We set $C = \max_{i,l}\{|h_{il}|\}$ and choose $\delta > 0$ such that

$$C\delta < \min\{|\gamma_1|, \ldots, |\gamma_m|\} \quad \text{and} \quad C\delta \leqslant \epsilon \min_{l \in L_-}\{\mathrm{ord}_{\Gamma_l}(D)\}.$$

Let $a'_1, \ldots, a'_n \in \mathbb{R}$ such that $|a'_1 - a_1| + \cdots + |a'_n - a_n| \leqslant \delta$. If we set $\gamma'_l = \sum_{i=1}^n a'_i h_{il}$, then $|\gamma'_l - \gamma_l| \leqslant C\delta$, so that $\{l \mid \gamma'_l > 0\} = L_+$ and $\{l \mid \gamma'_l < 0\} = L_-$. Further, for $l \in L_-$, we have

$$\begin{aligned}(1+\epsilon)\,\mathrm{ord}_{\Gamma_l}(D) + \gamma'_l &= (\mathrm{ord}_{\Gamma_l}(D) + \gamma_l) + (\epsilon\,\mathrm{ord}_{\Gamma_l}(D) + (\gamma'_l - \gamma_l)) \\ &\geqslant \epsilon\,\mathrm{ord}_{\Gamma_l}(D) - C\delta \geqslant 0.\end{aligned}$$

Therefore $(1+\epsilon)D + (s_1^{a'_1} \cdots s_n^{a'_n})$ is $\mathbb{R}$-effective because

$$\mathrm{ord}_{\Gamma_l}((1+\epsilon)D + (s_1^{a'_1} \cdots s_n^{a'_n})) = (1+\epsilon)\,\mathrm{ord}_{\Gamma_l}(D) + \gamma'_l$$

for $l \in \{1, \ldots, m\}$. □

2.5 Green functions

Let k be a field equipped with an absolute value $|\cdot|$, which is complete. If $|\cdot|$ is Archimedean, we assume that it is the usual absolute value on $\mathbb{R}$ or $\mathbb{C}$. Let X be an integral projective scheme over $\mathrm{Spec}\, k$.

2.5.1 Green functions of Cartier divisors

Let X^{an} be the Berkovich topological space associated with X. We denote by $C^0_{\mathrm{gen}}(X^{\mathrm{an}})$ the set of all continuous functions on a non-empty Zariski open subset of X^{an}, modulo the following equivalence relation

$$f \sim g \quad \overset{\text{def}}{\Longleftrightarrow} \quad f \text{ and } g \text{ coincide on a non-empty Zariski open subset.}$$

Note that the addition and the multiplication of functions induce a structure of $\mathbb{R}$-algebra on the set $C^0_{\mathrm{gen}}(X^{\mathrm{an}})$. Moreover, for any non-empty Zariski open subset U of X, we have a natural $\mathbb{R}$-algebra homomorphism from $C^0(U^{\mathrm{an}})$ to $C^0_{\mathrm{gen}}(X^{\mathrm{an}})$. Since U^{an} is dense in X^{an} (see [9] Corollary 3.4.5), we obtain that this homomorphism is injective. Moreover, the $\mathbb{R}$-algebra $C^0_{\mathrm{gen}}(X^{\mathrm{an}})$ is actually the colimit of the system $C^0(U^{\mathrm{an}})$ in the category of $\mathbb{R}$-algebras, where U runs over the set of all non-empty Zariski open subsets of X. We say that an element of $C^0_{\mathrm{gen}}(X^{\mathrm{an}})$ *extends to a continuous function on* U^{an} if it belongs to the image of the canonical homomorphism $C^0(U^{\mathrm{an}}) \to C^0_{\mathrm{gen}}(X^{\mathrm{an}})$.

Remark 2.5.1 Let f be an element of $C^0_{\mathrm{gen}}(X^{\mathrm{an}})$. If U is a non-empty Zariski open subset of X such that f extends to a continuous function on U^{an}, then, by the injectivity of the canonical homomorphism $C^0(U^{\mathrm{an}}) \to C^0_{\mathrm{gen}}(X^{\mathrm{an}})$ there exists a unique continuous function on U^{an} whose canonical image in $C^0_{\mathrm{gen}}(X^{\mathrm{an}})$ is f. Therefore, by gluing of continuous functions we obtain the existence of a largest Zariski open subset U_f of X such that f extends to a continuous function on U_f^{an}. The set U^{an} is called the *domain of definition* of the element f. By abuse of notation, we still use the expression f to denote the continuous function on U_f^{an} corresponding to the element $f \in C^0_{\mathrm{gen}}(X^{\mathrm{an}})$.

If f is a non-zero rational function on X, then it is an invertible regular function on a non-empty Zariski open subset U of X. Therefore $\ln|f|$ is a continuous function on U^{an}, which determines an element of $C^0_{\mathrm{gen}}(X^{\mathrm{an}})$. Note that this element does not depend on the choice of the non-empty Zariski open subset U. We still denote by $\ln|f|$ this element by abuse of notation.

Definition 2.5.2 Let D be a Cartier divisor on X. We call *Green function* of D any element g of $C^0_{\mathrm{gen}}(X^{\mathrm{an}})$ such that, for any local equation f of D over a non-empty Zariski open subset U, the element $g + \ln|f|$ of $C^0_{\mathrm{gen}}(X^{\mathrm{an}})$ extends to a continuous function on U^{an}.

Example 2.5.3 Let f be a non-zero rational function on X. Then $\mathrm{div}(f)$ is a Cartier divisor. By definition, $-\ln|f|$ is a Green function of $\mathrm{div}(f)$. More generally, let L be an invertible $\mathcal{O}_X$-module, equipped with a continuous metric φ. Let s a non-zero rational section of L. Then the function $-\ln|s|_\varphi$, which is well defined outside of the zero points and poles of the section s and is continuous, determines an element of $C^0_{\mathrm{gen}}(X^{\mathrm{an}})$. It is actually a Green function of the divisor $\mathrm{div}(s)$. In particular, we deduce from Remark 2.2.18 that, for any Cartier divisor D on X, there exists a Green function of D.

Remark 2.5.4 One can also construct a metrized invertible sheaf from a Cartier divisor equipped with a Green function. Let D be a Cartier divisor on X and g be a Green function of D. If f is a rational function of X which defines the divisor D on a non-empty Zariski open subset, then the element $f^{-1}s_D$ is a section of the invertible sheaf $\mathcal{O}_X(D)$ which trivialises the latter on U. Note that the element $-(g + \ln|f|)$ of $C^0_{\text{gen}}(X^{\text{an}})$ extends to a continuous function on U^{an}. We denote by $|f^{-1}s_D|_g$ the exponential of this function, which defines a continuous metric on the restriction of L to U. By gluing we obtain a continuous metric on L which we denote by φ_g. By definition one has $g = -\ln|s_D|_g$ in $C^0_{\text{gen}}(X^{\text{an}})$.

Proposition 2.5.5 (1) *An element g in $C^0_{\text{gen}}(X^{\text{an}})$ is a Green function of the trivial Cartier divisor if and only if it extends to a continuous function on X^{an}.*
(2) *Let D and D' be Cartier divisors on X and g, g' be Green functions of D and D', respectively. Then, for $(a, a') \in \mathbb{Z}^2$, $ag + a'g'$ is a Green function of $aD + a'D'$.*

Proof (1) follows from the definition.

(2) Let f and f' be local equations of D and D', respectively. Then $f^a f'^{a'}$ is a local equation of $aD + a'D'$. As $g + \ln|f|$ and $g' + \ln|f'|$ extend to continuous functions locally,

$$a(g + \ln|f|) + a'(g' + \ln|f'|) = ag + a'g' + \ln\left|f^a \cdot f'^{a'}\right|$$

is locally continuous, as required. □

We denote by $\widehat{\mathrm{Div}}(X)$ the set of all pairs of the form (D, g), where D is a Cartier divisor on X and g is a Green function of D. The above proposition shows that $\widehat{\mathrm{Div}}(X)$ forms a commutative group with the composition law

$$((D_1, g_1), (D_2, g_2)) \longmapsto (D_1 + D_2, g_1 + g_2).$$

One has a natural homomorphism of groups $\widehat{\mathrm{Div}}(X) \to \mathrm{Div}(X)$ sending (D, g) to D. The kernel of this homomorphism is $C^0(X^{\text{an}})$.

2.5.2 Green functions for $\mathbb{Q}$-Cartier and $\mathbb{R}$-Cartier divisors

Let $\mathbb{K}$ be either $\mathbb{Q}$ or $\mathbb{R}$. Let f be an element of $R(X)^\times \otimes_{\mathbb{Z}} \mathbb{K}$, that is,

$$f = f_1^{a_1} \cdots f_r^{a_r}, \quad (f_1, \ldots, f_r) \in (R(X)^\times)^r \text{ and } (a_1, \ldots, a_r) \in \mathbb{K}^r.$$

Then one can consider an element of $C^0_{\text{gen}}(X^{\text{an}})$ given by $a_1 \ln|f_1| + \cdots + a_r \ln|f_r|$, which dose not depend on the choice of the expression $f = f_1^{a_1} \cdots f_r^{a_r}$. Indeed, let $f = g_1^{b_1} \cdots g_l^{b_l}$ be another expression of f. Let us choose an affine open set $U = \mathrm{Spec}(A)$ such that $f_1, \ldots, f_r, g_1, \ldots, g_l$ belong to $A^\times$. For $x \in U^{\text{an}}$, as the seminorm $|\cdot|_x$ is multiplicative, $|\cdot|_x$ naturally extends to a map $|\cdot|_x : A^\times \otimes_{\mathbb{Z}} \mathbb{K} \to \mathbb{R}$, so that $|f_1|_x^{a_1} \cdots |f_r|_x^{a_r} = |g_1|_x^{b_1} \cdots |g_l|_x^{b_l}$. Therefore,

$$a_1 \ln|f_1| + \cdots + a_r \ln|f_r| = b_1 \ln|g_1| + \cdots + b_l \ln|g_l|$$

on U^{an}, which shows the assertion. We denote the above function by $\ln|f|$.

Definition 2.5.6 Let D be a $\mathbb{K}$-Cartier divisor on X. We say an element $g \in C^0_{\mathrm{gen}}(X^{\mathrm{an}})$ is a *D-Green function* or *Green function of D* if, for any point $x \in X^{\mathrm{an}}$ and any local equation f of D on a Zariski neighbourhood of $j(x)$, $g + \log|f|$ extends to a continuous function around x. We denote by $\widehat{\mathrm{Div}}_{\mathbb{K}}(X)$ the set of all pairs of the form (D, g), where D is a $\mathbb{K}$-Cartier divisor on X and g is a Green function of D. Note that $\widehat{\mathrm{Div}}_{\mathbb{K}}(X)$ is actually a vector space over $\mathbb{K}$, which is the quotient of $\widehat{\mathrm{Div}}(X) \otimes_{\mathbb{Z}} \mathbb{K}$ by the vector subspace generated by elements of the form $\lambda(D, g) - (\lambda D, \lambda g)$, where $(D, g) \in \widehat{\mathrm{Div}}(X)$ and $\lambda \in \mathbb{K}$.

Proposition 2.5.7 (1) *Let g be a Green function of the trivial $\mathbb{K}$-Cartier divisor. Then g extends to a continuous function on X^{an}.*
(2) *Let D and D' be $\mathbb{K}$-Cartier divisors on X and g, g' be Green functions of D and D', respectively. Then, for $a, a' \in \mathbb{K}$, $ag + a'g'$ is a Green function of $aD + a'D'$.*

Proof It can be proven in the same way as Proposition 2.5.5. □

Proposition 2.5.8 *Let $\mathbb{K}$ be either $\mathbb{Z}$ or $\mathbb{Q}$ or $\mathbb{R}$. Let D be an effective $\mathbb{K}$-Cartier divisor on X and g be a Green function of D. Then the element e^{-g} of $C^0_{\mathrm{gen}}(X)$ extends to a non-negative continuous function on X^{an}.*

Proof Locally on a Zariski open subset $U = \mathrm{Spec}(A)$ of X, the divisor D is defined by $f_1^{a_1} \cdots f_r^{a_r}$ (where $f_1, \ldots, f_r$ are elements of $A \setminus \{0\}$ and $a_1, \ldots, a_r$ are elements of $\mathbb{K}_{>0}$) and the element $g + \ln|f|$ of $C^0_{\mathrm{gen}}(X^{\mathrm{an}})$ extends to a continuous function on U^{an}. Hence $\mathrm{e}^{-g} = |f| \cdot \mathrm{e}^{-(g+\ln|f|)}$ extends to a continuous function on X^{an}, which is non-negative. □

Definition 2.5.9 Let $\mathbb{K}$ be either $\mathbb{Z}$ or $\mathbb{Q}$ or $\mathbb{R}$.

(1) For $f \in H^0_{\mathbb{K}}(D)$, $|f|\exp(-g)$ extends to a continuous function. Indeed, as $D+(f)$ is effective and $g - \ln|f|$ is a Green function of $D+(f)$, by the above proposition, $|f|\exp(-g) = \exp(-(g - \ln|f|))$ is a continuous function. We denote the function $|f|\exp(-g)$ by $|f|_g$. Moreover, $\sup\{|f|_g(x) \,:\, x \in X^{\mathrm{an}}\}$ is denoted by $\|f\|_g$.
(2) Let D be an effective $\mathbb{K}$-Cartier divisor on X and g be a Green function of X. By abuse of notation, we use the expression g to denote the map $-\ln(\mathrm{e}^{-g}) : X^{\mathrm{an}} \to \mathbb{R} \cup \{+\infty\}$, where e^{-g} is the non-negative continuous function on X^{an} described in Proposition 2.5.8. We say that an element (D, g) of $\widehat{\mathrm{Div}}(X)$ or $\widehat{\mathrm{Div}}_{\mathbb{R}}(X)$ is *effective* if D is effective and the map g takes non-negative values.
(3) Let $\overline{D} = (D, g)$ be an element of $\widehat{\mathrm{Div}}_{\mathbb{K}}(X)$. We define $\widehat{H}^0_{\mathbb{K}}(\overline{D})$ to be

$$\widehat{H}^0_{\mathbb{K}}(\overline{D}) := \{f \in R(X)^{\times} \,:\, \overline{D} + \widehat{(f)} \text{ is effective}\} \cup \{0\}.$$

Note that $\widehat{H}^0_{\mathbb{K}}(\overline{D}) = \{f \in H^0_{\mathbb{K}}(D) \,:\, \|f\|_g \leqslant 1\}$.

Remark 2.5.10 Let (D, g) be an element of $\widehat{\mathrm{Div}}_{\mathbb{K}}(X)$ and s be an element of $R(X)^{\times}$. Let $(D', g') = (D + \mathrm{div}(s), g - \ln|s|)$. Then the map $H^0_{\mathbb{K}}(D') \to H^0_{\mathbb{K}}(D)$ sending $f \in H^0_{\mathbb{K}}(D')$ to fs is a bijection. Moreover, for any $f \in H^0_{\mathbb{K}}(D')$ one has $\|f\|_{g'} = \|fs\|_g$.

2.5.3 Canonical Green functions with respect to endmorphisms

Given a polarised dynamical system on a projective variety over Spec K one can attach to the polarisation divisor a canonical Green function, which is closely related to the canonical local height function. We refer the readers to [116] for the original work of Néron in the Abelian variety case, and to [35, 155] for general dynamic systems in the setting of canonical local height and canonical metric respectively. See [95] for the non-Archimedean case.

Let X be an integral projective scheme over Spec k and $f : X \to X$ be a surjective endomorphism of X over k. Let D be a Cartier divisor on X. We assume that there are an integer d and $s \in R(X)^\times$ such that $d > 1$ and $f^*(D) = dD + \mathrm{div}(s)$. We fix a Green function g_0 of D. Then there exists a unique continuous function λ on X^{an} such that

$$(f^{\mathrm{an}})^*(g_0) = dg_0 - \log|s| + \lambda.$$

We set

$$h_0 = 0 \quad \text{and} \quad h_n = \sum_{i=0}^{n-1} \frac{1}{d^{i+1}}((f^{\mathrm{an}})^i)^*(\lambda) \quad (n \geqslant 1).$$

Proposition 2.5.11 (1) *If $n > m$, then one has*

$$\|h_n - h_m\|_{\sup} \leqslant \frac{\|\lambda\|_{\sup}}{d^m(d-1)}.$$

(2) *The sequence $\{h_n\}_{n\geqslant 1}$ of continuous functions on X^{an} converges uniformly to a continuous function h on X^{an}.*
(3) *For all $n \geqslant 1$, one has*

$$(f^{\mathrm{an}})^*(g_0 + h_{n-1}) = d(g_0 + h_n) - \log|s|.$$

(4) *There is a unique Green function g of D with $(f^{\mathrm{an}})^*(g) = dg - \log|s|$ on X^{an}.*
(5) *If $O_X(D)$ is semiample, then φ_g is semipositive.*

Proof (1) Note that

$$\begin{aligned}\|h_n - h_m\|_{\sup} &\leqslant \sum_{i=m}^{n-1} \frac{1}{d^{i+1}} \|((f^{\mathrm{an}})^i)^*(\lambda)\|_{\sup} = \frac{\|\lambda\|_{\sup}}{d^{m+1}} \sum_{i=0}^{n-m-1} \frac{1}{d^i}\\ &\leqslant \frac{\|\lambda\|_{\sup}}{d^{m+1}} \sum_{i=0}^{\infty} \frac{1}{d^i} = \frac{\|\lambda\|_{\sup}}{d^m(d-1)},\end{aligned}$$

as required.

(2) is a consequence of (1).

(3) In the case where $n = 1$, the assertion is obvious. If $n \geqslant 2$, then

$$
\begin{aligned}
(f^{\mathrm{an}})^*(g_0 + h_{n-1}) &= dg_0 - \log|s| + \lambda + \sum_{i=0}^{n-2} \frac{1}{d^{i+1}}((f^{\mathrm{an}})^{i+1})^*(\lambda) \\
&= d\left(g_0 + \frac{1}{d}\lambda + \sum_{i=0}^{n-2} \frac{1}{d^{i+2}}((f^{\mathrm{an}})^{i+1})^*(\lambda)\right) - \log|s| \\
&= d(g_0 + h_n) - \log|s|.
\end{aligned}
$$

(4) By (3), one has

$$
(f^{\mathrm{an}})^*(g_0 + h) = d(g_0 + h) - \log|s|,
$$

so that if we set $g = g_0 + h$, then the desired Green function is obtained.

Next we consider the uniqueness of g. Let g' be another Green function of D with $(f^{\mathrm{an}})^*(g') = dg' - \log|s|$ on X^{an}. Then $(f^{\mathrm{an}})^*(g' - g) = d(g' - g)$ on X^{an}. Note that there is a continuous function θ on X^{an} with $\theta = g' - g$, so that $(f^{\mathrm{an}})^*(\theta) = d\theta$. Here we consider the sup norm $\|\cdot\|_{\mathrm{sup}}$ of continuous functions. Then

$$
\|\theta\|_{\mathrm{sup}} = \|(f^{\mathrm{an}})^*(\theta)\|_{\mathrm{sup}} = \|d\theta\|_{\mathrm{sup}} = d\|\theta\|_{\mathrm{sup}},
$$

and hence $\|\theta\|_{\mathrm{sup}} = 0$. Therefore, $\theta = 0$.

(5) Since $\mathcal{O}_X(D)$ is semiample, there is a semipositive metric φ_0 of $\mathcal{O}_X(D)$ (see Remark 2.3.4). Let t be a non-zero rational section of $\mathcal{O}_X(D)$ such that $\mathrm{div}(t) = D$. We set $g_0 = -\ln|t|_{\varphi_0}$. By (3), $d\varphi_{g_0+h_n} = (f^{\mathrm{an}})^*(\varphi_{g_0+h_{n-1}}) + \varphi_{\log|s|}$. Note that $\varphi_{\log|s|}$ is a semipositive metric of $\mathcal{O}_X(-\mathrm{div}(s))$. Therefore, by induction on n together with Proposition 2.3.2 and Proposition 2.3.5, one can see that $\varphi_{g_0+h_n}$ is semipositive for $n \geqslant 0$, and hence $\varphi_g = \varphi_{g_0+h}$ is also semipositive by Proposition 2.3.3. □

Definition 2.5.12 A Green function g of D is called the *canonical Green function* of D with respect to f if $(f^{\mathrm{an}})^*(g) = dg - \log|s|$ on X^{an}.

Chapter 3
Adelic curves

The theory of adèles in the study of global fields was firstly introduced by Chevalley [49, Chapitre III] in the function field setting and by Weil [147] in the number field setting. This theory allows to consider all places of a global field in a unified way. It also leads to a uniform approach in the geometry of numbers in global fields, either via the adelic version of Minkowski's theorems and Siegel's lemma developed by McFeat [105], Bombieri-Vaaler [13], Thunder [142], Roy-Thunder [128], or via the study of adelic vector bundles developed by Gaudron [62], generalising the slope theory introduced by Stuhler [138], Grayson [71] and Bost [16,18]. The adelic point of view is also closely related to the Arakelov geometry approach to the height theory in arithmetic geometry. Recall that the Arakelov height theory has been developed by Arakelov [4,5], Szpiro [139], Faltings [60], Bost-Gillet-Soulé [23], (compare to the approach of Philippon [122], see also [136] for the comparison of these approaches). We refer the readers to [156] for an application of the Arakelov height theory in the adelic setting to the Bogomolov problem. The Arakelov height theory has been generalised by Moriwaki [106] to the setting of finitely generated fields over a number field (see also [108] for a panoramic view of this theory).

The purpose of this chapter is to develop a formal setting of adelic curves, which permits to include the above examples of global fields and finitely generated extensions of global fields, as well as less standard examples such as the trivial absolute value, polarised projective varieties and arithmetic varieties, and the combination of different adelic structures. More concretely, we consider a field equipped with a family of absolute values on the field, indexed by a measure space, which verifies a "*product formula*" (see Section 3.1 below). This construction is similar to that of M-field introduced by Gubler [79] (see [34] for the height theory of toric varieties in this setting, and the work of Ben Yaakov and Hrushovski in the model theory framework). Moreover, Gaudron and Rémond [65] have studied Siegel's lemma for fields of algebraic numbers with a similar point of view. However, our main concern is to establish a general setting with which we can develop not only the height theory but also the geometry of adelic vector bundles and birational Arakelov theory. Therefore our choice is different from the previous works. In particular, we require that the absolute values are well defined for all places (same as the setting of *globally valued*

H. Chen, A. Moriwaki, *Arakelov Geometry over Adelic Curves*, Lecture Notes in Mathematics 2258, https://doi.org/10.1007/978-981-15-1728-0_3

field of Ben Yaakov and Hrushovski, compare to [79, §2]) and we pay a particular attention to the algebraic coverings of adelic curves and the measurability properties (see Sections 3.3-3.4). We prove that, for any adelic curve S with underlying field K and any algebraic extension L of K, there exists a natural structure of adelic curve on L whose measure space is fibred over that of S with a disintegration kernel (compare to [65]). Curiously, even in the simplest case of finite separable extensions, this result is far from simple (see for example Theorem 3.3.4). The main subtleties appear in the proof of the measurability of the fibre integral, which is neither classic in the theory of disintegration of measures nor in the extension of absolute values in algebraic number theory. We combine the monotone class theorem (in a functional form) in measure theory with divers technics in algebra and number theory such as symmetric polynomials and Vandermonde matrix to resolve this problem.

The chapter is organized as follows. In Section 3.1 we give the definition of adelic curves and discuss several basic measurability properties concerning Archimedean absolute values. Various examples of adelic curves are presented in Section 1.2. In the subsequent two sections, we discuss algebraic extensions of adelic curves. The finite separable extension case is treated in Section 1.3, where we establish the measurability of fibre integrals (Theorem 3.3.4) and the construction of the extended adelic curve (Theorem 3.3.7). Section 3.4 is devoted to the generalisation of these results to arbitrary algebraic extensions case, where the compatibility of the construction in the situation of successive extensions (Theorem 3.4.12) is proved. These results will serve as the fundament for the geometry of adelic vector bundles and birational Arakelov geometry over adelic curves developed in further chapters.

3.1 Definition of Adelic curves

Let K be a commutative field and M_K be the set of all absolute values on K. We call *adelic structure on* K a measure space $(\Omega, \mathcal{A}, \nu)$ equipped with a map $\phi : \omega \mapsto |\cdot|_\omega$ from Ω to M_K satisfying the following properties:

(i) $\mathcal{A}$ is a σ-algebra on Ω and ν is a measure on $(\Omega, \mathcal{A})$;
(ii) for any $a \in K^\times := K \setminus \{0\}$, the function $\omega \mapsto \ln|a|_\omega$ is $\mathcal{A}$-measurable, integrable with respect to the measure ν.

The data $(K, (\Omega, \mathcal{A}, \nu), \phi)$ is called an *adelic curve*. Moreover, the space Ω and the map $\phi : \Omega \to M_K$ are called a *parameter space of* M_K and a *parameter map*, respectively. We do not require neither the injectivity nor the surjectivity of ϕ. Further, if the equality

$$\int_\Omega \ln|a|_\omega \,\nu(\mathrm{d}\omega) = 0 \tag{3.1}$$

holds for each $a \in K^\times$, then the adelic curve $(K, (\Omega, \mathcal{A}, \nu), \phi)$ is said to be *proper*. The equation (3.1) is called a *product formula*.

The set of all $\omega \in \Omega$ such that $|\cdot|_\omega$ is Archimedean (resp. non-Archimedean) is written as Ω_∞ (resp. Ω_{fin}). For any element $\omega \in \Omega$, let K_ω be the completion of

K with respect to the absolute value $|\cdot|_\omega$. Note that $|\cdot|_\omega$ extends by continuity to an absolute value on K_ω which we still denote by $|\cdot|_\omega$.

Proposition 3.1.1 *Let $(K, (\Omega, \mathcal{A}, \nu), \phi)$ be an adelic curve. The set Ω_∞ of all $\omega \in \Omega$ such that the absolute value $|\cdot|_\omega$ is Archimedean belongs to $\mathcal{A}$.*

Proof The result of the proposition is trivial if the characteristic of K is positive since in this case the set Ω_∞ is empty. In the following, we assume that the characteristic of K is zero. Let f be the function on Ω defined as $f(\omega) = \ln |2|_\omega$. Then $\Omega_\infty = \{\omega \in \Omega : f(\omega) > 0\}$. Hence Ω_∞ is a measurable set. □

In the case where $|\cdot|_\omega$ is Archimedean, the field K_ω is equal to $\mathbb{R}$ or $\mathbb{C}$. However, the absolute value $|\cdot|_\omega$ does not necessarily identify with the usual absolute value on K_ω. By Ostrowski's theorem (see [117] Chapter II, Theorem 4.2), there exists a number $\kappa(\omega) \in \,]0, 1]$ such that $|\cdot|_\omega$ equals $|\cdot|^{\kappa(\omega)}$ on $\mathbb{Q}$, where $|\cdot|$ denotes the usual absolute value on $\mathbb{C}$. Therefore one has $|\cdot|_\omega = |\cdot|^{\kappa(\omega)}$ on $K_\omega = \mathbb{R}$ or $\mathbb{C}$.

Proposition 3.1.2 *If we extend the domain of definition of the function κ to Ω by taking the value 0 on $\Omega \setminus \Omega_\infty$, then the function κ is $\mathcal{A}$-measurable and integrable with respect to ν. In particular, if the function κ is bounded from below on Ω_∞ by a positive number, then one has $\nu(\Omega_\infty) < +\infty$.*

Proof The result of the proposition is trivial if Ω_∞ is empty. In the following, we assume that Ω_∞ is non-empty. In this case the field K is of characteristic zero. One has

$$\forall\, \omega \in \Omega_\infty, \quad \ln |2|_\omega = \kappa(\omega) \ln(2),$$

so that

$$\forall\, \omega \in \Omega, \quad \kappa(\omega) = \frac{\max\{0, \ln |2|_\omega\}}{\ln(2)}.$$

Therefore the $\mathcal{A}$-measurability and ν-integrability of the function $\omega \mapsto \ln |2|_\omega$ imply the results of the proposition. □

Proposition 3.1.3 *Let $S = (K, (\Omega, \mathcal{A}, \nu), \phi)$ be an adelic curve. We assume that the field K is countable. Let Ω_0 be the set of points $\omega \in \Omega$ such that the absolute value $|\cdot|_\omega$ on K is trivial. Then Ω_0 belongs to $\mathcal{A}$.*

Proof By definition,

$$\Omega_0 = \bigcap_{a \in K^\times} \{\omega \in \Omega : |a|_\omega = 1\}.$$

Since the function $(\omega \in \Omega) \mapsto |a|_\omega$ is $\mathcal{A}$-measurable, the set $\{\omega \in \Omega : |a|_\omega = 1\}$ belongs to $\mathcal{A}$. Since $K^\times$ is a countable set, we obtain that Ω_0 also belongs to $\mathcal{A}$. □

3.2 Examples

We introduce several fundamental examples of proper adelic curves. Some of them are very classic objects in algebraic geometry or in arithmetic geometry.

3.2.1 Function fields

Let k be a field, C be a regular projective curve over $\operatorname{Spec} k$ and K be the field of all rational functions on C. We denote by Ω the set of all closed points of the curve C, equipped with the discrete σ-algebra $\mathcal{A}$ (namely $\mathcal{A}$ is the σ-algebra of all subsets of Ω). For any closed point x of C, the local ring $O_{C,x}$ is a discrete valuation ring whose fraction field is K. We let $\operatorname{ord}_x(\cdot) : K \to \mathbb{Z} \cup \{+\infty\}$ be the discrete valuation on K of valuation ring $O_{C,x}$ and $|\cdot|_x$ be the absolute value on K defined as

$$\forall\, a \in K^{\times}, \quad |a|_x = \mathrm{e}^{-\operatorname{ord}_x(a)}.$$

Let $n_x := [k(x) : k]$ be the degree of the residue field of x. Thus we obtain a map $\phi : \Omega \to M_K$ sending $x \in \Omega$ to $|\cdot|_x$. We equip the measurable space $(\Omega, \mathcal{A})$ with the measure ν such that $\nu(\{x\}) = n_x$. The relation

$$\forall\, a \in K^{\times}, \quad \sum_{x \in \Omega} n_x \operatorname{ord}_x(a) = 0$$

shows that the equality

$$\int_{\Omega} \ln |a|_x \, \nu(\mathrm{d}x) = 0$$

holds for any $a \in K^{\times}$. Therefore $(K, (\Omega, \mathcal{A}, \nu), \phi)$ is a proper adelic curve.

3.2.2 Number fields

Let K be a number field. Denote by Ω the set of all places of K, equipped with the discrete σ-algebra. For any $\omega \in \Omega$, let $|\cdot|_\omega$ be the absolute value on K in the equivalence class ω, which extends either the usual Archimedean absolute value on $\mathbb{Q}$ or one of the p-adic absolute values (such that the absolute value of p is $1/p$). Thus we obtain a map $\phi : \Omega \to M_K$ sending $\omega \in \Omega$ to $|\cdot|_\omega$. For each $\omega \in \Omega$, let n_ω be the local degree $[K_\omega : \mathbb{Q}_\omega]$, where K_ω and $\mathbb{Q}_\omega$ denote respectively the completion of K and $\mathbb{Q}$ with respect to the absolute value $|\cdot|_\omega$. Let ν be the measure on $(\Omega, \mathcal{A})$ such that $\nu(\{\omega\}) = n_\omega$ for any $\omega \in \Omega$. Note that the usual product formula (cf. [117] Chapter III, Proposition 1.3) asserts that

$$\forall\, a \in K^{\times}, \quad \prod_{\omega \in \Omega} |a|_\omega^{[K_\omega : \mathbb{Q}_\omega]} = 1,$$

which can also be written in the form

$$\forall\, a \in K^{\times}, \quad \int_{\Omega} \ln |a|_{\omega}\, \nu(\mathrm{d}\omega) = 0.$$

Therefore $(K, (\Omega, \mathcal{A}, \nu), \phi)$ is a proper adelic curve.

3.2.3 Copies of the trivial absolute value

Let K be any field and $(\Omega, \mathcal{A}, \nu)$ be an arbitrary measure space. For each $\omega \in \Omega$, let $|\cdot|_{\omega}$ be the trivial absolute value on K, namely one has $|a|_{\omega} = 1$ for any $a \in K^{\times}$. We denote by $\phi : \Omega \to M_K$ the map sending all elements of Ω to the trivial absolute value on K. Then the equality

$$\forall\, a \in K^{\times}, \quad \int_{\Omega} \ln |a|_{\omega}\, \nu(\mathrm{d}\omega) = 0$$

is trivially satisfied. Therefore the data $(K, (\Omega, \mathcal{A}, \nu), \phi)$ form a proper adelic curve.

3.2.4 Polarised varieties

Let k be a field and X be an integral and normal projective scheme of dimension $d \geqslant 1$ over $\operatorname{Spec} k$. Let $K = k(X)$ be the field of rational functions on X and $\Omega = X^{(1)}$ be the set of all prime divisors in X, equipped with the discrete σ-algebra $\mathcal{A}$. We also fix a family $\{D_i\}_{i=1}^{d-1}$ of ample divisors on X. Let ν be the measure on $(\Omega, \mathcal{A})$ such that

$$\forall\, Y \in \Omega = X^{(1)}, \quad \nu(\{Y\}) = \deg(D_1 \cdots D_{d-1} \cap [Y]).$$

Thus we obtain a measure space $(\Omega, \mathcal{A}, \nu)$.

For each $Y \in \Omega$, let $\mathcal{O}_{X,Y}$ be the local ring of X on the generic point of Y. It is a discrete valuation ring since it is a Noetherian normal domain of Krull dimension 1. Moreover, its fraction field is K. We denote by $\operatorname{ord}_Y(\cdot)$ the corresponding valuation on K and by $|\cdot|_Y$ the absolute value on K with $|\cdot|_Y := \mathrm{e}^{-\operatorname{ord}_Y(\cdot)}$. Thus we obtain a map ϕ from Ω to the set of all absolute values on K, sending $Y \in \Omega$ to $|\cdot|_Y$.

For any rational function $f \in K^{\times}$, let (f) be the principal divisor associated with f, which is

$$(f) := \sum_{Y \in \Omega} \operatorname{ord}_Y(f) \cdot Y.$$

Therefore, the relation $\deg(D_1 \cdots D_{d-1} \cdot (f)) = 0$ can be written as

$$\int_{\Omega} \ln |f|_Y\, \nu(\mathrm{d}Y) = 0.$$

Hence $(K, (\Omega, \mathcal{A}, \nu), \phi)$ is a proper adelic curve.

3.2.5 Function field over $\mathbb{Q}$

Let $K = \mathbb{Q}(T)$ be the field of rational functions of one variable T and with coefficients in $\mathbb{Q}$. We consider K as the field of all rational functions on $\mathbb{P}^1_{\mathbb{Q}}$. Any closed point $x \in \mathbb{P}^1_{\mathbb{Q}}$ defines a discrete valuation on K which we denote by $\mathrm{ord}_x(\cdot)$. Let ∞ be the rational point of $\mathbb{P}^1_{\mathbb{Q}}$ such that

$$\mathrm{ord}_\infty(f/g) = \deg(g) - \deg(f)$$

for polynomials f and g in $\mathbb{Q}[T]$ such that $g \neq 0$. Then the open subscheme $\mathbb{P}^1_{\mathbb{Q}} \setminus \{\infty\}$ is isomorphic to $\mathbb{A}^1_{\mathbb{Q}}$. Therefore any closed point x of $\mathbb{P}^1_{\mathbb{Q}}$ different from ∞ corresponds to an irreducible polynomial F_x in $\mathbb{Q}[T]$ (up to dilation by a scalar in $\mathbb{Q}^\times$). By convention we assume that $F_x \in \mathbb{Z}[T]$ and that the coefficients of F_x are coprime. Let $H(x)$ be the Mahler measure of the polynomial F_x, defined as

$$H(x) := \exp\left(\int_0^1 \ln |F_x(\mathrm{e}^{2\pi i t})| \,\mathrm{d}t\right).$$

Note that if the polynomial F_x is written in the form

$$F_x(T) = a_d T^d + \cdots + a_1 T + a_0 = a_d(T - \alpha_1) \cdots (T - \alpha_d),$$

then one has (by Jensen's formula, see [90])

$$H(x) = |a_d| \prod_{j=1}^{d} \max\{1, |\alpha_j|\} \geqslant 1.$$

Let $|\cdot|_x$ be the absolute value on $\mathbb{Q}(T)$ such that

$$\forall\, \varphi \in \mathbb{Q}(T), \quad |\varphi|_x = H(x)^{-\mathrm{ord}_x(\varphi)}.$$

For any prime number p, let $|\cdot|_p$ be the natural extension to $\mathbb{Q}(T)$ of the p-adic absolute value on $\mathbb{Q}$ constructed as follows. For any

$$f = a_d T^d + \cdots + a_1 T + a_0 \in \mathbb{Q}[T]$$

let

$$|f|_p := \max_{j \in \{0,\ldots,d\}} |a_j|_p.$$

Note that one has $|fg|_p = |f|_p \cdot |g|_p$ for f and g in $\mathbb{Q}[T]$ (see [12] Lemma 1.6.3 for example) and thus the function $|\cdot|_p$ on $\mathbb{Q}[T]$ extends in a unique way to a multiplicative function on $\mathbb{Q}(T)$. Moreover, the extended function satisfies the strong triangle inequality and therefore defines a non-Archimedean absolute value on $\mathbb{Q}(T)$.

Denote by $[0,1]_*$ the set of $t \in [0,1]$ such that $\mathrm{e}^{2\pi i t}$ is transcendental. For any $t \in [0,1]_*$, let $|\cdot|_t$ be the absolute value on $\mathbb{Q}(T)$ such that

$$\forall\, \varphi \in \mathbb{Q}(T), \quad |\varphi|_t := |\varphi(\mathrm{e}^{2\pi i t})|,$$

where $|\cdot|$ denotes the usual absolute value of $\mathbb{C}$. The absolute value $|\cdot|_t$ is Archimedean.

Denote by Ω the disjoint union $\Omega_h \amalg \mathcal{P} \amalg [0,1]_*$, where Ω_h is the set of all closed points of $\mathbb{P}^1_{\mathbb{Q}} \setminus \{\infty\}$, and $\mathcal{P}$ denotes the set of all prime numbers. Let $\phi : \Omega \to M_K$ be the map sending $\omega \in \Omega$ to $|\cdot|_\omega$. We equip Ω_h and $\mathcal{P}$ with the discrete σ-algebras, and $[0,1]_*$ with the restriction of the Borel σ-algebra on $[0,1]$. Let $\mathcal{A}$ be the σ-algebra on Ω generated by the above σ-algebras on Ω_h, $\mathcal{P}$ and $[0,1]_*$ respectively. Let ν be the measure on Ω such that $\nu(\{x\}) = 1$ for $x \in \Omega_h$, that $\nu(\{p\}) = 1$ for any prime number p and that the restriction of ν on $[0,1]_*$ coincides with the Lebesgue measure. Then for any $f \in K[T] \setminus \{0\}$ one has

$$\int_\Omega \ln|f|_\omega\, \nu(\mathrm{d}\omega) = \sum_{x\in\Omega_h} \ln|f|_x + \sum_{p\in\mathcal{P}} \ln|f|_p + \int_{[0,1]_*} \ln|f(\mathrm{e}^{2\pi i t})|\,\mathrm{d}t.$$

Since $[0,1] \setminus [0,1]_*$ is negligible with respect to the Lebesgue measure, we obtain that

$$\int_{[0,1]_*} \ln|f(\mathrm{e}^{2\pi i t})|\,\mathrm{d}t = \int_0^1 \ln|f(\mathrm{e}^{2\pi i t})|\,\mathrm{d}t$$

is equal to the logarithm of the Mahler measure of the polynomial f. In particular, if we write the polynomial f in the form

$$f = aF_{x_1}^{r_1} \cdots F_{x_n}^{r_n},$$

where $x_1, \ldots, x_n$ are distinct closed points of $\mathbb{P}^1_{\mathbb{Q}} \setminus \{\infty\}$, and $a \in \mathbb{Q}^\times$. Then one has

$$\int_{[0,1]_*} \ln|f(\mathrm{e}^{2\pi i t})|\,\mathrm{d}t = \ln|a| + \sum_{j=1}^n r_j \ln H(x_j).$$

Therefore one has

$$\int_\Omega \ln|f|_\omega\, \nu(\mathrm{d}\omega) = \sum_{j=1}^n (-r_j)\ln H(x_j) + \sum_{p\in\mathcal{P}} \ln|a|_p + \int_{[0,1]_*} \ln|f(\mathrm{e}^{2\pi i t})|\,\mathrm{d}t = 0.$$

Hence $(K, (\Omega, \mathcal{A}, \nu), \phi)$ is a proper adelic curve.

This example of proper adelic curve is much less classic and may looks artificial. However, it is actually very natural from the Arakelov geometry point of view. In fact, one can consider $\mathbb{Q}(T)$ as the field of the rational functions on the arithmetic variety $\mathbb{P}^1_{\mathbb{Z}} := \mathrm{Proj}(\mathbb{Z}[X,Y])$ with $T = X/Y$. Then the relation

$$\forall\, \varphi \in K^{\times}, \qquad \int_{\Omega} \ln|\varphi|_{\omega}\, \nu(\mathrm{d}\omega) = 0$$

can be interpreted as

$$\widehat{\deg}(\widehat{c}_1(\mathcal{O}_{\mathbb{P}^1_{\mathbb{Z}}}(1), \|\cdot\|) \cdot \widehat{(\varphi)}) = 0,$$

where $\|\cdot\|$ is the continuous Hermitian metric of $\mathcal{O}_{\mathbb{P}^1_{\mathbb{Z}}}(1)$ given by

$$\|aX + bY\|(\xi_1 : \xi_2) = \frac{|a\xi_1 + b\xi_2|}{\max\{|\xi_1|, |\xi_2|\}}$$

and $\widehat{(\varphi)}$ is the arithmetic divisor associated with the rational function φ. The integrals of $\ln|\varphi|_{\omega}$ on Ω_h, $\mathcal{P}$ and I_* correspond to the horizontal, vertical and Archimedean contributions in the arithmetic intersection product, respectively.

3.2.6 Polarised arithmetic variety

The previous example treated in Subsection 3.2.5 can be considered as a very particular case of adelic structures arising from polarised arithmetic varieties. Let K be a finitely generated field over $\mathbb{Q}$ and d be its transcendental degree over $\mathbb{Q}$. Let k be the set of all algebraic elements of K over $\mathbb{Q}$. Note that k is a finite extension over $\mathbb{Q}$. A normal model of K over $\mathbb{Q}$ means an integral and normal projective scheme X over $\mathbb{Q}$ such that the rational function field of X is K.

For simplicity, the set of all $\mathbb{C}$-valued points of $\mathrm{Spec}(K)$ is denoted by $K(\mathbb{C})$, that is, $K(\mathbb{C})$ is the set of all embeddings of K into $\mathbb{C}$. Let X be a *normal projective model* of K over $\mathbb{Q}$, namely X is an integral normal projective $\mathbb{Q}$-scheme, whose field of rational functions identifies with K. Let $\mathrm{Spec}(K) \to X$ be the canonical morphism. Considering the composition

$$\mathrm{Spec}(\mathbb{C}) \longrightarrow \mathrm{Spec}(K) \longrightarrow X,$$

we may treat $K(\mathbb{C})$ as a subset of $X(\mathbb{C})$. Note that

$$K(\mathbb{C}) = X(\mathbb{C}) \setminus \bigcup_{Y \subsetneq X} Y(\mathbb{C}),$$

where Y runs over all prime divisors on X. Indeed, "$\subseteq$" is obvious. Conversely, let $x \in X(\mathbb{C}) \setminus \bigcup_{Y \subsetneq X} Y(\mathbb{C})$. Then, for any $f \in K^{\times}$, f has no zero and pole at x as a rational function on $X(\mathbb{C})$, so that we have a homomorphism $K \to \mathbb{C}$ given by $f \mapsto f(x)$, as required. Note that the restriction to $K(\mathbb{C})$ of the Zariski topology on $X(\mathbb{C})$ does not depend on the choice of X. In fact, for any non-empty Zariski open set U of X, $K(\mathbb{C})$ is a subset of $U(\mathbb{C})$, so that if X' is another normal model of K over $\mathbb{Q}$ and U is a common open set of X and X', then $K(\mathbb{C})$ is a subset of $U(\mathbb{C})$. For $x \in K(\mathbb{C})$, we set $|\cdot|_x := |\sigma_x(\cdot)|$, where σ_x is the corresponding embedding $K \hookrightarrow \mathbb{C}$.

Let O_k be the ring of integers in k. For any maximal ideal $\mathfrak{p}$ of O_k, let $v_\mathfrak{p}$ be the absolute value of k given by

$$v_\mathfrak{p}(\cdot) = \#(O_k/\mathfrak{p})^{-\mathrm{ord}_\mathfrak{p}(\cdot)}.$$

Let $k_\mathfrak{p}$ be the completion of k with respect to $v_\mathfrak{p}$. By abuse of notation, the natural extension of $v_\mathfrak{p}$ to $k_\mathfrak{p}$ is also denoted by $v_\mathfrak{p}$. Let X be a normal projective model of K, $X_\mathfrak{p} := X \times_{\mathrm{Spec}(k)} \mathrm{Spec}(k_\mathfrak{p})$ and $h : X_\mathfrak{p} \to X$ be the natural projection. Let $X_\mathfrak{p}^{\mathrm{an}}$ be the analytification of $X_\mathfrak{p}$ in the sense of Berkovich [9]. For $x \in X_\mathfrak{p}^{\mathrm{an}}$, the associated scheme point of $X_\mathfrak{p}$ is denoted by p_x. We say that x is a *generic point* of $X_\mathfrak{p}^{\mathrm{an}}$ if $h(p_x)$ is the generic point of X. The set of all generic points of $X_\mathfrak{p}^{\mathrm{an}}$ is denoted by $K_\mathfrak{p}^{\mathrm{an}}$. If U is a non-empty Zariski open set of X, then $K_\mathfrak{p}^{\mathrm{an}} \subseteq U_\mathfrak{p}^{\mathrm{an}}$, so that $K_\mathfrak{p}^{\mathrm{an}}$ and the Berkovich topology of $K_\mathfrak{p}^{\mathrm{an}}$ do not depend on the choice of the model X. Moreover, as before, we can see that

$$K_\mathfrak{p}^{\mathrm{an}} = X_\mathfrak{p}^{\mathrm{an}} \setminus \bigcup_{Y \subsetneq X} Y_\mathfrak{p}^{\mathrm{an}},$$

where Y runs over all prime divisors on X. For $x \in K_\mathfrak{p}^{\mathrm{an}}$, the corresponding seminorm $|\cdot|_x$ induces an absolute value of K because K is contained in the residue field of $X_\mathfrak{p}$ at p_x. By abuse of notation, it is also denoted by $|\cdot|_x$.

The Zariski-Riemann space $\mathrm{ZR}(K/k)$ of K over k is defined by the set of all discrete valuation rings $\mathcal{O}$ such that $k \subseteq \mathcal{O} \subseteq K$ and the fraction field of $\mathcal{O}$ is K. For $\mathcal{O} \in \mathrm{ZR}(K/k)$, the associated valuation of K is denoted by $\mathrm{ord}_\mathcal{O}$. We set $|\cdot|_\mathcal{O} := \exp(-\mathrm{ord}_\mathcal{O}(\cdot))$.

Let $\Omega_{\mathrm{geom}}^{\mathrm{fin}} := \mathrm{ZR}(K/k)$, $\Omega_\mathfrak{p}^{\mathrm{fin}} := K_\mathfrak{p}^{\mathrm{an}}$, $\Omega^\infty := K(\mathbb{C})$ and

$$\Omega := \Omega_{\mathrm{geom}}^{\mathrm{fin}} \amalg \coprod_{\mathfrak{p} \in \mathrm{Max}(O_k)} \Omega_\mathfrak{p}^{\mathrm{fin}} \amalg \Omega^\infty,$$

where $\mathrm{Max}(O_k)$ is the set of all maximal ideals of O_k. Let $\phi : \Omega \to M_K$ be the map $\omega \mapsto |\cdot|_\omega$. Here we consider the σ-algebra $\mathcal{A}$ on Ω generated by the discrete σ-algebra on $\Omega_{\mathrm{geom}}^{\mathrm{fin}}$, the Borel σ-algebra on $\Omega_\mathfrak{p}^{\mathrm{fin}}$ with respect to the topology of $K_\mathfrak{p}^{\mathrm{an}}$ for each $\mathfrak{p} \in \mathrm{Max}(O_k)$, and the Borel σ-algebra on Ω^∞ with respect to the topology of $K(\mathbb{C})$. In order to introduce a measure on $(\Omega, \mathcal{A})$, let us fix a normal model X of K and nef adelic arithmetic $\mathbb{R}$-Cartier divisors

$$\overline{D}_1 = (D_1, g_1), \ldots, \overline{D}_d = (D_d, g_d)$$

of C^0-type on X (for details of adelic arithmetic $\mathbb{R}$-Cartier divisors, see [113]). The collection $(X; \overline{D}_1, \ldots, \overline{D}_d)$ is called a *polarisation of* K. Let $X^{(1)}$ be the set of all prime divisors on X. The Radon measure on $X_\mathfrak{p}^{\mathrm{an}}$ given by

$$\varphi \longmapsto \widehat{\deg}_\mathfrak{p}((D_1, g_{1,\mathfrak{p}}) \cdots (D_d, g_{d,\mathfrak{p}}); \varphi)$$

is denoted by $\mu_{(D_1,g_{1,\mathfrak{p}}),\ldots,(D_d,g_{d,\mathfrak{p}})}$. A measure ν on Ω is defined as follows: ν on $\Omega_{\mathrm{geom}}^{\mathrm{fin}}$ is a discrete measure given by

$$\nu(\{O\}) = \begin{cases} \widehat{\deg}\left(\overline{D}_1 \cdots \overline{D}_d \cdot (\Gamma, 0)\right) & \text{if } O = O_{X,\Gamma} \text{ for some } \Gamma \in X^{(1)}, \\ 0 & \text{otherwise,} \end{cases}$$

ν on $\Omega_{\mathfrak{p}}^{\mathrm{fin}}$ is the restriction of $2\mu_{(D_1,g_{1,\mathfrak{p}}),\ldots,(D_d,g_{d,\mathfrak{p}})}$ to $K_{\mathfrak{p}}^{\mathrm{an}}$, and ν on Ω^∞ is given by $2c_1(D_1, g_{1,\infty}) \wedge \ldots \wedge c_1(D_d, g_{d,\infty})$. Then $(K, (\Omega, \mathcal{A}, \nu))$ yields a proper adelic structure of K. Indeed, for each $f \in K^\times$, the product formula can be checked as follows:

$$\begin{aligned} \int_\Omega \ln |f|_\omega \, \nu(d\omega) &= \sum_{\Gamma \in X^{(1)}} -\operatorname{ord}_\Gamma(f) \widehat{\deg}\left(\overline{D}_1 \cdots \overline{D}_d \cdot (\Gamma, 0)\right) \\ &\quad + \sum_{\mathfrak{p} \in \operatorname{Max}(O_k)} \int_{K_{\mathfrak{p}}^{\mathrm{an}}} \ln |f|^2 d\mu_{(D_1,g_{1,\mathfrak{p}}),\ldots,(D_d,g_{d,\mathfrak{p}})} \\ &\quad + \int_{K(\mathbb{C})} \ln |f|^2 c_1(D_1, g_{1,\infty}) \wedge \ldots \wedge c_1(D_d, g_{d,\infty}). \end{aligned}$$

For a proper subvariety of Y of X, $Y_{\mathfrak{p}}^{\mathrm{an}}$ and $Y(\mathbb{C})$ are null sets with respect to the measures $\mu_{(D_1,g_{1,\mathfrak{p}}),\ldots,(D_d,g_{d,\mathfrak{p}})}$ and $c_1(D_1, g_{1,\infty}) \wedge \ldots \wedge c_1(D_d, g_{d,\infty})$, respectively. In addition, we have only countably many prime divisors on X. Therefore, the above equation implies

$$\begin{aligned} \int_\Omega \ln |f|_\omega \, \nu(d\omega) &= -\widehat{\deg}\left(\overline{D}_1 \cdots \overline{D}_d \cdot ((f), 0)\right) \\ &\quad + \sum_{\mathfrak{p} \in \operatorname{Max}(O_k)} \int_{X_{\mathfrak{p}}^{\mathrm{an}}} \ln |f|^2 d\mu_{(D_1,g_{1,\mathfrak{p}}),\ldots,(D_d,g_{d,\mathfrak{p}})} \\ &\quad + \int_{X(\mathbb{C})} \ln |f|^2 c_1(D_1, g_{1,\infty}) \wedge \ldots \wedge c_1(D_d, g_{d,\infty}). \end{aligned}$$

On the other hand,

$$\begin{aligned} 0 &= \widehat{\deg}\left(\overline{D}_1 \cdots \overline{D}_d \cdot \widehat{(f)}\right) \\ &= \widehat{\deg}\left(\overline{D}_1 \cdots \overline{D}_d \cdot \left((f), \sum_{\mathfrak{p} \in \operatorname{Max}(O_k)} -\ln |f|^2[\mathfrak{p}] - \ln |f|^2[\infty]\right)\right) \\ &= \widehat{\deg}\left(\overline{D}_1 \cdots \overline{D}_d \cdot ((f), 0)\right) \\ &\quad - \sum_{\mathfrak{p} \in \operatorname{Max}(O_k)} \int_{X_{\mathfrak{p}}^{\mathrm{an}}} \ln |f|^2 d\mu_{(D_1,g_{1,\mathfrak{p}}),\ldots,(D_d,g_{d,\mathfrak{p}})} \\ &\quad - \int_{X(\mathbb{C})} \ln |f|^2 c_1(D_1, g_{1,\infty}) \wedge \ldots \wedge c_1(D_d, g_{d,\infty}), \end{aligned}$$

as desired.

This proper adelic structure is denoted by $S(X; \overline{D}_1, \ldots, \overline{D}_d)$.

3.2.7 Amalgamation of adelic structures

Let K be a field,

$$\big((\Omega,\mathcal{A},\nu),\phi:\Omega\to M_K\big)\ \text{ and }\ \big((\Omega',\mathcal{A}',\nu'),\phi':\Omega'\to M_K\big)$$

be two adelic structures on K. Then the disjoint union of measure spaces

$$(\Omega,\mathcal{A},\nu)\amalg(\Omega',\mathcal{A}',\nu')$$

together with the map $\Phi:\Omega\amalg\Omega'\to M_K$ extending both ϕ and ϕ' form also an adelic structure on K. If S and S' denote the adelic curves $(K,(\Omega,\mathcal{A},\nu),\phi)$ and $(K,(\Omega',\mathcal{A}',\nu'),\phi')$ respectively, then we use the expression $S\amalg S'$ to denote the adelic curve

$$\big(K,(\Omega,\mathcal{A},\nu)\amalg(\Omega',\mathcal{A}',\nu'),\Phi\big),$$

called the *amalgamation* of the adelic curves S and S'. Similarly, one can define the amalgamation for any finite family of adelic arithmetic structures on the field K. Note that if S and S' are proper, then $S\amalg S'$ is also proper. In fact, for any $a\in K^{\times}$ one has

$$\int_{\Omega}\ln|a|_\omega\,\nu(d\omega)+\int_{\Omega'}\ln|a|_\omega\,\nu'(d\omega)=0.$$

3.2.8 Restriction of adelic structure to a subfield

Let $S=(K,(\Omega,\mathcal{A},\nu),\phi)$ be an adelic curve and let K_0 be a subfield of K. Let $\phi_0:\Omega\to M_{K_0}$ be the map sending $\omega\in\Omega$ to the restriction of $|\cdot|_\omega$ to K_0. Then ϕ_0 defines an adelic structure on K_0, called the *restriction* of the adelic structure of S to K_0. If S is proper, then its restriction to K_0 is also proper.

3.2.9 Restriction of adelic structure to a measurable subset

Let $S=(K,(\Omega,\mathcal{A},\nu),\phi)$ be an adelic curve and Ω_0 be an element of $\mathcal{A}$. Let $\mathcal{A}_0$ be the restriction of the σ-algebra $\mathcal{A}$ to Ω_0 and ν_0 be the restriction of the measure to Ω_0. Then $(K,(\Omega_0,\mathcal{A}_0,\nu_0),\phi|_{\Omega_0})$ is an adelic curve, called the *restriction of S to Ω_0*. Note that this adelic curve is not necessarily proper, even if the adelic curve S is proper.

3.3 Finite separable extensions

Let $S = (K, (\Omega_K, \mathcal{A}_K, \nu_K), \phi_K)$ be an adelic curve. Let K'/K be a finite and separable extension. For each $\omega \in \Omega_K$, let $M_{K',\omega}$ be the set of all absolute values on K' which extend the absolute value $|\cdot|_\omega$ on K. Let $\Omega_{K'}$ be the disjoint union

$$\coprod_{\omega\in\Omega_K} M_{K',\omega}.$$

One has a natural projection $\pi_{K'/K} : \Omega_{K'} \to \Omega_K$ which sends the elements of $M_{K',\omega}$ to ω. Let $\phi_{K'} : \Omega_{K'} \to M_{K'}$ be the map induced by the inclusion maps $M_{K',\omega} \to M_{K'}$. If x is an element of $\Omega_{K'}$, we also use the expression $|\cdot|_x$ to denote the corresponding absolute value. Note that the following diagram is commutative

$$\begin{array}{ccc} \Omega_{K'} & \xrightarrow{\pi_{K'/K}} & \Omega_K \\ {\scriptstyle \phi_{K'}}\downarrow & & \downarrow{\scriptstyle \phi_K} \\ M_{K'} & \xrightarrow[\varpi_{K'/K}]{} & M_K \end{array}$$

and identifies $\Omega_{K'}$ with the fibre product of $M_{K'}$ and Ω_K over M_K in the category of sets, where $\varpi_{K'/K}$ sends any absolute value on K' to its restriction to K. We equip the set $\Omega_{K'}$ with the σ-algebra $\mathcal{A}_{K'}$ generated by $\pi_{K'/K}$ and all real-valued functions of the form $(x \in \Omega_{K'}) \mapsto |a|_x$, where a runs over K'. Namely it is the smallest σ-algebra on $\Omega_{K'}$ which makes these maps measurable[1], where we consider the σ-algebra $\mathcal{A}_K$ on Ω_K and the Borel σ-algebra on $\mathbb{R}$.

We aim to construct a measure $\nu_{K'}$ on the measurable space $(\Omega_{K'}, \mathcal{A}_{K'})$ such that the direct image of $\nu_{K'}$ by $\pi_{K'/K}$ coincides with ν_K. Note that on each fibre $M_{K',\omega}$ of $\pi_{K'/K}$ there is a natural probability measure $\mathbb{P}_{K',\omega}$ such that

$$\forall\, x \in M_{K',\omega}, \quad \mathbb{P}_{K',\omega}(\{x\}) = \frac{[K'_x : K_\omega]}{[K' : K]}. \tag{3.2}$$

We refer to [117] Chapter II, Corollary 8.4 for a proof of the equality

$$\sum_{x\in M_{K',\omega}} \frac{[K'_x : K_w]}{[K' : K]} = 1. \tag{3.3}$$

Intuitively the family of probability measures $\{\mathbb{P}_{K',\omega}\}_{\omega\in\Omega_K}$ should form the disintegration of the measure $\nu_{K'}$ with respect to ν_K. However, as we will show below, the construction of the measure $\nu_{K'}$ relies actually on a subtil application of the monotone class theorem and the properties of extensions of absolute values.

---

[1] A map $f : X' \to X$ of measurable spaces $(X', \mathcal{A}')$ and $(X, \mathcal{A})$ is said to be measurable if $f^{-1}(B) \in \mathcal{A}'$ for all $B \in \mathcal{A}$.

3.3.1 Integration along fibres

If f is a function on $\Omega_{K'}$ valued in $\mathbb{R}$, we define $I_{K'/K}(f)$ to be the function on Ω_K which sends $\omega \in \Omega_K$ to

$$\sum_{x \in M_{K',\omega}} \frac{[K'_x : K_\omega]}{[K' : K]} f(x).$$

This is an $\mathbb{R}$-linear operator from the vector space of all real-valued functions on $\Omega_{K'}$ to that of all real-valued functions on Ω_K. The equality (3.3) shows that $I_{K'/K}$ sends the constant function 1 on $\Omega_{K'}$ to that on Ω_K. The following properties of the linear operator $I_{K'/K}$ are straightforward.

Proposition 3.3.1 *Let f be a real-valued function on $\Omega_{K'}$ and φ be a real-valued function on Ω_K. Let $\widetilde{\varphi} = \varphi \circ \pi_{K'/K}$. Then*

$$I_{K'/K}(\widetilde{\varphi} f) = \varphi I_{K'/K}(f).$$

Proof By definition, for any $\omega \in \Omega_K$ one has

$$\begin{aligned}\big(I_{K'/K}(\widetilde{\varphi} f)\big)(\omega) &= \sum_{x \in M_{K',\omega}} \frac{[K'_x : K_\omega]}{[K' : K]} \widetilde{\varphi}(x) f(x) \\ &= \sum_{x \in M_{K',\omega}} \frac{[K'_x : K_\omega]}{[K' : K]} \varphi(\omega) f(x).\end{aligned}$$

Proposition 3.3.2 *Let $K''/K'/K$ be successive finite separable extensions of fields. Let f be a real-valued function on $\Omega_{K'}$ and $\widetilde{f} = f \circ \pi_{K''/K'}$, where $\pi_{K''/K'} : \Omega_{K''} \to \Omega_{K'}$ sends any absolute value in $\Omega_{K'',\omega}$ to its restriction to K', viewed as an element in $\Omega_{K',\omega}$ ($\omega \in \Omega_K$). Then one has*

$$I_{K''/K}(\widetilde{f}) = I_{K'/K}(f). \tag{3.4}$$

Proof For any $\omega \in \Omega_K$, one has

$$\begin{aligned}\big(I_{K''/K}(\widetilde{f})\big)(\omega) &= \sum_{y \in M_{K'',\omega}} \frac{[K''_y : K_\omega]}{[K'' : K]} \widetilde{f}(y) \\ &= \sum_{x \in M_{K',\omega}} \sum_{\substack{y \in M_{K'',\omega} \\ y|_{K'} = x}} \frac{[K''_y : K'_x]}{[K'' : K']} \cdot \frac{[K'_x : K_\omega]}{[K' : K]} f(x),\end{aligned}$$

which can also be written as

$$\sum_{x \in M_{K',\omega}} \frac{[K'_x : K_\omega]}{[K' : K]} f(x) \sum_{\substack{y \in M_{K'',\omega} \\ y|_{K'} = x}} \frac{[K''_y : K'_x]}{[K'' : K']}.$$

Therefore the desired equality follows from the relation

$$\sum_{\substack{y\in M_{K'',\omega}\\ y|_{K'}=x}}\frac{[K''_y:K'_x]}{[K'':K']}=1.$$

Corollary 3.3.3 *Let φ be an $\mathcal{A}_K$-measurable function on Ω_K. One has*

$$I_{K'/K}(\varphi\circ\pi_{K'/K})=\varphi.$$

Proof It suffices to apply the previous proposition to the successive extensions $K'/K/K$ and then use the fact that $I_{K/K}$ is the identity map to obtain the result. □

3.3.2 Measurability of fibre integrals

The following theorem shows that the operator $I_{K'/K}$ sends an $\mathcal{A}_{K'}$-measurable function to an $\mathcal{A}_K$-measurable function. This result is fundamental in the construction of a suitable measure on the measurable space $(\Omega_{K'},\mathcal{A}_{K'})$.

Theorem 3.3.4 *For any real-valued $\mathcal{A}_{K'}$-measurable function f, the function $I_{K'/K}(f)$ is $\mathcal{A}_K$-measurable.*

Proof *Step 1:* We first prove that, if a is a primitive element of the finite separable extension K'/K (namely $K'=K(a)$) and if f_a is the function on $\Omega_{K'}$ sending $x\in\Omega_{K'}$ to $|a|_x$, then the function $I_{K'/K}(f_a)$ is $\mathcal{A}_K$-measurable.

Let K^{ac} be an algebraic closure of K containing K'. For each $\omega\in\Omega_K$, we extend the absolute value $|\cdot|_\omega$ to K^{ac} via an embedding of K^{ac} into an algebraic closure K_ω^{ac} of K_ω. We still denote by $|\cdot|_\omega$ the extended absolute value on K^{ac} by abuse of notation.

Lemma 3.3.5 *Let $d\in\mathbb{N}_{\geqslant 1}$ and $\{\alpha_1,\ldots,\alpha_d\}$ be a finite family of distinct elements in K^{ac}. For any $\omega\in\Omega_K$, one has*

$$\max_{j\in\{1,\ldots,d\}}|\alpha_j|_\omega=\limsup_{N\to+\infty}\bigg|\sum_{i=1}^d\alpha_i^N\bigg|_\omega^{\frac{1}{N}}. \tag{3.5}$$

Moreover, for any separable element $c\in K^{\mathrm{ac}}$, the function

$$(\omega\in\Omega_K)\longmapsto\max_{\tau\in\mathrm{Aut}_K(K^{\mathrm{ac}})}|\tau(c)|_\omega$$

is $\mathcal{A}_K$-measurable. □

Proof First of all, by the triangle inequality one has

$$\left|\sum_{i=1}^{d}\alpha_i^N\right|_{\omega}^{\frac{1}{N}} \leqslant d^{1/N}\max\{|\alpha_1|_{\omega},\ldots,|\alpha_d|_{\omega}\}.$$

Therefore

$$\max\{|\alpha_1|_{\omega},\ldots,|\alpha_d|_{\omega}\} \geqslant \limsup_{N\to+\infty}\left|\sum_{i=1}^{d}\alpha_i^N\right|_{\omega}^{\frac{1}{N}}.$$

Without loss of generality, we can assume that

$$|\alpha_1|_{\omega} = \ldots = |\alpha_\ell|_{\omega} > |\alpha_{\ell+1}|_{\omega} \geqslant \ldots \geqslant |\alpha_d|_{\omega},$$

where $\ell \in \{1,\ldots,d\}$. For $i \in \{1,\ldots,\ell\}$, let $\beta_i = \alpha_i/\alpha_1$. One has $\beta_1 = 1$ and

$$|\beta_1|_{\omega} = \ldots = |\beta_\ell|_{\omega} = 1.$$

For any integer $N \geqslant 1$, one has

$$\begin{pmatrix} \beta_1^N + \cdots + \beta_\ell^N \\ \vdots \\ \beta_1^{N+\ell-1} + \cdots + \beta_\ell^{N+\ell-1} \end{pmatrix} = \begin{pmatrix} 1 & \cdots & 1 \\ \beta_1^1 & \cdots & \beta_\ell^1 \\ \vdots & \ddots & \vdots \\ \beta_1^{\ell-1} & \cdots & \beta_\ell^{\ell-1} \end{pmatrix} \begin{pmatrix} \beta_1^N \\ \vdots \\ \beta_\ell^N \end{pmatrix}.$$

Let $\widehat{K^{\mathrm{ac}}}$ be the completion of K^{ac} with respect to $|\cdot|_{\omega}$. We equip the vector space $(\widehat{K^{\mathrm{ac}}})^{\ell}$ with the following norm

$$\forall\,(z_1,\ldots,z_\ell) \in (\widehat{K^{\mathrm{ac}}})^{\ell}, \quad \|(z_1,\ldots,z_\ell)\| := \max_{i\in\{1,\ldots,\ell\}} |z_i|_{\omega}.$$

Then the vector $(\beta_1^N,\ldots,\beta_\ell^N)$ has norm 1 with respect to $\|\cdot\|$. Moreover, the Vandermonde matrix above is invertible since $\beta_1,\ldots,\beta_\ell$ are distinct. Therefore the norm of the vector

$$\left(\sum_{i=1}^{\ell}\beta_i^N,\ldots,\sum_{i=1}^{\ell}\beta_i^{N+\ell-1}\right)$$

is bounded from below by a positive constant which does not depend on N, which shows that the sequence

$$\left|\sum_{i=1}^{d}\left(\frac{\alpha_i}{\alpha_1}\right)^N\right|_{\omega}, \quad N \in \mathbb{N},\ N \geqslant 1$$

does not converge to zero when $N \to +\infty$. This implies that

$$\limsup_{N\to+\infty}\left|\sum_{i=1}^{d}\alpha_i^N\right|_{\omega}^{\frac{1}{N}} \geqslant |\alpha_1|_{\omega} = \max\{|\alpha_1|_{\omega},\ldots,|\alpha_d|_{\omega}\}.$$

The equality (3.5) is thus proved.

We now proceed with the proof of the second statement. Let

$$T^d - \lambda_1 T^{d-1} + \cdots + (-1)^d \lambda_d \in K[T]$$

be the minimal polynomial of c over K, and $\alpha_1, \ldots, \alpha_d$ be its roots in K^{ac}. Since c is a separable element, these roots are distinct. By definition, for $k \in \{1, \ldots, d\}$ one has

$$\lambda_k = \sum_{\substack{(i_1,\ldots,i_k)\in\{1,\ldots,d\}^k \\ i_1<\ldots<i_k}} \alpha_{i_1} \cdots \alpha_{i_k}.$$

By the fundamental theorem on symmetric polynomials (see for example [55, §10-11]), if F is a polynomial in $K[X_1, \ldots, X_d]$ which is invariant by the action of the symmetric group $\mathfrak{S}_d$ by permuting the variables, then there exists a polynomial $G \in K[T_1, \ldots, T_d]$ such that

$$F(\alpha_1, \ldots, \alpha_d) = G(\lambda_1, \ldots, \lambda_d).$$

In particular, one has $F(\alpha_1, \ldots, \alpha_d) \in K$ and hence the function

$$(\omega \in \Omega_K) \longmapsto |F(\alpha_1, \ldots, \alpha_d)|_\omega$$

is $\mathcal{A}_K$-measurable. For any $N \in \mathbb{N}$, $N \geqslant 1$, the sum $\sum_{i=1}^d \alpha_i^N$ can be written as a symmetric polynomial evaluated at $(\alpha_1, \ldots, \alpha_d)$, thus the function

$$(\omega \in \Omega_K) \longmapsto \Big|\sum_{i=1}^d \alpha_i^N\Big|_\omega$$

is $\mathcal{A}_K$-measurable. Combining this observation with the equality (3.5), we obtain that the function

$$(\omega \in \Omega_K) \longmapsto \max_{\tau \in \mathrm{Aut}_K(K^{\mathrm{ac}})} |\tau(c)|_\omega = \max_{i\in\{1,\ldots,d\}} |\alpha_i|_\omega$$

is $\mathcal{A}_K$-measurable. □

We now continue with the proof of the statement that the function $I_{K'/K}$ is $\mathcal{A}_K$-measurable. Let $\{\gamma_1, \ldots, \gamma_n\}$ be the orbit of a under the action of $\mathrm{Aut}_K(K^{\mathrm{ac}})$. For any $\omega \in \Omega_K$, let $(s_1(\omega), \ldots, s_n(\omega))$ be the array $(|\gamma_1|_\omega, \ldots, |\gamma_n|_\omega)$ sorted in the decreasing order. Let k be an arbitrary element of $\{1, \ldots, n\}$. For any $\omega \in \Omega_K$, one has

$$s_1(\omega) \cdots s_k(\omega) = \max_{\substack{(i_1,\ldots,i_k)\in\{1,\ldots,n\}^k \\ i_1<\ldots<i_k}} \max_{\tau \in \mathrm{Aut}_K(K^{\mathrm{ac}})} |\tau(\gamma_{i_1} \cdots \gamma_{i_k})|_\omega.$$

By Lemma 3.3.5, we obtain that the function $s_1 \cdots s_k$ is $\mathcal{A}_K$-measurable. Therefore all the functions $s_1, \ldots, s_n$ on Ω_K are $\mathcal{A}_K$-measurable. In particular, if $f_a : \Omega_{K'} \to \mathbb{R}$ is the function sending $x \in \Omega_{K'}$ to $|a|_x$, where a is the primitive element of the finite separable extension K'/K fixed in the beginning of the step, then for any $\omega \in \Omega_K$ one has (we refer the readers to [117, page 163] for the second equality)

$$(I_{K'/K}(f_a))(\omega) = \sum_{x \in M_{K',\omega}} \frac{[K'_x : K_\omega]}{[K' : K]} f_a(x) = \frac{1}{n}\sum_{i=1}^{n} |\gamma_i|_\omega = \frac{1}{n}\sum_{i=1}^{n} s_i(\omega),$$

which implies that $I_{K'/K}(f_a)$ is $\mathcal{A}_K$-measurable.

Step 2: We then prove that, for any element $b \in K'$, the function $I_{K'/K}(f_b)$ on Ω_K is $\mathcal{A}_K$-measurable, where f_b denotes the function on $\Omega_{K'}$ sending $x \in \Omega_{K'}$ to $|b|_x$.

We consider the sub-extension $K(b)/K$ of K'/K. It is a finite and separable extension of K and b is a primitive element. Let g be the function on $\Omega_{K(b)}$ sending $y \in \Omega_{K(b)}$ to $|b|_y$. One has

$$f_b = g \circ \pi_{K'/K(b)},$$

where the map $\pi_{K'/K(b)} : \Omega_{K'} \to \Omega_{K(b)}$ is defined as in Proposition 3.3.2. By (3.4), one obtains

$$I_{K'/K}(f_b) = I_{K(b)/K}(g).$$

By the result obtained in Step 1, the function $I_{K(b)/K}(g)$ on Ω_K is $\mathcal{A}_K$-measurable. This proves the measurability of the function $I_{K'/K}(f_b)$.

Step 3: We are now able to apply the monotone class theorem to prove the announced measurability property.

Let $\mathcal{H}$ be the set of all non-negative and bounded functions f on $\Omega_{K'}$ such that the function $I_{K'/K}(f)$ on Ω_K is $\mathcal{A}_K$-measurable. Note that the constant function 1 on $\Omega_{K'}$ belongs to $\mathcal{H}$ since $I_{K'/K}(1)$ coincides with the constant function 1 on Ω_K. If f and g are two functions in $\mathcal{H}$ such that $f \geqslant g$, then $f - g \in \mathcal{H}$ since $I_{K'/K}(f-g) = I_{K'/K}(f) - I_{K'/K}(g)$ is $\mathcal{A}_K$-measurable. Moreover, Proposition 3.3.1 shows that, if f and g are two functions in $\mathcal{H}$, and φ and ψ are two non-negative and bounded $\mathcal{A}_K$-measurable functions on Ω_K, $\widetilde{\varphi} = \varphi \circ \pi_{K'/K}$ and $\widetilde{\psi} = \psi \circ \pi_{K'/K}$, then the function $\widetilde{\varphi} f + \widetilde{\psi} g$ belongs to $\mathcal{H}$ since

$$I_{K'/K}(\widetilde{\varphi} f + \widetilde{\psi} g) = \varphi I_{K'/K}(f) + \psi I_{K'/K}(g)$$

is $\mathcal{A}_K$-measurable. Finally, the operator $I_{K'/K}$ preserves pointwise limit. Therefore, if $\{f_n\}_{n\in\mathbb{N}}$ is a uniformly bounded sequence of functions in $\mathcal{H}$ which converges pointwisely to a function f, then one has $f \in \mathcal{H}$. These properties show that $\mathcal{H}$ is a λ-family (see Definition A.1.1) on $\Omega_{K'}$.

Let C be the set of all non-negative and bounded functions on $\Omega_{K'}$ which can be written in the form $f_b\widetilde{\varphi}$, where b is an element of K', φ is a non-negative and bounded $\mathcal{A}_K$-measurable function on Ω_K, and $\widetilde{\varphi} = \varphi \circ \pi_{K'/K}$. Note that if b_1 and b_2 are two elements of K' then one has

$$f_{b_1 b_2} = f_{b_1} f_{b_2}.$$

Therefore the family C is stable by multiplication. By the result obtained in Step 2 and Proposition 3.3.1, we obtain that $C \subseteq \mathcal{H}$. The monotone class theorem (Theorem A.1.3) then implies that the family $\mathcal{H}$ contains all non-negative and bounded $\sigma(C)$-measurable functions. Finally, any non-negative $\sigma(C)$-measurable function on $\Omega_{K'}$ can be written as the limit of an increasing sequence of non-negative and *bounded*

$\sigma(C)$-measurable functions, and any real-valued $\sigma(C)$-measurable function is the difference of two non-negative $\sigma(C)$-measurable functions. Therefore, for any real-valued $\sigma(C)$-measurable function f, the function $I_{K'/K}(f)$ is $\mathcal{A}_K$-measurable.

Step 4: It remains to prove that the σ-algebras $\sigma(C)$ and $\mathcal{A}_{K'}$ are the same. Clearly one has $\sigma(C) \subseteq \mathcal{A}_{K'}$ since any function in C is $\mathcal{A}_{K'}$-measurable. To prove the equality it suffices to show that any function of the form f_b with $b \in K'$ is $\sigma(C)$-measurable. Let

$$T^m + \mu_1 T^{m-1} + \cdots + \mu_m \in K[T]$$

be the minimal polynomial of b over K. Then for any $\omega \in \Omega_K$ and any $x \in M_{K',\omega}$, one has

$$|b|_x \leqslant m \cdot \max\{1, |\mu_1|_\omega, \ldots, |\mu_m|_\omega\} \tag{3.6}$$

since otherwise one should have

$$1 > \frac{|\mu_1|_\omega}{|b|_x} + \cdots + \frac{|\mu_m|_\omega}{|b|_x} \geqslant \sum_{i=1}^{m} \frac{|\mu_i|_\omega}{|b|_x^i},$$

which contradicts the equality

$$b^m = -\mu_1 b^{m-1} - \cdots - \mu_m.$$

For any $N \in \mathbb{N}$, let A_N be the set

$$\{\omega \in \Omega_K \,:\, \max\{|\mu_1|_\omega, \ldots, |\mu_m|_\omega\} \leqslant N\} \in \mathcal{A}_K.$$

The relation (3.6) shows that the function $f_b \cdot (\mathbb{1}_{A_N} \circ \pi_{K'/K})$ is non-negative and bounded. Hence it belongs to C. Finally, since

$$f_b = \lim_{N \to +\infty} f_b \cdot (\mathbb{1}_{A_N} \circ \pi_{K'/K}),$$

we obtain that the function f_b is $\sigma(C)$-measurable. The theorem is thus proved. □

3.3.3 Construction of the measure

In this subsection, we describe the construction of a suitable measure on the measurable space $(\Omega_{K'}, \mathcal{A}_{K'})$ to form an adelic structure on K' and prove some compatibility results.

Definition 3.3.6 We denote by $\nu_{K'} : \mathcal{A}_{K'} \to \mathbb{R}_+ \cup \{+\infty\}$ the map defined as follows:

$$\forall\, A \in \mathcal{A}_{K'}, \quad \nu_{K'}(A) := \int_{\Omega_K} I_{K'/K}(\mathbb{1}_A)\,\mathrm{d}\nu_K.$$

By Theorem 3.3.4, this map is well defined.

Theorem 3.3.7 (1) *The map $\nu_{K'}$ is a measure on the measurable space $(\Omega_{K'}, \mathcal{A}_{K'})$ such that, for any non-negative $\mathcal{A}_{K'}$-measurable function f on $\Omega_{K'}$ one has*

$$\int_{\Omega_{K'}} f \,\mathrm{d}\nu_{K'} = \int_{\Omega_K} I_{K'/K}(f)\,\mathrm{d}\nu_K. \tag{3.7}$$

(2) *A real-valued $\mathcal{A}_{K'}$-measurable function f on $\Omega_{K'}$ is integrable with respect to $\nu_{K'}$ if and only if $I_{K'/K}(|f|)$ is integrable with respect to ν_K. Moreover, the equality* (3.7) *also holds for all real-valued $\mathcal{A}'_K$-measurable functions on $\Omega_{K'}$ which are integrable with respect to $\nu_{K'}$.*

(3) *The direct image of the measure $\nu_{K'}$ by the measurable map $\pi_{K'/K}$ coincides with ν_K, namely for any real-valued $\mathcal{A}_K$-measurable function φ on Ω_K which is non-negative (resp. integrable with respect to ν_K), the function $\varphi \circ \pi_{K'/K}$ is non-negative (resp. integrable with respect to $\nu_{K'}$), and one has*

$$\int_{\Omega_{K'}} (\varphi \circ \pi_{K'/K})\,\mathrm{d}\nu_{K'} = \int_{\Omega_K} \varphi \,\mathrm{d}\nu_K. \tag{3.8}$$

(4) *$S' = (K', (\Omega_{K'}, \mathcal{A}_{K'}, \nu_{K'}), \phi_{K'})$ is an adelic curve.*

(5) *For $b \in K' \setminus \{0\}$, one has*

$$[K' : K] \int_{\Omega_{K'}} \ln |b|_x \,\mathrm{d}\nu_{K'} = \int_{\Omega_K} \ln |N_{K'/K}(b)|_\omega \,\mathrm{d}\nu_K, \tag{3.9}$$

where $N_{K'/K}(b)$ is the norm of b with respect to the extension K'/K. In particular, if S is proper, then S' is also proper.

Proof (1) The operator $I_{K'/K}$ preserves pointwise limits. Therefore, if $\{A_n\}_{n\in\mathbb{N}}$ is a countable family of disjoint sets in $\mathcal{A}_{K'}$ and if $A = \bigcup_{n\in\mathbb{N}} A_n$, one has

$$\nu_{K'}(A) = \int_{\Omega_K} I_{K'/K}(\mathbb{1}_A)\,\mathrm{d}\nu_K = \int_{\Omega_K} \sum_{n\in\mathbb{N}} I_{K'/K}(\mathbb{1}_{A_n})\,\mathrm{d}\nu_K = \sum_{n\in\mathbb{N}} \nu_{K'}(A_n),$$

where the last equality comes from the monotone convergence theorem.

The set of all non-negative and bounded $\mathcal{A}_{K'}$-measurable functions f which verify the equality (3.7) forms a λ-family. Moreover, this λ-family contains the set of all functions of the form $\mathbb{1}_A$ ($A \in \mathcal{A}_{K'}$), which is stable by multiplication. By Theorem A.1.3, we obtain that the equality (3.7) actually holds for all non-negative and bounded $\mathcal{A}_{K'}$-measurable functions, and hence holds for general non-negative $\mathcal{A}_{K'}$-measurable functions by the monotone convergence theorem again.

(2) The equality (3.7) clearly implies that a real-valued $\mathcal{A}_{K'}$-measurable function f on $\Omega_{K'}$ is integrable with respect to $\nu_{K'}$ if and only if $I_{K'/K}(|f|)$ is integrable with respect to ν_K. Moreover, if f is a real-valued $\mathcal{A}_{K'}$-measurable function on $\Omega_{K'}$ which is integrable with respect to $\nu_{K'}$, then the equality (3.7) applied to $\max(f, 0)$ and $-\min(f, 0)$ shows that

$$\int_{\Omega_{K'}} \max(f,0)\,\mathrm{d}\nu_{K'} = \int_{\Omega_K} I_{K'/K}(\max(f,0))\,\mathrm{d}\nu_K$$

and

$$\int_{\Omega_{K'}} (-\min(f,0))\,\mathrm{d}\nu_{K'} = \int_{\Omega_K} I_{K'/K}(-\min(f,0))\,\mathrm{d}\nu_K.$$

Since these numbers are finite, the difference of the above two equalities leads to

$$\int_{\Omega_{K'}} f\,\mathrm{d}\nu_{K'} = \int_{\Omega_K} I_{K'/K}(f)\,\mathrm{d}\nu_K.$$

(3) By the two assertions proved above, one has

$$\int_{\Omega_{K'}} (\varphi\circ\pi_{K'/K})\mathrm{d}\nu_{K'} = \int_{\Omega_K} I_{K'/K}(\varphi\circ\pi_{K'/K})\,\mathrm{d}\nu_K.$$

By Corollary 1.2.33, one has

$$I_{K'/K}(\varphi\circ\pi_{K'/K}) = \varphi.$$

Thus we obtain (3.8).

(4) Let b be an element in $K'\setminus\{0\}$ and f_b be the function on $\Omega_{K'}$ sending $x\in\Omega_{K'}$ to $|b|_x$. Let $\lambda = N_{K'/K}(b)$ be the norm of b with respect to the extension K'/K. For any $\omega\in\Omega_K$ one has (see [117] Chapter II, Corollary 8.4 and page 161)

$$\prod_{x\in M_{K',\omega}} |b|_x^{[K'_x:K_\omega]} = |\lambda|_\omega,$$

which implies that

$$I_{K/K'}(\ln f_b) = \frac{1}{[K':K]}\ln f_\lambda, \tag{3.10}$$

where f_λ is the function on Ω_K which sends $\omega\in\Omega_K$ to $|\lambda|_\omega$.

Let $\Omega_{K,\infty}$ be the set of all $\omega\in\Omega_K$ such that $|\cdot|_\omega$ is an Archimedean absolute value. By Proposition 3.1.1, this is an element of $\mathcal{A}_K$. Similarly, let $\Omega_{K',\infty}$ be the set of all $x\in\Omega_{K'}$ such that the absolute value $|\cdot|_x$ is Archimedean. One has $\Omega_{K',\infty} = \pi_{K'/K}^{-1}(\Omega_{K,\infty})$. We will prove the integrability of $\ln f_b$ on $\Omega_{K'}\setminus\Omega_{K',\infty}$ and on $\Omega_{K',\infty}$, respectively. For this purpose we use a refinement of the method in the Step 4 of the proof of Theorem 3.3.4.

Let

$$T^m + \mu_1 T^{m-1} + \cdots + \mu_m \in K[T]$$

be the minimal polynomial of b over K. Then for any $\omega\in\Omega_K\setminus\Omega_{K,\infty}$ and any $x\in M_{K',\omega}$, one has

$$|b|_x \leqslant \max\{1, |\mu_1|_\omega,\ldots,|\mu_m|_\omega\}.$$

Otherwise one should have

$$1 > \max_{i\in\{1,\ldots,m\}} \frac{|\mu_i|_\omega}{|b|_x} \geqslant \max_{i\in\{1,\ldots,m\}} \frac{|\mu_i|_\omega}{|b|_x^i}.$$

However, the equality

$$b^m = -\mu_1 b^{m-1} - \cdots - \mu_m.$$

implies that

$$|b|_x^m \leqslant \max_{i\in\{1,\ldots,m\}} |\mu_i|_\omega \cdot |b|_x^{m-i},$$

which leads to a contradiction. Therefore, if we denote by g the function

$$\omega \longmapsto \max\{0, \ln|\mu_1|_\omega, \ldots \ln|\mu_m|_\omega\} \tag{3.11}$$

on Ω_K, then $\ln f_b$ is bounded from above by $g \circ \pi_{K'/K}$ on $\Omega_K \setminus \Omega_{K,\infty}$. Moreover, by the definition of adelic curves, the functions $\omega \mapsto \ln|\mu_i|_\omega$ is integrable for any $i \in \{1,\ldots,m\}$, hence also is the function g. The function

$$(g \circ \pi_{K'/K} - \ln f_b)1\!\!1_{\Omega_{K'}\setminus\Omega_{K',\infty}}$$

is non-negative, and

$$\begin{aligned}
&I_{K'/K}((g\circ\pi_{K'/K} - \ln f_b)1\!\!1_{\Omega_{K'}\setminus\Omega_{K',\infty}})\\
&= (I_{K'/K}(g\circ\pi_{K'/K}) - I_{K'/K}(\ln f_b))1\!\!1_{\Omega_K\setminus\Omega_{K,\infty}}\\
&= \Big(g - \frac{1}{[K':K]}\ln f_\lambda\Big)1\!\!1_{\Omega_K\setminus\Omega_{K,\infty}}
\end{aligned}$$

is an integrable function with respect to ν_K, where the first equality comes from Proposition 3.3.1 and the fact that $I_{K'/K}$ is a linear operator, and the second equality comes from Corollary 1.2.33 and (3.10). By the second assertion of the theorem, the function $(g \circ \pi_{K'/K} - \ln f_b)1\!\!1_{\Omega_{K'}\setminus\Omega_{K',\infty}}$ is integrable, and hence also is $(\ln f_b)1\!\!1_{\Omega_{K'}\setminus\Omega_{K',\infty}}$.

We now consider the Archimedean case. We assume that Ω_∞ is non-empty. Then the characteristic of the field K is zero. In particular, it contains $\mathbb{Q}$ as its prime field. Moreover, for any $x \in M_{K',\omega}$, one has

$$|b|_x \leqslant m^{\kappa(\omega)} \cdot \max\{1, |\mu_1|_\omega, \ldots, |\mu_m|_\omega\}, \tag{3.12}$$

where $\kappa(\omega)$ is the exponent of $|\cdot|_\omega$ as a power of the usual absolute value on $K_\omega = \mathbb{R}$ or $\mathbb{C}$. Otherwise one should have

$$1 > \frac{|\mu_1|_\omega^{1/\kappa(\omega)}}{|b|_x^{1/\kappa(\omega)}} + \cdots + \frac{|\mu_m|_\omega^{1/\kappa(\omega)}}{|b|_x^{1/\kappa(\omega)}} \geqslant \sum_{i=1}^m \frac{|\mu_i|_\omega^{1/\kappa(\omega)}}{|b|_x^{i/\kappa(\omega)}}.$$

However, the equality

$$b^m = -\mu_1 b^{m-1} - \cdots - \mu_m.$$

implies that

$$|b|_x^{m/\kappa(\omega)} \leqslant |\mu_1|_\omega^{1/\kappa(\omega)}|b|_x^{(m-1)/\kappa(\omega)} + \cdots + |\mu_m|_\omega^{1/\kappa(\omega)}$$

since $|\cdot|_\omega^{1/\kappa(\omega)}$ and $|\cdot|_x^{1/\kappa(\omega)}$ are absolute values on K' and K respectively (which extend the usual absolute value on $\mathbb{Q}$). Therefore, the function $\ln f_b$ is bounded from above by $(\ln(m)\kappa + g) \circ \pi_{K'/K}$ on $\Omega_{K,\infty}$, where g is the function defined in (3.11), and we have extended the function κ on Ω_K by taking the value 0 on $\Omega_K \setminus \Omega_{K,\infty}$. Since the function $\ln(m)\kappa + g$ is integrable, by the same argument as in the non-Archimedean case, we obtain the integrability of the function $\ln f_b$ on $\Omega_{K',\infty}$.

(5) follows from (3.7) and (3.10).

3.4 General algebraic extensions

Let $S = (K, (\Omega_K, \mathcal{A}_K, \nu_K), \phi_K)$ be an adelic curve. In this section, we consider the construction of adelic curves from S whose underlying field are general algebraic extensions of K.

3.4.1 Finite extension

Let K'' be a finite extension of K and K' be the separable closure of K in the field K''. By the result of the previous section, one can construct an adelic structure on the field K' which we denote by $((\Omega_{K'}, \mathcal{A}_{K'}, \nu_{K'}), \phi_{K'})$.

Note that K'' is a purely inseparable extension of K' (see [29] Chapter V, §7, no.7, Proposition 13.a). If q is the degree of the extension K''/K', then for any $\alpha \in K''$ one has $\alpha^q \in K'$ (see [29] Chapter V, §5, no.1, Proposition 1). In particular, any absolute value $|\cdot|$ on K' extends in a unique way to K'' and one has

$$\forall\, \alpha \in K'', \quad |\alpha| = |\alpha^q|^{1/q}, \tag{3.13}$$

where $|\alpha|$ denotes the extended absolute value on K'' evaluated on α, and $|\alpha^q|$ denotes the initial absolute value on K' evaluated on α^q. In other words, the sets $M_{K'}$ and $M_{K''}$ are in canonical bijection. This observation permits to construct, for any $\alpha \in K'' \setminus \{0\}$ the function

$$\Omega_{K'} \to \mathbb{R}, \quad (x \in \Omega_{K'}) \mapsto \ln|\alpha|_x$$

This function is clearly $\mathcal{A}_{K'}$-measurable since one has

$$\forall\, x \in \Omega_{K'}, \quad \ln|\alpha|_x = \frac{1}{q}\ln|\alpha^q|_x.$$

Moreover, it is also integrable with respect to $\nu_{K'}$ and one has

$$\int_{\Omega_{K'}} \ln|\alpha|_x \, \nu_{K'}(\mathrm{d}x) = \frac{1}{q}\int_{\Omega_{K'}} \ln|\alpha^q|_x \, \nu_{K'}(dx). \tag{3.14}$$

This fact shows that

$$(K'', (\Omega_{K'}, \mathcal{A}_{K'}, \nu_{K'}), \phi_{K'})$$

is actually an adelic curve, where we identify $M_{K'}$ with $M_{K''}$. Note that the relation (3.13) shows that $\mathcal{A}_{K'}$ is also the smallest σ-algebra on $\Omega_{K'}$ which makes the canonical projection map $\Omega_{K'} \to \Omega_K$ and the functions $(x \in \Omega_{K'}) \mapsto |\alpha|_x$ measurable, where $\alpha \in K''$.

Definition 3.4.1 We denote by $S \otimes_K K''$ the adelic curve

$$(K'', (\Omega_{K'}, \mathcal{A}_{K'}, \nu_{K'}), \phi_{K'})$$

constructed as above, called the *finite extension* of S induced by the extension of fields K''/K. We also use the expression $\pi_{K''/K}$ to denote the projection map $\pi_{K'/K} : \Omega_{K'} \to \Omega_K$ described in the previous section. Similarly, we also use the expression $I_{K''/K}$ to denote the operator $I_{K'/K}$. Note that Ω'_K identifies also with the fibre product of Ω_K and $M_{K''}$ over M_K in the category of sets since $M_{K'}$ and $M_{K''}$ are the same. Similarly, $\phi_{K'}$ identifies with the projection map from $\Omega_{K'} = \Omega_K \times_{M_K} M_{K''}$ to $M_{K''} = M_{K'}$. Note that if S is proper, then $S \otimes_K K''$ is also proper (cf. (3.14) and Theorem 3.3.7, (5)).

In the following, we will prove that the above construction of adelic curves is compatible with successive finite extensions. The lemma below is important for the proof.

Lemma 3.4.2 *Let L/K be a finite extension of fields, $|\cdot|_v$ be an absolute value on K, and $|\cdot|_w$ be an absolute value on L extending $|\cdot|_v$. Let K^{sc} be the separable closure of K in L. Then the completion K^{sc}_w of K^{sc} with respect to the absolute value $|\cdot|_w$ identifies with the separable closure of K_v in L_w.*

Proof The case where $|\cdot|_v$ is Archimedean is trivial since the characteristic of the field K is then zero and hence $K^{\mathrm{sc}} = L$. In the following, we assume that $|\cdot|_v$ is non-Archimedean. We first prove that the extension K^{sc}_w/K_v is separable. Let $\alpha \in K^{\mathrm{sc}}$ be a primitive element (see [29] Chapter V, §7, no.4 Theorem 1 for its existence) of the separable extension K^{sc}/K and let F be its minimal polynomial. Assume that F is decomposed in $K^{\mathrm{sc}}_v[T]$ into the product of distinct irreducible polynomials as $F = F_1 \cdots F_m$. Since F is a separable polynomial, the same are the polynomials $F_1, \ldots, F_m$. For any extension $|\cdot|_w$ of the absolute value $|\cdot|_v$ to K^{sc}, there exists an index $i \in \{1, \ldots, m\}$ such that $K^{\mathrm{sc}}_w \cong K_v[T]/(F_i)$ (see [117] Chapter II, Propositions 8.2 and 8.3). Therefore the extension K^{sc}_w/K_v is separable.

In the following, we prove that the extension L_w/K^{sc}_w is purely inseparable. Note that $L_w = K_v(L) = K^{\mathrm{sc}}_w(L)$. Since the extension L/K^{sc} is purely inseparable, we obtain that the extension L_w/K^{sc}_w is also purely inseparable since it is generated by purely inseparable elements (see [29] Chapter V, §7, no.2, the corollary of Proposition 2). By [29] Chapter V, §7, no.7, Proposition 13.c, we obtain that K^{sc}_w is the separable closure of K_v in L_w since K_w^{sc}/K_v is separable and L_w/K_w^{sc} is purely inseparable. □

Remark 3.4.3 Let K''/K be a finite extension of fields and denote by

$$(K'', (\Omega'', \mathcal{A}'', \nu''), \phi'')$$

the adelic curve $S \otimes_K K''$. The above lemma allows to write the operator $I_{K''/K}$ in the following form: for any real-valued $\mathcal{A}''$-measurable function f on Ω''

$$\forall\, \omega \in \Omega_K, \quad (I_{K''/K}(f))(\omega) = \sum_{x \in M_{K'',\omega}} \frac{[K''_x : K_\omega]_s}{[K'' : K]_s} f(x),$$

where for any finite extension L_2/L_1 of fields, the expression $[L_2 : L_1]_s$ denotes the separable degree of the extension L_2/L_1.

Proposition 3.4.4 *Let $K_2/K_1/K$ be successive finite extensions of fields. If*

$$((\Omega_K, \mathcal{A}_K, \nu_K), \phi_K)$$

is an adelic structure on K and S is the corresponding adelic curve, then one has

$$(S \otimes_K K_1) \otimes_{K_1} K_2 = S \otimes_K K_2.$$

Moreover, one has

$$\pi_{K_1/K} \circ \pi_{K_2/K_1} = \pi_{K_2/K}, \quad I_{K_1/K} \circ I_{K_2/K_1} = I_{K_2/K}, \tag{3.15}$$

where we have used the conventions of notation described in Definition 3.4.1.

Proof We denote by $(K_1, (\Omega_1, \mathcal{A}_1, \nu_1), \phi_1)$ and $(K_2, (\Omega_2, \mathcal{A}_2, \nu_2), \phi_2)$ the adelic curves $S \otimes_K K_1$ and $S \otimes_K K_2$ respectively. First of all, set-theoretically one has

$$\Omega_2 = \coprod_{\omega \in \Omega_K} M_{K_2,\omega} = \coprod_{\omega \in \Omega_K} \coprod_{x \in M_{K_1,\omega}} M_{K_2,x} = \coprod_{x \in \Omega_1} M_{K_2,x},$$

and hence

$$\pi_{K_2/K} = \pi_{K_1/K} \circ \pi_{K_2/K_1}$$

Moreover, if f is a real-valued function on Ω_2, by Lemma 3.4.2 and Remark 3.4.3, for any $\omega \in \Omega_K$ one has

$$\begin{aligned}(I_{K_2/K}(f))(\omega) &= \sum_{y \in M_{K_2,\omega}} \frac{[K_{2,y} : K_\omega]_s}{[K_2 : K]_s} f(y)\\ &= \sum_{x \in M_{K_1,\omega}} \frac{[K_{1,x} : K_\omega]_s}{[K_1 : K]_s} \sum_{y \in M_{K_2,x}} \frac{[K_{2,y} : K_{1,x}]_s}{[K_2 : K_1]_s} f(y) = I_{K_1/K}(I_{K_2/K_1}(f))(\omega),\end{aligned}$$

where in the second equality we have used the multiplicativity of the separable degree (see [29] Chapter V, §6, no.5).

We then show that the map π_{K_2/K_1} is $\mathcal{A}_2$-measurable. Since the σ-algebra $\mathcal{A}_1$ is generated by $\pi_{K_1/K}$ and functions of the form $f_a : (x \in \Omega_1) \mapsto |a|_x$ with $a \in K_1$, it suffices to prove that the maps $\pi_{K_1/K} \circ \pi_{K_2/K_1}$ and $f_a \circ \pi_{K_2/K_1}$ are $\mathcal{A}_2$-measurable. We have shown that $\pi_{K_1/K} \circ \pi_{K_2/K_1} = \pi_{K_2/K}$, which is clearly $\mathcal{A}_2$-measurable by the definition of the adelic curve $S \otimes_K K_2$. Moreover, if a is an element in K_1, then the map $f_a \circ \pi_{K_2/K_1}$ sends $y \in \Omega_2$ to $|a|_y$. Hence it is also $\mathcal{A}_2$-measurable. In particular, the σ-algebra $\mathcal{A}_2$ contains the σ-algebra $\mathcal{A}_2'$ in the adelic structure of $(S \otimes_K K_1) \otimes_{K_1} K_2$, namely the smallest σ-algebra which makes the map π_{K_2/K_1} and all functions of the form $(y \in \Omega_K) \mapsto |\alpha|_y$ measurable, where α runs over K_2.

From the equality $\pi_{K_1/K} \circ \pi_{K_2/K_1} = \pi_{K_2/K}$ we also obtain that $\pi_{K_2/K}^{-1}(\mathcal{A}_K)$ is contained in $\pi_{K_2/K_1}^{-1}(\mathcal{A}_1)$, and thus is contained in $\mathcal{A}_2'$ since π_{K_2/K_1} is $\mathcal{A}_2'$-measurable. Since $\mathcal{A}_2$ is the smallest σ-algebra on Ω_2 which makes the map $\pi_{K_2/K}$ and all functions of the form $(y \in \Omega_K) \mapsto |\alpha|_y$ measurable, we obtain that $\mathcal{A}_2 \subseteq \mathcal{A}_2'$. Combining with the result obtained above, we obtain that the σ-algebras $\mathcal{A}_2$ and $\mathcal{A}_2'$ coincide.

Finally, the relation $I_{K_1/K} \circ I_{K_2/K_1} = I_{K_2/K}$ shows that the measure ν_2 coincides with the measure in the adelic structure of the adelic curve $(S \otimes_K K_1) \otimes_{K_1} K_2$. The proposition is thus proved. □

3.4.2 General algebraic extensions

We now consider an algebraic extension L of K which is not necessarily finite. Let $\mathscr{E}_{L/K}$ be the set of all finite extensions of K which are contained in L. This set is ordered by the relation of inclusion. Moreover, it is also filtered in the sense that, if K_1 and K_2 are two finite extensions of K which are contained in L, then there exists a finite extension $K_3 \in \mathscr{E}_{L/K}$ such that $K_3 \supseteq K_1 \cup K_2$.

By the result obtained in the previous subsection, for each element K'' in $\mathscr{E}_{L/K}$, we can equipped K'' with a natural adelic structure induced from the adelic structure of S, as described in Definition 3.4.1. We denote by $((\Omega_{K''}, \mathcal{A}_{K''}, \nu_{K''}), \phi_{K''})$ this adelic structure. Moreover, Proposition 3.4.4 shows that, for successive finite extensions $K_2/K_1/K$ of the field K which are contained in L, there exist a natural projection

$$\pi_{K_2/K_1} : (\Omega_{K_2}, \mathcal{A}_{K_2}) \longrightarrow (\Omega_{K_1}, \mathcal{A}_{K_1})$$

together with a disintegration operator I_{K_2/K_1} from the vector space of all real-valued $\mathcal{A}_{K_2}$-measurable functions on Ω_{K_2} to that of all real-valued $\mathcal{A}_{K_1}$-measurable functions on Ω_{K_1}, which sends ν_{K_2}-integrable functions to ν_{K_1}-integrable functions. These data actually define a functor from a filtered ordered set to the category of measure spaces. Intuitively one can define an adelic structure on L whose measure space part is the projective limit of this functor. However, the projective limite in the category of measure spaces does not exist in general (the product of infinitely many measures need not make sense). Therefore more careful treatment is needed for our setting of projective system of finite extensions of adelic curves. Our strategy is to construct the fibres as projective limits of probability spaces, which always exist.

For any $\omega \in \Omega_K$, let $M_{L,\omega}$ be the set of all absolute values on L which extend $|\cdot|_\omega$. Let Ω_L be the disjoint union of all $M_{L,\omega}$ with ω runs through Ω_K. In other words, Ω_L is the fibre product of Ω_K and M_L over M_K. The inclusion maps $M_{L,\omega} \to M_L$ define a map from Ω_L to M_L which we denote by ϕ_L. Moreover, for any extension $K'' \in \mathscr{E}_{L/K}$ and any $\omega \in \Omega_K$, one has a natural map from $M_{L,\omega}$ to $M_{K'',\omega}$ defined by restriction of absolute values. These maps induce a map from Ω_L to $\Omega_{K''}$ which we denote by $\pi_{L/K''}$. If x is an element of $\Omega_{K''}$, we denote by $M_{L,x}$ the set of absolute values on L which extends $|\cdot|_x$. It identifies with the inverse image of $\{x\}$ by $\pi_{L/K''}$. If $K_1 \subseteq K_2$ are extensions in $\mathscr{E}_{L/K}$, then one has

$$\pi_{L/K_1} = \pi_{K_2/K_1} \circ \pi_{L/K_2}. \tag{3.16}$$

Proposition 3.4.5 *For any $K'' \in \mathscr{E}_{L/K}$, the map $\pi_{L/K''} : \Omega_L \to \Omega_{K''}$ is surjective.*

Proof Let x be an element in $\Omega_{K''}$. The absolute value $|\cdot|_\omega$ extends in a unique way to the algebraic closure $(K''_x)^{\mathrm{ac}}$ of K''_x (see [117] Chapter II, Theorem 4.8). Therefore, if we choose an embedding of L into $(K''_x)^{\mathrm{ac}}$, then we obtain an absolute value on L which extends $|\cdot|_x$. □

Similarly to [65, Lemma 2.1], the set Ω_L described above gives an explicit construction of the projective limit of the projective system $\{\Omega_{K''}\}_{K''\in\mathscr{E}_{L/K}}$ in the category of sets, where $\{\pi_{L/K''}\}_{K''\in\mathscr{E}_{L/K}}$ are universal maps. In fact, any absolute value on L is uniquely determined by its restrictions on the subfields in $\mathscr{E}_{L/K}$. We equip Ω_L with the σ-algebra $\mathcal{A}_L$ generated by the maps $\pi_{L/K''}$ (namely the smallest σ-algebra which makes all maps $\pi_{L/K''}$ measurable) where K'' runs over $\mathscr{E}_{L/K}$. Thus $(\Omega_L, \mathcal{A}_L)$ identifies with the projective limit of the projective system $\{(\Omega_{K''}, \mathcal{A}_{K''})\}_{K''\in\mathscr{E}_{L/K}}$ in the category of mesurable spaces.

Let ω be an element in Ω_K. We equip $M_{L,\omega}$ with the smallest σ-algebra $\mathcal{A}_{L,\omega}$ such that the restriction of $\pi_{L/K''}$ to $M_{L,\omega}$ is measurable for any $K'' \in \mathscr{E}_{L/K}$, where we consider the discrete σ-algebra on $M_{K'',\omega} = \pi^{-1}_{K''/K}(\{\omega\})$. Let $V_{L,\omega}$ be the set of all real-valued functions on Ω_L which can be written in the form $f \circ (\pi_{L/K''}|_{M_{L,\omega}})$, where K'' is an element of $\mathscr{E}_{L/K}$, and f is a function on $M_{K'',\omega}$. Let

$$I_{L/K,\omega} : V_{L,\omega} \longrightarrow \mathbb{R}$$

be the map which sends any function of the form $f \circ (\pi_{L/K''}|_{M_{L,\omega}})$ to the integral

$$\int_{M_{K'',\omega}} f \, \mathrm{d}\mathbb{P}_{K',\omega}, \tag{3.17}$$

where K' is the separable closure of K in K'', and $\mathbb{P}_{K',\omega}$ is the probability measure on $M_{K',\omega}$ defined in (3.2). Similarly to (3.15), the fibre integral is compatible with successive finite extensions of the field K and the map $I_{L/K,\omega}$ is well defined since the value of the integral (3.17) does not depend on the choice of the field K'' upon which we write the function in $V_{L,\omega}$ as the composition of a function on $M_{K'',\omega}$ with $\pi_{L/K''}|_{M_{L,\omega}}$.

Proposition 3.4.6 *The set $V_{L,\omega}$ forms an algebra over $\mathbb{R}$ with respect to the composition laws of addition and multiplication of functions, and the map $I_{L/K,\omega} : V_{L,\omega} \to \mathbb{R}$ is an $\mathbb{R}$-linear operator. Moreover, it induces a probability measure on the measurable space $(M_{L,\omega}, \mathcal{A}_{L,\omega})$.*

Proof The first assertion comes from the fact that the set $\mathscr{E}_{L/K}$ is filtered, which implies that any finite collection of functions in $V_{L,\omega}$ descend on the same space $M_{K'',\omega}$, where $K'' \in \mathscr{E}_{L/K}$. In particular, the family $\mathcal{D}$ of subsets $A \subseteq M_{L,\omega}$ such that $1\!\!1_A \in V_{L,\omega}$ is an algebra (of sets), which generates $\mathcal{A}_{L,\omega}$ as a σ-algebra. Moreover, the map $\mathbb{P}_{L,\omega} : \mathcal{D} \to \mathbb{R}_+$ which sends $A \in \mathcal{D}$ to $I_{L/K,\omega}(1\!\!1_A)$ is an additive functional. Clearly it sends $M_{L,\omega}$ to 1.

The σ-algebra $\mathcal{A}_{L,\omega}$ is actually the Borel algebra of the projective limit topology on $M_{L,\omega}$ (namely the most coarse topology on $M_{L,\omega}$ which makes all maps $\pi_{L/K''}$ continuous, where $K'' \in \mathscr{E}_{L/K}$). This topology also identifies with the induced topology on $M_{L,\omega}$ viewed as a subset of $\prod_{F\in\mathscr{E}_{L/K}} M_{F,\omega}$ (equipped with the product topology), where on each set $M_{F,\omega}$ we consider the discrete topology. Note that $M_{L,\omega}$ is actually a closed subset of this product space since it is the intersection of closed subsets of the form

$$W_{K''} := \Big\{(x_F)_{F\in\mathscr{E}_{L/K}} \in \prod_{F\in\mathscr{E}_{L/K}} M_{F,\omega} \;:\; \pi_{K''/F}(x_{K''}) = x_F \text{ for } F \subseteq K''\Big\}.$$

Therefore, by Tychonoff's theorem, we obtain that $M_{L,\omega}$ is actually a compact topological space. Moreover, any set in $\mathcal{D}$ is open and closed since it is the inverse image of a discrete set by a continuous map. Therefore, the sets in $\mathcal{D}$ are open and compact. As a consequence, if $\{A_n\}_{n\in\mathbb{N}}$ is a sequence of disjoint sets in $\mathcal{D}$ whose union also lies in $\mathcal{D}$, then for sufficiently large n one has $A_n = \varnothing$. Hence the function $\mathbb{P}_{L,\omega} : \mathcal{D} \to \mathbb{R}_+$ is actually σ-additive. By Carathéodory's extension theorem, the function $\mathbb{P}_{L,\omega}$ extends to a Borel probability measure on $(M_{L,\omega}, \mathcal{A}_{L,\omega})$ such that

$$I_{L/K,\omega}(f) = \int_{M_{L,\omega}} f \,\mathrm{d}\mathbb{P}_{L,\omega}.$$

The proposition is thus proved. □

Remark 3.4.7 Let V_L be the vector space of all real-valued functions f on Ω_L which can be written as $g \circ \pi_{L/K''}$, where K''/K is a finite extension which is contained in L. Then the above construction leads to a linear operator $I_{L/K}$ from V_L to the vector space of all real-valued functions on Ω, sending $f \in V_L$ to the function

$$(\omega \in \Omega) \longmapsto I_{L/K,\omega}(f|_{M_{L,\omega}}).$$

Clearly, if g is a real-valued function on Ω, then $I_{L/K}(g \circ \pi_{L/K}) = g$.

The above proposition allows to define the fibre integrals for non-negative $\mathcal{A}_L$-measurable functions on Ω_L.

Proposition 3.4.8 *Let f be a non-negative $\mathcal{A}_L$-measurable function on Ω_L. For any $\omega \in \Omega_K$, the restriction of f to $M_{L,\omega}$ is $\mathcal{A}_{L,\omega}$-measurable. Moreover, the map $I_{L/K}(f)$ from Ω_K to $[0, +\infty]$ which sends $\omega \in \Omega_K$ to $\int_{M_{L,\omega}} f(x)\,\mathbb{P}_{L,\omega}(\mathrm{d}x)$ is $\mathcal{A}_K$-measurable.*

Proof Let $\mathcal{H}$ be the set of all bounded non-negative functions f on Ω_K such that $f|_{M_{L,\omega}}$ is $\mathcal{A}_{L,\omega}$-measurable for any $\omega \in \Omega_K$ and that the map

$$(\omega \in \Omega_K) \longmapsto \int_{M_{L,\omega}} f(x)\,\mathbb{P}_{L,\omega}(\mathrm{d}x)$$

is $\mathcal{A}_K$-measurable. Then $\mathcal{H}$ is a λ-family of non-negative functions on Ω_L (see Definition A.1.1). Moreover, the set $\mathcal{H}$ contains the subset C of all bounded non-negative functions of the form $g \circ \pi_{L/K''}$, where K'' is an element in $\mathscr{E}_{L/K}$ and g is an $\mathcal{A}_{K''}$-measurable function on $\Omega_{K''}$. In fact, for any $\omega \in \Omega_K$, one has

$$\int_{M_{L,\omega}} g(\pi_{L/K''}(x))\,\mathbb{P}_{L,\omega}(\mathrm{d}x) = I_{K''/K}(g)(\omega).$$

Since C is stable under multiplication, by the monotone class theorem A.1.3 we obtain that the family $\mathcal{H}$ actually contains all non-negative, bounded and $\sigma(C)$-measurable functions. By definition, $\mathcal{A}_L$ is the σ-algebra generated by the maps $\pi_{L/K''}$ with $K'' \in \mathscr{E}_{L/K}$. Therefore one has $\mathcal{A}_L = \sigma(C)$. Thus we obtain the result of the proposition for bounded non-negative $\mathcal{A}_L$-measurable functions. For general non-negative $\mathcal{A}_L$-measurable function f, we can apply the assertion of the proposition to the functions $\{\min(f, n)\}_{n\in\mathbb{N}}$ which form an increasing sequence converging to f. Passing to limit when n goes to the infinity, we obtain the result for f. □

The above proposition allows to construct a measure ν_L on the measurable space $(\Omega_L, \mathcal{A}_L)$ such that, for any subset A of $\mathcal{A}_L$, one has

$$\nu_L(A) = \int_{\Omega_K} \Big(\int_{M_{L,\omega}} \mathbb{1}_A(x)\,\mathbb{P}_{L,\omega}(\mathrm{d}x)\Big)\nu_K(\mathrm{d}\omega).$$

For any non-negative $\mathcal{A}_L$-measurable function f on Ω_L, one has

$$\int_{\Omega_L} f(x)\,\nu_L(\mathrm{d}x) = \int_{\Omega_K} \Big(\int_{M_{L,\omega}} f(x)\,\mathbb{P}_{L,\omega}(\mathrm{d}x)\Big)\nu_K(\mathrm{d}\omega). \tag{3.18}$$

We denote by $I_{L/K}(f)$ the map from Ω_K to $[0, +\infty]$ which sends $\omega \in \Omega_K$ to $\int_{M_{L,\omega}} f(x)\,\mathbb{P}_{L,\omega}(\mathrm{d}x)$. More generally, for any $\mathcal{A}_L$-measurable function f such that $I_{L/K}(|f|)$ is a real-valued function, we define $I_{L/K}(f)$ as the real-valued function

$$I_{L/K}(\max(f, 0)) - I_{L/K}(-\min(f, 0)).$$

Note that, if f is of the form $g\circ\pi_{L/K''}$ where $K''\in\mathscr{E}_{L/K}$ and g is an $\mathcal{A}_{K''}$-measurable function, then $I_{L/K}(f)$ is always well defined, and one has

$$I_{L/K}(g\circ\pi_{L/K''})=I_{K''/K}(g). \tag{3.19}$$

With this notation, the equality (3.18) can also be written as

$$\int_{\Omega_L} f\,\mathrm{d}\nu_L=\int_{\Omega_K} I_{L/K}(f)\,\mathrm{d}\nu_K. \tag{3.20}$$

Thus we obtain the following result.

Proposition 3.4.9 *An $\mathcal{A}_L$-measurable function f is ν_L-integrable if and only if $I_{L/K}(|f|)$ is ν_K-integrable. Moreover, $I_{L/K}$ defines a continuous linear operator from $\mathscr{L}^1(\Omega_L,\mathcal{A}_L,\nu_L)$ to $\mathscr{L}^1(\Omega_K,\mathcal{A}_K,\nu_K)$, and the equality* (3.20) *also holds for ν_L-integrable functions.*

Note that the relations (3.19) and (3.20) also imply that, if g is a function in $\mathscr{L}^1(\Omega_K,\mathcal{A}_K,\nu_K)$, then $g\circ\pi_{L/K}$ belongs to $\mathscr{L}^1(\Omega_L,\mathcal{A}_L,\nu_L)$, and one has

$$\int_{\Omega_L}(g\circ\pi_{L/K})\,\mathrm{d}\nu_L=\int_{\Omega_K} g\,\mathrm{d}\nu_K. \tag{3.21}$$

The following proposition shows that $(L,(\Omega_L,\mathcal{A}_L,\nu_L),\phi_L)$ forms an adelic curve.

Proposition 3.4.10 *For any non-zero element $a\in L$, the function*

$$(z\in\Omega_L)\longmapsto \ln|a|_z$$

is $\mathcal{A}_L$-measurable. Moreover, if $S=(K,(\Omega_K,\mathcal{A}_K,\nu_K),\phi_K)$ is proper, then

$$(L,(\Omega_L,\mathcal{A}_L,\nu_L),\phi_L)$$

is also proper.

Proof Denote by g the function on Ω_L such that $g(z)=\ln|a|_z$. We choose a finite extension $K''\in\mathscr{E}_{L/K}$ which contains a. Let $f:\Omega_{K''}\to\mathbb{R}$ be the function which sends $x\in\Omega_{K''}$ to $\ln|a|_x$. Then f is an $\mathcal{A}_{K''}$-measurable function on $\Omega_{K''}$. Since the function g identifies with the composition $f\circ\pi_{L/K''}$, we obtain that g is $\mathcal{A}_L$-measurable.

We assume that S is proper. For any $\omega\in\Omega_K$, one has

$$I_{L/K}(g)(\omega)=I_{L/K,\omega}(g)=I_{K''/K}(f)(\omega).$$

Therefore we obtain

$$\int_{\Omega_L} g\,\mathrm{d}\nu_L=\int_{\Omega_K} I_{K''/K}(f)\,\mathrm{d}\nu_K=0,$$

where the second equality comes from Theorem 3.3.7, (1) and (5). □

Definition 3.4.11 Let $S = (K, (\Omega_K, \mathcal{A}_K, \nu_K), \phi_K)$ be an adelic curve and L/K be an algebraic extension. The adelic curve $(L, (\Omega_L, \mathcal{A}_L, \nu_L), \phi_L)$ is called an *algebraic extension* of S, denoted by $S \otimes_K L$.

The following result, which is similar to Proposition 3.4.4, shows the compatibility property of the algebraic extensions of adelic curves.

Theorem 3.4.12 *Let $S = (K, (\Omega_K, \mathcal{A}_K, \nu_K), \phi_K)$ be an adelic curve and $L_2/L_1/K$ be successive algebraic extensions of fields. Then one has*

$$(S \otimes_K L_1) \otimes_{L_1} L_2 = S \otimes_K L_2. \tag{3.22}$$

Moreover, the following relations hold

$$\pi_{L_1/K} \circ \pi_{L_2/L_1} = \pi_{L_2/K}, \quad I_{L_1/K} \circ I_{L_2/L_1} = I_{L_2/K}. \tag{3.23}$$

Proof Let $(L_1, (\Omega_1, \mathcal{A}_1, \nu_1), \phi_1)$ and $(L_2, (\Omega_2, \mathcal{A}_2, \nu_2), \phi_2)$ be the adelic curves $S \otimes_K L_1$ and $S \otimes_K L_2$ respectively. First of all, set-theoretically one has

$$\Omega_2 = \coprod_{\omega \in \Omega_K} M_{L_2,\omega} = \coprod_{\omega \in \Omega_K} \coprod_{x \in M_{L_1,\omega}} M_{L_2,x} = \coprod_{x \in \Omega_1} M_{L_2,x},$$

and hence

$$\pi_{L_2/K} = \pi_{L_1/K} \circ \pi_{L_2/L_1}.$$

Moreover, for any extension $K_1 \in \mathscr{E}_{L_1/K}$ the map π_{L_2/K_1} is $\mathcal{A}_2$-measurable. Therefore $\pi_{L_2/L_1} : \Omega_2 \to \Omega_1$ is an $\mathcal{A}_2$-measurable map since the σ-algebra $\mathcal{A}_1$ is generated by the maps π_{L_1/K_1} with $K_1 \in \mathscr{E}_{L_1/K}$.

We now proceed with the proof of the equalities (3.22) and (3.23) with the supplementary assumption that the extension L_2/L_1 is *finite*. We first show that the σ-algebra $\mathcal{A}_2$ coincides with that in the adelic structure of $(S \otimes_K L_1) \otimes L_2$, namely the smallest σ-algebra $\mathcal{A}_2'$ on Ω_2 such that π_{L_2/L_1} and all functions of the form $(y \in \Omega_2) \mapsto |a|_y$ are $\mathcal{A}_2'$-measurable, where $a \in L_2$. We have already shown that the map π_{L_2/L_1} is measurable. Hence by proposition 3.4.10, we obtain that $\mathcal{A}_2' \subseteq \mathcal{A}_2$. Conversely, for any extension $K_1 \in \mathscr{E}_{L_1/K}$ one has

$$\pi_{L_2/K_1} = \pi_{L_1/K_1} \circ \pi_{L_2/L_1},$$

and hence

$$\pi_{L_2/K_1}^{-1}(\mathcal{A}_{K_1}) = \pi_{L_2/L_1}^{-1}(\pi_{L_1/K_1}^{-1}(\mathcal{A}_{K_1})) \subseteq \pi_{L_2/L_1}^{-1}(\mathcal{A}_1) \subseteq \mathcal{A}_2'.$$

If K_2 is an extension in $\mathscr{E}_{L_2/K}$, then $K_1 := K_2 \cap L_1 \in \mathscr{E}_{L_1/K}$. Moreover, the σ-algebra $\mathcal{A}_{K_2}$ is generated by π_{K_2/K_1} and the functions of the form $x \mapsto |a|_x$ on Ω_{K_2}, where $a \in K_2$. Note that $\pi_{L_2/K_1} = \pi_{L_1/K_1} \circ \pi_{L_2/L_1}$ is $\mathcal{A}_2'$-measurable, and for any $a \in K_2$, the composition of the function $x \mapsto |a|_x$ on Ω_{K_2} with π_{L_2/K_2}, which identifies with the function $y \mapsto |a|_y$ on Ω_{L_2}, is also $\mathcal{A}_2'$-measurable, we obtain that the map π_{L_2/K_2} is actually $\mathscr{A}'_2$-measurable. Since $\mathscr{A}_2$ is the smallest σ-algebra which makes all π_{L_2/K_2} measurable, where $K_2 \in \mathscr{E}_{L_2/K}$, we obtain $\mathscr{A}_2 \subseteq \mathscr{A}'_2$. Therefore one has $\mathscr{A}_2 = \mathscr{A}'_2$.

It remains to establish the relation $I_{L_1/K}(I_{L_2/L_1}(f)) = I_{L_2/K}(f)$ for any non-negative $\mathscr{A}_2$-measurable function on Ω_2. By induction it suffices to treat the case where $[L_2 : L_1]$ is a prime number. Moreover, similarly to the proof of Proposition 3.4.8, by using the monotone class theorem, we only need to verify the equality $I_{L_1/K}(I_{L_2/L_1}(f)) = I_{L_2/K}(f)$ for functions f of the form $g \circ \pi_{L_2/K_2}$, where $K_2 \in \mathscr{E}_{L_2/K}$ and g is a non-negative $\mathscr{A}_{K_2}$-measurable function on Ω_{K_2}. If K_2 belongs to $\mathscr{E}_{L_1/K}$, one has $I_{L_2/L_1}(f) = g \circ \pi_{L_1/K_2}$, and therefore

$$I_{L_1/K}(I_{L_2/L_1}(f)) = I_{K_2/K}(g) = I_{L_2/K}(f),$$

where the second equality comes from (3.19). Otherwise one has $[K_2 : K_1] = [L_2 : L_1]$ since $[K_2 : K_1]$ divides $[L_2 : L_1]$ which is a prime number, where K_1 denotes the intersection of K_2 with L_1. Moreover, there exists an element $a \in K_2$ such that $K_2 = K_1(a)$ and $L_2 = L_1(a)$. If a is totally inseparable over K_1, then it is also totally inseparable over L_1. In this case $I_{L_2/L_1}(f) = f = g \circ \pi_{L_1/K_1}$ and therefore the equality $I_{L_1/K}(I_{L_2/L_1}(f)) = I_{L_2/K}(f)$ also holds in this case.

In the following, we assume that the element a is separable over K_1. Let

$$P(T) = T^p + b_1 T^{p-1} + \cdots + b_p \in K_1[T]$$

be the minimal polynomial of a over K_1. Since p is a prime number, and $a \notin L_1$, we obtain that it is also the minimal polynomial of a over L_1. In particular, the element a is also separable over L_1. Let y be an element in Ω_1 and $x = \pi_{L_1/K_1}(y)$. Assume that

$$P = P_1 \cdots P_r$$

is the splitting of the polynomial P in the ring $K_{1,x}[T]$ into the product of distinct irreducible polynomials. Then the polynomials $P_1, \ldots, P_r$ correspond to points $x_1, \ldots, x_r$ which form the set $\pi_{K_2/K_1}^{-1}(\{x\})$. Moreover, one has (see (3.2) for the definition of $\mathbb{P}_{K_2,x}$)

$$\mathbb{P}_{K_2,x}(\{x_i\}) = \frac{\deg(P_i)}{p}.$$

Assume that each P_i splits in $L_{1,y}$ into the product of distinct irreducible polynomials as

$$P_i = Q_{i,1} \cdots Q_{i,n_i}.$$

Then each factor $Q_{i,j}$ corresponds to a point $y_{i,j}$ in $\pi_{L_2/L_1}^{-1}(\{y\})$ and one has

$$\mathbb{P}_{L_2,y}(x_{i,j}) = \frac{\deg(Q_{i,j})}{p}.$$

Therefore, if a non-negative function f on Ω_2 is of the form $g \circ \pi_{L_2/K_2}$, where g is a $\mathscr{A}_{K_2}$-measurable function on Ω_{K_2}, then one has

$$(I_{L_2/L_1}(f))(y) = \sum_{i=1}^{r}\sum_{j=1}^{n_i} \frac{\deg(Q_{i,j})}{p} g(x) = \frac{\deg(P_i)}{p} g(x) = (I_{K_2/K_1}(g))(x),$$

which shows that

$$I_{L_2/L_1}(f) = I_{K_2/K_1}(g) \circ \pi_{L_1/K_1}.$$

Therefore one has

$$\begin{aligned} I_{L_1/K}(I_{L_2/L_1}(f)) &= I_{L_1/K}(I_{K_2/K_1}(g) \circ \pi_{L_1/K_1}) \\ &= I_{K_1/K}(I_{K_2/K_1}(g)) = I_{K_2/K}(g) = I_{L_2/K}(f), \end{aligned}$$

where the second and the last equalities come from (3.19). Thus we have established the second equality in (3.23), which implies that the measure in the adelic structure of $(S \otimes_K L_1) \otimes_{L_1} L_2$ coincides with the measure ν_2 in the adelic structure of $S \otimes_K L_2$. The theorem is then established in the particular case where $[L_2 : L_1]$ is finite.

In the following, we will prove the general case of the theorem. Note that the previously proved case actually implies that, for any finite extension L'' of L_1, the map $\pi_{L_2/L''}$ is $\mathcal{A}_2$-measurable since the σ-algebra $\mathcal{A}_{L''}$ in the adelic structure of $(S \otimes_K L_1) \otimes L''$ coincides with that in the adelic structure of $S \otimes_K L''$, which is generated by the maps $\pi_{L''/K''}$ with $K'' \in \mathscr{E}_{L''/K}$. In particular, if we denote by $\mathcal{A}_2'$ the σ-algebra in the adelic structure of $(S \otimes_K L_1) \otimes_{L_1} L_2$, then one has $\mathcal{A}_2' \subseteq \mathcal{A}_2$. Conversely, for any $K_2 \in \mathscr{E}_{L_2/K}$, one has $\pi_{L_2/K_2} = \pi_{L''/K_2} \circ \pi_{L_2/L''}$, where $L'' = L_1K_2$ is an element in $\mathscr{E}_{L_2/L_1}$. Hence π_{L_2/K_2} is $\mathcal{A}_2'$-measurable. Since $K_2 \in \mathscr{E}_{L_2/L}$ is arbitrary, we obtain that $\mathcal{A}_2 \subseteq \mathcal{A}_2'$ and hence $\mathcal{A}_2 = \mathcal{A}_2'$.

Again it remains to establish the equality $I_{L_1/K}(I_{L_2/L_1}(f)) = I_{L_2/K}(f)$ for any non-negative $\mathcal{A}_2$-measurable function f on Ω_2, which can be written in the form $g \circ \pi_{L_2/K_2}$, where $K_2 \in \mathscr{E}_{L_2/K}$. Let $L'' = K_2L_1$. One has $L'' \in \mathscr{E}_{L_2/L_1}$ and $g \circ \pi_{L_2/K_2} = g \circ \pi_{L''/K_2} \circ \pi_{L_2/L''}$. Therefore (3.19) implies

$$I_{L_2/K}(g \circ \pi_{L_2/K_2}) = I_{L''/K}(g \circ \pi_{L''/K_2}) = I_{K_2/K}(g).$$

Moreover, also by (3.19) one obtains

$$\begin{aligned} I_{L_1/K}(I_{L_2/L_1}(g \circ \pi_{L_2/K_2})) &= I_{L_1/K}(I_{L_2/L_1}(g \circ \pi_{L''/K_2} \circ \pi_{L_2/L''}))) \\ &= I_{L_1/K}(I_{L''/L_1}(g \circ \pi_{L''/K_2})) = I_{L''/K}(g \circ \pi_{L''/K_2}) = I_{K_2/K}(g), \end{aligned}$$

where the third equality comes from the proved case of finite extensions. Thus we establish the relation $I_{L_1/K} \circ I_{L_2/L_1} = I_{L_2/K}$ which implies that the measure ν_2 identifies with that in the adelic structure of $(S \otimes_K L_1) \otimes_{L_1} L_2$. The theorem is thus proved. □

Proposition 3.4.13 *Let $S = (K, (\Omega_K, \mathcal{A}_K, \nu_K), \phi_K)$ be an adelic curve and L/K be an algebraic extension. Let $(L, (\Omega_L, \mathcal{A}_L, \nu_L), \phi_L)$ be the adelic curve $S \otimes_K L$. Then $\mathcal{A}_L$ is the smallest σ-algebra making the canonical projection map $\pi_{L/K} : \Omega_L \to \Omega_K$ and the functions $(x \in \Omega_L) \mapsto |a|_x$ measurable for all $a \in L$.*

Proof By definition the projection map $\pi_{L/K}$ is $\mathcal{A}_L$-measurable. Moreover, by Proposition 3.4.10, for any $a \in L$, the function $(x \in \Omega_L) \mapsto |a|_x$ is $\mathcal{A}_L$-measurable.

Suppose that F is a map from a measurable space $(E, \mathcal{E})$ to Ω_L such that the composed map $\pi_{L/K} \circ F$ and the functions $(y \in E) \mapsto |a|_{F(y)}$ are measurable, where $a \in L$. We will show that F is measurable if we consider the σ-algebra $\mathcal{A}_L$ on Ω_L. This implies that $\mathcal{A}_L$ is contained in the smallest σ-algebra making the canonical projection $\Omega_L \to \Omega_K$ and the functions $(x \in \Omega_L) \mapsto |a|_x$ measurable, where $a \in L$.

Recall that $\mathcal{A}_L$ is the smallest σ-algebra making the projection maps $\pi_{L/K''}$: $\Omega_L \to \Omega_{K''}$ measurable, where K''/K runs over the set of finite extensions contained in L. To show the measurability of F it suffices to verify the measurability of $\pi_{L/K''} \circ F$ for any finite extension K''/K contained in L. Moreover, since $\mathcal{A}_{K''}$ is the smallest σ-algebra making the projection map $\pi_{K''/K} : \Omega_{K''} \to \Omega_K$ and the functions $(x \in \Omega_{K''}) \mapsto |a|_x$ measurable, where $a \in K''$, we are reduced to verify the measurability of $\pi_{K''/K} \circ \pi_{L/K''} \circ F = \pi_{L/K} \circ F$ and

$$(y \in E) \longmapsto |a|_{\pi_{L/K''}(F(y))}, \quad \text{where } a \in K''. \tag{3.24}$$

By the assumption on F, the map $\pi_{L/K} \circ F$ is measurable. Moreover, since $a \in K''$, one has

$$|a|_{\pi_{L/K''}(F(y))} = |a|_{F(y)}.$$

Hence the function in (3.24) is also measurable. The proposition is thus proved. □

3.5 Height function and Northcott property

Let $S = (K, (\Omega, \mathcal{A}, \nu), \phi)$ be a *proper* adelic curve and K^{ac} be an algebraic closure of K. Let $S \otimes_K K^{\mathrm{ac}} = (K^{\mathrm{ac}}, (\Omega_{K^{\mathrm{ac}}}, \mathcal{A}_{K^{\mathrm{ac}}}, \nu_{K^{\mathrm{ac}}}), \phi_{K^{\mathrm{ac}}})$ be the algebraic extension of S by K^{ac}.

Definition 3.5.1 For $(a_0, a_1, \ldots, a_n) \in (K^{\mathrm{ac}})^{n+1} \setminus \{(0, \ldots, 0)\}$, we define the invariant $h_S(a_0, \ldots, a_n)$ to be

$$h_S(a_0, \ldots, a_n) := \int_{\Omega_{K^{\mathrm{ac}}}} \ln\left(\max\{|a_0|_\chi, \ldots, |a_n|_\chi\}\right) \nu_{K^{\mathrm{ac}}}(d\chi).$$

By the product formula, $h_S(\lambda a_0, \ldots, \lambda a_n) = h_S(a_0, \ldots, a_n)$ for all $\lambda \in K^{\mathrm{ac}} \setminus \{0\}$, so that there is a map $\mathbb{P}^n(K^{\mathrm{ac}}) \to \mathbb{R}$ such that the following diagram is commutative:

$$\begin{array}{ccc} (K^{\mathrm{ac}})^{n+1} \setminus \{(0, \ldots, 0)\} & \xrightarrow{\;h_S\;} & \mathbb{R} \\ \downarrow & \nearrow & \\ \mathbb{P}^n(K^{\mathrm{ac}}) & & \end{array}$$

By abuse of notation, the map $\mathbb{P}^n(K^{\mathrm{ac}}) \to \mathbb{R}$ is also denoted by h_S. For $x \in \mathbb{P}^n(K^{\mathrm{ac}})$, the value $h_S(x)$ is called the *height* of x with respect to the adelic curve S.

Definition 3.5.2 We say that S has the Northcott property if the set

$$\{a \in K \,:\, h_S(1:a) \leqslant C\}$$

is finite for any $C \geqslant 0$. In the cases of Example 3.2.2, Example 3.2.5 and Example 3.2.6, the Northcott property holds (for details, see [106] and [108]).

The purpose of this section is to prove the following theorem:

Theorem 3.5.3 (Northcott's theorem) *If S has the Northcott property, then the set $\{x \in \mathbb{P}^n(K^{\mathrm{ac}}) \,:\, h_S(x) \leqslant C,\ [K(x):K] \leqslant \delta\}$ is finite for any C and δ.*

Before starting the proof of Theorem 3.5.3, we need to prepare two lemmas.

Lemma 3.5.4 *Let K' be a finite normal extension of K. Then $h_S(1:\sigma(\alpha)) = h_S(1:\alpha)$ for all $\alpha \in K'$ and $\sigma \in \mathrm{Aut}_K(K')$.*

Proof Let K'' be the separable closure of K in K' and $q = [K' : K'']$. Then $\sigma|_{K''} \in \mathrm{Aut}_K(K'')$ and $a^q \in K''$. If the assertion holds for the extension K''/K, then

$$qh_S(1:\sigma(\alpha)) = h_S(1:\sigma(\alpha^q)) = h_S(1:\alpha^q) = qh_S(1:\alpha),$$

so that we may assume that the extension K'/K is separable.

For $\chi \in \pi_{K'/K}^{-1}(\{\omega\})$ and $\tau \in \mathrm{Gal}(K'/K)$, a map $(\beta \in K') \mapsto |\tau(\beta)|_\chi$ gives rise to an element of $\pi_{K'/K}^{-1}(\{\omega\})$, which is denoted by χ^τ. In this way, one has an action $\mathrm{Gal}(K'/K) \times \pi_{K'/K}^{-1}(\{\omega\}) \to \pi_{K'/K}^{-1}(\{\omega\})$ given by $(\tau,\chi) \mapsto \chi^\tau$. Note that the action is transitive (cf. [117, Chapter II, Proposition 9.1]) and $\mathrm{Gal}(K'_\chi/K_\omega) = \mathrm{Stab}_{\mathrm{Gal}(K'/K)}(\chi)$ (cf. [117, Chapter II, Proposition 9.6]). In particular, $[K'_\chi : K_\omega] = [K'_{\chi'} : K_\omega]$ for all $\chi,\chi' \in \pi_{K'/K}^{-1}(\{\omega\})$. Therefore,

$$\begin{aligned}
I_{K'/K}(\ln(\max\{1,|\sigma(\alpha)|\}))(\omega) &= \sum_{\chi\in\pi_{K'/K}^{-1}(\{\omega\})} \frac{[K'_\chi : K_\omega]}{[K':K]} \ln(\max\{1,|\sigma(\alpha)|_\chi\}) \\
&= \sum_{\chi\in\pi_{K'/K}^{-1}(\{\omega\})} \frac{[K'_{\chi^\sigma} : K_\omega]}{[K':K]} \ln(\max\{1,|\alpha|_{\chi^\sigma}\}) \\
&= I_{K'/K}(\ln(\max\{1,|\alpha|\}))(\omega),
\end{aligned}$$

and hence the assertion follows. □

For a polynomial $F = a_nX^n + \cdots + a_1X + a_0 \in K^{\mathrm{ac}}[X] \setminus \{0\}$, we define $h_S(F)$ to be $h_S(F) := h_S(a_n : \cdots : a_1 : a_0)$. If we set $\|F\|_\chi := \max\{|a_n|_\chi, \ldots, |a_0|_\chi\}$ for $\chi \in \Omega_{K^{\mathrm{ac}}}$ as in Subsection 1.1.15, then

$$h_S(F) = \int_{\Omega_{K^{\mathrm{ac}}}} \ln(\|F\|_\chi)\, \nu_{K^{\mathrm{ac}}}(\mathrm{d}\chi).$$

Lemma 3.5.5 *For $F, G \in K^{\mathrm{ac}}[X] \setminus \{0\}$, one has*

$$h_S(FG) \leqslant h_S(F) + h_S(G) + \ln \min\{\deg(F)+1, \deg(G)+1\} \int_{\Omega_\infty} \nu(\mathrm{d}\omega).$$

Proof By Proposition 1.1.72,

$$\begin{aligned} h_S(FG) &= \int_{\Omega_{K^{\mathrm{ac}}}} \ln(\|FG\|_\chi)\nu_{K^{\mathrm{ac}}}(\mathrm{d}\chi) \\ &\leqslant \int_{\Omega_{K^{\mathrm{ac}}}} \big(\ln(\|F\|_\chi) + \ln(\|G\|_\chi)\big)\, \nu_{K^{\mathrm{ac}}}(\mathrm{d}\chi) \\ &\qquad + \int_{\Omega_\infty} \ln(\min\{\deg(F)+1, \deg(G)+1\})\nu(\mathrm{d}\omega), \end{aligned}$$

so that the assertion follows. □

Proof (Proof of Theorem 3.5.3) Clearly we may assume that $C \geqslant 0$ and $\delta \geqslant 1$. Let us begin with the following special case:

Claim 3.5.6 *The set $\{\alpha \in K^{\mathrm{ac}} \,:\, h_S(1:\alpha) \leqslant C,\ [K(\alpha):K] \leqslant \delta\}$ is finite.* □

Proof Let F be the minimal monic polynomial of α over K. We set $F = X^n + a_{n-1}X^{n-1} + \cdots + a_1X + a_0 = (X-\alpha_1)\cdots(X-\alpha_n)$ and $K' = K(\alpha_1, \ldots, \alpha_n)$, where $\alpha_1 = \alpha$. Then, by Lemma 3.5.4 and Lemma 3.5.5,

$$\begin{aligned} h_S(F) &\leqslant \sum_{i=1}^n h_S(X-\alpha_i) + (n-1)\ln(2)\int_{\Omega_\infty} \nu(\mathrm{d}\omega) \\ &= \sum_{i=1}^n h_S(1:\alpha_i) + (n-1)\ln(2)\int_{\Omega_\infty} \nu(\mathrm{d}\omega) \\ &= nh_S(1:\alpha) + (n-1)\ln(2)\int_{\Omega_\infty} \nu(\mathrm{d}\omega) \\ &\leqslant \delta C + (\delta-1)\ln(2)\int_{\Omega_\infty} \nu(\mathrm{d}\omega). \end{aligned}$$

Note that $h_S(1:a_i) \leqslant h_S(F)$ and $a_i \in K$ for all $i = 0, \ldots, n-1$, so that one can see that there are finitely many possibilities of F because S has the Northcott property. Therefore the assertion of the claim follows. □

Let us go back to the proof of Theorem 3.5.3. For $i = 0, \ldots, n$, let

$$\Upsilon_i := \{x = (x_0 : \cdots : x_n) \in \mathbb{P}^n(K^{\mathrm{ac}}) \,:\, h_S(x) \leqslant C,\ [K(x):K] \leqslant \delta,\ x_i \neq 0\}.$$

It is sufficient to show that $\#(\Upsilon_i) < \infty$ for all i. Without loss of generality, we may assume that $i = 0$. Then

$$\Upsilon_0 = \{a = (a_1, \ldots, a_n) \in (K^{\mathrm{ac}})^n \,:\, h_S(1,a) \leqslant C,\ [K(a):K] \leqslant \delta\}.$$

Note that $[K(a_i) : K] \leqslant [K(a_1, \ldots, a_n) : K]$ and $h_S(1 : a_i) \leqslant h_S(1, a_1, \ldots, a_n)$ for all $i = 1, \ldots, n$. Thus the assertion is a consequence of the above special case. □

Corollary 3.5.7 *We assume that S has the Northcott property. Let K' be a finite extension of K. Then $S \otimes_K K'$ has also the Northcott property.*

Remark 3.5.8 Theorem 3.5.3 can be generalised to the case of an adelic vector bundle. For details, see Proposition 6.2.3.

3.6 Measurability of automorphism actions

Let $S = (K, (\Omega, \mathcal{A}, \nu), \phi)$ be an adelic curve and L/K be an algebraic extension. We denote by $\mathrm{Aut}_K(L)$ the group of field automorphisms of L which are K-linear. The group $\mathrm{Aut}_K(L)$ acts on M_L as follows: for any $\tau \in \mathrm{Aut}_K(L)$ and any $x \in M_L$, one has

$$\forall\, a \in L, \quad |a|_{\tau(x)} = |\tau(a)|_x.$$

Moreover, by definition the restrictions of the absolute values $|\cdot|_x$ and $|\cdot|_{\tau(x)}$ on K are the same. Therefore we obtain an action of the K-linear automorphism group $\mathrm{Aut}_K(L)$ on the set $\Omega_L = \Omega \times_{M_K} M_L$ (where we consider trivial actions of $\mathrm{Aut}_K(L)$ on Ω and on M_K).

Proposition 3.6.1 *Let $S = (K, (\Omega_K, \mathcal{A}_K, \nu_K), \phi_K)$ be an adelic curve and L/K be an algebraic extension. For any $\tau \in \mathrm{Aut}_K(L)$, the action of τ on Ω_L is measurable, where on Ω_L we consider the σ-algebra $\mathcal{A}_L$ in the adelic structure of $S \otimes_K L$.*

Proof By Proposition 3.4.13 the σ-algebra $\mathcal{A}_L$ is the smallest σ-algebra which makes measurable the canonical projection map $\pi_{L/K} : \Omega_L \to \Omega_K$ and the functions $(x \in \Omega_L) \mapsto |a|_x$, where $a \in L$. Let $\tau \in \mathrm{Aut}_K(L)$. To show the measurability of the action of τ on Ω_L, it suffices to verify the measurability of the map $\pi_{L/K} \circ \tau$ and the functions $(x \in \Omega_L) \mapsto |a|_{\tau(x)}$ with $a \in L$. Note that by definition $\pi_{L/K} \circ \tau = \pi_{L/K}$ and $|\alpha|_{\tau(x)} = |\tau(\alpha)|_x$. The proposition is thus proved. □

Proposition 3.6.2 *Let $S = (K, (\Omega_K, \mathcal{A}_K, \nu_K), \phi_K)$ be an adelic curve and L/K be a finite extension. Let $F : \Omega_L \to \mathbb{R}$ be an $\mathcal{A}_L$-measurable function. For any $\omega \in \Omega$, let*

$$f(\omega) = \max_{x \in \pi_{L/K}^{-1}(\{\omega\})} F(x).$$

Then the function $f : \Omega_K \to \mathbb{R}$ is $\mathcal{A}_K$-measurable.

Proof We first assume that the extension L/K is normal. By [27], Chapitre VI, §8, n°6, Proposition 7, for any $\omega \in \Omega_K$, the action of the K-linear automorphism group $\mathrm{Aut}_K(L)$ on $M_{L,\omega}$ is transitive. As a consequence, if we denote by $\widetilde{F}$ the function

$$\max_{\tau \in \mathrm{Aut}_K(L)} F \circ \tau,$$

then for each $\omega \in \Omega_K$, the restriction of $\widetilde{F}$ to $\pi_{L/K}^{-1}(\{\omega\})$ is constante, the value of which is equal to $f(\omega)$. By Proposition 3.6.1, for any $\tau \in \mathrm{Aut}_K(L)$, the action of τ on Ω_L is measurable and hence the function $F \circ \tau$ is $\mathcal{A}_L$-measurable. Since $\mathrm{Aut}_K(L)$ is a finite set, we deduce that the function $\widetilde{F}$ is also $\mathcal{A}_L$-measurable. By Proposition 3.4.8, the function $f = I_{L/K}(\widetilde{F})$ is $\mathcal{A}_K$-measurable.

In the general case, we pick a finite normal extension L_1/K which contains L. By applying the proved result to the function $F \circ \pi_{L_1/L}$, we still obtain the measurability of the function f. The proposition is thus proved. □

3.7 Morphisms of adelic curves

In this section, we consider morphism of adelic curves.

Definition 3.7.1 Let $S = (K, (\Omega, \mathcal{A}, \nu), \phi)$ and $S' = (K', (\Omega', \mathcal{A}', \nu'), \phi')$ be two adelic curves, we call *morphism from S' to S* any triplet $\alpha = (\alpha^{\#}, \alpha_{\#}, I_\alpha)$, where

(a) $\alpha^{\#} : K \longrightarrow K'$ is a field homomorphism,
(b) $\alpha_{\#} : (\Omega', \mathcal{A}') \longrightarrow (\Omega, \mathcal{A})$ is a measurable map such that the following diagram is commutative

$$\begin{array}{ccc} \Omega' & \xrightarrow{\alpha_{\#}} & \Omega \\ {\scriptstyle \phi'}\downarrow & & \downarrow{\scriptstyle \phi} \\ M_{K'} & \xrightarrow[-\circ\alpha^{\#}]{} & M_K \end{array}$$

and that the direct image of ν' by $\alpha_{\#}$ coincides with ν, namely, for any function $f \in \mathscr{L}^1(\Omega, \mathcal{A}, \nu)$, one has

$$\int_{\Omega'} f \circ \alpha_{\#} \, \mathrm{d}\nu' = \int_{\Omega} f \, \mathrm{d}\nu.$$

(c) $I_\alpha : L^1(\Omega', \mathcal{A}', \nu') \to L^1(\Omega, \mathcal{A}, \nu)$ is a disintegration kernel of $\alpha_{\#}$, namely I_α is a linear map such that, for any element $g \in L^1(\Omega', \mathcal{A}', \nu')$, one has

$$\int_{\Omega} I_\alpha(g) \, \mathrm{d}\nu = \int_{\Omega'} g \, \mathrm{d}\nu',$$

and for any function $f \in \mathscr{L}^1(\Omega, \mathcal{A}, \nu)$ one has I_α sends the equivalence class of $f \circ \alpha_{\#}$ to that of f.

Naturally, if S, S' and S'' are adelic curves and if $\alpha = (\alpha^{\#}, \alpha_{\#}, I_\alpha) : S' \to S$ and $\beta = (\beta^{\#}, \beta_{\#}, I_\beta) : S'' \to S$ are morphisms of adelic curves, then $\alpha \circ \beta := (\beta^{\#} \circ \alpha^{\#}, \alpha_{\#} \circ \beta_{\#}, I_\alpha \circ I_\beta)$ forms a morphism of adelic curves from S'' to S. Thus the adelic curves and their morphisms form a category.

Example 3.7.2 Let $S = (K, (\Omega, \mathcal{A}, \nu), \phi)$ be an adelic curve.

(1) If $S \otimes_K K' = (K', (\Omega_{K'}, \mathcal{A}_{K'}, \nu_{K'}), \phi_{K'})$ is an algebraic extension of S, then the triplet $(K \hookrightarrow K', \pi_{K'/K}, \overline{I}_{K'/K})$ is a morphism of adelic curves from $S \otimes_K K'$ to S, where $\overline{I}_{K'/K} : L^1(\Omega_{K'}, \mathcal{A}_{K'}, \nu_{K'}) \to L^1(\Omega, \mathcal{A}, \nu)$ is the linear map induced by $I_{K'/K} : \mathscr{L}^1(\Omega_{K'}, \mathcal{A}_{K'}, \nu_{K'}) \to \mathscr{L}^1(\Omega, \mathcal{A}, \nu)$.

(2) Assume that K_0 is a subfield of K. Let S_0 be the field K_0 equipped with the restriction to K_0 of the adelic structure of S (see Subsection 3.2.8). Then the triplet $(K_0 \hookrightarrow K, \mathrm{Id}_\Omega, \mathrm{Id}_{L^1(\Omega,\mathcal{A},\nu)})$ forms a morphism of adelic curves from S to S_0.

(3) Let $K = \mathbb{Q}(T)$ be the field of rational functions of one variable T with coefficients in $\mathbb{Q}$. Let $S = (K, (\Omega, \mathcal{A}, \nu), \phi)$ constructed in Subsection 3.2.5. Recall that $(\Omega, \mathcal{A}, \nu)$ is written as a disjoint union $\Omega_h \coprod \mathcal{P} \coprod [0,1]_*$, where Ω_h is the set of closed points of $\mathbb{P}^1_{\mathbb{Q}}$, $\mathcal{P}$ is the set of prime numbers, and $[0,1]_*$ is the subset of $[0,1]$ of t such that $\mathrm{e}^{2\pi i t}$ is transcendental. Let $S_{\mathbb{Q}} = (\mathbb{Q}, (\Omega_{\mathbb{Q}}, \mathcal{A}_{\mathbb{Q}}, \nu_{\mathbb{Q}}), \phi_{\mathbb{Q}})$ be the adelic curve defined in Subsection 3.2.2 and $\widetilde{S}_{\mathbb{Q}} = (\mathbb{Q}, (\widetilde{\Omega}_{\mathbb{Q}}, \widetilde{\mathcal{A}}_{\mathbb{Q}}, \widetilde{\nu}_{\mathbb{Q}}), \widetilde{\phi}_{\mathbb{Q}})$ be the adelic curve consisting of the filed $\mathbb{Q}$ equipped with the amalgamation of the adelic structure of $S_{\mathbb{Q}}$ and a family of copies of the trivial absolute value on $\mathbb{Q}$ indexed by Ω_h. We can also write $\widetilde{\Omega}$ as the disjoint union of three subsets $\Omega_h \coprod \mathcal{P} \coprod \{\infty\}$, where 0 denotes the trivial absolute value on $\mathbb{Q}$ and ∞ denotes the infinite place of $\mathbb{Q}$. Let $\alpha^\# : \mathbb{Q} \to \mathbb{Q}(T)$ be the inclusion map. Let $\alpha_\# : \Omega \to \widetilde{\Omega}_{\mathbb{Q}}$ be the map which sends any element of $\Omega_h \coprod \mathcal{P}$ to itself and send any element of $[0,1]_*$ identically to ∞. Finally, let $I_\alpha : L^1(\Omega, \mathcal{A}, \nu) \to L^1(\widetilde{\Omega}_{\mathbb{Q}}, \widetilde{\mathcal{A}}_{\mathbb{Q}}, \widetilde{\nu}_{\mathbb{Q}})$ be the linear map sending the equivalence class of any function $f \in \mathscr{L}^1(\Omega, \mathcal{A}, \nu)$ to that of the function $I_\alpha(f)$ sending $\omega \in \Omega_h \coprod \mathcal{P}$ to $f(\omega)$ and ∞ to $\int_{[0,1]_*} f(t)\,\mathrm{d}t$. Then the triplet $(\alpha^\#, \alpha_\#, I_\alpha)$ forms a morphism of adelic curves from S to $\widetilde{S}_{\mathbb{Q}}$.

Chapter 4
Vector bundles on adelic curves: global theory

The purpose of this chapter is to study the geometry of adelic curves, notably the divisors and vector bundles.

4.1 Norm families

Let $S = (K, (\Omega, \mathcal{A}, \nu), \phi)$ be an adelic curve (see §3.1). Recall that, for any $\omega \in \Omega$, we denote by $|\cdot|_\omega$ the absolute value of K indexed by ω. Note that, in the case where $|\cdot|_\omega$ is Archimedean, there exists a constant $\kappa(\omega)$, $0 < \kappa(\omega) \leqslant 1$, such that $|\cdot|_\omega = |\cdot|^{\kappa(\omega)}$, where $|\cdot|$ denotes the usual absolute value on $\mathbb{R}$ or $\mathbb{C}$. For simplicity, we assume that $\kappa(\omega) = 1$ for any $\omega \in \Omega_\infty$, namely $|\cdot|_\omega$ identifies with the usual absolute value on $\mathbb{R}$ or $\mathbb{C}$. Note that this assumption is harmless for the generality of the theory since in general case we can replace the absolute values $\{|\cdot|_\omega\}_{\omega\in\Omega_\infty}$ by the usual ones and consider the measure $d\widetilde{\nu} = (1\!\!1_{\Omega\setminus\Omega_\infty} + \kappa 1\!\!1_{\Omega_\infty})\,d\nu$ instead.

4.1.1 Definition and algebraic constructions

Let E be a vector space of finite dimension over K. We denote by $\mathcal{N}_E$ the set of norm families $\{\|\cdot\|_\omega\}_{\omega\in\Omega}$, where each $\|\cdot\|_\omega$ is a norm on $E_{K_\omega} := E \otimes_K K_\omega$. We say that a norm family $\xi = \{\|\cdot\|_\omega\}_{\omega\in\Omega}$ is *ultrametric on* $\Omega \setminus \Omega_\infty$ if the norm $\|\cdot\|_\omega$ is ultrametric for any $\omega \in \Omega \setminus \Omega_\infty$. We say that a norm family $\{\|\cdot\|_\omega\}_{\omega\in\Omega}$ in $\mathcal{N}_E$ is *Hermitian* if the following conditions are satisfied:

(a) for any $\omega \in \Omega \setminus \Omega_\infty$, the norm $\|\cdot\|_\omega$ is ultrametric (namely the norm family is ultrametric on $\Omega \setminus \Omega_\infty$);
(b) for any $\omega \in \Omega_\infty$, the norm $\|\cdot\|_\omega$ is induced by an inner product (see §1.2.1), namely there exists an inner product $\langle\ ,\ \rangle_\omega$ on E_{K_ω} such that $\|\ell\|_\omega = \langle \ell, \ell\rangle_\omega^{1/2}$ for any $\ell \in E_{K_\omega}$.

H. Chen, A. Moriwaki, *Arakelov Geometry over Adelic Curves*, Lecture Notes in Mathematics 2258, https://doi.org/10.1007/978-981-15-1728-0_4

We denote by $\mathcal{H}_E$ the subset of $\mathcal{N}_E$ consisting of all Hermitian norm families. In the following, we describe some algebraic constructions of norm families.

4.1.1.1 Multiplication by a numerical function

Let E be a finite-dimensional vector space over K, $\xi = \{\|\cdot\|_\omega\}_{\omega\in\Omega} \in \mathcal{N}_E$ and $f : \Omega \to]0, +\infty[$ be a positive function on Ω. We denote by $f\xi$ the norm family $\{f(\omega)\|\cdot\|_\omega\}_{\omega\in\Omega}$ in $\mathcal{N}_E$.

4.1.1.2 Restrict and quotient norm families

Let E be a finite-dimensional vector space over K and $\xi = \{\|\cdot\|_{E,\omega}\}_{\omega\in\Omega}$ be a norm family in $\mathcal{N}_E$. Let F be a vector subspace of E. For any $\omega \in \Omega$, let $\|\cdot\|_{F,\omega}$ be the restriction of the norm $\|\cdot\|_{E,\omega}$ to F_{K_ω} (see Definition 1.1.2). Then $\{\|\cdot\|_{F,\omega}\}_{\omega\in\Omega}$ forms a norm family in $\mathcal{N}_F$, called the *restriction* of ξ to F. Similarly, if G is a quotient vector space of E, then each norm $\|\cdot\|_{E,\omega}$ induces by quotient a norm $\|\cdot\|_{G,\omega}$ on G_{K_ω} (see §1.1.3). Thus we obtain a norm family $\{\|\cdot\|_{G,\omega}\}_{\omega\in\Omega}$ in $\mathcal{N}_G$, called the *quotient* of ξ on G. Note that, if the norm family ξ is ultrametric on $\Omega \setminus \Omega_\infty$ (resp. Hermitian), then all its restrictions and quotients are also ultrametric on $\Omega \setminus \Omega_\infty$ (resp. Hermitian).

4.1.1.3 Direct sums

Let $\mathscr{S}$ be the set of all convex and continuous functions $f : [0, 1] \to [0, 1]$ such that $\max\{t, 1-t\} \leqslant f(t)$ for any $t \in [0, 1]$. If E and F are finite-dimensional vector spaces over K and if $\xi_E = \{\|\cdot\|_{E,\omega}\}_{\omega\in\Omega}$ and $\xi_F = \{\|\cdot\|_{F,\omega}\}_{\omega\in\Omega}$ are respectively norm families in $\mathcal{N}_E$ and $\mathcal{N}_F$, for any family $\psi = \{\psi_\omega\}_{\omega\in\Omega}$ of elements of $\mathscr{S}$ we define a norm family $\xi_E \oplus_\psi \xi_F = \{\|\cdot\|_\omega\}_{\omega\in\Omega}$ in $\mathcal{N}_{E\oplus F}$ (see Subsection 1.1.10) such that, for any $(x, y) \in E_{K_\omega} \oplus F_{K_\omega}$,

$$\|(x, y)\|_\omega := (\|x\|_{E,\omega} + \|y\|_{F,\omega})\psi_\omega\Big(\frac{\|x\|_{E,\omega}}{\|x\|_{E,\omega} + \|y\|_{F,\omega}}\Big).$$

We call $\xi_E \oplus_\psi \xi_F$ the *ψ-direct sum* of ξ_E and ξ_F. If both norm families ξ_E and ξ_F are Hermitian, and if

$$\psi_\omega(t) = \begin{cases} \max\{t, 1-t\}, & \omega \in \Omega \setminus \Omega_\infty, \\ (t^2 + (1-t)^2)^{1/2}, & \omega \in \Omega_\infty, \end{cases}$$

then the direct sum $\xi_E \oplus_\psi \xi_F$ belongs to $\mathcal{H}_{E\oplus F}$. We call it the *orthogonal direct sum* of ξ_E and ξ_F.

4.1.1.4 Dual norm family

Let E be a vector space of finite dimension over K and $\xi = \{\|\cdot\|_\omega\}_{\omega\in\Omega}$ be a norm family in $\mathcal{N}_E$. The dual norms (see §1.1.5) $\{\|\cdot\|_{\omega,*}\}_{\omega\in\Omega}$ form a norm family in $\mathcal{N}_{E^\vee}$, called the *dual* of ξ, denoted by $\xi^\vee$. Note that $\xi^\vee$ is always ultrametric on $\Omega\setminus\Omega_\infty$, and it is Hermitian if ξ is. Moreover, if for any $\omega \in \Omega\setminus\Omega_\infty$ the norm $\|\cdot\|_\omega$ is ultrametric, then one has $(\xi^\vee)^\vee = \xi$, where we identify E with its double dual space $E^{\vee\vee}$ (see Proposition 1.1.18 and Corollary 1.2.12). In particular, if E is a K-vector space of dimension 1, then one has $(\xi^\vee)^\vee = \xi$.

4.1.1.5 Tensor products

Let $\{E_i\}_{i=1}^n$ be a family of finite-dimensional vector spaces over K. For any $i \in \{1,\ldots,n\}$, let $\xi_i = \{\|\cdot\|_{i,\omega}\}_{\omega\in\Omega}$ be an element of $\mathcal{N}_{E_i}$. We denote by $\xi_1 \otimes_\pi \cdots \otimes_\pi \xi_n$ the norm family $\{\|\cdot\|_{\omega,\pi}\}_{\omega\in\Omega}$ in $\mathcal{N}_{E_1\otimes\cdots\otimes E_n}$, where $\|\cdot\|_{\omega,\pi}$ is the π-tensor product of the norms $\|\cdot\|_{i,\omega}$, $i \in \{1,\ldots,n\}$ (see §1.1.11). The norm family $\xi_1 \otimes_\pi \cdots \otimes_\pi \xi_n$ is called the *π-tensor product* of $\xi_1,\ldots,\xi_n$. Similarly, we denote by $\xi_1 \otimes_\varepsilon \cdots \otimes_\varepsilon \xi_n$ the norm family in $\mathcal{N}_{E_1\otimes\cdots\otimes E_n}$ consisting of ε-tensor products (see §1.1.11) of the norms $\|\cdot\|_{1,\omega},\ldots,\|\cdot\|_{n,\omega}$. We call $\xi_1 \otimes_\varepsilon \cdots \otimes_\varepsilon \xi_n$ the *ε-tensor product* of $\xi_1,\ldots,\xi_n$. We also introduce the following mixed version of ε-tensor product and π-tensor product. We denote by $\xi_1 \otimes_{\varepsilon,\pi} \cdots \otimes_{\varepsilon,\pi} \xi_n$ consisting of the norms $\|\cdot\|_{\omega,\varepsilon}$ if $\omega \in \Omega\setminus\Omega_\infty$ and $\|\cdot\|_{\omega,\pi}$ if $\omega \in \Omega_\infty$. This norm family is called the ε,π-tensor product of $\xi_1,\ldots,\xi_n$. Note that the ε-tensor product and the ε,π-tensor product are both ultrametric on $\Omega\setminus\Omega_\infty$.

If all norm families $\xi_1,\ldots,\xi_n$ are Hermitian, we denote by $\xi_1 \otimes \cdots \otimes \xi_n$ the norm family $\{\|\cdot\|_\omega\}_{\omega\in\Omega}$ in $\mathcal{H}_{E_1\otimes\cdots\otimes E_n}$, where for each $\omega \in \Omega\setminus\Omega_\infty$, the norm $\|\cdot\|_\omega$ is the ε-tensor product of $\{\|\cdot\|_{i,\omega}\}_{i\in\{1,\ldots,n\}}$, and for each $\omega \in \Omega_\infty$, the norm $\|\cdot\|_\omega$ is the orthogonal tensor product (see §1.2.9) of $\{\|\cdot\|_{i,\omega}\}_{i\in\{1,\ldots,n\}}$. We call $\xi_1 \otimes \cdots \otimes \xi_n$ the *orthogonal tensor product* of $\xi_1,\ldots,\xi_n$.

4.1.1.6 Exterior powers

Let E be a finite-dimensional vector space over K and $\xi = \{\|\cdot\|_{E,\omega}\}_{\omega\in\Omega}$ be a norm family in $\mathcal{N}_E$. Let i be a non-negative integer. We equip $E^{\otimes i}$ with the ε,π-tensor power of the norm family ξ, which induces by quotient a norm family on the exterior power $\Lambda^i(E)$ which we denote by $\Lambda^i\xi$.

4.1.1.7 Determinant

Let E be a finite-dimensional vector space over K and $\xi = \{\|\cdot\|_{E,\omega}\}_{\omega\in\Omega}$ be a norm family in $\mathcal{N}_E$. Each norm $\|\cdot\|_{E,\omega}$ induces a determinant norm $\|\cdot\|_{\det(E),\omega}$ on $\det(E)\otimes_K K_\omega \cong \det(E_{K_\omega})$ (see §1.1.13). The norm family $\{\|\cdot\|_{\det(E),\omega}\}_{\omega\in\Omega}$ is called the *determinant* of ξ, denoted by $\det(\xi)$. By Proposition 1.2.15 we obtain that $\det(\xi)$ coincides with $\Lambda^r\xi$, where r is the dimension of E over K.

4.1.1.8 Extension of scalars

Let E be a vector space of finite dimension over K and $\xi = \{\|\cdot\|_\omega\}_{\omega\in\Omega}$ be a norm family in $\mathcal{N}_E$. Let L/K be an algebraic extension of the field K and $((\Omega_L, \mathcal{A}_L, \nu_L), \phi_L)$ be the adelic structure of the adelic curve $S \otimes_K L$. We construct a norm family $\xi_L = \{\|\cdot\|_x\}_{x\in\Omega_L} \in \mathcal{N}_{E\otimes_K L}$ as follows: for any $x \in \Omega_L$ whose canonical image in Ω is ω, if $|\cdot|_\omega$ is non-Archimedean, $\|\cdot\|_x$ is the norm $\|\cdot\|_{\omega,L_x,\varepsilon}$ on $(E_{K_\omega})\otimes_{K_\omega} L_x$ induced by $\|\cdot\|_\omega$ by ε-extension of scalars; otherwise $\|\cdot\|_x$ is the norm $\|\cdot\|_{\omega,L_x,\pi}$ on $(E_{K_\omega})\otimes_{K_\omega} L_x$ induced by $\|\cdot\|_\omega$ by π-extension of scalars (see §1.3). By Proposition 1.3.20 (1), if the dimension of E over K is 1, then the norm family $(\xi^\vee)_L$ identifies with the dual norm family of ξ_L. Moreover, by Corollary 1.3.15, if $L_2/L_1/K$ are successive algebraic extensions, then one has $(\xi_{L_1})_{L_2} = \xi_{L_2}$, where we identify $E \otimes_K L_2$ with $(E \otimes_K L_1) \otimes_{L_1} L_2$.

Let $\xi = \{\|\cdot\|_\omega\}_{\omega\in\Omega} \in \mathcal{H}_E$ be a Hermitian norm family, where for $\omega \in \Omega$, the norm $\|\cdot\|_\omega$ is induced by an inner product $\langle\,,\rangle_\omega$. We denote by $\xi_L^H = \{\|\cdot\|_x\}_{x\in\Omega_L} \in \mathcal{H}_{E\otimes_K L}$ the Hermitian norm family defined as follows. For any $x \in \Omega_L \setminus \Omega_{L,\infty}$ over $\omega \in \Omega \setminus \Omega_\infty$, one has $\|\cdot\|_x = \|\cdot\|_{\omega,L_x,\varepsilon}$; for any $x \in \Omega_{L,\infty}$ over $\omega \in \Omega_\infty$, $\|\cdot\|_x$ is the norm $\|\cdot\|_{\omega,L_x,\mathrm{HS}}$ induced by the inner product $\langle\,,\rangle_{\omega,L_x}$ on E_{L_x} which extends $\langle\,,\rangle_\omega$ on E_{K_ω} (namely $\|\cdot\|_x$ is the orthogonal tensor product norm of $\|\cdot\|_\omega$ and $|\cdot|_x$ if we identify E_{L_x} with $E_{K_\omega} \otimes_{K_\omega} L_x$, see Remark 1.3.2). One has $(\xi^\vee)_L^H = (\xi_L^H)^\vee$ (see Proposition 1.3.20 (1) for the ultrametric part and Remark 1.3.2 for the inner product part).

4.1.1.9 Comparison of norm families

Let E be a finite-dimensional vector space over K, and $\xi = \{\|\cdot\|_\omega\}_{\omega\in\Omega}$ and $\xi' = \{\|\cdot\|'_\omega\}_{\omega\in\Omega}$ be two elements of $\mathcal{N}_E$. We say that ξ is *smaller than* ξ' if for any $\omega \in \Omega$ one has $\|\cdot\|_\omega \leqslant \|\cdot\|'_\omega$. The condition "$\xi$ is smaller than ξ" is denoted by $\xi \leqslant \xi'$ or $\xi' \geqslant \xi$.

4.1.1.10 Local distance

Let E be a finite-dimensional vector space over K, and $\xi = \{\|\cdot\|_\omega\}_{\omega\in\Omega}$ and $\xi' = \{\|\cdot\|'_\omega\}_{\omega\in\Omega}$ be two norm families in $\mathcal{N}_E$. For any $\omega \in \Omega$, let

$$d_\omega(\xi, \xi') := \sup_{s\in E_{K_\omega}\setminus\{0\}} \Big|\ln\|s\|_\omega - \ln\|s\|'_\omega\Big|.$$

We call $d_\omega(\xi,\xi')$ the *local distance* on ω of the norm families ξ and ξ'. By Proposition 1.1.43, one has

$$d_\omega(\xi^\vee,(\xi')^\vee) \leqslant d_\omega(\xi,\xi'), \tag{4.1}$$

and the equality holds if $\omega \in \Omega_\infty$ or if $\|\cdot\|_\omega$ and $\|\cdot\|'_\omega$ are both ultrametric.

4.1.2 Dominated norm families

Let E be a finite-dimensional vector space over K and $\xi = \{\|\cdot\|_\omega\}_{\omega\in\Omega}$ be a norm family in $\mathcal{N}_E$.

Definition 4.1.1 We say that the norm family ξ is *upper dominated* if, for any non-zero element $s \in E$, there exists a ν-integrable function $A(\cdot)$ on Ω such that $\ln\|s\|_\omega \leqslant A(\omega)$ ν-almost everywhere. Note that the upper dominancy is equivalent to

$$\forall s \in E\setminus\{0\},\qquad \overline{\int_\Omega} \ln\|s\|_\omega\,\nu(\mathrm{d}\omega) < +\infty$$

with the notation of Definition A.4.1. Similarly, we say that the norm family ξ is *lower dominated* if, for any non-zero element $s \in E$, there exists a ν-integrable function $B(\cdot)$ on Ω such that $B(\omega) \leqslant \ln\|s\|_\omega$ ν-almost everywhere. Note that the lower dominancy is equivalent to

$$\forall s \in E\setminus\{0\},\qquad \underline{\int_\Omega} \ln\|s\|_\omega\,\nu(\mathrm{d}\omega) > -\infty.$$

Definition 4.1.2 We say that ξ is *dominated* if ξ and $\xi^\vee$ are both upper dominated. Note that the upper dominancy of ξ and $\xi^\vee$ implies the lower dominancy of $\xi^\vee$ and ξ, respectively, because (see Proposition A.4.7) $\ln\|\alpha\|_{\omega,*} + \ln\|s\|_\omega \geqslant 0$ for all $s \in E$ and $\alpha \in E^\vee$ with $\alpha(s) = 1$, so that if ξ is dominated, then ξ and $\xi^\vee$ are upper and lower dominated.

Remark 4.1.3 If ξ is a dominated norm family, then also is $\xi^\vee$. In fact, for any $\omega \in \Omega$ one has $\|\cdot\|_{\omega,**} \leqslant \|\cdot\|_\omega$ (see (1.5)). Therefore the upper dominancy of ξ implies that of $\xi^{\vee\vee}$. The converse is true when $\|\cdot\|_\omega$ is ultrametric for $\omega \in \Omega\setminus\Omega_\infty$ since in this case one has $\|\cdot\|_{\omega,**} = \|\cdot\|_\omega$ for any $\omega \in \Omega$ (see Proposition 1.1.18 and Corollary 1.2.12).

Remark 4.1.4 It is *not* true that if ξ is upper and lower dominated then it is dominated. Consider the following example. Let K be an infinite field. We equip K with the discrete σ-algebra $\mathcal{A}$ and let ν be the atomic measure on K such that $\nu(\{a\}) = 1$ for any $a \in K$. For any $a \in K$, let $|\cdot|_a$ be the trivial absolute value on K. Then $S = (K,(K,\mathcal{A},\nu),\{|\cdot|_a\}_{a\in K})$ forms an adelic curve. Consider now the vector space $E = K^2$ over K. For any $a \in K$ let $\|\cdot\|_a$ be the norm on K^2 such that

$$\|(x,y)\|_a = \begin{cases} 0, & \text{if } x = y = 0, \\ 1/2, & \text{if } y = ax,\ x \neq 0, \\ 1, & \text{else.} \end{cases}$$

Then for any vector $s \in K^2$, $s \neq 0$, one has $\|s\|_a = 1$ for all except at most one $a \in K$. Therefore the function $(a \in K) \mapsto \ln \|s\|_a$ on K is integrable, and in particular dominated. Therefore the norm family $\xi = \{\|\cdot\|_a\}_{a \in K}$ is upper dominated and lower dominated. However, for any $a \in K$, the dual norm $\|\cdot\|_{a,*}$ on K^2 (we identify K^2 with the dual vector space of itself in the canonical way) satisfies

$$\|(x,y)\|_{a,*} = \begin{cases} 0, & \text{if } x = y = 0, \\ 1, & \text{if } x = -ay,\ y \neq 0 \\ 2, & \text{else.} \end{cases}$$

Therefore, for any non-zero element $s \in K^2$, one has $\ln \|s\|_{a,*} = \ln(2)$ for all except at most one element $a \in K$. The dual norm family $\xi^\vee$ is thus not upper dominated.

Example 4.1.5 A fundamental example of dominated norm family is that arising from a basis. Let E be a vector space of finite dimension r over K, and $\boldsymbol{e} = \{e_1, \ldots, e_r\}$ be a basis of E over K. For any algebraic extension L/K and any $x \in \Omega_L$, let $\|\cdot\|_{\boldsymbol{e},x}$ be the norm on $E \otimes_K L_x$ such that, for any $(\lambda_1, \ldots, \lambda_r) \in L_x^r$,

$$\|\lambda_1 e_1 + \cdots + \lambda_r e_r\|_{\boldsymbol{e},x} := \begin{cases} \max_{i \in \{1,\ldots,r\}} |\lambda_i|_x, & \text{if } x \in \Omega_L \setminus \Omega_{L,\infty} \\ |\lambda_1|_x + \cdots + |\lambda_r|_x, & \text{if } x \in \Omega_{L,\infty}, \end{cases}$$

where $\Omega_{L,\infty}$ denotes the set of all $x \in \Omega_L$ such that the absolute value $|\cdot|_x$ is Archimedean. Let $\xi_{\boldsymbol{e}}$ be the norm family $\{\|\cdot\|_{\boldsymbol{e},\omega}\}_{\omega \in \Omega}$. Note that one has $\xi_{\boldsymbol{e},L} = \{\|\cdot\|_{\boldsymbol{e},x}\}_{x \in \Omega_L}$ for any algebraic extension L/K. Moreover, for any non-zero vector $s = a_1 e_1 + \cdots + a_r e_r \in E \otimes_K L$, with $(a_1, \ldots, a_r) \in L^r$, one has

$$\forall\, x \in \Omega_L, \quad \ln \|s\|_{\boldsymbol{e},x} \leqslant \max_{\substack{i \in \{1,\ldots,r\} \\ a_i \neq 0}} \ln |a_i|_x + \ln(r) \mathbb{1}_{\Omega_{L,\infty}}(x).$$

Since the functions $x \mapsto \ln |a|_x$ are ν_L-integrable for all $a \in L \setminus \{0\}$ and since $\nu_L(\Omega_{L,\infty}) < +\infty$ (see Proposition 3.1.2), we obtain that the function $(x \in \Omega_L) \mapsto \ln \|s\|_{\boldsymbol{e},x}$ is ν_L-integrable. If we denote by $\{e_1^\vee, \ldots, e_r^\vee\}$ the dual basis of $\boldsymbol{e}$, then for any $\alpha = a_1 e_1^\vee + \cdots + a_r e_r^\vee \in E^\vee \otimes_K L$ with $(a_1, \ldots, a_r) \in L^r$ and any $x \in \Omega_L$ one has

$$\|\alpha\|_{\boldsymbol{e},x,*} = \max\{|a_1|_x, \ldots, |a_r|_x\}.$$

Therefore the function $(x \in \Omega_L) \mapsto \|\alpha\|_{\boldsymbol{e},x,*}$ is $\mathcal{A}$-measurable. If $\alpha \neq 0$, then the function $(x \in \Omega_L) \mapsto \ln \|\alpha\|_{\boldsymbol{e},x,*}$ is ν_L-integrable. Hence the norm family $\xi_{\boldsymbol{e},L}$ is dominated. Note that (see page 209 for the definition of the local distance function)

$$\forall\, x \in \Omega_L, \quad d_x((\xi_{\boldsymbol{e},L})^\vee, \xi_{\boldsymbol{e}^\vee,L}) \leqslant \ln(r) \mathbb{1}_{\Omega_{L,\infty}}(x), \tag{4.2}$$

where $e^\vee$ denotes the dual basis of e.

Proposition 4.1.6 *Let E be a finite-dimensional vector space over K, ξ_1 and ξ_2 be norm families in $\mathcal{N}_E$. We assume that ξ_1 is dominated. If the local distance function $(\omega \in \Omega) \mapsto d_\omega(\xi_1, \xi_2)$ is ν-dominated (see Definition A.4.9), then the norm family ξ_2 is dominated. In particular, if there exists a basis $\mathbf{e}$ of E over K such that the function $(\omega \in \Omega) \mapsto d_\omega(\xi_{\mathbf{e}}, \xi_2)$ is ν-dominated, then the norm family ξ_2 is dominated.*

Proof Assume that ξ_i is of the form $\{\|\cdot\|_{i,\omega}\}_{\omega\in\Omega}$, $i \in \{1, 2\}$. Let s be a non-zero element in E. For any $\omega \in \Omega$, one has,

$$\ln \|s\|_{2,\omega} - \ln \|s\|_{1,\omega} \leqslant d_\omega(\xi_1, \xi_2) \quad \nu\text{-almost everywhere.}$$

Moreover, since the norm family ξ_1 is dominated, one has

$$\overline{\int_\Omega} \ln \|s\|_{1,\omega}\, \nu(\mathrm{d}\omega) < +\infty;$$

since the local distance function $d(\xi_1, \xi_2)$ is dominated, one has

$$\overline{\int_\Omega} d_\omega(\xi_1, \xi_2)\, \nu(\mathrm{d}\omega) < +\infty.$$

Therefore by Proposition A.4.4 one has

$$\overline{\int_\Omega} \ln \|s\|_{2,\omega}\, \nu(\mathrm{d}\omega) \leqslant \overline{\int_\Omega} \ln \|s\|_{1,\omega}\, \nu(\mathrm{d}\omega) + \overline{\int_\Omega} d_\omega(\xi_1, \xi_2)\, \nu(\mathrm{d}\omega) < +\infty.$$

By (4.1), one has $d_\omega(\xi_1^\vee, \xi_2^\vee) \leqslant d_\omega(\xi_1, \xi_2)$ for any ω. Hence the same argument as above applied to the dual norm families shows that

$$\forall\, \alpha \in E^\vee \setminus \{0\}, \qquad \overline{\int_\Omega} \ln \|\alpha\|_{2,\omega,*}\, \nu(\mathrm{d}\omega) < +\infty.$$

Therefore, the norm family ξ_2 is dominated. To establish the last assertion, it suffices to apply the obtained result to the case where $\xi_1 = \xi_{\boldsymbol{e}}$ (see Example 4.1.5 for the fact that the norm family $\xi_{\boldsymbol{e}}$ is dominated). □

Proposition 4.1.7 *Let E be a vector space of finite dimension over K and $\xi = \{\|\cdot\|_\omega\}_{\omega\in\Omega}$ be an element of $\mathcal{N}_E$ which is dominated. Then for any basis $\boldsymbol{e} = \{e_1, \ldots, e_r\}$ of E, there exists a ν-integrable function $A_{\boldsymbol{e}}$ on Ω such that, for any algebraic extension L/K and any $x \in \Omega_L$ one has (note that $\xi_K = \xi^{\vee\vee}$ in the case where $L = K$)*

$$d_x(\xi_L, \xi_{\boldsymbol{e},L}) \leqslant A_{\boldsymbol{e}}(\pi_{L/K}(x)). \tag{4.3}$$

In particular, the local distance function $(x \in \Omega_L) \mapsto d_x(\xi_L, \xi_{\boldsymbol{e},L})$ is ν_L-dominated.

Proof Let x be an element in Ω_L. Assume that s is a non-zero vector of E_{L_x} which is written as $s = a_1e_1 + \cdots + a_re_r$ with $(a_1, \ldots, a_r) \in L_x^r$. Then one has

$$\|s\|_x \leqslant \begin{cases} \max_{i\in\{1,\ldots,r\}} (|a_i|_x \cdot \|e_i\|_x), & \text{if } x \in \Omega_L \setminus \Omega_{L,\infty}, \\ \sum_{i=1}^r |a_i|_x \cdot \|e_i\|_x, & \text{if } x \in \Omega_{L,\infty}. \end{cases}$$

Thus $\ln \|s\|_x \leqslant \ln \|s\|_{e,x} + \max_{i\in\{1,\ldots,r\}} \ln\|e_i\|_x$. Moreover, one can also interpret

$$\sup_{0\neq s\in E_{L_x}} \frac{\|s\|_{e,x}}{\|s\|_x}$$

as the operator norm of the L_x-linear map

$$\mathrm{Id}_{E_{L_x}} : (E_{L_x}, \|\cdot\|_x) \longrightarrow (E_{L_x}, \|\cdot\|_{e,x}),$$

which is equal to the operator norm of the dual L_x-linear map

$$\mathrm{Id}_{E^\vee_{L_x}} : (E^\vee_{L_x}, \|\cdot\|_{e,x,*}) \longrightarrow (E^\vee_{L_x}, \|\cdot\|_{x,*})$$

since the norm $\|\cdot\|_{e,x}$ is reflexive (see Proposition 1.1.22). Let $\{e_i^\vee\}_{i=1}^r$ be the dual basis of $\boldsymbol{e}$. For any $\alpha = b_1 e_1^\vee + \cdots + b_r e_r^\vee \in E^\vee_{L_x}$, one has

$$\|\alpha\|_{x,*} \leqslant \begin{cases} \max_{i\in\{1,\ldots,r\}} (|b_i|_x \cdot \|e_i^\vee\|_{x,*}), & \text{if } x \in \Omega_L \setminus \Omega_{L,\infty}, \\ \sum_{i=1}^r |b_i|_x \cdot \|e_i^\vee\|_{x,*}, & \text{if } x \in \Omega_{L,\infty}. \end{cases}$$

Thus we obtain

$$\ln \|\alpha\|_{x,*} - \ln \|\alpha\|_{e,x,*} \leqslant \max_{i\in\{1,\ldots,r\}} \ln \|e_i^\vee\|_{x,*} + \ln(r)1\!\!1_{\Omega_{L,\infty}}(x).$$

Therefore, for any $s \in E_L$, one has

$$-\max_{i\in\{1,\ldots,r\}} \ln\|e_i\|_x \leqslant \ln \|s\|_{e,x} - \ln \|s\|_x \leqslant \max_{i\in\{1,\ldots,r\}} \ln \|e_i^\vee\|_{x,*} + \ln(r)1\!\!1_{\Omega_{L,\infty}}(x). \quad (4.4)$$

Note that, if $\omega = \pi_{L/K}(x)$, then one has (see Proposition 1.3.1)

$$\forall\, i \in \{1, \ldots, r\}, \quad \|e_i\|_x = \|e_i\|_{\omega,**}.$$

Moreover, if $\omega = \pi_{L/K}(x)$ belongs to $\Omega \setminus \Omega_\infty$, then

$$\|e_i^\vee\|_{x,*} = \|e_i^\vee\|_{\omega,L_x,\varepsilon,*} = \|e_i^\vee\|_{\omega,*,L_x,\varepsilon} = \|e_i^\vee\|_{\omega,*},$$

where the second equality comes from Proposition 1.3.20 (1) and the last one comes from Proposition 1.3.1. If $\omega = \pi_{L/K}(x) \in \Omega_\infty$, then

$$\|e_i^\vee\|_{x,*} = \|e_i^\vee\|_{\omega,L_x,\pi,*} = \|e_i^\vee\|_{\omega,*,L_x,\varepsilon} = \|e_i^\vee\|_{\omega,*},$$

where the second equality comes from Proposition 1.3.20 (2), and the last one comes from Proposition 1.3.1. Since the norm family ξ is dominated, there exists a ν-integrable function A on Ω such that (see Remark 4.1.3)

$$\max_{i\in\{1,\ldots,r\}} \max\{\ln \|e_i\|_{\omega,**}, \ln \|e_i^\vee\|_{\omega,*}\} \leqslant A(\omega) \quad \nu\text{-almost everywhere.}$$

Therefore, by (4.4), we obtain

$$d_x(\xi_L, \xi_{e,L}) \leqslant A_e(\pi_{L/K}(x)) \quad \nu\text{-almost everywhere,}$$

with

$$\forall\, \omega \in \Omega, \quad A_e(\omega) := A(\omega) + \ln(r)\mathbb{1}_{\Omega_\infty}(\omega).$$

Note that the function A_e is ν-integrable on $(\Omega, \mathcal{A})$. The proposition is thus proved.□

Corollary 4.1.8 *Let E be a vector space of finite dimension over K, ξ_1 and ξ_2 be norm families in $\mathcal{N}_E$ which are dominated and ultrametric on $\Omega\setminus\Omega_\infty$. Then the local distance function $(\omega \in \Omega) \mapsto d_\omega(\xi_1, \xi_2)$ is ν-dominated.*

Proof Let e be a basis of E over K. By Proposition 4.1.7, the local distance functions $(\omega \in \Omega) \mapsto d_\omega(\xi_1, \xi_e)$ and $(\omega \in \Omega) \mapsto d_\omega(\xi_2, \xi_e)$ are both ν-dominated. Since for any $\omega \in \Omega$ one has

$$d_\omega(\xi_1, \xi_2) \leqslant d_\omega(\xi_1, \xi_e) + d_\omega(\xi_2, \xi_e),$$

by Propositions A.4.2 and A.4.4 the function $(\omega \in \Omega) \mapsto d_\omega(\xi_1, \xi_2)$ is ν-dominated.□

Remark 4.1.9 The assertion of Corollary 4.1.8 does not necessarily hold without the condition that the norm families are ultrametric on $\Omega \setminus \Omega_\infty$. Consider the following counter-example. Let K be an infinite field, $\mathcal{A}$ be the discrete σ-algebra on K and ν be the atomic measure on K such that $\nu(\{a\}) = 1$ for any $a \in K$. For any $a \in K$, let $|\cdot|_a$ be the trivial absolute value on K. We consider the adelic curve $S = (K, (K, \mathcal{A}, \nu), \{|\cdot|_a\}_{a\in K})$. For any $a \in K$, let $\|\cdot\|_a$ be the norm on K^2 such that

$$\|(x, y)\|_a = \begin{cases} 0, & \text{if } x = y = 0,\\ 2, & \text{if } y = ax,\ x \neq 0,\\ 1, & \text{else.}\end{cases}$$

Note that for any $s \in K^2 \setminus \{(0,0)\}$, one has $\|s\|_a = 1$ for all except at most one $a \in K$. Therefore the norm family $\xi = \{\|\cdot\|_a\}_{a\in K}$ is upper dominated. If we identify K^2 with the dual vector space of itself in the canonical way, then for any $a \in K$ one has

$$\forall\, (x, y) \in K^2, \quad \|(x, y)\|_{a,*} = \begin{cases} 0, & \text{if } x = y = 0,\\ 1, & \text{else.}\end{cases}$$

Hence the dual norm family $\xi^\vee$ is also upper dominated. Now let $e = \{(1, 0), (0, 1)\}$ be the canonical basis of K^2. For any $a \in K$ one has

$$\forall\,(x,y)\in K^2,\quad \|(x,y)\|_{e,a}=\begin{cases}0, & \text{if } x=y=0,\\ 1, & \text{else.}\end{cases}$$

Therefore one has $d_a(\xi,\xi_e)=\ln(2)$ for any $a\in K$. Since K is an infinite set, the local distance function $(a\in K)\mapsto d_a(\xi,\xi_e)$ is clearly not upper dominated.

Corollary 4.1.10 *Let E be a finite-dimensional vector space over K and $\xi=\{\|\cdot\|_\omega\}_{\omega\in\Omega}$ be a norm family in $\mathcal{N}_E$. The following assertions are equivalent:*

(1) *the norm family ξ is dominated and the local distance function $(\omega\in\Omega)\mapsto d_\omega(\xi,\xi^{\vee\vee})$ is ν-dominated;*
(2) *for any basis $\boldsymbol{e}$ of E, the local distance function $(\omega\in\Omega)\mapsto d_\omega(\xi,\xi_{\boldsymbol{e}})$ is ν-dominated;*
(3) *there exists a basis $\boldsymbol{e}$ of E such that the local distance function $(\omega\in\Omega)\mapsto d_\omega(\xi,\xi_{\boldsymbol{e}})$ is ν-dominated.*

Proof "(1)$\Longrightarrow$(2)": Note that the norm family $\xi^{\vee\vee}$ is ultrametric on $\Omega\setminus\Omega_\infty$. Moreover it is dominated since ξ is dominated (see Remark 4.1.3). By Proposition 4.1.7, we obtain that, for any basis $\boldsymbol{e}$ of E, the local distance function $(\omega\in\Omega)\mapsto d_\omega(\xi^{\vee\vee},\xi_{\boldsymbol{e}})$ is ν-dominated. By the assumption that the function $(\omega\in\Omega)\mapsto d_\omega(\xi,\xi^{\vee\vee})$ is ν-dominated, we deduce that the function $(\omega\in\Omega)\mapsto d_\omega(\xi,\xi_{\boldsymbol{e}})$ is also ν-dominated.

"(2)$\Longrightarrow$(3)" is trivial.

"(3)$\Longrightarrow$(1)": By Proposition 4.1.6, the norm family ξ is ν-dominated. By Proposition 4.1.7, the function $(\omega\in\Omega)\mapsto d_\omega(\xi^{\vee\vee},\xi_{\boldsymbol{e}})$ is ν-dominated. Since $d_\omega(\xi,\xi^{\vee\vee})\leqslant d_\omega(\xi,\xi_{\boldsymbol{e}})+d_\omega(\xi^{\vee\vee},\xi_{\boldsymbol{e}})$, we obtain that the function $(\omega\in\Omega)\mapsto d_\omega(\xi,\xi^{\vee\vee})$ is ν-dominated. □

Definition 4.1.11 Let E be a finite-dimensional vector space over K and ξ be a norm family on E. We say that ξ is *strongly dominated* if it is dominated and if the function $(\omega\in\Omega)\mapsto d_\omega(\xi,\xi^{\vee\vee})$ is ν-dominated (or equivalently, (E,ξ) satisfies any of the assertions in Corollary 4.1.10).

Remark 4.1.12 Let E be a finite-dimensional vector space over K and ξ be a norm family on E. If the norm family ξ is ultrametric on $\Omega\setminus\Omega_\infty$, then the function $(\omega\in\Omega)\mapsto d_\omega(\xi,\xi^{\vee\vee})$ is identically zero (and hence is ν-dominated). Therefore, in this case ξ is dominated if and only if it is strongly dominated. In particular, in the case where E is of dimension 1 over K, the norm family ξ is dominated if and only if it is strongly dominated. Moreover, if ξ is a dominated norm family on a finite-dimensional vector space, then the dual norm family ξ is strongly dominated since it is dominated (see Remark 4.1.3) and ultrametric on $\Omega\setminus\Omega_\infty$.

Corollary 4.1.13 *Let E be a vector space of finite dimension over K and $\xi=\{\|\cdot\|_\omega\}_{\omega\in\Omega}$ be a norm family in $\mathcal{N}_E$. If ξ is dominated, then for any algebraic extension L/K the norm family ξ_L is strongly dominated. Conversely, if there exists an algebraic extension L/K such that the norm family ξ_L is dominated and if the function $(\omega\in\Omega)\mapsto d_\omega(\xi,\xi^{\vee\vee})$ is ν-dominated, then the norm family ξ is also dominated.*

Proof Assume that the norm family ξ is dominated. By Proposition 4.1.7, for any basis $\boldsymbol{e}$ of E, the local distance function $(x \in \Omega_L) \mapsto d_x(\xi_L, \xi_{\boldsymbol{e},L})$ is ν_L-dominated. Therefore, by Proposition 4.1.6 we obtain that the norm family ξ_L is strongly dominated.

Conversely, we assume that L/K is an algebraic extension and ξ_L is dominated. Let $\boldsymbol{e}$ be a basis of E over K. Since ξ_L is dominated and since the norms in the family ξ_L corresponding to non-Archimedean absolute values are ultrametric, the function $f : (x \in \Omega_L) \mapsto d_x(\xi_L, \xi_{\boldsymbol{e},L})$ is ν_L-dominated. Moreover, by Proposition 1.3.7, one has $f = g \circ \pi_{L/K}$, where g sends $\omega \in \Omega$ to $d_\omega(\xi^{\vee\vee}, \xi_{\boldsymbol{e}})$ (one has $\xi_{\boldsymbol{e}} = \xi_{\boldsymbol{e}}^{\vee\vee}$ since it is ultrametric on $\Omega \setminus \Omega_\infty$). Since the function f is ν_L-dominated, there exists a ν_L-integrable function A on Ω_L such that $|f| \leqslant A$ almost everywhere. By Proposition 3.4.9, the function $I_{L/K}(A)$ is ν-integrable. Moreover, one has $I_{L/K}(|f|) = |g|$ since $|f| = |g| \circ \pi_{L/K}$ (see Remark 3.4.7). Therefore, the function g is ν-dominated by the ν-integrable function $I_{L/K}(A)$, which implies that the norm family $\xi^{\vee\vee}$ is dominated. Finally, by Proposition 4.1.6 and the assumption that the function $(\omega \in \Omega) \mapsto d_\omega(\xi, \xi^{\vee\vee})$ is ν-dominated, we obtain that the norm family ξ is dominated. □

Remark 4.1.14 Let E be a finite-dimensional vector space over K. Corollary 4.1.10 implies that there exist Hermitian norm families on E which are dominated. In fact, let $\boldsymbol{e} = \{e_i\}_{i=1}^r$ be a basis of E over K. Consider the following norm family $\xi = \{\|\cdot\|_\omega\}_{\omega \in \Omega}$ with

$$\|a_1 e_1 + \cdots + a_r e_r\|_\omega = \begin{cases} \max_{i \in \{1,\ldots,r\}} |a_i|_\omega, & \text{if } \omega \in \Omega \setminus \Omega_\infty, \\ \left(\sum_{i=1}^r |a_i|_\omega^2\right)^{1/2}, & \text{if } \omega \in \Omega_\infty, \end{cases}$$

for all $(a_1, \ldots, a_r) \in K_\omega^r$. It is a Hermitian norm family on E. Note that one has $d_\omega(\xi, \xi_{\boldsymbol{e}}) \leqslant \frac{1}{2}\ln(r)1\!\!1_{\Omega_\infty}(\omega)$. Therefore $(\omega \in \Omega) \mapsto d_\omega(\xi, \xi_{\boldsymbol{e}})$ is a ν-dominated function on Ω. By Corollary 4.1.10, we obtain that ξ is a dominated norm family.

Let ξ be a Hermitian norm family on E. If ξ is dominated, then for any algebraic extension L/K, the norm family ξ_L^H is dominated. In fact, by Corollary 4.1.10, the norm family ξ_L is dominated. By Proposition 1.3.1 (3), the local distance function $(x \in \Omega_L) \mapsto d_x(\xi_L, \xi_L^H)$ is bounded from above by $\frac{1}{2}\ln(2)1\!\!1_{\Omega_{L,\infty}}$. By Proposition 4.1.6, we obtain that the norm family ξ_L^H is dominated.

Proposition 4.1.15 *Let E be a finite-dimensional vector space over K and L/K be an algebraic extension of fields. For any $\omega \in \Omega$, we fix an extension $|\cdot|_{L,\omega}$ on L of the absolute value $|\cdot|_\omega$ and denote by L_ω the completion of L with respect to the extended absolute value. Let $\boldsymbol{e} = \{e_i\}_{i=1}^r$ be a basis of $E \otimes_K L$. For any $\omega \in \Omega$, let $\|\cdot\|'_\omega$ be the norm on $E \otimes_K L_\omega$ defined as*

$$\forall\, (\lambda_1, \ldots, \lambda_r) \in L_\omega^r, \quad \|\lambda_1 e_1 + \cdots + \lambda_r e_r\|'_\omega = \max_{i \in \{1,\ldots,r\}} |\lambda_i|_{L,\omega}$$

and let $\|\cdot\|_\omega$ be the restriction of $\|\cdot\|'_\omega$ to $E \otimes_K K_\omega$. Then the norm family $\xi = \{\|\cdot\|_\omega\}_{\omega \in \Omega}$ in $\mathcal{N}_E$ is strongly dominated.

Proof We first prove that, for any element $b \in L$, the function

$$(\omega \in \Omega) \longmapsto \ln |b|_{L,\omega}$$

is bounded from above by a ν-integrable function. Let

$$F(X) = X^n + a_{n-1}X^{n-1} + \cdots + a_0 \in K[X]$$

be the minimal polynomial of b. By the same argument as in the proof of Theorem 3.3.7 (4), we obtain that

$$\ln |b|_\omega \leqslant 1\!\!1_{\Omega_\infty}(\omega)\ln(n) + \max\{0, \ln |a_0|_\omega, \ldots, \ln |a_{n-1}|_\omega\}.$$

By Proposition 3.1.2, the function $1\!\!1_{\Omega_\infty}$ is ν-integrable. Moreover, by the definition of adelic curves, for any $i \in \{0, \ldots, n-1\}$ such that $a_i \neq 0$, the function $(\omega \in \Omega) \mapsto \ln |a_i|_\omega$ is also ν-integrable, we thus obtain the assertion.

Let $\boldsymbol{f} = \{f_i\}_{i=1}^r$ be a basis of E over K and $A = (a_{ij})_{(i,j)\in\{1,\ldots,r\}^2} \in M_{r\times r}(L)$ be the transition matrix between $\boldsymbol{e}$ and $\boldsymbol{f}$, namely

$$\forall\, i \in \{1, \ldots, r\}, \quad f_i = \sum_{j=1}^r a_{ij} e_j.$$

Let $(b_{ij})_{(i,j)\in\{1,\ldots,r\}^2} \in M_{r\times r}(L)$ be the inverse matrix of A. Then one has

$$\forall\, i \in \{1, \ldots, r\}, \quad e_i = \sum_{j=1}^r b_{ij} f_j.$$

By the above assertion, there exists a ν-integrable function g on Ω such that

$$\forall\, \omega \in \Omega, \quad \max_{(i,j)\in\{1,\ldots,r\}^2} \max\{\ln |a_{ij}|_\omega, \ln |b_{ij}|_\omega\} \leqslant g(\omega).$$

We will prove that the local distance function $d(\xi, \xi_{\boldsymbol{f}})$ is ν-dominated. Let $\omega \in \Omega$ and $x = \lambda_1 f_1 + \cdots + \lambda_r f_r$ be an element of $E \otimes_K K_\omega$. One has

$$x = \sum_{i=1}^r \lambda_i \sum_{j=1}^r a_{ij} e_j = \sum_{j=1}^r \Big(\sum_{i=1}^r a_{ij}\lambda_i\Big) e_j.$$

Therefore

$$\begin{aligned}\ln \|x\|_\omega &= \max_{j\in\{1,\ldots,r\}} \ln \Big|\sum_{i=1}^r a_{ij}\lambda_i\Big|_{L,\omega} \leqslant \max_{i\in\{1,\ldots,r\}} \ln |\lambda_i|_\omega + g(\omega) + \ln(r)1\!\!1_{\Omega_\infty}(\omega)\\ &\leqslant \ln \|x\|_{\boldsymbol{f},\omega} + g(\omega) + \ln(r)1\!\!1_{\Omega_\infty}(\omega).\end{aligned}$$

Similarly, if we write x as $x = \mu_1 e_1 + \cdots + \mu_r e_r$, with $(\mu_1, \ldots, \mu_r) \in L_\omega^r$, one has

$$x = \sum_{i=1}^r \mu_i \sum_{j=1}^r b_{ij} f_j = \sum_{j=1}^r \Big(\sum_{i=1}^r b_{ij}\mu_i\Big) f_j.$$

Namely $\lambda_j = \sum_{i=1}^r b_{ij}\mu_i$ for any $j \in \{1,\ldots,r\}$. If $\omega \in \Omega \setminus \Omega_\infty$ then

$$\begin{aligned}\ln\|x\|_{f,\omega} &= \ln\Big(\max_{j\in\{1,\ldots,r\}} |\lambda_j|_\omega\Big) = \ln\Big(\max_{j\in\{1,\ldots,r\}} \Big|\sum_{i=1}^r b_{ij}\mu_i\Big|_\omega\Big)\\ &\leqslant \ln\Big(\max_{i\in\{1,\ldots,r\}} |\mu_i|_{L,\omega}\Big) + g(\omega) = \ln\|x\|_\omega + g(\omega).\end{aligned}$$

If $\omega \in \Omega_\infty$, then

$$\begin{aligned}\ln\|x\|_{f,\omega} &= \ln\Big(\sum_{j=1}^r |\lambda_j|_\omega\Big) = \ln\Big(\sum_{j=1}^r \Big|\sum_{i=1}^r b_{ij}\mu_i\Big|_\omega\Big)\\ &\leqslant \ln\Big(\max_{i\in\{1,\ldots,r\}} |\mu_i|_{L,\omega}\Big) + g(\omega) + \ln(r^2) = \ln\|x\|_\omega + g(\omega) + \ln(r^2).\end{aligned}$$

Therefore, one has

$$\forall\, \omega \in \Omega, \quad d_\omega(\xi, \xi_f) \leqslant g(\omega) + 2\ln(r)\mathbb{1}_{\Omega_\infty}(\omega),$$

which implies that the local distance function $d(\xi, \xi_f)$ is ν-dominated. By Corollary 4.1.10, we obtain that the norm family ξ is dominated. □

The following proposition is a criterion of the dominance property in the case where the vector space is of dimension 1.

Proposition 4.1.16 *Let E be a vector space of dimension 1 over K and $\xi = \{\|\cdot\|_\omega\}_{\omega\in\Omega}$ be a norm family on E. Then the following conditions are equivalent:*

(1) *the norm family ξ is dominated;*
(2) *for any non-zero element $s \in E$, the function $(\omega \in \Omega) \mapsto \ln\|s\|_\omega$ is ν-dominated;*
(3) *there exists a non-zero element $s \in E$ such that the function $(\omega \in \Omega) \mapsto \ln\|s\|_\omega$ is ν-dominated.*

Proof "(1)⟹(2)⟹(3)" are trivial. In the following, we prove "(3)⟹(1)". If s' is a non-zero element of E, then we can write it in the form $s' = as$, where a is a non-zero element of K. Then one has

$$\forall\, \omega \in \Omega, \quad \ln\|s'\|_\omega = \ln|a|_\omega + \ln\|s\|_\omega = \ln|a|_\omega + \ln\|s\|_\omega.$$

Since the function $(\omega \in \Omega) \mapsto \ln\|s\|_\omega$ is ν-dominated and the function $(\omega \in \Omega) \mapsto \ln|a|_\omega$ is ν-integrable, we obtain that the function $(\omega \in \Omega) \mapsto \ln\|s'\|_\omega$ is ν-dominated. Moreover, if we denote by $s^\vee$ the dual element of s in $E^\vee$, then one has

$$\ln\|s^\vee\|_{\omega,*} = -\ln\|s\|_\omega \tag{4.5}$$

for any $\omega \in \Omega$. By the same argument as above, we obtain that, for any non-zero element $\alpha \in E^\vee$, the function $(\omega \in \Omega) \mapsto \ln\|\alpha\|_{\omega,*}$ is ν-dominated. Therefore the norm family ξ is dominated. □

Let K' be a finite extension field of K and let

$$S_{K'} = S \otimes_K K' = (K', (\Omega_{K'}, \mathcal{A}_{K'}, \nu_{K'}), \phi_{K'})$$

be the algebraic extension of S by K'. Let E be a finite-dimensional vector space over K and $E_{K'} = E \otimes_K K'$. Note that, for any $\omega \in \Omega$ and any $\omega' \in \Omega_{K'}$ such that $\pi_{K'/K}(\omega') = \omega$, the vector space $E \otimes_K K_\omega$ can be naturally considered as a K_ω-vector subspace of $E_{K'} \otimes_{K'} K'_{\omega'}$.

Proposition 4.1.17 *Let* $\xi = \{\|\cdot\|_\omega\}_{\omega\in\Omega}$ *and* $\xi' = \{\|\cdot\|'_{\omega'}\}_{\omega'\in\Omega_{K'}}$ *be norm familes of* E *and* $E_{K'}$*, respectively, such that*

$$\forall\, \omega \in \Omega,\ \forall\, \omega' \in \pi_{K'/K}^{-1}(\{\omega\}),\ \forall s \in E \otimes_K K_\omega, \quad \|s\|_\omega = \|s\|'_{\omega'}. \tag{4.6}$$

If ξ' *is dominated (resp. strongly dominated), then* ξ *is also dominated (resp. strongly dominated).*

Proof Assume that ξ' is dominated. Let s be a non-zero element of E. By the assumption (4.6) one has

$$\overline{\int_\Omega} \ln\|s\|_\omega\, \nu(\mathrm{d}\omega) = \overline{\int_{\Omega_{K'}}} \ln\|s\|'_{\omega'}\, \nu_{K'}(\mathrm{d}\omega') < +\infty.$$

Hence the norm family ξ is upper dominated. Let α be a non-zero element in $E^\vee$. For any $\omega \in \Omega$ and any $\omega' \in \pi_{K'/K}^{-1}(\{\omega\})$, one has

$$\|\alpha\|_{\omega,*} = \sup_{s\in(E\otimes_K K_\omega)\setminus\{0\}} \frac{|\alpha(s)|_\omega}{\|s\|_\omega} = \sup_{s\in(E\otimes_K K_\omega)\setminus\{0\}} \frac{|\alpha(s)|_\omega}{\|s\|'_{\omega'}} \leqslant \|\alpha\|'_{\omega',*}.$$

Since $(\xi')^\vee$ is upper dominated, we deduce that $\xi^\vee$ is also upper dominated.

Assume that ξ' is strongly dominated. Let $\boldsymbol{e} = \{e_i\}_{i=1}^n$ be a basis of E. Then, for $\omega \in \Omega$ and $\omega' \in \pi_{K'/K}^{-1}(\omega)$,

$$\begin{aligned}
d_{\omega'}(\xi', \xi_{\boldsymbol{e},K'}) &= \sup_{s'\in(E_{K'}\otimes_{K'} K'_{\omega'})\setminus\{0\}} \Big|\ln\|s'\|'_{\omega'} - \ln\|s'\|_{\boldsymbol{e},K',\omega'}\Big| \\
&\geqslant \sup_{s\in(E\otimes_K K_\omega)\setminus\{0\}} \Big|\ln\|s\|'_{\omega'} - \ln\|s\|_{\boldsymbol{e},K',\omega'}\Big| \\
&= \sup_{s\in(E\otimes_K K_\omega)\setminus\{0\}} \Big|\ln\|s\|_\omega - \ln\|s\|_{\boldsymbol{e},\omega}\Big| = d_\omega(\xi, \xi_{\boldsymbol{e}}).
\end{aligned}$$

By our assumption together with Corollary 4.1.10, the function $(\omega' \in \Omega_{K'}) \mapsto d_{\omega'}(\xi', \xi_{\boldsymbol{e},K'})$ is $\nu_{K'}$-dominated, that is, there is an integrable function A' on $\Omega_{K'}$ such that $d_{\omega'}(\xi', \xi_{\boldsymbol{e},K'}) \leqslant A'(\omega')$ for all $\omega' \in \Omega_{K'}$, so that the above estimate implies that $d_\omega(\xi, \xi_{\boldsymbol{e}}) \leqslant I_{K'/K}(A')(\omega)$ for all $\omega \in \Omega$. By Proposition 3.4.9, one has

$$\int_{\Omega_{K'}} A'(\omega')\, \nu_{K'}(\mathrm{d}\omega') = \int_\Omega I_{K'/K}(A')(\omega)\, \nu(\mathrm{d}\omega)$$

and hence ξ is strongly dominated by Corollary 4.1.10 again. □

Corollary 4.1.18 *Let $f : X \to \operatorname{Spec} K$ be a geometrically reduced projective K-scheme and L be an invertible $\mathcal{O}_X$-module. Let $X_{K'} := X \times_{\operatorname{Spec} K} \operatorname{Spec} K'$ and $L_{K'} := L \otimes_K K'$. For each $\omega \in \Omega$ and $\omega' \in \Omega_{K'}$, X_ω, L_ω, $X_{K',\omega'}$ and $L_{K',\omega'}$ are defined by*

$$\begin{cases} X_\omega := X \times_{\operatorname{Spec}(K)} \operatorname{Spec}(K_\omega), & L_\omega := L \otimes_K K_\omega, \\ X_{K',\omega'} := X_{K'} \times_{\operatorname{Spec}(K')} \operatorname{Spec}(K'_{\omega'}), & L_{K',\omega'} := L_{K'} \otimes_{K'} K'_{\omega'}. \end{cases}$$

Moreover, for each $\omega \in \Omega$ and $\omega' \in \pi_{K'/K}^{-1}(\omega)$, let φ_ω be a metric of L_ω on X_ω, and $\varphi_{K',\omega'}$ be the metric of $L_{K'}$ obtained by φ_ω by the extension of scalars (cf. Definition 2.1.18). Let $\|\cdot\|_{\varphi_\omega}$ and $\|\cdot\|_{\varphi_{K',\omega'}}$ be the sup norms on $H^0(X_\omega, L_\omega)$ and $H^0(X_{K',\omega'}, L_{K',\omega'})$ associated with the metrics φ_ω and $\varphi_{K',\omega'}$, respectively. If $\xi_{K'} = \{\|\cdot\|_{\varphi_{K',\omega'}}\}_{\omega' \in \Omega_{K'}}$ on $H^0(X_{K'}, L_{K'})$ is dominated, then $\xi = \{\|\cdot\|_{\varphi_\omega}\}_{\omega \in \Omega}$ on $H^0(X, L)$ is also dominated.

Proof For $\omega \in \Omega$, $\omega' \in \pi_{K'/K}^{-1}(\omega)$ and $s \in H(X_\omega, L_\omega)$, one has $\|s\|_{\varphi_{K',\omega'}} = \|s\|_{\varphi_\omega}$ (see Proposition 2.1.19), so that the assertion follows from Proposition 4.1.17. □

The following proposition shows that the dominance property is actually preserved by most of the algebraic constructions on norm families.

Proposition 4.1.19 (1) *Let E be a finite-dimensional vector space over K and ξ be a dominated (resp. strongly dominated) norm family on E. Then the restriction of ξ to any vector subspace of E is a dominated (resp. strongly dominated) norm family.*

(2) *Let E be a finite-dimensional vector space over K and ξ be a dominated (resp. strongly dominated) norm family on E. Then the quotient norm family of ξ on any quotient vector space of E is a dominated (resp. strongly dominated) norm family.*

(3) *Let E be a finite-dimensional vector space over K and ξ be an element of $\mathcal{N}_E$. If ξ is dominated, then the norm family $\xi^\vee$ is strongly dominated.*

(4) *Let E and F be finite-dimensional vector spaces over K, and ξ_E and ξ_F be elements of $\mathcal{N}_E$ and $\mathcal{N}_F$, respectively. Let $\psi : \Omega \to \mathscr{S}$ be a map such that $\psi = \psi_0$ outside of a measurable subset Ω' of Ω with $\nu(\Omega') < +\infty$, where ψ_0 denotes the function in $\mathscr{S}$ sending $t \in [0, 1]$ to $\max\{t, 1 - t\}$. If both norm families ξ_E and ξ_F are dominated (resp. strongly dominated), then the ψ-direct sum $\xi_E \oplus_\psi \xi_F$ is also dominated (resp. strongly dominated).*

(5) *Let E and F be finite-dimensional vector spaces over K, and ξ_E and ξ_F be elements of $\mathcal{N}_E$ and $\mathcal{N}_F$, respectively. Assume that both norm families ξ_E and ξ_F are dominated. Then the ε-tensor product $\xi_E \otimes_\varepsilon \xi_F$ and the ε,π-tensor product $\xi_E \otimes_{\varepsilon,\pi} \xi_F$ are strongly dominated. If in addition both norm families ξ_E and ξ_F are Hermitian, then the orthogonal tensor product $\xi_E \otimes \xi_F$ is strongly dominated.*

(6) *Let E be a finite-dimensional vector space over K and ξ be an element of $\mathcal{N}_E$. Assume that ξ is dominated. Then, for any $i \in \mathbb{N}$, the exterior power norm family $\Lambda^i \xi$ is strongly dominated. In particular, the determinant norm family* $\det(\xi)$ *is strongly dominated.*

Proof (1) and (2) in the dominated case: We first show the following claim: if $\xi = \{\|\cdot\|_\omega\}_{\omega\in\Omega}$ is an upper dominated norm family, then all its restrictions and quotients are also upper dominated. Let F be a vector space of E and $\xi_F = \{\|\cdot\|_\omega\}_{\omega\in\Omega}$ be the restriction of ξ to F. For any $s \in F\setminus\{0\}$ and any $\omega \in \Omega$ one has $\|s\|_{F,\omega} = \|s\|_\omega$. Since the norm family ξ is upper dominated, the function $(\omega \in \Omega) \mapsto \ln\|s\|_{F,\omega}$ is upper dominated. Let G be a quotient vector space of E and $\xi_G = \{\|\cdot\|_{G,\omega}\}_{\omega\in\Omega}$ be the quotient of ξ on G. For any $t \in G \setminus \{0\}$ and any $s \in E$ which represents the class t in G, one has $\|t\|_{G,\omega} \leqslant \|s\|_\omega$ for any $\omega \in \Omega$. Therefore the function $(\omega \in \Omega) \mapsto \ln\|t\|_{G,\omega}$ is upper dominated.

Let $\xi = \{\|\cdot\|_\omega\}_{\omega\in\Omega}$ be a norm family on E such that the dual norm family $\xi^\vee$ is upper dominated. Let G be a quotient vector space of E. We identify $G^\vee$ with a vector subspace of $E^\vee$. By Proposition 1.1.20, if ξ_G denotes the quotient norm family of ξ on G, then $\xi_G^\vee$ identifies with the restriction of $\xi^\vee$ to $G^\vee$. By the claim proved above, we obtain that $\xi_G^\vee$ is upper dominated. Similarly, if F is a vector subspace of E and if $\xi_F = \{\|\cdot\|_{F,\omega}\}_{\omega\in\Omega}$ is the restriction of ξ to F, then, for any $\omega \in \Omega$, $\|\cdot\|_{F,\omega,*}$ is bounded from above by the quotient of the norm $\|\cdot\|_{\omega,*}$ on $F^\vee$ (viewed as a quotient vector space of $E^\vee$). Therefore, by the claim proved above, we obtain that the norm family $\xi_F^\vee$ is upper dominated.

(1) in the strongly dominated case: Let F be a vector subspace of E. Let $\boldsymbol{f}$ be a basis of F. We complete it into a basis $\boldsymbol{e}$ of E. For any $\omega \in \Omega$ one has $d_\omega(\xi_F, \xi_{\boldsymbol{f}}) \leqslant d_\omega(\xi, \xi_{\boldsymbol{e}})$. Since the function $(\omega \in \Omega) \mapsto d_\omega(\xi, \xi_{\boldsymbol{e}})$ is ν-dominated, also is the function $(\omega \in \Omega) \mapsto d_\omega(\xi_F, \xi_{\boldsymbol{f}})$. By Corollary 4.1.10, we obtain that the norm family ξ_F is strongly dominated.

(2) in the strongly dominated case: Let $\boldsymbol{g} = \{g_i\}_{i=1}^m$ be a basis of G. For any $i \in \{1, \ldots, m\}$, we choose a vector e_i in E such that the canonical image of e_i in G is g_i. We complete the family $\{e_i\}_{i=1}^m$ into a basis $\boldsymbol{e}$ of E. Then for any $\omega \in \Omega$ one has $d_\omega(\xi_G, \xi_{\boldsymbol{g}}) \leqslant d_\omega(\xi, \xi_{\boldsymbol{e}})$. Therefore, the function $(\omega \in \Omega) \mapsto d_\omega(\xi_G, \xi_{\boldsymbol{g}})$ is ν-dominated, which implies that ξ_G is strongly dominated.

(3) has already been shown in Remark 4.1.3, see also Remark 4.1.12 for the strong dominancy.

(4) in the dominated case: We first show the following claim: if both norm families ξ_E and ξ_F are upper dominated, then also is the direct sum $\xi_E \oplus_\psi \xi_F$. In fact, if (s, t) is an element in $E \oplus F$, for $\omega \in \Omega \setminus \Omega'$ one has

$$\|(s,t)\|_\omega = \max\{\|s\|_{E,\omega}, \|t\|_{F,\omega}\},$$

and for $\omega \in \Omega'$, one has

$$\|(s,t)\|_\omega \leqslant \|s\|_{E,\omega} + \|t\|_{F,\omega} \leqslant 2\max\{\|s\|_{E,\omega}, \|t\|_{F,\omega}\},$$

where $\|\cdot\|_\omega$ denotes the norm indexed by ω in $\xi_E \oplus_\psi \xi_F$. Therefore the function $(\omega \in \Omega) \mapsto \|(s,t)\|_\omega$ is upper dominated.

Let ψ' be the map from Ω to $\mathscr{S}$ sending any $\omega \in \Omega \setminus \Omega_\infty$ to ψ_0 and any $\omega \in \Omega_\infty$ to $\psi(\omega)_*$ (see Definition 1.1.48). By Proposition 1.1.49 we obtain that $(\xi_E \oplus_\psi \xi_F)^\vee$ identifies with $\xi_E^\vee \oplus_{\psi'} \xi_F^\vee$. By the claim proved above, if $\xi_E^\vee$ and $\xi_F^\vee$ are upper dominated, then also is $(\xi_E \oplus_\psi \xi_F)^\vee$.

(4) in the strongly dominated case: Let $\boldsymbol{e}'$ and $\boldsymbol{e}''$ be bases of E and F respectively, and $\boldsymbol{e}$ be the disjoint union of $\boldsymbol{e}'$ and $\boldsymbol{e}''$, viewed as a basis of $E \oplus F$. Since ξ_E and ξ_F are both dominated, by Corollary 4.1.10 there exist ν-integrable functions A' and A'' such that

$$d_\omega(\xi_E, \xi_{\boldsymbol{e}'}) \leqslant A'(\omega), \quad d_\omega(\xi_F, \xi_{\boldsymbol{e}''}) \leqslant A''(\omega) \quad \nu\text{-almost everywhere.}$$

Moreover, if (s,t) is an element in $E \oplus F$, for $\omega \in \Omega \setminus \Omega'$ one has

$$\|(s,t)\|_\omega = \max\{\|s\|_{E,\omega}, \|t\|_{F,\omega}\},$$

and for $\omega \in \Omega'$, one has

$$\max\{\|s\|_{E,\omega}, \|t\|_{F,\omega}\} \leqslant \|(s,t)\|_\omega \leqslant \|s\|_{E,\omega} + \|t\|_{F,\omega} \leqslant 2\max\{\|s\|_{E,\omega}, \|t\|_{F,\omega}\},$$

where $\|\cdot\|_\omega$ denotes the norm indexed by ω in $\xi_E \oplus_\psi \xi_F$. Therefore

$$d_\omega(\xi_E \oplus_\psi \xi_F, \xi_{\boldsymbol{e}}) \leqslant \max\{A'(\omega), A''(\omega)\} + \ln(2)\mathbb{1}_{\Omega'}(\omega) \quad \nu\text{-almost everywhere.}$$

Note that the function $(\omega \in \Omega) \mapsto \max\{A'(\omega), A''(\omega)\} + \ln(2)\mathbb{1}_{\Omega'}(\omega)$ is ν-integrable. Hence the norm family $\xi_E \oplus_\psi \xi_F$ is strongly dominated (by Corollary 4.1.10).

(5) By (3), the norm families $\xi_E^{\vee\vee}$ and $\xi_F^{\vee\vee}$ are both dominated. Therefore, without loss of generality, we may assume that $\|\cdot\|_{E,\omega}$ and $\|\cdot\|_{F,\omega}$ are ultrametric norms for $\omega \in \Omega \setminus \Omega_\infty$ (see Definition 1.1.52, see also Proposition 1.2.14). Let $\boldsymbol{e} = \{e_i\}_{i=1}^n$ and $\boldsymbol{f} = \{f_j\}_{j=1}^m$ be bases of E and F over K, and let $\boldsymbol{e} \otimes \boldsymbol{f} = \{e_i \otimes f_j\}_{(i,j)\in\{1,\ldots,n\}\times\{1,\ldots,m\}}$. Note that $\boldsymbol{e} \otimes \boldsymbol{f}$ is a basis of $E \otimes F$. Moreover, for any $\omega \in \Omega$, the norm $\|\cdot\|_{\boldsymbol{e}\otimes\boldsymbol{f},\omega}$ identifies with the ε-tensor product of the norms $\|\cdot\|_{E,\omega}$ and $\|\cdot\|_{F,\omega}$. Since the norm families ξ_E and ξ_F are dominated, there exist ν-integrable functions A_E and A_F on Ω such that

$$d_\omega(\xi_E, \xi_{\boldsymbol{e}}) = \sup_{0 \neq s \in E} \Big| \ln \|s\|_{E,\omega} - \ln \|s\|_{\boldsymbol{e},\omega} \Big| \leqslant A_E(\omega) \quad \nu\text{-almost everywhere,} \tag{4.7}$$

and

$$d_\omega(\xi_F, \xi_{\boldsymbol{f}}) = \sup_{0 \neq t \in F} \Big| \ln \|t\|_{F,\omega} - \ln \|t\|_{\boldsymbol{f},\omega} \Big| \leqslant A_F(\omega) \quad \nu\text{-almost everywhere.}$$

By (4.1), we obtain

$$d_\omega(\xi_E^\vee, \xi_{\boldsymbol{e}}^\vee) = \sup_{0 \neq \alpha \in E^\vee} \Big| \ln \|\alpha\|_{E,\omega,*} - \ln \|\alpha\|_{\boldsymbol{e},\omega,*} \Big| \leqslant A_E(\omega) \quad \nu\text{-almost everywhere,}$$

which implies (see (4.2))

$$d_\omega(\xi_E^\vee, \xi_{e^\vee}) \leqslant A_E(\omega) + \ln(n)1\!\!1_{\Omega_\infty}(\omega) \quad \nu\text{-almost everywhere.}$$

Therefore, for any $\omega \in \Omega$ and any non-zero tensor $\varphi \in \mathrm{Hom}_K(E^\vee, F) \cong E \otimes_K F$, one has (see Remark 1.1.53)

$$\begin{aligned}\Big|\ln \|\varphi\|_{\varepsilon,\omega} - \ln \|\varphi\|_{e\otimes f,\omega}\Big| &= \Big|\sup_{0\neq\alpha\in E^\vee} \ln\frac{\|\varphi(\alpha)\|_{F,\omega}}{\|\alpha\|_{E,\omega,*}} - \sup_{0\neq\alpha\in E^\vee}\ln\frac{\|\varphi(\alpha)\|_{f,\omega}}{\|\alpha\|_{e^\vee,\omega}}\Big| \\ &\leqslant A_E(\omega) + A_F(\omega) + \ln(n)1\!\!1_{\Omega_\infty}(\omega)\end{aligned}$$

ν-almost everywhere, where $\|\cdot\|_{\varepsilon,\omega}$ denotes the ε-tensor product of $\|\cdot\|_{E,\omega}$ and $\|\cdot\|_{F,\omega}$. Therefore the ε-tensor product norm family $\xi_E \otimes_\varepsilon \xi_F$ is dominated (and hence is strongly dominated since it is ultrametric on $\Omega \setminus \Omega_\infty$, see Remark 4.1.12). By using the fact that $\xi_E \otimes_{\varepsilon,\pi} \xi_F = (\xi_E^\vee \otimes_\varepsilon \xi_F^\vee)^\vee$ (see Corollary 1.2.20 and Proposition 1.1.57 for the non-Archimedean and the Archimedean cases respectively), we deduce the dominance property of $\xi_E \otimes_{\varepsilon,\pi} \xi_F$ from the above result and the assertion (3) of the Proposition. Finally, by Propositions 1.2.59 and 4.1.6, we deduce that the orthogonal tensor product norm family $\xi_E \otimes \xi_F$ is also strongly dominated, provided that ξ_E and ξ_F are both Hermitian.

(6) is a direct consequence of (2) and (5) since $\Lambda^i\xi$ is a quotient norm family of the i-th ε, π-tensor power of ξ. □

Remark 4.1.20 Let E be a finite-dimensional vector space over K. We denote by $\mathcal{D}_E$ the subset of $\mathcal{N}_E$ of all strongly dominated norm families $\xi = \{\|\cdot\|_\omega\}_{\omega\in\Omega}$ on E. By Corollary 4.1.8, we obtain that, for any pair (ξ, ξ') of norm families in $\mathcal{D}_E$, the local distance function $(\omega \in \Omega) \mapsto d_\omega(\xi, \xi')$ is ν-dominated. This observation allows to construct a function $\mathrm{dist}(\cdot,\cdot)$ on $\mathcal{D}_E \times \mathcal{D}_E$, defined as (see §A.4 for the definition of the upper integral $\overline{\int_\Omega} h(\omega)\,\nu(\mathrm{d}\omega)$)

$$\mathrm{dist}(\xi, \xi') := \overline{\int_\Omega} d_\omega(\xi, \xi')\,\nu(\mathrm{d}\omega).$$

Clearly this function is symmetric with respect to its two variables, and verifies the triangle inequality, where the latter assertion follows from the triangle inequality of the local distance function and Proposition A.4.4. Therefore, $\mathrm{dist}(\cdot,\cdot)$ is actually a pseudometric on $\mathcal{D}_E$. Moreover, for any pair (ξ, ξ') of elements of $\mathcal{D}_E$, $\mathrm{dist}(\xi, \xi') = 0$ if and only if $\xi_\omega = \xi'_\omega$ ν-almost everywhere (see Proposition A.4.10). Therefore, the pseudometric $\mathrm{dist}(\cdot,\cdot)$ induces a metric on the quotient space of $\mathcal{D}_E$ modulo the equivalence relation

$$\xi \sim \xi' \Longleftrightarrow \xi_\omega = \xi'_\omega \quad \nu\text{-almost everywhere.}$$

This quotient metric space is actually complete. In fact, assume that $\{\xi_n\}_{n\in\mathbb{N}}$ is a Cauchy sequence in $\mathcal{D}_E$. Then we can pick a subsequence $\{\xi_{n_k}\}_{k\in\mathbb{N}}$ such that

$$\forall\, k \in \mathbb{N}, \qquad \overline{\int_{\Omega}} \left(1\!\!1_{\{d(\xi_{n_k}, \xi_{n_{k+1}}) \geqslant 2^{-k}\}} \right) \nu(\mathrm{d}\omega) \leqslant 2^{-k}.$$

The set of $\omega \in \Omega$ such that $\{\xi_{n_k,\omega}\}_{k\in\mathbb{N}}$ is not a Cauchy sequence (with respect the the metric defined in §1.1.9) is a ν-negligible set. Let ξ be a norm family such that $\{\xi_{n_k,\omega}\}_{k\in\mathbb{N}}$ converges to ξ_ω ν-almost everywhere (see Remark 1.1.41 for the local completeness). Then, by the same argument as in the proof of Proposition A.4.14, we obtain that $\operatorname{dist}(\xi_n, \xi)$ converges to 0 when n goes to the infinity.

4.1.3 Measurability of norm families

Let E be a vector space of finite dimension over K and $\xi = \{\|\cdot\|_\omega\}_{\omega\in\Omega}$ be a norm family in $\mathcal{N}_E$. We say that the norm family ξ is *$\mathcal{A}$-measurable* (or simply *measurable* when there is no ambiguity on the σ-algebra $\mathcal{A}$) if for any $s \in E$ the function $(\omega \in \Omega) \mapsto \|s\|_\omega$ is $\mathcal{A}$-measurable. By definition, if the norm family ξ is $\mathcal{A}$-measurable on Ω, then also is its restriction to a vector subspace of E. The following proposition shows that measurable direct sums preserve the measurability of norm families.

Proposition 4.1.21 *Let E and F be finite-dimensional vector spaces over K and $\xi_E = \{\|\cdot\|_{E,\omega}\}_{\omega\in\Omega}$, $\xi_F = \{\|\cdot\|_{F,\omega}\}_{\omega\in\Omega}$ be respectively norm families in $\mathcal{N}_E$ and $\mathcal{N}_F$, which are both $\mathcal{A}$-measurable. For any map $\psi : \Omega \to \mathscr{S}$ which is $\mathcal{A}$-measurable, where we consider the Borel σ-algebra on $\mathscr{S}$ induced by the topology of uniform convergence, the direct sum $\xi_E \oplus_\psi \xi_F = \{\|\cdot\|_{\psi,\omega}\}_{\omega\in\Omega}$ is $\mathcal{A}$-measurable.*

Proof Consider the map $g : \mathscr{S} \times [0, +\infty[^2 \to \mathbb{R}$

$$g(\eta, a, b) \longrightarrow \begin{cases} 0, & a + b = 0, \\ (a+b)\eta(a/(a+b)), & a + b \neq 0. \end{cases}$$

We claim that the map g is continuous. Let $\{(\eta_n, a_n, b_n)\}_{n\in\mathbb{N}}$ be a sequence in $\mathscr{S} \times [0, +\infty[^2$ which converges to $(\eta, a, b) \in \mathscr{S} \times [0, +\infty[^2$. If $a + b \neq 0$, then $a_n/(a_n + b_n)$ converges to $a/(a + b)$, and therefore

$$\begin{aligned} &\big|\eta_n(a_n/(a_n + b_n)) - \eta(a/(a + b))\big| \\ &\leqslant \big|\eta_n(a_n/(a_n + b_n)) - \eta(a_n/(a_n + b_n))\big| + \big|\eta(a_n/(a_n + b_n)) - \eta(a/(a + b))\big| \\ &\leqslant \|\eta_n - \eta\|_{\sup} + \big|\eta(a_n/(a_n + b_n)) - \eta(a/(a + b))\big| \end{aligned}$$

converges to 0 when n tends to the infinity. We then deduce that

$$\lim_{n\to+\infty} g(\eta_n, a_n, b_n) = g(\eta, a, b).$$

If $a + b = 0$, then

$$\lim_{n\to+\infty} g(\eta_n, a_n, b_n) = 0 = g(\eta, a, b)$$

since the sequence of functions $\{\eta_n\}_{n\in\mathbb{N}}$ is uniformly bounded and $a_n + b_n$ converges to 0 when n tends to the infinity.

Note that $\mathscr{S}$ is a closed subset of $C^0([0,1])$, the space of all continuous real functions on $[0,1]$. Since $C^0([0,1])$ admits a countable topological basis (see [32] Chapter X, §3.3, Theorem 1), also is $\mathscr{S}$. Being a metric space, the topological space $\mathscr{S}$ is thus separable. Therefore, the Borel σ-algebra of the product topological space $\mathscr{S} \times [0,+\infty[^2$ coincides with the product σ-algebra of the Borel σ-algebras of $\mathscr{S}$ and $[0,+\infty[^2$ (see [93] Lemma 1.2). In particular, the function F is measurable with respect to the product σ-algebra. If (s,t) is an element in $E \oplus F$, then one has

$$\|(s,t)\|_{\psi,\omega} = g(\psi(\omega), \|s\|_{E,\omega}, \|t\|_{F,\omega}),$$

which is an $\mathcal{A}$-measurable function since the maps ψ, $\omega \mapsto \|s\|_{E,\omega}$ and $\omega \mapsto \|t\|_{F,\omega}$ are all $\mathcal{A}$-measurable. □

Proposition 4.1.22 (1) *Let E be a vector space of dimension 1 over K and ξ be a norm family in $\mathcal{N}_E$. Then ξ is $\mathcal{A}$-measurable if and only if there exists an element $s \in E \setminus \{0\}$ such that the function $(\omega \in \Omega) \mapsto \|s\|_\omega$ is $\mathcal{A}$-measurable.*

(2) *Let E be a vector space of dimension 1 over K and ξ be a norm family in $\mathcal{N}_E$ which is $\mathcal{A}$-measurable. Then the dual norm family $\xi^\vee$ is also $\mathcal{A}$-measurable.*

(3) *Let E_1 and E_2 be vector spaces of dimension 1 over K, and ξ_1 and ξ_2 be norm families in $\mathcal{N}_{E_1}$ and $\mathcal{N}_{E_2}$ respectively. We assume that both norm families ξ_1 and ξ_2 are $\mathcal{A}$-measurable. Then the tensor product $\xi_1 \otimes \xi_2$ (which is also equal to $\xi_1 \otimes_\varepsilon \xi_2$ and $\xi_1 \otimes_\pi \xi_2$) is also $\mathcal{A}$-measurable.*

(4) *Let E be a vector space of dimension 1 over K and ξ be a norm family in $\mathcal{N}_E$. Assume that there exists an integer $n \geqslant 1$ such that $\xi^{\otimes n}$ is $\mathcal{A}$-measurable, then the norm family ξ is also measurable.*

Proof (1) The necessity follows from the definition. For the sufficiency, we assume that there exists $s \in E \setminus \{0\}$ such that the function $(\omega \in \Omega) \mapsto \|s\|_\omega$ is $\mathcal{A}$-measurable. If s' is a general element in E, there exists $a \in K$ such that $s' = as$. Note that for any $\omega \in \Omega$ one has

$$\|s'\|_\omega = |a|_\omega \cdot \|s\|_\omega.$$

Since the function $(\omega \in \Omega) \mapsto |a|_\omega$ is $\mathcal{A}$-measurable, we obtain that the function $(\omega \in \Omega) \to \|s'\|_\omega$ is $\mathcal{A}$-measurable.

(2) Let s be a non-zero element of E and α be the element of $E^\vee$ such that $\alpha(s) = 1$. For any $\omega \in \Omega$ one has $\|\alpha\|_\omega = \|s\|_\omega^{-1}$. Since the function $(\omega \in \Omega) \mapsto \|s\|_\omega$ is $\mathcal{A}$-measurable, we obtain that the function $(\omega \in \Omega) \mapsto \|\alpha\|_\omega$ is also $\mathcal{A}$-measurable. Therefore, by (1) we obtain that the norm family $\xi^\vee$ is $\mathcal{A}$-measurable on Ω.

(3) Let s_1 and s_2 be non-zero elements of E_1 and E_2 respectively. Then $s_1 \otimes s_2$ is a non-zero element of $E_1 \otimes E_2$. Moreover, for any $\omega \in \Omega$ one has $\|s_1 \otimes s_2\|_\omega = \|s_1\|_\omega \cdot \|s_2\|_\omega$. Since the functions $(\omega \in \Omega) \mapsto \|s_1\|_\omega$ and $(\omega \in \Omega) \mapsto \|s_2\|_\omega$ are $\mathcal{A}$-measurable, we obtain that the function $(\omega \in \Omega) \mapsto \|s_1 \otimes s_2\|_\omega$ is $\mathcal{A}$-measurable. By (1), the norm family $\xi_1 \otimes \xi_2$ is $\mathcal{A}$-measurable.

(4) Let s be a non-zero element of E. For any $\omega \in \Omega$, one has $\|s^{\otimes n}\|_\omega = \|s\|_\omega^n$. Hence $\|s\|_\omega = \|s^{\otimes n}\|_\omega^{1/n}$. Since $\xi^{\otimes n}$ is $\mathcal{A}$-measurable, the function $(\omega \in \Omega) \mapsto$

$\|s^{\otimes n}\|_\omega$ is $\mathcal{A}$-measurable. As a consequence, the function $(\omega \in \Omega) \mapsto \|s\|_\omega$ is also $\mathcal{A}$-measurable. By (1), we obtain that the norm family ξ is $\mathcal{A}$-measurable. □

Remark 4.1.23 It is not clear that other algebraic constructions of norm families preserve the $\mathcal{A}$-measurability. We consider the following counter-example. Let $K = \mathbb{R}$ and $(\Omega, \mathcal{A}, \nu)$ be the set $\mathbb{R}$ equipped with the Borel σ-algebra and the Lebesgue measure. Let $\phi : \Omega \to M_{\mathbb{R}}$ be the constant map which sends any point of Ω to the trivial absolute value. Then $(K, (\Omega, \mathcal{A}, \nu), \phi)$ is an adelic curve. Let $f : \mathbb{R} \to]0, 1]$ be a map which is not Borel measurable. Let E be a vector space of dimension 2 over $\mathbb{R}$ and $\{e_1, e_2\}$ be a basis of E. For each $t \in \Omega = \mathbb{R}$, let $\|\cdot\|_t$ be the norm on E such that $\|\lambda(e_1 + te_2)\|_t = f(t)$ for $\lambda \in \mathbb{R} \setminus \{0\}$, and $\|s\|_t = 1$ if $s \in E \setminus \mathbb{R}(e_1 + te_2)$. Then $\xi = \{\|\cdot\|_t\}_{t\in\mathbb{R}}$ is an element of $\mathcal{N}_E$. Note that, for any $s \in E$ the function $t \mapsto \|s\|_t$ is Borel measurable on $\mathbb{R}$ since it is constant except at most one point of $\mathbb{R}$. However, if we denote by G the quotient space $E/\mathbb{R}e_2$ and by $\xi_G = \{\|\cdot\|_{G,t}\}_{t\in\mathbb{R}}$ the quotient norm family of ξ. The one has

$$\forall\, t \in \mathbb{R},\ \big\|[e_1]\big\|_{G,t} = f(t).$$

Therefore the quotient norm family ξ_G is not $\mathcal{A}$-measurable on Ω.

The following results show that, at least in the particular case where K is a countable set, the algebraic constructions of norm families defined in the previous subsection preserve the $\mathcal{A}$-measurability of norm families.

Proposition 4.1.24 *We assume that, either $\mathcal{A}$ is discrete, or the field K admits a countable subfield K_0 which is dense in K_ω for any $\omega \in \Omega$.*

(1) *Let E be a vector space of finite dimension over K and $\xi = \{\|\cdot\|_\omega\}_{\omega\in\Omega}$ be a norm family in $\mathcal{N}_E$ which is $\mathcal{A}$-measurable. Then*
   (1.a) *for any quotient space G of E, the quotient norm family $\xi_G = \{\|\cdot\|_{G,\omega}\}_{\omega\in\Omega}$ on G of ξ is $\mathcal{A}$-measurable;*
   (1.b) *the dual norm family $\xi^\vee = \{\|\cdot\|_{\omega,*}\}_{\omega\in\Omega}$ is $\mathcal{A}$-measurable;*
   (1.c) *for any algebraic extension L/K, the norm family $\xi_L = \{\|\cdot\|_x\}_{x\in\Omega_L}$ on $E_L := E \otimes_K L$ is $\mathcal{A}_L$-measurable. If in addition ξ is Hermitian, then the norm family ξ_L^H is $\mathcal{A}_L$-mesurable.*

(2) *Let E and F be finite-dimensional vector spaces over K, $\xi_E = \{\|\cdot\|_{E,\omega}\}_{\omega\in\Omega}$ and $\xi_F = \{\|\cdot\|_{F,\omega}\}_{\omega\in\Omega}$ be respectively norm families in $\mathcal{N}_E$ and $\mathcal{N}_F$. We assume that ξ_E and ξ_F are both $\mathcal{A}$-measurable. Then*
   (2.a) *the π-tensor product $\xi_E \otimes_\pi \xi_F$, the ε-tensor product $\xi_E \otimes_\varepsilon \xi_F$ and the ε, π-tensor product $\xi_E \otimes_{\varepsilon,\pi} \xi_F$ are all $\mathcal{A}$-measurable;*
   (2.b) *if in addition the norm families ξ_E and ξ_F are Hermitian, the orthogonal tensor product $\xi_E \otimes \xi_F = \{\|\cdot\|_{E\otimes F,\omega}\}_{\omega\in\Omega}$ is $\mathcal{A}$-measurable.*

(3) *Let E be a finite-dimensional vector space over K and $\xi_E = \{\|\cdot\|_{E,\omega}\}_{\omega\in\Omega}$ be a norm family in $\mathcal{N}_E$ which is $\mathcal{A}$-measurable. Then the exterior product norm family $\Lambda^i\xi$ is $\mathcal{A}$-measurable for any $i \in \mathbb{N}$. In particular, the determinant norm family $\det(\xi)$ is $\mathcal{A}$-measurable.*

(4) *Let E be a finite-dimensional vector space over K, and $\xi = \{\|\cdot\|_\omega\}_{\omega\in\Omega}$ and $\xi' = \{\|\cdot\|'_\omega\}_{\omega\in\Omega}$ be two $\mathcal{A}$-measurable norm families in $\mathcal{N}_E$. Then the local distance function $\omega \mapsto d_\omega(\xi,\xi')$ is $\mathcal{A}$-measurable.*

Proof It suffices to treat the case where K admits a countable subfield which is dense in all K_ω.

(1.a) Let $p : E \to G$ be the projection map and F be its kernel. Let F_0 be a finite-dimensional K_0-vector subspace of F which generates F as a vector space over K. Note that for any $\omega \in \Omega$ the set F_0 is dense in F_{K_ω}. For any $\ell \in G$ and any $\omega \in \Omega$, one has

$$\|\ell\|_{G,\omega} = \inf_{s\in E,\, p(s)=\ell} \|s\|_\omega = \inf_{t\in F_0} \|s_0 + t\|_\omega,$$

where s_0 is an element of E such that $p(s_0) = \ell$. As the norm family ξ_E is $\mathcal{A}$-measurable on Ω, the function $(\omega \in \Omega) \mapsto \|s\|_\omega$ is $\mathcal{A}$-measurable for any $s \in E$. Hence the function $\omega \mapsto \|\ell\|_{G,\omega}$ is also $\mathcal{A}$-measurable since it is the infimum of a countable family of $\mathcal{A}$-measurable functions.

(1.b) Let E_0 be a finite-dimensional K_0-vector subspace of E which generates E as a vector space over K. For any $\alpha \in E^\vee$ and any $\omega \in \Omega$, one has

$$\|\alpha\|_{\omega,*} = \sup_{s\in E\setminus\{0\}} \frac{|\alpha(s)|_\omega}{\|s\|_\omega} = \sup_{s\in E_0\setminus\{0\}} \frac{|\alpha(s)|_\omega}{\|s\|_\omega}$$

since $E_0 \setminus \{0\}$ is dense in $E_{K_\omega} \setminus \{0\}$. As the norm family ξ_E is $\mathcal{A}$-measurable on Ω, the function $(\omega \in \Omega) \mapsto \|s\|_\omega$ is $\mathcal{A}$-measurable for any $s \in E$. Moreover, $\alpha(s)$ belongs to K, and thus the function $\omega \mapsto |\alpha(s)|_\omega$ is $\mathcal{A}$-measurable. Hence the function $(\omega \in \Omega) \mapsto \|\alpha\|_{\omega,*}$ is $\mathcal{A}$-measurable since it is the supremum of a countable family of $\mathcal{A}$-measurable functions.

(1.c) Let H_0 be a finite-dimensional K_0-vector subspace of $E^\vee$ which generates $E^\vee$ as a vector space over K. Then $H_0 \setminus \{0\}$ is dense in $E^\vee_{K_\omega} \setminus \{0\}$ for any $\omega \in \Omega$. Let s be an element in E_L. For any $x \in \Omega_L$, let

$$\|s\|_{\omega,L_x,\varepsilon} = \sup_{\varphi\in E^\vee_{K_\omega}\setminus\{0\}} \frac{|\varphi(s)|_x}{\|\varphi\|_{\omega,*}} = \sup_{\varphi\in H_0\setminus\{0\}} \frac{|\varphi(s)|_x}{\|\varphi\|_{\omega,*}},$$

where $\omega = \pi_{L/K}(x)$, and $\|\cdot\|_{\omega,*}$ denotes the dual norm of $\|\cdot\|_\omega$. We have seen in (1.b) that the dual norm family $\xi^\vee = \{\|\cdot\|_{\omega,*}\}_{\omega\in\Omega}$ is $\mathcal{A}$-mesurable on Ω. Therefore the function

$$(x \in \Omega_L) \longmapsto \|\varphi\|_{\pi_{L/K}(x),*},$$

which is the composition of the $\mathcal{A}$-measurable function $\omega \mapsto \|\varphi\|_{\omega,*}$ with $\pi_{L/K}$, is $\mathcal{A}_L$-measurable. Moreover, the function $x \mapsto |\varphi(s)|_x$ on Ω_L is $\mathcal{A}_L$-mesurable. Therefore, the function $x \mapsto \|s\|_{\omega,L_x,\varepsilon}$ on Ω_L, which is the supremum of a countable family of measurable functions, is also measurable. Therefore the norm family $\{\|\cdot\|'_x\}_{x\in\Omega_L}$ is measurable. This result applied to $\xi^\vee$ shows that the norm family $\{\|\cdot\|_{\omega,*,L_x,\varepsilon}\}_{x\in\Omega_L}$ is measurable. By Proposition 1.3.20 (1), (2), the norm family ξ_L identifies with the dual of $\{\|\cdot\|_{\omega,*,L_x,\varepsilon}\}_{x\in\Omega_L}$, and hence is measurable.

Assume that the norm family ξ is Hermitian. Let s be an element of E_L, which is written as $s_1 \otimes \lambda_1 + \cdots + s_n \otimes \lambda_n$, where $(s_1, \ldots, s_n) \in E^n$ and $(\lambda_1, \ldots, \lambda_n) \in L^n$. For any $x \in \Omega_L$ one has

$$\|s\|_x^2 = \sum_{i=1}^{n} \sum_{j=1}^{n} \langle s_i, s_j \rangle_{\pi_{L/K}(x)} \langle \lambda_i, \lambda_j \rangle_x.$$

Note that the function

$$(\omega \in \Omega_\infty) \mapsto \langle s_i, s_j \rangle_\omega = \frac{1}{2}\big(\|s_i + s_j\|_\omega^2 - \|s_i\|_\omega^2 - \|s_j\|_\omega^2\big)$$

is $\mathcal{A}|_{\Omega_\infty}$-measurable (hence its composition with $\pi_{L/K}|_{\Omega_{L,\infty}}$ is $\mathcal{A}_L|_{\Omega_{L,\infty}}$-measurable) and the function

$$(x \in \Omega_{L,\infty}) \longmapsto \langle \lambda_i, \lambda_j \rangle_x = \frac{1}{2}\big(|\lambda_i + \lambda_j|_x^2 - |\lambda_i|_x^2 - |\lambda_j|_x^2\big)$$

is $\mathcal{A}_L|_{\Omega_{L,\infty}}$-measurable. Therefore the function $(x \in \Omega_L) \mapsto \|s\|_x^2$ is $\mathcal{A}_L$-measurable on $\Omega_{L,\infty}$. Moreover, by the measurability of ξ_L proved above, this function is also $\mathcal{A}_L$-measurable on $\Omega_L \setminus \Omega_{L,\infty}$. Hence it is $\mathcal{A}_L$-measurable.

(2.a) Let s be an element of $E \otimes F$, φ be the K-linear map from $E^\vee$ to F which corresponds to s, and r be the rank of φ. Let $\{\varphi_i\}_{i=1}^n$ be a basis of $E^\vee$ such that $\varphi_{r+1}, \ldots, \varphi_n$ belong to the kernel of f and let $\{e_i\}_{i=1}^n$ be the dual basis of $\{\varphi_i\}_{i=1}^n$. For $i \in \{1, \ldots, r\}$, let f_i be the image of α_i by φ. We complete the family $\{f_i\}_{i=1}^r$ to a basis $\{f_j\}_{j=1}^m$ of F. One has

$$s = e_1 \otimes f_1 + \cdots + e_r \otimes f_r.$$

Let E_0 and F_0 be K_0-vector subspaces of E and F generated by $\{e_i\}_{i=1}^n$ and $\{f_j\}_{j=1}^m$ respectively.

By definition, for $\omega \in \Omega$ one has

$$\|s\|_{\pi,\omega} = \inf\left\{\sum_{i=1}^{N} \|x_i\|_{E,\omega} \cdot \|y_i\|_{F,\omega} \;:\; \begin{array}{l} s = x_1 \otimes y_1 + \cdots + x_N \otimes y_N \\ \text{for some } N \in \mathbb{N}, (x_1, \ldots, x_N) \in E_{K_\omega}^N \\ \text{and } (y_1, \ldots, y_N) \in F_{K_\omega}^N \end{array}\right\}.$$

We claim that $\|s\|_{\pi,\omega}$ is eqal to

$$\|s\|'_{\pi,\omega} := \inf\left\{\sum_{i=1}^{N} \|x_i\|_{E,\omega} \cdot \|y_i\|_{F,\omega} \;:\; \begin{array}{l} s = x_1 \otimes y_1 + \cdots + x_N \otimes y_N \\ \text{for some } N \in \mathbb{N}, (x_1, \ldots, x_N) \in E_0^N \\ \text{and } (y_1, \ldots, y_N) \in F_0^N \end{array}\right\}. \quad (4.8)$$

Clearly $\|s\|_{\pi,\omega}$ is bounded from above by $\|s\|'_{\pi,\omega}$. We will show that $\|s\|_{\pi,\omega} \geqslant \|s\|'_{\pi,\omega}$. By Proposition 1.1.11, there exists $\alpha \in \,]0, 1]$ such that the bases $\{e_i\}_{i=1}^n$ and $\{f_j\}_{j=1}^m$ of E_{K_ω} and F_{K_ω} are both α-orthogonal (see Definition 1.2.4). Assume that s is written in the form

$$s = x_1 \otimes y_1 + \cdots + x_N \otimes y_N,$$

where $(x_1, \ldots, x_N) \in E_{K_\omega}^N$ and $(y_1, \ldots, y_N) \in F_{K_\omega}^N$. For any $\epsilon > 0$, there exist $(x'_1, \ldots, x'_N) \in E_0^N$ and $(y'_1, \ldots, y'_N) \in F_0^N$ such that

$$\sup_{\ell \in \{1,\ldots,N\}} \max\{\|x_\ell - x'_\ell\|_{E,\omega}, \|y_\ell - y'_\ell\|_{F,\omega}\} \leqslant \epsilon. \tag{4.9}$$

We write $x_\ell - x'_\ell$ into the form

$$x_\ell - x'_\ell = \sum_{i=1}^n a_{\ell,i} e_i, \quad \text{where } (a_{\ell,1}, \ldots, a_{\ell,n}) \in K_\omega^n.$$

Since the basis $\{e_i\}_{i=1}^n$ is α-orthogonal, one has

$$\epsilon \geqslant \|x_\ell - x'_\ell\|_{E,\omega} \geqslant \alpha \max_{i \in \{1,\ldots,n\}} |a_{\ell,i}|_\omega \cdot |e_i\|_{E,\omega}. \tag{4.10}$$

Similarly, if we write $y_\ell - y'_\ell$ as

$$\sum_{j=1}^m b_{\ell,j} f_j, \quad (a_{\ell,1}, \ldots, a_{\ell,m}) \in K_\omega^m,$$

one has

$$\epsilon \geqslant \|y_\ell - y'_\ell\|_{F,\omega} \geqslant \alpha \max_{j \in \{1,\ldots,m\}} |b_{\ell,j}|_\omega \cdot \|f_j\|_{F,\omega}. \tag{4.11}$$

Let

$$M = \sup_{\ell \in \{1,\ldots,N\}} \max\{\|x'_\ell\|_{E,\omega}, \|y_\ell\|_{F,\omega}\}.$$

If we write x'_ℓ and y_ℓ into linear combinations of $\{e_i\}_{i=1}^n$ and $\{f_j\}_{j=1}^m$ respectively:

$$x'_\ell = \sum_{i=1}^n c_{\ell,i} e_i, \quad y_\ell = \sum_{j=1}^m d_{\ell,j} f_j,$$

one has

$$M \geqslant \alpha \max_{i \in \{1,\ldots,n\}} |c_{\ell,i}|_\omega \cdot \|e_i\|_{E,\omega} \text{ and } M \geqslant \alpha \max_{j \in \{1,\ldots,m\}} |d_{\ell,j}|_\omega \cdot \|f_j\|_{F,\omega}. \tag{4.12}$$

Note that

$$\begin{aligned} s - x'_1 \otimes y'_1 - \cdots - x'_N \otimes y'_N &= \sum_{\ell=1}^N (x_\ell \otimes y_\ell - x'_\ell \otimes y'_\ell) \\ &= \sum_{\ell=1}^N ((x_\ell - x'_\ell) \otimes y_\ell + x'_\ell \otimes (y_\ell - y'_\ell)) \end{aligned}$$

$$= \sum_{i=1}^{n}\sum_{j=1}^{m}\Big(\sum_{\ell=1}^{N} a_{\ell,i}d_{\ell,j} + c_{\ell,i}b_{\ell,j}\Big)e_i \otimes f_j.$$

For $(i,j) \in \{1,\ldots,n\} \times \{1,\ldots,m\}$, let

$$A_{i,j} = \sum_{\ell=1}^{N} a_{\ell,i}d_{\ell,j} + c_{\ell,i}b_{\ell,j}.$$

Since $\{e_i \otimes f_j\}_{(i,j)\in\{1,\ldots,n\}\times\{1,\ldots,m\}}$ is a basis of $E_0 \otimes_{K_0} F_0$ over K_0 and a basis of $E_{K_\omega} \otimes_{K_\omega} F_{K_\omega}$ over K_ω, one obtains that $A_{i,j} \in K_0$ for any $(i,j) \in \{1,\ldots,n\} \times \{1,\ldots,m\}$. Therefore one has

$$\begin{aligned}
\|s\|'_{\pi,\omega} &\leqslant \sum_{\ell=1}^{N} \|x'_\ell\|_{E,\omega} \cdot \|y'_\ell\|_{F,\omega} + \sum_{i=1}^{n}\sum_{j=1}^{m} |A_{i,j}|_\omega \cdot \|e_i\|_{E,\omega} \cdot \|f_i\|_{F,\omega} \\
&\leqslant \sum_{\ell=1}^{N} (\|x_\ell\|_{E,\omega} + \epsilon)(\|y_\ell\|_{F,\omega} + \epsilon) + \sum_{i=1}^{n}\sum_{j=1}^{m} |A_{i,j}|_\omega \cdot \|e_i\|_{E,\omega} \cdot \|f_i\|_{F,\omega} \\
&\leqslant \sum_{\ell=1}^{N} (\|x_\ell\|_{E,\omega} + \epsilon)(\|y_\ell\|_{F,\omega} + \epsilon) + \alpha^{-2}\epsilon mnMN,
\end{aligned}$$

where the first inequality comes from the definition (4.8) of $\|\cdot\|'_{\pi,\omega}$, the second inequality results from (4.9), and the third inequality comes from (4.10), (4.11) and (4.12). Since ϵ is arbitrary, we obtain $\|s\|'_{\pi,\omega} \leqslant \sum_{\ell=1}^{N} \|x_\ell\| \cdot \|y_\ell\|$, which leads to $\|s\|'_{\pi,\omega} \leqslant \|s\|_{\pi,\omega}$ since the writing $s = x_1 \otimes y_1 + \cdots + x_N \otimes y_N$ is arbitrary.

As the set

$$\bigcup_{N\in\mathbb{N}} \{(x_1,\ldots,x_N,y_1,\ldots,y_N) \in E_0^N \times F_0^N \;:\; s = x_1 \otimes y_1 + \cdots + x_N \otimes y_N\}$$

is countable, we obtain that the function $(\omega \in \Omega) \mapsto \|s\|_{\pi,\omega} = \|s\|_{\pi',\omega}$ is $\mathcal{A}$-measurable. Therefore the norm family $\xi_E \otimes_\pi \xi_F$ is $\mathcal{A}$-measurable.

By Proposition 1.1.57 one has $\xi_E \otimes_\varepsilon \xi_F = (\xi_E^\vee \otimes_\pi \xi_F^\vee)^\vee$. Hence by the (1.b) of the proposition established above, we obtain that the norm family $\xi_E \otimes_\varepsilon \xi_F$ is also $\mathcal{A}$-measurable. Finally, by Corollary 1.2.20 and Proposition 1.1.57 one has $\xi_E \otimes_{\varepsilon,\pi} \xi_F = (\xi_E^\vee \otimes_\varepsilon \xi_F^\vee)^\vee$. Therefore the norm family $\xi_E \otimes_{\varepsilon,\pi} \xi_F$ is also $\mathcal{A}$-measurable.

(2.b) We proceed with the measurability of the orthogonal tensor product norm family in assuming that both norm families ξ_E and ξ_F are Hermitian. In the first step, we treat a particular case where $E = F^\vee$ and $\xi_E = \xi_F^\vee$. In this case the tensor product space $E \otimes_K F$ is isomorphic to the space $\mathrm{End}_K(F)$ of K-linear endomorphisms of F, and the ε-tensor product norm family $\xi_F^\vee \otimes_\varepsilon \xi_F$ identifies with the family of operator norms. Let $\{x_i\}_{i=1}^n$ be a basis of F over K and F_0 be the K_0-vector subspace of F generated by $\{x_i\}_{i=1}^n$. By using the basis $\{x_i\}_{i=1}^n$ one can identify $\mathrm{End}_{K_0}(F_0)$ with $M_{n\times n}(K_0)$, the space of all matrices of size $n\times n$ with coefficients in K_0. Similarly, for any $\omega \in \Omega$, one can identify $\mathrm{End}_{K_\omega}(F_{K_\omega})$ with the space $M_{n\times n}(K_\omega)$ of all matrices of size $n \times n$ with coefficients in K_ω. In particular, $\mathrm{End}_{K_0}(F_0)$ is dense in $\mathrm{End}_{K_\omega}(F_{K_\omega})$.

For any $f \in \mathrm{End}_K(F)$ and any $\omega \in \Omega_\infty$, let $\|f\|_{\mathrm{HS},\omega}$ be the Hilbert-Schmidt norm of f. By Proposition 1.2.61, one has

$$\|f\|_{\mathrm{HS},\omega} = \bigg(\sum_{i=1}^{r} \inf_{\substack{g\in\mathrm{End}_{K_\omega}(F_{K_\omega})\\ \mathrm{rk}(g)\leqslant i-1}} \|f-g\|_\omega^2\bigg)^{1/2},$$

where $\|\cdot\|_\omega$ denotes the operator norm on $\mathrm{End}_{K_\omega}(F_{K_\omega})$. Note that the set

$$\{g \in \mathrm{End}_{K_0}(F_0) \;:\; \mathrm{rk}(g) \leqslant i-1\}$$

is dense in

$$\{g \in \mathrm{End}_{K_\omega}(F_{K_\omega}) \;:\; \mathrm{rk}(g) \leqslant i-1\}.$$

Hence

$$\|f\|_{\mathrm{HS},\omega} = \bigg(\sum_{i=1}^{r} \inf_{\substack{g\in\mathrm{End}_{K_0}(F_0)\\ \mathrm{rk}(g)\leqslant i-1}} \|f-g\|_\omega^2\bigg)^{1/2}.$$

By the result of (2.a) on the measurability of ε-tensor product, the function

$$(\omega \in \Omega_\infty) \longmapsto \|f-g\|_\omega$$

is $\mathcal{A}|_{\Omega_\infty}$-measurable. Hence we deduce the $\mathcal{A}|_{\Omega_\infty}$-measurability of the function $(\omega \in \Omega_\infty) \mapsto \|f\|_{\mathrm{HS},\omega}$. Finally, since the norm $\|\cdot\|_{E\otimes F,\omega}$ identifies with the ε-tensor product $\|\cdot\|_{\varepsilon,\omega}$ for $\omega \in \Omega \setminus \Omega_\infty$, by the result of (2.a), the function $(\omega \in \Omega \setminus \Omega_\infty) \mapsto \|f\|_{E\otimes F,\omega}$ is $\mathcal{A}|_{\Omega\setminus\Omega_\infty}$-measurable. Thus we obtain the $\mathcal{A}$-measurability of the norm family $\xi_E \otimes \xi_F$.

We now consider the general case. Let T be a tensor vector in $E \otimes_K F$, viewed as a linear map from $E^\vee$ to F. Let G be the direct sum $E^\vee \oplus F$ and ξ_G be the orthogonal direct sum of $\xi_E^\vee$ and ξ_F. By Proposition 4.1.21 and the result obtained in (1.b), the norm family ξ_G is $\mathcal{A}$-measurable. Moreover, the linear map $T : E^\vee \to F$ induces a K-linear endomorphism $f = \begin{pmatrix} 0 & 0 \\ T & 0 \end{pmatrix}$ of $E^\vee \oplus F$. For any $\omega \in \Omega_\infty$, the Hilbert-Schmidt norm of T with respect to $\|\cdot\|_{E,\omega}$ and $\|\cdot\|_{F,\omega}$ identifies with the Hilbert-Schmidt norm of f with respect to the orthogonal direct sum norm $\|\cdot\|_{E^\vee\oplus F,\omega}$. By the particular case proved above, we obtain the measurability of the function $(\omega \in \Omega_\infty) \mapsto \|T\|_{E\otimes F,\omega}$. Combined with the measurability of $\xi_E \otimes_\varepsilon \xi_F$ proved in (2.a), which implies the measurability of the function $(\omega \in \Omega\setminus\Omega_\infty) \mapsto \|T\|_{E\otimes F,\omega}$, we obtain that the function $(\omega \in \Omega) \mapsto \|T\|_{E\otimes F,\omega}$ is $\mathcal{A}$-measurable. The assertion is thus proved.

(3) We equip $E^{\otimes i}$ with the i-th ε, π-tensor power of ξ. By (2.a), this norm family is $\mathcal{A}$-measurable. The exterior power norm family is its quotient. By (1.a) we obtain that the norm family $\Lambda^i\xi$ is $\mathcal{A}$-measurable.

(4) Let E_0 be a finite-dimensional K_0-vector subspace of E, which generates E as a K-vector space. For any $\omega \in \Omega$, E_0 is dense in E_{K_ω}. Therefore one has

$$d_\omega(\xi,\xi') = \sup_{0\neq s\in E_0} \left|\ln\|s\|_\omega - \ln\|s\|'_\omega\right|.$$

Since E_0 is a countable set and since the functions $(\omega\in\Omega)\mapsto\|s\|_\omega$ and $(\omega\in\Omega)\mapsto\|s\|'_\omega$ are both measurable, we obtain that the function $(\omega\in\Omega)\mapsto d_\omega(\xi,\xi')$ is also measurable. □

Remark 4.1.25 We assume that K is a number field. Let $S=(K,(\Omega,\mathcal{A},\nu),\phi)$ be the standard adelic curve as in Subsection 3.2.2. Let E be a finite-dimensional vector space over K and $\xi=\{\|\cdot\|_\omega\}_{\omega\in\Omega}$ be a norm family of E. Since $\mathcal{A}$ is the disctere σ-algebra, every function on Ω is measurable, so that (E,ξ) is measurable.

Theorem 4.1.26 *We assume that K admits a countable subfield $\widetilde{K}$ which is dense in K_ω for any $\omega\in\Omega_\infty$. Let E be a finite-dimensional vector space over K and $\xi=\{\|\cdot\|_\omega\}_{\omega\in\Omega}$ be a measurable norm family on E which is ultrametric on $\Omega\setminus\Omega_\infty$. For any $\epsilon>0$, there exists a* measurable *Hermitian norm family $\xi^H=\{\|\cdot\|^H_\omega\}_{\omega\in\Omega}$ on E such that $\|\cdot\|^H_\omega=\|\cdot\|_\omega$ for any $\omega\in\Omega\setminus\Omega_\infty$ and that*

$$\|\cdot\|_\omega \leqslant \|\cdot\|^H_\omega \leqslant (r+\epsilon)^{1/2}\|\cdot\|_\omega$$

for any $\omega\in\Omega_\infty$, where r denotes the dimension of E over K.

Proof The assertion is trivial when Ω_∞ is empty. In what follows, we assume that Ω_∞ is non-empty. In this case the field K is of characteristic 0. We divide the proof of the theorem into two steps.

Step 1: In this step, we show that there is a family $\{\varphi_\omega\}_{\omega\in\Omega_\infty}$ of embeddings $\varphi_\omega: K\to\mathbb{C}$ such that $|\cdot|_\omega=|\varphi_\omega(\cdot)|$ for all $\omega\in\Omega_\infty$ and, for any $a\in\widetilde{K}$, the function $(\omega\in\Omega_\infty)\mapsto(\varphi_\omega(a)\in\mathbb{C})$ is $\mathcal{A}|_{\Omega_\infty}$-measurable, where $|\cdot|$ denotes the usual absolute value on $\mathbb{C}$.

For each $\omega\in\Omega_\infty$, we fix an embedding $\widetilde{\varphi}_\omega: K\to\mathbb{C}$ such that $|\cdot|_\omega=|\widetilde{\varphi}_\omega(\cdot)|$. We denote by $f_\omega(a)$ and $g_\omega(a)$ the real part and the imaginary part of $\widetilde{\varphi}_\omega(a)$, respectively, that is, $\widetilde{\varphi}_\omega(a)=f_\omega(a)+\sqrt{-1}g_\omega(a)$ for $a\in K$. We claim that, for any $a\in K$, the function $(\omega\in\Omega_\infty)\mapsto f_\omega(a)$ is $\mathcal{A}|_{\Omega_\infty}$-measurable. In fact, we can write $f_\omega(a)$ as $|a+\frac{1}{2}|^2_\omega-|a|^2_\omega-\frac{1}{4}$. Therefore the claim follows from the definition of adelic curve. Moreover, for any $a\in K$, the function $(\omega\in\Omega_\infty)\mapsto g_\omega(a)^2$ is measurable since we can write $g_\omega(a)^2$ as $|a|^2_\omega-f_\omega(a)^2$. In particular, the function $(\omega\in\Omega_\infty)\mapsto|g_\omega(a)|$ is measurable. As a consequence, for any couple of elements (a,b) in K, the function $(\omega\in\Omega_\infty)\mapsto g_\omega(a)g_\omega(b)$ is measurable because

$$g_\omega(a)g_\omega(b)=\tfrac{1}{2}(g_\omega(a+b)^2-g_\omega(a)^2-g_\omega(b)^2).$$

Claim 4.1.27 *Let a be an element of K. Assume that $s:\Omega_\infty\to\{1,-1\}$ is a map such that the function $(\omega\in\Omega_\infty)\mapsto s(\omega)g_\omega(a)$ is $\mathcal{A}|_{\Omega_\infty}$-measurable. Then for any function $\eta:\Omega_\infty\to\mathbb{R}$ such that the function $(\omega\in\Omega_\infty)\mapsto g_\omega(a)\eta(\omega)$ is $\mathcal{A}|_{\Omega_\infty}$-measurable, the function $(\omega\in\Omega_\infty)\mapsto s(\omega)\eta(\omega)\mathbb{1}_{g_\omega(a)\neq 0}$ is $\mathcal{A}|_{\Omega_\infty}$-measurable. In particular, for any $b\in K$, the function $(\omega\in\Omega_\infty)\mapsto s(\omega)g_\omega(b)\mathbb{1}_{g_\omega(a)\neq 0}$ is $\mathcal{A}|_{\Omega_\infty}$-measurable.* □

Proof Let $h_{a,s} : \Omega_\infty \to \mathbb{R}$ be the function defined by

$$h_{a,s}(\omega) := \begin{cases} s(\omega)g_\omega(a)^{-1} = (s(\omega)g_\omega(a))^{-1}, & \text{if } g_\omega(a) \neq 0, \\ 0, & \text{if } g_\omega(a) = 0. \end{cases}$$

As the function $(\omega \in \Omega_\infty) \mapsto s(\omega)g_\omega(a)$ is $\mathcal{A}|_{\Omega_\infty}$-measurable, also is $h_{a,s}$. Since the function $(\omega \in \Omega_\infty) \mapsto g_\omega(a)\eta(\omega)$ is measurable, we deduce that the function

$$(\omega \in \Omega_\infty) \longmapsto h_{a,s}(\omega)g_\omega(a)\eta(\omega) = s(\omega)\eta(\omega)\mathbb{1}_{g_\omega(a)\neq 0}$$

is $\mathcal{A}|_{\Omega_\infty}$-measurable, which proves the first assertion. The second assertion follows from the first one and the fact that the function $(\omega \in \Omega_\infty) \mapsto g_\omega(a)g_\omega(b)$ is $\mathcal{A}|_{\Omega_\infty}$-measurable. □

In the following, we show that there exists a map $s : \Omega_\infty \to \{1, -1\}$ such that, for any $a \in \widetilde{K}$, the function $(\omega \in \Omega_\infty) \mapsto s(\omega)g_\omega(a)$ is $\mathcal{A}|_{\Omega_\infty}$-measurable. Since $\widetilde{K}$ is countable, we can write its elements in a sequence $\{a_n\}_{n\in\mathbb{N}}$. We will construct by induction a decreasing sequence of functions $s_n : \Omega_\infty \to \{1, -1\}$ $(n \in \mathbb{N})$ which satisfy the following conditions:

(1) for any $n \in \mathbb{N}$ and any $i \in \{0, \ldots, n\}$, the function $(\omega \in \Omega_\infty) \mapsto s_n(\omega)g_\omega(a_i)$ is $\mathcal{A}|_{\Omega_\infty}$-measurable,
(2) for any $n \in \mathbb{N}$, one has

$$\{\omega \in \Omega_\infty \,:\, s_n(\omega) = 1\} \supseteq \{\omega \in \Omega_\infty \,:\, g_\omega(a_0) = \cdots = g_\omega(a_n) = 0\}.$$

In the case where $n = 0$, we just choose

$$s_0(\omega) = \begin{cases} 1, & \text{if } g_\omega(a_0) \geqslant 0, \\ -1, & \text{if } g_\omega(a_0) < 0. \end{cases}$$

Then one has $s_0(\omega)g_\omega(a_0) = |g_\omega(a_0)|$. Therefore the function

$$(\omega \in \Omega_\infty) \mapsto s_0(\omega)g_\omega(a_i)$$

is $\mathcal{A}|_{\Omega_\infty}$-measurable. Moreover, by definition one has

$$\{\omega \in \Omega_\infty \,:\, s_0(\omega) = 1\} \supseteq \{\omega \in \Omega_\infty \,:\, g_\omega(a_0) = 0\}.$$

Assume that the functions $s_0, \ldots, s_n$ have been constructed, which satisfy the conditions above. We choose

$$s_{n+1}(\omega) = \begin{cases} -s_n(\omega), & \text{if } g_\omega(a_0) = \cdots = g_\omega(a_n) = 0 \text{ and } g_\omega(a_{n+1}) < 0, \\ s_n(\omega), & \text{otherwise.} \end{cases}$$

Clearly, if $g_\omega(a_0) = \cdots = g_\omega(a_n) = g_\omega(a_{n+1}) = 0$, then $s_{n+1}(\omega) = s_n(\omega) = 1$, so that the above condition (2) for s_{n+1} is satisfied. Moreover, if $g_\omega(a_0) = \cdots = g_\omega(a_n) = 0$

and $g_\omega(a_{n+1}) < 0$, then $s_n(\omega) = 1$ and $s_{n+1}(\omega) = -s_n(\omega) = -1$. Hence we always have $s_{n+1}(\omega) \leqslant s_n(\omega)$ for any $\omega \in \Omega_\infty$. For any $i \in \{0, \ldots, n\}$ and any $\omega \in \Omega_\infty$ one has

$$s_{n+1}(\omega)g_\omega(a_i) = s_n(\omega)g_\omega(a_i)$$

since $s_{n+1}(\omega) = s_n(\omega)$ once $g_\omega(a_i) \neq 0$. Hence by the induction hypothesis the function $(\omega \in \Omega_\infty) \mapsto s_{n+1}(\omega)g_\omega(a_i)$ is $\mathcal{A}|_{\Omega_\infty}$-measurable.

In what follows, we show that the function $(\omega \in \Omega_\infty) \mapsto s_{n+1}(\omega)g_\omega(a_{n+1})$ is also $\mathcal{A}|_{\Omega_\infty}$-measurable. For any $i \in \{0, \ldots, n\}$, the set

$$\{\omega \in \Omega_\infty \,:\, g_\omega(a_i) = 0\} = \{\omega \in \Omega_\infty \,:\, g_\omega(a_i)^2 = 0\}$$

belongs to $\mathcal{A}$ since the function $(\omega \in \Omega_\infty) \mapsto g_\omega(a_i)^2$ is $\mathcal{A}|_{\Omega_\infty}$-measurable. By Claim 4.1.27, the function

$$(\omega \in \Omega_\infty) \longmapsto s_{n+1}(\omega)g_\omega(a_{n+1})\mathbb{1}_{g_\omega(a_0)\neq 0} \tag{4.13}$$

is $\mathcal{A}|_{\Omega_\infty}$-measurable. Similarly, for any $i \in \{1, \ldots, n\}$ we deduce from the $\mathcal{A}|_{\Omega_\infty}$-measurability of the function

$$(\omega \in \Omega) \longmapsto g_\omega(a_i)g_\omega(a_{n+1})\mathbb{1}_{g_\omega(a_0)=0} \cdots \mathbb{1}_{g_\omega(a_{i-1})=0}$$

that the function

$$(\omega \in \Omega) \longmapsto s_{n+1}(\omega)g_\omega(a_{n+1})\mathbb{1}_{g_\omega(a_0)=0} \cdots \mathbb{1}_{g_\omega(a_{i-1})=0}\mathbb{1}_{g_\omega(a_i)\neq 0} \tag{4.14}$$

is $\mathcal{A}|_{\Omega_\infty}$-measurable. Moreover, the function

$$(\omega \in \Omega_\infty) \longmapsto s_{n+1}(\omega)g_\omega(a_{n+1})\mathbb{1}_{g_\omega(a_0)=\cdots=g_\omega(a_n)=0} \tag{4.15}$$

is $\mathcal{A}|_{\Omega_\infty}$-measurable since $s_{n+1}(\omega)g_\omega(a_{n+1}) = |g_\omega(a_{n+1})|$ once the condition $g_\omega(a_0) = \cdots = g_\omega(a_n) = 0$ holds and since the set

$$\begin{aligned}\{\omega \in \Omega_\infty \,:\, g_\omega(a_0) = \cdots = g_\omega(a_n) = 0\}\\ = \{\omega \in \Omega_\infty \,:\, g_\omega(a_0)^2 = \cdots = g_\omega(a_n)^2 = 0\}\end{aligned}$$

belongs to $\mathcal{A}$. We then obtain the $\mathcal{A}|_{\Omega_\infty}$-measurability of the function $(\omega \in \Omega_\infty) \mapsto s_{n+1}(\omega)g_\omega(a_{n+1})$ since we can write the function as the sum of (4.13), (4.15) and (4.14) for $i \in \{1, \ldots, n\}$. Let s be the limit of the decreasing sequence of functions $\{s_n\}_{n\in\mathbb{N}}$. For any $n \in \mathbb{N}$, the function $(\omega \in \Omega_\infty) \mapsto s_m(\omega)g_\omega(a_n)$ is $\mathcal{A}|_{\Omega_\infty}$-measurable for any integer $m \geqslant n$. By passing to limit when $m \to +\infty$, we obtain that the function $(\omega \in \Omega_\infty) \mapsto s(\omega)g_\omega(a_n)$ is $\mathcal{A}|_{\Omega_\infty}$-measurable.

Here we define $\varphi_\omega : K \to \mathbb{C}$ to be

$$\varphi_\omega := \begin{cases} \widetilde{\varphi}_\omega & \text{if } s(\omega) = 1, \\ \text{the complex conjugation of } \widetilde{\varphi}_\omega & \text{if } s(\omega) = -1. \end{cases}$$

By the measurability result proved above, we obtain that, for any $a \in \widetilde{K}$, the function $(\omega \in \Omega_\infty) \mapsto \varphi_\omega(a)$ is $\mathcal{A}|_{\Omega_\infty}$-measurable.

Step 2: Let $\Omega_{\infty,c}$ be the set of $\omega \in \Omega_\infty$ such that $K_\omega = \mathbb{C}$. Then one has

$$\Omega_\infty \setminus \Omega_{\infty,c} = \bigcap_{a \in \widetilde{K}} \{\omega \in \Omega_\infty \,:\, \varphi_\omega(a) \in \mathbb{R}\}.$$

Therefore, the sets $\Omega_{\infty,c}$ and $\Omega_\infty \setminus \Omega_{\infty,c}$ belong to $\mathcal{A}$ (by Proposition 3.1.1, the set Ω_∞ belongs to $\mathcal{A}$).

Let $\{e_i\}_{i=1}^r$ be a basis of E over K. We consider $\mathbb{C}^{r\times r}$ as the space of complex matrices of size $r \times r$ and equip it with the Euclidean topology and Borel σ-algebra.

Let $r' = r + \varepsilon$. For any $\omega \in \Omega_{\infty,c}$, let $G(\omega)$ be the set of all matrices $A \in \mathbb{C}^{r\times r}$ such that, for any $(\lambda_1, \ldots, \lambda_r) \in \widetilde{K}^r$, if we note $(x_1, \ldots, x_r) = (\varphi_\omega(\lambda_1), \ldots, \varphi_\omega(\lambda_r))$, then one has

$$\|x_1e_1 + \cdots + x_re_r\|_\omega^2 \leqslant (\overline{x}_1, \ldots, \overline{x}_r)A^*A\begin{pmatrix} x_1 \\ \vdots \\ x_r \end{pmatrix} \leqslant r'\|x_1e_1 + \cdots + x_re_r\|_\omega^2,$$

where $A^* = {}^t\overline{A}$. For any $\omega \in \Omega_\infty \setminus \Omega_{\infty,c}$, let $G(\omega)$ be the set of all matrices $A \in \mathbb{C}^{r\times r}$ such that, for any $(\lambda_1, \ldots, \lambda_r) \in \widetilde{K}^r$ and any $(\lambda_1', \ldots, \lambda_r') \in \widetilde{K}^r$, if we note

$$(z_1, \ldots, z_r) := (\varphi_\omega(\lambda_1) + \sqrt{-1}\varphi_\omega(\lambda_1'), \ldots, \varphi_\omega(\lambda_r) + \sqrt{-1}\varphi_\omega(\lambda_r')),$$

then

$$\|z_1e_1 + \cdots + z_re_r\|_{\omega,\mathbb{C}}^2 \leqslant (\overline{z}_1, \ldots, \overline{z}_r)A^*A\begin{pmatrix} z_1 \\ \vdots \\ z_r \end{pmatrix} \leqslant r'\|z_1e_1 + \cdots + z_re_r\|_{\omega,\mathbb{C}}^2,$$

where the norm $\|\cdot\|_{\omega,\mathbb{C}}$ is defined as follows

$$\forall\,(x, y) \in E_{K_\omega}^2, \quad \|x + \sqrt{-1}y\|_{\omega,\mathbb{C}} := (\|x\|_\omega^2 + \|y\|_\omega^2)^{-1/2}.$$

For any $A \in \mathbb{C}^{r\times r}$, let $E(A)$ be the subset of $\mathbb{C}^r$ defined as

$$\{x \in \mathbb{C}^r \,:\, {}^t\overline{x}A^*Ax \leqslant 1\}.$$

For any $\omega \in \Omega_\infty$, let

$$B_\omega := \{(x_1, \ldots, x_r) \in \mathbb{C}^r \,:\, \|x_1e_1 + \cdots + x_re_r\|_{\omega,\mathbb{C}} \leqslant 1\}.$$

Then a matrix A belongs to $G(\omega)$ if and only if the following conditions hold

$$(r')^{-1/2}B_\omega \subseteq E(A) \subseteq B_\omega.$$

In fact, this relation is equivalent to, for any $(z_1, \ldots, z_r) \in \mathbb{C}^r$,

$$\|z_1e_1 + \cdots + z_re_r\|_{\omega,\mathbb{C}}^2 \leqslant (\overline{z}_1, \ldots, \overline{z}_r)A^*A\begin{pmatrix} z_1 \\ \vdots \\ z_r \end{pmatrix} \leqslant r'\|z_1e_1 + \cdots + z_re_r\|_{\omega,\mathbb{C}}^2. \quad (4.16)$$

Hence it implies that $A \in G(\omega)$. The converse implication is also true since $\varphi_\omega(\widetilde{K}) + \sqrt{-1}\varphi_\omega(\widetilde{K})$ is dense in $\mathbb{C}$ if $K_\omega = \mathbb{R}$, and $\varphi_\omega(\widetilde{K})$ is dense in $\mathbb{C}$ if $K_\omega = \mathbb{C}$.

Let $\delta > 0$ such that $(1+\delta)\sqrt{r} < \sqrt{r'}$. By Theorem 1.2.54 (see also Remark 1.2.55), there exists a Hermitian matrix A_0 such that

$$(1+\delta)^{-1}r^{-1/2}B_\omega \subseteq E(A_0) \subseteq (1+\delta)^{-1}B_\omega.$$

This shows that A_0 belongs to the interior of $G(\omega)$ and hence the interior $G(\omega)^\circ$ of $G(\omega)$ is not empty. We denote by $F(\omega)$ the closure of $G(\omega)^\circ$. If U is an open subset of $\mathbb{C}^{r\times r}$, one has

$$\begin{aligned}\{\omega \in \Omega_\infty : U \cap F(\omega) \neq \varnothing\} &= \{\omega \in \Omega_\infty : U \cap G(\omega)^\circ \neq \varnothing\} \\ &= \bigcup_{A \in \overline{\mathbb{Q}}^{r\times r} \cap U} \{\omega \in \Omega_\infty : A \in G(\omega)^\circ\},\end{aligned}$$

where $\overline{\mathbb{Q}}$ denotes the set of algebraic numbers in $\mathbb{C}$. For any matrix $A \in \mathbb{C}^{r\times r}$

$$\{\omega \in \Omega_{\infty,c} : A \in G(\omega)^\circ\} = \bigcup_{\mu \in \mathbb{Q}\cap(0,1)} \bigcap_{\lambda=(\lambda_1,\ldots,\lambda_r)\in\widetilde{K}^r} \Omega_{\infty,c}(\mu,\lambda),$$

where $\Omega_{\infty,c}(\mu, \lambda_1, \ldots, \lambda_r)$ is the set of $\omega \in \Omega_{\infty,c}$ such that

$$\begin{aligned}(1+\mu)\|\lambda_1e_1 + \cdots + \lambda_re_r\|_\omega^2 &\leqslant (\overline{\varphi_\omega(\lambda_1)}, \ldots, \overline{\varphi_\omega(\lambda_r)})A^*A\begin{pmatrix} \varphi_\omega(\lambda_1) \\ \vdots \\ \varphi_\omega(\lambda_r) \end{pmatrix} \\ &\leqslant (1-\mu)r'\|\lambda_1e_1 + \cdots + \lambda_re_r\|_\omega^2.\end{aligned}$$

Note that $\Omega_\infty(\mu, \lambda_1, \ldots, \lambda_r)$ belongs to $\mathcal{A}$ since the functions

$$(\omega \in \Omega_\infty) \longmapsto \|\lambda_1e_1 + \cdots + \lambda_re_r\|_\omega^2$$

and

$$(\omega \in \Omega_\omega) \longmapsto \varphi_\omega(\lambda_i) \quad (i \in \{1, \ldots, r\})$$

are $\mathcal{A}|_{\Omega_\infty}$-measurable. We then deduce that $\{\omega \in \Omega_{\infty,c} : A \in G(\omega)^\circ\}$ belongs to $\mathcal{A}$. Similarly,

$$\{\omega \in \Omega_\infty \setminus \Omega_{\infty,c} \;:\; A \in G(\omega)^\circ\} = \bigcup_{\mu \in \mathbb{Q}\cap(0,1)} \bigcap_{\substack{\lambda=(\lambda_1,\ldots,\lambda_r)\in\widetilde{K}^r \\ \lambda'=(\lambda'_1,\ldots,\lambda'_r)\in\widetilde{K}^r}} (\Omega_\infty \setminus \Omega_{\infty,c})(\mu,\lambda,\lambda'),$$

where $(\Omega_\infty \setminus \Omega_{\infty,c})(\mu,\lambda,\lambda')$ is the set of $\omega \in \Omega_{\infty,c}$ such that

$$\begin{aligned}(1+\mu)\Big(\|\lambda_1 e_1 + \cdots + \lambda_r e_r\|_\omega^2 + \|\lambda_1 e_1 + \cdots + \lambda_r e_r\|_\omega^2\Big) \\ \leqslant (\overline{z}_1(\omega),\ldots,\overline{z}_r(\omega))A^*A\begin{pmatrix} z_1(\omega) \\ \vdots \\ z_n(\omega)\end{pmatrix} \\ \leqslant (1-\mu)r'\Big(\|\lambda_1 e_1 + \cdots + \lambda_r e_r\|_\omega^2 + \|\lambda'_1 e_1 + \cdots + \lambda'_r e_r\|_\omega^2\Big).\end{aligned}$$

Hence $\{\omega \in \Omega_\infty \setminus \Omega_{\infty,c} \;:\; A \in G(\omega)^\circ\}$ belongs to $\mathcal{A}$.

Gathering the results we obtained, one can conclude that

$$\{\omega \in \Omega_\infty \;:\; U \cap F(\omega) \neq \varnothing\}$$

belongs to $\mathcal{A}$, so that by the measurable selection theorem of Kuratowski and Ryll-Nardzweski (see A.2.1), we obtain that there exists an $\mathcal{A}|_{\Omega_\infty}$-measurable map $\alpha : \Omega_\infty \to \mathbb{C}^{r\times r}$ such that $\alpha(\omega)$ belongs to $F(\omega)$ for any $(\omega \in \Omega)$. Finally, for any $\omega \in \Omega_\infty$ and any $(\lambda_1,\ldots,\lambda_r) \in K_\omega^r$ we let (where we extend φ_ω by continuity to $K_\omega \to \mathbb{C}$)

$$\|\lambda_1 e_1 + \cdots + \lambda_r e_r\|_\omega^H := \left[(\overline{\varphi_\omega(\lambda_1)},\ldots,\overline{\varphi_\omega(\lambda_r)})\alpha(\omega)^*\alpha(\omega)\begin{pmatrix}\varphi_\omega(\lambda_1) \\ \vdots \\ \varphi_\omega(\lambda_r)\end{pmatrix}\right]^{1/2}.$$

Then $\|\cdot\|_\omega^H$ is a Hermitian norm which satisfies

$$\|\cdot\|_\omega \leqslant \|\cdot\|_\omega^H \leqslant (r+\varepsilon)^{1/2}\|\cdot\|_\omega.$$

For $\omega \in \Omega \setminus \Omega_\infty$, let $\|\cdot\|_\omega^H := \|\cdot\|_\omega$. Then by the measurability of the map $\alpha(\cdot)$ we obtain that the norm family $\xi^H := \{\|\cdot\|_\omega^H\}_{\omega\in\Omega}$ is measurable. The theorem is thus proved. □

4.1.4 Adelic vector bundles

In this section, we introduce the notion of adelic vector bundles on an adelic curve $S = (K, (\Omega, \mathcal{A}, \nu), \phi)$. An adelic vector bundle is a finite-dimensional vector space E over K equipped with a family of norms indexed by Ω, which satisfies some measurability and dominance conditions so that the height of non-zero vectors is well defined (see Definition 4.1.28). In the classic setting of global fields, the notion of adelic vector bundles was defined differently in the literature (see for example [62]): one requires that almost all norms come from a common integral model of E. However, in our setting it is not relevant to consider integral models. The readers will discover the link between our definition and the classic one via the dominance property described in Proposition 4.1.7.

Definition 4.1.28 Let E be a finite-dimensional vector space over K, and ξ be a norm family in $\mathcal{N}_E$. If both norm families ξ and $\xi^\vee$ are $\mathcal{A}$-measurable on Ω and if ξ is dominated (resp. strongly dominated), we say that the couple (E, ξ) is *an adelic vector bundle* (resp. a *strongly adelic vector bundle*) on S. The *rank* of an adelic vector bundle or a strongly adelic vector bundle (E, ξ) is defined as the dimension of E over K.

If the norm family ξ is Hermitian (in this case (E, ξ) is necessarily a strongly adelic vector bundle), we say that (E, ξ) is a *Hermitian adelic vector bundle* on S. If the dimension of E is 1 (in this case ξ is necessarily Hermitian), we say that (E, ξ) is an *adelic line bundle* on S.

Proposition 4.1.29 *Let E be a vector space of dimension* 1 *over K and ξ be a norm family in $\mathcal{N}_E$. If ξ is $\mathcal{A}$-measurable and dominated, then (E, ξ) is an adelic line bundle on S.*

Proof Since E is of dimension 1 over K, any dominated norm family is strongly dominated (see Remark 4.1.12). Moreover, by Proposition 4.1.22, if ξ is $\mathcal{A}$-measurable, then also is $\xi^\vee$. □

Proposition 4.1.30 *Let (E, ξ) be an adelic line bundle on S. Then $(E^\vee, \xi^\vee)$ is an adelic line bundle on S.*

Proof By definition, the norm family $\xi^\vee$ is $\mathcal{A}$-measurable on Ω. Moreover, by Proposition 4.1.19 (3), the norm family $\xi^\vee$ is dominated. By Proposition 4.1.29, we obtain that $(E^\vee, \xi^\vee)$ is an adelic line bundle on S. □

The following proposition is fundamental in the height theory of rational points in a projective space over an adelic curve.

Proposition 4.1.31 *Let (E, ξ) be an adelic vector bundle on S.*

(1) *Any vector subspace of dimension* 1 *of E equipped with the restriction of the norm family ξ forms an adelic line bundle on S.*
(2) *Any quotient vector space of dimension* 1 *of E equipped with the quotient norm family of ξ forms an adelic line bundle on S.*

Proof (1) Let F be a vector subspace of rank 1 of E and ξ_F be the restriction of ξ to F. Clearly ξ_F is $\mathcal{A}$-measurable. Moreover, by Proposition 4.1.19 (1), the norm family ξ_F is dominated. By Proposition 4.1.29, (F, ξ_F) is an adelic line bundle on S.

(2) Let G be a quotient space of dimension 1 of E and ξ_G be the quotient of the norm family ξ on G. Then $G^\vee$ identifies with a vector subspace of dimension 1 of $E^\vee$

and $\xi_G^\vee$ identifies with the restriction of $\xi^\vee$ to $G^\vee$ (see Proposition 1.1.20). Therefore $(G^\vee, \xi_G^\vee)$ is an adelic line bundle on S. Finally, by Proposition 4.1.30 and the fact that $\xi_G = \xi_G^{\vee\vee}$ (where we identify G with $G^{\vee\vee}$), we obtain that (G, ξ_G) is an adelic line bundle on S. □

Proposition 4.1.32 (1) *Let (E, ξ) be an adelic vector bundle (resp. a strongly adelic vector bundle) on S, F be a vector subspace of E and ξ_F be the restriction of ξ to F. If the norm family $\xi_F^\vee$ is $\mathcal{A}$-measurable, then (F, ξ_F) is an adelic vector bundle (resp. a strongly adelic vector bundle) on S.*

(2) *Let (E, ξ) be an adelic vector bundle (resp. a strongly adelic vector bundle) on S, G be a quotient vector space of E and ξ_G be the quotient norm family of ξ. If the norm family ξ_G is $\mathcal{A}$-measurable, then (G, ξ_G) is an adelic vector bundle (resp. a strongly adelic vector bundle) on S.*

(3) *Let (E, ξ) be an adelic vector bundle on S. Assume that the norm family $\xi^{\vee\vee}$ is measurable. Then $(E^\vee, \xi^\vee)$ is a strongly adelic vector bundle on S.*

(4) *Let (E, ξ_E) and (F, ξ_F) be adelic vector bundles on S. If the norm families $\xi_E \otimes_\varepsilon \xi_F$ and $(\xi_E \otimes_\varepsilon \xi_F)^\vee$ are $\mathcal{A}$-measurable, then $(E \otimes F, \xi_E \otimes_\varepsilon \xi_F)$ is a strongly adelic vector bundle on S. Similarly, $(E \otimes F, \xi_E \otimes_{\varepsilon,\pi} \xi_F)$ is a strongly adelic vector bundle on S provided that the both norm families $\xi_E \otimes_{\varepsilon,\pi} \xi_F$ and $(\xi_E \otimes_{\varepsilon,\pi} \xi_F)^\vee$ are measurable. If in addition ξ_E and ξ_F are both Hermitian, and if both norm families $\xi_E \otimes \xi_F$ and $(\xi_E \otimes \xi_F)^\vee$ are $\mathcal{A}$-measurable, then the orthogonal tensor product $(E \otimes F, \xi_E \otimes \xi_F)$ is a Hermitian adelic vector bundle on S.*

(5) *Let (E, ξ) be an adelic vector bundle on S. If $\det(\xi)$ is $\mathcal{A}$-measurable then $(\det(E), \det(\xi))$ is an adelic line bundle on S.*

Proof These assertions are direct consequences of Proposition 4.1.19. We just mention below some particular points. For the assertion (1), since ξ is $\mathcal{A}$-measurable, by definition ξ_F is also measurable. For the assertion (2), since $\xi^\vee$ is $\mathcal{A}$-measurable, and $\xi_G^\vee$ identifies with the restriction of $\xi^\vee$ to $G^\vee$, it is also $\mathcal{A}$-measurable. For the last assertion, since $\det(E)$ is of dimension 1 on K, the $\mathcal{A}$-measurability of $\det(\xi)$ implies that of its dual. □

Corollary 4.1.33 *Let (E_1, ξ_1) and (E_2, ξ_2) be adelic line bundles on S. Then the tensor product $(E_1 \otimes E_2, \xi_1 \otimes \xi_2)$ is also an adelic line bundle on S.*

Proof By Proposition 4.1.22 (3), the tensor product norm family $\xi_1 \otimes \xi_2$ is $\mathcal{A}$-measurable. By Proposition 4.1.32 (4), we obtain that $(E_1 \otimes E_2, \xi_1 \otimes \xi_2)$ is an adelic line bundle on S. □

Remark 4.1.34 By using the measurability results obtained in the previous subsection (notably Proposition 4.1.24), we obtain that the assertions of Proposition 4.1.32 remain true without supplementary measurability assumptions, if the σ-algebra $\mathcal{A}$ is discrete, or if the field K admits a countable subfield which is dense in all completions K_ω, $\omega \in \Omega$.

In the case of direct sums, the measurability result in Proposition 4.1.21 leads to the following criterion (without any condition on K).

Proposition 4.1.35 *Let (E, ξ_E) and (F, ξ_F) be adelic vector bundles (resp. a strongly adelic vector bundle) on S, and $\psi : (\omega \in \Omega) \mapsto \psi_\omega \in \mathscr{S}$ be an $\mathcal{A}$-measurable map. We assume that there exists a measurable subset Ω' of Ω such that $\nu(\Omega') < +\infty$ and that $\psi_\omega = \psi_0$ on $\Omega \setminus \Omega'$, where ψ_0 denotes the function in $\mathscr{S}$ sending $t \in [0, 1]$ to $\max\{t, 1-t\}$. Then $(E \oplus F, \xi_E \oplus_\psi \xi_F)$ is an adelic vector bundle (resp. a strongly adelic vector bundle) on S.*

Proof Since (E, ξ_E) and (F, ξ_F) are adelic vector bundles (resp. strongly adelic vector bundles) on S, the norm families ξ_E, ξ_F, $\xi_E^\vee$ and $\xi_F^\vee$ are all $\mathcal{A}$-measurable, and the norm families ξ_E and ξ_F are dominated (resp. strongly dominated).

By Proposition 4.1.21, the ψ-direct sum $\xi_E \oplus_\psi \xi_F$ is also $\mathcal{A}$-measurable. Let $\psi' = \{\psi'_\omega\}_{\omega\in\Omega}$ be the family in $\mathscr{S}$ such that $\psi_\omega = \psi_0$ on $\Omega \setminus \Omega_\infty$ and $\psi'_\omega = \psi_{\omega,*}$ (see Definition 1.1.48) on Ω_∞, then one has

$$(\xi_E \oplus_\psi \xi_F)^\vee = \xi_E^\vee \oplus_{\psi'} \xi_F^\vee.$$

Note that the map from $\mathscr{S}$ to itself sending $\varphi \in \mathscr{S}$ to φ_* is continuous. This is a consequence of (1.11) and Proposition 1.1.43. Therefore, the map ψ' is also $\mathcal{A}$-measurable. Still by Proposition 4.1.21, we obtain that the norm family $(\xi_E \oplus \xi_F)^\vee$ is $\mathcal{A}$-measurable.

By Proposition 4.1.19 (4), the norm family $\xi_E \oplus_\psi \xi_F$ is dominated (resp. strongly dominated). Therefore $(E \oplus F, \xi_E \oplus_\psi \xi_F)$ is an adelic vector bundle (resp. strongly adelic vector bundle) on S. □

4.1.5 Examples

In this subsection, we present several examples of adelic vector bundles, which include most classic constructions.

Torsion free coherent sheaves

Let k be a field and X be a normal projective scheme of dimension $d \geqslant 1$ over $\operatorname{Spec} k$, equipped with a family $\{D_i\}_{i=1}^{d-1}$ of ample divisors on X. Let $K = k(X)$ be the field of rational functions on X and $\Omega = X^{(1)}$, equipped with the discrete σ-algebra. We have seen in §3.2.4 that $S = (K, (\Omega, \mathcal{A}, \nu), \phi)$ is an adelic curve, where the measure ν is defined as

$$\forall\, Y \in \Omega = X^{(1)}, \quad \nu(\{Y\}) = \deg(D_1 \cdots D_{d-1} \cap [Y])$$

and the map $\phi : \Omega \to M_K$ sends $Y \in \Omega$ to $|\cdot|_Y = \mathrm{e}^{-\mathrm{ord}_Y(\cdot)}$.

Let $\mathcal{E}$ be a torsion-free (namely the canonical homomorphism $\mathcal{E} \to \mathcal{E}^{\vee\vee}$ is injective) coherent sheaf on X and $E := \mathcal{E} \otimes_{\mathcal{O}_X} K$. The latter is a finite-dimensional vector space over K. Moreover, for any $Y \in \Omega$, $\mathcal{E} \otimes_{\mathcal{O}_X} \mathcal{O}_{X,Y}$ is a torsion-free module of finite type over the discrete valuation ring $\mathcal{O}_{X,Y}$ (the local ring of X at the generic point of Y), hence is a free $\mathcal{O}_{X,Y}$-module of finite rank. We define a norm $\|\cdot\|_Y$ on $E \otimes_K K_Y$ as follows

$$\forall\, s \in E \otimes_K K_Y, \quad \|s\|_Y := \inf\{|a|_Y \;:\; a \in K_Y^{\times},\; a^{-1}s \in \mathcal{E} \otimes_{\mathcal{O}_X} \widehat{\mathcal{O}}_{X,Y}\},$$

where $\widehat{\mathcal{O}}_{X,Y}$ is the completion of $\mathcal{O}_{X,Y}$, which identifies with the valuation ring of K_Y (the completion of K with respect to $|\cdot|_Y$). This norm is clearly ultrametric. Thus we obtain a Hermitian norm family in $\mathcal{N}_E$, which we denote by $\xi_{\mathcal{E}}$. Note that the dual norm family $\xi_{\mathcal{E}}^{\vee}$ identifies with $\xi_{\mathcal{E}^{\vee}}$, where $\mathcal{E}^{\vee}$ denotes the dual $\mathcal{O}_X$-module of $\mathcal{E}$.

Since torsion-free coherent sheaves are locally free on codimension 1, we obtain that, for any basis $\boldsymbol{e}$ of E, the norms $\|\cdot\|_Y$ and $\|\cdot\|_{\boldsymbol{e},Y}$ are identical for all but a finite number of $Y \in \Omega$. Therefore, the couple $(E, \xi_{\mathcal{E}})$ is a strongly adelic vector bundle on S.

Hermitian vector bundles on an arithmetic curve

Let K be a number field and $\mathcal{O}_K$ be the ring of algebraic integers in K. Recall that a Hermitian vector bundle on $\operatorname{Spec} \mathcal{O}_K$ is by definition a couple $(\mathcal{E}, \{\|\cdot\|_\sigma\}_{\sigma:K\to\mathbb{C}})$, where $\mathcal{E}$ is a projective $\mathcal{O}_K$-module of finite rank, and for any embedding $\sigma : K \to \mathbb{C}$, $\|\cdot\|_\sigma$ is a Hermitian norm on $\mathcal{E}\otimes_{\mathcal{O}_K,\sigma}\mathbb{C}$. We also require that the norms $\{\|\cdot\|_\sigma\}_{\sigma:K\to\mathbb{C}}$ are invariant under the complex conjugation, namely for $s_1, \ldots, s_n$ in $\mathcal{E}$, $\lambda_1, \ldots, \lambda_n$ in $\mathbb{C}$, and $\sigma : K \to \mathbb{C}$, one has

$$\|\lambda_1 \otimes s_1 + \cdots + \lambda_n \otimes s_n\|_\sigma = \|\overline{\lambda}_1 \otimes s_1 + \cdots + \overline{\lambda}_n \otimes s_n\|_{\overline{\sigma}}.$$

We let $E := \mathcal{E} \otimes_{\mathcal{O}_K} K$.

Let $S = (K, (\Omega, \mathcal{A}, \nu), \phi)$ be the adelic curve associated with the number field K, as described in §3.2.2. Recall that Ω is the set of all places of K, $\mathcal{A}$ is the discrete σ-algebra on Ω and $\nu(\{\omega\}) = [K_\omega : \mathbb{Q}_\omega]$.

Recall that any finite place of K is determined by a maximal ideal $\mathfrak{p}$ of $\mathcal{O}_K$. Let $\widehat{\mathcal{O}}_{K,\mathfrak{p}}$ be the completion of the local ring $\mathcal{O}_{K,\mathfrak{p}}$, which is also the valuation ring of $K_{\mathfrak{p}}$. We construct a norm $\|\cdot\|_{\mathfrak{p}}$ as follows

$$\forall\, s \in E \otimes_K K_{\mathfrak{p}}, \quad \|s\|_{\mathfrak{p}} := \inf\{|a|_{\mathfrak{p}} \;:\; a \in K_{\mathfrak{p}}^{\times},\; a^{-1}s \in \mathcal{E} \otimes_{\mathcal{O}_K} \widehat{\mathcal{O}}_{K,\mathfrak{p}}\}.$$

Let v be an Archimedean place of K. Then v corresponds to an embedding σ of K into $\mathbb{C}$, we let $\|\cdot\|_v$ be the restriction of $\|\cdot\|_\sigma$ to $E \otimes_K K_v$. Note that the condition that the norms $\{\|\cdot\|_\sigma\}_{\sigma:K\to\mathbb{C}}$ are invariant under the complex conjugation ensures that the norm $\|\cdot\|_v$ does not depend on the choice of the embedding $\sigma : K \to \mathbb{C}$ corresponding to v. Thus we obtain a norm family $\xi = \{\|\cdot\|_v\}_{v\in\Omega}$ in $\mathcal{N}_E$. Since $\mathcal{E}$ is a locally free sheaf, we obtain that, for any basis $\boldsymbol{e}$ of E over K, one has $\|\cdot\|_v = \|\cdot\|_{\boldsymbol{e},v}$ for all but a finite number of v.

Ultrametrically normed vector space over a trivially valued field

Let K be an arbitrary field and Ω be the one point set $\{\omega\}$. Let $|\cdot|_\omega$ be the trivial absolute value on K. We then obtain an adelic curve S by taking the discrete σ-algebra $\mathcal{A}$ on Ω and the measure ν on $(\Omega, \mathcal{A})$ such that $\nu(\{\omega\}) = 1$. Then any ultrametrically normed finite-dimensional vector space over K is a strongly adelic vector bundle on S.

Remark 4.1.36 Let K be a number field and E be a finite-dimensional vector space over K. In [62, §3], a structure of adelic vector bundle on E has been defined as a norm family $\xi \in \mathcal{N}_E$ such that, for all but finitely many $\omega \in \Omega$ (where Ω denotes the set of all places of K), the norm ξ_ω is induced by a projective O_K-module of finite type $\mathcal{E}$. Clearly such a structure of adelic vector bundle is a dominated norm family. We denote by $\mathcal{D}_E^\circ$ the subset of $\mathcal{D}_E$ consisting of all structures of adelic vector bundles in the sense of [62]. We claim that $\mathcal{D}_E^\circ$ is dense in $\mathcal{D}_E$ (with respect to the metric $\mathrm{dist}(\cdot,\cdot)$ defined in Remark 4.1.20). In other words, given a dominated norm family ξ on E, there exists a sequence $\{\xi_n\}_{n\in\mathbb{N}}$ in $\mathcal{D}_E^\circ$ which converges to ξ. In fact, we can choose an arbitrary element ξ_0 in $\mathcal{D}_E^\circ$. The main point is that, if we modify finitely many norms in the family ξ_0, we still obtain a norm family in $\mathcal{D}_E^\circ$. Since the local distance function $d(\xi, \xi_0)$ is ν-dominated, we can construct a sequence $\{\Omega_n\}_{n\geqslant 1}$ of subsets of Ω, such that $\Omega \setminus \Omega_n$ is a finite set and that

$$\lim_{n\to+\infty} \overline{\int_\Omega} 1\!\!1_{\Omega_n}(\omega) d_\omega(\xi, \xi_0)\, \nu(d\omega) = 0. \tag{4.17}$$

We then let ξ_n be the norm family such that

$$\xi_{n,\omega} = \begin{cases} \xi_{0,\omega}, & \omega \in \Omega_n \\ \xi_\omega, & \omega \in \Omega \setminus \Omega_n. \end{cases}$$

Then the sequence $\{\xi_n\}_{n\in\mathbb{N}}$ is contained in $\mathcal{D}_E^\circ$ and converges to ξ (see (4.17)). Combined with the completeness of the space $\mathcal{D}_E$ explained in Remark 4.1.20, we obtain that $\mathcal{D}_E$ is actually the completion of the metric space $\mathcal{D}_E^\circ$.

4.2 Adelic divisors

Let $S = (K, (\Omega, \mathcal{A}, \nu), \phi)$ be an adelic curve. We call *adelic divisor* on S any element ζ in the vector space $L^1(\Omega, \mathcal{A}, \nu)$ (see Section A.5). For the reason of customs of arithmetic geometry, we use the notation $\widehat{\mathrm{Div}}_{\mathbb{R}}(S)$ to denote the vector space $L^1(\Omega, \mathcal{A}, \nu)$.

If ζ is an adelic divisor on S, we define its *Arakelov degree* as

$$\widehat{\deg}(\zeta) := \int_\Omega \zeta(\omega)\, \nu(d\omega) \in \mathbb{R}. \tag{4.18}$$

The function deg is a continuous linear form on $\widehat{\mathrm{Div}}_{\mathbb{R}}(S)$. If a is an element of $K^{\times}$, we denote by $\widehat{(a)}$ the adelic divisor represented by the function which sends $\omega \in \Omega$ to $-\ln|a|_{\omega}$, called the *adelic divisor associated with a*. The map

$$\widehat{(\cdot)} : K^{\times} \longrightarrow \widehat{\mathrm{Div}}_{\mathbb{R}}(S)$$

is additive and hence extends to an $\mathbb{R}$-linear homomorphism from $K^{\times}\otimes_{\mathbb{Z}}\mathbb{R}$ to $\widehat{\mathrm{Div}}_{\mathbb{R}}(S)$, which we denote by $\widehat{(\cdot)}_{\mathbb{R}}$. The closure of the image of this map is denoted by $\widehat{\mathrm{PDiv}}_{\mathbb{R}}(S)$ and the elements of this vector space are called *principal adelic divisors*. We denote by $\widehat{\mathrm{Cl}}_{\mathbb{R}}(S)$ the quotient space $\widehat{\mathrm{Div}}_{\mathbb{R}}(S)/\widehat{\mathrm{PDiv}}_{\mathbb{R}}(S)$. Note that it forms actually a Banach space with respect to the quotient norm. Two adelic divisors lying in the same equivalent class in $\widehat{\mathrm{Cl}}_{\mathbb{R}}(S)$ are said to be $\mathbb{R}$*-linearly equivalent*.

We say that an adelic divisor ζ on S is *effective* if ζ is ν-almost everywhere non-negative. Denote by $\widehat{\mathrm{Div}}_{\mathbb{R}}(S)_{+}$ the cone of all effective adelic divisors on S. Clearly, if ζ is effective, then $\deg(\zeta) \geqslant 0$.

Let $S' = (K', (\Omega', \mathcal{A}', \nu'), \phi')$ be another adelic curve and $\alpha = (\alpha^{\#}, \alpha_{\#}, I_{\alpha}) : S' \to S$ be a morphism of adelic curves (see Section 3.7). If ζ is an adelic divisor on S, which is represented by an element $f \in \mathscr{L}^1(\Omega, \mathcal{A}, \nu)$, we denote by $\alpha^*(\zeta)$ the adelic divisor on S' represented by the function $f \circ \alpha_{\#}$ (the equivalence class of $f \circ \alpha_{\#}$ does not depend on the choice of the representative f since ν identifies with the direct image of ν' by $\alpha_{\#}$). If ζ' is an adelic divisor on S', we denote by $\alpha_*(\zeta')$ the adelic divisor $I_{\alpha}(\zeta')$ on S. Since I_{α} is a disintegration kernel of ν' over ν, one has $\alpha_*(\alpha^*(\zeta)) = \zeta$ for any adelic divisor ζ on S.

We assume that S is proper. Then one has $\deg(\zeta) = 0$ if ζ is a principal adelic divisor. This is a direct consequence of the product formula and the fact that $\deg(\cdot)$ is a continuous linear operator. Therefore the $\mathbb{R}$-linear map $\deg(\cdot)$ induces by passing to quotient a continuous $\mathbb{R}$-linear map from $\widehat{\mathrm{Cl}}_{\mathbb{R}}(S)$ to $\mathbb{R}$ which sends any class $[\zeta]$ to $\deg(\zeta)$. We still denote this linear map by $\deg(\cdot)$ by abuse of notation.

4.3 Arakelov degree and slopes

The purpose of this section is to generalise the theory of Arakelov degree and slopes to the setting of adelic vector bundles over adelic curves. Throughout the section, let $S = (K, (\Omega, \mathcal{A}, \nu), \phi)$ be a *proper* adelic curve. For all subsections except the first one, we assume in addition that, either the σ-algebra $\mathcal{A}$ is discrete, or there exists a countable subfield K_0 of K which is dense in all K_{ω}, $\omega \in \Omega$.

4.3.1 Arakelov degree of adelic line bundles

Definition 4.3.1 Let E be a finite-dimensional vector space over K and ξ be a dominated and measurable norm family on E. If s is a non-zero vector of E, then the function $(\omega \in \Omega) \mapsto \ln\|s\|_\omega$ is ν-dominated and $\mathcal{A}$-measure, hence is ν-integrable. We denote $\widehat{\mathrm{div}}_\xi(s)$ by the adelic divisor defined by this function, which is called the *adelic divisor* of s with respect to ξ. We define the *Arakelov degree* of s with respect to ξ as the Arakelov degree of $\widehat{\mathrm{div}}_\xi(s)$, that is,

$$\widehat{\deg}_\xi(s) := \deg\left(\widehat{\mathrm{div}}_\xi(s)\right) = -\int_\Omega \ln\|s\|_\omega\, \nu(\mathrm{d}\omega).$$

Moreover, by the product formula (3.1) we obtain that, for any $a \in K^\times$, one has

$$\widehat{\deg}_\xi(as) = \widehat{\deg}_\xi(s). \tag{4.19}$$

Remark 4.3.2 We assume $E = K^n$ and $\xi = \{\|\cdot\|_\omega\}_{\omega\in\Omega}$ is given by

$$\|(a_1, \ldots, a_n)\|_\omega = \max\{|a_1|_\omega, \ldots, |a_n|_\omega\}$$

for each $\omega \in \Omega$. Then $h_S(s) = -\widehat{\deg}_\xi(s)$ for all $s \in E \setminus \{0\}$ (for the definition of h_S, see Defintion 3.5.1).

Remark 4.3.3 Let E be a finite-dimensional vector space over K and $\xi = \{\|\cdot\|_\omega\}_{\omega\in\Omega}$ be a family of seminorms, where $\|\cdot\|_\omega$ is a seminorm on $E \otimes_K K_\omega$. If s is an element of E such that the function $(\omega \in \Omega) \mapsto \|s\|_\omega$ is measurable and upper dominated, we denote by $\widehat{\deg}_\xi(s)$ the value

$$-\overline{\int_\Omega} \ln\|s\|_\omega\, \nu(\mathrm{d}\omega) \in\,]-\infty, +\infty].$$

Since the adelic curve S is proper, for any $a \in K^\times$, one has $\widehat{\deg}_\xi(as) = \widehat{\deg}_\xi(s)$. This notation will be used in Chapter 7 on families of supremum seminorms.

Definition 4.3.4 Let (E, ξ) be an adelic line bundle on S. We call *Arakelov degree* of (E, ξ) the number $\widehat{\deg}_\xi(s)$, where s is a non-zero element of E. Note that the relation (4.19) shows that the definition does not depend on the choice of the non-zero element s of E. We denote the Arakelov degree of (E, ξ) by $\widehat{\deg}(E, \xi)$.

Proposition 4.3.5 *Let (E, ξ) be an adelic line bundle on S. Then $(E^\vee, \xi^\vee)$ is also an adelic line bundle on S. Moreover, one has*

$$\widehat{\deg}(E^\vee, \xi^\vee) = -\widehat{\deg}(E, \xi). \tag{4.20}$$

Proof By Proposition 4.1.30, the couple $(E^\vee, \xi^\vee)$ is also an adelic line bundle, so that the Arakelov degree $\widehat{\deg}(E^\vee, \xi^\vee)$ is well defined. If α is a non-zero element of $E^\vee$ and s is a non-zero element of E then one has

$$\forall\, \omega \in \Omega, \quad |\alpha(s)|_\omega = \|\alpha\|_{\omega,*} \cdot \|s\|_\omega.$$

By the product formula

$$\int_{\Omega} \ln |\alpha(s)|_\omega \, \nu(\mathrm{d}\omega) = 0,$$

we obtain the equality (4.20). □

Proposition 4.3.6 *Let (E_1, ξ_1) and (E_2, ξ_2) be adelic line bundles on S. Let $E = E_1 \otimes_K E_2$ and $\xi = \xi_1 \otimes_\varepsilon \xi_2$ (which is also equal to $\xi_1 \otimes \xi_2$ and $\xi_1 \otimes_\pi \xi_2$). Then one has*

$$\widehat{\deg}(E_1 \otimes E_2, \xi_1 \otimes \xi_2) = \widehat{\deg}(E_1, \xi_1) + \widehat{\deg}(E_2, \xi_2). \tag{4.21}$$

Proof Let s_1 and s_2 be non-zero elements of E_1 and E_2, respectively. For any $\omega \in \Omega$, one has

$$\ln \|s_1 \otimes s_2\|_\omega = \ln \|s_1\|_\omega + \ln \|s_2\|_\omega. \tag{4.22}$$

By taking the integral with respect to ν, we obtain the equality (4.21). □

4.3.2 Arakelov degree of adelic vector bundles

From now on and until the end of the section, we assume that, either the σ-algebra $\mathcal{A}$ is discrete, or there exists a countable subfield K_0 of K which is dense in each K_ω, where $\omega \in \Omega$.

Definition 4.3.7 Let (E, ξ) be an adelic vector bundle on S. By Proposition 4.1.32, $(\det(E), \det(\xi))$ is an adelic line bundle on S. We define the *Arakelov degree* of (E, ξ) as

$$\widehat{\deg}(E, \xi) := \widehat{\deg}(\det(E), \det(\xi)).$$

Note that, the Arakelov degree of the zero adelic vector bundle is 0. By Proposition 1.2.15, one has $\det(\xi) = \det(\xi^{\vee\vee})$. Therefore

$$\widehat{\deg}(E, \xi) = \widehat{\deg}(E, \xi^{\vee\vee}). \tag{4.23}$$

Proposition 4.3.8 *Let (E, ξ) be a* Hermitian *adelic vector bundle on S. One has*

$$\widehat{\deg}(E, \xi) = -\widehat{\deg}(E^\vee, \xi^\vee). \tag{4.24}$$

Proof The determinant of $E^\vee$ is canonically isomorphic to the dual vector space of $\det(E)$, and the norm family $\det(\xi^\vee)$ identifies with $\det(\xi)^\vee$ under this isomorphism (see Proposition 1.2.47), provided that ξ is Hermitian. Therefore, by Proposition 4.3.5 we obtain the equalities. □

Definition 4.3.9 Let (E, ξ) be an adelic vector bundle on S. We denote by $\delta(\xi)$ the function on Ω sending $\omega \in \Omega$ to $\delta_\omega(\xi) := \delta(E_{K_\omega}, \|\cdot\|_\omega)$ (see §1.2.7). Recall that the function $\delta(\xi)$ is identically 1 on $\Omega \setminus \Omega_\infty$ (see Proposition 1.2.47), and takes value in $[1, r^{r/2}]$ on Ω_∞ (see Proposition 1.2.46 and the inequalities (1.44) and (1.47)), where r is the dimension of E over K. In particular, the function $\ln \delta(\xi)$ is ν-dominated since it is bounded and vanishes outside a set of finite measure.

Similarly, we denote by $\Delta(\xi)$ the function on Ω sending $\omega \in \Omega$ to $\Delta_\omega(\xi) := \Delta(E_{K_\omega}, \|\cdot\|_\omega)$ (see Definition 1.2.41). This function is bounded from below by the constant function 1. Moreover, it identifies with the constant function 1 if ξ is Hermitian.

Proposition 4.3.10 *Let $(E, \xi = \{\|\cdot\|_\omega\}_{\omega\in\Omega})$ be an adelic vector bundle on S. Then the function $(\omega \in \Omega) \mapsto \delta_\omega(\xi)$ is $\mathcal{A}$-measurable and its logarithm is integrable with respect to ν. It is a constant function of value 1 when $\|\cdot\|_\omega$ is induced by an inner product once $\omega \in \Omega_\infty$. Moreover, the following relations hold*

$$\begin{aligned} 0 &\leqslant \widehat{\deg}(E,\xi) + \widehat{\deg}(E^\vee,\xi^\vee) \\ &= \int_\Omega \ln(\delta_\omega(\xi))\,\nu(d\omega) \leqslant \frac{1}{2}\dim_K(E)\ln(\dim_K(E))\nu(\Omega_\infty). \end{aligned} \tag{4.25}$$

In particular, one has $\widehat{\deg}(E,\xi) + \widehat{\deg}(E^\vee,\xi^\vee) = 0$ if for any $\omega \in \Omega_\infty$ the norm $\|\cdot\|_\omega$ is induced by an inner product.

Proof By Proposition 4.1.32, we obtain that both couples $(\det(E), \det(\xi))$ and $(\det(E^\vee), \det(\xi^\vee))$ are adelic line bundles on S. Let η be a non-zero element in $\det(E)$ and $\eta^\vee$ be its dual element in $\det(\xi)$. By (1.38), one has

$$(-\ln\|\eta\|_{\omega,\det}) + (-\ln\|\eta^\vee\|_{\omega,*,\det}) = \ln\delta_\omega(\xi).$$

Therefore the function $(\omega \in \Omega) \mapsto \delta_\omega(\xi)$ is $\mathcal{A}$-measurable. Moreover, by Proposition 1.2.47 we obtain that $\delta_\omega(\xi) = 1$ if $\omega \in \Omega \setminus \Omega_\infty$ or if $\omega \in \Omega_\infty$ and the norm $\|\cdot\|_\omega$ is induced by an inner product. Therefore (4.25) follows from the inequalities

$$0 \leqslant \ln\delta(\xi) \leqslant \frac{1}{2}\dim_K(E)\ln(\dim_K(E))1\!\!1_{\Omega_\infty},$$

which also implies the ν-integrability of $\ln\delta(\xi)$. □

Proposition 4.3.11 *Let (E, ξ) be a strongly adelic vector bundle on S. The function $\ln\Delta(\xi)$ is ν-dominated.*

Proof Let $\boldsymbol{e} = \{e_i\}_{i=1}^r$ be a basis of E over K. By Corollary 4.1.10, the local distance function $(\omega \in \Omega) \mapsto d_\omega(\xi, \xi_{\boldsymbol{e}})$ is ν-dominated. We write the norm families ξ and $\xi_{\boldsymbol{e}}$ in the form of $\xi = \{\|\cdot\|_\omega\}_{\omega\in\Omega}$ and $\xi_{\boldsymbol{e}} = \{\|\cdot\|_{\boldsymbol{e},\omega}\}_{\omega\in\Omega}$, respectively. Let $\omega \in \Omega$. If $\|\cdot\|_{h,\omega}$ is a norm on E_{K_ω} bounded from below by $\|\cdot\|_{\boldsymbol{e},\omega}$, which is either ultrametric or induced by an inner product, then the norm $\mathrm{e}^{d_\omega(\xi,\xi_{\boldsymbol{e}})}\|\cdot\|_{h,\omega}$ is bounded from below by $\|\cdot\|_\omega$. This norm is also ultrametric or induced by an inner product. Therefore we obtain that

$$\ln\Delta_\omega(\xi) \leqslant \ln\Delta_\omega(\xi_{\boldsymbol{e}}) + d_\omega(\xi,\xi_{\boldsymbol{e}}) \leqslant \frac{1}{2}r\ln(r)1\!\!1_{\Omega_\infty}(\omega) + d_\omega(\xi,\xi_{\boldsymbol{e}}),$$

where the second inequality comes from (1.44). Since $\nu(\Omega_\infty) < +\infty$ (see Proposition 3.1.2), we obtain that the function $\ln\Delta(\xi)$ is ν-dominated. □

Definition 4.3.12 Let $\overline{E} = (E, \xi)$ be an adelic vector bundle on S. We denote by $\delta(\overline{E})$ the integral $\int_\Omega \ln(\delta_\omega(\xi))\,\nu(\mathrm{d}\omega)$, which is also equal to $\widehat{\deg}(E, \xi) + \widehat{\deg}(E^\vee, \xi^\vee)$ (see Proposition 4.3.10). We denote by $\Delta(\overline{E})$ the lower integral $\underline{\int_\Omega} \ln(\Delta_\omega(\xi))\,\nu(\mathrm{d}\omega)$, which takes value in $[0, +\infty]$. It is finite once the function $(\omega \in \Omega) \to d_\omega(\xi, \xi^{\vee\vee})$ is ν-dominated (see Proposition 4.3.11), namely $\overline{E}$ is a strongly adelic vector bundle.

Let E, F and G be vector spaces of finite dimension over K, and ξ_E, ξ_F and ξ_G be norm families in $\mathcal{N}_E$, $\mathcal{N}_F$ and $\mathcal{N}_G$ respectively. We say that a diagram

$$0 \longrightarrow (F, \xi_F) \xrightarrow{\ f\ } (E, \xi_E) \xrightarrow{\ g\ } (G, \xi_G) \longrightarrow 0$$

is an *exact sequence* if

$$0 \longrightarrow F \xrightarrow{\ f\ } E \xrightarrow{\ g\ } G \longrightarrow 0$$

is an exact sequence of vector spaces over K and if the norm family ξ_F is the restriction of ξ_E to F, and the norm family ξ_G is the quotient of the norm family of ξ_E on G (see §4.1.1, page 206). Note that if ξ_E is Hermitian, then both norm families ξ_F and ξ_G are Hermitian.

Proposition 4.3.13 *Let (E, ξ) be an adelic vector bundle over S and*

$$0 = E_0 \subsetneq E_1 \subsetneq \ldots \subsetneq E_n = E$$

be a flag of vector subspaces of E. For any $i \in \{1, \ldots, n\}$, let ξ_i be the restriction of ξ to E_i and let η_i be the quotient norm family of ξ_i on E_i/E_{i-1}. Then one has

$$\sum_{i=1}^n \widehat{\deg}(E_i/E_{i-1}, \eta_i) \leqslant \widehat{\deg}(E, \xi) \tag{4.26}$$

and

$$\widehat{\deg}(E, \xi) - \Delta(E, \xi) \leqslant \sum_{i=1}^n \Big(\widehat{\deg}(E_i/E_{i-1}, \eta_i) - \Delta(E_i/E_{i-1}, \eta_i)\Big). \tag{4.27}$$

If in addition ξ is ultrametric on $\Omega \setminus \Omega_\infty$, then

$$\widehat{\deg}(E, \xi) - \delta(E, \xi) \leqslant \sum_{i=1}^n \Big(\widehat{\deg}(E_i/E_{i-1}, \eta_i) - \delta(E_i/E_{i-1}, \eta_i)\Big). \tag{4.28}$$

In particular, if ξ is Hermitian, then one has

$$\widehat{\deg}(E, \xi) = \sum_{i=1}^n \widehat{\deg}(E_i/E_{i-1}, \eta_i). \tag{4.29}$$

Proof For any $i \in \{1, \ldots, n\}$, we have an exact sequence

$$0 \longrightarrow (E_{i-1}, \xi_{i-1}) \longrightarrow (E_i, \xi_i) \longrightarrow (E_i/E_{i-1}, \eta_i) \longrightarrow 0 .$$

In particular, one has a canonical isomorphism (see [28] Chapter III, §7, no.7, Proposition 10)

$$\det(E_i) \cong \det(E_{i-1}) \otimes \det(E_i/E_{i-1}).$$

For any $i \in \{1, \ldots, n\}$, we pick a non-zero element $\alpha_i \in \det(E_i/E_{i-1})$ and let $\beta_i = \alpha_1 \otimes \cdots \otimes \alpha_i$, viewed as an element in $\det(E_i)$. By convention, let β_0 be the element $1 \in \det(E_0) \cong K$. By Corollary 1.1.68 and Proposition 1.2.43, one has

$$\ln \|\alpha_i\|_\omega + \ln \|\beta_{i-1}\|_\omega + \ln \frac{\Delta_\omega(\xi_{i-1})\Delta_\omega(\eta_i)}{\Delta_\omega(\xi_i)} \leqslant \ln \|\beta_i\|_\omega \leqslant \ln \|\alpha_i\|_\omega + \ln \|\beta_{i-1}\|_\omega.$$

If ξ is ultrametric on $\Omega \setminus \Omega_\infty$, then by Proposition 1.2.51, one has

$$\ln \|\alpha_i\|_\omega + \ln \|\beta_{i-1}\|_\omega + \ln \frac{\delta_\omega(\xi_{i-1})\delta_\omega(\eta_i)}{\delta_\omega(\xi_i)} \leqslant \ln \|\beta_i\|_\omega,$$

Taking the sum with respect to i, we obtain

$$\sum_{i=1}^n - \ln \|\alpha_i\|_\omega \leqslant - \ln \|\beta_n\|_\omega \leqslant \Big(\sum_{i=1}^n - \ln \|\alpha_i\|_\omega\Big) + \ln \Delta_\omega(\xi) - \Big(\sum_{i=1}^n \ln \Delta_\omega(\eta_i)\Big) \tag{4.30}$$

and, in the case where ξ is ultrametric on $\Omega \setminus \Omega_\infty$,

$$- \ln \|\beta_n\|_\omega \leqslant \Big(\sum_{i=1}^n - \ln \|\alpha_i\|_\omega\Big) + \ln \delta_\omega(\xi) - \Big(\sum_{i=1}^n \ln \delta_\omega(\eta_i)\Big).$$

By taking the integrals with respect to ν, we obtain the inequalities (4.26) and (4.28). Moreover, (4.30) leads to

$$- \ln\|\beta_n\|_\omega + \sum_{i=1}^n \ln \Delta_\omega(\eta_i) \leqslant \Big(\sum_{i=1}^n - \ln \|\alpha_i\|_\omega\Big) + \ln \Delta_\omega(\xi).$$

Taking the lower integrals, by Proposition A.4.5 and the inequality (A.4) we obtain (4.27).

If ξ is a Hermitian norm family, then each η_i is a Hermitian norm family, $i \in \{1, \ldots, n\}$. By Proposition 4.3.10, all functions $\ln \delta(\xi)$ and $\ln \delta(\eta_i)$ vanish. Therefore the equality (4.29) holds. □

Proposition 4.3.14 *Let (E, ξ) be an adelic vector bundle on S. If L/K is an algebraic extension of fields, then one has* $\det(\xi_L) = \det(\xi)_L$. *In particular,* $\widehat{\deg}(E, \xi) = \widehat{\deg}(E_L, \xi_L)$.

Proof The relation $\det(\xi_L) = (\det \xi)_L$ comes from (1) (for the non-Archimedean case) and (2) (for the Archimedean case) in Proposition 1.3.19.

Let α be a non-zero element of $\det(E)$. For any $x \in \Omega_L$ and $\omega = \pi_{L/K}(x)$, one has $\ln\|\alpha\|_x = \ln\|\alpha\|_\omega$. By taking the integral with respect to ν_L, we obtain $\widehat{\deg}(E,\xi) = \widehat{\deg}(E_L,\xi_L)$. □

Definition 4.3.15 Let (E,ξ_E) and (F,ξ_F) be adelic vector bundles on S. Let $f : E \to F$ be a K-linear map. If f is non-zero, we define the *local height function* of f as the real-valued function on Ω sending $\omega \in \Omega$ to $\ln\|f_{K_\omega}\|$, where f_{K_ω} is the K_ω-linear map $E_{K_\omega} \to F_{K_\omega}$ induced by f, and $\|f_{K_\omega}\|$ is its operator norm.

Proposition 4.3.16 *Let (E,ξ_E) and (F,ξ_F) be adelic vector bundles, and $f : E \to F$ be a K-linear map. The local height function of f is $\mathcal{A}$-measurable. If (E,ξ_E) and (F,ξ_F) are strongly adelic vector bundles, then the local height function of f is ν-dominated.*

Proof We first prove the measurability of the local height function. The statement is trivial when the σ-algebra $\mathcal{A}$ is discrete. In the following, we prove the measurability under the hypothesis that there exists a countable subfield K_0 of K which is dense in each K_ω. In this case there exists a countable sub-K_0-module E_0 of E which generates E as a vector space over K. For any $\omega \in \Omega$, viewed as a subset of E_{K_ω}, E_0 is dense. Therefore one has

$$\|f_{K_\omega}\| = \sup_{x \in E_0 \setminus \{0\}} \frac{\|f(x)\|_{F,\omega}}{\|x\|_{E,\omega}}.$$

Since (E,ξ_E) and (F,ξ_F) are adelic vector bundles, for any $x \in E_0$, the functions $(\omega \in \Omega) \mapsto \|x\|_{E,\omega}$ and $(\omega \in \Omega) \mapsto \|f(x)\|_{F,\omega}$ are $\mathcal{A}$-measurable. Therefore the function $(\omega \in \Omega) \mapsto \ln\|f_{K_\omega}\|$ is $\mathcal{A}$-measurable.

We now proceed with the proof of the dominancy of the function. We consider f as an element of $E^\vee \otimes F$ and equip this vector space with the norm family $\xi_E^\vee \otimes_\varepsilon \xi_F$ denoted by $\{\|\cdot\|_{\omega,\varepsilon}\}_{\omega\in\Omega}$. By Remark 1.1.53, the norm of

$$f_{K_\omega} : (E_{K_\omega}, \|\cdot\|_{E,\omega,**}) \to (F_{K_\omega}, \|\cdot\|_{F,\omega,**})$$

identifies with $\|f\|_{\omega,\varepsilon}$. By Proposition 4.1.19 (3) and (5), the norm family $\xi_E^\vee \otimes_\varepsilon \xi_F$ is dominated. Hence the function $(\omega \in \Omega) \mapsto \ln\|f\|_{\omega,\varepsilon}$ is ν-dominated. Note that one has

$$\left|\ln\|f_{K_\omega}\| - \ln\|f\|_{\omega,\varepsilon}\right| \leqslant d_\omega(\xi_E, \xi_E^{\vee\vee}) + d_\omega(\xi_F, \xi_F^{\vee\vee}),$$

where $\|f_{K_\omega}\|$ denotes the operator norm of $f_{K_\omega} : (E_{K_\omega}, \|\cdot\|_{E,\omega}) \to (F_{K_\omega}, \|\cdot\|_{F,\omega})$. As the local distance functions $(\omega \in \Omega) \mapsto d_\omega(\xi_E, \xi_E^{\vee\vee})$ and $(\omega \in \Omega) \mapsto d_\omega(\xi_F, \xi_F^{\vee\vee})$ are ν-dominated, we obtain that the function $(\omega \in \Omega) \mapsto \ln\|f_{K_\omega}\|$ is ν-dominated. The proposition is thus proved. □

Definition 4.3.17 Let (E,ξ_E) and (F,ξ_F) be adelic vector bundles, and $f : E \to F$ be a K-linear map. We define the *height* $h(f)$ of f as the lower integral

$$\underline{\int_\Omega} \ln\|f_{K_\omega}\|\,\nu(d\omega).$$

By Remark 1.1.53, in the case where ξ_E and ξ_F are ultrametric on $\Omega\setminus\Omega_\infty$, one has

$$h(f) = -\widehat{\deg}_{\xi_E^\vee\otimes_\varepsilon\xi_F}(f).$$

Proposition 4.3.18 *Let (E_1,ξ_1) and (E_2,ξ_2) be adelic vector bundles on S and $f: E_1\to E_2$ be a K-linear isomorphism. One has*

$$\widehat{\deg}(E_1,\xi_1)-\widehat{\deg}(E_2,\xi_2)=h(\det(f)). \tag{4.31}$$

In particular,

$$\widehat{\deg}(E_1,\xi_1)\leqslant\widehat{\deg}(E_2,\xi_2)+rh(f). \tag{4.32}$$

Proof By definition one has

$$\begin{aligned} h(\det(f)) &= -\widehat{\deg}\big(\det(E_1)^\vee\otimes\det(E_2),\det(\xi_1)^\vee\otimes\det(\xi_2)\big)\\ &= -\widehat{\deg}(E_1,\xi_1)+\widehat{\deg}(E_2,\xi_2), \end{aligned} \tag{4.33}$$

where the second equality comes from Propositions 4.3.5 and 4.3.6. Finally the inequality (4.32) is a consequence of (4.31) and Proposition 1.1.64. □

4.3.3 Arakelov degree of tensor adelic vector bundles

Let $\overline{E}=(E,\xi_E)$ and $\overline{F}=(F,\xi_F)$ be adelic vector bundles on S. We denote by $\overline{E}\otimes_{\varepsilon,\pi}\overline{F}$ the couple $(E\otimes_K F,\xi_E\otimes_{\varepsilon,\pi}\xi_F)$, called the ε,π-tensor product of $\overline{E}$ and $\overline{F}$. By Proposition 4.1.32 (see also Remark 4.1.34), $\overline{E}\otimes_{\varepsilon,\pi}\overline{F}$ is an adelic vector bundle on S. If both $\overline{E}$ and $\overline{F}$ are Hermitian adelic vector bundles, we denote by $\overline{E}\otimes\overline{F}$ the couple $(E\otimes_K F,\xi_E\otimes\xi_F)$, called the *orthogonal tensor product* of $\overline{E}$ and $\overline{F}$. By Proposition 4.1.32, $\overline{E}\otimes\overline{F}$ is a Hermitian adelic vector bundle on S.

Proposition 4.3.19 *Let $\overline{E}=(E,\xi_E)$ and $\overline{F}=(F,\xi_F)$ be adelic vector bundles on S. One has*

$$\widehat{\deg}(\overline{E}\otimes_{\varepsilon,\pi}\overline{F})=\dim_K(F)\,\widehat{\deg}(\overline{E})+\dim_K(E)\,\widehat{\deg}(\overline{F}). \tag{4.34}$$

If $\overline{E}$ and $\overline{F}$ are Hermitian adelic vector bundles, then one has

$$\widehat{\deg}(\overline{E}\otimes\overline{F})=\dim_K(F)\,\widehat{\deg}(\overline{E})+\dim_K(E)\,\widehat{\deg}(\overline{F}). \tag{4.35}$$

Proof Let n and m be the dimensions of E and F over K respectively. By Propositions 1.1.69 and 1.2.39, under the canonical isomorphism

$$\det(E)^{\otimes m}\otimes\det(F)^{\otimes n}\cong\det(E\otimes_K F),$$

the norm family $\det(\xi_E)^{\otimes m}\otimes\det(\xi_F)^{\otimes n}$ identifies with $\det(\xi_E\otimes_{\varepsilon,\pi}\xi_F)$. Therefore the equality (4.34) results from Proposition 4.3.6.

The equality (4.35) can be proved in a similar way by using Propositions 1.2.63 and 1.2.39. □

4.3.4 Positive degree

Let (E, ξ) be an adelic vector bundle on S. We define the *positive degree* of (E, ξ) as

$$\widehat{\deg}_+(E, \xi) := \sup_{F \subseteq E} \widehat{\deg}(F, \xi_F),$$

where F runs over the set of all vector subspaces of E, and ξ_F denotes the restriction of ξ to F. Clearly one has $\widehat{\deg}(E, \xi) \geqslant 0$.

Proposition 4.3.20 *Let (E, ξ_E) be an adelic vector bundle on S, F be a vector subspace of E and G be the quotient space of E by F. Let ξ_F be the restriction of ξ_E to F and ξ_G be the quotient of ξ_E on G. Then one has*

$$\widehat{\deg}_+(F, \xi_F) \leqslant \widehat{\deg}_+(E, \xi_E) \leqslant \widehat{\deg}_+(F, \xi_F) + \widehat{\deg}_+(G, \xi_G) + \Delta(E, \xi_E). \qquad (4.36)$$

If in addition (E, ξ_E) is ultrametric on $\Omega \setminus \Omega_\infty$, then

$$\widehat{\deg}_+(F, \xi_F) \leqslant \widehat{\deg}_+(E, \xi_E) \leqslant \widehat{\deg}_+(F, \xi_F) + \widehat{\deg}_+(G, \xi_G) + \delta(E, \xi_E). \qquad (4.37)$$

Proof The first inequality of (4.36) follows directly from the definition of positive degree. In the following, we prove the second inequality of (4.36).

Let E_1 be a vector subspace of E, $F_1 = F \cap E_1$ and G_1 be the canonical image of E_1 in G. Then we obtain the following short exact sequence of adelic vector bundles:

$$0 \longrightarrow (F_1, \xi_{F_1}) \longrightarrow (E_1, \xi_{E_1}) \longrightarrow (G_1, \xi_{G_1}) \longrightarrow 0,$$

where ξ_{E_1} is the restriction of the norm family ξ_E, ξ_{F_1} is the restriction of ξ_{E_1} to F_1 and ξ_{G_1} is the quotient norm family of ξ_{E_1} on G_1. Note that the norm family ξ_{F_1} coincides with the restricted norm family of ξ_F induced by the inclusion map $F_1 \to F$. Moreover, if we denote by ξ'_{G_1} the restricted norm family induced by the inclusion map $G_1 \to G$, then the identity map $(G_1, \xi_{G_1}) \to (G_1, \xi'_{G_1})$ has norm $\leqslant 1$ (see Proposition 1.1.14 (2.b)) on any $\omega \in \Omega$. In particular, by Proposition 4.3.18 one has

$$\widehat{\deg}(G_1, \xi_{G_1}) \leqslant \widehat{\deg}(G_1, \xi'_{G_1}) \leqslant \widehat{\deg}_+(G, \xi_G).$$

Therefore, by Proposition 4.3.13, one has

$$\begin{aligned}\widehat{\deg}(E_1, \xi_{E_1}) &\leqslant \widehat{\deg}(F_1, \xi_{F_1}) + \widehat{\deg}(G_1, \xi_{G_1}) + \Delta(E_1, \xi_{E_1}) \\ &\leqslant \widehat{\deg}_+(F, \xi_F) + \widehat{\deg}_+(G, \xi_G) + \Delta(E_1, \xi_{E_1}) \\ &\leqslant \widehat{\deg}_+(F, \xi_F) + \widehat{\deg}_+(G, \xi_G) + \Delta(E, \xi_E),\end{aligned}$$

where the last inequality comes from Corollary 1.2.44. Similarly, by Propositions 4.3.13 and 1.2.51, in the case where ξ_E is ultrametric on $\Omega \setminus \Omega_\infty$ we have

$$\widehat{\deg}(E_1, \xi_{E_1}) \leqslant \widehat{\deg}_+(F, \xi_F) + \widehat{\deg}_+(G, \xi_G) + \delta(E, \xi_E).$$

Since $E_1 \subseteq E$ is arbitrary, we obtain (4.37). □

Proposition 4.3.21 *Let $(E, \xi = \{\|\cdot\|_\omega\}_{\omega\in\Omega})$ be an adelic vector bundle on S. Then we have the following:*

(1) *Let (F, η) be an adelic vector bundle on S such that F is a vector subspace of E and $\eta \geqslant \xi_F$ on F. Then $\widehat{\deg}_+(F,\eta) \leqslant \widehat{\deg}_+(F,\xi_F) \leqslant \widehat{\deg}_+(E,\xi)$.*

(2) *Let φ be an integrable function on Ω. Then*

$$\begin{cases} \widehat{\deg}_+(E,\exp(-\varphi)\xi) \leqslant \widehat{\deg}_+(E,\xi) + \dim_K(E)\left|\displaystyle\int_\Omega \varphi(\omega)\,\nu(\mathrm{d}\omega)\right|, \\ \widehat{\deg}_+(E,\exp(\varphi)\xi) \geqslant \widehat{\deg}_+(E,\xi) - \dim_K(E)\left|\displaystyle\int_\Omega \varphi(\omega)\,\nu(\mathrm{d}\omega)\right|. \end{cases}$$

Moreover, if $\displaystyle\int_\Omega \varphi\,\nu(\mathrm{d}\omega) \geqslant 0$, then

$$\widehat{\deg}_+(E,\xi) \leqslant \widehat{\deg}_+(E,\exp(-\varphi)\xi) \quad \textit{and} \quad \widehat{\deg}_+(E,\exp(\varphi)\xi) \leqslant \widehat{\deg}_+(E,\xi).$$

Proof (1) The inequality $\widehat{\deg}_+(F,\xi_F) \leqslant \widehat{\deg}_+(E,\xi)$ has been proved in Proposition 4.3.20. For $\epsilon > 0$, there is a vector subspace W of F such that

$$\widehat{\deg}_+(F,\eta) - \epsilon \leqslant \widehat{\deg}(W,\eta_W),$$

so that

$$\widehat{\deg}_+(F,\eta) - \varepsilon \leqslant \widehat{\deg}(W,\eta_W) \leqslant \widehat{\deg}(W,\xi_W) \leqslant \widehat{\deg}_+(F,\xi_F),$$

as required.

(2) Let F be a vector subspace of E over K. Then

$$\widehat{\deg}(F,\exp(-\varphi)\xi_F) = \widehat{\deg}(F,\xi_F) + \dim_K(F)\int_\Omega \varphi(\omega)\,\nu(\mathrm{d}\omega), \tag{4.38}$$

so that if $\displaystyle\int_\Omega \varphi\nu(\mathrm{d}\omega) \geqslant 0$, then $\widehat{\deg}(F,\exp(-\varphi)\xi_F) \geqslant \widehat{\deg}(F,\xi_F)$, which leads to the third inequality. Moreover, by (4.38),

$$\widehat{\deg}(F,\exp(-\varphi)\xi_F) \leqslant \widehat{\deg}_+(E,\exp(-\varphi)\xi) + \dim_K(E)\left|\int_\Omega \varphi(\omega)\,\nu(\mathrm{d}\omega)\right|,$$

and hence the first inequality follows.

If we set $\xi' = \exp(\varphi)\xi$, then the first and third inequalities imply the second and fourth inequalities, respectively. □

4.3.5 Riemann-Roch formula

Here we consider a Riemann-Roch formula of an adelic vector bundle on S.

Proposition 4.3.22 *Let $\overline{V} = (V, \xi)$ be an adelic vector bundle on S. Then one has*

$$0 \leqslant \widehat{\deg}(\overline{V}) - \left(\widehat{\deg}_+(\overline{V}^{\vee\vee}) - \widehat{\deg}_+(\overline{V}^{\vee})\right) \leqslant \delta(\overline{V}). \tag{4.39}$$

If ξ is ultrametric on $\Omega \setminus \Omega_\infty$, then one has

$$0 \leqslant \widehat{\deg}(\overline{V}) - \left(\widehat{\deg}_+(\overline{V}) - \widehat{\deg}_+(\overline{V}^{\vee})\right) \leqslant \delta(\overline{V}). \tag{4.40}$$

Proof Let $\epsilon > 0$. We choose a vector subspace W of V such that $\widehat{\deg}(W, \xi_W) \geqslant \widehat{\deg}_+(\overline{V}) - \epsilon$, where ξ_W is the restriction of ξ to W. Let $\xi^\vee$ be the dual norm family of ξ on $V^\vee$, $\xi_{V/W}$ be the quotient norm family of ξ on V/W, and $\xi^\vee_{V/W}$ be the dual norm family of $\xi_{V/W}$ on $(V/W)^\vee$. If we consider $(V/W)^\vee$ as a vector subspace of $V^\vee$, then $\xi^\vee_{V/W}$ identifies with the restriction of $\xi^\vee$ to $(V/W)^\vee$ by Proposition 1.1.20, so that

$$\widehat{\deg}((V/W)^\vee, \xi^\vee_{V/W}) \leqslant \widehat{\deg}_+(V^\vee, \xi^\vee).$$

On the other hand, one has

$$\widehat{\deg}((V/W)^\vee, \xi^\vee_{V/W}) + \widehat{\deg}(V/W, \xi_{V/W}) \geqslant 0$$

by Proposition 4.3.10 and

$$\widehat{\deg}(W, \xi_W) + \widehat{\deg}(V/W, \xi_{V/W}) \leqslant \widehat{\deg}(V, \xi)$$

by Proposition 4.3.13. Therefore,

$$\begin{aligned}\widehat{\deg}_+(V^\vee, \xi^\vee) &\geqslant \widehat{\deg}((V/W)^\vee, \xi^\vee_{V/W}) \geqslant -\widehat{\deg}(V/W, \xi_{V/W}) \\ &\geqslant \widehat{\deg}(W, \xi_W) - \widehat{\deg}(V, \xi) \geqslant \widehat{\deg}_+(V, \xi) - \widehat{\deg}(V, \xi) - \epsilon\end{aligned}$$

and hence

$$\widehat{\deg}_+(\overline{V}) - \widehat{\deg}_+(\overline{V}^\vee) \leqslant \widehat{\deg}(\overline{V}). \tag{4.41}$$

Replacing $\overline{V}$ by $\overline{V}^\vee$ in (4.41), we obtain

$$\widehat{\deg}_+(\overline{V}^\vee) - \widehat{\deg}_+(\overline{V}^{\vee\vee}) \leqslant \widehat{\deg}(\overline{V}^\vee), \tag{4.42}$$

which, by Proposition 4.3.10, implies the second inequality of (4.39). Replacing $\overline{V}$ by $\overline{V}^{\vee\vee}$ in (4.41), by the fact that $\|\cdot\|_{\omega,**,*} = \|\cdot\|_{\omega,*}$ for any $\omega \in \Omega$ (see Proposition 1.2.14 (1)) and the equality (4.23) we obtain

$$\widehat{\deg}_+(\overline{V}^{\vee\vee}) - \widehat{\deg}_+(\overline{V}^\vee) \leqslant \widehat{\deg}(\overline{V}), \tag{4.43}$$

which leads to the first inequality of (4.39).

In the case where ξ is ultrametric on $\Omega\setminus\Omega_\infty$, one has $\|\cdot\|_\omega = \|\cdot\|_{\omega,**}$ for all $\omega\in\Omega$ by Proposition 1.1.18 (Archimedean case) and Corollary 1.2.12. Thus (4.40) follows from (4.39). □

4.3.6 Comparison of $\widehat{\deg}_+$ and $\widehat{h}^0$ in the classic setting

In this subsection, we compare $\widehat{\deg}_+$ with the invariant $\widehat{h}^0$ in the classic settings of vector bundles on a regular projective curve and Hermitian vector bundles on an arithmetic curve.

4.3.6.1 Function field case

Let k be a field, C be a regular projective curve over $\operatorname{Spec} k$ and $K = k(C)$ be the field of rational functions on C. Let Ω be the set of all closed points of the curve C, equipped with the discrete σ-algebra $\mathcal{A}$ and the measure ν such that $\nu(\{x\}) = [k(x):k]$ for any $x\in\Omega$. Let $\phi:\Omega\to M_K$ be the map sending x to $|\cdot|_x = \mathrm{e}^{-\operatorname{ord}_x(\cdot)}$. We have seen in §3.2.1 that $S=(K,(\Omega,\mathcal{A},\nu),\phi)$ is an adelic curve.

Let $\mathcal{E}$ be a locally free $\mathcal{O}_C$-module of finite type and $E=\mathcal{E}_K:=\mathcal{E}\otimes_{\mathcal{O}_C}K$ be its generic fibre. For any $x\in\Omega$, let $\|\cdot\|_x$ be the norm on $E\otimes_K K_x$ defined as

$$\forall\, s\in E\otimes_K K_x,\quad \|s\|_x = \inf\{|a|_x \,:\, a\in K_x^{\times},\ a^{-1}s\in\mathcal{E}\otimes_{\mathcal{O}_{C,x}}\widehat{\mathcal{O}}_{C,x},\}$$

where $\widehat{\mathcal{O}}_{C,x}$ is the completion of $\mathcal{O}_{C,x}$, which identifies with the valuation ring of K_x. Then $\xi_{\mathcal{E}}=\{\|\cdot\|_x\}_{x\in\Omega}$ forms a Hermitian norm family on E and $(E,\xi_{\mathcal{E}})$ is an adelic vector bundle on Ω. Note that the Arakelov degree of $(E,\xi_{\mathcal{E}})$ identifies with the degree of the locally free $\mathcal{O}_C$-module $\mathcal{E}$, namely

$$\widehat{\deg}(E,\xi_{\mathcal{E}}) = \deg(c_1(\mathcal{E})\cap[C]).$$

The Harder-Narasimhan flag of $(E,\xi_{\mathcal{E}})$ is also related to the classic construction of Harder-Narasimhan filtration of $\mathcal{E}$. In fact, there exists a unique flag of locally free $\mathcal{O}_C$-modules

$$0=\mathcal{E}_0\subsetneq\mathcal{E}_1\subsetneq\ldots\subsetneq\mathcal{E}_n=\mathcal{E}$$

such that each sub-quotient $\mathcal{E}_i/\mathcal{E}_{i-1}$ is a locally free $\mathcal{O}_C$-module which is semistable and that

$$\mu(\mathcal{E}_1/\mathcal{E}_0)>\mu(\mathcal{E}_2/\mathcal{E}_1)>\ldots>\mu(\mathcal{E}_n/\mathcal{E}_{n-1}).$$

Then the Harder-Narasimhan flag of the Hermitian adelic vector bundle $(E,\xi_{\mathcal{E}})$ is given by

$$0=\mathcal{E}_{0,K}\subsetneq\mathcal{E}_{1,K}\subsetneq\ldots\subsetneq\mathcal{E}_{n,K}=E.$$

The notion of positive degree for locally free $\mathcal{O}_C$-modules of finite rank has been proposed in [42] and compared with the dimension (over k) of the vector space of global sections, by using the Riemann-Roch formula on curves. For any locally free $\mathcal{O}_C$-module of finite rank $\mathcal{E}$, we denote by $h^0(\mathcal{E})$ the dimension of $H^0(C,\mathcal{E})$ over k.

Theorem 4.3.23 *Let $g(C)$ be the genus of C relatively to the field k (namely $g(C) = h^0(\omega_{C/k})$, $\omega_{C/k}$ being the relative dualising sheaf of C over* Spec k*). For any locally free $\mathcal{O}_C$-module of finite rank $\mathcal{E}$, one has*

$$|h^0(\mathcal{E}) - \widehat{\deg}_+(E,\xi_{\mathcal{E}})| \leqslant \dim_K(E)\max(g(C)-1,1). \tag{4.44}$$

We refer the readers to [42, §2] for a proof.

4.3.6.2 Number field case

Let K be a number field and Ω be the set of all places of K, equipped with the discrete σ-algebra $\mathcal{A}$. For each $\omega \in \Omega$, we denote by $|\cdot|_\omega$ the absolute value on K extending either the usual absolute value on $\mathbb{Q}$ or one of the p-adic absolute values (with $|p|_\omega = p^{-1}$ in the latter case). We let K_ω (resp. $\mathbb{Q}_\omega$) be the completion of K (resp. $\mathbb{Q}$) with respect to the absolute value $|\cdot|_\omega$. Let ν be the measure on the measurable space $(\Omega,\mathcal{A})$ such that $\nu(\{\omega\}) = [K_\omega : \mathbb{Q}_\omega]$. Then $S = (K,(\Omega,\mathcal{A},\nu),\phi : \omega \mapsto |\cdot|_\omega)$ forms an adelic curve.

Let $\mathcal{O}_K$ be the ring of algebraic integers in K. Recall that a Hermitian vector bundle on Spec $\mathcal{O}_K$ is by definition the data $\overline{\mathcal{E}} = (\mathcal{E},\{\|\cdot\|_\sigma\}_{\sigma\in\Omega_\infty})$ of a projective $\mathcal{O}_K$-module of finite type $\mathcal{E}$ together with a family of norms, where $\|\cdot\|_\sigma$ is a norm on the vector space $\mathcal{E}\otimes_{\mathcal{O}_K} K_\omega$ which is induced by an inner product. Similarly to the function field case, the $\mathcal{O}_K$-module structure of $\mathcal{E}$ induces, for each non-Archimedean place $\mathfrak{p} \in \Omega\setminus\Omega_\infty$, an ultrametric norm $\|\cdot\|_{\mathfrak{p}}$ on $\mathcal{E}\otimes_{\mathcal{O}_K} K_{\mathfrak{p}}$ as follows :

$$\forall\, s \in \mathcal{E}\otimes_{\mathcal{O}_K} K_{\mathfrak{p}},\quad \|s\|_{\mathfrak{p}} = \inf\{|a|_{\mathfrak{p}} \;:\; a \in K_{\mathfrak{p}}^{\times},\ a^{-1}s \in \mathcal{E}\otimes_{\mathcal{O}_K} \mathcal{O}_{\mathfrak{p}}\},$$

where $\mathcal{O}_{\mathfrak{p}}$ is the valuation ring of $K_{\mathfrak{p}}$. Let E be $\mathcal{E}\otimes_{\mathcal{O}_K} K$ and let $\xi_{\overline{\mathcal{E}}}$ be the norm family $\{\|\cdot\|_\omega\}_{\omega\in\Omega}$. Then the couple $(E,\xi_{\overline{\mathcal{E}}})$ forms an adelic vector bundle on S, which is said to be *induced by* $\overline{\mathcal{E}}$.

Recall that the space $\widehat{H}^0(\overline{\mathcal{E}})$ of "global sections" of $\overline{\mathcal{E}}$ is defined as

$$\widehat{H}^0(\overline{\mathcal{E}}) := \{s \in \mathcal{E} \;:\; \sup_{\sigma\in\Omega_\infty}\|s\|_\sigma \leqslant 1\} = \{s \in E \;:\; \sup_{\omega\in\Omega}\|s\|_\omega \leqslant 1\}.$$

This is a finite set. However it does not possess a natural vector space structure over a base field. We define (compare to the case of function field of a regular projective curve over a finite field) $\widehat{h}^0(\overline{\mathcal{E}})$ to be $\ln\operatorname{card}(\widehat{H}^0(\overline{\mathcal{E}}))$. The invariants $\widehat{h}^0(\overline{\mathcal{E}})$ and $\widehat{\deg}_+(E,\xi_{\overline{\mathcal{E}}})$ have been compared in [42, §6] (see also [37]). We denote by

- B_n the unit ball in $\mathbb{R}^n$, where $n \in \mathbb{N}$,
- $\operatorname{vol}(B_n)$ the Lebesgue measure of B_n, which is equal to $\pi^{n/2}/\Gamma(n/2+1)$,
- $r_1(K)$ the number of real places of K,

– $r_2(K)$ the number of complex places of K.

Theorem 4.3.24 *Let $\overline{\mathcal{E}}$ be a Hermitian vector bundle on* $\operatorname{Spec} \mathcal{O}_K$*, and $(E, \xi_{\overline{\mathcal{E}}})$ be the adelic vector bundle on S induced by $\overline{\mathcal{E}}$. then*

$$|\widehat{h}^0(\overline{\mathcal{E}}) - \widehat{\deg}_+(E, \xi_{\overline{\mathcal{E}}})| \leqslant \dim_K(E) \ln |\mathfrak{D}_K| + C(K, \dim_K(E)),$$

where $\mathfrak{D}_K$ is the discriminant of K over $\mathbb{Q}$, and for any $n \in \mathbb{N}$,

$$\begin{aligned} C(K,n) := n[K:\mathbb{Q}]\ln(3) + n(r_1(K)+r_2(K))\ln(2) + \frac{n}{2}\ln|\mathfrak{D}_K| \\ - r_1 \ln(\operatorname{vol}(B_n)n!) - r_2 \ln(V(B_{2n})(2n)!) + \ln(([K:\mathbb{Q}]n)!). \end{aligned}$$

4.3.7 Slopes and slope inequalities

Let (E, ξ) be an adelic vector bundle on S such that $E \neq \{0\}$. We define the *slope* of (E, ξ) as

$$\widehat{\mu}(E, \xi) := \frac{\widehat{\deg}(E, \xi)}{\dim_K(E)}.$$

We define the *maximal slope* of (E, ξ) as

$$\widehat{\mu}_{\max}(E, \xi) := \sup_{0 \neq F \subseteq E} \widehat{\mu}(F, \xi_F),$$

where F runs over the set of all non-zero vector subspaces of E and ξ_F denotes the restriction of ξ to F. Similarly, we define the *minimal slope* of (E, ξ) as

$$\widehat{\mu}_{\min}(E, \xi) = \inf_{E \twoheadrightarrow G \neq \{0\}} \widehat{\mu}_{\max}(G, \xi_G), \tag{4.45}$$

where G runs over the set of all non-zero quotient spaces of E, and ξ_G denotes the quotient norm family of ξ. By definition one has $\widehat{\mu}_{\min}(\overline{E}) \leqslant \widehat{\mu}_{\max}(\overline{E})$ and $\widehat{\mu}(\overline{E}) \leqslant \widehat{\mu}_{\max}(\overline{E})$ (note that here the vector space E has been assumed to be non-zero). If $\overline{E} = \overline{\mathbf{0}}$ is the zero adelic vector bundle, we define by convention

$$\widehat{\mu}_{\max}(\overline{\mathbf{0}}) := -\infty, \quad \widehat{\mu}(\overline{\mathbf{0}}) := 0, \quad \widehat{\mu}_{\min}(\overline{\mathbf{0}}) := +\infty.$$

Proposition 4.3.25 *Let $\overline{E} = (E, \xi_E)$ be a non-zero adelic vector bundle on S. One has*

$$\widehat{\mu}_{\min}(\overline{E}) + \widehat{\mu}_{\max}(\overline{E}^\vee) \geqslant 0, \tag{4.46}$$

provided that $\widehat{\mu}_{\max}(\overline{E}^\vee) < +\infty$ (we will show in Proposition 4.3.30 that this condition is always satisfied, and, as a consequence of the current proposition, that one has $\widehat{\mu}_{\min}(\overline{E}) > -\infty$).

Proof Let G be a non-zero quotient vector space of E and ξ_G be the quotient norm family of ξ_E. Note that $G^\vee$ identifies with a vector subspace of $E^\vee$ and by

Proposition 1.1.20, the dual norm family $\xi_G^\vee$ identifies with the restriction of $\xi_E^\vee$ to $G^\vee$. By Proposition 4.3.10, one has

$$0 \leqslant \widehat{\mu}(G, \xi_G) + \widehat{\mu}(G^\vee, \xi_G^\vee) \leqslant \widehat{\mu}_{\max}(G, \xi_G) + \widehat{\mu}_{\max}(E^\vee, \xi_E^\vee).$$

Since G is arbitrary, we obtain the inequality (4.46). □

Classically in the setting of vector bundles over a regular projective curve or in that of Hermitian vector bundle over an arithmetic curve, the minimal slope is rather defined as the minimal value of slopes of quotient bundles. A direct analogue would replace $\widehat{\mu}_{\max}$ by $\widehat{\mu}$ in (4.45) for the definition of the minimal slope. However, it can be shown that the two definitions are actually equivalent.

Proposition 4.3.26 *Let $\overline{E}$ be a non-zero adelic vector bundle on S. One has*

$$\widehat{\mu}_{\min}(\overline{E}) = \inf_{E \twoheadrightarrow G \neq \{0\}} \widehat{\mu}(\overline{G}),$$

where G runs over the set of non-zero quotient vector spaces of E, and in $\overline{G}$ we consider the quotient norm family of that in $\overline{E}$.

Proof Clearly $\widehat{\mu}_{\min}(\overline{E}) \geqslant \inf_{E \twoheadrightarrow G \neq \{0\}} \widehat{\mu}(\overline{G})$, so that we assume by contradiction that

$$\widehat{\mu}_{\min}(\overline{E}) > \inf_{E \twoheadrightarrow G \neq \{0\}} \widehat{\mu}(\overline{G}).$$

We fix $\varepsilon \in \mathbb{R}_{>0}$ such that

$$\inf_{E \twoheadrightarrow G \neq \{0\}} \widehat{\mu}(\overline{G}) < \widehat{\mu}_{\min}(\overline{E}) - \varepsilon.$$

Among the non-zero quotient vector spaces of E of slope bounded from above by $\widehat{\mu}_{\min}(\overline{E}) - \varepsilon$, we choose a G having the least dimension over K, that is, $\widehat{\mu}(\overline{G}) \leqslant \widehat{\mu}_{\min}(\overline{E}) - \varepsilon$ and if Q is a non-zero quotient vector space of E such that $\dim_K(Q) < \dim_K(G)$, then $\widehat{\mu}_{\min}(\overline{E}) - \varepsilon < \widehat{\mu}(\overline{Q})$. As $\widehat{\mu}_{\min}(\overline{E}) - \varepsilon < \widehat{\mu}_{\max}(\overline{G})$, one can find a non-zero subspace G' such that $\widehat{\mu}_{\min}(\overline{E}) - \varepsilon < \widehat{\mu}(\overline{G}')$. Note that $G' \subsetneq G$. We have a short exact sequence

$$0 \longrightarrow G' \longrightarrow G \longrightarrow G/G' \longrightarrow 0,$$

which leads to (by Proposition 4.3.13)

$$\widehat{\deg}(\overline{G'}) + \widehat{\deg}(\overline{G/G'}) \leqslant \widehat{\deg}(\overline{G}). \tag{4.47}$$

On the other hand, since $0 < \dim_K(G/G') < \dim_K(G)$, one has

$$\begin{aligned}\widehat{\deg}(\overline{G'}) + \widehat{\deg}(\overline{G/G'}) &= \dim_K G' \widehat{\mu}(\overline{G'}) + \dim_K(G/G')\widehat{\mu}(\overline{G/G'}) \\ &> \dim_K(G')(\widehat{\mu}_{\min}(\overline{E}) - \varepsilon) + \dim_K(G/G')(\widehat{\mu}_{\min}(\overline{E}) - \varepsilon) \\ &= \dim_K(G)(\widehat{\mu}_{\min}(\overline{E}) - \varepsilon),\end{aligned}$$

and hence $\widehat{\mu}(\overline{G}) > \widehat{\mu}_{\min}(\overline{E}) - \varepsilon$ by (4.47). This is a contradiction.

Corollary 4.3.27 *Let $\overline{E} = (E, \xi_E)$ be a non-zero adelic vector bundle on S. One has*

$$\widehat{\mu}_{\min}(\overline{E}^\vee) + \widehat{\mu}_{\max}(\overline{E}) \leqslant \frac{1}{2}\ln(\dim_K(E))\,\nu(\Omega_\infty). \tag{4.48}$$

Moreover, one has $\widehat{\mu}_{\min}(\overline{E}^\vee) + \widehat{\mu}_{\max}(\overline{E}) = 0$ if $\overline{E}$ is Hermitian.

Proof Let F be a non-zero vector subspace of E and $\xi_F = \{\|\cdot\|_{F,\omega}\}_{\omega\in\Omega}$ be the restriction of $\xi_E = \{\|\cdot\|_{E,\omega}\}_{\omega\in\Omega}$ to F. For any $\omega \in \Omega$, $\|\cdot\|_{F,\omega,*}$ is bounded from above by the quotient norm of $\|\cdot\|_{E,\omega,*}$ by the canonical surjective map $E^\vee_{K_\omega} \to F^\vee_{K_\omega}$. Hence by Proposition 4.3.26, one has $\widehat{\mu}(\overline{F}^\vee) \geqslant \widehat{\mu}_{\min}(\overline{E}^\vee)$. Therefore,

$$\begin{aligned}\widehat{\mu}(\overline{F}) + \widehat{\mu}_{\min}(\overline{E}^\vee) \leqslant \widehat{\mu}(\overline{F}) + \widehat{\mu}(\overline{F}^\vee) &= \frac{1}{\dim_K(F)}\int_\Omega \ln(\delta_\omega(\xi_F))\,\nu(\mathrm{d}\omega)\\ &\leqslant \frac{1}{2}\ln(\mathrm{rk}_K(F))\nu(\Omega_\infty),\end{aligned}$$

where the equality follows from Proposition 4.3.10 and the last inequality comes from Remark 1.2.55. Since F is arbitrary, we obtain (4.48).

If $\overline{E}$ is Hermitian, then for any non-zero vector subspace F of E one has

$$\widehat{\mu}(\overline{F}) + \widehat{\mu}_{\min}(\overline{E}^\vee) \leqslant \widehat{\mu}(\overline{F}) + \widehat{\mu}(\overline{F}^\vee) = 0,$$

which leads to $\widehat{\mu}_{\min}(\overline{E}^\vee) + \widehat{\mu}_{\max}(\overline{E}) \leqslant 0$. As we have seen that $\widehat{\mu}_{\min}(\overline{E}^\vee) + \widehat{\mu}_{\max}(\overline{E}) \geqslant 0$ in Proposition 4.3.25 (note that $\overline{E}^{\vee\vee} = \overline{E}$ when $\overline{E}$ is Hermitian), the equality $\widehat{\mu}_{\min}(\overline{E}^\vee) + \widehat{\mu}_{\max}(\overline{E}) = 0$ holds. □

4.3.8 Finiteness of slopes

Let $\overline{E} = (E, \xi)$ be an adelic vector bundle on S such that ξ is Hermitian. We assume that the vector space E is non-zero and we denote by $\Theta(E)$ the set of all K-vector subspaces of E. For any $F \in \Theta(E)$, the vector subspace F equipped with the restricted norm family forms a Hermitian adelic vector bundle on S (see Proposition 4.1.32). We denote by $\overline{F}$ this Hermitian adelic vector bundle. Note that the dimension and the Arakelov degree define two functions on $\Theta(E)$, which satisfy the following relations: for any pair (E_1, E_2) of elements in $\Theta(E)$, one has

$$\dim_K(E_1 \cap E_2) + \dim_K(E_1 + E_2) = \dim_K(E_1) + \dim_K(E_2), \tag{4.49}$$

$$\widehat{\deg}(\overline{E_1 \cap E_2}) + \widehat{\deg}(\overline{E_1 + E_2}) \geqslant \widehat{\deg}(\overline{E}_1) + \widehat{\deg}(\overline{E}_2), \tag{4.50}$$

where the inequality (4.50) comes from Corollary 1.2.52.

Proposition 4.3.28 *Let E be a non-zero vector space of finite dimension over K and $\Theta(E)$ be the set of all vector subspaces of E. Assume given two functions $r : \Theta(E) \to \mathbb{R}_+$ and $d : \Theta(E) \to \mathbb{R}$ which verify the following conditions:*

(1) *the function $r(\cdot)$ takes value 0 on the zero vector subspace of E and takes positive values on non-zero vector subspaces of E;*

(2) *for any couple (E_1, E_2) of elements in $\Theta(E)$ one has*

$$r(E_1 \cap E_2) + r(E_1 + E_2) = r(E_1) + r(E_2)$$

and

$$d(E_1 \cap E_2) + d(E_1 + E_2) \geqslant d(E_1) + d(E_2);$$

(3) $d(\{0\}) \leqslant 0$.

Then the function $\mu = d/r$ attains its maximal value $\mu_{\max}$ on the set $\Theta^*(E)$ of all non-zero vector subspaces of E. Moreover, there exists a non-zero vector subspace E_{des} of E such that $\mu(E_{\mathrm{des}}) = \mu_{\max}$ and which contains all non-zero vector subspaces F of E such that $\mu(F) = \mu_{\max}$.

Proof The first relation in the condition (2) implies that, if $L_1, \ldots, L_n$ are vector subspaces of dimension 1 of E, which are linearly independent, then one has

$$r(L_1 + \cdots + L_n) = r(L_1) + \cdots + r(L_n).$$

In particular, if L and L' are different vector subspaces of dimension 1 in E then one has $r(L) = r(L')$. In fact, if s and s' are non-zero vectors of L and L' respectively, and L'' is the vector subspace of E generated by $s + s'$ (which is of dimension 1), then one has

$$r(L) + r(L'') = r(L + L') = r(L') + r(L'').$$

Therefore the function $r(\cdot)$ is proportional to the dimension function. Without loss of generality, we may assume that the function $r(\cdot)$ identifies with the dimension function.

We prove the proposition by induction on the dimension of the vector space E. The case where the $r(E) = 1$ is trivial. In the following, we assume that $r(E) \geqslant 2$ and that the proposition has been proved for vector spaces of dimension $< r(E)$. If for any non-zero vector subspace F of E one has $\mu(F) \leqslant \mu(E)$, then there is nothing to prove since $\mu(E) = \mu_{\max}$ and $E = E_{\mathrm{des}}$. Otherwise there exists a non-zero vector subspace E' of E such that $\mu(E') > \mu(E)$. Moreover, we can choose E' such that $r(E')$ is maximal (among the non-zero vector subspaces of E having this property). Clearly one has $r(E') < r(E)$. Hence by the induction hypothesis the restriction of the function $\mu(\cdot)$ to $\Theta^*(E')$ attains its maximum, and among the non-zero vector subspaces of E' on which the restriction of the function $\mu(\cdot)$ to $\Theta(E')$ attains the maximal value there exists a greatest one E'_{des} with respect to the relation of inclusion. Let $E_{\mathrm{des}} := E'_{\mathrm{des}}$ be this vector space. We claim that E_{des} verifies the properties announced in the proposition.

Let F be a non-zero vector subspace of E. If $F \subseteq E'$, then clearly one has $\mu(F) \leqslant \mu(E_{\mathrm{des}})$. Otherwise the dimension of $F \cap E'$ is smaller than $r(F)$ and the dimension of $F + E'$ is greater than $r(E')$. Moreover, since $F \cap E' \subsetneq E'$, one has (here we use the condition that $d(\{0\}) \leqslant 0$ to treat the case where $F \cap E' = \{0\}$)

$$d(F \cap E') \leqslant \mu(E_{\mathrm{des}})r(F \cap E');$$

since $F + E' \supsetneq E'$, one has $\mu(F + E') \leqslant \mu(E) < \mu(E')$. Therefore

$$\begin{aligned} d(F \cap E') + d(F + E') &= \mu(F \cap E')r(F \cap E') + \mu(F + E')r(F + E') \\ &< \mu(E_{\mathrm{des}})r(F \cap E') + \mu(E')r(F + E'). \end{aligned}$$

Combining this relation with the inequality in the condition (2) of the proposition, we obtain

$$\mu(E_{\mathrm{des}})r(F \cap E') + \mu(E')r(F + E') > \mu(E')r(E') + \mu(F)r(F).$$

By the equality in the condition (2), we deduce

$$\mu(F)r(F) < \mu(E_{\mathrm{des}})r(F \cap E') + \mu(E')(r(F) - r(F \cap E')) \leqslant \mu(E_{\mathrm{des}})r(F).$$

Therefore, the function $\mu(\cdot)$ attains its maximal value $\mu_{\max}$ at E_{des}. Moreover, if F is a non-zero vector subspace of E such that $\mu(F) = \mu(E_{\mathrm{des}})$, then one should have $F \subseteq E'$, and hence $F \subseteq E_{\mathrm{des}}$ by the induction hypothesis. The proposition is thus proved. □

Definition 4.3.29 Let $\overline{E}$ be a non-zero Hermitian adelic vector bundle on S. We can apply the above proposition to the functions of dimension and of Arakelov degree to obtain the existence of a (unique) non-zero vector subspace E_{des} of E such that

$$\widehat{\mu}(\overline{E}_{\mathrm{des}}) = \widehat{\mu}_{\max}(\overline{E}) = \sup_{0 \neq F \in \Theta(E)} \widehat{\mu}(\overline{F})$$

and which contains all non-zero vector subspaces of E on which the function $\widehat{\mu}$ attains the maximal slope of $\overline{E}$. The vector subspace E_{des} is called the *destabilising vector subspace* of the Hermitian adelic vector bundle $\overline{E}$. If $\overline{E}_{\mathrm{des}} = \overline{E}$, we say that the Hermitian adelic vector bundle $\overline{E}$ is *semistable*. In particular, for any non-zero Hermitian adelic vector bundle $\overline{E}$ on S, the Hermitian adelic vector bundle $\overline{E}_{\mathrm{des}}$ is always semistable.

Proposition 4.3.30 *Let (E, ξ) be a non-zero adelic vector bundle on S. Then one has $\widehat{\mu}_{\max}(E, \xi) < +\infty$ and $\widehat{\mu}_{\min}(E, \xi) > -\infty$.*

Proof Let r be the dimension of E over K. We first prove the inequality $\widehat{\mu}_{\max}(E, \xi) < +\infty$ in the particular case where ξ is ultrametric on $\Omega \setminus \Omega_\infty$. By Theorem 4.1.26, there exists a measurable Hermitian norm family ξ^H on E such that

$$\forall\, \omega \in \Omega, \quad d_\omega(\xi, \xi^H) \leqslant \frac{1}{2}\ln(r + 1)\mathbb{1}_{\Omega_\infty}(\omega).$$

Therefore, for any non-zero vector subspace F of E one has

$$\left|\widehat{\mu}(F,\xi_F)-\widehat{\mu}(F,\xi_F^H)\right| \leqslant \frac{1}{2}\ln(r+1)\nu(\Omega_\infty).$$

Moreover, by Proposition 4.3.28, the maximal slope $\widehat{\mu}_{\max}(E,\xi^H)$ is finite. Therefore one has $\widehat{\mu}_{\max}(E,\xi) < +\infty$.

We now proceed with the proof of the relation $\widehat{\mu}_{\max}(E,\xi) < +\infty$ in the general case. We write ξ in the form $\{\|\cdot\|_\omega\}_{\omega\in\Omega}$. Note that for any $\omega \in \Omega$ one has $\|\cdot\|_{\omega,**} \leqslant \|\cdot\|_\omega$. Therefore $\widehat{\mu}_{\max}(E,\xi^{\vee\vee}) \geqslant \widehat{\mu}_{\max}(E,\xi)$. The norm family $\xi^{\vee\vee}$ is ultrametric on $\Omega \setminus \Omega_\infty$, and $(E,\xi^{\vee\vee})$ is an adelic vector bundle (see Proposition 4.1.32 (1)). By the particular case proved above, one has $\widehat{\mu}_{\max}(E,\xi^{\vee\vee}) < +\infty$. Thus we obtain $\widehat{\mu}_{\max}(E,\xi) < +\infty$.

Applying the above proved result to $(E^\vee,\xi^\vee)$ we obtain $\widehat{\mu}_{\max}(E^\vee,\xi^\vee) < +\infty$. Therefore, by Proposition 4.3.25 we obtain $\widehat{\mu}_{\min}(E,\xi) > -\infty$.

4.3.9 Some slope estimates

The following proposition is a natural generalisation of the slope inequalities to the setting of adelic curves. We refer the readers to [18, §4.1] for this theory in the classic setting of Hermitian vector bundles over an algebraic integer ring.

Proposition 4.3.31 *Let (E_1,ξ_1) and (E_2,ξ_2) be adelic vector bundles on S, and $f : E_1 \to E_2$ be a K-linear map.*

(1) If f is injective, then one has $\widehat{\mu}_{\max}(E_1,\xi_1) \leqslant \widehat{\mu}_{\max}(E_2,\xi_2) + h(f)$.
(2) If f is surjective, then one has $\widehat{\mu}_{\min}(E_1,\xi_1) \leqslant \widehat{\mu}_{\min}(E_2,\xi_2) + h(f)$.
(3) If f is non-zero, then one has $\widehat{\mu}_{\min}(E_1,\xi_1) \leqslant \widehat{\mu}_{\max}(E_2,\xi_2) + h(f)$.

Proof (1) The assertion is trivial if f is the zero map since in this case $E_1 = \{0\}$ and $\widehat{\mu}_{\max}(E_1,\xi_1) = -\infty$ by convention. In the following, we assume that the linear map f is non-zero. Let F_1 be a non-zero vector subspace of E_1 and F_2 be its image in E_2. Let $g : F_1 \to F_2$ be the restriction of f to F_1. It is an isomorphism of vector spaces. Moreover, if we equip F_1 and F_2 with induced norm families, by Proposition 4.3.18 one has

$$\widehat{\mu}(F_1,\xi_{F_1}) \leqslant \widehat{\mu}(F_2,\xi_{F_2}) + h(g) \leqslant \widehat{\mu}_{\max}(E_2,\xi_2) + h(f),$$

where ξ_{F_1} and ξ_{F_2} are restrictions of ξ_1 and ξ_2 to F_1 and F_2, respectively. Since F_1 is arbitrary, we obtain the inequality $\widehat{\mu}_{\max}(E_1,\xi_1) \leqslant \widehat{\mu}_{\max}(E_2,\xi_2) + h(f)$.

(2) The assertion is trivial if f is the zero map since in this case $E_2 = \{0\}$ and $\widehat{\mu}_{\min}(E_2,\xi_2) = +\infty$ by convention. In the following, we assume that the linear map f is non-zero. Let G_2 be a non-zero quotient vector space of E_2 and $\widetilde{f}$ be the composition of f with the quotient map $E_2 \to G_2$. Let F_1 be the kernel of $\widetilde{f}$, G_1 be the quotient space E_1/F_1, and $g : G_1 \to G_2$ be the K-linear map induced by $\widetilde{f}$. It is a K-linear isomorphism. By (1), one has

$$\widehat{\mu}_{\min}(E_1,\xi_1) \leqslant \widehat{\mu}_{\max}(G_1,\xi_{G_1}) \leqslant \widehat{\mu}_{\max}(G_2,\xi_{G_2}) + h(g) \leqslant \widehat{\mu}_{\max}(G_2,\xi_{G_2}) + h(f),$$

where ξ_{G_1} and ξ_{G_2} are the quotient norm family of ξ_1 and ξ_2, respectively. Since G_2 is arbitrary, one obtains $\widehat{\mu}_{\min}(E_1,\xi_1) \leqslant \widehat{\mu}_{\min}(E_2,\xi_2) + h(f)$.

(3) Let G be the image of E_1 by f, which is non-zero since f is non-zero. We equip G with the restriction ξ_G of ξ_2 to G. As G is non-zero, one has $\widehat{\mu}_{\min}(G,\xi_G) \leqslant \widehat{\mu}_{\max}(G,\xi_G) \leqslant \widehat{\mu}_{\max}(E_2,\xi_2)$. By (2), one has $\widehat{\mu}_{\min}(E_1,\xi_1) \leqslant \widehat{\mu}_{\min}(G,\xi_G) + h(f)$. Hence $\widehat{\mu}_{\min}(E_1,\xi_1) \leqslant \widehat{\mu}_{\max}(E_2,\xi_2) + h(f)$. □

Proposition 4.3.32 *Let (E',ξ') and (E,ξ) be adelic vector bundles on S, and $f : E' \to E$ be an injective K-linear map. Let E'' be the quotient vector space $E/f(E')$ and ξ'' be the quotient norm family of ξ on E''. Then the following inequality holds*

$$\widehat{\mu}_{\min}(E,\xi) \geqslant \min(\widehat{\mu}_{\min}(E',\xi') - h(f), \widehat{\mu}_{\min}(E'',\xi'')). \tag{4.51}$$

If in addition $\widehat{\mu}_{\min}(E',\xi') - h(f) \geqslant \widehat{\mu}_{\min}(E'',\xi'')$, then the equality $\widehat{\mu}_{\min}(E,\xi) = \widehat{\mu}_{\min}(E'',\xi'')$ holds.

Proof The inequality (4.51) is trivial if $E = \{0\}$ since in this case one has $\widehat{\mu}_{\min}(E,\xi) = +\infty$ by convention. Moreover, one has $E'' = \{0\}$ since E'' is a quotient vector space of E. Therefore the equality $\widehat{\mu}_{\min}(E,\xi) = \widehat{\mu}_{\min}(E'',\xi'')$ holds.

In the following, we assume that $E \neq \{0\}$. Let Q be a quotient vector space of E and ξ_Q be the quotient norm family of ξ. Let $\pi : E \to Q$ be the quotient map. If the composed map $\pi \circ f$ is non-zero, by Proposition 4.3.31 (3), one has

$$\widehat{\mu}_{\min}(E',\xi') \leqslant \widehat{\mu}_{\max}(Q,\xi_Q) + h(\pi \circ f) \leqslant \widehat{\mu}_{\max}(Q,\xi_Q) + h(f).$$

Otherwise the quotient map $\pi : E \to Q$ factorises through E'' and by Proposition 4.3.31 (2) one has

$$\widehat{\mu}_{\min}(E'',\xi'') \leqslant \widehat{\mu}_{\min}(Q,\xi_Q) \leqslant \widehat{\mu}_{\max}(Q,\xi_Q).$$

Since Q is arbitrary, the inequality (4.51) is true.

If $\widehat{\mu}_{\min}(E',\xi') - h(f) \geqslant \widehat{\mu}_{\min}(E'',\xi'')$, then (4.51) implies

$$\widehat{\mu}_{\min}(E,\xi) \geqslant \widehat{\mu}_{\min}(E'',\xi'').$$

Moreover, by Proposition 4.3.31 (2) one has

$$\widehat{\mu}_{\min}(E,\xi) \leqslant \widehat{\mu}_{\min}(E'',\xi'').$$

Hence the equality $\widehat{\mu}_{\min}(E,\xi) = \widehat{\mu}_{\min}(E'',\xi'')$ holds. □

Proposition 4.3.33 *Let $\{(E_i,\xi_{E_i})\}_{i=1}^n$ be a family of adelic vector bundles, where $n \in \mathbb{N}$, $n \geqslant 2$. Assume that*

$$E_0 := \{0\} \xrightarrow{\alpha_1} E_1 \xrightarrow{\alpha_2} E_2 \xrightarrow{\alpha_3} E_3 \longrightarrow \cdots \xrightarrow{\alpha_{n-1}} E_{n-1} \xrightarrow{\alpha_n} E_n \tag{4.52}$$

is a sequence of injective K-linear maps. For any $i \in \{1,\ldots,n\}$, let $\beta_i = \alpha_n \circ \cdots \circ \alpha_{i+1}$, where by convention $\beta_n := \mathrm{Id}_{E_n}$. For any $i \in \{1,\ldots,n\}$, let Q_i be the quotient space $E_i/\alpha_i(E_{i-1})$ and ξ_{Q_i} be the quotient norm family of ξ_{E_i}. Then one has

$$\widehat{\mu}_{\min}(E_n,\xi_{E_n}) \geqslant \min_{i\in\{1,\ldots,n\}} \big(\widehat{\mu}_{\min}(Q_i,\xi_{Q_i}) - h(\beta_i)\big)$$

Proof The case where $n=2$ was proved in Proposition 4.3.32. In the following, we assume that $n \geqslant 3$ and that the proposition has been proved for the case of $n-1$ adelic vector bundles. For any $i \in \{2,\ldots,n\}$, let E_i' be the cokernel of the composed linear map

$$\alpha_i \circ \cdots \circ \alpha_2 : E_1 \longrightarrow E_i$$

and let $\xi_{E_i'}$ be the quotient norm family of ξ_{E_i} on E_i'. Let $E_1' = \{0\}$. Then the sequence (4.52) induces a sequence of K-linear maps

$$E_1' := \{0\} \xrightarrow{\alpha_2'} E_2' \xrightarrow{\alpha_3'} E_3' \longrightarrow \cdots \xrightarrow{\alpha_{n-1}'} E_{n-1}' \xrightarrow{\alpha_n'} E_n' . \tag{4.53}$$

For any $i \in \{2,\ldots,n\}$, let $\beta_i' = \alpha_n' \circ \cdots \circ \alpha_{i+1}'$, where by convention $\beta_n' = \mathrm{Id}_{E_n'}$. For any $i \in \{2,\ldots,n\}$, let Q_i' be the quotient space $E_i'/\alpha_i'(E_{i-1}')$ and $\xi_{Q_i'}$ be the quotient norm family of $\xi_{E_i'}$. Note that Q_i' is canonically isomorphic to Q_i, and under the canonical isomorphism $Q_i \cong Q_i'$, the norm family ξ_{Q_i} identifies with $\xi_{Q_i'}$ (see Proposition 1.1.14). Therefore one has $\widehat{\mu}_{\min}(Q_i,\xi_{Q_i}) = \widehat{\mu}_{\min}(Q_i',\xi_{Q_i'})$ for any $i \in \{2,\ldots,n\}$. Applying the induction hypothesis to (4.53) we obtain

$$\begin{aligned}\widehat{\mu}_{\min}(E_n',\xi_{E_n'}) &\geqslant \min_{i\in\{2,\ldots,n\}} \big(\widehat{\mu}_{\min}(Q_i,\xi_{Q_i}) - h(\beta_i')\big)\\ &\geqslant \min_{i\in\{2,\ldots,n\}} \big(\widehat{\mu}_{\min}(Q_i,\xi_{Q_i}) - h(\beta_i)\big),\end{aligned}$$

where the second inequality comes from Proposition 1.1.14. Finally, by Proposition 4.3.32 one has

$$\widehat{\mu}_{\min}(E_n,\xi_{E_n}) \geqslant \min\big\{\widehat{\mu}_{\min}(E_n',\xi_{E_n'}), \widehat{\mu}_{\min}(E_1,\xi_{E_1}) - h(\beta_1)\big\}.$$

The proposition is thus proved. □

Proposition 4.3.34 *Let $\overline{E}$ and $\overline{F}$ be adelic vector bundles on S. One has*

$$\widehat{\mu}(\overline{E} \otimes_{\varepsilon,\pi} \overline{F}) = \widehat{\mu}(\overline{E}) + \widehat{\mu}(\overline{F}).$$

If $\overline{E}$ and $\overline{F}$ are both Hermitian, then

$$\widehat{\mu}(\overline{E} \otimes \overline{F}) = \widehat{\mu}(\overline{E}) + \widehat{\mu}(\overline{F}).$$

Proof These equalities are direct consequences of Proposition 4.3.19. □

Proposition 4.3.35 *Let $\overline{E} = (E,\xi_E)$ and $\overline{F} = (F,\xi_F)$ be adelic vector bundles on S. One has*

$$\widehat{\mu}_{\max}(\overline{E} \otimes_{\varepsilon,\pi} \overline{F}) \geqslant \widehat{\mu}_{\max}(\overline{E}) + \widehat{\mu}_{\max}(\overline{F}). \tag{4.54}$$

If $\overline{E}$ and $\overline{F}$ are Hermitian adelic vector bundles, then

$$\widehat{\mu}_{\max}(\overline{E} \otimes \overline{F}) \geqslant \widehat{\mu}_{\max}(\overline{E}) + \widehat{\mu}_{\max}(\overline{F}). \tag{4.55}$$

Proof Let E_1 and F_1 be vector subspaces of E and F respectively. Let ξ_{E_1} and ξ_{F_1} be the restrictions of ξ_E and ξ_F to E_1 and F_1 respectively. By Proposition 4.3.34, one has

$$\widehat{\mu}(\overline{E}_1 \otimes_{\varepsilon,\pi} \overline{F}_1) = \widehat{\mu}(\overline{E}_1) + \widehat{\mu}(\overline{F}_1).$$

If ξ_E and ξ_F are both Hermitian, then

$$\widehat{\mu}(\overline{E}_1 \otimes \overline{F}_1) = \widehat{\mu}(\overline{E}_1) + \widehat{\mu}(\overline{F}_1).$$

By Proposition 1.1.60, if we denote by ξ the restriction of $\xi_E \otimes_{\varepsilon,\pi} \xi_F$ to $E_1 \otimes F_1$, then the identity map from $\overline{E}_1 \otimes_{\varepsilon,\pi} \overline{F}_1$ to $(E_1 \otimes_k F_1, \xi)$ has height $\leqslant 0$ and therefore

$$\widehat{\mu}(\overline{E}_1) + \widehat{\mu}(\overline{F}_1) = \widehat{\mu}(\overline{E}_1 \otimes_{\varepsilon,\pi} \overline{F}_1) \leqslant \widehat{\mu}(E_1 \otimes_k F_1, \xi) \leqslant \widehat{\mu}_{\max}(\overline{E} \otimes_{\varepsilon,\pi} \overline{F}).$$

Similarly, if both norm families ξ_E and ξ_F are Hermitian, then by Proposition 1.2.58 the restriction of $\xi_E \otimes \xi_F$ to $E_1 \otimes_K F_1$ identifies with $\xi_{E_1} \otimes \xi_{F_1}$. Hence

$$\widehat{\mu}(\overline{E}_1) + \widehat{\mu}(\overline{F}_1) = \widehat{\mu}(\overline{E}_1 \otimes \overline{F}_1) \leqslant \widehat{\mu}_{\max}(\overline{E} \otimes \overline{F}).$$

Since E_1 and F_1 are arbitrary, we obtain the inequalities (4.54) and (4.55). □

Lemma 4.3.36 *Let (E, ξ) be an adelic vector bundle over S. Then we have the following:*

(1) *Let ψ be an integrable function on Ω. Then*

$$\widehat{\mu}_{\max}(E, e^{\psi}\xi) = \widehat{\mu}_{\max}(E, \xi) - \int_{\Omega} \psi\, \nu(d\omega).$$

(2) *If $\widehat{\mu}_{\max}(E, \xi) \leqslant 0$, then $\widehat{\deg}_+(E, \xi) = 0$.*

Proof (1) Let F be a non-zero vector subspace of E. Then, as

$$\widehat{\mu}(F, e^{\psi}\xi_F) = \widehat{\mu}(F, \xi_F) - \int_{\Omega} \psi\, \nu(d\omega),$$

we obtain (1).

(2) Let F be a non-zero vector subspace of E. By our assumption, $\widehat{\mu}(F, \xi_F) \leqslant 0$, that is, $\widehat{\deg}(F, \xi_F) \leqslant 0$, so that the assertion follows. □

4.3.10 Harder-Narasimhan filtration: Hermitian case

It had been discovered by Stuhler [138] (generalised by Grayson [71]) that the Euclidean lattices and vector bundles on projective algebraic curves share some common constructions and properties such as slopes and Harder-Narasimhan filtration etc. Later Bost has developed the slope theory of Hermitian vector bundles over spectra of algebraic integer rings, see [16, Appendice] (see also [137] and [18, §4.1] for more details, and [62, 22, 65] for further generalisations).

The Hermitian adelic vector bundles on S form a category in which a theory of Hader-Narasimhan filtration can be developed in a functorial way. We refer the readers to [2,41] for more details. In this subsection, we adopt a more direct approach as in [22].

Let $\overline{E}$ be a non-zero Hermitian adelic vector bundle on S. We can construct in a recursive way a flag

$$0 = E_0 \subsetneq E_1 \subsetneq \ldots \subsetneq E_n = E$$

of vector subspaces of E such that $\overline{E_i/E_{i-1}} = \overline{(E/E_{i-1})}_{\mathrm{des}}$, called the *Harder-Narasimhan flag* of $\overline{E}$, where E/E_{i-1} is equipped with the quotient norm family, and E_i/E_{i-1} is equipped with the subquotient norm family (namely the restriction of the norm family of $\overline{E/E_{i-1}}$ to E_i/E_{i-1}).

Proposition 4.3.37 *Let $\overline{E}$ be a non-zero Hermitian adelic vector bundle on S and*

$$0 = E_0 \subsetneq E_1 \subsetneq \ldots \subsetneq E_n = E$$

be the Harder-Narasimhan flag of $\overline{E}$. Then each subquotient $\overline{E_i/E_{i-1}}$ is a semistable Hermitian adelic vector bundle. Moreover, if we let $\mu_i = \widehat{\mu}(\overline{E_i/E_{i-1}})$ for $i \in \{1, \ldots, n\}$, then one has $\mu_1 > \ldots > \mu_n$.

Proof We reason by induction on the length n of the Harder-Narasimhan flag. When $n = 1$, the assertion is trivial. In the following, we suppose that $n \geqslant 2$. By definition

$$0 = E_1/E_1 \subsetneq E_2/E_1 \subsetneq \ldots \subsetneq E_n/E_1$$

is the Harder-Narasimhan flag of $\overline{E/E_1}$. Therefore the induction hypothesis leads to $\mu_2 > \ldots > \mu_n$. It remains to establish $\mu_1 > \mu_2$. Since E_1 is the destabilising vector subspace of E and E_2 contains strictly E_1, one has

$$\mu_1 = \widehat{\mu}(\overline{E}_1) > \widehat{\mu}(\overline{E}_2). \tag{4.56}$$

Moreover,

$$0 \longrightarrow \overline{E}_1 \longrightarrow \overline{E}_2 \longrightarrow \overline{E_2/E_1} \longrightarrow 0$$

forms an exact sequence of adelic vector bundles on S. Therefore one has

$$\widehat{\deg}(\overline{E}_2) = \widehat{\deg}(\overline{E}_1) + \widehat{\deg}(\overline{E_2/E_1}) = \mu_1 \dim_K(E_1) + \mu_2 \dim_K(E_2/E_1).$$

By (4.56) we obtain

$$\mu_1 \dim_K(E_1) + \mu_2 \dim_K(E_2/E_1) < \mu_1 \dim(E_2)$$

and hence $\mu_1 > \mu_2$. The proposition is thus proved. □

The Harder-Narasimhan flag and the slopes of the successive subquotients in the previous proposition permit to construct an $\mathbb{R}$-filtration $\mathcal{F}_{\mathrm{hn}}$ on the vector space E, called the *Harder-Narasimhan* $\mathbb{R}$*-filtration* as follows:

$$\forall\, t \in \mathbb{R}, \quad \mathcal{F}^t_{\mathrm{hn}}(\overline{E}) := E_i \text{ if } \mu_{i+1} < t \leqslant \mu_i, \tag{4.57}$$

where by convention $\mu_0 = +\infty$ and $\mu_{n+1} = -\infty$. If $\overline{E}$ is the zero Hermitian adelic vector bundle, by convention its Harder-Narasimhan $\mathbb{R}$-filtration is defined as the only $\mathbb{R}$-filtration of the zero vector space: for any $t \in \mathbb{R}$ one has $\mathcal{F}^t(\overline{E}) = \{0\}$. Note that the Harder-Narasimhan $\mathbb{R}$-filtration is locally constant on the left, namely $\mathcal{F}^{t-\varepsilon}_{\mathrm{hn}}(\overline{E}) = \mathcal{F}^t_{\mathrm{hn}}(\overline{E})$ if $\varepsilon > 0$ is sufficiently small. Moreover, each subquotient

$$\mathrm{Sq}^t_{\mathrm{hn}}(\overline{E}) := \mathcal{F}^t_{\mathrm{hn}}(\overline{E})/\mathcal{F}^{t+}_{\mathrm{hn}}(\overline{E})$$

with $\mathcal{F}^{t+}_{\mathrm{hn}}(\overline{E}) := \bigcup_{\varepsilon>0} \mathcal{F}^{t+\varepsilon}_{\mathrm{hn}}(\overline{E})$, viewed as a Hermitian adelic vector bundle in considering the *subqutient norm family*, namely the quotient of the restricted norm family on $\mathcal{F}^t_{\mathrm{hn}}(\overline{E})$, is either zero or a semistable Hermitian adelic vector bundle of slope t. The following proposition shows that the Harder-Narasimhan $\mathbb{R}$-filtration is actually characterised by these properties.

Proposition 4.3.38 *Let $\overline{E}$ be a non-zero Hermitian adelic vector bundle on S and $\mathcal{F}$ be a decreasing $\mathbb{R}$-filtration of E which is separated, exhaustive*[1] *and locally constant on the left. Assume that each subquotient $\mathrm{Sq}^t(\overline{E}) := \mathcal{F}^t(E)/\mathcal{F}^{t+}(E)$ equipped with the subquotient norm family, is either zero or a semistable Hermitian adelic vector bundle of slope t. Then the $\mathbb{R}$-filtration $\mathcal{F}$ coincides with the Harder-Narasimhan $\mathbb{R}$-filtration of $\overline{E}$.*

Proof We will prove an alternative statement as follows. Let

$$0 = F_0 \subsetneq F_1 \subsetneq \ldots \subsetneq F_m \subsetneq E \tag{4.58}$$

be a flag of vector subspaces of E. We will prove that, if each subquotient $\overline{F_i/F_{i-1}}$ ($i \in \{1, \ldots, m\}$) is a semistable Hermitian adelic vector bundle and if the relations

$$\widehat{\mu}(\overline{F_1/F_0}) > \ldots > \widehat{\mu}(\overline{F_m/F_{m-1}})$$

hold, then (4.58) identifies with the Harder-Narasimhan flag of $\overline{E}$. This alternative statement is actually equivalent to the form announced in the proposition. In fact, the data of an $\mathbb{R}$-filtration of E is equivalent to that of a flag (of vector subspaces

[1] Let E be a vector space over K and $\{\mathcal{F}^t(E)\}_{t\in\mathbb{R}}$ be a decreasing $\mathbb{R}$-filtration of E. We say that the filtration $\mathcal{F}$ is *separated* if $\mathcal{F}^t(E) = \{0\}$ for sufficiently positive t. We say that the filtration $\mathcal{F}$ is *exhaustive* if $\mathcal{F}^t(E) = E$ for sufficiently negative t.

figuring in the $\mathbb{R}$-filtration) and a decreasing sequenc of real numbers indicating the indices where the $\mathbb{R}$-filtration has jumps, see §1.1.8, notably Remark 1.1.40.

We will prove the statement by induction on the dimension of E over K. The case where $\dim_K(E) = 1$ is trivial. In the following, we assume that $\dim_K(E) \geqslant 2$ and that the alternative assertion has been proved for any Hermitian adelic vector bundle of rank $< \dim_K(E)$. Let

$$0 = E_0 \subsetneq E_1 \subsetneq \ldots \subsetneq E_n = E$$

be the Harder-Narasimhan flag of $\overline{E}$. We begin by showing that $F_1 = E_1$. Since E_1 is the destabilising vector subspace of $\overline{E}$, one has $\widehat{\mu}(\overline{F}_1) \leqslant \widehat{\mu}(\overline{E}_1)$. Moreover one has

$$0 = F_0 \cap E_1 \subseteq F_1 \cap E_1 \subseteq F_2 \cap E_1 \subseteq \ldots F_m \cap E_1 = E_1.$$

Note that each subquotient $(F_i \cap E_1)/(F_{i-1} \cap E_1)$ identifies with a vector subspace of F_i/F_{i-1}. Since $\overline{F_i/F_{i-1}}$ is semistable, one has

$$\begin{aligned}\widehat{\deg}\big(\overline{(F_i \cap E_1)/(F_{i-1} \cap E_1)}\big) &\leqslant \widehat{\mu}(\overline{F_i/F_{i-1}}) \dim_K \big((F_i \cap E_1)/(F_{i-1} \cap E_1)\big) \\ &\leqslant \widehat{\mu}(\overline{F}_1) \dim_K \big((F_i \cap E_1)/(F_{i-1} \cap E_1)\big),\end{aligned}$$

where the second inequality is strict if $i > 1$ and if $(F_i \cap E_1)/(F_{i-1} \cap E_1)$ is non-zero. Therefore we obtain

$$\widehat{\deg}(\overline{E}_1) = \sum_{i=1}^{m} \widehat{\deg}\big(\overline{(F_i \cap E_1)/(F_{i-1} \cap E_1)}\big) \leqslant \widehat{\mu}(\overline{F}_1) \dim_K(E_1). \tag{4.59}$$

Combining with the inequality $\widehat{\mu}(\overline{F}_1) \leqslant \widehat{\mu}(\overline{E}_1) = \widehat{\mu}_{\max}(\overline{E})$, we deduce that the inequality (4.59) is actually an equality, which also implies that $(F_i \cap E_1)/(F_{i-1} \cap E_1) = \{0\}$ once $i > 1$. Therefore one has $F_1 = E_1$, which also leads to the alternative assertion in the particular case where $\overline{E}$ is semistable.

In the case where $\overline{E}$ is not semistable, namely $n \geqslant 2$, note that

$$0 = E_1/E_1 \subsetneq E_2/E_1 \subsetneq \ldots \subsetneq E_n/E_1 = E/E_1 \tag{4.60}$$

is the Harder-Narasimhan flag of $\overline{E/E_1} = \overline{E/F_1}$. By the induction hypothesis applied to $\overline{E/F_1}$, we obtain that the flag

$$0 = F_1/F_1 \subsetneq F_2/F_1 \subsetneq \ldots \subsetneq F_m/F_1 = E/F_1$$

coincides with (4.60). The proposition is thus proved. □

Definition 4.3.39 Let $\overline{E}$ be a non-zero Hermitian adelic vector bundle on S, and

$$0 = E_0 \subsetneq E_1 \subsetneq \ldots \subsetneq E_n = E$$

be its Harder-Narasimhan flag. For any $i \in \{1, \ldots, \dim_K(E)\}$, there exists a unique $j \in \{1, \ldots, n\}$ such that $\dim_K(E_{j-1}) < i \leqslant \dim_K(E_j)$. We let $\widehat{\mu}_i(\overline{E})$ be the slope

$\widehat{\mu}(\overline{E_j/E_{j-1}})$, called the *i-th slope* of $\overline{E}$. Clearly one has

$$\widehat{\mu}_1(\overline{E}) \geqslant \ldots \geqslant \widehat{\mu}_r(\overline{E}),$$

where r is the dimension of E over K. Moreover, by definition $\widehat{\mu}_1(\overline{E})$ coincides with the maximal slope of $\overline{E}$.

Remark 4.3.40 As in the classic case of vector bundles on projective curves or Hermitian vector bundles over algebraic integer rings, one can naturally construct Harder-Narasimhan polygones associated with Hermitian adelic vector bundles on adelic curves. Let $\overline{E}$ be a non-zero Hermitian adelic vector bundle on the adelic curve S. We consider the convex hull $C_{\overline{E}}$ in $\mathbb{R}^2$ of the points $(\dim_K(F), \widehat{\deg}(\overline{F}))$, where F runs over the set of all vector subspaces of E. The upper boundary of this convex set identifies with the graph of a concave function $P_{\overline{E}}$ on $[0, \dim_K(E)]$ which is affine on each interval $[i-1, i]$ with $i \in \{1, \ldots, \dim_K(E)\}$. This function is called the *Harder-Narasimhan polygon* of $\overline{E}$. If

$$0 = E_0 \subsetneq E_1 \subsetneq \ldots \subsetneq E_n = E$$

is the Harder-Narasimhan flag of $\overline{E}$, then the abscissae on which the Harder-Narasimhan polygon $P_{\overline{E}}$ changes slopes are exactly $\dim_K(E_i)$ for $i \in \{0, \ldots, n\}$. Moreover, the value of $P_{\overline{E}}$ on $\dim_K(E_i)$ is $\widehat{\deg}(\overline{E}_i)$.

Proposition 4.3.41 *Let $\overline{E}$ be a non-zero Hermitian adelic vector bundle on S and r be the dimension of E over K. One has*

$$\widehat{\mu}_r(\overline{E}) = -\widehat{\mu}_1(\overline{E}^\vee). \tag{4.61}$$

In particular, $\widehat{\mu}_r(\overline{E})$ is equal to $\widehat{\mu}_{\min}(\overline{E})$.

Proof Let

$$0 = E_0 \subsetneq E_1 \subsetneq \ldots \subsetneq E_n = E$$

be the Harder-Narasimhan flag of $\overline{E}$. Note that

$$0 = (E/E_n)^\vee \subsetneq (E/E_{n-1})^\vee \subsetneq \ldots \subsetneq (E/E_0)^\vee = E^\vee \tag{4.62}$$

is a flag of vector subspaces of $E^\vee$, and for $i \in \{1, \ldots, n\}$ one has

$$(E/E_{i-1})^\vee/(E/E_i)^\vee \cong (E_i/E_{i-1}).$$

By Proposition 4.3.38 (notably the alternative form stated in the proof), we obtain that (4.62) is actually the Harder-Narasimhan flag of $\overline{E}^\vee$. Therefore

$$\widehat{\mu}_1(\overline{E}^\vee) = \widehat{\mu}(\overline{(E/E_{n-1})^\vee}) = -\widehat{\mu}(\overline{E/E_{n-1}}) = -\widehat{\mu}_r(\overline{E}),$$

where the second equality comes from Proposition 4.3.10.

Note that $\overline{E/E_{n-1}}$ is a non-zero quotient Hermitian adelic bundle of $\overline{E}$ which is semistable (so that $\widehat{\mu}(\overline{E/E_{n-1}}) = \widehat{\mu}_{\max}(\overline{E/E_{n-1}})$). Therefore one has $\widehat{\mu}_{\min}(\overline{E}) \leqslant$

$\widehat{\mu}_r(\overline{E})$. Conversely, if F is a vector subspace of E such that $F \subsetneq E$ and G is the quotient space E/F. Then $G^\vee$ identifies with a non-zero vector subspace of $E^\vee$. Moreover, by Proposition 1.1.20 the dual of the quotient norm family of $\overline{G}$ identifies with the restriction of the dual norm family in the adelic vector bundle structure of $\overline{E}^\vee$. Hence one has

$$\widehat{\mu}(\overline{G}^\vee) \leqslant \widehat{\mu}_1(\overline{E}^\vee) = -\widehat{\mu}_r(\overline{E}).$$

Still by Proposition 4.3.10, one obtains

$$\widehat{\mu}_{\max}(\overline{G}) \geqslant \widehat{\mu}(\overline{G}) \geqslant \widehat{\mu}_r(\overline{E}).$$

The equality $\widehat{\mu}_{\min}(\overline{E}) = \widehat{\mu}_r(\overline{E})$ is thus proved. □

The following proposition, which results from the slope inequalities, proves the functoriality of Harder-Narasimhan $\mathbb{R}$-filtration (see [41] for the meaning of the functoriality of Harder-Narasimhan $\mathbb{R}$-filtration).

Proposition 4.3.42 *Let $\overline{E}$ and $\overline{F}$ be two Hermitian adelic vector bundles on S, and $f : E \to F$ be a non-zero K-linear map. For any $t \in \mathbb{R}$ one has*

$$f(\mathcal{F}^t_{\mathrm{hn}}(\overline{E})) \subseteq \mathcal{F}^{t-h(f)}_{\mathrm{hn}}(\overline{F}).$$

Proof We will actually show by contradiction that the composition of maps

$$\mathcal{F}^t_{\mathrm{hn}}(\overline{E}) \xrightarrow{\ f\ } F \longrightarrow F/\mathcal{F}^{t-h(f)}_{\mathrm{hn}}(\overline{F})$$

is zero. If this map is not zero, then by Proposition 4.3.31 (3) one obtains

$$\widehat{\mu}_{\min}(\mathcal{F}^t_{\mathrm{hn}}(\overline{E})) \leqslant \widehat{\mu}_{\max}(\overline{F}/\mathcal{F}^{t-h(f)}_{\mathrm{hn}}(\overline{F})) + h(f).$$

By (4.57) we obtain

$$t \leqslant \widehat{\mu}_{\min}(\mathcal{F}^t_{\mathrm{hn}}(\overline{E})) \leqslant \widehat{\mu}_{\max}(\overline{F}/\mathcal{F}^{t-h(f)}_{\mathrm{hn}}(\overline{F})) + h(f) < t - h(f) + h(f) = t,$$

which leads to a contradiction. □

Corollary 4.3.43 *Let $\overline{E}$ be a non-zero Hermitian adelic vector bundle on S. One has*

$$\mathcal{F}^t_{\mathrm{hn}}(\overline{E}) = \sum_{\substack{0\neq F\in\Theta(E)\\ \widehat{\mu}_{\min}(\overline{F})\geqslant t}} F,$$

where F runs over the set $\Theta(E)$ of all non-zero vector subspaces of E with minimal slope $\geqslant t$. In other words, $\mathcal{F}^t_{\mathrm{hn}}(\overline{E})$ is the largest vector subspace of E whose minimal slope is bounded from below by t.

Proof By the definition of the Harder-Narasimhan $\mathbb{R}$-filtration (see (4.57)), for any $t \in \mathbb{R}$ one has $\widehat{\mu}_{\min}(\mathcal{F}^t_{\mathrm{hn}}(\overline{E})) \geqslant t$. Moreover, if F is a non-zero vector subspace of

E, then one has $\mathcal{F}_{\mathrm{hn}}^t(\overline{F}) = F$ provided that $t \leqslant \widehat{\mu}_{\min}(\overline{F})$. Therefore the proposition 4.3.42 applied to the inclusion map $F \longrightarrow E$ leads to $F \subseteq \mathcal{F}_{\mathrm{hn}}^t(\overline{E})$. □

Proposition 4.3.44 *Let $\overline{E}$ be a non-zero Hermitian adelic vector bundle on S and r be the dimension of E over K. The following equalities hold:*

$$\widehat{\deg}(\overline{E}) = \sum_{i=1}^{r} \widehat{\mu}_i(\overline{E}) = -\int_{\mathbb{R}} t\,\mathrm{d}\big(\dim_K(\mathcal{F}_{\mathrm{hn}}^t(\overline{E}))\big), \tag{4.63}$$

$$\widehat{\deg}_+(\overline{E}) = \sum_{i=1}^{r} \max\{\widehat{\mu}_i(\overline{E}), 0\} = \int_0^{+\infty} \dim_K(\mathcal{F}_{\mathrm{hn}}^t(\overline{E}))\,\mathrm{d}t. \tag{4.64}$$

Proof By definition the sum of Dirac measures

$$\sum_{i=1}^{r} \delta_{\widehat{\mu}_i(\overline{E})}$$

identifies with the derivative $-\mathrm{d}\big(\dim_K(\mathcal{F}_{\mathrm{hn}}^t(\overline{E}))\big)$ in the sense of distribution. Therefore, the second equality in (4.63) is true, and the second equality in (4.64) follows from the relation

$$\sum_{i=1}^{r} \max\{\widehat{\mu}_i(\overline{E}), 0\} = -\int_0^{+\infty} t\,\mathrm{d}\big(\dim_K(\mathcal{F}_{\mathrm{hn}}^t(\overline{E}))\big)$$

and the formula of integration by part.

Let

$$0 = E_0 \subsetneq E_1 \subsetneq \ldots \subsetneq E_n = E$$

be the Harder-Narasimhan flag of $\overline{E}$. By Proposition 4.3.13 one has

$$\widehat{\deg}(\overline{E}) = \sum_{j=1}^{n} \widehat{\deg}(\overline{E_j/E_{j-1}}) = \sum_{j=1}^{n} \widehat{\mu}(\overline{E_j/E_{j-1}}) \dim_K(E_j/E_{j-1}) = \sum_{i=1}^{r} \widehat{\mu}_i(\overline{E}),$$

which proves (4.63).

Let ℓ be the largest element of $\{1, \ldots, n\}$ such that $\widehat{\mu}(\overline{E}_\ell/\overline{E}_{\ell-1}) \geqslant 0$. If $\widehat{\mu}(\overline{E_j/E_{j-1}}) < 0$ for any $j \in \{1, \ldots, n\}$, by convention we let $\ell = 0$. Then by (4.63) one has

$$\widehat{\deg}(\overline{E}_\ell) = \sum_{i=1}^{r} \max\{\widehat{\mu}_i(\overline{E}), 0\}.$$

Hence we obtain

$$\sum_{i=1}^{r} \max\{\widehat{\mu}_i(\overline{E}), 0\} \leqslant \widehat{\deg}_+(\overline{E}).$$

Conversely, if F is a non-zero vector subspace of E and m is its dimension over K, by Proposition 4.3.42 one has $\widehat{\mu}_i(\overline{F}) \leqslant \widehat{\mu}_i(\overline{E})$ for any $i \in \{1, \ldots, m\}$. Therefore by

(4.63) one obtains

$$\widehat{\deg}(\overline{F}) = \sum_{i=1}^{m} \widehat{\mu}_i(\overline{F}) \leqslant \sum_{i=1}^{m} \widehat{\mu}_i(\overline{E}) \leqslant \sum_{i=1}^{r} \max\{\widehat{\mu}_i(\overline{E}), 0\}.$$

4.3.11 Harder-Narasimhan filtration: general case

Inspired by Corollary 4.3.43, we extend the definition of Harder-Narasimhan $\mathbb{R}$-filtration to the setting of general adelic vector bundles.

Definition 4.3.45 Let $\overline{E}$ be a non-zero adelic vector bundle on S. For any $t \in \mathbb{R}$, let

$$\mathcal{F}^t_{\mathrm{hn}}(\overline{E}) := \bigcap_{\epsilon>0} \sum_{\substack{\{0\}\neq F\in\Theta(E) \\ \widehat{\mu}_{\min}(\overline{F})\geqslant t-\epsilon}} F, \tag{4.65}$$

where $\Theta(E)$ denotes the set of vector subspaces of E. By the finiteness of maximal and minimal slopes proved in Proposition 4.3.30, we obtain that $\mathcal{F}^t_{\mathrm{hn}}(\overline{E}) = E$ when t is sufficiently negative, and $\mathcal{F}^t_{\mathrm{hn}}(\overline{E}) = \{0\}$ when t is sufficiently positive. By convention we let $\mathcal{F}^{+\infty}_{\mathrm{hn}}(\overline{E}) = \{0\}$ and $\mathcal{F}^{-\infty}_{\mathrm{hn}} = E$.

Proposition 4.3.46 *Let $\overline{E}$ be a non-zero adelic vector bundle on S. For any $t \in \mathbb{R}$, the vector space $\mathcal{F}^t_{\mathrm{hn}}(\overline{E})$ equipped with the induced norm family has a minimal slope $\geqslant t$. In particular, one has*

$$\forall\, t \in \mathbb{R}, \quad \mathcal{F}^t_{\mathrm{hn}}(\overline{E}) = \sum_{\substack{\{0\}\neq F\in\Theta(E) \\ \widehat{\mu}_{\min}(\overline{F})\geqslant t}} F$$

and

$$\widehat{\mu}_{\min}(\overline{E}) = \max\{t \in \mathbb{R} \,:\, \mathcal{F}^t_{\mathrm{hn}}(\overline{E}) = E\}.$$

Proof Let $t \in \mathbb{R}$. For sufficiently small $\varepsilon > 0$, one has

$$\mathcal{F}^t_{\mathrm{hn}}(\overline{E}) = \sum_{\substack{\{0\}\neq F\in\Theta(E) \\ \widehat{\mu}_{\min}(\overline{F})\geqslant t-\epsilon}} F$$

Let M be a non-zero quotient vector space of $\mathcal{F}^t_{\mathrm{hn}}(\overline{E})$. By definition, for any $\epsilon > 0$ there exists a vector subspace F_ϵ of E such that $\widehat{\mu}_{\min}(\overline{F_\epsilon}) \geqslant t-\epsilon$ and that the composed map $F_\epsilon \to \mathcal{F}^t_{\mathrm{hn}}(\overline{E}) \to M$ is non-zero. By the slope inequality (see Proposition 4.3.31) we have $t - \epsilon \leqslant \widehat{\mu}_{\min}(\overline{F_\epsilon}) \leqslant \widehat{\mu}_{\max}(\overline{M})$, which leads to $\widehat{\mu}_{\max}(\overline{M}) \geqslant t$ since $\epsilon > 0$ is arbitrary. As M is arbitrary, we obtain the first statement.

By the first statement of the proposition, for any $t \in \mathbb{R}$ such that $\mathcal{F}^t_{\mathrm{hn}}(\overline{E}) = E$, one has $\widehat{\mu}_{\min}(\overline{E}) \geqslant t$. Conversely, by definition, if t is a real number such that $\widehat{\mu}_{\min}(\overline{E}) \geqslant t$, then $E \subseteq \mathcal{F}^t_{\mathrm{hn}}(\overline{E})$ and hence $E = \mathcal{F}^t_{\mathrm{hn}}(\overline{E})$. Therefore, the equality $\widehat{\mu}_{\min}(\overline{E}) = \max\{t \in \mathbb{R} : \mathcal{F}^t_{\mathrm{hn}}(\overline{E}) = E\}$ holds. □

Definition 4.3.47 Let $\overline{E}$ be a non-zero adelic vector bundle on S. For any $i \in \{1, \ldots, \dim_K(E)\}$, we let

$$\widehat{\mu}_i(\overline{E}) := \sup\{t \in \mathbb{R} : \dim_K(\mathcal{F}^t_{\mathrm{hn}}(\overline{E})) \geqslant i\}.$$

The number $\widehat{\mu}_i(\overline{E})$ is called the *i-th slope* of $\overline{E}$. Proposition 4.3.46 shows that the last slope of $\overline{E}$ identifies with the minimal slope $\widehat{\mu}_{\min}(\overline{E})$ of $\overline{E}$.

Remark 4.3.48 Let $\overline{E}$ be a non-zero adelic vector bundle. In general the first slope $\widehat{\mu}_1(\overline{E})$ does not coincide with $\widehat{\mu}_{\max}(\overline{E})$ and we only have an inequality $\widehat{\mu}_1(\overline{E}) \leqslant \widehat{\mu}_{\max}(\overline{E})$. Moreover, if the norm family of $\overline{E}$ is ultrametric on $\Omega \setminus \Omega_\infty$, then one has $\widehat{\mu}_{\max}(\overline{E}) \leqslant \widehat{\mu}_1(\overline{E}) + \frac{1}{2}\ln(\mathrm{rk}_K(E))\nu(\Omega_\infty)$. This follows from (4.69) and (4.25).

With the extended definition, the statement of Proposition 4.3.42 still holds for general adelic vector bundles.

Proposition 4.3.49 *Let $\overline{E}$ and $\overline{F}$ be adelic vector bundles on S, and $f : E \to F$ be a non-zero K-linear map. For any $t \in \mathbb{R}$ one has*

$$f(\mathcal{F}^t_{\mathrm{hn}}(\overline{E})) \subseteq \mathcal{F}^{t-h(f)}_{\mathrm{hn}}(\overline{F}).$$

Proof Let M be a non-zero vector subspace of E such that $\widehat{\mu}_{\min}(\overline{M}) \geqslant t$. By Proposition 4.3.31 (2), one has

$$\widehat{\mu}_{\min}(\overline{M}) \leqslant \widehat{\mu}_{\min}(\overline{f(M)}) + h(f|_M) \leqslant \widehat{\mu}_{\min}(\overline{f(M)}) + h(f).$$

Therefore $f(M) \subseteq \mathcal{F}^{t-h(f)}_{\mathrm{hn}}(F)$. □

Proposition 4.3.50 *Let $\overline{E}$ be a non-zero adelic vector bundle on S. Let r be the dimension of E over K. Then the following inequalities hold:*

$$\widehat{\deg}(\overline{E}) \geqslant \sum_{i=1}^r \widehat{\mu}_i(\overline{E}) = -\int_{\mathbb{R}} t \,\mathrm{d}\big(\dim_K(\mathcal{F}^t_{\mathrm{hn}}(\overline{E}))\big), \tag{4.66}$$

$$\widehat{\deg}_+(\overline{E}) \geqslant \sum_{i=1}^r \max\{\widehat{\mu}_i(\overline{E}), 0\} = \int_0^{+\infty} \dim_K(\mathcal{F}^t_{\mathrm{hn}}(\overline{E}))\,\mathrm{d}t. \tag{4.67}$$

Proof For $i \in \{1, \ldots, r\}$, let E_i be $\mathcal{F}^{\widehat{\mu}_i(\overline{E})}_{\mathrm{hn}}(\overline{E})$. Let $E_0 = \{0\}$. Then for each $i \in \{1, \ldots, r\}$, such that $E_i \supsetneq E_{i-1}$ one has

$$\widehat{\mu}(\overline{E_i/E_{i-1}}) \geqslant \widehat{\mu}_{\min}(\overline{E}_i) = \widehat{\mu}_i(\overline{E}),$$

where the last equality comes from Proposition 4.3.46 and the fact that the restriction of the $\mathbb{R}$-filtration $\mathcal{F}_{\mathrm{hn}}$ to E_i coincides with the Harder-Narasimhan $\mathbb{R}$-filtration of $\overline{E}_i$. Therefore, by Proposition 4.3.13 one has

$$\widehat{\deg}(\overline{E}) \geqslant \sum_{\substack{i\in\{1,\ldots,r\}\\ E_i\supsetneq E_{i-1}}} \widehat{\deg}(\overline{E_i/E_{i-1}}) \geqslant \sum_{\substack{i\in\{1,\ldots,r\}\\ E_i\supsetneq E_{i-1}}} \dim_K(E_i/E_{i-1})\widehat{\mu}_i(\overline{E}) = \sum_{i=1}^{r}\widehat{\mu}_i(\overline{E}),$$

which proves (4.66). Finally, if we let j be the largest index in $\{1,\ldots,r\}$ such that $\widehat{\mu}_j(\overline{E}) \geqslant 0$. Then by what we have proved

$$\widehat{\deg}(\overline{E}_j) \geqslant \sum_{i=1}^{j}\widehat{\mu}_i(\overline{E}) = \sum_{i=1}^{r}\max(\widehat{\mu}_i(\overline{E}),0).$$

Therefore, the inequality (4.67) holds. □

Proposition 4.3.51 *Let $\overline{E}=(E,\xi)$ be a non-zero adelic vector bundle on S. Let r be the dimension of E over K. Then the following inequalities hold:*

$$\widehat{\deg}(\overline{E}) \leqslant \sum_{i=1}^{r}\widehat{\mu}_i(\overline{E}) + \Delta(\overline{E}). \tag{4.68}$$

If in addition ξ is ultrametric on $\Omega\setminus\Omega_\infty$, then one has

$$\widehat{\deg}(\overline{E}) \leqslant \sum_{i=1}^{r}\widehat{\mu}_i(\overline{E}) + \delta(\overline{E}). \tag{4.69}$$

Proof We reason by induction on the dimension of E over K. The case where $\dim_K(E)=1$ is trivial since in this case $\overline{E}$ is Hermitian. In the following, we assume that $\dim_K(E)>1$ and that the proposition has been proved for adelic vector bundles of rank $<\dim_K(E)$.

The Harder-Narasimhan $\mathbb{R}$-filtration corresponds to an increasing flag

$$0=E_1\subsetneq E_2\subsetneq\ldots\subsetneq E_n=E$$

and a decreasing sequence of numbers $\mu_1>\ldots>\mu_n$ representing the points of jump of the $\mathbb{R}$-filtration. By Proposition 4.3.46, the minimal slope of $\overline{E}$ is equal to μ_n.

Let ϵ be a positive number such that $\epsilon<\mu_{n-1}-\mu_n$ and E' be a vector subspace of E such that $E'\subsetneq E$ and $\widehat{\mu}_{\max}(\overline{E/E'})\leqslant\widehat{\mu}_{\min}(\overline{E})+\epsilon=\mu_n+\epsilon$. By Proposition 4.3.26, one has

$$\widehat{\mu}(\overline{E/E'})\geqslant\widehat{\mu}_{\min}(\overline{E})=\mu_n.$$

Therefore, one has

$$\mu_n\leqslant\widehat{\mu}(\overline{E/E'})\leqslant\widehat{\mu}_{\max}(\overline{E/E'})\leqslant\mu_n+\epsilon.$$

Moreover, by Proposition 4.3.46, one has

$$\widehat{\mu}_{\min}(\overline{E}_{n-1}) \geqslant \mu_{n-1} > \mu_n + \epsilon.$$

By Proposition 4.3.31, we obtain that the composed map $E_{n-1} \to E \to E/E'$ is zero, or equivalently, E_{n-1} is contained in E'. Note that for any vector subspace F of E' such that $F \supsetneq E_{n-1}$ one has $\widehat{\mu}_{\min}(\overline{F}) \leqslant \mu_n$, otherwise the Harder-Narasimhan $\mathbb{R}$-filtration of $\overline{E}$ could not correspond to the flag $0 = E_1 \subsetneq E_2 \subsetneq \ldots \subsetneq E_n = E$ and the decreasing sequence $\mu_1 > \ldots > \mu_n$. Therefore, the Harder-Narasimhan $\mathbb{R}$-filtration of $\overline{E}'$ corresponds to a flag of the form ($\ell \in \mathbb{N}$)

$$0 = E_1 \subsetneq \ldots \subsetneq E_{n-1} \subsetneq E'_n \subsetneq \ldots \subsetneq E'_{n-1+\ell} = E'$$

together with a decreasing sequence $\mu_1 > \ldots > \mu_{n-1} > \mu'_n > \ldots > \mu'_{n-1+\ell}$, where $\mu'_n \leqslant \mu_n$ whenever $\ell \geqslant 1$. We apply the induction hypothesis to $\overline{E}'$ and obtain

$$\widehat{\deg}(\overline{E}') \leqslant \sum_{i=1}^{n-1} \mu_i \dim_K(E_i/E_{i-1}) + \sum_{i=n}^{n-1+\ell} \mu'_i \dim_K(E'_i/E_{i-1'}) + \Delta(\overline{E}'),$$

with the convention $E_{n-1} = E'_{n-1}$. By the condition that $\mu'_n \leqslant \mu_n$ whenever $\ell \geqslant 1$ we obtain

$$\widehat{\deg}(\overline{E}') \leqslant \sum_{i=1}^{n-1} \mu_i \dim_K(E_i/E_{i-1}) + \mu_n \dim_K(E'/E_{n-1}) + \Delta(\overline{E}').$$

Finally, by Proposition 4.3.13 (notably the inequality (4.27)) one obtains

$$\begin{aligned}\widehat{\deg}(\overline{E}) &\leqslant \widehat{\deg}(\overline{E}') + \widehat{\deg}(\overline{E/E'}) + \Delta(\overline{E}) - \Delta(\overline{E}') - \Delta(\overline{E/E'}) \\ &\leqslant \sum_{i=1}^{n-1} \mu_i \dim_K(E_i/E_{i-1}) + \mu_n \dim_K(E'/E_{n-1}) \\ &\qquad + (\mu_n + \epsilon)\dim_K(E/E') + \Delta(\overline{E}) \\ &= \sum_{j=1}^{r} \widehat{\mu}_j(\overline{E}) + \epsilon \dim_K(E/E') + \Delta(\overline{E}) \leqslant \sum_{j=1}^{r} \widehat{\mu}_j(\overline{E}) + \epsilon \dim_K(E) + \Delta(\overline{E}).\end{aligned}$$

Since ϵ is arbitrary, we obtain the inequality (4.68). In the case where ξ is ultrametric on $\Omega \setminus \Omega_\infty$, as above we deduce (4.69) from (4.28). The proposition is thus proved.□

Corollary 4.3.52 *Let $\overline{E} = (E, \xi)$ be a non-zero adelic vector bundle on S and r be the dimension of E over K. One has*

$$\widehat{\deg}_+(\overline{E}) \leqslant \sum_{i=1}^{r} \max\{\widehat{\mu}_i(\overline{E}), 0\} + \Delta(\overline{E}). \tag{4.70}$$

If in addition ξ is ultrametric on $\Omega \setminus \Omega_\infty$, then one has

$$\widehat{\deg}_+(\overline{E}) \leqslant \sum_{i=1}^{r} \max\{\widehat{\mu}_i(\overline{E}), 0\} + \delta(\overline{E}). \tag{4.71}$$

Proof Let F be a non-zero vector subspace of E and m be the dimension of F over K. By (4.68) one has

$$\widehat{\deg}(\overline{F}) \leqslant \sum_{j=1}^{m} \widehat{\mu}_j(\overline{F}) + \Delta(\overline{F}) \leqslant \sum_{j=1}^{m} \max\{\widehat{\mu}_j(\overline{F}), 0\} + \Delta(\overline{F}).$$

Note that

$$\sum_{j=1}^{m} \max\{\widehat{\mu}_j(\overline{F}), 0\} = -\int_{\mathbb{R}} \max\{t, 0\}\, \mathrm{d}(\dim_K(\mathcal{F}_{\mathrm{hn}}^t(\overline{F}))) = \int_0^{+\infty} \dim_K(\mathcal{F}_{\mathrm{hn}}^t(\overline{F}))\, \mathrm{d}t.$$

Moreover, by Proposition 4.3.49, for any $t \in \mathbb{R}$, one has

$$\dim_K(\mathcal{F}_{\mathrm{hn}}^t(\overline{F})) \leqslant \dim_K(\mathcal{F}_{\mathrm{hn}}^t(\overline{E})).$$

Therefore,

$$\widehat{\deg}(\overline{F}) \leqslant \int_0^{+\infty} \dim_K(\mathcal{F}_{\mathrm{hn}}^t(\overline{E}))\, \mathrm{d}t + \Delta(\overline{F}) = \sum_{i=1}^{r} \max\{\widehat{\mu}_i(\overline{E}), 0\} + \Delta(\overline{F}).$$

Note that $\Delta(\overline{F}) \leqslant \Delta(\overline{E})$ (see Corollary 1.2.44). By taking the supremum with respect to F, we obtain the inequality (4.70).

The proof of the inequality (4.71) is quite similar, where we combine the above argument with the inequality (4.69). □

Remark 4.3.53 Let (E, ξ) be an adelic vector bundle on S. Then one has the following inequality: if $\widehat{\deg}_+(E, \xi) > 0$, then

$$\widehat{\deg}_+(E, \xi) \leqslant \dim_K(E)\, \widehat{\mu}_{\max}(E, \xi). \tag{4.72}$$

As a consequence, we obtain

$$\widehat{\deg}_+(E, \xi) \leqslant \dim_K(E) \max\{\widehat{\mu}_{\max}(E, \xi), 0\} \tag{4.73}$$

in general. The inequality (4.73) is weaker than (4.70) and (4.71), but it holds without an error term. Moreover, the inequality (4.72) can be proved as follows: for any $\epsilon \in\,]0, \widehat{\deg}_+(E, \xi)[$, one can find a non-zero vector subspace F of E such that $0 \leqslant \widehat{\deg}_+(E, \xi) - \epsilon \leqslant \widehat{\deg}(F, \xi_F)$, so that

$$0 < \frac{\widehat{\deg}_+(E, \xi) - \epsilon}{\dim_K(E)} \leqslant \frac{\widehat{\deg}(F, \xi_F)}{\dim_K(E)} \leqslant \widehat{\mu}(F, \xi_F) \leqslant \widehat{\mu}_{\max}(E, \xi),$$

which implies (4.72).

Definition 4.3.54 Let $\overline{E}$ be an adelic vector bundle on S and r be the dimension of E over K. We denote by $\widetilde{\deg}(\overline{E})$ the sum $\widehat{\mu}_1(\overline{E}) + \cdots + \widehat{\mu}_r(\overline{E})$. If $\overline{E}$ is the zero adelic vector bundle on S, then by convention $\widetilde{\deg}(\overline{E})$ is defined to be 0. If $\overline{E}$ is non-zero, we define $\widetilde{\mu}(\overline{E})$ to be the quotient $\widetilde{\deg}(\overline{E})/\mathrm{rk}_K(E)$.

Proposition 4.3.55 *Let $\overline{E}$ and $\overline{F}$ be non-zero adelic vector bundles on S and $f : E \to F$ be a K-linear map.*

(1) *Suppose that f is a bijection. Then one has*

$$\widetilde{\deg}(\overline{E}) \leqslant \widetilde{\deg}(\overline{F}) + \dim_K(F) \cdot h(f). \tag{4.74}$$

(2) *Suppose that f is injective. Then $\widehat{\mu}_1(\overline{E}) \leqslant \widehat{\mu}_1(\overline{F}) + h(f)$.*

Proof (1) By Proposition 4.3.49, for any $t \in \mathbb{R}$ one has

$$f(\mathcal{F}^t_{\mathrm{hn}}(\overline{E})) \subseteq \mathcal{F}^{t-h(f)}_{\mathrm{hn}}(\overline{F}).$$

Therefore the inequality (4.74) follows from Proposition 1.1.39.

(2) Let $\lambda = \widehat{\mu}_1(\overline{E})$. Then $\mathcal{F}^{\lambda}_{\mathrm{hn}}(\overline{E}) \neq \{0\}$. Since f is injective, by Proposition 4.3.49, this implies that $\mathcal{F}^{\lambda-h(f)}_{\mathrm{hn}}(\overline{E}) \neq \{0\}$ and hence $\lambda - h(f) \leqslant \widehat{\mu}_1(\overline{F})$. □

Proposition 4.3.56 *Let $\overline{E} = (E, \xi)$ be an adelic vector bundle on S and*

$$0 = E_0 \subseteq E_1 \subseteq \ldots \subseteq E_n$$

be a flag of vector subspaces of E. One has

$$\widehat{\deg}(\overline{E}) - \Delta(\overline{E}) \leqslant \sum_{i=1}^{n} \widetilde{\deg}(\overline{E_i/E_{i-1}}) \leqslant \widehat{\deg}(\overline{E}) \tag{4.75}$$

If in addition ξ is ultrametric on $\Omega \setminus \Omega_\infty$, one has

$$\widehat{\deg}(\overline{E}) - \delta(\overline{E}) \leqslant \sum_{i=1}^{n} \widetilde{\deg}(\overline{E_i/E_{i-1}}) \leqslant \widehat{\deg}(\overline{E}) \tag{4.76}$$

Proof By Propositions 4.3.50 and 4.3.51 (notably the inequality (4.68)), for any $i \in \{1, \ldots, n\}$, one has

$$\widehat{\deg}(\overline{E_i/E_{i-1}}) - \Delta(\overline{E_i/E_{i-1}}) \leqslant \widetilde{\deg}(\overline{E_i/E_{i-1}}) \leqslant \widehat{\deg}(\overline{E_i/E_{i-1}}).$$

Taking the sum with respect to i, by Proposition 4.3.13, we obtain

$$\widehat{\deg}(\overline{E}) - \Delta(\overline{E}) \leqslant \sum_{i=1}^{n} \widetilde{\deg}(\overline{E_i/E_{i-1}}) \leqslant \widehat{\deg}(\overline{E}).$$

In the case where ξ is ultrametric on $\Omega \setminus \Omega_\infty$, the above argument combined with (4.69) leads to (4.76). □

Definition 4.3.57 Let $\overline{E}$ be a non-zero adelic vector bundle on S. We say that $\overline{E}$ is *semistable* if its Harder-Narasimhan $\mathbb{R}$-filtration only has one jump point, namely one has $\widehat{\mu}_1(\overline{E}) = \cdots = \widehat{\mu}_r(\overline{E})$ with $r = \dim_K(E)$. By definition, the following conditions are equivalent:

(1) $\overline{E}$ is semistable;
(2) for any non-zero vector subspace F of E, one has $\widehat{\mu}_{\min}(\overline{F}) \leqslant \widehat{\mu}_{\min}(\overline{E})$;
(3) $\widetilde{\mu}(\overline{E}) = \widehat{\mu}_{\min}(\overline{E})$.

Theorem 4.3.58 *Let $\overline{E}$ be a non-zero adelic vector bundle on S. We assume that the Harder-Narasimhan $\mathbb{R}$-filtration corresponds to the flag*

$$0 = E_0 \subsetneq E_1 \subsetneq \ldots \subsetneq E_n = E \tag{4.77}$$

and the decreasing sequence $\mu_1 > \ldots > \mu_n$ of real numbers. Then each subquotient $\overline{E_i/E_{i-1}}$ is semistable and $\widetilde{\mu}(\overline{E_i/E_{i-1}}) = \mu_i$, $i \in \{1, \ldots, n\}$. Moreover, (4.77) *is the only flag of vector subspaces of E such that each subquotient $\overline{E_i/E_{i-1}}$ is semistable and*

$$\widetilde{\mu}(\overline{E_1/E_0}) > \ldots > \widetilde{\mu}(\overline{E_n/E_{n-1}}).$$

Proof We begin with showing that each subquotient $\overline{E_i/E_{i-1}}$ is semistable and that $\widehat{\mu}(\overline{E_i/E_{i-1}}) = \mu_i$. The case where $i = 1$ results from the definition of Harder-Narasimhan $\mathbb{R}$-filtration. In what follows, we suppose that $i \geqslant 2$.

By Proposition 4.3.46, for any $j \in \{1, \ldots, n\}$ one has $\widehat{\mu}_{\min}(\overline{E}_j) \geqslant \mu_j$. Moreover, by definition of the Harder-Narasimhan $\mathbb{R}$-filtration, one has $\widehat{\mu}_{\min}(\overline{E}_j) \leqslant \mu_j$. Hence we obtain the equality $\widehat{\mu}_{\min}(\overline{E}_j) = \mu_j$.

We claim that any vector subspace G' of E_i/E_{i-1} has a minimal slope $\leqslant \mu_i$. Let $\pi : E_i \longrightarrow E_i/E_{i-1}$ be the canonical quotient map and E_i' be the preimage of G' by the quotient map π. Since E_i' contains strictly E_{i-1}, one has $\widehat{\mu}_{\min}(\overline{E_i'}) \leqslant \mu_i$. For any $\epsilon > 0$ there exists a quotient vector space H' of E_i' such that $\widehat{\mu}_{\max}(\overline{H}') \leqslant \mu_i + \epsilon$. If $\epsilon < \mu_{i-1} - \mu_i$, then $\widehat{\mu}_{\min}(\overline{E}_{i-1}) = \mu_{i-1} > \mu_i + \epsilon$. By Proposition 4.3.31 (3), we obtain that the composed map $E_{i-1} \longrightarrow E_i' \longrightarrow H'$ is zero, or equivalently, H' is actually a quotient vector space of $E_i'/E_{i-1} = G'$. Hence we obtain $\widehat{\mu}_{\min}(\overline{G}') \leqslant \mu_i$. Therefore one has $\widehat{\mu}_1(\overline{E_i/E_{i-1}}) \leqslant \mu_i \leqslant \mu_{\min}(\overline{E_i/E_{i-1}})$, which implies that $\overline{E_i/E_{i-1}}$ is semistable and $\widetilde{\mu}(\overline{E_i/E_{i-1}}) = \mu_i$.

We now proceed with the proof of the uniqueness by induction on the dimension of E over K. The case where $\dim_K(E) = 1$ is trivial. In the following, we assume that the assertion has been proved for non-zero adelic vector bundles of rank $< \dim_K(E)$. We still denote by

$$0 = E_0 \subsetneq E_1 \subsetneq \ldots \subsetneq E_n = E$$

the Harder-Narasimhan flag of $\overline{E}$. Let

$$0 = F_0 \subsetneq F_1 \subsetneq \ldots \subsetneq F_m = E$$

be a flag of vector subspaces of E such that each subquotient $\overline{F_j/F_{j-1}}$ is semistable and that

$$\widetilde{\mu}(\overline{F_1/F_0}) > \ldots > \widetilde{\mu}(\overline{F_m/F_{m-1}}).$$

Since the subquotients $\overline{F_j/F_{j-1}}$ are semistable, we can rewrite these inequalities as

$$\widehat{\mu}_{\min}(\overline{F_1/F_0}) > \ldots > \widehat{\mu}_{\min}(\overline{F_m/F_{m-1}}). \tag{4.78}$$

We claim that E_1 is actually contained in F_1. Assume that i is the smallest index in $\{1, \ldots, m\}$ such that $E_1 \subset F_i$. We identifie $E_1/(E_1 \cap F_{i-1})$ with a vector subspace of F_i/F_{i-1}. Since $\overline{F_i/F_{i-1}}$ is semistable, one has

$$\widehat{\mu}_{\min}(\overline{E_1/(E_1 \cap F_{i-1})}) \leqslant \widehat{\mu}_{\min}(\overline{F_i/F_{i-1}}).$$

If $i > 1$, then by (4.78) one has $\widehat{\mu}_{\min}(\overline{F_i/F_{i-1}}) < \widehat{\mu}_{\min}(\overline{F_1}) \leqslant \widehat{\mu}_{\min}(\overline{E_1})$, which leads to a contradiction since $\widehat{\mu}_{\min}(\overline{E_1/(E_1 \cap F_{i-1})}) \geqslant \widehat{\mu}_{\min}(\overline{E}_1)$. Therefore one has $E_1 \subseteq F_1$. If the inclusion is strict, then by the definition of Harder-Narasimhan filtration one has $\widehat{\mu}_{\min}(\overline{F_1}) < \widehat{\mu}_{\min}(\overline{E_1})$. This contradicts the semi-stability of $\overline{F_1}$. Therefore we have $E_1 = F_1$. Moreover, for any vector subspace M of E which contains strictly E_1, one has $\widehat{\mu}_{\min}(\overline{M}) < \widehat{\mu}_{\min}(\overline{E}_1)$. Hence, by Proposition 4.3.32 one has $\widehat{\mu}_{\min}(\overline{M}) = \widehat{\mu}_{\min}(\overline{M/E_1})$. Therefore, if E/E_1 is non-zero, then

$$0 = E_1/E_1 \subsetneq \ldots \subsetneq E_n/E_1 = E/E_1$$

is the Harder-Narasimhan flag of E/E_1. By the induction hypothesis one has $n = m$ and $E_i = F_i$ for any $i \in \{2, \ldots, n\}$. The uniqueness is thus proved. □

Proposition 4.3.59 *Let $\overline{E}$ be a non-zero adelic vector bundle on S. The following assertions are equivalent:*

(1) *$\overline{E}$ is semistable,*
(2) *for any non-zero vector subspace F of E, one has $\widetilde{\mu}(\overline{F}) \leqslant \widetilde{\mu}(\overline{E})$*
(3) *for any non-zero quotient vector space G of E, one has $\widetilde{\mu}(\overline{G}) \geqslant \widetilde{\mu}(\overline{E})$.*

Proof Let r be the dimension of E over K. Assume that $\overline{E}$ is semistable. Then one has $\widehat{\mu}_1(\overline{E}) = \cdots = \widehat{\mu}_r(\overline{E}) = \widetilde{\mu}(\overline{E})$. If F is a non-zero vector subspace of E, then by Proposition 4.3.49 we obtain that, for any $t \in \mathbb{R}$, one has $\mathcal{F}^t_{\mathrm{hn}}(\overline{F}) \subseteq \mathcal{F}^t_{\mathrm{hn}}(\overline{E})$. Therefore, for any $i \in \{1, \ldots, \dim_K(F)\}$ one has $\widehat{\mu}_i(\overline{F}) \leqslant \widetilde{\mu}(\overline{E})$, which implies $\widetilde{\mu}(\overline{F}) \leqslant \widetilde{\mu}(\overline{E})$. Similarly, if G is a non-zero quotient vector space of E and $\pi : E \to G$ is the quotient map, then, by Proposition 4.3.49, one has $\pi(\mathcal{F}^t_{\mathrm{hn}}(\overline{E})) \subseteq \mathcal{F}^t_{\mathrm{hn}}(\overline{G})$ for any $t \in \mathbb{R}$. Therefor for any $i \in \{1, \ldots, \dim_K(G)\}$ one has $\widehat{\mu}_i(\overline{G}) \geqslant \widetilde{\mu}(\overline{E})$, which implies that $\widetilde{\mu}(\overline{G}) \geqslant \widetilde{\mu}(\overline{E})$. Hence we have proved the implications (1)⇒(2) and (1)⇒(3).

We will prove the converse implications by contraposition. Suppose that $\overline{E}$ is not semistable and its Harder-Narasimhan $\mathbb{R}$-filtration corresponds to the flag

$$0 = E_0 \subsetneq E_1 \subsetneq \ldots \subsetneq E_n = E,$$

and the successive jump points $\mu_1 < \ldots < \mu_n$, where $n \in \mathbb{N}$, $n \geqslant 1$. Then one has

$$\widetilde{\mu}(\overline{E}) = \frac{1}{\dim_K(E)} \sum_{i=1}^{n} \mu_i \dim_K(E_i/E_{i-1}).$$

By Theorem 4.3.58 we obtain that $\widetilde{\mu}(\overline{E}_1) = \mu_1 > \widetilde{\mu}(\overline{E})$ and $\widetilde{\mu}(\overline{E_n/E_{n-1}}) = \mu_n < \widetilde{\mu}(\overline{E})$. The proposition is thus proved. □

Remark 4.3.60 Consider the particular case where the adelic curve consists of exactly one copy of the trivial absolute value on K (of measure 1 with respect to ν). In this case an adelic vector bundle on S is just a finite-dimensional vector space E over K equipped with a norm $\|\cdot\|$ (which is not necessary ultrametric), where we consider the trivial absolute value on K. We have shown in §1.1.8 that ultrametric norms on a finite-dimensional vector space over K correspond bijectively to $\mathbb{R}$-filtrations on the same vector space. In particular, if $(E, \|\cdot\|)$ is a Hermitian adelic vector bundle on S, then the $\mathbb{R}$-filtration on E corresponding to $\|\cdot\|$ identifies with the Harder-Narasimhan $\mathbb{R}$-filtration of $(E, \|\cdot\|)$.

Proposition 4.3.61 *We equip K with the trivial absolute value. Let $(E, \|\cdot\|)$ be a finite-dimensional normed vector space over K, which is also considered as an adelic vector bundle as in Remark 4.3.60. The adelic vector bundle $(E, \|\cdot\|)$ is semistable if and only if the double dual norm $\|\cdot\|_{**}$ is constant on $E \setminus \{0\}$. Moreover, in this case one has*

$$-\ln\|x\|_{**} = \widehat{\mu}(E, \|\cdot\|) = \widehat{\mu}_{\min}(E, \|\cdot\|)$$

for any $x \in E \setminus \{0\}$.

Proof First we assume that $(E, \|\cdot\|)$ is semistable. Let $\{e_i\}_{i=1}^r$ be an α-orthogonal basis of $(E, \|\cdot\|)$, where $\alpha \in \,]0, 1[$. Without loss of generality, we assume that $\|e_1\| \leqslant \ldots \leqslant \|e_r\|$. Moreover, by Proposition 1.2.23 one has

$$\widehat{\deg}(E, \|\cdot\|) \leqslant -r\ln(\alpha) - \sum_{i=1}^{r} \ln\|e_i\|.$$

In particular, if $\|e_r\|/\|e_1\| > \alpha^{-r}$, that is,

$$-\frac{1}{r}\ln\|e_1\| > -\frac{1}{r}\ln\|e_r\| - \ln(\alpha),$$

then

$$\begin{aligned}-\ln\|e_1\| = -\frac{r-1}{r}\ln\|e_1\| - \frac{1}{r}\ln\|e_1\| &> -\frac{1}{r}\sum_{i=1}^{r-1}\ln\|e_i\| - \frac{1}{r}\ln\|e_r\| - \ln(\alpha)\\ &\geqslant \widehat{\mu}(E, \|\cdot\|) \geqslant \widehat{\mu}_{\min}(E, \|\cdot\|),\end{aligned}$$

which shows that $(E, \|\cdot\|)$ is not semistable, so that $\|e_r\|/\|e_1\| \leqslant \alpha^{-r}$. This observation shows that, for any α-orthogonal basis $\{e_i\}_{i=1}^r$ of E, one has

$$\max_{(i,j)\in\{1,\ldots,r\}^2} \Big|\ln\|e_i\| - \ln\|e_j\|\Big| \leqslant -r\ln(\alpha). \tag{4.79}$$

Note that $\{e_i\}_{i=1}^r$ is also an α-orthogonal basis of $(E, \|\cdot\|_{**})$ (see Proposition 1.2.11). Moreover, we deduce from (4.79) and (1.25) that

$$\max_{(i,j)\in\{1,\ldots,r\}^2} \Big| \ln\|e_i\|_{**} - \ln\|e_j\|_{**} \Big| \leqslant -(r+1)\ln(\alpha). \tag{4.80}$$

Note that one has $\widehat{\deg}(E, \|\cdot\|) = \widehat{\deg}(E, \|\cdot\|_{**})$ (see Proposition 1.2.15). Moreover, by Propositions 1.1.66 and 1.2.23 one has

$$-\sum_{i=1}^r \ln\|e_i\|_{**} \leqslant \widehat{\deg}(E, \|\cdot\|) \leqslant -r\ln(\alpha) - \sum_{i=1}^r \ln\|e_i\|_{**}.$$

Combining this estimate with (4.80) we obtain

$$\max_{i\in\{1,\ldots,r\}} \Big| \ln\|e_i\|_{**} - \widehat{\mu}(E, \|\cdot\|) \Big| \leqslant -(r+2)\ln(\alpha).$$

In particular, for any $(\lambda_1, \ldots, \lambda_r) \in K^r \setminus \{(0, \ldots, 0)\}$, one has

$$\Big| \ln\|\lambda_1 e_1 + \cdots + \lambda_r e_r\|_{**} - \widehat{\mu}(E, \|\cdot\|) \Big| \leqslant -(r+3)\ln(\alpha)$$

Since $(E, \|\cdot\|)$ admits an α-orthogonal basis for any $\alpha \in \,]0,1[$ (see Corollary 1.2.9), we obtain that the restriction of $\ln\|\cdot\|_{**}$ to $E \setminus \{0\}$ is constant (which is equal to $-\widehat{\mu}(E, \|\cdot\|)$).

Assume now that the double dual norm $\|\cdot\|_{**}$ is constant on $E \setminus \{0\}$. Since $\|\cdot\|_{**}$ and $\|\cdot\|$ induce the same dual norm on $E^\vee$, we obtain that the restriction of $\ln\|\cdot\|_*$ to $E^\vee \setminus \{0\}$ is constant and takes $-\widehat{\mu}(E^\vee, \|\cdot\|_*)$ as its value. Note that one has $-\widehat{\mu}(E^\vee, \|\cdot\|_*) = \widehat{\mu}(E, \|\cdot\|)$ by Proposition 4.3.10. We will show that $(E, \|\cdot\|)$ is semistable. First we show that $\widehat{\mu}_{\min}(E, \|\cdot\|) = \widehat{\mu}(E, \|\cdot\|)$. Let G be a non-zero quotient vector space of E and $\|\cdot\|_G$ be the quotient norm of $\|\cdot\|$ on G. By Proposition 1.1.20, $\|\cdot\|_{G,*}$ coincides with the restriction of $\|\cdot\|_*$ to $G^\vee$. Since the function $\ln\|\cdot\|_*$ takes constant value $\widehat{\mu}(E, \|\cdot\|)$ on $E^\vee \setminus \{0\}$ we obtain that

$$\widehat{\mu}(G, \|\cdot\|_G) = -\widehat{\mu}(G^\vee, \|\cdot\|_{G,*}) = \widehat{\mu}(E, \|\cdot\|).$$

Therefore $\widehat{\mu}_{\min}(E, \|\cdot\|) = \widehat{\mu}(E, \|\cdot\|)$. Now for any non-zero vector subspace F of E one has

$$\widehat{\mu}_{\min}(F, \|\cdot\|_F) \leqslant \widehat{\mu}(F, \|\cdot\|_F) \leqslant \widehat{\mu}(E, \|\cdot\|_E),$$

where $\|\cdot\|_F$ denotes the restriction of $\|\cdot\|$ to F. In fact, $\ln\|\cdot\|_F$ is bounded from below by the restriction of $\ln\|\cdot\|_{**}$ to F, which is constant on $F \setminus \{0\}$ of value $-\widehat{\mu}(E, \|\cdot\|)$. Therefore $(E, \|\cdot\|)$ is semistable. □

Remark 4.3.62 We keep the notation of the previous proposition. Note that the norms $\|\cdot\|$ and $\|\cdot\|_{**}$ induce the same dual norm on $E^\vee$ (see Proposition 1.2.14 (1)), so that we obtain that the adelic vector bundle $(E, \|\cdot\|)$ is semistable if and only if the restriction of the function $\|\cdot\|_*$ on $E^\vee \setminus \{0\}$ is constant. Moreover, in this case one has (see Proposition 1.2.47)

$$\forall\,\varphi \in E^\vee \setminus \{0\}, \quad -\ln\|\varphi\|_* = -\widehat{\deg}(E, \|\cdot\|).$$

Remark 4.3.63 In the case where $\|\cdot\|$ is ultrmetric, the normed vector space $(E, \|\cdot\|)$ corresponds to a sequence

$$0 = E_0 \subsetneq E_1 \subsetneq \ldots \subsetneq E_n = E$$

of vector subspaces of E and a decreasing sequence $\mu_1 > \ldots > \mu_n$ of real numbers (see Remark 1.1.40). Note that, for any $i \in \{1, \ldots, n\}$ the restriction of the subquotient norm $\|\cdot\|_{E_i/E_{i-1}}$ to $(E_i/E_{i-1}) \setminus \{0\}$ is constant and takes $\mathrm{e}^{-\mu_i}$ as its value. Therefore Proposition 4.3.61 implies that $(E_i/E_{i-1}, \|\cdot\|_{E_i/E_{i-1}})$ is semistable and admits μ_i as its minimal slope. Therefore Theorem 4.3.58 shows that

$$0 = E_0 \subsetneq E_1 \subsetneq \ldots \subsetneq E_n = E$$

is the Harder-Narasimhan flag of the adelic vector bundle $(E, \|\cdot\|)$.

Proposition 4.3.64 *We equip K with the trivial absolute value. Consider a finite-dimensional non-zero normed vector space $(E, \|\cdot\|)$ over K. The Harder-Narasimhan flags of $(E, \|\cdot\|)$ and $(E, \|\cdot\|_{**})$ are the same. Moreover, for any $i \in \{1, \ldots, \dim_K(E)\}$ one has $\widehat{\mu}_i(E, \|\cdot\|) = \widehat{\mu}_i(E, \|\cdot\|_{**})$.*

Proof Let n be the dimension of E over K. We reason by induction on n. First of all, if $(E, \|\cdot\|)$ is semistable, then by Proposition 4.3.61 (see also its proof), the function $-\ln\|\cdot\|_{**}$ is constant on $E \setminus \{0\}$ and takes $\widehat{\mu}(E, \|\cdot\|) = \widehat{\mu}_{\min}(\|\cdot\|)$ as its value. Therefore the assertion of the proposition holds in this case, and in particular the assertion is true when $n = 1$. In the following we suppose that $(E, \|\cdot\|)$ is not semistable (hence $n \geqslant 2$) and that the proposition has been proved for normed vector spaces of dimension $\leqslant n - 1$.

For any $i \in \{1, \ldots, n\}$, let $\mu_i = \widehat{\mu}_i(E, \|\cdot\|_{**})$ and E_i be the ball of radius $\mathrm{e}^{-\mu_i}$ in $(E, \|\cdot\|_{**})$ centered at the origin. Let

$$\beta = \min\{\mu_i - \mu_{i-1} \,:\, i \in \{2, \ldots, n\},\ \mu_i > \mu_{i-1}\}$$

and α be an element of $]0, 1[$ such that $\alpha > \mathrm{e}^{-\beta/n}$. Let $\{e_i\}_{i=1}^n$ be an α-orthogonal bases of $(E, \|\cdot\|)$. By Proposition 1.2.11, it is also an α-orthgonal basis of $(E, \|\cdot\|_{**})$. By Proposition 1.2.26, $\{e_i\}_{i=1}^n$ is an orthogonal basis of $(E, \|\cdot\|_{**})$. Without loss of generality, we may assume that

$$\{e_i\}_{i=1}^n \cap E_1 = \{e_1, \ldots, e_m\},$$

where m is the dimension of E_1 over K (see Proposition 1.2.26 (1)). Let $\{e_i^\vee\}_{i=1}^n$ be the dual basis of $\{e_i\}_{i=1}^n$ and $\|\cdot\|_1$ be the restriction of the norm $\|\cdot\|$ to E_1. For any $i \in \{1, \ldots, m\}$, let φ_i be the restriction of $e_i^\vee$ to E_1. Then $\{\varphi_i\}_{i=1}^m$ forms a basis of $E_1^\vee$, which is the dual basis of $\{e_i\}_{i=1}^m$. By lemma 1.2.10, $\{\varphi_i\}_{i=1}^m$ is an α-orthogonal basis of $E_1^\vee$, and one has

$$\forall\, i \in \{1,\ldots,m\},\quad \|e_i\|^{-1} \leqslant \|\varphi_i\|_{1,*} \leqslant \alpha^{-1}\|e_i\|^{-1}.$$

By Proposition 1.2.11, one has

$$\forall\, i \in \{1,\ldots,m\},\quad \alpha\|e_i\| \leqslant \|e_i\|_{**} = \mathrm{e}^{-\mu_1} \leqslant \|e_i\|.$$

Therefore one obtains

$$\forall\, i \in \{1,\ldots,m\},\quad \alpha\mathrm{e}^{\mu_1} \leqslant \|\varphi_i\|_{1,*} \leqslant \alpha^{-1}\mathrm{e}^{\mu_1}.$$

As a consequence, for a general non-zero element φ of $E_1^\vee$, which is written in the form $\lambda_1\varphi_1 + \cdots + \lambda_m\varphi_m$, one has

$$\|\varphi\|_{1,*} \leqslant \max_{\substack{i\in\{1,\ldots,m\}\\ \lambda_i\neq 0}} \|\varphi_i\|_{1,*} \leqslant \alpha^{-1}\mathrm{e}^{\mu_1}$$

and

$$\|\varphi\|_{1,*} \geqslant \alpha \max_{\substack{i\in\{1,\ldots,m\}\\ \lambda_i\neq 0}} \|\varphi_i\|_{1,*} \geqslant \alpha^2\mathrm{e}^{\mu_1}.$$

Since $(E,\|\cdot\|)$ admits an α-orthogonal basis for any $\alpha \in\,]0,1[$, we obtain that the restriction of $\|\cdot\|_{1,*}$ to $E_1 \setminus \{0\}$ is constant and takes e^{μ_1} as its value. Therefore, Proposition 4.3.61 (see also Remark 4.3.62), we obtain that $(E_1,\|\cdot\|)$ is semistable and admits μ_1 as its minimal slope.

By Proposition 1.1.20, one has (see Definition 1.1.2 and Subsection 1.1.3 for notation)

$$\|\cdot\|_{*,(E/E_1)^\vee\hookrightarrow E^\vee} = \|\cdot\|_{E\twoheadrightarrow E/E_1,*}.$$

Moreover, since $\|\cdot\|_*$ is ultrametric, by Proposition 1.2.35 one has

$$\|\cdot\|_{**,E\twoheadrightarrow E/E_1} = \|\cdot\|_{*,(E/E_1)^\vee\hookrightarrow E^\vee,*} = \|\cdot\|_{E\twoheadrightarrow E/E_1,**}.$$

Applying the induction hypothesis to $(E/E_1,\|\cdot\|_{E\twoheadrightarrow E/E_1})$ we obtain that the Harder-Narasimhan flags and the successive slopes of

$$(E/E_1,\|\cdot\|_{E\twoheadrightarrow E/E_1}) \quad\text{and}\quad (E/E_1,\|\cdot\|_{**,E\twoheadrightarrow E/E_1})$$

are the same. Therefore, by Theorem 4.3.58 we obtain that the Harder-Narasimhan flag and the successive slopes of $(E,\|\cdot\|)$ coincides with those of $(E,\|\cdot\|_{**})$. The proposition is thus proved. □

Remark 4.3.65 In the framework of linear code, Randriambololona [125] has proposed a Harder-Narasimhan theory based on semimodular degree functions on the modular lattice of vector subspaces. Note that our approach, which relies on the Arakelov degree function of quotient vector spaces, has a very different nature from the classic method (due to the fact that the equality (4.29) and the inequality (4.50) fail in general for non-Hermitian adelic vector bundles). It is an intriguing question to compare the Harder-Narasimhan filtrations constructed in our setting and in [125].

4.3.12 Absolute positive degree and absolute maximal slope

We have seen in Proposition 4.3.14 that the Arakelov degree is preserved by extension of scalars. In this subsection, we discuss the behaviour of the maximal slope and the positive degree under extension of scalars to the algebraic closure of K. We denote by K^{ac} the algebraic closure of the field K.

Definition 4.3.66 Let (E,ξ) be an adelic vector bundle on S. We denote by $\widehat{\deg}_+^{\mathrm{a}}(E,\xi)$ the positive degree of $(E_{K^{\mathrm{ac}}},\xi_{K^{\mathrm{ac}}})$, called the *absolute positive degree* of (E,ξ). If E is non-zero, we denote by $\widehat{\mu}_{\max}^{\mathrm{a}}(E,\xi)$ the maximal slope of $(E_{K^{\mathrm{ac}}},\xi_{K^{\mathrm{ac}}})$, called the *absolute maximal slope* of (E,ξ).

Proposition 4.3.67 *Let (E,ξ) be an adelic vector bundle on S. One has*

$$\widehat{\deg}_+(E,\xi)\leqslant\widehat{\deg}_+^{\mathrm{a}}(E,\xi)\ \ \textit{and}\ \ \widehat{\mu}_{\max}(E,\xi)\leqslant\widehat{\mu}_{\max}^{\mathrm{a}}(E,\xi).$$

Moreover, for any algebraic extension L of K, one has

$$\widehat{\deg}_+^{\mathrm{a}}(E,\xi)=\widehat{\deg}_+^{\mathrm{a}}(E_L,\xi_L)\ \ \textit{and}\ \ \widehat{\mu}_{\max}^{\mathrm{a}}(E,\xi)=\widehat{\mu}_{\max}^{\mathrm{a}}(E_L,\xi_L).$$

Proof Let F be a vector subspace of E and ξ_F be the restriction of ξ to F. Let $\xi_{F_{K^{\mathrm{ac}}}}$ be the restriction of $\xi_{K^{\mathrm{ac}}}$ to $F_{K^{\mathrm{ac}}}$. By Proposition 1.3.17 (1), (2), the identity map $(F_{K^{\mathrm{ac}}},\xi_{F,K^{\mathrm{ac}}})\to(F_{K^{\mathrm{ac}}},\xi_{F_{K^{\mathrm{ac}}}})$ has norm $\leqslant 1$ on any $\omega\in\Omega$. By Proposition 4.3.18, one has

$$\begin{aligned}\widehat{\deg}(F,\xi_F)=\widehat{\deg}(F_{K^{\mathrm{ac}}},\xi_{F,K^{\mathrm{ac}}})&\leqslant\widehat{\deg}(F_{K^{\mathrm{ac}}},\xi_{F_{K^{\mathrm{ac}}}})\\&\leqslant\widehat{\deg}_+(E_{K^{\mathrm{ac}}},\xi_{K^{\mathrm{ac}}})=\widehat{\deg}_+^{\mathrm{a}}(E,\xi),\end{aligned}$$

where the first equality comes from Proposition 4.3.14. Similarly, if F is non-zero, one has

$$\begin{aligned}\widehat{\mu}(F,\xi_F)\leqslant\widehat{\mu}(F_{K^{\mathrm{ac}}},\xi_{F,K^{\mathrm{ac}}})&\leqslant\widehat{\mu}(F_{K^{\mathrm{ac}}},\xi_{F_{K^{\mathrm{ac}}}})\\&\leqslant\widehat{\mu}_{\max}(F_{K^{\mathrm{ac}}},\xi_{F_{K^{\mathrm{ac}}}})=\widehat{\mu}_{\max}^{\mathrm{a}}(F,\xi_F).\end{aligned}$$

Since F is arbitrary, we obtain

$$\widehat{\deg}_+(E,\xi)\leqslant\widehat{\deg}_+^{\mathrm{a}}(E,\xi)\quad\text{and}\quad\widehat{\mu}_{\max}(E,\xi)\leqslant\widehat{\mu}_{\max}^{\mathrm{a}}(E,\xi).$$

By Corollary 1.3.15, if L is an algebraic extension of K, then one has $(\xi_L)_{K^{\mathrm{ac}}}=\xi_{K^{\mathrm{ac}}}$. Therefore $\widehat{\deg}_+^{\mathrm{a}}(E,\xi)=\widehat{\deg}_+^{\mathrm{a}}(E_L,\xi_L)$, and $\widehat{\mu}_{\max}^{\mathrm{a}}(E,\xi)=\widehat{\mu}_{\max}^{\mathrm{a}}(E_L,\xi_L)$. □

Proposition 4.3.68 *Assume that the field K is perfect. Let (E,ξ) be a* Hermitian *adelic vector bundle on S. Then one has*

$$\widehat{\deg}_+(E,\xi)=\widehat{\deg}_+^{\mathrm{a}}(E,\xi)\quad\textit{and}\quad\widehat{\mu}_{\max}(E,\xi)=\widehat{\mu}_{\max}^{\mathrm{a}}(E,\xi).$$

Proof Without loss of generality, we may assume that the vector space E is non-zero. Let

$$\{0\} = \widetilde{E}_0 \subsetneq \widetilde{E}_1 \subsetneq \ldots \subsetneq \widetilde{E}_n = E_{K^{\mathrm{a}}}$$

be the Harder-Narasimhan flag of $(E_{K^{\mathrm{ac}}}, \xi_{K^{\mathrm{ac}}})$. By the uniqueness of Harder-Narasimhan filtration (see Proposition 4.3.38), for any K-automorphism τ of K^{ac} and any $i \in \{1, \ldots, n\}$, the vector space $\widetilde{E}_i$ is stable by τ. Since the filed K is perfect, by Galois descent (see [29], Chapter V, §10, no.4, Corollary of Proposition 6), there exists a flag

$$\{0\} = E_0 \subsetneq E_1 \subsetneq \ldots \subsetneq E_n = E \tag{4.81}$$

such that $\widetilde{E}_i = E_{i,K^{\mathrm{ac}}}$ for any $i \in \{1, \ldots, n\}$. Moreover, by Propositions 1.3.17 (1), (2) (here we use the hypothesis that ξ is Hermitian), if we denote by ξ_i the restriction of ξ to E_i, then $\xi_{i,K^{\mathrm{ac}}}$ coincides with the restriction $\widetilde{\xi}_i$ of $\xi_{K^{\mathrm{ac}}}$ to $\widetilde{E}_i$. Therefore by Proposition 4.3.14 one has $\widehat{\deg}(E_i, \xi_i) = \widehat{\deg}(\widetilde{E}_i, \widetilde{\xi}_i)$. We then deduce that the slopes of E_i/E_{i-1} and $\widetilde{E}_i/\widetilde{E}_{i-1}$ (equipped with subquotient norm families) are the same. Hence by Proposition 4.3.38 we obtain that (4.81) is the Harder-Narasimhan flag of (E, ξ). Therefore,

$$\widehat{\mu}_{\max}(E, \xi) = \widehat{\mu}(E_1, \xi_1) = \widehat{\mu}(\widetilde{E}_1, \widetilde{\xi}_1) = \widehat{\mu}^{\mathrm{a}}_{\max}(E, \xi)$$

and

$$\widehat{\deg}_+(E, \xi) = \max_{i \in \{0,\ldots,n\}} \widehat{\deg}(E_i, \xi_i) = \max_{i \in \{0,\ldots,n\}} \widehat{\deg}(\widetilde{E}_i, \widetilde{\xi}_i) = \widehat{\deg}^{\mathrm{a}}_+(E, \xi).$$

4.3.13 Successive minima

The successive minima are classic invariants of Hermitian vector bundles on an arithmetic curve. In this subsection, we extend their construction (more precisely, the construction of successive minima of Roy-Thunder [128]) to the setting of adelic vector bundles on an adelic curve.

Definition 4.3.69 Let (E, ξ) be an adelic vector bundle on S and r be the dimension of E over K. For any $i \in \{1, \ldots, r\}$, let

$$\nu_i(E, \xi) := \sup\{t \in \mathbb{R} \,:\, \dim_K(\mathrm{Vect}_K(\{s \in E_K \,:\, \widehat{\deg}_\xi(s) \geqslant t\})) \geqslant i\},$$

called the ith *(logarithmic) minimum* of (E, ξ). In other words, $\nu_i(E, \xi)$ is the supremum of the set of real numbers t such that there exist at least i linearly independent vectors of Arakelov degree $\geqslant t$. Clearly one has

$$\nu_1(E, \xi) \geqslant \ldots \geqslant \nu_r(E, \xi).$$

The first minimum $\nu_1(E,\xi)$ is also denoted by $\nu_{\max}(E,\xi)$, and the last minimum $\nu_r(E,\xi)$ is also denoted by $\nu_{\min}(E,\xi)$. For any $t\in\mathbb{R}$, let

$$\mathcal{F}_{\mathrm{m}}^t(E,\xi) := \bigcap_{\varepsilon>0} \mathrm{Vect}_K(\{s\in E\setminus\{0\}\ :\ \widehat{\deg}_\xi(s)\geqslant t-\varepsilon\}).$$

By definition

$$\nu_i(E,\xi)=\sup\{t\in\mathbb{R}\ :\ \dim_K(\mathcal{F}_{\mathrm{m}}^t(\overline{E}))\geqslant i\}.$$

If E is the zero vector space, then by convention we define

$$\nu_{\max}(E,\xi):=-\infty \quad \text{and} \quad \nu_{\min}(E,\xi):=+\infty.$$

We also define the absolute version of the successive minima as follows. For any $i\in\{1,\ldots,r\}$, let $\nu_i^{\mathrm{a}}(E,\xi):=\nu_i(E_{K^{\mathrm{ac}}},\xi_{K^{\mathrm{ac}}})$, where K^{ac} denotes the algebraic closure of (E,ξ). Similarly, we let

$$\nu_{\max}^{\mathrm{a}}(E,\xi):=\nu_{\max}(E_{K^{\mathrm{ac}}},\xi_{K^{\mathrm{ac}}}) \quad \text{and} \quad \nu_{\min}^{\mathrm{a}}(E,\xi):=\nu_{\min}(E_{K^{\mathrm{ac}}},\xi_{K^{\mathrm{ac}}}).$$

Proposition 4.3.70 *Let $\overline{E}=(E,\xi)$ be a non-zero adelic vector bundle on S. For any $t\in\mathbb{R}$ one has*

$$\mathcal{F}_{\mathrm{m}}^t(\overline{E})=\bigcap_{\varepsilon>0}\ \sum_{\substack{0\neq F\subseteq E\\ \nu_{\min}(\overline{F})\geqslant t-\varepsilon}} F, \tag{4.82}$$

where F runs over the set of all non-zero vector subspaces of E, and in the structure of adelic vector bundle of $\overline{F}$ we consider the restricted norm family.

Proof Let $\varepsilon>0$ and F be a non-zero vector subspace of E such that $\nu_{\min}(\overline{F})>t-\varepsilon$. There exists a basis $\{s_i\}_{i=1}^n$ of F over K such that

$$\min_{i\in\{1,\ldots,n\}}\widehat{\deg}_\xi(s_i)\geqslant t-2\varepsilon.$$

Therefore one has

$$\sum_{\substack{0\neq F\subseteq E\\ \widehat{\mu}_{\min}(F)\geqslant t-\varepsilon}} F\subseteq \mathrm{Vect}_K(\{s\in E\setminus\{0\}\ :\ \widehat{\deg}_\xi(s)\geqslant t-2\varepsilon\}).$$

Conversely, for any $t\in\mathbb{R}$ such that $\mathcal{F}_{\mathrm{m}}^t(\overline{E})\neq\{0\}$ and any $\varepsilon>0$, there exist elements $u_1,\ldots,u_r$ in E which generate $\mathcal{F}_{\mathrm{m}}^t(\overline{E})$ as vector space over K and such that

$$\forall\, j\in\{1,\ldots,r\},\quad \widehat{\deg}_\xi(u_j)\geqslant t-\varepsilon.$$

Therefore $\nu_{\min}(\mathcal{F}_{\mathrm{m}}^t(\overline{E}))\geqslant t$. □

Proposition 4.3.71 *Let (E,ξ) be an adelic vector bundle on S and r be the dimension of E over K. For any $i\in\{1,\ldots,r\}$, one has $\nu_i(E,\xi)\leqslant\nu_i^{\mathrm{a}}(E,\xi)$. Moreover, for any algebraic extension L of K and any $i\in\{1,\ldots,r\}$ one has $\nu_i^{\mathrm{a}}(E,\xi)=\nu_i^{\mathrm{a}}(E_L,\xi_L)$.*

Proof Let $\{s_j\}_{j=1}^i$ be a linearly independent family in E. Then it is also a linearly independent family in $E_{K^{\mathrm{ac}}}$. Moreover, by the same argument as in the proof of Proposition 4.3.67, for any $j \in \{1, \ldots, i\}$ one has $\widehat{\deg}_\xi(s_j) \leqslant \widehat{\deg}_{\xi_{K^{\mathrm{ac}}}}(s_j)$. Therefore $\nu_i(E,\xi) \leqslant \nu_i(E_{K^{\mathrm{ac}}}, \xi_{K^{\mathrm{ac}}}) = \nu_i^{\mathrm{a}}(E,\xi)$. The equality $\nu_i^{\mathrm{a}}(E_L,\xi_L) = \nu_i^{\mathrm{a}}(E,\xi)$ comes from the relation $(\xi_L)_{K^{\mathrm{ac}}} = \xi_{K^{\mathrm{ac}}}$, which is a consequence of Corollary 1.3.15. □

The following proposition is straightforward from the definition of the (absolute) first minimum and the (absolute) maximal slope.

Proposition 4.3.72 *If (E,ξ) is an adelic vector bundle on S, then one has*

$$\nu_1(E,\xi) \leqslant \widehat{\mu}_{\max}(E,\xi) \quad \text{and} \quad \nu_1^{\mathrm{a}}(E,\xi) \leqslant \widehat{\mu}_{\max}^{\mathrm{a}}(E,\xi). \tag{4.83}$$

4.3.14 Minkowski property

Definition 4.3.73 Let $S = (K, (\Omega, \mathcal{A}, \nu), \phi)$ be a proper adelic curve. Let C be a non-negative real number. We say that the adelic curve S satisfies the *Minkowski property of level* $\geqslant C$ if, for any adelic vector bundle (E,ξ) on S such that ξ is ultrametric on $\Omega \setminus \Omega_\infty$, one has

$$\nu_1(E,\xi) \geqslant \widehat{\mu}_{\max}(E,\xi) - C\ln(\dim_K(E)).$$

We say that S satisfies the *absolute Minkowski property of level* $\geqslant C$ if, for any adelic vector bundle (E,ξ) on S, one has

$$\nu_1^{\mathrm{a}}(E,\xi) \geqslant \widehat{\mu}_{\max}^{\mathrm{a}}(E,\xi) - C\ln(\dim_K(E)).$$

Remark 4.3.74 Let $\overline{V}$ be an Euclidean lattice. The first theorem of Minkowski can be stated as (see [24, §3.2] for more details)

$$\nu_1(\overline{V}) \geqslant \widehat{\mu}_{\max}(\overline{V}) - \frac{1}{2}\ln(\dim_K(V)).$$

Hence the Minkowski property should be considered as an analogue in the general setting of adelic curve of the statement of the first theorem of Minkowski. For general number fields, it has been shown in [66, §5] that, for any adelic vector bundle $\overline{E}$ of rank n over a number field K, one has

$$\nu_1(\overline{E}) \geqslant \widehat{\mu}_{\max}(\overline{E}) - \frac{1}{2}\ln(n) - \frac{1}{2}\ln|\mathfrak{D}_{K/\mathbb{Q}}|$$

and

$$\nu_1^{\mathrm{a}}(\overline{E}) \geqslant \widehat{\mu}_{\max}(\overline{E}) - \frac{1}{2}\sum_{\ell=2}^{n}\frac{1}{\ell} = \widehat{\mu}_{\max}^{\mathrm{a}}(\overline{E}) - \frac{1}{2}\sum_{\ell=2}^{n}\frac{1}{\ell}$$

where $\mathfrak{D}_{K/\mathbb{Q}}$ is the discriminant of K over $\mathbb{Q}$. Therefore the adelic curve corresponding to a number field satisfies the Minkowski property of level $\geqslant \frac{1}{2} + \frac{1}{2}\ln|\mathfrak{D}_{K/\mathbb{Q}}|$ and the absolute Minkowski property of level $\geqslant \frac{1}{2}$.

In the function field case, given a regular projective curve (over a base field), by Riemann-Roch formula there exists a constant $A > 0$ which only depends on the curve, such that

$$\nu_1(E) \geqslant \mu_{\max}(E) - A$$

for any vector bundle E on the curve (see [42, Remark 8.3]). Therefore the Minkowski property of level $\geqslant A/\ln(2)$ is satisfied in this case. Moreover, if the base field is of characteristic zero, then it has been shown in [22] that the absolute Minkowski property of level $\geqslant 0$ is satisfied.

The Minkowski property may fail for general adelic curves. Consider the adelic curve $S = (\mathbb{Q}, (\mathbb{Q}, \mathcal{A}, \nu), \phi)$ consisting of the field of rational numbers, the measure space of $\mathbb{Q}$ equipped with the discrete σ-algebra and the atomic measure such that $\nu(\{\omega\}) = 1$ for any $\omega \in \mathbb{Q}$, together with the map ϕ sending any $\omega \in \mathbb{Q}$ to the trivial absolute value on $\mathbb{Q}$. We write the rational numbers into a sequence $\{q_n\}_{n\in\mathbb{N}}$. For any $n \in \mathbb{N}_{\geqslant 2}$, consider the following adelic vector bundle $\overline{E}_n$ on S. Let $E_n = K^2$. For $m \in \mathbb{N}$ such that $m < n$ let $\|\cdot\|_{q_m}$ be the norm on E_n defined as

$$\|(x,y)\|_{q_m} = \begin{cases} \mathrm{e}^{-1}, & \text{if there exists } a \in K^{\times} \text{ such that } (x,y) = a(1,q_m), \\ 0, & \text{if } (x,y) = (0,0), \\ 1, & \text{else.} \end{cases}$$

For $m \in \mathbb{N}$ such that $m \geqslant n$, let $\|\cdot\|_{q_m}$ be the norm on K^2 such that $\|(x,y)\| = 1$ for any $(x,y) \in K^2 \setminus \{(0,0)\}$. Then by definition one has $\widehat{\deg}(\overline{E}_n) = n$ and hence $\widehat{\mu}(\overline{E}_n) = n/2$. Moreover, for any vector subspace F of E, either there exists $m \in \{0,\ldots,n-1\}$ such that $F = K(1,q_m)$ and thus $\widehat{\mu}(\overline{F}) = 1$, or one has $\widehat{\mu}(\overline{F}) = 0$. Since $n \geqslant 2$, we obtain that the adelic vector bundle $\overline{E}_n$ is semistable and of slope $n/2$. Moreover the first minimum of $\overline{E}_n$ is 1. Therefore it is not possible to find a constant C only depending on S such that $\widehat{\mu}_{\max}(\overline{E}_n)$ is bounded from above by $\nu_1(\overline{E}_n) + C\ln(2)$.

In the literature, the absolute Minkowski property is closely related to the semistability of tensor vector bundles and the estimation of the maximal slope of them. We refer the readers to [3, 22, 64] for more detailed discussions. In the following, we prove several slope estimates in assuming the Minkowski property.

Proposition 4.3.75 *Let $\overline{E} = (E, \xi_E)$ and $\overline{F} = (F, \xi_F)$ be adelic vector bundles on S. We assume that ξ_E and ξ_F are ultrametric on $\Omega \setminus \Omega_\infty$. One has*

$$\nu_1(\overline{E} \otimes_{\varepsilon} \overline{F}) \leqslant \widehat{\mu}_{\max}(\overline{E}) + \widehat{\mu}_{\max}(\overline{F}). \tag{4.84}$$

Proof Let f be a non-zero element of $E \otimes_K F$, viewed as a K-linear map from $E^{\vee}$ to F. By Proposition 4.3.31 (3), one has

$$\widehat{\mu}_{\min}(\overline{E}^{\vee}) \leqslant \widehat{\mu}_{\max}(\overline{F}) + h(f) = \widehat{\mu}_{\max}(\overline{F}) - \widehat{\deg}_{\xi_E \otimes_{\varepsilon} \xi_F}(f).$$

By Proposition 4.3.25, we obtain

$$0 \leqslant \widehat{\mu}_{\max}(\overline{E}) + \widehat{\mu}_{\max}(\overline{F}) - \widehat{\deg}_{\xi_E \otimes_{\varepsilon} \xi_F}(f).$$

Since f is arbitrary, we obtain the inequality 4.84. □

Corollary 4.3.76 *Let C be a non-negative real number. We assume that the adelic curve S satisfies the Minkowski property of level $\geqslant C$. Let $\overline{E} = (E, \xi_E)$ and $\overline{F} = (F, \xi_F)$ be adelic vector bundles on S.*

(1) *Assume that ξ_E and ξ_F are ultrametric on $\Omega \setminus \Omega_\infty$. Then*

$$\widehat{\mu}_{\max}(\overline{E} \otimes_\varepsilon \overline{F}) \leqslant \widehat{\mu}_{\max}(\overline{E}) + \widehat{\mu}_{\max}(\overline{F}) + C \ln(\dim_K(E) \cdot \dim_K(F)), \tag{4.85}$$

(2) *One has*

$$\widehat{\mu}_{\min}(\overline{E} \otimes_{\varepsilon,\pi} \overline{F}) \geqslant \widehat{\mu}_{\min}(\overline{E}) + \widehat{\mu}_{\min}(\overline{F}) - (C + \tfrac{1}{2}\nu(\Omega_\infty)) \ln(\dim_K(E) \cdot \dim_K(F)). \tag{4.86}$$

Proof By the assumption of Minkowski property, we have

$$\nu_1(\overline{E} \otimes_\varepsilon \overline{F}) \geqslant \widehat{\mu}_{\max}(\overline{E} \otimes_\varepsilon \overline{F}) - C \ln(\dim_K(E) \cdot \dim_K(F)).$$

Hence (4.85) follows from (4.84).

If we apply the inequality (4.85) to $\overline{E}^\vee$ and $\overline{F}^\vee$ (note that $\xi_E^\vee$ and $\xi_F^\vee$ are always ultrametric on $\Omega \setminus \Omega_\infty$), we obtain

$$\widehat{\mu}_{\max}(\overline{E}^\vee \otimes_\varepsilon \overline{F}^\vee) \leqslant \widehat{\mu}_{\max}(\overline{E}^\vee) + \widehat{\mu}_{\max}(\overline{F}^\vee) + C \ln(\dim_K(E) \cdot \dim_K(F)).$$

By Proposition 4.3.25 we deduce that

$$\widehat{\mu}_{\min}(\overline{E} \otimes_{\varepsilon,\pi} \overline{F}) \geqslant -\widehat{\mu}_{\max}(\overline{E}^\vee) - \widehat{\mu}_{\max}(\overline{F}^\vee) - C \ln(\dim_K(E) \cdot \dim_K(F)).$$

Finally, by Corollary 4.3.27 we obtain (4.86). □

Proposition 4.3.77 *Let $\overline{E}$ be a non-zero adelic vector bundle on S. One has*

$$\nu_{\min}(\overline{E}) \leqslant \widehat{\mu}_{\min}(\overline{E}).$$

Proof Let r be the dimension of E over K. Assume that $\overline{E}$ is of the form $\overline{E} = (E, \xi)$, with $\xi = \{\|\cdot\|_\omega\}_{\omega\in\Omega}$. Let t be a real number and $\{s_i\}_{i=1}^r$ be a basis of E over K such that $\widehat{\deg}_\xi(s_i) \geqslant t$ for any $i \in \{1, \ldots, r\}$.

Let G be a quotient vector space of E and $\xi_G = \{\|\cdot\|_{G,\omega}\}_{\omega\in\Omega}$ be the quotient norm family of ξ on G. For any $i \in \{1, \ldots, r\}$, let α_i be the canonical image of s_i in G. Without loss of generality, we may assume that $\{\alpha_1, \ldots, \alpha_n\}$ forms a basis of G over K. For any $\omega \in \Omega$, one has

$$\|\alpha_1 \wedge \cdots \wedge \alpha_n\|_{G,\omega,\det} \leqslant \prod_{i=1}^n \|\alpha_i\|_{G,\omega} \leqslant \prod_{i=1}^n \|s_i\|_\omega,$$

where the first inequality comes from Proposition 1.1.63. Therefore one has

$$\widehat{\deg}(\overline{G}) \geqslant \sum_{i=1}^{n} \widehat{\deg}_{\xi}(s_i) \geqslant nt,$$

which implies $\widehat{\mu}(\overline{G}) \geqslant t$. Therefore we obtain $\nu_{\min}(\overline{E}) \leqslant \widehat{\mu}_{\min}(\overline{E})$. □

Corollary 4.3.78 *Let $\overline{E} = (E, \xi)$ be a non-zero adelic vector bundle on S. For any $i \in \{1, \ldots, \dim_K(E)\}$ one has $\nu_i(\overline{E}) \leqslant \widehat{\mu}_i(\overline{E})$.*

Proof By the relations (4.82) and (4.65), Proposition 4.3.77 leads to $\mathcal{F}^t_{\mathrm{m}}(\overline{E}) \subseteq \mathcal{F}^t_{\mathrm{hn}}(\overline{E})$ for any $t \in \mathbb{R}$. Therefore, by Proposition 1.1.39 we obtain that $\nu_i(\overline{E}) \leqslant \widehat{\mu}_i(\overline{E})$ for any $i \in \{1, \ldots, \dim_K(E)\}$. □

Definition 4.3.79 Let $S = (K, (\Omega, \mathcal{A}, \nu), \phi)$ be an adelic curve. Let C be a non-negative real number. We say that S satisfies the *strong Minkowski property of level* $\geqslant C$ if for any adelic vector bundle (E, ξ) on S such that ξ is ultrametric on $\Omega \setminus \Omega_\infty$ one has

$$\nu_{\min}(E, \xi) \geqslant \widehat{\mu}_{\min}(E, \xi) - C \ln(\dim_K(E)). \tag{4.87}$$

Proposition 4.3.80 *Assume that the adelic curve S satisfies the strong Minkowski property of level $\geqslant C$. For any non-zero adelic vector bundle $\overline{E} = (E, \xi)$ on S one has*

$$\widehat{\mu}_i(\overline{E}) \leqslant \nu_i(\overline{E}) + C \ln(\dim_K(E)). \tag{4.88}$$

Proof Since the adelic curve S satisfies the strong Minkowski property of level $\geqslant C$, by the relation (4.87) we obtain that

$$\forall\, t \in \mathbb{R}, \quad \mathcal{F}^t_{\mathrm{hn}}(\overline{E}) \subseteq \mathcal{F}^{t - C\ln(\dim_K(E))}_{\mathrm{m}}(\overline{E}).$$

Therefore, by Proposition 1.1.39 we obtain the inequality (4.88). □

Remark 4.3.81 Proposition 4.3.80 shows that, if the adelic curve S satisfies the strong Minkowski property of level $\geqslant C$, then it also satisfies Minkowski property of level $\geqslant C$. Moreover, the transference theorem of Gaudron [63, Theorem 36] shows that, for any *Hermitian* adelic vector bundle $\overline{E}$ of rank n over a number field K, one has

$$\begin{aligned}\nu_{\min}(\overline{E}) - \widehat{\mu}_{\min}(\overline{E}) = \nu_{\min}(\overline{E}) + \widehat{\mu}_{\max}(\overline{E}^\vee) &\geqslant \nu_{\min}(\overline{E}) + \widehat{\nu}_{\max}(\overline{E}^\vee) \\ &\geqslant \ln(n) + \ln|\mathfrak{D}_{K/\mathbb{Q}}|,\end{aligned}$$

where $\mathfrak{D}_{K/\mathbb{Q}}$ is the discriminant of K over $\mathbb{Q}$. We then deduce that, if $\overline{E}$ is a general adelic vector bundle of dimension n over K, which is not necessarily Hermitian, one has (by Theorem 4.1.26)

$$\nu_{\min}(\overline{E}) \geqslant \widehat{\mu}_{\min}(\overline{E}) - \Big(1 + \frac{1}{2}[K : \mathbb{Q}]\Big) \ln(n) + \ln|\mathfrak{D}_{K/\mathbb{Q}}|.$$

Therefore the adelic curve corresponding to a number field K satisfies the strong Minkowski property of level $1 + \frac{1}{2}[K : \mathbb{Q}] \ln|\mathfrak{D}_{K/\mathbb{Q}}|$.

4.4 Adelic vector bundles over number fields

Throughout this section, we fix a number field K and the standard adelic curve $S = (K, (\Omega, \mathcal{A}, \nu), \phi)$ of K as in Subsection 3.2.2. Note that S is proper. Denote by Ω_{fin} the set $\Omega \setminus \Omega_\infty$ of finite places of K, and by $\mathfrak{o}_K$ the ring of algebraic integers in K. Note that the absolute value $|\cdot|_\omega$ at ω is given by

$$\forall\, x \in K_\omega, \quad |x|_\omega = \begin{cases} \text{the standard absolute value of } x \text{ in either } \mathbb{R} \text{ or } \mathbb{C} & \text{if } \omega \in \Omega_\infty, \\ \exp\left(\dfrac{-\ln(p_\omega)\operatorname{ord}_\omega(x)}{\operatorname{ord}_\omega(p_\omega)}\right) & \text{if } \omega \in \Omega_{\text{fin}}, \end{cases}$$

where p_ω is the characteristic of the residue field of the valuation ring of K_ω. Further, for $\omega \in \Omega_{\text{fin}}$, let $\mathfrak{o}_{K,\omega}$ be the localisation of $\mathfrak{o}_K$ at ω and $\mathfrak{o}_\omega$ be the valuation ring of the completion K_ω of K with respect to ω, that is,

$$\mathfrak{o}_{K,\omega} = \{a \in K \,:\, |a|_\omega \leqslant 1\} \quad \text{and} \quad \mathfrak{o}_\omega = \{a \in K_\omega \,:\, |a|_\omega \leqslant 1\}.$$

Moreover, $\nu(\{\omega\}) = [K_\omega : \mathbb{Q}_\omega]$ for $\omega \in \Omega$ and $\sum_{\omega \in \Omega_\infty} \nu(\{\omega\}) = [K : \mathbb{Q}]$.

Let E be a finite-dimensional vector space over K and $\xi = \{\|\cdot\|_\omega\}_{\omega \in \Omega}$ be a norm family of E over S. In this section, we always assume that $\|\cdot\|_\omega$ is ultrametric for every $\omega \in \Omega_{\text{fin}}$. For $\omega \in \Omega_{\text{fin}}$, we set

$$E_\omega := E \otimes_K K_\omega \quad \text{and} \quad \mathscr{E}_\omega := \{x \in E_\omega \mid \|x\|_\omega \leqslant 1\}.$$

By Propositions 1.1.25 and 1.1.30, $\mathscr{E}_\omega$ is a free $\mathfrak{o}_\omega$-module of rank $\dim_K E$ and $\mathscr{E}_\omega \otimes_{\mathfrak{o}_\omega} K_\omega = E_\omega$.

Remark 4.4.1 As in the next subsection, let $(E, \xi)^\omega_{\leqslant 1} := \{x \in E \,:\, \|x\|_\omega \leqslant 1\}$. Then one can see the following:

(1) $(E, \xi)^\omega_{\leqslant 1}$ is a free $\mathfrak{o}_{K,\omega}$-module.
(2) $(E, \xi)^\omega_{\leqslant 1} \otimes_{\mathfrak{o}_{K,\omega}} \mathfrak{o}_\omega = \mathscr{E}_\omega$.

(1) Fix a basis of $\{x_i\}_{i=1}^r$ of E. We consider a norm $\|\cdot\|'_\omega$ on E_ω given by

$$\forall\, (\lambda_1, \ldots, \lambda_r) \in K_\omega^r, \quad \|\lambda_1 x_1 + \cdots + \lambda_r x_r\|'_\omega = \max\{|\lambda_1|_\omega, \ldots, |\lambda_r|_\omega\}.$$

By Proposition 1.1.11, there is a positive integer n such that $|\varpi_\omega|^n \|\cdot\|'_\omega \leqslant \|\cdot\|_\omega$, where ϖ_ω is a uniformizing parameter of $\mathfrak{o}_{K,\omega}$. Therefore,

$$(E, \xi)^\omega_{\leqslant 1} \subseteq \mathfrak{o}_{K,\omega} e_1 \varpi_\omega^{-n} + \cdots + \mathfrak{o}_{K,\omega} e_r \varpi_\omega^{-n},$$

as required.

(2) Obviously $(E, \xi)^\omega_{\leqslant 1} \otimes_{\mathfrak{o}_{K,\omega}} \mathfrak{o}_\omega \subseteq \mathscr{E}_\omega$. Let $\{e_i\}_{i=1}^r$ be a free basis of $(E, \xi)^\omega_{\leqslant 1}$ over $\mathfrak{o}_{K,\omega}$. For $x \in \mathscr{E}_\omega$, we choose $(a_1, \ldots, a_r) \in K_\omega^r$ such that $x = a_1 e_1 + \cdots + a_r e_r$. One can find $(a'_1, \ldots, a'_r) \in K^r$ such that

$$|a_i - a'_i|_\omega \leqslant \frac{1}{2}|a_i|_\omega \text{ for any } i \in \{1, \ldots, r\} \text{ and } \|x - (a'_1 e_1 + \cdots + a'_r e_r)\|_\omega \leqslant \frac{1}{2}\|x\|_\omega.$$

If we set $x' = a'_1 e_1 + \cdots + a'_r e_r$, then $\|x\|_\omega = \|x'\|_\omega$ and $|a_i|_\omega = |a'_i|_\omega$ for all i. In particular $x' \in (E, \xi)^\omega_{\leqslant 1}$, so that $a'_i \in \mathfrak{o}_{K,\omega}$, and hence $|a_i|_\omega = |a'_i|_\omega \leqslant 1$. Therefore, $x \in (E, \xi)^\omega_{\leqslant 1} \otimes_{\mathfrak{o}_{K,\omega}} \mathfrak{o}_\omega$.

4.4.1 Coherency for a norm family

Let E be a finite-dimensional vector space over K. Let $\xi = \{\|\cdot\|_\omega\}_{\omega \in \Omega}$ be a norm family of E over S. We define $(E, \xi)^{\mathrm{fin}}_{\leqslant 1}$ and $(E, \xi)^\omega_{\leqslant 1}$ ($\omega \in \Omega_{\mathrm{fin}}$) to be

$$\begin{cases} (E, \xi)^{\mathrm{fin}}_{\leqslant 1} := \{x \in E \,:\, \|x\|_\omega \leqslant 1 \text{ for all } \omega \in \Omega_{\mathrm{fin}}\}, \\ (E, \xi)^\omega_{\leqslant 1} := \{x \in E \,:\, \|x\|_\omega \leqslant 1\}. \end{cases}$$

Note that $(E, \xi)^{\mathrm{fin}}_{\leqslant 1}$ and $(E, \xi)^\omega_{\leqslant 1}$ are an $\mathfrak{o}_K$-module and an $\mathfrak{o}_{K,\omega}$-module, respectively. Furthermore, by Remark 4.4.1, $(E,\xi)^\omega_{\leqslant 1}$ is a free $\mathfrak{o}_{K,\omega}$-module and $(E, \xi)^\omega_{\leqslant 1} \otimes_{\mathfrak{o}_{K,\omega}} \mathfrak{o}_\omega = \mathscr{E}_\omega$. Let us begin with the following proposition.

Proposition 4.4.2 *The following are equivalent:*

(1) *For any* $v \in E$, $\|v\|_\omega \leqslant 1$ *except finitely many* $\omega \in \Omega_{\mathrm{fin}}$.
(2) $(E, \xi)^{\mathrm{fin}}_{\leqslant 1} \otimes_{\mathfrak{o}_K} \mathfrak{o}_{K,\omega} = (E, \xi)^\omega_{\leqslant 1}$ *for all* $\omega \in \Omega_{\mathrm{fin}}$.
(3) $(E, \xi)^{\mathrm{fin}}_{\leqslant 1} \otimes_{\mathfrak{o}_K} \mathfrak{o}_{K,\omega} = (E, \xi)^\omega_{\leqslant 1}$ *for some* $\omega \in \Omega_{\mathrm{fin}}$.
(4) $(E, \xi)^{\mathrm{fin}}_{\leqslant 1} \otimes_{\mathfrak{o}_K} K = E$.

Moreover, under the above equivalent conditions, $(E, \xi)^{\mathrm{fin}}_{\leqslant 1} \otimes_{\mathbb{Z}} \mathbb{Q} = E$.

Proof First of all, let us see the following claim:

Claim 4.4.3 (a) *Let S be a finite subset of Ω_{fin}. Then there is $f \in \mathfrak{o}_K \setminus \{0\}$ such that*

$$\operatorname{ord}_\omega(f) \begin{cases} > 0 & \text{if } \omega \in S, \\ = 0 & \text{if } \omega \notin S. \end{cases}$$

(b) $(E, \xi)^\omega_{\leqslant 1} \otimes_{\mathfrak{o}_{K,\omega}} K = E$ *for all* $\omega \in \Omega_{\mathrm{fin}}$. □

Proof (a) Let us consider the ideal given by $\mathfrak{I} = \prod_{\mathfrak{p} \in S} \mathfrak{p}$. As the class group of K is finite, there are a positive integer a and $f \in \mathfrak{o}_K$ such that $f\mathfrak{o}_K = \mathfrak{I}^a$, as required.

(b) Obviously $(E, \xi)^\omega_{\leqslant 1} \otimes_{\mathfrak{o}_{K,\omega}} K \subseteq E$. For $v \in E$, there is $a \in \mathfrak{o}_{K,\omega} \setminus \{0\}$ such that $av \in (E, \xi)^\omega_{\leqslant 1}$, which shows the converse inclusion. □

(1) $\Longrightarrow$ (2): Clearly $(E, \xi)^{\mathrm{fin}}_{\leqslant 1} \otimes_{\mathfrak{o}_K} \mathfrak{o}_{K,\omega} \subseteq (E, \xi)^\omega_{\leqslant 1}$. Conversely, for $v \in (E, \xi)^\omega_{\leqslant 1}$, as $S = \{\omega' \in \Omega_{\mathrm{fin}} \,:\, \|v\|_{\omega'} > 1\}$ is finite, there is $f \in \mathfrak{o}_K \setminus \{0\}$ such that $|f|_{\omega'} < 1$

for $\omega' \in S$ and $|f|_{\omega'} = 1$ for $\omega' \in \Omega_{\mathrm{fin}} \setminus S$ by the above claim (a). Thus, there is a positive integer n such that $f^n v \in (E, \xi)^{\mathrm{fin}}_{\leqslant 1}$. Note that $f \in \mathfrak{o}_{K,\omega}^{\times}$. Thus the converse inclusion holds.

"(2) $\Longrightarrow$ (3)" is obvious and "(3) $\Longrightarrow$ (4)" follows from (b) in the claim. Let us see that "(4) $\Longrightarrow$ (1)". For $v \in E$, there is $a \in \mathfrak{o}_K \setminus \{0\}$ such that $av \in (E, \xi)^{\mathrm{fin}}_{\leqslant 1}$, that is, $|a|_\omega \|v\|_\omega \leq 1$ for all $\omega \in \Omega_{\mathrm{fin}}$. Note that $|a|_\omega = 1$ except finitely many ω, so that one has (1).

Note that $(E, \xi)^{\mathrm{fin}}_{\leqslant 1} \otimes_{\mathfrak{o}_K} K$ and $(E, \xi)^{\mathrm{fin}}_{\leqslant 1} \otimes_{\mathbb{Z}} \mathbb{Q}$ are the localisations of $(E, \|\cdot\|)^{\mathrm{fin}}_{\leqslant 1}$ with respect to $\mathfrak{o}_K \setminus \{0\}$ and $\mathbb{Z} \setminus \{0\}$, respectively. Therefore, for the last assertion, it is sufficient to show that, for $\alpha \in \mathfrak{o}_K \setminus \{0\}$, there is $\alpha' \in \mathfrak{o}_K \setminus \{0\}$ such that $\alpha\alpha' \in \mathbb{Z} \setminus \{0\}$. Indeed, one can find $(a_1, \ldots, a_n) \in \mathbb{Z}^n$ such that $\alpha^n + a_1\alpha^{n-1} + \cdots + a_{n-1}\alpha + a_n = 0$. We may assume that $a_n \neq 0$. Thus $\alpha(\alpha^{n-1} + a_1\alpha^{n-2} + \cdots + a_{n-1}) = -a_n \in \mathbb{Z} \setminus \{0\}$. □

Definition 4.4.4 We say that (E, ξ) is *coherent* if the equivalent conditions of Proposition 4.4.2 are satisfied.

Proposition 4.4.5 *If there are a non-empty open set U of* $\mathrm{Spec}(\mathfrak{o}_K)$ *and a locally free $\mathfrak{o}_U$-module $\mathscr{E}$ such that $\mathscr{E} \otimes_{\mathfrak{o}_U} K = E$ and $\|\cdot\|_\omega = \|\cdot\|_{\mathscr{E} \otimes_{\mathfrak{o}_U} \mathfrak{o}_\omega}$ for all $\omega \in U \cap \Omega_{\mathrm{fin}}$, then (E, ξ) is coherent and dominated, where $\mathfrak{o}_U$ is the ring of regular functions on the open set U.*

Proof For $s \in E \setminus \{0\}$, we can find a non-empty open set $U' \subseteq U$ such that $s \in \mathscr{E} \otimes_{\mathfrak{o}_U} \mathfrak{o}_\omega$ and $\mathscr{E} \otimes_{\mathfrak{o}_U} \mathfrak{o}_\omega / \mathfrak{o}_\omega s$ is torsion free for all $\omega \in U' \cap \Omega_{\mathrm{fin}}$, so that

$$\|s\|_\omega = \|s\|_{\mathscr{E} \otimes_{\mathfrak{o}_U} \mathfrak{o}_\omega} = 1.$$

In particular, (E, ξ) is upper dominated and coherent. Let $\mathscr{E}^\vee$ be the dual of $\mathscr{E}$ over U. Note that $\mathscr{E}^\vee \otimes_{\mathfrak{o}_U} \mathfrak{o}_\omega = (\mathscr{E} \otimes_{\mathfrak{o}_U} \mathfrak{o}_\omega)^\vee$, so that by Propsotion 1.1.34, $\|\cdot\|_{\omega,*} = \|\cdot\|_{\mathscr{E}^\vee \otimes_{\mathfrak{o}_U} \mathfrak{o}_\omega}$ for all $\omega \in U \cap \Omega_{\mathrm{fin}}$. Therefore, in the same way as above, one can see that $(E^\vee, \xi^\vee)$ is upper dominated, and hence (E, ξ) is dominated. □

4.4.2 Finite generation of a dominated vector bundle over S

Let E be a finite-dimensional vector space over K and $\xi = \{\|\cdot\|_\omega\}_{\omega \in \Omega}$ be a norm family of E over S. The purpose of this subsection is to prove the following proposition.

Proposition 4.4.6 *If ξ is dominated, then*

$$(E, \xi)^{\mathrm{fin}}_{\leqslant 1} := \{x \in E \;:\; \|x\|_\omega \leqslant 1 \textit{ for all } \omega \in \Omega_{\mathrm{fin}}\}$$

is a finitely generated $\mathfrak{o}_K$-module.

Proof First we assume that $\dim_K(E) = 1$. Fix $x \in E \setminus \{0\}$. For each $\omega \in \Omega_{\mathrm{fin}}$, let a_ω be the smallest integer a with $a \geqslant -\ln\|x\|_\omega/\ln|\varpi_\omega|_\omega$, where ϖ_ω is a local parameter of $\mathfrak{o}_{K,\omega}$.

As ξ is lower dominated, there is an integrable function $A(\omega)$ on Ω such that

$$\forall\,\omega \in \Omega,\quad -\ln\|x\|_\omega \leqslant A(\omega).$$

Here we assume that there are infinitely many $\omega \in \Omega_{\mathrm{fin}}$ with $a_\omega \leqslant -1$. As $a_\omega \leqslant -1$ implies $-\ln|\varpi_\omega|_\omega \leqslant -\ln\|x\|_\omega$, one has

$$\begin{aligned} A(\omega)\nu(\{\omega\}) &\geqslant -\ln\|x\|_\omega\nu(\{\omega\}) \\ &\geqslant -\ln|\varpi_\omega|_\omega\nu(\{\omega\}) = \ln\operatorname{card}(\mathfrak{o}_K/\mathfrak{p}_\omega) \geqslant \ln p_\omega, \end{aligned}$$

where $\mathfrak{p}_\omega$ is the maximal ideal of $\mathfrak{o}_K$ and p_ω is the characteristic of the residue field $\mathfrak{o}_K/\mathfrak{p}_\omega$, which gives a contradiction to the integrability of the function $A(\cdot)$. Therefore, $a_\omega \geqslant 0$ except finitely many $\omega \in \Omega_{\mathrm{fin}}$.

Note that

$$\begin{aligned} \|ax\|_\omega = |a|_\omega\|x\|_\omega \leqslant 1 &\iff |\varpi_\omega|_\omega^{\mathrm{ord}_\omega(a)+\ln\|x\|_\omega/\ln|\varpi_\omega|_\omega} \leqslant 1 \\ &\iff \mathrm{ord}_\omega(a) \geqslant -\ln\|x\|_\omega/\ln|\varpi_\omega|_\omega \\ &\iff \mathrm{ord}_\omega(a) \geqslant a_\omega. \end{aligned}$$

Therefore

$$\begin{aligned} &\{a \in K \;:\; \|ax\|_\omega \leqslant 1 \text{ for all } \omega \in \Omega_{\mathrm{fin}}\} \\ &\qquad = \{a \in K \;:\; \mathrm{ord}_\omega(a) - a_\omega \geqslant 0 \text{ for all } \omega \in \Omega_{\mathrm{fin}}\} \end{aligned}$$

is finitely generated over $\mathfrak{o}_K$ by Lemma 4.4.7. Thus one has the assertion in the case where $\dim_K(E) = 1$.

In general, we prove the theorem by induction on $\dim_K(E)$. By the previous observation, we may assume $\dim_K(E) \geqslant 2$. Fix $x \in E \setminus \{0\}$. We set $E' = Kx$ and $E'' = E/E'$. Let ξ' be the norm family on E' given by the restriction of ξ, and ξ'' be the norm family on E'' given by the quotient of ξ. Then ξ' and ξ'' are dominated by Proposition 4.1.19, so that, by the hypothesis of induction, $(E',\xi')^{\mathrm{fin}}_{\leqslant 1}$ and $(E'',\xi'')^{\mathrm{fin}}_{\leqslant 1}$ are finitely generated over $\mathfrak{o}_K$. Note that $\beta((E,\xi)^{\mathrm{fin}}_{\leqslant 1}) \subseteq (E'',\xi'')^{\mathrm{fin}}_{\leqslant 1}$, where β is the canonical homomorphism $E \to E''$. In particular, $\beta((E,\xi)^{\mathrm{fin}}_{\leqslant 1})$ is finitely generated over $\mathfrak{o}_K$ because $\mathfrak{o}_K$ is Noetherian. Therefore, one has the exact sequence

$$0 \longrightarrow (E',\xi')^{\mathrm{fin}}_{\leqslant 1} \longrightarrow (E,\xi)^{\mathrm{fin}}_{\leqslant 1} \longrightarrow \beta((E,\xi)^{\mathrm{fin}}_{\leqslant 1}) \longrightarrow 0,$$

and hence the assertion follows. □

Lemma 4.4.7 *Let $\{b_\omega\}_{\omega\in\Omega_{\mathrm{fin}}}$ be a family of integers indexed by Ω_{fin}. Then*

$$\mathfrak{o}_K(\{b_\omega\}_{\omega\in\Omega_{\mathrm{fin}}}) := \{a \in K \;:\; \mathrm{ord}_\omega(a) + b_\omega \geqslant 0 \textit{ for all } \omega \in \Omega_{\mathrm{fin}}\}$$

is finitely generated over $\mathfrak{o}_K$ *if and only if either* $b_\omega \leqslant 0$ *except finitely many* ω*, or* $b_\omega < 0$ *for infinitely many* ω.

Proof We set $S = \{\omega \in \Omega_{\rm fin} \,:\, b_\omega \geqslant 1\}$ and $T = \{\omega \in \Omega_{\rm fin} \,:\, b_\omega \leqslant -1\}$.

First we assume that $b_\omega \leqslant 0$ except finitely many ω, that is, $\mathrm{card}(S) < \infty$. Then one can choose $f \in \mathfrak{o}_K \setminus \{0\}$ such that $\mathrm{ord}_\omega(f) \geqslant b_\omega$ for all $\omega \in \Omega_{\rm fin}$. Note that $\{a \in K \,:\, \mathrm{ord}_\omega(a) \geqslant 0 \text{ for all } \omega \in \Omega_{\rm fin}\} = \mathfrak{o}_K$, so that $\mathfrak{o}_K(\{b_\omega\}_{\omega\in\Omega_{\rm fin}})f \subseteq \mathfrak{o}_K$. Thus $\mathfrak{o}_K(\{b_\omega\}_{\omega\in\Omega_{\rm fin}})f$ is finitely generated over $\mathfrak{o}_K$ because $\mathfrak{o}_K$ is Noetherian. Therefore $\mathfrak{o}_K(\{b_\omega\}_{\omega\in\Omega_{\rm fin}})$ is also finitely generated over $\mathfrak{o}_K$.

Next we assume that $b_\omega < 0$ for infinitely many ω, that is, $\mathrm{card}(T) = \infty$. In this case, $\mathfrak{o}_K(\{b_\omega\}_{\omega\in\Omega_{\rm fin}}) = \{0\}$. Indeed, if $x \in \mathfrak{o}_K(\{b_\omega\}_{\omega\in\Omega_{\rm fin}}) \setminus \{0\}$, then $\mathrm{ord}_\omega(x) \geq 1$ for all $\omega \in T$, which is a contradiction.

Finally we assume that $\mathrm{card}(S) = \infty$ and $\mathrm{card}(T) < \infty$. In this case, we need to show that $\mathfrak{o}_K(\{b_\omega\}_{\omega\in\Omega_{\rm fin}})$ is not finitely generated over $\mathfrak{o}_K$. We set $S = \{\omega_1, \omega_2, \ldots, \omega_n, \ldots\}$. For each positive integer N, let us consider a family $\{b_{N,\omega}\}_{\omega\in\Omega_{\rm fin}}$ of integers given by

$$b_{N,\omega} = \begin{cases} 0 & \text{if } \omega \in \{\omega_n \,:\, n \geqslant N+1\}, \\ b_\omega & \text{otherwise.} \end{cases}$$

Then one has a strictly increasing sequence of finitely generated $\mathfrak{o}_K$-modules:

$$\mathfrak{o}_K(\{b_{1,\omega}\}_{\omega\in\Omega_{\rm fin}}) \subsetneq \mathfrak{o}_K(\{b_{2,\omega}\}_{\omega\in\Omega_{\rm fin}}) \subsetneq \cdots \subsetneq \mathfrak{o}_K(\{b_{N,\omega}\}_{\omega\in\Omega_{\rm fin}}) \subsetneq \cdots$$

such that $\bigcup_{N=1}^{\infty} \mathfrak{o}_K(\{b_{N,\omega}\}_{\omega\in\Omega_{\rm fin}}) = \mathfrak{o}_K(\{b_\omega\}_{\omega\in\Omega_{\rm fin}})$. Therefore $\mathfrak{o}_K(\{b_\omega\}_{\omega\in\Omega_{\rm fin}})$ is not finitely generated over $\mathfrak{o}_K$. □

Example 4.4.8 Let $\{b_\omega\}_{\omega\in\Omega_{\rm fin}}$ be a family of integers indexed by $\Omega_{\rm fin}$. To each $\omega \in \Omega_{\rm fin}$, we assign a norm $\|\cdot\|_\omega$ of K_ω given by

$$\|x\|_\omega = \exp\left(\frac{-b_\omega \ln(p_\omega)}{\mathrm{ord}_\omega(p_\omega)}\right)|x|_\omega$$

for $x \in K_\omega$. Moreover, for $\omega \in \Omega_\infty$ let $\|\cdot\|_\omega$ be the standard absolute value of either $\mathbb{R}$ or $\mathbb{C}$. Then $\xi := \{\|\cdot\|_\omega\}_{\omega\in\Omega}$ yields a norm family on K. Note that, for $\omega \in \Omega_{\rm fin}$, $\|x\|_\omega \leqslant 1$ if and only if $\mathrm{ord}_\omega(x) + b_\omega \geqslant 0$ for $x \in K$, that is,

$$(K, \xi)^{\rm fin}_{\leqslant 1} = \mathfrak{o}_K(\{b_\omega\}_{\omega\in\Omega_{\rm fin}}).$$

For example, if we set $b_\omega = 1$ for all $\omega \in \Omega_{\rm fin}$, then $(K, \xi)^{\rm fin}_{\leqslant 1}$ is not finitely generated over $\mathfrak{o}_K$ by Lemma 4.4.7.

4.4.3 Invariants λ and σ

Let $(E, \xi = \{\|\cdot\|_\omega\}_{\omega\in\Omega})$ be an adelic vector bundle on S. Let $\mathscr{E} := (E,\xi)^{\mathrm{fin}}_{\leqslant 1} = \{x \in E : \|x\|_\omega \leqslant 1 \text{ for all } \omega \in \Omega_{\mathrm{fin}}\}$. If ξ is coherent and dominated, then, by Proposition 4.4.2, 4.4.6 and Remark 4.4.1, $\mathscr{E}$ is a finitely generated $\mathfrak{o}_K$-module, $\mathscr{E} \otimes_{\mathfrak{o}_K} K = E$ and $\mathscr{E} \otimes_{\mathfrak{o}_K} \mathfrak{o}_\omega = \mathscr{E}_\omega$ for all $\omega \in \Omega_{\mathrm{fin}}$.

We define $\|\cdot\|_\infty$ to be

$$\forall\, x \in E,\quad \|x\|_\infty := \max_{\omega\in\Omega_\infty} \{\|\iota_\omega(x)\|_\sigma\},$$

where ι_ω is the canonical homomorphism $E \to E_\omega$. Under the assumption that ξ is coherent and dominated, the invariant $\lambda(E,\xi)$ is defined to be

$$\lambda(E,\xi) := \begin{cases} \infty & \text{if } E = \{0\}, \\ \sup\left\{ \lambda \in \mathbb{R} \;:\; \begin{array}{l} \text{There is a basis } e_1,\ldots,e_r \text{ of } E \\ \text{over } K \text{ such that } e_1,\ldots,e_r \in \mathscr{E} \\ \text{and } \max\{\|e_1\|_\infty,\ldots,\|e_r\|_\infty\} \leqslant e^{-\lambda} \end{array} \right\} & \text{otherwise.} \end{cases}$$

By Proposition 1.1.30,

$$0 \leqslant \sup_{x\in E_\omega\setminus\{0\}} \ln\left(\frac{\|x\|_{\mathscr{E}_\omega}}{\|x\|_\omega}\right) \leqslant -\ln|\varpi_\omega|_\omega \tag{4.89}$$

for any $\omega \in \Omega_{\mathrm{fin}}$, where $\|\cdot\|_{\mathscr{E}_\omega}$ is the norm arising from $\mathscr{E}_\omega$ (cf. Subsection 1.1.7). The *impurity* $\sigma(E,\xi)$ of (E,ξ) is defined to be

$$\sigma(E,\xi) := \sum_{\omega\in\Omega_{\mathrm{fin}}} \sup_{x\in E_\omega\setminus\{0\}} \ln\left(\frac{\|x\|_{\mathscr{E}_\omega}}{\|x\|_\omega}\right) \nu(\{\omega\}) \in [0,\infty].$$

Note that $\sigma(E,\xi) = 0$ if and only if $\|\cdot\|_\omega = \|\cdot\|_{\mathscr{E}_\omega}$ for all $\omega \in \Omega_{\mathrm{fin}}$. Moreover, if ξ is coherent and dominated, then, by Proposition 4.4.5, $\xi' = \{\|\cdot\|_{\mathscr{E}_\omega}\}_{\omega\in\Omega_{\mathrm{fin}}} \cup \{\|\cdot\|_\omega\}_{\omega\in\Omega_\infty}$ is also coherent and dominated, so that $\sigma(E,\xi) < \infty$ by Corollary 4.1.10.

Proposition 4.4.9 *We assume that ξ is coherent. Then the following are equivalent:*

(1) *ξ is dominated.*
(2) *$\mathscr{E}$ is finitely generated over $\mathfrak{o}_K$ and $\sigma(E,\xi) < \infty$.*

Proof It is sufficient to see that (2) $\Longrightarrow$ (1). If we set

$$\xi' = \{\|\cdot\|_{\mathscr{E}_\omega}\}_{\omega\in\Omega_{\mathrm{fin}}} \cup \{\|\cdot\|_\omega\}_{\omega\in\Omega_\infty},$$

then ξ' is dominated by Proposition 4.4.5 together with Proposition 4.4.2 and Remark 4.4.1. Therefore the assertion follows from the assumption $\sigma(E,\xi) < \infty$. □

Proposition 4.4.10 *We assume that ξ is coherent and dominated. There is a constant c_K depending only on K such that*

$$[K:\mathbb{Q}]\lambda(E,\xi) \leqslant \nu_{\min}(E,\xi) \leqslant [K:\mathbb{Q}]\lambda(E,\xi) + \sigma(E,\xi) + c_K.$$

Proof First we consider the inequality $[K:\mathbb{Q}]\lambda(E,\xi) \leqslant \nu_{\min}(E,\xi)$. We set $\lambda = \lambda(E,\xi)$. For $\epsilon > 0$, there is a basis $\{e_1,\ldots,e_r\}$ of E over K such that $e_i \in \mathscr{E}$ and $\|e_i\|_\infty \leqslant e^{-\lambda+\epsilon}$ for all i. On the other hand,

$$\begin{aligned}\widehat{\deg}_\xi(e_i) &= \sum_{\omega\in\Omega} -\ln\|e_i\|_\omega \nu(\{\omega\}) \geqslant \sum_{\omega\in\Omega_\infty} -\ln\|e_i\|_\omega \nu(\{\omega\}) \\ &\geqslant \sum_{\omega\in\Omega_\infty} (\lambda-\epsilon)\nu(\{\omega\}) = [K:\mathbb{Q}](\lambda-\epsilon),\end{aligned}$$

so that the assertion follows.

Next let us see the second inequality

$$\nu_{\min}(E,\xi) \leqslant [K:\mathbb{Q}]\lambda(E,\xi) + \sigma(E,\xi) + c_K.$$

For $\epsilon > 0$, there is a basis $\{e'_1,\ldots,e'_r\}$ of E over K such that $\widehat{\deg}(e'_i) \geq \nu_{\min}(E,\xi) - \epsilon$ for all i. We set $E_i = Ke'_i$ and $\mathscr{E}_i = \mathscr{E} \cap E_i$. By Lemma 4.4.11 below, there is an $e''_i \in \mathscr{E}_i$ such that

$$\operatorname{card}(\mathscr{E}_i/\mathfrak{o}_K e''_i) \leqslant C'_K,$$

where C'_K is a constant depending only on the number field K. Therefore $\nu_{\min}(E,\xi) - \epsilon$ is bounded from above by

$$\begin{aligned}\widehat{\deg}_\xi(e'_i) = \widehat{\deg}_\xi(e''_i) &= \sum_{\omega\in\Omega_{\rm fin}} -\ln\|e''_i\|_\omega\nu(\{\omega\}) + \sum_{\omega\in\Omega_\infty} -\ln\|e''_i\|_\omega\nu(\{\omega\}) \\ &= \sum_{\omega\in\Omega_{\rm fin}} -\ln\|e''_i\|_{\mathscr{E}_\omega}\nu(\{\omega\}) + \sum_{\omega\in\Omega_{\rm fin}} \ln\left(\frac{\|e''_i\|_{\mathscr{E}_\omega}}{\|e''_i\|_\omega}\right)\nu(\{\omega\}) \\ &\qquad + \sum_{\omega\in\Omega_\infty} -\ln\|e''_i\|_\omega\nu(\{\omega\}) \\ &\leqslant \ln\operatorname{card}(\mathscr{E}_i/\mathfrak{o}_K e''_i) + \sum_{\omega\in\Omega_{\rm fin}} \ln(\|\mathrm{Id}_{E_\omega}\|_\omega^{\rm op})\nu(\{\omega\}) + \sum_{\omega\in\Omega_\infty} -\ln\|e''_i\|_\omega\nu(\{\omega\}) \\ &\leqslant \ln C'_K + \sigma(E,\xi) + \sum_{\omega\in\Omega_\infty} -\ln\|e''_i\|_\omega\nu(\{\omega\}).\end{aligned}$$

If we set

$$A = \frac{1}{[K:\mathbb{Q}]}\sum_{\omega\in\Omega_\infty} \ln\|e''_i\|_\omega\nu(\{\omega\}),$$

then $\sum_{\omega\in\Omega_\infty}(\ln\|e''_i\|_\omega - A)\nu(\{\omega\}) = 0$. Let $\{u_1,\ldots,u_s\}$ be a free basis of $\mathfrak{o}_K^\times$ modulo the torsion subgroup. Then, by Dirichlet's unit theorem, there are $a'_{i1},\ldots,a'_{is} \in \mathbb{R}$ such that

$$\ln\|e''_i\|_\omega - A = \sum_{j=1}^s a'_{ij}\ln|u_j|_\omega$$

for all $\omega \in \Omega_\infty$. Let a_{ij} be the round-up of a'_{ij}. Then

$$\sum_{j=1}^{s}(a'_{ij} - a_{ij}) \ln |u_j|_\omega \leqslant \sum_{j=1}^{s} |a'_{ij} - a_{ij}| \cdot \big| \ln |u_j|_\omega \big| \leqslant \sum_{j=1}^{s} \big| \ln |u_j|_\omega \big| \leqslant C''_K,$$

where $C''_K = \sum_{\omega \in \Omega_\infty} \sum_{j=1}^{s} | \ln |u_j|_\omega |$. Therefore,

$$\begin{aligned} -A &= \sum_{j=1}^{s} a'_{ij} \ln |u_j|_\omega - \ln \|e''_i\|_\omega \leqslant C''_K + \sum_{j=1}^{s} a_{ij} \ln |u_j|_\omega - \ln \|e''_i\|_\omega \\ &= C''_K - \ln \|v_i e''_i\|_\omega, \end{aligned}$$

where $v_i = \prod_{j=1}^{s} u_j^{-a_{ij}}$, and hence, if we set $e_i = v_i e''_i$, then $e_i \in \mathscr{E}$ and

$$\nu_{\min}(E, \xi) - \epsilon \leqslant \ln C'_K + \sigma(E, \xi) + [K : \mathbb{Q}] C''_K - [K : \mathbb{Q}] \ln \|e_i\|_\omega,$$

that is, there is a constant c_K depending only on K such that

$$\nu_{\min}(E, \xi) - \epsilon \leqslant c_K + \sigma(E, \xi) - [K : \mathbb{Q}] \ln \|e_i\|_\omega$$

for all i and $\omega \in \Omega_\infty$. We choose i and ω such that $\max\{\|e_1\|_\omega, \ldots, \|e_r\|_\omega\} = \|e_i\|_\omega$. Then, as $e^{-\lambda(E,\xi)} \leqslant \|e_i\|_\omega$, that is, $-\ln \|e_i\|_\omega \leqslant \lambda(E, \xi)$,

$$\nu_{\min}(E, \xi) - \epsilon \leqslant c_K + \sigma(E, \xi) + [K : \mathbb{Q}] \lambda(E, \xi),$$

and hence the assertion follows. □

Lemma 4.4.11 *There is a constant e_K depending only on K such that, for any invertible $\mathfrak{o}_K$-module $\mathscr{L}$, we can find $l \in \mathscr{L} \setminus \{0\}$ such that* $\operatorname{card}(\mathscr{L}/\mathfrak{o}_K l) \leqslant e_K$.

Proof Since the class group is finite, there are finitely many invertible $\mathfrak{o}_K$-modules $\mathscr{L}_1, \ldots, \mathscr{L}_h$ such that, for any invertible $\mathfrak{o}_K$-module $\mathscr{L}$, there is $\mathscr{L}_i$ such that $\mathscr{L}_i \simeq \mathscr{L}$. For each $i \in \{1, \ldots, h\}$, fix $l_i \in \mathscr{L}_i \setminus \{0\}$. Let $\mathscr{L}$ be an invertible $\mathfrak{o}_K$-module. Then there are $\mathscr{L}_i$ and an isomorphism $\varphi : \mathscr{L}_i \to \mathscr{L}$. If we set $l = \varphi(l_i)$, then $\mathscr{L}_i/\mathfrak{o}_K l_i \simeq \mathscr{L}/\mathfrak{o}_K l$, as required. □

Chapter 5
Slopes of tensor product

The purpose of this chapter is to study the minimal slope of the tensor product of a finite family of adelic vector bundles on an adelic curve. More precisely, give a family $\overline{E}_1, \ldots, \overline{E}_d$ of adelic vector bundles over a proper adelic curve S, we give a lower bound of $\widehat{\mu}_{\min}(\overline{E}_1 \otimes_{\varepsilon,\pi} \cdots \otimes_{\varepsilon,\pi} \overline{E}_d)$ in terms of the sum of the minimal slopes of $\overline{E}_i$ minus a term which is the product of three half of the measure of the infinite places and the sum of $\ln(\dim_K(E_i))$, see Corollary 5.6.2 for details. This result, whose form is similar to the main results of [64, 22, 38], does not rely on the comparison of successive minima and the height proved in [155], which des not hold for general adelic curves. Our method inspires the work of Totaro [143] on p-adic Hodge theory and relies on the geometric invariant theory on Grassmannian. The chapter is organised as follows. In the first section, we regroup several fundamental properties of $\mathbb{R}$-filtrations. We then recall in the second section some basic notions and results of the geometric invariant theory, in particular the Hilbert-Mumford criterion of the semistability. In the third section we give an estimate for the slope of a quotient adelic vector bundle of the tensor product adelic vector bundle, under the assumption that the underlying quotient space, viewed as a rational point of the Grassmannian (with the Plücker coordinates), is semistable in the sense of geometric invariant theory. In the fifth section, we prove a non-stability criterion which generalises [143, Proposition 1]. Finally, we prove in the sixth section the lower bound of the minimal slope of the tensor product adelic vector bundle in the general case.

5.1 Reminder on $\mathbb{R}$-filtrations

Let K be a field. We equip K with the trivial absolute value $|\cdot|$ such that $|a| = 1$ for any $a \in K \setminus \{0\}$. Note that K equipped with the trivial absolute value forms a proper adelic curve whose underlying measure space is a one point set equipped with the counting measure (which is a probability measure), see §3.2.3. Moreover, any finite-dimensional normed vector space over $(K, |\cdot|)$ can be considered as an adelic vector bundle on S. In fact, if V is a finite-dimensional vector space over K, any

H. Chen, A. Moriwaki, *Arakelov Geometry over Adelic Curves*, Lecture Notes in Mathematics 2258, https://doi.org/10.1007/978-981-15-1728-0_5

norm on V can be considered as a norm family indexed by the one point set. This norm family is clearly measurable. It is also dominated since all norms on V are equivalent (see Corollaries 1.1.13 and 4.1.10).

Let V be a finite-dimensional vector space over K. Recall that the set of ultrametric norms on V are canonically in bijection with the set of $\mathbb{R}$-filtrations on V (see Remark 1.1.40). If $\|\cdot\|$ is an ultrametric norm on V, then the balls centered at the origin are vector subspaces of V, and $\{(V, \|\cdot\|)_{\leqslant e^{-t}}\}_{t\in\mathbb{R}}$ is the corresponding $\mathbb{R}$-filtration. Conversely, given an $\mathbb{R}$-filtration $\mathcal{F}$ on V, we define a function $\lambda_{\mathcal{F}} : V \to \mathbb{R} \cup \{+\infty\}$ as follows

$$\forall\, x \in V, \quad \lambda_{\mathcal{F}}(x) := \sup\{t \in \mathbb{R} \,:\, x \in \mathcal{F}^t(V)\}.$$

Then the ultrametric norm $\|\cdot\|_{\mathcal{F}}$ corresponding to the $\mathbb{R}$-filtration $\mathcal{F}$ is given by

$$\forall\, x \in V, \quad \|x\|_{\mathcal{F}} = \mathrm{e}^{-\lambda_{\mathcal{F}}(x)}.$$

Definition 5.1.1 Let V be a finite-dimensional vector space over K and $\mathcal{F}$ be an $\mathbb{R}$-filtration on V. For any $t \in \mathbb{R}$, we denote by $\mathrm{sq}^t_{\mathcal{F}}(V)$ the quotient vector space

$$\mathcal{F}^t(V)\Big/\bigcup_{\varepsilon>0}\mathcal{F}^{t+\varepsilon}(V).$$

Clearly, if $\mathcal{F}$ corresponds to the flag

$$0 = V_0 \subsetneq V_1 \subsetneq \ldots \subsetneq V_n = V$$

of vector subspaces of V together with the sequence

$$\mu_1 > \ldots > \mu_n$$

in $\mathbb{R}$, then

$$\forall\, i \in \{1, \ldots, n\}, \quad \mathrm{sq}^{\mu_i}_{\mathcal{F}}(V) = V_i/V_{i-1},$$

and $\mathrm{sq}^t_{\mathcal{F}}(V) = \{0\}$ if $t \notin \{\mu_1, \ldots, \mu_n\}$.

Proposition 5.1.2 *Let $(V, \|\cdot\|)$ be a finite-dimensional ultrametrically normed vector space over K. The following assertions hold.*

(1) *The normed vector space $(V, \|\cdot\|)$ admits an orthogonal basis.*
(2) *If $\boldsymbol{e} = \{e_i\}_{i=1}^r$ is an orthogonal basis of $(V, \|\cdot\|)$, then the Arakelov degree of $(V, \|\cdot\|)$ is equal to $\lambda_{\mathcal{F}}(e_1) + \cdots + \lambda_{\mathcal{F}}(e_r)$.*
(3) *A basis $\boldsymbol{e} = \{e_i\}_{i=1}^r$ of V is orthogonal if and only if it is compatible with the $\mathbb{R}$-filtration $\mathcal{F}$, namely $\mathrm{card}(\mathcal{F}^t(V) \cap \boldsymbol{e}) = \dim_K(\mathcal{F}^t(V))$ for any $t \in \mathbb{R}$.*
(4) *Assume that the vector space V is non-zero. The adelic vector bundle $(V, \|\cdot\|)$ is semistable if and only if the function $\|\cdot\|$ is constant on $V \setminus \{0\}$.*
(5) *The Harder-Narasimhan $\mathbb{R}$-filtration of $(V, \|\cdot\|)$ identifies with $\mathcal{F}$.*
(6) *Let $\boldsymbol{e} = \{e_i\}_{i=1}^r$ be an orthogonal basis of $(V, \|\cdot\|)$. Then the sequence of successive slopes of $(V, \|\cdot\|)$ identifies with the sorted sequence of $\{\lambda_{\mathcal{F}}(e_i)\}_{i=1}^r$.*
(7) *Let*

$$0 = V_0 \subsetneq V_1 \subsetneq \ldots \subsetneq V_r = V$$

be a complete *flag of vector subspaces of* V. *For any* $i \in \{1, \ldots, r\}$, *let* $\|\cdot\|_i$ *be the subquotient norm of* $\|\cdot\|$ *on the vector space* V_i/V_{i-1}. *Then the sequence of successive slopes of* $(V, \|\cdot\|)$ *identifies with the sorted sequence of*

$$\left\{\widehat{\deg}(V_i/V_{i-1}, \|\cdot\|_i)\right\}_{i=1}^r.$$

Proof (1) Note that the valued field $(K, |\cdot|)$ is locally compact. By Proposition 1.2.30, there exists an orthogonal basis of $(V, \|\cdot\|)$.

(2) Let $\boldsymbol{e} = \{e_i\}_{i=1}^r$ be an orthogonal basis of $(V, \|\cdot\|)$. By Proposition 1.2.23, it is a Hadamard basis, namely

$$\|e_1 \wedge \cdots \wedge e_r\| = \prod_{i=1}^r \|e_i\|.$$

Therefore one has

$$\widehat{\deg}(V, \|\cdot\|) = -\ln \|e_1 \wedge \cdots \wedge e_r\| = -\sum_{i=1}^r \ln \|e_i\| = \sum_{i=1}^r \lambda_{\mathcal{F}}(e_i)$$

(3) Assume that the $\mathbb{R}$-filtration $\mathcal{F}$ corresponds to the flag

$$0 = V_0 \subsetneq V_1 \subsetneq \ldots \subsetneq V_n = V$$

together with the sequence

$$\mu_1 > \ldots > \mu_n$$

(cf. Remark 1.1.40). Let $\boldsymbol{e} = \{e_i\}_{i=1}^r$ be a basis of V. Then $\boldsymbol{e}$ is compatible with the $\mathbb{R}$-filtration $\mathcal{F}$ if and only if $\operatorname{card}(\boldsymbol{e} \cap V_j) = \dim_K(V_j)$ for any $j \in \{1, \ldots, n\}$. By Proposition 1.2.26 (1), this condition is also equivalent to the orthogonality of the basis $\boldsymbol{e}$.

(4) follows directly from Proposition 4.3.61 since $\|\cdot\| = \|\cdot\|_{**}$ (see Corollary 1.2.12).

(5) The $\mathbb{R}$-filtration corresponds to an increasing flag

$$0 = V_0 \subsetneq V_1 \subsetneq \ldots \subsetneq V_n = V$$

of vector subspaces of V, together with a decreasing sequence of real numbers

$$\mu_1 > \ldots > \mu_n.$$

Note that for any $i \in \{1, \ldots, n\}$ and any $x \in V_i \setminus V_{i-1}$ one has $\lambda_{\mathcal{F}}(x) = \mu_i$. In particular, the subquotient norm $\|\cdot\|_i$ on V_i/V_{i-1} induced by $\|\cdot\|$ takes constant value $\mathrm{e}^{-\mu_i}$ on $(V_i/V_{i-1}) \setminus \{0\}$. Therefore, by (4) the adelic vector bundle $(V_i/V_{i-1}, \|\cdot\|_i)$ is semistable of slope μ_i. By Proposition 4.3.38, we obtain that $\mathcal{F}$ is the Harder-Narasimhan $\mathbb{R}$-filtration of $(V, \|\cdot\|)$.

(6) Assume that the $\mathbb{R}$-filtration $\mathcal{F}$ corresponds to the flag

$$0 = V_0 \subsetneq V_1 \subsetneq \ldots \subsetneq V_n = V$$

and the sequence

$$\mu_1 > \ldots > \mu_n.$$

By definition, the value μ_i appears exactly $\dim_K(V_i/V_{i-1})$ times in the successive slopes of $(V, \|\cdot\|)$. Moreover, a basis $\boldsymbol{e}$ is orthogonal if and only if it is compatible with the flag

$$0 = V_0 \subsetneq V_1 \subsetneq \ldots \subsetneq V_n = V,$$

or equivalently, for any $i \in \{1, \ldots, n\}$, the set $\boldsymbol{e} \cap (V_i \setminus V_{i-1})$ contains exactly $\dim_K(V_i/V_{i-1})$ elements. Since the function $\lambda_{\mathcal{F}}(\cdot)$ takes the constant value μ_i on $V_i \setminus V_{i-1}$, we obtain the assertion.

(7) By Proposition 1.2.30, there exists an orthogonal basis $\boldsymbol{e} = \{e_i\}_{i=1}^r$ which is compatible with the flag

$$0 = V_0 \subsetneq V_1 \subsetneq \ldots \subsetneq V_r = V.$$

Without loss of generality, we may assume that $e_i \in V_i \setminus V_{i-1}$ for any $i \in \{1, \ldots, r\}$ Since the basis $\boldsymbol{e} = \{e_i\}_{i=1}^r$ is orthogonal, the image of e_i in V_i/V_{i-1} has norm $\|e_i\|$. In fact, any element x in $e_i + V_{i-1}$ can be written in the form

$$e_i + \sum_{j=1}^{i-1} a_j e_j$$

and hence $\|x\|_i \geqslant \|e_i\|$. Therefore one has

$$\widehat{\deg}(V_i/V_{i-1}, \|\cdot\|_i) = -\ln \|e_i\| = \lambda_{\mathcal{F}}(e_i).$$

Proposition 5.1.3 *Let V be a finite-dimensional vector space over K and $\|\cdot\|$ and $\|\cdot\|'$ be two ultrametric norms on V. Then there exists a basis $\boldsymbol{e}$ of V which is orthogonal with respect to $\|\cdot\|$ and $\|\cdot\|'$ simultaneously.*

Proof Let $\mathcal{F}$ be the $\mathbb{R}$-filtration on V associated with the norm $\|\cdot\|$, which corresponds to a flag

$$0 = V_0 \subsetneq V_1 \subsetneq \ldots \subsetneq V_n = V$$

together with a sequence $\mu_1 > \ldots > \mu_n$. By Proposition 1.2.30, there exists an orthogonal basis $\boldsymbol{e}$ of $(V, \|\cdot\|')$ which is compatible with the the flag

$$0 = V_0 \subsetneq V_1 \subsetneq \ldots \subsetneq V_n = V.$$

By Proposition 5.1.2 (3), we obtain that $\boldsymbol{e}$ is also orthogonal with respect to $\|\cdot\|$. □

Corollary 5.1.4 *Let V be a finite-dimensional vector space over K, and*

$$0 = V_0 \subsetneq V_1 \subsetneq \ldots \subsetneq V_n = V \quad and \quad 0 = W_0 \subsetneq W_1 \subsetneq \ldots \subsetneq W_m = V$$

be two flags of vector subspaces of V. There exists a basis $\boldsymbol{e}$ of V which is compatible with the two flags simultaneously.

Proof By choosing two decreasing sequences of real numbers $\mu_1 > \ldots > \mu_n$ and $\lambda_1 > \ldots > \lambda_m$ we obtain two $\mathbb{R}$-filtrations on V, which correspond to two ultrametric norms on V. By Proposition 5.1.3, there exists a basis of V which is orthogonal with respect to the two norms simultaneously. By Proposition 5.1.2 (3), this basis is compatible with the two flags simultaneously. □

Definition 5.1.5 Let $d \in \mathbb{N}_{\geqslant 2}$ and $\{E_j\}_{j=1}^d$ be a family of finite-dimensional vector spaces over K. For any $j \in \{1, \ldots, d\}$, let $\mathcal{F}_j$ be an $\mathbb{R}$-filtration on E_j, which corresponds to an ultrametric norm $\|\cdot\|_j$ on E_j. Let $\|\cdot\|_\varepsilon$ be the ε-tensor product of the norms $\|\cdot\|_1, \ldots, \|\cdot\|_d$. The $\mathbb{R}$-filtration on the tensor product space

$$E_1 \otimes_K \cdots \otimes_K E_d$$

corresponding to $\|\cdot\|_\varepsilon$ is called the *tensor product* of the $\mathbb{R}$-filtrations $\mathcal{F}_1, \ldots, \mathcal{F}_d$, which is denoted by $\mathcal{F}_1 \otimes \cdots \otimes \mathcal{F}_d$.

Remark 5.1.6 We keep the notation of Definition 5.1.5. For any $j \in \{1, \ldots, d\}$, let $\boldsymbol{e}^{(j)} = \{e_i^{(j)}\}_{i=1}^{n_j}$ be an orthogonal base of $(E_j, \|\cdot\|_j)$. By Proposition 1.2.19 together with Remark 1.1.56, one has the following:

(i) $\{e_{i_1}^{(1)} \otimes \cdots \otimes e_{i_d}^{(d)}\}_{(i_1,\ldots,i_d) \in \prod_{j=1}^d \{1,\ldots,n_j\}}$ forms an orthogonal basis of the vector space $E_1 \otimes_K \cdots \otimes_K E_d$ with respect to $\|\cdot\|_\varepsilon$.

(ii) $\|e_{i_1}^{(1)} \otimes \cdots \otimes e_{i_d}^{(d)}\|_\varepsilon = \prod_{j=1}^d \|e_{i_j}^{(j)}\|_j$ for any $(i_1, \ldots, i_d) \in \prod_{j=1}^d \{1, \ldots, n_j\}$.

Therefore, if we denote by $\mathcal{F}$ the tensor product $\mathbb{R}$-filtration $\mathcal{F}_1 \otimes \cdots \otimes \mathcal{F}_d$, then the vector space $\mathcal{F}^t(E_1 \otimes_K \cdots \otimes_K E_d)$ is generated by the vectors $e_{i_1}^{(1)} \otimes \cdots \otimes e_{i_d}^{(d)}$ such that

$$\lambda_{\mathcal{F}_1}(e_{i_1}^{(1)}) + \cdots + \lambda_{\mathcal{F}_d}(e_{i_d}^{(d)}) \geqslant t.$$

Therefore, one has

$$\begin{aligned}\mathcal{F}^t(E_1 \otimes_K \cdots \otimes_K E_d) &= \sum_{t_1 + \cdots + t_d \geqslant t} \mathcal{F}_1^{t_1}(E_1) \otimes_K \cdots \otimes_K \mathcal{F}_d^{t_d}(E_d) \\ &= \sum_{t_1 + \cdots + t_d = t} \mathcal{F}_1^{t_1}(E_1) \otimes_K \cdots \otimes_K \mathcal{F}_d^{t_d}(E_d).\end{aligned}$$

Furthermore, by (i), (ii) and Proposition 5.1.2 (6), if $(E_j, \|\cdot\|_j)$ are all semistable, where $j \in \{1, \ldots, d\}$, then $(E_1 \otimes_K \cdots \otimes_K E_d, \|\cdot\|_\varepsilon)$ is also semistable.

5.2 Reminder on geometric invariant theory

Let K be a field. By *group scheme over* $\operatorname{Spec} K$ or by *K-group scheme*, we mean a K-scheme $\pi : G \to \operatorname{Spec} K$ equipped with a K-morphism $m_G : G \times_K G \to G$ (called the *group scheme structure of* G) such that, for any K-scheme $f : S \to \operatorname{Spec} K$, the set $G(S)$ of K-morphisms from S to G equipped with the composition law $m_G(S) : G(S) \times G(S) \to G(S)$ forms a group. Note that the maps of inverse $\iota_G(S) : G(S) \to G(S)$ and the maps of unity

$$e_G(S) : (\operatorname{Spec} K)(S) = \{S \xrightarrow{f} \operatorname{Spec} K\} \longrightarrow G(S)$$

actually define K-morphisms $\iota_G : G \to G$ and $e_G : \operatorname{Spec} K \to G$, which make the following diagrams commutative:

$$\begin{array}{ccc} G \times_K G \times_K G & \xrightarrow{m_G \times \mathrm{Id}_G} & G \times_K G \\ {\scriptstyle \mathrm{Id}_G \times m_G}\downarrow & & \downarrow{\scriptstyle m_G} \\ G \times_K G & \xrightarrow{m_G} & G \end{array}$$

$$\begin{array}{ccccc} G & \xrightarrow{(e_G\pi, \mathrm{Id}_G)} & G \times_K G & \xleftarrow{(\mathrm{Id}_G, e_G\pi)} & G \\ & {\scriptstyle \mathrm{Id}_G}\searrow & \downarrow{\scriptstyle m_G} & \swarrow{\scriptstyle \mathrm{Id}_G} & \\ & & G & & \end{array}$$

$$\begin{array}{ccc} G & \xrightarrow{(\mathrm{Id}_G, \iota_G)} & G \times_K G \\ {\scriptstyle (\iota_G, \mathrm{Id}_G)}\downarrow & \searrow^{e_G\pi} & \downarrow{\scriptstyle m_G} \\ G \times_K G & \xrightarrow{m_G} & G \end{array}$$

Let G and H be group schemes over $\operatorname{Spec} K$. We call *morphism of K-group schemes* from G to H any K-morphism $f : G \to H$ such that, for any K-scheme S, le morphism f induces a morphism of groups $f(S) : G(S) \to H(S)$.

Example 5.2.1 Let V be a finite-dimensional vector space over K. We denote by $\mathbb{GL}(V)$ the open subscheme of the affine K-scheme $\mathbb{A}(\operatorname{End}(V)^\vee)$ defined by the non-vanishing of the determinant. For any K-scheme $\pi : S \to \operatorname{Spec} K$, one has

$$\mathbb{GL}(V)(S) = \operatorname{Aut}_{\mathcal{O}_S}(\pi^*(V)).$$

The set $\mathbb{GL}(V)(S)$ is canonically equipped with a structure of group. The group structures of $\mathbb{GL}(V)(S)$ where S runs over the set of K-schemes define a K-morphism $\mathbb{GL}(V) \times_K \mathbb{GL}(V) \to \mathbb{GL}(V)$, which makes $\mathbb{GL}(V)$ a group scheme over K. The group scheme $\mathbb{GL}(V)$ is called the *general linear group scheme associated with* V.

Definition 5.2.2 Let G be a group scheme over $\operatorname{Spec} K$ and X be a scheme over $\operatorname{Spec} K$. As *action* of G on X, we refer to a K-morphism $f : G \times_K X \to X$ such that, for any K-scheme S, the map

$$f(S) : G(S) \times X(S) \longrightarrow X(S)$$

defines an action of the group $G(S)$ on $X(S)$.

Example 5.2.3 Let V be a finite-dimensional vector space over K and X be the projective space $\mathbb{P}(V)$. Recall that, for any K-scheme $p : S \to \operatorname{Spec} K$, $\mathbb{P}(V)(S)$ identifies with the set of all invertible quotient modules of $p^*(V)$. Note that any automorphism of $p^*(V)$ acts naturally on the set $\mathbb{P}(V)(S)$ of invertible quotient modules of $p^*(V)$. Hence we obtain an action of the general linear group scheme $\mathbb{GL}(V)$ on the projective space $\mathbb{P}(V)$.

More generally, let G be a group scheme over $\operatorname{Spec} K$. By (finite-dimensional) *linear representation* of G we refer to a morphism of K-group schemes from G to certain $\mathbb{GL}(V)$, where V is a finite-dimensional vector space over K. Note that such a morphism induces an action of G on the projective space $\mathbb{P}(V)$. This action is said to be *linear*.

Let G be a group scheme over $\operatorname{Spec} K$ which acts on a K-scheme X. We denote by $f : G \times_K X \to X$ the action of G on X and by $\mathrm{pr}_2 : G \times_K X \to X$ the projection to the second coordinate. Let L be an invertible $\mathcal{O}_X$-module. We call *G-linear structure on L* any isomorphism η of $\mathcal{O}_{G\times_K X}$-modules from $f^*(L)$ to $\mathrm{pr}_2^*(L)$ such that the following diagram commutes

$$\begin{array}{ccc}
(\mathrm{pr}_2 \circ (\mathrm{Id}_G \times f))^*(L) & = & (f \circ \mathrm{pr}_{23})^*(L) \\
\uparrow{\scriptstyle (\mathrm{Id}_G \times f)^*(\eta)} & & \downarrow{\scriptstyle \mathrm{pr}_{23}^*(\eta)} \\
(f \circ (\mathrm{Id}_G \times f))^*(L) & & (\mathrm{pr}_2 \circ \mathrm{pr}_{23})^*(L) \\
\| & & \| \\
(f \circ (m_G \times \mathrm{Id}_X))^*(L) & \xrightarrow[(m_G \times \mathrm{Id}_X)^*(\eta)]{} & (\mathrm{pr}_2 \circ (m \times \mathrm{Id}_X))^*(L)
\end{array} \tag{5.1}$$

where $\mathrm{pr}_{23} : G \times_K G \times_K X \to G \times_K X$ is the projection to the second and the third coordinates, and $m_G : G \times_K G \to G$ is the group scheme structure of G. The couple (L, η) is called a *G-linearised invertible $\mathcal{O}_X$-module*.

Note that, if $\eta : f^*(L) \to \mathrm{pr}_2^*(L)$ is a G-linear structure on L, then $\eta^\vee : f^*(L^\vee) \to \mathrm{pr}_2^*(L^\vee)$ is a G-linear structure on $L^\vee$. Moreover, if (L_1, η_1) and (L_2, η_2) are G-linearised invertible $\mathcal{O}_X$-modules, then

$$\eta_1 \otimes \eta_2 : f^*(L_1) \otimes f^*(L_2) \cong f^*(L_1 \otimes L_2) \longrightarrow \mathrm{pr}_2^*(L_1 \otimes L_2) \cong \mathrm{pr}_2^*(L_1) \otimes \mathrm{pr}_2^*(L_2)$$

is a G-linear structure on $L_1 \otimes L_2$.

Example 5.2.4 Let G be a group scheme over K and V be a finite-dimensional vector space over K. A linear action of G on $\mathbb{P}(V)$ defines canonically a G-linear structure on the universal invertible sheaf $\mathcal{O}_V(1)$. Let $f : G \times_K \mathbb{P}(V) \to \mathbb{P}(V)$ be a linear action of G on $\mathbb{P}(V)$. Let $\pi : \mathbb{P}(V) \to \operatorname{Spec} K$ be the structural morphism and $\beta : \pi^*(V) \to \mathcal{O}_V(1)$ be the tautological invertible quotient sheaf of $\pi^*(V)$. Note that the morphism

$$G \times_K \mathbb{P}(V) \xrightarrow{(\mathrm{pr}_1, f)} G \times_K \mathbb{P}(V)$$

is an isomorphism of K-schemes, the inverse of which is given by the composed K-morphism

$$G \times_K \mathbb{P}(V) \xrightarrow{(\mathrm{pr}_1, \iota_G \,\mathrm{pr}_1, \mathrm{pr}_2)} G \times_K G \times_K \mathbb{P}(V) \xrightarrow{(\mathrm{pr}_1, f\,\mathrm{pr}_{23})} G \times_K \mathbb{P}(V)\,.$$

Moreover, one has $\mathrm{pr}_2 \circ (\mathrm{pr}_1, f) = f$. Therefore $((\mathrm{pr}_1, f)^*)^{-1}$ defines an isomorphism from $f^*(\mathcal{O}_V(1))$ to $\mathrm{pr}_2^*(\mathcal{O}_V(1))$ as invertible quotient modules of $f^*(\pi^*(V)) \cong \mathrm{pr}_2^*(\pi^*(V))$. The fact that the action of G on $\mathbb{P}(V)$ is linear shows that this isomorphism actually defines a G-linear structure on $\mathcal{O}_V(1)$.

Definition 5.2.5 We denote by $\mathbb{G}_{\mathrm{m},K} = \operatorname{Spec} K[T, T^{-1}]$ the multiplicative group scheme over $\operatorname{Spec} K$ (recall that for any K-scheme S one has $\mathbb{G}_{\mathrm{m},K}(S) = \mathcal{O}_S(S)^\times$). If G is a group scheme over $\operatorname{Spec} K$, by *one-parameter subgroup* of G any morphism of K-group schemes from $\mathbb{G}_{\mathrm{m},K}$ to G.

Let G be a group scheme over K, which acts on a K-scheme X. Denote by $f : G \times_K X \to X$ the action. If $\varphi : \mathbb{G}_{\mathrm{m},K} \to G$ is a one-parameter subgroup of G, then f and φ induce an action of $\mathbb{G}_{\mathrm{m},K}$ on X, denoted by f_φ. Note that f_φ is the composed morphism

$$\mathbb{G}_{\mathrm{m},K} \times_K X \xrightarrow{\varphi \times \mathrm{Id}_X} G \times_K X \xrightarrow{f} X\,.$$

Let $g : \mathbb{G}_{\mathrm{m},K} \times_K X \to X$ be an action of the multiplicative group $\mathbb{G}_{\mathrm{m},K}$ on a *proper* K-scheme. Suppose that $x : \operatorname{Spec} K \to X$ is a rational point of X. The *orbit* of x by the action of $\mathbb{G}_{\mathrm{m},K}$ is by definition the following composed morphism orb_x

$$\mathbb{G}_{\mathrm{m},K} \cong \mathbb{G}_{\mathrm{m},K} \times_K \operatorname{Spec} K \xrightarrow{\mathrm{Id}_{\mathbb{G}_{\mathrm{m},K}} \times x} \mathbb{G}_{\mathrm{m},K} \times_K X \xrightarrow{g} X.$$

Since X is proper over $\operatorname{Spec} K$, by the valuative criterion of properness, the morphism $\mathrm{orb}_x : \mathbb{G}_{\mathrm{m},K} \to X$ extends in a unique way to a K-morphism $\widetilde{\mathrm{orb}}_x : \mathbb{A}^1_K = \operatorname{Spec} K[T] \to X$. The image by $\widetilde{\mathrm{orb}}_x$ of the rational point of $\mathbb{A}^1_K$ corresponding to the prime ideal (T) is denoted by $\widetilde{x}_g$. Note that $\widetilde{x}_g$ is a rational point of X which is invariant by the action of $\mathbb{G}_{\mathrm{m},K}$.

Assume that L is a $\mathbb{G}_{\mathrm{m},K}$-linearised invertible $\mathcal{O}_X$-module. Since $\widetilde{x}_g$ is a fixed rational point of the action g, the $\mathbb{G}_{\mathrm{m},K}$-linear structure corresponds to an action of $\mathbb{G}_{\mathrm{m},K}$ on $\widetilde{x}_g^*(L)$, which is induced by an endomorphism of the K-group scheme $\mathbb{G}_{\mathrm{m},K}$. Note that any endomorphism of the K-group scheme $\mathbb{G}_{\mathrm{m},K}$ is of the form $t \mapsto t^n$, where the exponent n is an integer. We denote by $\mu(x, L)$ the *opposite* of the exponent corresponding to the action of $\mathbb{G}_{\mathrm{m},K}$ on $\widetilde{x}_g^*(L)$. Note that our choice of the constant $\mu(x, L)$ conforms with that of the book [115].

More generally, if G is a K-group scheme, $f : G \times_K X \to X$ is an action of G on a proper K-scheme X and if $\varphi : \mathbb{G}_{\mathrm{m},K} \to G$ is a one-parameter subgroup of G, for any $x \in X(K)$ we denote by $\mu(x, \varphi, L)$ the exponent corresponding to the action of $\mathbb{G}_{\mathrm{m},K}$ on $\widetilde{x}_{f_\varphi}^*(L)$.

Example 5.2.6 Consider the one-parameter subgroups of the general linear group. Let E be a finite-dimensional vector space over K and $\varphi : \mathbb{G}_{\mathrm{m},K} \to \mathbb{GL}(E)$ be a one-parameter subgroup. By [53, II.§2, n°2, 2.5], we can decompose the vector space E as a direct sum of K-vector subspaces $E_1, \ldots, E_n$ which are invariant by the action of $\mathbb{G}_{\mathrm{m},K}$, and integers $(a_1, \ldots, a_n)$ such that the action of $\mathbb{G}_{\mathrm{m},K}$ on E_i is given by $t \mapsto t^{a_i} \mathrm{Id}_{E_i}$. Therefore the one-parameter subgroup φ determines an $\mathbb{R}$-filtration $\mathcal{F}_\varphi$ on E such that

$$\mathcal{F}_\varphi^t(E) = \bigoplus_{\substack{i \in \{1,\ldots,n\} \\ -a_i \geqslant t}} E_i.$$

We now consider the canonical action of $\mathbb{GL}(E)$ on the projective space $\mathbb{P}(E)$ (see Example 5.2.3). Let x be a rational point of $\mathbb{P}(E)$ and $\pi_x : E \to K$ be the one dimensional quotient vector space of E corresponding to x. Then the morphism $\mathrm{orb}_x : \mathbb{G}_{m,K} = \mathrm{Spec}\, K[T, T^{-1}] \to \mathbb{P}(E)$ is represented by the surjective $K[T, T^{-1}]$-linear map $p_x : E \otimes_K K[T, T^{-1}] \twoheadrightarrow K[T, T^{-1}]$ sending $v_i \otimes 1$ to $\pi_x(v_i)T^{a_i}$ for any $v_i \in E_i$. The extended morphism $\widetilde{\mathrm{ord}}_x : \mathbb{A}_K^1 \to \mathbb{P}(E)$ corresponds to the surjective $K[T]$-linear map

$$\widetilde{p}_x : E \otimes_K K[T] \longrightarrow K[T] \cdot T^{-\mu(x,\varphi,\mathcal{O}_E(1))}$$

given by the restriction of p_x to $E \otimes_K K[T]$, where $\mathcal{O}_E(1)$ denotes the universal invertible sheaf. In particular, one has

$$\mu(x, \varphi, \mathcal{O}_E(1)) = -\min\{a_i \;:\; i \in \{1, \ldots, n\},\ \pi_x(E_i) \neq \{0\}\}.$$

Therefore, we can interpret the constant $\mu(x, \varphi, \mathcal{O}_E(1))$ via the $\mathbb{R}$-filtration $\mathcal{F}_\varphi$. In fact, the $\mathbb{R}$-filtration $\mathcal{F}_\varphi$ induces by the surjective map $\pi_x : E \to K$ an $\mathbb{R}$-filtration on K (viewed as a one-dimensional vector space over K), which corresponds to the quotient norm of $\|\cdot\|_{\mathcal{F}_\varphi}$. Then the number $\mu(x, \varphi, \mathcal{O}_E(1))$ is equal to the jump point of this quotient $\mathbb{R}$-filtration.

The following theorem of Hilbert-Mumford relates the positivity of the function $\mu(x, \cdot, L)$ to the non-vanishing of a global section invariant by the action of the K-group scheme.

Theorem 5.2.7 *We assume that the field K is perfect. Let G be a reductive K-group scheme acting on a projective K-scheme X, L be a G-linearised* ample *invertible $\mathcal{O}_X$-module. For any rational point $x \in X(K)$, the following two conditions are equivalent:*

(1) *for any one-parameter subgroup* $\lambda : \mathbb{G}_{\mathrm{m},K} \to G$ *of* G, $\mu(x, \lambda, L) \geqslant 0$;
(2) *there exists an integer* $n \in \mathbb{N}_{\geqslant 1}$ *and a section* $s \in H^0(X, L^{\otimes n})$ *which is invariant under the action of* $G(K)$ *such that* x *lies outside of the vanishing locus of* s.

We just explain why the condition (2) implies the positivity of $\mu(x, \lambda, L)$ for any one-parameter group. Let

$$\widetilde{\mathrm{orb}}_x : \mathbb{A}^1_K = \mathrm{Spec}\, K[T] \longrightarrow X$$

be the extended orbit of the rational point x by the action of $\mathbb{G}_{\mathrm{m},K}$ via λ. Then the pull-back of L by $\widetilde{\mathrm{orb}}_x$ corresponds to a free $K[T]$-module of rank 1, which is equipped with a linear action of $\mathbb{G}_{\mathrm{m},K}$. This action corresponds to an invertible element of the tensorial algebra

$$K[t, t^{-1}] \otimes_K K[T] \cong K[t, t^{-1}, T],$$

where t and T are variables. Moreover, the compatibility condition (5.1) shows that $\eta(t, T)$ satisfies the following relation

$$\eta(t, T)\eta(u, T) = \eta(tu, T) \text{ in } K[t, t^{-1}, u, u^{-1}, T],$$

where t, u, and T are variables. Therefore $\eta(t, T)$ is of the form $t^{-\mu(x,\lambda,L)}$. We fix a section m of $\widetilde{\mathrm{orb}}_x^*(L)$ which trivialises this invertible sheaf. Note that the pull-back of the section s is an element of this free $K[T]$-module which is invariant by the action of $\mathbb{G}_{\mathrm{m},K}(K) = K^{\times}$. We write s in the form $P(T)m$, where $P \in k[T]$. Then the action of an element $a \in K^{\times}$ on s gives the section $P(aT)a^{-\mu(x,\lambda,L)}m$. Hence P is a homogeneous polynomial and $\mu(x, \lambda, L)$ is equal to the degree of P, which is a non-negative integer. We refer the readers to [115, §2.1] for a proof of the above theorem. See also [96] and [127].

Definition 5.2.8 Under the assumption and with the notation of Theorem 5.2.7, if $x \in X(K)$ satisfies the equivalent conditions of the theorem, we say that the point x is *semistable* with respect to the G-linearised invertible $\mathcal{O}_X$-module L.

Remark 5.2.9 Let $d \in \mathbb{N}_{\geqslant 2}$ and $\{E_j\}_{j=1}^d$ be a family of finite-dimensional non-zero vector spaces over K. Any one-parameter subgroup

$$\lambda : \mathbb{G}_{\mathrm{m},K} \longrightarrow \mathbb{GL}(E_1) \times_K \cdots \times_K \mathbb{GL}(E_d)$$

can be written in the form $(\lambda_1, \cdots, \lambda_d)$, where $\lambda_j : \mathbb{G}_{\mathrm{m},K} \to \mathbb{GL}(E_j)$ is a one-parameter subgroup of $\mathbb{GL}(E_j)$, $j \in \{1, \ldots, d\}$. We can then decompose the vector space E_j into the direct sum of eigenspaces of the action λ_j as follows:

$$E_j = E_{j,1} \oplus \cdots \oplus E_{j,n_j},$$

where each $E_{j,i}$ is stable by the action of λ_j, and on $E_{j,i}$ the action of $\mathbb{G}_{\mathrm{m},K}$ is given by $t \mapsto t^{a_{j,i}}\mathrm{Id}_{E_{j,i}}$, $i \in \{1, \ldots, n_j\}$. Note that the one-parameter subgroup λ induces an action of $\mathbb{G}_{m,K}$ on the tensor product space $E_1 \otimes_K \cdots \otimes_K E_d$ via the canonical morphisme of K-group schemes

$$\mathbb{GL}(E_1) \times_K \cdots \times_K \mathbb{GL}(E_d) \longrightarrow \mathbb{GL}(E_1 \otimes_K \cdots \otimes_K E_d).$$

For any $(i_1, \ldots, i_d) \in \prod_{j=1}^d \{1, \ldots, n_j\}$, the vector subspace $E_{1,i_1} \otimes_K \cdots \otimes_K E_{d,i_d}$ of $E_1 \otimes_K \cdots \otimes_K E_d$ is invariant by the action of $\mathbb{G}_{m,K}$, and on $E_{1,i_1} \otimes_K \cdots \otimes_K E_{d,i_d}$ the action of $\mathbb{G}_{m,K}$ is given by

$$t \longmapsto t^{a_{1,i_1}+\cdots+a_{d,i_d}} \mathrm{Id}_{E_{1,i_1} \otimes_K \cdots \otimes_K E_{d,i_d}}.$$

We construct an $\mathbb{R}$-filtration $\mathcal{F}_\lambda$ on $E_1 \otimes_K \cdots \otimes_K E_d$ as follows

$$\mathcal{F}_\lambda^t(E_1 \otimes_K \cdots \otimes_K E_d) := \sum_{\substack{(i_1,\ldots,i_d) \in \prod_{j=1}^d \{1,\ldots,n_j\} \\ -a_{1,i_1}-\cdots-a_{d,i_d} \geqslant t}} E_{1,i_1} \otimes_K \cdots \otimes_K E_{d,i_d}. \tag{5.2}$$

Moreover, if we denote by $\mathcal{F}_{\lambda_j}$ the $\mathbb{R}$-filtrations on E_j defined as

$$\mathcal{F}_{\lambda_j}^t(E_j) = \sum_{\substack{i \in \{1,\ldots,n_j\} \\ -a_i \geqslant t}} E_{j,i},$$

then the $\mathbb{R}$-filtration $\mathcal{F}_\lambda$ defined in (5.2) identifies with the tensor product of $\mathcal{F}_{\lambda_1}, \ldots, \mathcal{F}_{\lambda_d}$ (see Definition 5.1.5, see also Remark 5.1.6). Conversely, for any $\mathbb{R}$-filtration $\mathcal{F}_j$ with integral jump points on the vector spaces E_j, there exists a one-parameter subgroup $\lambda_j : \mathbb{G}_{m,K} \to \mathbb{GL}(E_j)$ such that $\mathcal{F}_{\lambda_j} = \mathcal{F}_j$. This comes from Proposition 1.2.30 which allows us to construct the actions of $\mathbb{G}_{m,K}$ on E_j diagonally with respect to an orthogonal basis.

More generally, for any integer $r \geqslant 1$, any one-parameter subgroup

$$\lambda = (\lambda_1, \ldots, \lambda_d) : \mathbb{G}_{m,K} \longrightarrow \mathbb{GL}(E_1) \times_K \cdots \times_K \mathbb{GL}(E_d)$$

induces an action of $\mathbb{G}_{m,K}$ on the K-vector space

$$(E_1 \otimes_K \cdots \otimes_K E_d)^{\otimes r}.$$

Again the $\mathbb{R}$-filtration on $(E_1 \otimes_K \cdots \otimes_K E_d)^{\otimes r}$ corresponding to the eigenspace decomposition of the action of $\mathbb{G}_{m,K}$ identifies with

$$(\mathcal{F}_{\lambda_1} \otimes \cdots \otimes \mathcal{F}_{\lambda_d})^{\otimes r}.$$

For any $j \in \{1, \ldots, d\}$, let a_j be the dimension of E_j over K. Consider a non-zero quotient vector space V of $E_1 \otimes_K \cdots \otimes_K E_d$. Let r be the dimension of V over K. The canonical surjective map $E_1 \otimes_K \cdots \otimes_K E_d \to V$ determines a rational point x of

$$\pi : P = \mathbb{P}((E_1 \otimes_K \cdots \otimes E_d)^{\otimes r}) \longrightarrow \operatorname{Spec} K,$$

which corresponds the composed map

$$(E_1 \otimes_K \cdots \otimes_K E_d)^{\otimes r} \longrightarrow V^{\otimes r} \longrightarrow \det(V).$$

We consider the semistability of the point x with respect to the $\mathbb{GL}(E_1) \times_K \cdots \times_K \mathbb{GL}(E_d)$-invertible sheaf

$$L := \mathcal{O}_P(a_1 \cdots a_d) \otimes \pi^*(\det(E_1^\vee)^{\otimes r b_1} \otimes \cdots \otimes \det(E_d^\vee)^{\otimes r b_d}),$$

where for any $j \in \{1, \ldots, d\}$,

$$b_j := \frac{a_1 \cdots a_d}{a_j}.$$

Let

$$\lambda = (\lambda_1, \ldots, \lambda_d) : \mathbb{G}_{m,K} \longrightarrow \mathbb{GL}(E_1) \times_K \cdots \times_K \mathbb{GL}(E_d)$$

be a one-parameter subgroup, which determines, for each $j \in \{1, \ldots, d\}$, an $\mathbb{R}$-filtration $\mathcal{F}_{\lambda_j}$ on E_j. We let $\|\cdot\|_j$ be the ultrametric norm on E_j corresponding to $\mathcal{F}_{\lambda_j}$, where we consider the trivial absolute value on K. We equip $E_1 \otimes_K \cdots \otimes_K E_d$ with the ε-tensor product of the norms $\|\cdot\|_j$ and equip V with the quotient norm. By Example 5.2.6 and Proposition 1.2.15, we obtain that

$$\mu(x, \lambda, \mathcal{O}_P(1)) = \widehat{\deg}(\overline{V}).$$

Moreover, by definition, for any $j \in \{1, \ldots, d\}$ one has

$$\mu(x, \lambda_j, \pi^*(\det(E_j^\vee))) = -\widehat{\deg}(E_j, \|\cdot\|_j),$$

which leads to

$$\mu(x, \lambda, L) = a_1 \cdots a_d \, \widehat{\deg}(\overline{V}) - r \sum_{j=1}^{d} b_j \, \widehat{\deg}(\overline{E}_j). \tag{5.3}$$

Therefore we deduce from the Hilbert-Mumford criterion the following result.

Corollary 5.2.10 *We assume that K is perfect and equip K with the trivial absolute value. Let $\{E_j\}_{j=1}^d$ be a finite family of finite-dimensional non-zero vector spaces over K, and V be a non-zero quotient vector space of $E_1 \otimes_K \cdots \otimes_K E_d$. Let r be the dimension of V over K, and, for any $j \in \{1, \ldots, d\}$, let a_j the dimension of E_j over K. Let*

$$P = \mathbb{P}((E_1 \otimes_K \cdots \otimes_K E_d)^{\otimes r}),$$

$\pi : P \to \operatorname{Spec} K$ be the canonical morphism and

$$L = \mathcal{O}_P(a_1 \cdots a_d) \otimes \pi^*(\det(E_1^\vee)^{\otimes r b_1} \otimes \cdots \otimes \det(E_d^\vee)^{\otimes r b_d}),$$

where

$$\forall\, j \in \{1, \ldots, d\}, \quad b_j = \frac{a_1 \cdots a_d}{a_j}.$$

Then the composed surjective map

$$(E_1 \otimes_K \cdots \otimes_K E_d)^{\otimes r} \longrightarrow V^{\otimes r} \longrightarrow \det(V), \tag{5.4}$$

viewed as a rational point x of P, is semistable with respect to the $\mathbb{GL}(E_1) \times_K \cdots \times_K \mathbb{GL}(E_d)$-linearised invertible sheaf L if and only if, for all ultrametric norms $\|\cdot\|_j$ on E_j, $j \in \{1, \ldots, d\}$, if we equip V with the quotient norm of the ε-tensor product of $\|\cdot\|_1, \ldots, \|\cdot\|_d$, then one has

$$\widehat{\mu}(\overline{V}) \geqslant \sum_{j=1}^{d} \widehat{\mu}(\overline{E}_j). \tag{5.5}$$

Proof Assume that the inequality (5.5) holds for any choice of norms $\|\cdot\|_j$. By (5.3), for any one-parameter subgroup $\lambda : \mathbb{G}_{m,K} \to \mathbb{GL}(E_1) \times_K \cdots \times_K \mathbb{GL}(E_d)$, one has

$$\mu(x, \lambda, L) = a_1 \cdots a_d r \Big(\widehat{\mu}(\overline{V}) - \sum_{j=1}^{d} \widehat{\mu}(\overline{E}_j) \Big) \geqslant 0.$$

Hence the rational point x of P defined by (5.4) is semistable with respect to L.

Conversely, by Remark 5.2.9 the semi-stability of the rational points x implies that the inequality (5.5) holds for any choice of ultrametric norms $\|\cdot\|_j$ such that $\ln\|E_j \setminus \{0\}\|_j \subseteq \mathbb{Z}$. As a consequence the inequality (5.5) holds for any choice of ultrametric norms $\|\cdot\|_j$ such that $\ln\|E_j \setminus \{0\}\|_j \subseteq \mathbb{Q}$. In fact, in this case there exists $n \in \mathbb{N}_{>0}$ such that the (finite) set $\ln\|E_j \setminus \{0\}\|_j$ is contained in $\frac{1}{n}\mathbb{Z}$ for any $j \in \{1, \ldots, d\}$. Note that the nth power of the function $\|\cdot\|_j$ forms a norm on E_j. If we denote by $\|\cdot\|_V$ the quotient norm of the ε-tensor product of $\|\cdot\|_1, \ldots, \|\cdot\|_d$, then the quotient norm of the ε-tensor product of $\|\cdot\|_1^n, \ldots, \|\cdot\|_d^n$ is $\|\cdot\|_V^n$. Note that

$$\forall\, j \in \{1, \ldots, d\}, \quad \ln\|E_j \setminus \{0\}\|_j^n = n \ln\|E_j \setminus \{0\}\|_j \subseteq \mathbb{Z}$$

and hence

$$n\widehat{\mu}(V, \|\cdot\|_V) = \widehat{\mu}(V, \|\cdot\|_V^n) \geqslant \sum_{j=1}^{d} \widehat{\mu}(E_j, \|\cdot\|_j^n) = n \sum_{j=1}^{d} \widehat{\mu}(E_j, \|\cdot\|_j).$$

Finally the general case follows from a limite procedure by using Proposition 4.3.18.□

5.3 Estimate for the minimal slope under semi-stability assumption

In this section, we fix a proper adelic curve $S = (K, (\Omega, \mathcal{A}, \nu), \phi)$ such that, either the σ-algebra $\mathcal{A}$ is discrete, or K admits a countable subfield which is dense in each

K_ω, where $\omega \in \Omega$. We assume in addition that the field K is of characteristic 0. We fix an integer $d \geqslant 2$ and we let $\{\overline{E}_j = (E_j, \xi_j)\}_{j=1}^d$ be a family of non-zero adelic vector bundles on S. Let V be a quotient vector space of $E_1 \otimes_K \cdots \otimes_K E_d$ and r be the dimension of V over K. For any $j \in \{1, \ldots, d\}$, let a_j be the dimension of E_j over K. We equip V with the quotient norm family of $\xi_1 \otimes_{\varepsilon,\pi} \cdots \otimes_{\varepsilon,\pi} \xi_d$. Note that the quotient map $E_1 \otimes_K \cdots \otimes_K E_d \to V$ induces a surjective map

$$\Lambda^r(E_1 \otimes_K \cdots \otimes_K E_d) \longrightarrow \Lambda^r V = \det(V).$$

Consider the composed map

$$(E_1 \otimes_K \cdots \otimes_K E_d)^{\otimes r} \longrightarrow \Lambda^r(E_1 \otimes_K \cdots \otimes_K E_d) \longrightarrow \det(V),$$

which permits to consider $\det(V)$ as a rational point of $P = \mathbb{P}((E_1 \otimes_K \cdots \otimes_K E_d)^{\otimes r})$. Denote by $\pi : P \to \operatorname{Spec} K$ the structural morphism and by L the invertible sheaf

$$\mathcal{O}_P(a_1 \cdots a_d) \otimes \pi^*(\det(E_1^\vee)^{\otimes r b_1} \otimes \cdots \otimes \det(E_d^\vee)^{r \otimes b_d}), \tag{5.6}$$

where

$$\forall\, j \in \{1, \ldots, d\}, \quad b_j = \frac{a_1 \cdots a_d}{a_j}.$$

We equip L with its natural $\mathbb{GL}(E_1) \times_K \cdots \times_K \mathbb{GL}(E_d)$-linear structure. Note that L and $\mathcal{O}_P(a_1 \cdots a_d)$ are isomorphic as invertible $\mathcal{O}_P$-modules, however the natural $\mathbb{GL}(E_1) \times_K \cdots \times_K \mathbb{GL}(E_d)$-linear structures on these two invertible sheaves are different.

Our purpose is to estimate $\widehat{\mu}(\overline{V})$ under the additional assumption that, as a rational point of $P = \mathbb{P}((E_1 \otimes_K \cdots \otimes_K E_d)^{\otimes r})$, the determinant line $\det(V)$ is semistable with respect to the $\mathbb{GL}(E_1) \times_K \cdots \times_K \mathbb{GL}(E_d)$-linearised invertible sheaf L.

Proposition 5.3.1 *We equip V with the quotient norm family ξ_V of $\xi_1 \otimes_{\varepsilon,\pi} \cdots \otimes_{\varepsilon,\pi} \xi_d$. Assume that, as a rational point of the K-scheme P, $\det(V)$ is semistable with respect to the $\mathbb{GL}(E_1) \times_K \cdots \times_K \mathbb{GL}(E_d)$-linearised invertible sheaf L defined in (5.6). Then the following inequality holds:*

$$\widehat{\mu}(V, \xi_V) \geqslant \sum_{j=1}^d \Big(\widehat{\mu}(E_j, \xi_j) - \nu(\Omega_\infty) \ln(\dim_K(E_j))\Big). \tag{5.7}$$

Proof For any integer $m \in \mathbb{N}_{\geqslant 1}$, let $\mathfrak{S}_m$ be the symmetric group of $\{1, \ldots, m\}$. Let $A = a_1 \cdots a_d$. By the first principal theorem of the classic invariant theory (see [148, Chapter III] and [6, Appendix 1], see also [38, Theorem 3.3]), there exist an integer $n \geqslant 1$ and an element $(\sigma_1, \ldots, \sigma_d) \in \mathfrak{S}_{nrA}^d$ such that the composed map

$$\begin{array}{ccc} \det(V)^{\vee\otimes nA} & \longrightarrow & E_1^{\vee\otimes nrA}\otimes\cdots\otimes E_d^{\vee\otimes nrA} \\ & & \Big\downarrow \sigma_1\otimes\cdots\otimes\sigma_d \\ \det(E_1^\vee)^{\otimes nrb_1}\otimes\cdots\otimes\det(E_d^\vee)^{\otimes nrb_d} & \xleftarrow[\varpi]{} & E_1^{\vee\otimes nrA}\otimes\cdots\otimes E_d^{\vee\otimes nrA} \end{array} \tag{5.8}$$

is non-zero. Since ξ_V is the quotient norm family of $\xi_1 \otimes_{\varepsilon,\pi} \cdots \otimes_{\varepsilon,\pi} \xi_d$, the determinant norm family $\det(\xi_V)$ is the quotient norm family of the ε,π-tensor product norm family $\xi_1^{\otimes_{\varepsilon,\pi} r} \otimes_{\varepsilon,\pi} \cdots \otimes_{\varepsilon,\pi} \xi_d^{\otimes_{\varepsilon,\pi} r}$ by the following composed map (this is a consequence of Propositions 1.1.14 (1), 1.1.58 and 1.2.39)

$$E_1^{\otimes r} \otimes_K \cdots \otimes_K E_d^{\otimes r} \longrightarrow \Lambda^r(E_1 \otimes_K \cdots \otimes_K E_d) \longrightarrow \Lambda^r(V) = \det(V).$$

By passing to the dual vector space, we obtain that the dual of the determinant norm family $\det(\xi_V)^\vee$ identifies with the restrict norm family of $\xi_1^{\vee\otimes_\varepsilon r} \otimes_\varepsilon \cdots \otimes_\varepsilon \xi_d^{\vee\otimes_\varepsilon r}$. This is a consequence of Proposition 1.1.57, Corollary 1.2.20 and Proposition 1.1.20.

By Proposition 1.2.18, the height of the K-linear map ϖ in (5.8) is bounded from above by

$$\sum_{j=1}^d nrb_j \ln(a_j!),$$

where we consider the norm family $\xi_1^{\vee\otimes_\varepsilon nrA} \otimes_\varepsilon \cdots \otimes_\varepsilon \xi_d^{\vee\otimes_\varepsilon nrA}$ on $E_1^{\vee\otimes nrA} \otimes_K \cdots \otimes_K E_d^{\vee\otimes nrA}$, and the norm family $\det(\xi_1)^{\vee\otimes nrb_1} \otimes \cdots \otimes \det(\xi_d)^{\vee\otimes nrb_d}$ on $\det(E_1^\vee)^{\otimes nrb_1} \otimes_K \cdots \otimes_K \det(E_d^\vee)^{\otimes nrb_d}$.

Therefore by the slope inequality we obtain

$$-nA\,\widehat{\deg}(V,\xi_V) \leqslant -\sum_{j=1}^d \Big(nrb_j\,\widehat{\deg}(E_j,\xi_j) - \nu(\Omega_\infty) nrb_j \ln(a_j!)\Big),$$

which leads to

$$\begin{aligned} \widehat{\mu}(V,\xi_V) &\geqslant \sum_{j=1}^d \Big(\widehat{\mu}(E_j,\xi_j) - \frac{1}{a_j}\ln(a_j!)\nu(\Omega_\infty)\Big) \\ &\geqslant \sum_{j=1}^d \Big(\widehat{\mu}(E_j,\xi_j) - \ln(a_j)\nu(\Omega_\infty)\Big). \end{aligned}$$

Remark 5.3.2 Assume that the norm families $\xi_1, \ldots, \xi_d$ are Hermitian. If we equip V with the quotient norm family ξ_V^H of the orthogonal tensor product $\xi_1 \otimes \cdots \otimes \xi_d$, then a similar argument as above leads to the following inequality (where we use Proposition 1.2.62 to compute the height of ϖ)

$$\widehat{\mu}(V,\xi_V^H) \geqslant \sum_{j=1}^d \Big(\widehat{\mu}(E_j,\xi_j) - \frac{1}{2}\nu(\Omega_\infty)\ln(\dim_K(E_j))\Big). \tag{5.9}$$

5.4 An interpretation of the geometric semistability

Let K be a perfect field, E be a finite-dimensional non-zero vector space over K and r be the dimension of E over K. We denote by $\mathbf{Fil}(E)$ the set of $\mathbb{R}$-filtrations $\mathcal{E}$ on E. Let $\mathbf{Fil}_0(E)$ be the subset of $\mathbf{Fil}(E)$ of $\mathbb{R}$-filtrations $\mathcal{E}$ such that $\widehat{\deg}(E, \|\cdot\|_{\mathcal{E}}) = 0$, where $\|\cdot\|_{\mathcal{E}}$ is the norm on E associated with the $\mathbb{R}$-filtration $\mathcal{E}$ (here we consider the trivial absolute value on K). In other words,

$$\forall\, x \in E, \quad \|x\|_{\mathcal{E}} = \exp(-\sup\{t \in \mathbb{R} \,:\, x \in \mathcal{E}^t(E)\}).$$

Let $\mathcal{E}_1$ and $\mathcal{E}_2$ be two elements of $\mathbf{Fil}(E)$. By Proposition 5.1.3, there exists a basis $\boldsymbol{e} = \{e_i\}_{i=1}^r$ of E which is orthogonal with respect to the norms $\|\cdot\|_{\mathcal{E}_1}$ and $\|\cdot\|_{\mathcal{E}_2}$ simultaneously. We denote by $\langle \mathcal{E}_1, \mathcal{E}_2 \rangle$ the number

$$\frac{1}{r}\sum_{i=1}^{r}(-\ln\|e_i\|_{\mathcal{E}_1})(-\ln\|e_i\|_{\mathcal{E}_2}). \tag{5.10}$$

As shown by the following proposition, this number actually does not depend on the choice of the basis $\boldsymbol{e}$.

Proposition 5.4.1 *Let E be a finite-dimensional non-zero vector space over K, and $\mathcal{E}_1$ and $\mathcal{E}_2$ be $\mathbb{R}$-filtrations on E. If $\boldsymbol{e} = \{e_i\}_{i=1}^r$ is a basis of E which is compatible with the $\mathbb{R}$-filtrations $\mathcal{E}_1$ and $\mathcal{E}_2$ simultaneously, then the following equality holds*

$$\sum_{i=1}^{r}\lambda_{\mathcal{E}_1}(e_i)\lambda_{\mathcal{E}_2}(e_i) = \sum_{t\in\mathbb{R}} t\,\widehat{\deg}(\mathrm{sq}^t_{\mathcal{E}_1}(E), \|\cdot\|_{\mathcal{E}_2,\mathrm{sq}^t_{\mathcal{E}_1}(E)}), \tag{5.11}$$

where $\|\cdot\|_{\mathcal{E}_2,\mathrm{sq}^t_{\mathcal{E}_1}(E)}$ denotes the subquotient norm induced by $\|\cdot\|_{\mathcal{E}_2}$ on the vector space $\mathrm{sq}^t_{\mathcal{E}_1}(E)$.

Proof We assume that the $\mathbb{R}$-filtration $\mathcal{E}_1$ corresponds to the flag

$$0 = V_0 \subsetneq V_1 \subsetneq \ldots \subsetneq V_n = V$$

together with the sequence

$$\mu_1 > \ldots > \mu_n.$$

Then the right hand side of the formula can be written as

$$\sum_{j=1}^{n}\mu_j\,\widehat{\deg}(V_j/V_{j-1}, \|\cdot\|_{\mathcal{E}_2,j}),$$

where $\|\cdot\|_{\mathcal{E}_2,j}$ is the subquotient norm on V_j/V_{j-1} induced by $\|\cdot\|_{\mathcal{E}_2}$. By Proposition 5.1.2 (3) the basis $\boldsymbol{e}$ is compatible with respect to the flag

$$0 = V_0 \subsetneq V_1 \subsetneq \ldots \subsetneq V_n = V.$$

By Proposition 1.2.6, the canonical image of $\boldsymbol{e} \cap (V_j \setminus V_{j-1})$ in V_j/V_{j-1} forms an orthogonal basis of $(V_j/V_{j-1}, \|\cdot\|_{\mathcal{E}_2,j})$. Moreover, for any $x \in \boldsymbol{e} \cap (V_j \setminus V_{j-1})$ one has

$$\|x\|_{\mathcal{E}_2} = \|[x]\|_{\mathcal{E}_2,j}$$

and $\|x\|_{\mathcal{E}_1} = \mathrm{e}^{-\mu_j}$ Therefore,

$$\begin{aligned}\sum_{i=1}^{r} \lambda_{\mathcal{E}_1}(e_i)\lambda_{\mathcal{E}_2}(e_i) &= \sum_{j=1}^{n} \sum_{x \in \boldsymbol{e}\cap(V_j\setminus V_{j-1})} \lambda_{\mathcal{E}_1}(x)\lambda_{\mathcal{E}_2}(x)\\ &= \sum_{j=1}^{n} \mu_j \sum_{x \in \boldsymbol{e}\cap(V_j\setminus V_{j-1})} (-\ln\|[x]\|_{\mathcal{E}_2,j}) = \sum_{j=1}^{n} \mu_j \widehat{\deg}(V_j/V_{j-1}, \|\cdot\|_{\mathcal{E}_2,j}),\end{aligned}$$

where the last equality comes from Proposition 5.1.2 (2). The equality (5.11) is thus proved. □

We say that an $\mathbb{R}$-filtration $\mathcal{E} \in \mathbf{Fil}(E)$ is *trivial* if the function $Z_{\mathcal{E}}$ (see Definition 1.1.38) is constantly zero, or equivalently, $\langle \mathcal{E}, \mathcal{E} \rangle = 0$.

Lemma 5.4.2 *Let V be a finite-dimensional non-zero vector space over $\mathbb{R}$, equipped with an inner product $\langle , \rangle$. Let $\{\ell_i\}_{i=1}^n$ be a finite family of linear forms on V, where $n \in \mathbb{N}$, $n \geqslant 1$. Let $\theta : V \setminus \{0\} \to \mathbb{R}$ be the function defined by*

$$\theta(x) = \max_{i \in \{1,\ldots,n\}} \frac{\ell_i(x)}{\|x\|},$$

where $\|\cdot\|$ is the norm induced by the inner product $\langle , \rangle$. Then the function θ attains its minimal value on $V \setminus \{0\}$. Moreover, if c is the minimal value of θ and if x_0 is a point of $V \setminus \{0\}$ minimising the function θ, then for any $x \in \mathbb{R}^n$ one has

$$\theta(x) \geqslant c \frac{\langle x, x_0 \rangle}{\|x\| \cdot \|x_0\|}. \tag{5.12}$$

Proof Note that the function θ is invariant by positive dilatations, namely for any $x \in V \setminus \{0\}$ and any $\lambda > 0$ one has $\theta(\lambda x) = \theta(x)$. Moreover, the function θ is clearly continuous. Hence it attains its minimal value, which is equal to $\min_{x \in V,\ \|x\|=1} \theta(x)$.

To show the inequality (5.12), we may assume without loss of generality that $\|x\| = \|x_0\| = 1$. Note that for any $t \in [0, 1]$ one has

$$\begin{aligned}c\|tx + (1-t)x_0\| &\leqslant \|tx + (1-t)x_0\|\theta(tx + (1-t)x_0)\\ &= \max_{i \in \{1,\ldots,n\}} \ell_i(tx + (1-t)x_0) \leqslant t\theta(x) + (1-t)c.\end{aligned}$$

Note that when $t = 0$ one has

$$c\|tx + (1-t)x_0\| = c = t\theta(x) + (1-t)c.$$

Therefore the right derivative at $t = 0$ of the convex function

$$t \in [0,1] \longmapsto c\|tx + (1-t)x_0\|$$

is bounded from above by $\theta(x) - c$, which leads to

$$c\frac{t\|x\|^2 - (1-t)\|x\|^2 + (1-2t)\langle x, x_0\rangle}{\|tx + (1-t)x_0\|}\Big|_{t=0} = c(\langle x, x_0\rangle - 1) \leqslant \theta(x) - c,$$

namely $\theta(x) \geqslant c\langle x, x_0\rangle$. □

Theorem 5.4.3 *Let $d \in \mathbb{N}_{\geqslant 2}$, $\{E_j\}_{j=1}^d$ be a family of finite-dimensional non-zero vector spaces over K, and V be a quotient vector space of $E_1 \otimes_K \cdots \otimes_K E_d$. Let r be the dimension of V over K. For any $j \in \{1, \ldots, d\}$, let a_j be the dimension of E_j over K. Then the following conditions are equivalent.*

(1) *The rational point x of*

$$\pi : P = \mathbb{P}((E_1 \otimes_K \cdots \otimes_K E_d)^{\otimes r}) \longrightarrow \operatorname{Spec} K$$

corresponding to $\det(V)$ *is not semistable with respect to the $\mathbb{GL}(E_1) \times \cdots \times \mathbb{GL}(E_d)$-linearised invertible sheaf*

$$L := \mathcal{O}_P(a_1 \cdots a_d) \otimes \pi^*(\det(E_1^\vee)^{\otimes rb_1} \otimes \cdots \det(E_d^\vee)^{\otimes rb_d}),$$

where $b_j = a_1 \cdots a_d / a_j$ for any $j \in \{1, \ldots, d\}$.

(2) *Let $\mathbf{S}$ be the subset of $\mathbf{Fil}_0(E_1) \times \cdots \times \mathbf{Fil}_0(E_d)$ consisting of the filtrations $(\mathcal{F}_1, \ldots, \mathcal{F}_d)$ which are not simultaneously trivial. For each*

$$\mathcal{F} = (\mathcal{F}_1, \ldots, \mathcal{F}_d) \in \mathbf{Fil}(E_1) \times \cdots \times \mathbf{Fil}(E_d),$$

let $\|\cdot\|_{\mathcal{F},V}$ be the quotient norm on V of the ε-tensor product of $\|\cdot\|_{\mathcal{F}_1}, \ldots, \|\cdot\|_{\mathcal{F}_d}$. Then the function $\Theta : \mathbf{S} \to \mathbb{R}$ defined as

$$\forall\, \mathcal{F} = (\mathcal{F}_1, \ldots, \mathcal{F}_d) \in \mathbf{S}, \quad \Theta(\mathcal{F}) = \frac{\widehat{\mu}(V, \|\cdot\|_{\mathcal{F},V})}{(\langle \mathcal{F}_1, \mathcal{F}_1\rangle + \cdots + \langle \mathcal{F}_d, \mathcal{F}_d\rangle)^{1/2}}$$

attains its minimal value, which is negative.

Moreover, if the above conditions are satisfied and if $\mathcal{E} = (\mathcal{E}_1, \ldots, \mathcal{E}_d)$ is a minimal point of the function Θ, then for any $\mathcal{F} = (\mathcal{F}_1, \ldots, \mathcal{F}_d) \in \mathbf{Fil}(E_1) \times \cdots \times \mathbf{Fil}(E_d)$ one has

$$\widehat{\mu}(V, \|\cdot\|_{\mathcal{F},V}) \geqslant \sum_{j=1}^d \widehat{\mu}(E_j, \|\cdot\|_{\mathcal{F}_j}) + c\frac{\langle \mathcal{E}_1, \mathcal{F}_1\rangle + \cdots + \langle \mathcal{E}_d, \mathcal{F}_d\rangle}{(\langle \mathcal{E}_1, \mathcal{E}_1\rangle + \cdots + \langle \mathcal{E}_d, \mathcal{E}_d\rangle)^{1/2}}, \tag{5.13}$$

where c is the minimal value of Θ.

Proof Assume that the condition (2) holds, then there exists $\mathcal{F} = (\mathcal{F}_1, \ldots, \mathcal{F}_d) \in \mathbf{S}$ such that

$$\widehat{\mu}(V, \|\cdot\|_{\mathcal{F},V}) < 0 = \widehat{\mu}(E_1, \|\cdot\|_{\mathcal{F}_1}) + \cdots + \widehat{\mu}(E_d, \|\cdot\|_{\mathcal{F}_d}).$$

By Corollary 5.2.10, the point x is not semistable with respect to L. Conversely, if the condition (1) holds, then there exists

$$\mathcal{F}' = (\mathcal{F}_1', \ldots, \mathcal{F}_d') \in \mathbf{Fil}(E_1) \times \cdots \times \mathbf{Fil}(E_d)$$

such that

$$\widehat{\mu}(V, \|\cdot\|_{\mathcal{F}',V}) < \sum_{j=1}^{d} \widehat{\mu}(E, \|\cdot\|_{\mathcal{F}_j'}).$$

For $j \in \{1, \ldots, d\}$, let $\mathcal{F}_j$ be $\mathbb{R}$-filtrations on E_j such that

$$\|\cdot\|_{\mathcal{F}_j} = \exp(\widehat{\mu}(E, \|\cdot\|_{\mathcal{F}_j'}))\|\cdot\|_{\mathcal{F}_j'}.$$

Then one has $\mathcal{F}_j \in \mathbf{Fil}_0(E_j)$ for any $j \in \{1, \ldots, d\}$. Moreover, if we denote by $\mathcal{F}$ the vector $(\mathcal{F}_1, \ldots, \mathcal{F}_d)$, then one has

$$\widehat{\mu}(V, \|\cdot\|_{\mathcal{F},V}) = \widehat{\mu}(V, \|\cdot\|_{\mathcal{F}',V}) - \sum_{j=1}^{d} \widehat{\mu}(E, \|\cdot\|_{\mathcal{F}_j'}) < 0.$$

In particular, the $\mathbb{R}$-filtrations $\mathcal{F}_j$ are not simultaneously trivial since otherwise we should have $\widehat{\mu}(V, \|\cdot\|_{\mathcal{F},V}) = 0$. Therefore one has $\mathcal{F} \in \mathbf{S}$, which implies that the function Θ takes at least a negative value.

In the following, we show that the function Θ attains its minimal value. For any $j \in \{1, \ldots, d\}$, let $\mathscr{B}_j$ be the set of bases of E_j. For $n \in \mathbb{N}$, let Δ_n be the vector subspace of $\mathbb{R}^n$ of vectors $(z_1, \ldots, z_n)$ such that $z_1 + \cdots + z_n = 0$. For any

$$\boldsymbol{e} = (\boldsymbol{e}^{(1)}, \ldots, \boldsymbol{e}^{(d)}) \in \mathscr{B}_1 \times \cdots \times \mathscr{B}_d,$$

let

$$\Psi_{\boldsymbol{e}} : \Delta_{a_1} \times \cdots \times \Delta_{a_d} \to \mathbf{Fil}_0(E_1) \times \cdots \times \mathbf{Fil}_0(E_d)$$

be the map sending $(\boldsymbol{y}^{(1)}, \ldots, \boldsymbol{y}^{(d)})$ to the vector of $\mathbb{R}$-filtrations $(\mathcal{F}_1, \ldots, \mathcal{F}_d)$ such that, for any $j \in \{1, \ldots, d\}$, $\boldsymbol{e}^{(j)} = \{e_i^{(j)}\}_{i=1}^{a_j}$ forms an orthogonal basis of $\|\cdot\|_{\mathcal{F}_j}$ with

$$\boldsymbol{y}^{(j)} = (\lambda_{\mathcal{F}_j}(e_1^{(j)}), \ldots, \lambda_{\mathcal{F}_j}(e_{a_j}^{(j)})).$$

For $j \in \{1, \ldots, d\}$, let $\boldsymbol{y}^{(j)} = (y_1^{(j)}, \ldots, y_{a_j}^{(j)})$ be an element of Δ_{a_j}. If

$$\mathcal{F} = (\mathcal{F}_1, \ldots, \mathcal{F}_d) = \Psi_{\boldsymbol{e}}(\boldsymbol{y}^{(1)}, \ldots, \boldsymbol{y}^{(d)}),$$

then $\widehat{\mu}(V, \|\cdot\|_{\mathcal{F},V})$ is equal to the maximal value of

$$\sum_{j=1}^{d} y_{i_1^{(j)}}^{(j)} + \cdots + y_{i_r^{(j)}}^{(j)}$$

for those $(i_1^{(j)},\ldots,i_r^{(j)})\in\{1,\ldots,a_j\}^r$ such that the image of

$$(e^{(1)}_{i_1^{(1)}}\otimes\cdots\otimes e^{(1)}_{i_r^{(1)}})\otimes\cdots\otimes(e^{(d)}_{i_1^{(d)}}\otimes\cdots\otimes e^{(d)}_{i_r^{(d)}})$$

by the canonical composed surjective map

$$E_1^{\otimes r}\otimes_K\cdots\otimes_K E_d^{\otimes r}\cong(E_1\otimes_K\cdots\otimes_K E_d)^{\otimes r}\longrightarrow V^{\otimes r}\longrightarrow\det(V)$$

is non-zero. Moreover, for any $j\in\{1,\ldots,d\}$ one has

$$\langle\mathcal{F}_j,\mathcal{F}_j\rangle=\sum_{i=1}^{a_j}(y_i^{(j)})^2.$$

Therefore, the composition of Θ with the restriction of $\Psi_{\boldsymbol{e}}$ on

$$(\Delta_{a_1}\times\ldots\times\Delta_{a_d})\setminus\{(0,\ldots,0)\}$$

defines a continuous function on $(\Delta_{a_1}\times\ldots\times\Delta_{a_d})\setminus\{(0,\ldots,0)\}$ which is invariant by dilatation by elements in $\mathbb{R}_{>0}$. It hence attains its minimal value. Moreover, although $\mathscr{B}_1\times\cdots\times\mathscr{B}_d$ may contain infinitely many elements, from the expression of the value $\widehat{\mu}(V,\|\cdot\|_{\mathcal{E}\otimes_\varepsilon\mathcal{F},V})$ as above we obtain that there are only finitely many (at most $2^{(a_1\cdots a_d)^r}$) choices for the composed function

$$\Theta\circ(\Psi_{\boldsymbol{e}}|_{\Delta_{a_1}\times\cdots\times\Delta_{a_d}\setminus\{(0,\ldots,0)\}}).$$

Therefore, the function Θ attains its minimal value, which is negative since Θ takes at least one negative value.

In the following, we prove the inequality (5.13). Let $\mathcal{E}=(\mathcal{E}_1,\ldots,\mathcal{E}_d)$ be an element of $\mathbf{Fil}_0(E_1)\times\cdots\times\mathbf{Fil}_0(E_d)$ which minimise the function Θ. Let $\mathcal{F}=(\mathcal{F}_1,\ldots,\mathcal{F}_d)$ be an element of $\mathbf{Fil}(E_1)\times\cdots\times\mathbf{Fil}(E_d)$. Note that, if $\mathcal{F}_j'$ is the $\mathbb{R}$-filtration of E_j such that

$$\|\cdot\|_{\mathcal{F}_j'}=\exp(\widehat{\deg}(E_j,\|\cdot\|_{\mathcal{F}_j}))\|\cdot\|_{\mathcal{F}_j},$$

then one has $\mathcal{F}_j'\in\mathbf{Fil}_0(E_j)$ and $\langle\mathcal{E}_j,\mathcal{F}_j'\rangle=\langle\mathcal{E}_j,\mathcal{F}_j\rangle$. Moreover, if we denote by $\mathcal{F}'$ the vector $(\mathcal{F}_1',\ldots,\mathcal{F}_d')$, then

$$\widehat{\mu}(V,\|\cdot\|_{\mathcal{F}',V})=\widehat{\mu}(V,\|\cdot\|_{\mathcal{F},V})-\sum_{j=1}^d\widehat{\deg}(E_j,\mathcal{F}_j).$$

Therefore, to show the inequality (5.13), it suffices to treat the case where $\mathcal{F}\in\mathbf{Fil}_0(E_1)\times\cdots\times\mathbf{Fil}_0(E_d)$.

By Proposition 5.1.3, for any $j\in\{1,\ldots,d\}$, there exists a basis $\boldsymbol{e}^{(j)}$ of E which is orthogonal with respect to the norms $\|\cdot\|_{\mathcal{E}_j}$ and $\|\cdot\|_{\mathcal{F}_j}$ simultaneously. Therefore the inequality (5.13) follows from Lemma 5.4.2. □

5.5 Lifting and refinement of filtrations

Let K be a field. Let V be a finite-dimensional vector space over K and

$$0 = V_0 \subsetneq V_1 \subsetneq \ldots \subsetneq V_n = V \tag{5.14}$$

be a flag of vector subspaces of V. Suppose given, for any $i \in \{1, \ldots, n\}$, an $\mathbb{R}$-filtration $\mathcal{F}_i$ of the sub-quotient vector space V_i/V_{i-1}. We will construct an $\mathbb{R}$-filtration on V from the data of $\{\mathcal{F}_i\}_{i=1}^n$ as follows. For any $i \in \{1, \ldots, n\}$, let $\widetilde{\boldsymbol{e}}^{(i)}$ be a basis of V_i/V_{i-1} which is orthogonal with respect to the norm $\|\cdot\|_{\mathcal{F}_i}$, where we consider the trivial valuation on K. The basis $\widetilde{\boldsymbol{e}}^{(i)}$ gives rise to a linearly independent family $\boldsymbol{e}^{(i)}$ in V_i. Let $\boldsymbol{e} = \bigcup_{i=1}^n \boldsymbol{e}^{(i)}$ be the (disjoint) union of $\boldsymbol{e}^{(i)}$, $i \in \{1, \ldots, n\}$. Note that $\boldsymbol{e}$ forms actually a basis of V over K. We define an ultrametric norm $\|\cdot\|$ on V such that $\boldsymbol{e}$ is an orthogonal basis under this norm and that, for any $i \in \{1, \ldots, n\}$ and any $x \in \boldsymbol{e}^{(i)}$, the norm of x is $\|\widetilde{x}\|_{\mathcal{F}_i}$, where $\widetilde{x}$ denotes the class of x in V_i/V_{i-1}.

Remark 5.5.1 (1) For any $i \in \{1, \ldots, n\}$, the subquotient norm on V_i/V_{i-1} induced by $\|\cdot\|$ coincides with $\|\cdot\|_{\mathcal{F}_i}$. In particular, one has

$$\widehat{\deg}(V, \|\cdot\|) = \sum_{i=1}^n \widehat{\deg}(V_i/V_{i-1}, \|\cdot\|_{\mathcal{F}_i}).$$

(2) Let $\mathcal{F}$ be the $\mathbb{R}$-filtration corresponding to the ultrametric norm $\|\cdot\|$. Assume that (5.14) is the flag of vector subspaces of V defined by an $\mathbb{R}$-filtration $\mathcal{G}$ on V and that $\mu_1 > \ldots > \mu_n$ are jump points of the $\mathbb{R}$-filtration $\mathcal{G}$. Then we can compute $\langle \mathcal{F}, \mathcal{G} \rangle$ as follows:

$$\langle \mathcal{F}, \mathcal{G} \rangle = \frac{1}{\dim_K(V)} \sum_{i=1}^n \mu_i \, \widehat{\deg}(V_i/V_{i-1}, \|\cdot\|_{\mathcal{F}_i}).$$

Definition 5.5.2 The $\mathbb{R}$-filtration on V corresponding to the norm $\|\cdot\|$ constructed above is called a *lifting* of the family $\{\mathcal{F}_i\}_{i=1}^n$ (relatively to the basis $\boldsymbol{e}$). We emphasis that the lifting depends on the choice of the basis $\boldsymbol{e}$.

Definition 5.5.3 Let V be a finite-dimensional vector space over k and $\mathcal{F}$ be an $\mathbb{R}$-filtration on V. We call *refinement* of $\mathcal{F}$ any flag

$$0 = V_0 \subsetneq V_1 \subsetneq \ldots \subsetneq V_n = V$$

of vector subspaces of V together with a non-increasing sequence

$$t_1 \geqslant \ldots \geqslant t_n$$

such that, for any $i \in \{1, \ldots, n\}$ and any $x \in V_i \setminus V_{i-1}$, one has $\|x\|_{\mathcal{F}} = \mathrm{e}^{-t_i}$.

Remark 5.5.4 Let V be a finite-dimensional vector space over K and $\mathcal{F}$ be an $\mathbb{R}$-filtration on V. Recall that the $\mathbb{R}$-filtration $\mathcal{F}$ corresponds to a flag

$$0 = W_0 \subsetneq W_1 \subsetneq \ldots \subsetneq W_m = V$$

together with a decreasing sequence

$$\lambda_1 > \ldots > \lambda_m.$$

To choose a refinement of $\mathcal{F}$ is equivalent to specify, for any $j \in \{1, \ldots, m\}$, a flag

$$0 = V_j^{(0)}/W_{j-1} \subsetneq V_j^{(1)}/W_{j-1} \subsetneq \ldots \subsetneq V_j^{(n_j)}/W_{j-1} = W_j/W_{j-1}$$

of W_j/W_{j-1}. The corresponding refinement is given by the flag

$$0 = V_0 \subsetneq V_1^{(1)} \subsetneq \ldots \subsetneq V_1^{(n_1)} \subsetneq \ldots \subsetneq V_m^{(1)} \subsetneq \ldots \subsetneq V_m^{(n_m)} = V$$

and the non-increasing sequence

$$\underbrace{\lambda_1 = \cdots = \lambda_1}_{n_1 \text{ copies}} > \underbrace{\lambda_2 = \cdots = \lambda_2}_{n_2 \text{ copies}} > \ldots > \underbrace{\lambda_m = \cdots = \lambda_m}_{n_m \text{ copies}}.$$

Proposition 5.5.5 *Let $d \in \mathbb{N}_{\geqslant 2}$, $\{E_j\}_{j=1}^d$ be a family of finite-dimensional vector spaces over K, and $(\mathcal{F}_1, \ldots, \mathcal{F}_d) \in \mathbf{Fil}(E_1) \times \cdots \mathbf{Fil}(E_d)$. Let $G = E_1 \otimes_K \cdots \otimes_K E_d$ be the tensor product space and let $\mathcal{G}$ be the tensor product $\mathbb{R}$-filtration of $\mathcal{F}_1, \ldots, \mathcal{F}_d$ (namely the $\mathbb{R}$-filtration on G corresponding to the ε-tensor product of $\|\cdot\|_{\mathcal{F}_1}, \ldots, \|\cdot\|_{\mathcal{F}_d}$). Then there exists a refinement*

$$0 = G_0 \subsetneq G_1 \subsetneq \ldots \subsetneq G_n = G, \quad t_1 \geqslant \ldots \geqslant t_n$$

of the $\mathbb{R}$-filtration $\mathcal{G}$, such that, for any $i \in \{1, \ldots, n\}$, the subquotient G_i/G_{i-1} is canonically isomorphic to a tensor product of subquotients of the form

$$\mathrm{sq}_{\mathcal{F}_1}^{\lambda_{i,1}}(E_1) \otimes_K \cdots \otimes_K \mathrm{sq}_{\mathcal{F}_d}^{\lambda_{i,d}}(E_d)$$

with $\lambda_{i,1} + \cdots + \lambda_{i,d} = t_i$.

Proof Since $\mathcal{G}$ is the tensor product $\mathbb{R}$-filtration of $\mathcal{F}_1, \ldots, \mathcal{F}_d$, one has

$$\mathcal{G}^t(G) = \sum_{\mu_1 + \cdots + \mu_d \geqslant t} \mathcal{F}_1^{\mu_1}(E_1) \otimes_K \cdots \otimes_K \mathcal{F}_d^{\mu_d}(E_d).$$

Therefore,

$$\mathrm{sq}_{\mathcal{G}}^t(G) = \bigoplus_{\mu_1 + \cdots + \mu_d = t} \mathrm{sq}_{\mathcal{F}_1}^{\mu_1}(E_1) \otimes_K \cdots \otimes_K \mathrm{sq}_{\mathcal{F}_d}^{\mu_d}(E_d).$$

For any $t \in \mathbb{R}$ there exists clearly a flag of $\mathrm{sq}_{\mathcal{G}}^t(G)$ whose successive subquotients are of the form $\mathrm{sq}_{\mathcal{F}_1}^{\mu_1}(E_1) \otimes_K \cdots \otimes_K \mathrm{sq}_{\mathcal{F}_d}^{\mu_d}(E_d)$ with $\mu_1 + \cdots + \mu_d = t$. Hence we can construct a refinement of the $\mathbb{R}$-filtration $\mathcal{G}$ by using the construction in Remark 5.5.4. □

Remark 5.5.6 Consider a proper adelic curve $S = (K, (\Omega, \mathcal{A}, \nu), \phi)$ whose underlying vector space is K. We keep the notation of Proposition 5.5.5 and suppose that each E_j is equipped with a norm family ξ_j such that (E_j, ξ_j) forms an adelic vector bundle on S, and we equip G with the ε, π-tensor product norm family $\xi_G = \xi_1 \otimes_{\varepsilon,\pi} \cdots \otimes_{\varepsilon,\pi} \xi_d$. For any $t \in \mathbb{R}$ and any $j \in \{1, \ldots, d\}$, let ξ_j^t be the induced norm families of ξ_j on $\mathcal{F}_j^t(E_j)$ and $\xi_{j,\mathrm{sq}}^t$ be the quotient norm family of ξ_j^t on $\mathrm{sq}_{\mathcal{F}_j}^t(E_j)$. By Propositions 1.1.58 and 1.2.36, for any $(\mu_1, \ldots, \mu_d) \in \mathbb{R}^d$, the quotient norm family of $\xi_1^{\mu_1} \otimes_{\varepsilon,\pi} \cdots \otimes_{\varepsilon,\pi} \xi_d^{\mu_d}$ on $\mathrm{sq}_{\mathcal{F}_1}^{\mu_1}(E_1) \otimes_K \cdots \otimes_K \mathrm{sq}_{\mathcal{F}_d}^{\mu_d}(E_d)$ identifies with $\xi_{1,\mathrm{sq}}^{\mu_1} \otimes_{\varepsilon,\pi} \cdots \otimes_{\varepsilon,\pi} \xi_{d,\mathrm{sq}}^{\mu_d}$.

We consider a refinement

$$0 = G_0 \subsetneq G_1 \subsetneq \ldots \subsetneq G_n = G, \quad t_1 \geqslant \ldots \geqslant t_n$$

of the $\mathbb{R}$-filtration $\mathcal{G}$ such that each subquotient G_i/G_{i-1} is canonically isomorphic to a tensor product of the form $\mathrm{sq}_{\mathcal{F}_1}^{\lambda_{i,1}}(E_1) \otimes_K \cdots \otimes_K \mathrm{sq}_{\mathcal{F}_d}^{\lambda_{i,d}}(E_i)$ with $\lambda_{i,1} + \cdots + \lambda_{i,d} = t_i$. Note that the canonicity of the isomorphism means that the vector space G_i contains $\mathcal{F}^{\lambda_{i,1}}(E_1) \otimes_K \cdots \otimes_K \mathcal{F}^{\lambda_{i,d}}(E_d)$ and the composition

$$\mathcal{F}^{\lambda_{i,1}}(E_1) \otimes_K \cdots \otimes_K \mathcal{F}^{\lambda_{i,d}}(E_d) \longrightarrow G_i \longrightarrow G_i/G_{i-1}$$

of the inclusion map $\mathcal{F}^{\lambda_{i,1}}(E_1) \otimes_K \cdots \otimes_K \mathcal{F}^{\lambda_{i,d}}(E_d) \to G_i$ with the quotient map $G_i \to G_i/G_{i-1}$ induces an isomorphism

$$\varphi_i : \mathrm{sq}_{\mathcal{F}_1}^{\lambda_{i,1}}(E_1) \otimes_K \cdots \otimes_K \mathrm{sq}_{\mathcal{F}_d}^{\lambda_{i,d}}(E_d) \longrightarrow G_i/G_{i-1}.$$

We are interested in the comparison between $\xi_{1,\mathrm{sq}}^{\lambda_{i,1}} \otimes_{\varepsilon,\pi} \cdots \otimes_{\varepsilon,\pi} \xi_{d,\mathrm{sq}}^{\lambda_{i,d}}$ and the subquotient norm family of ξ_G on G_i/G_{i-1}. By Propositions 1.1.60, the restriction of ξ_G on $\mathcal{F}_1^{\lambda_{i,1}}(E_1) \otimes_K \cdots \otimes_K \mathcal{F}_d^{\lambda_{i,d}}(E_d)$ is bounded from above by $\xi_1^{\lambda_{i,1}} \otimes_{\varepsilon,\pi} \cdots \otimes_{\varepsilon,\pi} \xi_d^{\lambda_{i,d}}$. Therefore, for any $\omega \in \Omega$, the isomorphism $\varphi_{i,\omega}$ has an operator norm $\leqslant 1$.

Assume that the norm families $\xi_1, \ldots, \xi_d$ are Hermitian. Let $\widetilde{\xi}_G$ be the orthogonal tensor product of $\xi_1, \ldots, \xi_d$. If we equip $\mathrm{sq}_{\mathcal{F}_1}^{\lambda_{i,1}}(E_1) \otimes_K \cdots \otimes_K \mathrm{sq}_{\mathcal{F}_d}^{\lambda_{i,d}}(E_d)$ with the orthogonal product norm family $\xi_{1,\mathrm{sq}}^{\lambda_{i,1}} \otimes \cdots \otimes \xi_{d,\mathrm{sq}}^{\lambda_{i,d}}$ and G_i/G_{i-1} with the subquotient norm family of $\widetilde{\xi}_G$ on G_i/G_{i-1}, then, for any $\omega \in \Omega$, the operator norm of $\varphi_{i,\omega}$ is bounded from above by 1. This follows from the fact that the restriction of ξ_G on $\mathcal{F}_1^{\lambda_{i,1}}(E_1) \otimes_K \cdots \otimes_K \mathcal{F}_d^{\lambda_{i,d}}(E_d)$ identifies with $\xi_1^{\lambda_{i,1}} \otimes \cdots \otimes \xi_d^{\lambda_{i,d}}$ (see Proposition 1.2.58).

5.6 Estimation in general case

Let $S = (K, (\Omega, \mathcal{A}, \nu), \phi)$ be a proper adelic curve such that, either the σ-algebra $\mathcal{A}$ is discrete, or K admits a countable subfield which is dense in each K_ω, where $\omega \in \Omega$. We assume in addition that K is of characteristic 0. In this section, we establish the following result.

Theorem 5.6.1 *Let $d \in \mathbb{N}_{\geqslant 2}$, $\{(E_j, \xi_j)\}_{j=1}^d$ be a family of non-zero* Hermitian *adelic vector bundles on S, and V be a non-zero quotient vector space of $E_1 \otimes_K \cdots \otimes_K E_d$. For any $j \in \{1, \ldots, d\}$ let $\mathcal{H}_j$ be the Harder-Narasimhan $\mathbb{R}$-filtrations of (E_j, ξ_j), and $\|\cdot\|_V$ be the quotient norm of the ε-tensor product of $\|\cdot\|_{\mathcal{H}_1}, \ldots, \|\cdot\|_{\mathcal{H}_d}$, ξ_V be the quotient norm family of the ε, π-tensor product $\xi_1 \otimes_{\varepsilon,\pi} \cdots \otimes_{\varepsilon,\pi} \xi_d$, and $\widetilde{\xi}_V$ be the quotient norm family of the orthogonal tensor product $\xi_1 \otimes \cdots \otimes \xi_d$. Then the following inequalities hold*

$$\widehat{\mu}(V, \xi_V) \geqslant \widehat{\mu}(V, \|\cdot\|_V) - \nu(\Omega_\infty) \sum_{j=1}^d \ln(\dim_K(E_j)), \tag{5.15}$$

$$\widehat{\mu}(V, \widetilde{\xi}_V) \geqslant \widehat{\mu}(V, \|\cdot\|_V) - \frac{1}{2}\nu(\Omega_\infty) \sum_{j=1}^d \ln(\dim_K(E_j)). \tag{5.16}$$

In particular, if all adelic vector bundles (E_j, ξ_j) are semistable, then one has

$$\widehat{\mu}(V, \widetilde{\xi}_V) \geqslant \sum_{j=1}^d \Big(\widehat{\mu}(E_j, \xi_j) - \nu(\Omega_\infty) \ln(\dim_K(E_j))\Big), \tag{5.17}$$

$$\widehat{\mu}(V, \xi_V) \geqslant \sum_{j=1}^d \Big(\widehat{\mu}(E_j, \xi_j) - \frac{1}{2}\nu(\Omega_\infty) \ln(\dim_K(E_j))\Big). \tag{5.18}$$

Proof For any $j \in \{1, \ldots, d\}$, let a_j be the dimension of E_j over K. We reason by induction on $A = a_1 + \cdots + a_d$. The theorem is clearly true when $A = d$ (namely $\mathrm{rk}(E_j) = 1$ for any j). In the following, we assume that the theorem has been proved for any family of adelic vector bundles whose dimensions have a sum $< A$.

Step 1: In this step, we assume that the adelic vector bundles (E_j, ξ_j) are not simultaneously semistable, or equivalently, at least one of the $\mathbb{R}$-filtrations $\mathcal{H}_j$ has more than one jump point. Let G be the tensor product space $E_1 \otimes_K \cdots \otimes_K E_d$ and $\mathcal{G} \in \mathbf{Fil}(G)$ be the tensor product of the $\mathbb{R}$-filtrations $\mathcal{H}_1, \ldots, \mathcal{H}_d$, which corresponds to the ε-tensor product of the norms $\|\cdot\|_{\mathcal{F}_1}, \ldots \|\cdot\|_{\mathcal{F}_d}$. We choose a refinement

$$0 = G_0 \subsetneq G_1 \subsetneq \ldots \subsetneq G_n = G, \quad t_1 \geqslant \ldots \geqslant t_n$$

of the $\mathbb{R}$-filtration $\mathcal{G}$ such that, for any $i \in \{1, \ldots, n\}$, the subquotient G_i/G_{i-1} is canonically isomorphic to a tensor product of the form

$$\mathrm{sq}_{\mathcal{H}_1}^{\lambda_{i,1}}(E_1) \otimes_K \otimes \cdots \otimes_K \mathrm{sq}_{\mathcal{H}_d}^{\lambda_{i,d}}(E_d)$$

with $\lambda_{i,1} + \cdots + \lambda_{i,d} = t_i$ (see Proposition 5.5.5). The assumption that at least one of the $\mathbb{R}$-filtrations $\mathcal{F}_j$ has more than one jump point implies that

$$\dim_K(\mathrm{sq}_{\mathcal{H}_1}^{\lambda_{i,1}}(E_1)) + \cdots + \dim_K(\mathrm{sq}_{\mathcal{H}_d}^{\lambda_{i,d}}(E_d)) < a_1 + \cdots + a_d.$$

For any $i \in \{1, \ldots, n\}$ and any $j \in \{1, \ldots, d\}$, denoted by $\xi_{j,\mathrm{sq}}^{\lambda_{i,j}}$ the subquotient norm families of ξ_j on $\mathrm{sq}_{\mathcal{H}_j}^{\lambda_{i,j}}(E_j)$. For any $i \in \{1, \ldots, n\}$, let $\xi_{G_i/G_{i-1}}$ be the subquotient norm family of ξ_G on G_i/G_{i-1}, $\xi'_{V_i/V_{i-1}}$ be the quotient norm family of $\xi_{G_i/G_{i-1}}$ on V_i/V_{i-1}, $\xi_{V_i/V_{i-1}}$ be the subquotient norm family of ξ_V on V_i/V_{i-1}, and $\xi''_{V_i/V_{i-1}}$ be the quotient norm family of $\xi_{1,\mathrm{sq}}^{\lambda_{i,1}} \otimes_{\varepsilon,\pi} \cdots \otimes_{\varepsilon,\pi} \xi_{d,\mathrm{sq}}^{\lambda_{i,d}}$ on V_i/V_{i-1}, where we identify G_i/G_{i-1} with

$$\mathrm{sq}_{\mathcal{H}_1}^{\lambda_{i,1}}(E_1) \otimes_K \cdots \otimes_K \mathrm{sq}_{\mathcal{H}_d}^{\lambda_{i,d}}(E_d).$$

By Proposition 1.1.14 (2), one has

$$\widehat{\deg}(V_i/V_{i-1}, \xi_{V_i/V_{i-1}}) \geqslant \widehat{\deg}(V_i/V_{i-1}, \xi'_{V_i/V_{i-1}}). \tag{5.19}$$

By Remark 5.5.6, one has

$$\widehat{\deg}(V_i/V_{i-1}, \xi'_{V_i/V_{i-1}}) \geqslant \widehat{\deg}(V_i/V_{i-1}, \xi''_{V_i/V_{i-1}}). \tag{5.20}$$

Moreover, for any $(i, j) \in \{1, \ldots, n\} \times \{1, \ldots, d\}$, the Hermitian adelic vector bundle $(\mathrm{sq}_{\mathcal{H}_j}^{\lambda_{i,j}}(E_j), \xi_{j,\mathrm{sq}}^{\lambda_{i,j}})$ is semistable of slope $\lambda_{i,j}$. Therefore, by the induction hypothesis one has

$$\begin{aligned}\widehat{\deg}(V_i/V_{i-1}, \xi_{V_i/V_{i-1}}) &\geqslant \widehat{\deg}(V_i/V_{i-1}, \xi''_{V_i/V_{i-1}})\\ &\geqslant \dim_K(V_i/V_{i-1}) \sum_{j=1}^{d} \big(\lambda_{i,j} - \nu(\Omega_\infty)\ln(a_j)\big).\end{aligned}$$

Taking the sum with respect to $i \in \{1, \ldots, n\}$, we obtain

$$\begin{aligned}\widehat{\deg}(V, \xi_V) &= \sum_{i=1}^{n} \widehat{\deg}(V_i/V_{i-1}, \xi_{V_i/V_{i-1}})\\ &\geqslant \sum_{i=1}^{n} \dim_K(V_i/V_{i-1}) \sum_{j=1}^{d} \big(\lambda_{i,j} - \nu(\Omega_\infty)\ln(a_j)\big)\\ &= \sum_{i=1}^{n} \dim_K(V_i/V_{i-1}) t_i - \dim_K(V)\nu(\Omega_\infty) \sum_{j=1}^{d} \ln(a_j)\\ &= \widehat{\deg}(V, \|\cdot\|_V) - \dim_K(V)\nu(\Omega_\infty) \sum_{j=1}^{d} \ln(a_j),\end{aligned}$$

which leads to

$$\widehat{\mu}(V, \xi_V) \geqslant \widehat{\mu}(V, \|\cdot\|_V) - \nu(\Omega_\infty) \sum_{j=1}^{d} \ln(a_j).$$

Similarly, if we denote by $\widetilde{\xi}_{V_i/V_{i-1}}$ the subquotient norm family of $\widetilde{\xi}_V$, then the induction hypothesis gives

$$\widehat{\deg}(V_i/V_{i-1},\widetilde{\xi}_{V_i/V_{i-1}})\geqslant \dim_K(V_i/V_{i-1})\sum_{j=1}^d\Big(\lambda_{i,j}-\frac{1}{2}\nu(\Omega_\infty)\ln(a_j)\Big),$$

which leads to

$$\widehat{\mu}(V,\xi_V)\geqslant\widehat{\mu}(V,\|\cdot\|_V)-\frac{1}{2}\nu(\Omega_\infty)\sum_{j=1}^d\ln(a_j).$$

Step 2: In this step, we assume that all adelic vector bundles (E_j,ξ_j) are semistable. Note that the Harder-Narasimhan $\mathbb{R}$-filtration of (E_j,ξ_j) has then only one jump point. Therefore, it suffices to prove (5.17) and (5.18). Note that the case where $\det(V)$ is semistable as a rational point of

$$P:=\mathbb{P}((E_1\otimes_K\cdots\otimes_K E_d)^{\otimes\dim_K(V)})$$

has been proved in Proposition 5.3.1. In the following, we assume that $\det(V)$ is not semistable (in the sense of geometric invariant theory) as a rational point of P.

For each

$$\mathcal{F}=(\mathcal{F}_1,\ldots,\mathcal{F}_d)\in\mathbf{Fil}(E_1)\times\cdots\times\mathbf{Fil}(E_d),$$

let $\|\cdot\|_{\mathcal{F},V}$ be the quotient norm on V of the ε-tensor product of $\|\cdot\|_{\mathcal{F}_1},\ldots,\|\cdot\|_{\mathcal{F}_d}$. Let $\mathbf{S}$ be the subset of $\mathbf{Fil}_0(E_1)\times\cdots\times\mathbf{Fil}_0(E_d)$ consisting of vectors $(\mathcal{F}_1,\ldots,\mathcal{F}_d)$ such that the filtrations $\mathcal{F}_1,\ldots,\mathcal{F}_d$ are not simultaneously trivial. Then, by Theorem 5.4.3, the function $\Theta:\mathbf{S}\rightarrow\mathbb{R}$

$$\forall\,\mathcal{F}=(\mathcal{F}_1,\ldots,\mathcal{F}_d)\in\mathbf{S},\quad\Theta(\mathcal{F}):=\frac{\widehat{\mu}(V,\|\cdot\|_{\mathcal{F},V})}{(\langle\mathcal{F}_1,\mathcal{F}_1\rangle+\cdots+\langle\mathcal{F}_d,\mathcal{F}_d\rangle)^{1/2}}$$

attains its minimal value c, which is negative. In the following, we denote by $\mathcal{E}=(\mathcal{E}_1,\ldots,\mathcal{E}_d)$ a minimal point of the function Θ. Then, for any $\mathcal{F}=(\mathcal{F}_1,\ldots,\mathcal{F}_d)\in\mathbf{Fil}(E_1)\times\cdots\times\mathbf{Fil}(E_d)$, one has

$$\widehat{\mu}(V,\|\cdot\|_{\mathcal{F},V})\geqslant\sum_{j=1}^d\widehat{\mu}(E_j,\|\cdot\|_{\mathcal{F}_j})+c\frac{\langle\mathcal{E}_1,\mathcal{F}_1\rangle+\cdots+\langle\mathcal{E}_d,\mathcal{F}_d\rangle}{(\langle\mathcal{E}_1,\mathcal{E}_1\rangle+\cdots+\langle\mathcal{E}_d,\mathcal{E}_d\rangle)^{1/2}}.$$

In the following, for each $j\in\{1,\ldots,d\}$, we denote by $\mathcal{F}_j$ the $\mathbb{R}$-filtration on E which induces on each subquotient $\mathrm{sq}^t_{\mathcal{E}_j}(E_j)$ the Harder-Narasimhan filtration of this vector space equipped with the subquotient norm family $\xi^t_{j,\mathrm{sq},\mathcal{E}_j}$ of ξ_j. By (4.63), for any $t\in\mathbb{R}$, if we denote by $\|\cdot\|_{\mathcal{F}_j,\mathrm{sq},t}$ the subquotient norm of $\|\cdot\|_{\mathcal{F}_j}$ on $\mathrm{sq}^t_{\mathcal{E}_j}(E_j)$, then one has

$$\widehat{\deg}(\mathrm{sq}^t_{\mathcal{E}_j}(E_j),\|\cdot\|_{\mathcal{F}_j,\mathrm{sq},t})=\widehat{\deg}(\mathrm{sq}^t_{\mathcal{E}_j}(E_j),\xi^t_{j,\mathrm{sq},\mathcal{E}_j}).$$

Taking the sum with respect to t, by Proposition 4.3.13 and the assumption that ξ_j is Hermitian we obtain that

$$\widehat{\deg}(E_j, \|\cdot\|_{\mathcal{E}_j}) = \widehat{\deg}(E_j, \xi_j).$$

Moreover, by (4.63) one has

$$\langle \mathcal{E}_j, \mathcal{F}_j \rangle = \frac{1}{a_j} \sum_{t \in \mathbb{R}} t \sum_{i=1}^{r_j(t)} \widehat{\mu}_i(\mathrm{sq}^t_{\mathcal{E}_j}(E_j), \xi^t_{j,\mathrm{sq},\mathcal{E}_j}) = \frac{1}{a_j} \sum_{t \in \mathbb{R}} t\, \widehat{\deg}(\mathrm{sq}^t_{\mathcal{E}_j}(E_j), \xi^t_{j,\mathrm{sq},\mathcal{E}_j}),$$

where $r_j(t) = \dim_K(\mathrm{sq}^t_{\mathcal{E}_j}(E_j))$, and the second equality comes from (4.63). For any $j \in \{1, \ldots, d\}$ and any $u \in \mathbb{R}$, let

$$\Psi_j(u) = \sum_{t<u} \widehat{\deg}(\mathrm{sq}^t_{\mathcal{E}_j}(E_j), \xi^t_{j,\mathrm{sq},\mathcal{E}_j}) = \widehat{\deg}(\overline{E/\mathcal{E}^u_j(E_j)}),$$

where we consider the quotient norm family on $E/\mathcal{E}^u_j(E_j)$. Since (E_j, ξ_j) is semistable, one has

$$\widehat{\mu}(\overline{E/\mathcal{E}^u_j(E_j)}) \geqslant \widehat{\mu}_{\min}(E_j, \xi_j) = \widehat{\mu}(E_j, \xi_j)$$

and hence

$$\Psi_j(u) \geqslant \widehat{\mu}(E_j, \xi_j)\,\mathrm{rk}(E/\mathcal{E}^u_j(E_j)).$$

By Abel's summation formula we obtain

$$\langle \mathcal{E}_j, \mathcal{F}_j \rangle = \frac{1}{a_j} \int_{\mathbb{R}} t\, \mathrm{d}\Psi_j(t) = M_j \frac{\Psi_j(M_j)}{a_j} - \frac{1}{a_j} \int_{-\infty}^{M_j} \Psi_j(t)\, \mathrm{d}t,$$

where M_j is a sufficiently positive number such that $\mathcal{E}^{M_j}_j(E_j) = \{0\}$. Therefore one has

$$\begin{aligned}
\langle \mathcal{E}_j, \mathcal{F}_j \rangle &\leqslant M_j \frac{\widehat{\deg}(E_j, \xi_j)}{a_j} - \frac{1}{a_j} \widehat{\mu}(E_j, \xi_j) \int_{-\infty}^{M_j} \dim_K(E_j/\mathcal{E}^t_j(E_j))\, \mathrm{d}t \\
&= \frac{\widehat{\mu}(E_j, \xi_j)}{a_j} \int_{-\infty}^{M_j} t\, \mathrm{d}\big(\dim_K(E_j/\mathcal{E}^t_j(E_j))\big) = \frac{\widehat{\mu}(E_j, \xi_j)}{a_j} \widehat{\deg}(E_j, \|\cdot\|_{\mathcal{E}_j}) = 0.
\end{aligned}$$

Therefore we obtain

$$\widehat{\mu}(V, \|\cdot\|_{\mathcal{F},V}) \geqslant \sum_{j=1}^{d} \widehat{\mu}(E_j, \xi_j).$$

It remains to compare $\widehat{\mu}(V, \|\cdot\|_{\mathcal{F},V})$ with the slopes of (V, ξ_V) and $(V, \widetilde{\xi}_V)$. We choose a refinement

$$0 = G_0 \subsetneq G_1 \subsetneq \ldots \subsetneq G_n = E_1 \otimes_K \cdots \otimes_K E_d, \quad t_1 \geqslant \ldots \geqslant t_n$$

of the $\mathbb{R}$-filtration $\mathcal{E}_1 \otimes \cdots \otimes \mathcal{E}_d$ such that, for any $i \in \{1, \ldots, n\}$, the subquotient G_i/G_{i-1} is canonically isomorphic to a tensor product of the form

$$\mathrm{sq}_{\mathcal{E}_1}^{\lambda_{i,1}}(E_1)\otimes_K\cdots\otimes_K\mathrm{sq}_{\mathcal{E}_d}^{\lambda_{i,d}}(E_d)$$

with $\lambda_{i,1}+\cdots+\lambda_{i,d}=t_i$. For any $i\in\{1,\ldots,n\}$, let $\|\cdot\|_{\mathcal{F},G_i/G_{i-1}}$ be the subquotient norm on G_i/G_{i-1} of the ε-tensor product of $\|\cdot\|_{\mathcal{F}_1},\ldots,\|\cdot\|_{\mathcal{F}_d}$. By the construction of $\mathcal{F}_1,\ldots,\mathcal{F}_d$, the subquotient norm $\|\cdot\|_{\mathcal{F},G_i/G_{i-1}}$ on G_i/G_{i-1} corresponds to the tensor product of the Harder-Narasimhan $\mathbb{R}$-filtrations of

$$(\mathrm{sq}_{\mathcal{E}_1}^{\lambda_{i,1}}(E_1),\xi_{1,\mathrm{sq},\mathcal{E}_1}^t),\ldots,(\mathrm{sq}_{\mathcal{E}_d}^{\lambda_{i,d}}(E_d),\xi_{d,\mathrm{sq},\mathcal{E}_d}^t).$$

By the induction hypothesis, and the same argument showing (5.19) and (5.20), we obtain

$$\widehat{\deg}(V_i/V_{i-1},\xi_{V_i/V_{i-1}})\geqslant\widehat{\deg}(V_i/V_{i-1},\|\cdot\|_{\mathcal{F},V_i/V_{i-1}})-\dim_K(V_i/V_{i-1})\nu(\Omega_\infty)\sum_{j=1}^d\ln(a_j),$$

where $\xi_{V_i/V_{i-1}}$ is the subquotient norm family of ξ_V on V_i/V_{i-1}, $\|\cdot\|_{\mathcal{F},V_i/V_{i-1}}$ is the quotient norm of $\|\cdot\|_{\mathcal{F},G_i/G_{i-1}}$, which identifies with the subquotient norm of $\|\cdot\|_{F,V}$ since the flag $0=G_0\subsetneq G_1\subsetneq\ldots\subsetneq G_n$ is compatible with the $\mathbb{R}$-filtration $\mathcal{F}_1\otimes\cdots\otimes\mathcal{F}_d$. Taking the sum of the above formula with respect to $i\in\{1,\ldots,n\}$, we obtain

$$\widehat{\deg}(V,\xi_V)\geqslant\widehat{\deg}(V,\|\cdot\|_{\mathcal{F},V})-\dim_K(V)\nu(\Omega_\infty)\sum_{j=1}^d\ln(a_j),$$

which leads to

$$\widehat{\mu}(V,\xi_V)\geqslant\widehat{\mu}(V,\|\cdot\|_{\mathcal{F},V})-\nu(\Omega_\infty)\sum_{j=1}^d\ln(a_j).$$

Similarly, one has

$$\widehat{\mu}(V,\widetilde{\xi}_V)\geqslant\widehat{\mu}(V,\|\cdot\|_{\mathcal{F},V})-\frac{1}{2}\nu(\Omega_\infty)\sum_{j=1}^d\ln(a_j).$$

The theorem is thus proved. □

Corollary 5.6.2 *Let $d\in\mathbb{N}_{\geqslant2}$, $\{(E,\xi_j)\}_{j=1}^d$ be a family of adelic vector bundles on S and V be a non-zero quotient vector space of $E_1\otimes_K\cdots\otimes_K E_d$. Let ξ_V be the quotient norm families of $\xi_1\otimes_{\varepsilon,\pi}\cdots\otimes_{\varepsilon,\pi}\xi_d$. Then one has*

$$\widehat{\mu}_{\min}(V,\xi_V)\geqslant\sum_{j=1}^d\Big(\widehat{\mu}_{\min}(E_j,\xi_j)-\frac{3}{2}\nu(\Omega_\infty)\ln(\dim_K(E_j))\Big).\tag{5.21}$$

If all norm families $\xi_1,\ldots,\xi_d$ are Hermitian, then one has

$$\widehat{\mu}_{\min}(V,\xi_V) \geqslant \sum_{j=1}^{d}\Big(\widehat{\mu}_{\min}(E_j,\xi_j) - \nu(\Omega_\infty)\ln(\dim_K(E_j))\Big), \tag{5.22}$$

$$\widehat{\mu}_{\min}(V,\widetilde{\xi}_V) \geqslant \sum_{j=1}^{d}\Big(\widehat{\mu}_{\min}(E_j,\xi_j) - \frac{1}{2}\nu(\Omega_\infty)\ln(\dim_K(E_j))\Big), \tag{5.23}$$

where $\widetilde{\xi}_V$ is the quotient norm family of the orthogonal tensor product $\xi_1 \otimes \cdots \otimes \xi_d$.

Proof We begin with the proof of the Hermitian case. To establish (5.22) it suffices to prove weaker inequalities

$$\widehat{\mu}(V,\xi_V) \geqslant \sum_{j=1}^{d}\Big(\widehat{\mu}_{\min}(E_j,\xi_j) - \nu(\Omega_\infty)\ln(\dim_K(E_j))\Big), \tag{5.24}$$

$$\widehat{\mu}(V,\widetilde{\xi}_V) \geqslant \sum_{j=1}^{d}\Big(\widehat{\mu}_{\min}(E_j,\xi_j) - \frac{1}{2}\nu(\Omega_\infty)\ln(\dim_K(E_j))\Big) \tag{5.25}$$

for all non-zero quotient vector space V of $E_1 \otimes_K \cdots \otimes_K E_d$. For any $j \in \{1,\ldots,d\}$, let $\mathcal{H}_j$ be the Harder-Narasimhan $\mathbb{R}$-filtration of (E_j,ξ_j). Let $\|\cdot\|_V$ be the quotient norm of the ε-tensor product of $\|\cdot\|_{\mathcal{H}_1},\ldots,\|\cdot\|_{\mathcal{H}_d}$. By Theorem 5.6.1, one has

$$\widehat{\mu}(V,\xi_V) \geqslant \widehat{\mu}(V,\|\cdot\|_V) - \nu(\Omega_\infty)\sum_{j=1}^{d}\ln(\dim_K(E_j)), \tag{5.26}$$

$$\widehat{\mu}(V,\widetilde{\xi}_V) \geqslant \widehat{\mu}(V,\|\cdot\|_V) - \frac{1}{2}\nu(\Omega_\infty)\sum_{j=1}^{d}\ln(\dim_K(E_j)). \tag{5.27}$$

Moreover, since $\|\cdot\|_V$ is the quotient norm of the ε-tensor product of $\|\cdot\|_{\mathcal{H}_1},\ldots,\|\cdot\|_{\mathcal{H}_d}$, one has (see Remark 5.1.6)

$$\widehat{\mu}(V,\|\cdot\|_V) \geqslant \sum_{j=1}^{d}\widehat{\mu}_{\min}(E_j,\xi_j).$$

Therefore (5.24) follows from (5.26) and (5.25) follows from (5.27).

In the following, we proceed with the proof of (5.21) in the general (non-necessarily Hermitian) case. Note that one has

$$\xi_1^{\vee\vee} \otimes_{\varepsilon,\pi} \cdots \otimes_{\varepsilon,\pi} \xi_d^{\vee\vee} = \xi_1 \otimes_{\varepsilon,\pi} \cdots \otimes_{\varepsilon,\pi} \xi_d.$$

Moreover, since $\xi_j^{\vee\vee} \leqslant \xi_j$, one has

$$\widehat{\mu}_{\min}(E_j,\xi_j^{\vee\vee}) \geqslant \widehat{\mu}_{\min}(E_j,\xi_j)$$

for any $j \in \{1, \dots, d\}$. Therefore, by replacing ξ_j by $\xi_j^{\vee\vee}$ we may suppose without loss of generality that all ξ_j are non-Archimedean on $\Omega \setminus \Omega_\infty$.

Assume that ξ_j is of the form $\{\|\cdot\|_{j,\omega}\}_{\omega\in\Omega}$. By Theorem 4.1.26, for any $\epsilon > 0$ and any $j \in \{1, \dots, d\}$ there exist measurable Hermitian norm families $\xi_j^H = \{\|\cdot\|_{j,\omega}^H\}_{\omega\in\Omega}$ of E_j such that $\|\cdot\|_{j,\omega}^H = \|\cdot\|_{j,\omega}$ for any $\omega \in \Omega \setminus \Omega_\infty$ and

$$\|\cdot\|_{j,\omega} \leqslant \|\cdot\|_{j,\omega}^H \leqslant (\dim_K(E_j) + \epsilon)^{1/2} \|\cdot\|_{j,\omega}$$

for any $\omega \in \Omega_\infty$. By the slope inequality (see Proposition 4.3.31) one has

$$\widehat{\mu}_{\min}(E_j, \xi_j^H) \geqslant \widehat{\mu}(E_j, \xi_j) - \frac{1}{2}\nu(\Omega_\infty) \ln(\dim_K(E_j) + \epsilon). \tag{5.28}$$

Moreover, if we denote by ξ_V' the quotient norm family of $\xi_1^H \otimes_{\varepsilon,\pi} \cdots \otimes_{\varepsilon,\pi} \xi_d^H$ on V, one has

$$\widehat{\mu}_{\min}(V, \xi_V') \leqslant \widehat{\mu}_{\min}(V, \xi_V) \tag{5.29}$$

by the slope inequality. Applying the Hermitian case of the corollary to (E_j, ξ_j^H) ($j \in \{1, \dots, d\}$) and (V, ξ_V'), we obtain

$$\widehat{\mu}_{\min}(V, \xi_V') \geqslant \sum_{j=1}^{d} \Big(\widehat{\mu}_{\min}(E_j, \xi_j^H) - \nu(\Omega_\infty) \ln(\dim_K(E_j)) \Big).$$

Combining this inequality with (5.28) and (5.29), by passing to limite when ϵ tend to $0+$ we obtain (5.21). The corollary is thus proved. □

Remark 5.6.3 In the case where the adelic curve S comes from an arithmetic curve. The inequality (5.23) recovers essentially the second inequality of [64, Corollary 5.4], which strengthens [38, Theorem 1]. From the methodological point of view, the arguments in this chapter rely on the geometric invariant theory without using the theorem of successive minima of Zhang, which was a key argument in [64, 22] (see [155, Theorem 5.2], see also [64, §3]).

Chapter 6
Adelic line bundles on arithmetic varieties

In this chapter, we fix a proper adelic curve $S = (K, (\Omega, \mathcal{A}, \nu), \phi)$.

6.1 Metrised line bundles on an arithmetic variety

Let X be a projective scheme over $\operatorname{Spec} K$ and L be an invertible $\mathcal{O}_X$-module. For any $\omega \in \Omega$, we let X_ω be the fibre product $X \times_{\operatorname{Spec} K} \operatorname{Spec} K_\omega$ (recall that K_ω is the completion of K with respect to $|\cdot|_\omega$) and L_ω be the pull-back of L by the canonical projection morphism $X_\omega \to X$. By *metric family* on L, we refer to a family of continuous metrics $\varphi = \{\varphi_\omega\}_{\omega\in\Omega}$, where φ_ω is a continuous metric on L_ω. If $\varphi = \{\varphi_\omega\}_{\omega\in\Omega}$ and $\varphi' = \{\varphi'_\omega\}_{\omega\in\Omega}$ are two metric families on L, the *local distance* of φ and φ' at $\omega \in \Omega$ is defined as (see Definition 2.2.7)

$$d_\omega(\varphi, \varphi') := d(\varphi_\omega, \varphi'_\omega).$$

The global distance between φ and φ' is defined as the upper integral

$$\operatorname{dist}(\varphi, \varphi') := \overline{\int_\Omega} d_\omega(\varphi, \varphi')\, \nu(\mathrm{d}\omega). \tag{6.1}$$

If $\varphi = \{\varphi_\omega\}_{\omega\in\Omega}$ is a metric family on L, then the dual metrics $\{-\varphi_\omega\}_{\omega\in\Omega}$ form a metric family on $L^\vee$, denoted by $-\varphi$. If L and L' are invertible $\mathcal{O}_X$-modules, and $\varphi = \{\varphi_\omega\}_{\omega\in\Omega}$ and $\varphi' = \{\varphi'_\omega\}_{\omega\in\Omega}$ are metric families on L and L' respectively, then $\{\varphi_\omega + \varphi'_\omega\}_{\omega\in\Omega}$ is a metric family on $L \otimes L'$, denoted by $\varphi + \varphi'$. The metric family $\varphi + (-\varphi')$ on $L \otimes L'^\vee$ is also denoted by $\varphi - \varphi'$. Similarly, for any integer $n \geqslant 0$, $\{n\varphi_\omega\}_{\omega\in\Omega}$ is a metric family on $L^{\otimes n}$, denoted by $n\varphi$.

Definition 6.1.1 Let Y and X be projective schemes over $\operatorname{Spec} K$ and $f : Y \to X$ be a projective K-morphism. Let L be an invertible $\mathcal{O}_X$-module equipped with a metric family $\varphi = \{\varphi_\omega\}_{\omega\in\Omega}$. We denote by $f^*(\varphi)$ the metric family $\{f^*_\omega(\varphi_\omega)\}_{\omega\in\Omega}$ on $f^*(L)$, where for any $\omega \in \Omega$, $f_\omega : Y_\omega \to X_\omega$ is the K_ω-morphisme induced by

H. Chen, A. Moriwaki, *Arakelov Geometry over Adelic Curves*, Lecture Notes in Mathematics 2258, https://doi.org/10.1007/978-981-15-1728-0_6

f, and $f_\omega^*(\varphi_\omega)$ is defined in Definition 2.2.9. The norm family $f^*(\varphi)$ is called the *pull-back* of φ by f. In the particular case where Y is a closed subscheme of X and f is the canonical immersion, the norm family $f^*(\varphi)$ is also denoted by $\varphi|_Y$ and called the *restriction* of φ to Y.

The following properties are straightforward from the definition.

Proposition 6.1.2 *Let X be a projective scheme over* $\operatorname{Spec} K$ *and $f : Y \to X$ be a projective morphism of K-schemes.*

(1) *If L_1 and L_2 are two invertible $\mathcal{O}_X$-modules, and φ_1 and φ_2 are metric families on L_1 and L_2 respectively, then one has $f^*(\varphi_1 + \varphi_2) = f^*(\varphi_1) + f^*(\varphi_2)$ on $f^*(L_1 \otimes L_2) \cong f^*(L_1) \otimes f^*(L_2)$.*
(2) *For any invertible $\mathcal{O}_X$-module L and any metric family φ on L, one has $f^*(-\varphi) = -f^*(\varphi)$ on $f^*(L^\vee) \cong f^*(L)^\vee$.*

Remark 6.1.3 Let us consider the particular case where X is the spectrum of a finite extension K' of the field K. For any $\omega \in \Omega$, the Berkovich space of X_ω identifies with the discrete set of absolute values on K' extending $|\cdot|_\omega$ on K. Moreover, any invertible $\mathcal{O}_X$-module L could be considered as a vector space of dimension 1 over K', and any metric family on L is just a norm family with respect to the adelic curve $S \otimes_K K'$ (cf. Definition 3.4.1) if we consider L as a vector space over K'.

6.1.1 Quotient metric families

Let E be a finite-dimensional vector space over K and $\xi = \{\|\cdot\|_\omega\}_{\omega \in \Omega}$ be a norm family on E. Let $f : X \to \operatorname{Spec} K$ be a projective K-scheme and L be an invertible $\mathcal{O}_X$-module. Suppose given a surjective homomorphism $\beta : f^*(E) \to L$. For any $\omega \in \Omega$, the morphism $f : X \to \operatorname{Spec} K$ induces by base change a morphism f_ω from $X_\omega := X \times_{\operatorname{Spec} K} \operatorname{Spec} K_\omega$ to $\operatorname{Spec} K_\omega$. We denote by L_ω the pull-back of L on X_ω. The homomorphism β induces a surjective homomorphism $\beta_\omega : f_\omega^*(E) \to L_\omega$. Therefore, the norm $\|\cdot\|_\omega$ induces a quotient metric φ_ω on L_ω (see Definition 2.2.15). The family $\varphi = \{\varphi_\omega\}_{\omega \in \Omega}$ is called the *quotient metric family induced by (E, ξ) and β*.

Let $\boldsymbol{e} = \{e_i\}_{i=1}^r$ be a basis of E. For each $\omega \in \Omega$, let $\|\cdot\|_{\boldsymbol{e},\omega}$ be the norm on $E_\omega := E \otimes_K K_\omega$ given by

$$\|a_1 e_1 + \cdots + a_r e_r\|_{\boldsymbol{e},\omega} := \begin{cases} \max\{|a_1|_\omega, \ldots, |a_r|_\omega\} & \text{if } \omega \in \Omega \setminus \Omega_\infty, \\ |a_1|_\omega + \cdots + |a_r|_\omega & \text{if } \omega \in \Omega_\infty, \end{cases}$$

for all $a_1, \ldots, a_r \in K_\omega$, and let $\varphi_{\boldsymbol{e},\omega}$ be the metric of L_ω induced by $\|\cdot\|_{\boldsymbol{e},\omega}$ and the surjective homomorphism $E_\omega \otimes_{K_\omega} \mathcal{O}_{X_\omega} \to L_\omega$. Let $\xi_{\boldsymbol{e}} := \{\|\cdot\|_{\boldsymbol{e},\omega}\}_{\omega \in \Omega}$ and let $\varphi_{\boldsymbol{e}} := \{\varphi_{\boldsymbol{e},\omega}\}_{\omega \in \Omega}$. The metric family $\varphi_{\boldsymbol{e}}$ is called the *quotient metric family of L induced by β and $\boldsymbol{e}$*.

Remark 6.1.4 We keep the above notation. Given a fixed surjective homomorphism $\beta : f^*(E) \to L$, by Proposition 1.2.14 (see also Remark 1.3.2), the norm family ξ and the double dual norm family $\xi^{\vee\vee}$ induce the same quotient metric family on L.

Proposition 6.1.5 *Let* (E, ξ) *and* (E', ξ') *be finite-dimensional vector spaces equipped with* dominated *norm families. Let* $f : X \to \operatorname{Spec} K$ *be a projective scheme over* $\operatorname{Spec} K$, $\beta : f^*(E) \to L$ *and* $\beta' : f'^*(E') \to L$ *be two surjective homomorphisms inducing closed immersions* $i : X \to \mathbb{P}(E)$ *and* $i' : X \to \mathbb{P}(E')$, *and* φ *and* φ' *be quotient metric families induced by* (E, ξ) *and* β, *and by* (E', ξ') *and* β', *respectively. Then the local distance function* $(\omega \in \Omega) \mapsto d_\omega(\varphi, \varphi')$ *is* ν*-dominated.*

Proof We begin with the particular case where $E = E'$ and $\beta = \beta'$. By Proposition 2.2.20, for any $\omega \in \Omega$ one has

$$d_\omega(\varphi, \varphi') \leqslant d_\omega(\xi^{\vee\vee}, \xi'^{\vee\vee}).$$

Note that the norm families $\xi^{\vee\vee}$ and $\xi'^{\vee\vee}$ are strongly dominated (see Remark 4.1.12). By Corollary 4.1.10 and the triangle inequality of the local distance function, we obtain that the function $(\omega \in \Omega) \mapsto d_\omega(\xi^{\vee\vee}, \xi'^{\vee\vee})$ is ν-dominated and then deduce that the function $(\omega \in \Omega) \mapsto d_\omega(\varphi, \varphi')$ is ν-dominated.

In the general case, since i and i' are closed immersions, by Serre's vanishing therorem (cf. [85, Theorem 5.2, Chapter III]), there exists an integer $n \geqslant 1$ such that

$$\Gamma(\mathbb{P}(E), \mathcal{O}_{\mathbb{P}(E)}(n)) \longrightarrow \Gamma(X, L^{\otimes n}) \text{ and } \Gamma(\mathbb{P}(E'), \mathcal{O}_{\mathbb{P}(E')}(n)) \longrightarrow \Gamma(X, L^{\otimes n})$$

are surjective, so that the natural homomorphisms

$$E^{\otimes n} \longrightarrow \operatorname{Sym}^n(E) = \Gamma(\mathbb{P}(E), \mathcal{O}_{\mathbb{P}(E)}(n)) \longrightarrow \Gamma(X, L^{\otimes n})$$

and

$$E'^{\otimes n} \longrightarrow \operatorname{Sym}^n(E') = \Gamma(\mathbb{P}(E'), \mathcal{O}_{\mathbb{P}(E')}(n)) \longrightarrow \Gamma(X, L^{\otimes n})$$

are both surjective. Therefore the above surjective homomorphisms factorise through $f^*\Gamma(X, L^{\otimes n})$. Moreover, by Remark 2.2.19, if we equip $E^{\otimes n}$ and $E'^{\otimes n}$ with the ε, π-tensor power norm families (see §4.1.1) of ξ and ξ' respectively, then the corresponding quotient metric families are $n\varphi$ and $n\varphi'$ respectively. Note that the ε, π-tensor powers of ξ and ξ' are dominated (see Proposition 4.1.19 (5)). Therefore, by the special case proved above, we obtain that the function

$$(\omega \in \Omega) \longmapsto d_\omega(n\varphi, n\varphi') = n d_\omega(\varphi, \varphi')$$

is ν-dominated. The proposition is thus proved. □

6.1.2 Dominated metric families

Throughout this subsection, let $f : X \to \operatorname{Spec} K$ be a projective K-scheme.

Definition 6.1.6 Let L be a very ample invertible $\mathcal{O}_X$-module. We say that a metric family φ on L is *dominated* if there exist a finite-dimensional vector space E over K, a dominated norm family ξ on E, and a surjective homomorphism $\beta : f^*(E) \to L$ inducing a closed immersion $X \to \mathbb{P}(E)$, such that the quotient metric family φ' induced by (E, ξ) and β satisfies the following condition:

the local distance function $(\omega \in \Omega) \mapsto d_\omega(\varphi, \varphi')$ *is* ν*-dominated.*

Remark 6.1.7 With the above definition, Proposition 6.1.5 implies the following assertions. Let E be a finite-dimensional vector space over K equipped with a dominated norm family ξ. Let L be an invertible $\mathcal{O}_X$-module and $\beta : f^*(E) \to L$ be a surjective homomorphism inducing a closed immersion $X \to \mathbb{P}(E)$. Then the quotient metric family induced by (E, ξ) and β is dominated. Moreover, if φ_1 and φ_2 are two metric families on L which are dominated, then the local distance function $(\omega \in \Omega) \mapsto d_\omega(\varphi_1, \varphi_2)$ is ν-dominated.

Proposition 6.1.8 *Let L_1 and L_2 be very ample invertible $\mathcal{O}_X$-modules. Assume that φ_1 and φ_2 are dominated metric families on L_1 and L_2 respectively. Then $\varphi_1 + \varphi_2$ is a dominated metric family on $L_1 \otimes L_2$.*

Proof Since the metric families φ_1 and φ_2 are dominated, there exist finite-dimensional vector spaces E_1 and E_2 over K, dominated norm families ξ_1 and ξ_2 on E_1 and E_2 respectively, and surjective homomorphisms $\beta_1 : f^*(E_1) \to L_1$ and $\beta_2 : f^*(E_2) \to L_2$ inducing closed immersions $X \to \mathbb{P}(E_1)$ and $X \to \mathbb{P}(E_2)$ respectively, such that, if we denote by $\widetilde{\varphi}_1$ and $\widetilde{\varphi}_2$ the quotient metric families induced by (E_1, ξ_1) and β_1 and by (E_2, ξ_2) and β_2 respectively, then the local distance functions

$$(\omega \in \Omega) \longmapsto d_\omega(\varphi_1, \widetilde{\varphi}_1) \quad \text{and} \quad (\omega \in \Omega) \longmapsto d_\omega(\varphi_2, \widetilde{\varphi}_2)$$

are ν-dominated. Consider now the composed morphism

$$\iota : X \xrightarrow{(\iota_1, \iota_2)} \mathbb{P}(E_1) \times_K \mathbb{P}(E_2) \xrightarrow{\varsigma} \mathbb{P}(E_1 \otimes_K E_2),$$

where ι_1 and ι_2 are closed immersions corresponding to β_1 and β_2, and ς is the Segre embedding. Note that ι is the closed immersion corresponding to the surjective homomorphism

$$\beta_1 \otimes \beta_2 : f^*(E_1 \otimes_K E_2) \cong f^*(E_1) \otimes_{\mathcal{O}_X} f^*(E_2) \longrightarrow L_1 \otimes_{\mathcal{O}_X} L_2.$$

Moreover, if we equip $E_1 \otimes_K E_2$ with the ε, π-tensor product norm family of ξ_1 and ξ_2, then the quotient metric family on $L_1 \otimes L_2$ induced by $(E_1 \otimes_K E_2, \xi_1 \otimes_{\varepsilon,\pi} \xi_2)$ and $\beta_1 \otimes \beta_2$ identifies with $\widetilde{\varphi}_1 + \widetilde{\varphi}_2$. This is a consequence of Proposition 1.2.36 (for the non-Archimedean case) and Proposition 1.1.58 (for the Archimedean case). Since

$$\forall\, \omega \in \Omega, \quad d_\omega(\varphi_1 + \varphi_2, \widetilde{\varphi}_1 + \widetilde{\varphi}_2) \leqslant d_\omega(\varphi_1, \widetilde{\varphi}_1) + d_\omega(\varphi_2, \widetilde{\varphi}_2),$$

we obtain that the function $(\omega \in \Omega) \mapsto d_\omega(\varphi_1 + \varphi_2, \widetilde{\varphi}_1 + \widetilde{\varphi}_2)$ is ν-dominated. Therefore the metric family $\varphi_1 + \varphi_2$ is dominated. □

Definition 6.1.9 Let L be an invertible $\mathcal{O}_X$-module and φ be a metric family on L. We say that φ is *dominated* if there exist two very ample invertible $\mathcal{O}_X$-modules L_1 and L_2 together with dominated metric families φ_1 and φ_2 on L_1 and L_2 respectively, such that $L = L_2 \otimes L_1^\vee$ and $\varphi = \varphi_2 - \varphi_1$.

Remark 6.1.10 In the case where the invertible $\mathcal{O}_X$-module L is very ample, the condition of dominancy in Definition 6.1.9 is actually equivalent to that in Definition 6.1.6. In order to explain this fact (in avoiding confusions), in this remark we temporary say that a metric family φ on a very ample invertible $\mathcal{O}_X$-module L is *strictly dominated* if it satisfies the condition in Definition 6.1.6. Clearly, if φ is strictly dominated, then it is dominated (namely satisfies the condition in Definition 6.1.9) since we can write L as $L^{\otimes 2} \otimes L^\vee$ and φ as $2\varphi - \varphi$. Conversely, if φ is dominated, then there exist very ample invertible $\mathcal{O}_X$-modules L_1 and L_2 such that $L \cong L_2 \otimes L_1^\vee$, and strictly dominated metric families φ_1 and φ_2 on L_1 and L_2 such that $\varphi = \varphi_2 - \varphi_1$. We pick an arbitrary strictly dominated metric family φ' on L. By Proposition 6.1.8, we obtain that $\varphi' + \varphi_1$ is a strictly dominated metric family on L_2. Hence the local distance function

$$(\omega \in \Omega) \longmapsto d_\omega(\varphi_2, \varphi' + \varphi_1) = d_\omega(\varphi + \varphi_1, \varphi' + \varphi_1) = d_\omega(\varphi, \varphi')$$

is ν-dominated. Therefore the metric family φ is strictly dominated.

Proposition 6.1.11 *Let E be a finite-dimensional vector space over K equipped with a norm family ξ, $\beta : f^*(E) \to L$ be a surjective homomorphism (we do not assume that β induces a closed immersion), and φ be the quotient metric family induced by (E, ξ) and β. Suppose that ξ is a dominated norm family. Then φ is a dominated metric family.*

Proof Since X is a projective K-scheme, there exists a very ample invertible $\mathcal{O}_X$-module L'. Let E' be a finite-dimensional vector space over K and $\beta' : f^*(E') \to L'$ be a surjective homomorphism, which induces a closed embedding of X in $\mathbb{P}(E')$, which we denote by λ'. Let $\lambda : X \to \mathbb{P}(E)$ be the K-morphism induced by β. Then the tensor product homomorphism

$$\beta \otimes \beta' : f^*(E) \otimes_{\mathcal{O}_X} f^*(E') \cong f^*(E \otimes_K E') \longrightarrow L \otimes_{\mathcal{O}_X} L'$$

corresponds to the composed K-morphism

$$X \xrightarrow{(\lambda,\lambda')} \mathbb{P}(E) \times_K \mathbb{P}(E') \xrightarrow{\varsigma} \mathbb{P}(E \otimes_K E'),$$

where ς is the Segre embedding. Since X is separated over $\operatorname{Spec} K$ and λ' is a closed immersion, the morphism (λ, λ') is a closed immersion. Therefore, the morphism from X to $\mathbb{P}(E \otimes_K E')$ induced by $\beta \otimes \beta'$ is a closed embedding.

Let ξ' be a dominated norm family on E' and φ' be the metric family on L' induced by (E', ξ') and β'. By definition the metric family φ' is dominated (see Proposition

4.1.19 (5)). Moreover, the metric family $\varphi+\varphi'$ is induced by $(E\otimes E',\xi\otimes_{\varepsilon,\pi}\xi')$ and $\beta\otimes\beta'$ (see the proof of Proposition 6.1.8). As ξ and ξ' are dominated, we obtain that $\xi\otimes_{\varepsilon,\pi}\xi'$ is dominated. Therefore the metric families $\varphi+\varphi'$ is dominated. Hence the metric family φ is also dominated. □

Proposition 6.1.12 *Let L and L' be invertible $\mathcal{O}_X$-modules, and φ and φ' be metric families on L and L', respectively.*

(1) *If φ is dominated, then the dual metric family $-\varphi$ on $L^\vee$ is dominated.*
(2) *If φ and φ' are dominated, then the tensor product metric family $\varphi+\varphi'$ on $L\otimes L'$ is dominated.*
(3) *If $L=L'$ and φ and φ' are dominated, then the local distance function $(\omega\in\Omega)\mapsto d_\omega(\varphi,\varphi')$ is ν-dominated.*
(4) *If $L=L'$, φ' is dominated and the local distance function $(\omega\in\Omega)\mapsto d_\omega(\varphi,\varphi')$ is ν-dominated, then φ is dominated.*
(5) *If $r\varphi$ is dominated for some non-zero integer r, then φ is dominated.*
(6) *Let $g:Y\to X$ be a projective morphism of K-schemes. If φ is dominated, then $g^*(\varphi)$ is also dominated.*

Proof (1) Let L_1 and L_2 be very ample invertible $\mathcal{O}_X$-modules and φ_1 and φ_2 be dominated metric families on L_1 and L_2 respectively, such that $L\cong L_2\otimes L_1^\vee$ and that $\varphi=\varphi_2-\varphi_1$. Then one has $L^\vee\cong L_1\otimes L_2^\vee$ and $-\varphi=\varphi_1-\varphi_2$. Hence the metric family $-\varphi$ is dominated.

(2) Let L_1, L_2, L_1' and L_2' be very ample invertible $\mathcal{O}_X$-modules, $\varphi_1,\varphi_2,\varphi_1'$ and φ_2' be dominated metric families on L_1, L_2, L_1' and L_2' respectively, such that $L\cong L_2\otimes L_1^\vee$, $L'\cong L_2'\otimes L_1'^\vee$, $\varphi=\varphi_2-\varphi_1$ and $\varphi'=\varphi_2'-\varphi_1'$. Note that $L\otimes L'\cong(L_2\otimes L_2')\otimes(L_1\otimes L_1')^\vee$, and $\varphi+\varphi'=(\varphi_2+\varphi_2')-(\varphi_1+\varphi_1')$. By Proposition 6.1.8, the metric families $\varphi_2+\varphi_2'$ and $\varphi_1+\varphi_1'$ are dominated. Hence $\varphi+\varphi'$ is dominated.

(3) Let L_1 be a very ample invertible $\mathcal{O}_X$-module and φ_1 be a dominated metric family on L_1. Let $\varphi_2=\varphi+\varphi_1$ and $\varphi_2'=\varphi'+\varphi_1$. By (2), the metric families φ_2 and φ_2' are dominated. Since the invertible $\mathcal{O}_X$-module L_2 is very ample, by Proposition 6.1.5 we obtain that the local distance function $(\omega\in\Omega)\mapsto d_\omega(\varphi_2,\varphi_2')=d_\omega(\varphi,\varphi')$ is ν-dominated.

(4) First we assume that L is very ample. As φ' is dominated, there exist a finite-dimensional vector space E over K, a dominated norm family ξ on E, and a surjective homomorphism $\beta:f^*(E)\to L$ inducing a closed immersion $X\to\mathbb{P}(E)$ such that, if ψ is the quotient metric family induced by (E,ξ) and β, then the local distance function $(\omega\in\Omega)\mapsto d_\omega(\varphi',\psi)$ is ν-dominated. Note that

$$\forall\,\omega\in\Omega,\quad d_\omega(\varphi,\psi)\leqslant d_\omega(\varphi,\varphi')+d_\omega(\varphi',\psi).$$

Thus $(\omega\in\Omega)\mapsto d_\omega(\varphi,\psi)$ is ν-dominated, as required.

In general, there are very ample invertible $\mathcal{O}_X$-modules L_1 and L_2, and dominated metric families φ_1' and φ_2' on L_1 and L_2, respectively, such that $L=L_1\otimes L_2^\vee$ and $\varphi'=\varphi_1'-\varphi_2'$. We set $\varphi_1=\varphi+\varphi_2'$ and $\varphi_2=\varphi_2'$. Then $\varphi=\varphi_1-\varphi_2$, and φ_1 and φ_2 are metric families of L_1 and L_2, respectively. As

$$d_\omega(\varphi_1, \varphi_1') = d_\omega(\varphi_1 - \varphi_2, \varphi_1' - \varphi_2') = d_\omega(\varphi, \varphi'),$$

$(\omega \in \Omega) \mapsto d_\omega(\varphi_1, \varphi_1')$ is ν-dominated, so that φ_1 is dominated by the previous observation. Therefore, φ is also dominated.

(5) As $-r\varphi$ is dominated by (1), we may assume that $r > 0$. We choose a dominated metric family ψ on L. By (2), $r\psi$ is dominated, so that, by (3),

$$(\omega \in \Omega) \mapsto d_\omega(r\varphi, r\psi) = r d_\omega(\varphi, \psi)$$

is ν-dominated. Therefore $(\omega \in \Omega) \mapsto d_\omega(\varphi, \psi)$ is also ν-dominated, and hence the assertion follows from (4).

(6) First we assume that L is very ample. Then there exist a finite-dimensional vector space E over K, a dominated norm family ξ on E, a surjective homomorphism $\beta : f^*(E) \to L$ inducing a closed immersion $X \to \mathbb{P}(E)$ such that, if ψ denotes the quotient metric family induced by (E, ξ) and β, then the local distance function $(\omega \in \Omega) \mapsto d_\omega(\varphi, \psi)$ is ν-dominated. Note that $\beta : E \otimes_K \mathcal{O}_X \to L$ yields the surjective homomorphism $g^*(\beta) : E \otimes_K \mathcal{O}_Y \to g^*(L)$. Moreover, $g^*(\psi)$ coincides with quotient metric family induced by (E, ξ) and $g^*(\beta)$. By Proposition 6.1.11, $g^*(\psi)$ is a dominated metric family. Moreover, for any $\omega \in \Omega$ one has

$$d_\omega(g^*(\varphi), g^*(\psi)) \leqslant d_\omega(\psi, \varphi).$$

By (4) we obtain that $g^*(\varphi)$ is a dominated metric family.

In general, there are very ample invertible $\mathcal{O}_X$-modules L_1 and L_2 such that $L = L_1 \otimes L_2^\vee$. Let φ_1 be a dominated metric family on L_1. If we set $\varphi_2 = \varphi_1 - \varphi$, then φ_2 is dominated by (1) and (2). By the previous case, $g^*(\varphi_1)$ and $g^*(\varphi_2)$ are dominated, so that, by (1) and (2) again, $g^*(\varphi) = g^*(\varphi_1) - g^*(\varphi_2)$ is also dominated. □

Theorem 6.1.13 *Let $f : X \to \operatorname{Spec} K$ be a geometrically reduced projective K-scheme and L be an invertible $\mathcal{O}_X$-module, equipped with a dominated metric family $\varphi = \{\varphi_\omega\}_{\omega \in \Omega}$. For any $\omega \in \Omega$, let $\|\cdot\|_{\varphi_\omega}$ be the sup norm on $H^0(X, L) \otimes_K K_\omega$ corresponding to the metric φ_ω. Then the norm family $\xi = \{\|\cdot\|_{\varphi_\omega}\}_{\omega \in \Omega}$ on $H^0(X, L)$ is strongly dominated.*

Proof Let us begin with the following claim:

Claim 6.1.14 *If the assertion of the theorem holds under the assumption that X is geometrically integral, then it holds in general.* □

Proof One can find a finite extension K' of K and the irreducible decomposition $X_1 \cup \cdots \cup X_n$ of $X_{K'}$ such that $X_1, \ldots, X_n$ are geometrically integral. We use the same notation as in Corollary 4.1.18, which says that it is sufficient to see that $\xi_{K'}$ is dominated.

Let $\psi = \{\psi_\omega\}_{\omega \in \Omega}$ be another metric family of L. Then $d_{\omega'}(\varphi_{K'}, \psi_{K'}) = d_\omega(\varphi, \psi)$ for all $\omega \in \Omega$ and $\omega' \in \Omega_{K'}$ with $\pi_{K'/K}(\omega') = \omega$. Therefore, one can see that $\varphi_{K'} = \{\varphi_{K',\omega'}\}_{\omega' \in \Omega_{K'}}$ is dominated.

Let $\varphi_{K',i}$ be the restriction of $\varphi_{K'}$ to X_i. Then, by Proposition 6.1.12, (6), $\varphi_{K',i}$ is also dominated for all i. On the other hand, one has the natural injective homomorphism

$$H^0(X_{K',\omega'}, L_{K',\omega'}) \longrightarrow H^0(X_{1,\omega'}, L_{K',\omega'}) \oplus \cdots \oplus H^0(X_{n,\omega'}, L_{K',\omega'}).$$

Here we give a norm $\|\cdot\|_{\omega'}$ on $H^0(X_{1,\omega'}, L_{K',\omega'}) \oplus \cdots \oplus H^0(X_{n,\omega'}, L_{K',\omega'})$ given by

$$\|(x_1, \ldots, x_n)\|_{\omega'} = \max_{i \in \{1,\ldots,n\}} \left\{\|x_i\|_{\varphi_{K',i,\omega'}}\right\}.$$

By our assumption, $\xi_{K',i}$ is dominated for all i, so that $\{\|\cdot\|_{\omega'}\}_{\omega \in \Omega_{K'}}$ is also dominated by Proposition 4.1.19, (4). Note that $\|\cdot\|_{\varphi_{K',\omega'}}$ is the restriction of $\|\cdot\|_{\omega'}$, and hence $\xi_{K'} = \{\|\cdot\|_{\varphi_{K',\omega'}}\}_{\omega' \in \Omega_{K'}}$ is dominated by Proposition 4.1.19, (1), as desired. □

From now on, we assume that X is geometrically integral.

Claim 6.1.15 *If L is very ample, the assertion of the theorem holds.* □

Proof Let $E = H^0(X, L)$, $r = \dim_K(E)$ and $\beta : f^*(E) \to L$ be the canonical surjective homomorphism which induces a closed immersion of X in $\mathbb{P}(E)$. Note that any non-zero element of E can not identically vanish on X. Hence there exist a finite extension K' of K together with closed points $P_1, \ldots, P_r$ of X such that the residue filed $\kappa(P_i)$ at P_i is contained in K', and that we have a strictly decreasing sequence of K'-vector spaces

$$E_0 \supsetneq E_1 \supsetneq E_2 \supsetneq \cdots \supsetneq E_{r-1} \supsetneq E_r = \{0\},$$

where $E_0 = E \otimes_K K'$ and

$$E_i = \{s \in E \otimes_K K' \;:\; s(P_1) = \cdots = s(P_i) = 0\}$$

for $i \in \{1, \ldots, r\}$. In order to prove Claim 6.1.15, by virtue of Corollary 4.1.18, we may assume that $K' = K$.

Let $\omega_1, \ldots, \omega_r$ be local bases of L around $P_1, \ldots, P_r$, respectively. For each $i \in \{1, \ldots, r\}$, we define $\theta_i \in E^\vee$ to be

$$\forall\, s \in E, \quad \theta_i(s) = f_s(P_i) \qquad (s = f_s\omega_i \text{ around } P_i).$$

Note that $\theta_1, \ldots, \theta_r$ are linearly independent over K. Let $\boldsymbol{e} = \{e_i\}_{i=1}^r$ be the dual basis of $\{\theta_i\}_{i=1}^r$ over K, that is, $(e_1, \ldots, e_r) \in E^r$ and $\theta_i(e_j) = \delta_{ij}$. Here we define a norm $\|\cdot\|'_\omega$ as follows: for any element $s \in E_\omega$ written as $s = \lambda_1 e_1 + \cdots + \lambda_r e_r$ with $(\lambda_1, \ldots, \lambda_r) \in K_\omega^r$,

$$\|s\|'_\omega = \max_{i \in \{1,\ldots,r\}} |\lambda_i|_\omega.$$

Let ξ' be the norm family of E given by $\{\|\cdot\|'_\omega\}_{\omega \in \Omega}$. Since the measure of Ω_∞ is finite, ξ' is dominated by Corollary 4.1.10.

Let φ' be the metric family of L induced by (E, ξ') and β. Note that φ' is dominated. Let us see that $\|\cdot\|'_\omega = \|\cdot\|_{\varphi'_\omega}$ for any $\omega \in \Omega$. First of all, by Proposition 2.2.23,

$\|\cdot\|_{\varphi'_\omega} \leqslant \|\cdot\|'_\omega$. For any $(i,j) \in \{1,\ldots,r\}^2$, one has

$$|e_j|_{\varphi'_\omega}(P_i) = \begin{cases} 1, & \text{if } i = j, \\ 0, & \text{if } i \neq j. \end{cases}$$

Therefore, for any $(\lambda_1,\ldots,\lambda_r) \in K_\omega^r$,

$$\|\lambda_1 e_1 + \cdots + \lambda_r e_r\|_{\varphi'_\omega} \geqslant \max_{i\in\{1,\ldots,r\}} |\lambda_1 e_1 + \cdots + \lambda_r e_r|_{\varphi'_\omega}(P_i) = \max_{i\in\{1,\ldots,r\}} |\lambda_i|_\omega,$$

as required.

By the inequality (2.5) in Subsection 2.2.2, one has

$$d_\omega(\|\cdot\|_{\varphi_\omega}, \|\cdot\|'_\omega) = d_\omega(\|\cdot\|_{\varphi_\omega}, \|\cdot\|_{\varphi'_\omega}) \leqslant d_\omega(\varphi_\omega, \varphi'_\omega).$$

Therefore, $\omega \mapsto d_\omega(\|\cdot\|_{\varphi_\omega}, \|\cdot\|'_\omega)$ is ν-dominaited, and hence ξ is dominaited by Proposition 4.1.6. □

Finally let us consider the following claim:

Claim 6.1.16 *For any non-zero element $s \in H^0(X, L)$, the function $(\omega \in \Omega) \longmapsto \ln \|s\|_{\varphi_\omega}$ is ν-dominated.* □

Proof Fix a non-zero element $s \in H^0(X, L)$.

Let us construct a dominated metric family φ' of L such that the function $(\omega \in \Omega) \longmapsto \ln \|s\|_{\varphi'_\omega}$ is ν-dominated. Let L_1 be a very ample invertible $\mathcal{O}_X$-module such that $L_2 := L \otimes L_1$ is also very ample. The multiplication by the non-zero section s defines an injective K-linear map from $H^0(X, L_1)$ to $H^0(X, L_2)$. We choose a dominated norm family $\xi'_2 = \{\|\cdot\|'_{2,\omega}\}_{\omega\in\Omega}$ on $H^0(X, L_2)$ and let ξ'_1 be the restriction of ξ'_2 on $H^0(X, L_1)$ via this injective map. By Proposition 4.1.19 (1), the norm family ξ'_1 is also dominated. Let φ'_1 and φ'_2 be the quotient metric families induced by ξ'_1 and ξ'_2 respectively, where we consider the canonical surjective homomorphisms $f^*(H^0(X, L_1)) \to L_1$ and $f^*(H^0(X, L_2)) \to L_2$. We set $\varphi' = \varphi'_2 - \varphi'_1$. By Propositions 1.3.26 and 1.3.25, for all $\omega \in \Omega$ and $x \in X_\omega^{\mathrm{an}}$ such that $s(x) \neq 0$ and $\ell \in L_{1,\omega} \otimes \widehat{\kappa}(x) \setminus \{0\}$, one has

$$\begin{aligned} |\ell|_{\varphi'_{1,\omega}}(x) &= \inf_{\substack{u\in H^0(X,L_1),\ \lambda\in\widehat{\kappa}(x)^\times \\ u(x)=\lambda\ell}} |\lambda|_x^{-1} \cdot \|su\|'_{2,\omega} \\ &\geqslant \inf_{\substack{v\in H^0(X,L_2),\ \lambda\in\widehat{\kappa}(x)^\times \\ v(x)=\lambda s(x)\ell}} |\lambda|_x^{-1} \cdot \|v\|'_{2,\omega} \\ &= |s(x)\ell|_{\varphi'_{2,\omega}}(x) = |s(x)|_{\varphi'_\omega}(x) \cdot |\ell|_{\varphi'_{1,\omega}}(x), \end{aligned}$$

which leads to the inequality $\|s\|_{\varphi'_\omega} \leqslant 1$. Moreover, by Proposition 2.2.5, for any non-zero section $u \in H^0(X, L_1)$, one has

$$\|su\|_{\varphi'_{2,\omega}} \leqslant \|s\|_{\varphi'_\omega} \cdot \|u\|_{\varphi'_{1,\omega}}.$$

Therefore, the function $(\omega \in \Omega) \mapsto \ln \|s\|_{\varphi'_\omega}$ is non-positive and bounded from below by a ν-dominated function.

By the inequality (2.5) in Subsection 2.2.2, one has

$$d_\omega(\|\cdot\|_{\varphi_\omega}, \|\cdot\|_{\varphi'_\omega}) \leqslant d_\omega(\varphi_\omega, \varphi'_\omega).$$

On the other hand, by Proposition 6.1.12 (3), the function $(\omega \in \Omega) \longmapsto d_\omega(\varphi_\omega, \varphi'_\omega)$ is ν-dominated, so that the assertion follows. □

We now proceed with the proof of the theorem. Let L_1 be a very ample invertible $\mathcal{O}_X$-module such that $L_2 := L \otimes L_1$ is also very ample. We fix a non-zero global section t of L_1, which defines an injective K-linear map from $H^0(X, L)$ to $H^0(X, L_2)$. We choose a dominated metric family $\varphi_1 = \{\varphi_{1,\omega}\}_{\omega\in\Omega}$ on L_1 such that $\|t\|_{\varphi_{1,\omega}} \leqslant 1$ for any $\omega \in \Omega$, which is possible if we take a strongly dominated norm family $\xi_1 = \{\|\cdot\|_{1,\omega}\}_{\omega\in\Omega}$ on $H^0(X, L_1)$ such that $\|t\|_{1,\omega} \leqslant 1$ for any $\omega \in \Omega$, and choose φ_1 as the quotient metric family induced by $(H^0(X, L_1), \xi_1)$ and the canonical surjective homomorphism $f^*(H^0(X, L_1)) \to L_1$. Let $\varphi_2 = \{\varphi_{2,\omega}\}_{\omega\in\Omega}$ be the metric family $\varphi + \varphi_1$ on L_2. By Proposition 6.1.12 (2), the metric family φ_2 is dominated. Let ξ_2 be the norm family $\{\|\cdot\|_{\varphi_{2,\omega}}\}_{\omega\in\Omega}$ on $H^0(X, L_2)$. By Claim 6.1.15, the norm family ξ_2 is strongly dominated.

Let $\{s_1, \ldots, s_m\}$ be a basis of $H^0(X, L)$. For any $i \in \{1, \ldots, m\}$, let $t_i = ts_i \in H^0(X, L_2)$. We choose sections $t_{m+1}, \ldots, t_n$ in $H^0(X, L_2)$ such that $\{t_1, \ldots, t_n\}$ forms a basis of $H^0(X, L_2)$. Let $\xi_2^\circ = \{\|\cdot\|_{2,\omega}^\circ\}_{\omega\in\Omega}$ be the norm family on $H^0(X, L_2)$ such that, for any $\omega \in \Omega$ and any $(\lambda_1, \ldots, \lambda_n) \in K_\omega^n$,

$$\|\lambda_1 t_1 + \cdots + \lambda_n t_n\|_{2,\omega}^\circ = \begin{cases} \max_{i\in\{1,\ldots,n\}} |\lambda_i|_\omega, & \text{if } \omega \in \Omega \setminus \Omega_\infty, \\ |\lambda_1|_\omega + \cdots + |\lambda_n|_\omega, & \text{if } \omega \in \Omega_\infty. \end{cases}$$

For any $\omega \in \Omega$ and any $(\lambda_1, \ldots, \lambda_\omega) \in K_\omega^m$, one has

$$\|\lambda_1 s_1 + \cdots + \lambda_m s_m\|_{\varphi_\omega} \geqslant \|\lambda_1 t_1 + \cdots + \lambda_m t_m\|_{\varphi_{2,\omega}}$$

since $\|t\|_{\varphi_{1,\omega}} \leqslant 1$. As the norm family ξ_2 is strongly dominated, the local distance function $(\omega \in \Omega) \mapsto d_\omega(\xi_2, \xi_2^\circ)$ is ν-dominated (see Corollary 4.1.10). In particular, there exists a ν-dominated function A on Ω such that, for any $\omega \in \Omega$ and any $(\lambda_1, \ldots, \lambda_m) \in K_\omega^m$, one has

$$\ln \|\lambda_1 s_1 + \cdots + \lambda_m s_m\|_{\varphi_\omega} \geqslant \ln \|\lambda_1 t_1 + \cdots + \lambda_m t_m\|_{2,\omega}^\circ - A(\omega). \tag{6.2}$$

Moreover,

$$\|\lambda_1 s_1 + \cdots + \lambda_m s_m\|_{\varphi_\omega} \leqslant \|\lambda_1 t_1 + \cdots + \lambda_m t_m\|_{2,\omega}^\circ \cdot \max_{i\in\{1,\ldots,m\}} \|s_i\|_{\varphi_\omega}$$

By Claim 6.1.16, for any $i \in \{1, \ldots, m\}$, the function $(\omega \in \Omega) \mapsto \ln \|s_i\|_{\varphi_\omega}$ is ν-dominated. Therefore, there exists a ν-dominated function B on Ω such that, for any $\omega \in \Omega$ and any $(\lambda_1, \ldots, \lambda_m) \in K_\omega^m$, one has

$$\ln\|\lambda_1 s_1 + \cdots + \lambda_m s_m\|_{\varphi_\omega} \leqslant \ln\|\lambda_1 t_1 + \cdots + \lambda_m t_m\|^{\circ}_{2,\omega} + B(\omega). \tag{6.3}$$

The inequalities (6.2) and (6.3) imply that the local distance function $(\omega \in \Omega) \mapsto d_\omega(\xi, \xi^\circ)$ is ν-dominated, where ξ° is the restricted norm family of ξ_2° on $H^0(X, L)$. The strong dominancy of ξ then follows from Corollary 4.1.10. The theorem is thus proved. □

Remark 6.1.17 We keep the notation of Theorem 6.1.13. In the case where X is not geometrically integral, it is possible that $\|\cdot\|_{\varphi_\omega}$ is only a seminorm instead of a norm. However, the argument of Claim 6.1.16 shows that, for any $s \in H^0(X, L)$, the function

$$(\omega \in \Omega) \longrightarrow \ln\|s\|_\omega$$

is upper dominated.

Proposition 6.1.18 *Let K' be a finite extension of K and $X = \operatorname{Spec} K'$. Let L be an invertible $\mathcal{O}_X$-module. Then a metric family φ on L is dominated if and only if the corresponding norm family ξ_L on L relatively to the adelic curve $S \otimes_K K'$ (cf. Remark 6.1.3) is dominated.*

Proof First we assume that there exists a finite-dimensional vector space E over K, a surjective K'-linear map $\beta : E \otimes_K K' \to L$, and a dominated norm family ξ_E on E such that the local distance function $(\omega \in \Omega) \mapsto d_\omega(\varphi, \varphi')$ is ν-dominated, where φ' denotes the metric family on L induced by (E, ξ_E) and β. Denote by ξ'_L the norm families on L relatively to the adelic curve $S \otimes_K K'$, which correspond to the metric families φ'. By definition ξ'_L identifies with the quotient norm family of $\xi_{K'}$ induced by β. Since ξ is dominated, also is $\xi_{K'}$ (cf. Corollary 4.1.13). Hence the norm family ξ'_L is also dominated (cf. Proposition 4.1.19 (2)). By Proposition 4.1.6, we obtain that the norm family ξ_L is dominated.

Conversely, we assume that the norm family ξ_L is dominated. Let e be a generator of L as vector space over K' and let $\xi_e = \{\|\cdot\|_{e,\omega}\}_{\omega \in \Omega}$ be the norm family on Ke such that $\|\lambda e\|_{e,\omega} = |\lambda|_\omega$ for any $\omega \in \Omega$. Note that the norm family $\xi_{e,K'}$ on $(Ke) \otimes_K K' \cong L$ is dominated. By Corollary 4.1.10, the local distance function $(x \in \Omega_{K'}) \mapsto d_x(\xi_{e,K'}, \xi_L)$ is $\nu_{K'}$-dominated. Let φ' be the metric family induced by (Ke, ξ_e) and the canonical isomorphism $(Ke) \otimes_K K' \cong L$. For any $x \in \Omega_{K'}$ one has $d_x(\xi_{e,K'}, \xi_L) = d_x(\varphi', \varphi)$. By Proposition 6.1.12 (4), the metric family φ is dominated. □

6.1.3 Universally dense point families

In this subsection, we consider universally dense point families (cf. Lemma 6.1.19) and their consequences.

Lemma 6.1.19 *Let K be a field, X be a scheme locally of finite type over* Spec K *and K'/K be a field extension. Let $X_{K'}$ be the fibre product $X \times_{\operatorname{Spec} K} \operatorname{Spec} K'$ and*

$\pi : X_{K'} \to X$ be the morphism of projection. For any closed point P of X, the set $\pi^{-1}(\{P\})$ is finite and consists of closed points of $X_{K'}$. Moreover, if F is a set consisting of closed points of X, which is Zariski dense in X, then the subset $\pi^{-1}(F)$ of $X_{K'}$ is Zariski dense.

Proof Let P' be a point of $X_{K'}$ such that $P = \pi(P')$ is a closed point. Since X is locally of finite type over $\operatorname{Spec} K$, the residue field of P is a finite extension of K (this is a consequence of Zariski's lemma, see [154]). As the residue field of P' is a quotient ring of $\kappa(P) \otimes_K K'$, we obtain that it is a finite extension of K'. Moreover, since $\kappa(P) \otimes_K K'$ is an Artinian K'-algebra, the set $\pi^{-1}(\{P\})$ is finite.

In the following, we fix an algebraic closure K'^{ac} of the field K' and we denote by K^{ac} the algebraic closure of K in K'^{ac}. For any closed point P of X, we choose an arbitrary embedding of $\kappa(P)$ in K^{ac} so that we can consider the residue field $\kappa(P)$ as a subfield of K^{ac}. Similarly, for any $P' \in \pi^{-1}(\{P\})$, we choose an embedding of the residue field $\kappa(P')$ in K'^{ac} which extends the embedding $\kappa(P) \to K^{\mathrm{ac}}$.

To prove the lemma it suffices to verify that, for any affine open subset U of X, $\pi^{-1}(U \cap F)$ is Zariski dense in $U_{K'}$. Therefore we may assume without loss of generality that X is an affine scheme of finite type over K. We let A be the coordinate ring of X. Thus the coordinate ring of $X_{K'}$ is $A \otimes_K K'$. Let f be an element of $A \otimes_K K'$ and $\widetilde{f}$ be the canonical image of f in $A \otimes_K K'^{\mathrm{ac}}$. We write $\widetilde{f}$ as a linear combination

$$\widetilde{f} = a_1 g_1 + \cdots + a_n g_n,$$

where $g_1, \ldots, g_n$ are elements of $A \otimes_K K^{\mathrm{ac}}$, and $a_1, \ldots, a_n$ are elements of K'^{ac} which are linearly independent over K^{ac}. Let P be a closed point of X and $\mathfrak{m}_P$ be the maximal ideal of A corresponding to P. Assume that for any $P' \in \pi^{-1}(\{P\})$ one has $f(P') = 0$, where $f(P')$ denotes the image of f by the projection map $A \otimes_K K' \to (A \otimes_K K')/\mathfrak{m}_{P'}$, with $\mathfrak{m}_{P'}$ being the (maximal) ideal of $A \otimes_K K'$ corresponding to P'. Then the canonical image of f in $(A/\mathfrak{m}_P) \otimes_K K'$ is nilpotent, which implies that the canonical image of $\widetilde{f}$ in $(A/\mathfrak{m}_P) \otimes_K K'^{\mathrm{ac}}$ is nilpotent. In particular, the canonical image of $\widetilde{f}$ by the composed map

$$A \otimes_K K'^{\mathrm{ac}} \longrightarrow (A/\mathfrak{m}_P) \otimes_K K'^{\mathrm{ac}} = \kappa(P) \otimes_K K'^{\mathrm{ac}} \longrightarrow K'^{\mathrm{ac}}$$

is zero, where the last map in the above diagram is given by $\lambda \otimes \mu \mapsto \lambda\mu$ for any $\lambda \in \kappa(P) \subseteq K^{\mathrm{ac}}$ and $\mu \in K'^{\mathrm{ac}}$. In other words, one has

$$a_1 g_1(P) + \cdots + a_n g_n(P) = 0,$$

where for each $i \in \{1, \ldots, n\}$, $g_n(P)$ denotes the image of g_n by the composed map

$$A \otimes_K K^{\mathrm{ac}} \longrightarrow (A/\mathfrak{m}_P) \otimes_K K^{\mathrm{ac}} = \kappa(P) \otimes_K K^{\mathrm{ac}} \longrightarrow K^{\mathrm{ac}}.$$

Since $a_1, \ldots, a_n$ are linearly independent over K^{ac}, we obtain that $g_1(P) = \cdots = g_n(P) = 0$. Since this holds for any $P \in F$ and since F is Zariski dense in X, we obtain that $g_1, \ldots, g_n$ are nilpotent elements of $A \otimes_K K^{\mathrm{ac}}$. Therefore $\widetilde{f}$ is a nilpotent element of $A \otimes_K K'^{\mathrm{ac}}$. Since the extension $K'^{\mathrm{ac}}/K^{\mathrm{ac}}$ equips K'^{ac} with a structure of K^{ac}-algebra which is faithfully flat, the canonical map $A \otimes_K K^{\mathrm{ac}} \to A \otimes_K K'^{\mathrm{ac}}$ is injective (see [53], Chapitre I, §1, n°2, Lemme 2.7). Therefore, f is a nilpotent element in $A \otimes_K K'$. This shows that $\pi^{-1}(F)$ is Zariski dense in $X_{K'}$. □

Proposition 6.1.20 *Let $S = (K, (\Omega, \mathcal{A}, \nu), \phi)$ be an adelic curve. Let X be a projective K-scheme, L be an invertible $\mathcal{O}_X$-module, equipped with a metric family φ. We assume that, for any closed point P in X, the norm family $P^*(\varphi)$ (cf. Remark 6.1.3) on $P^*(L)$ is measurable. Then, for any $s \in H^0(X, L)$, the function*

$$(\omega \in \Omega \setminus \Omega_0) \longmapsto \|s\|_{\varphi_\omega}$$

is measurable, where Ω_0 denotes the set of $\omega \in \Omega$ such that $|\cdot|_\omega$ is the trivial absolute value, and we consider the restriction of the σ-algebra $\mathcal{A}$ on $\Omega \setminus \Omega_0$.

Proof As X is projective over K, considering the coefficients of defining homogeneous polynomials of X, we can find a subfield K_0 of K which is finitely generated over the prime field of K (and hence K_0 is countable) and a projective scheme X_0 over $\mathrm{Spec}\, K_0$ such that $X \cong X_0 \times_{\mathrm{Spec}\, K_0} \mathrm{Spec}\, K$. Let $\mathscr{P}$ be the set of closed points P in X whose canonical image in X_0 is a closed point. By Lemma 6.1.19, $\mathscr{P}$ is a Zariski dense and countable subset of X.

By the assumption of the proposition, for any closed point P of X, the function

$$(x \in \Omega_{\kappa(P)}) \longmapsto |s|_{\varphi_\omega}(P_x), \qquad (\omega = \pi_{\kappa(P)/K}(x))$$

is $\mathcal{A}_{\kappa(P)}$-measurable. By Proposition 3.6.2, we obtain that the function

$$(\omega \in \Omega) \longmapsto \max_{x \in \pi^{-1}_{\kappa(P)/K}(\{\omega\})} |s|_{\varphi_\omega}(P_x)$$

is $\mathcal{A}$-measurable. Therefore, the function

$$(\omega \in \Omega) \longmapsto \sup_{P \in \mathscr{P}} \max_{x \in \pi^{-1}_{\kappa(P)/K}(\{\omega\})} |s|_{\varphi_\omega}(x) \tag{6.4}$$

is $\mathcal{A}$-measurable since $\mathscr{P}$ is countable.

To obtain the conclusion of the proposition, it remains to show that the function coincides with $\omega \mapsto \|s\|_{\varphi_\omega}$ on $\Omega \setminus \Omega_0$. For this purpose it suffices to verify that, for any $\omega \in \Omega \setminus \Omega_0$, the set

$$F_\omega = \left\{P_x \;:\; P \in \mathscr{P} \text{ and } x \in \pi^{-1}_{\kappa(P)/K}(\{\omega\})\right\}$$

is dense in X_ω^{an}, where $X_\omega := X \times_{\mathrm{Spec}\, K} \mathrm{Spec}\, K_\omega$. Let $j_\omega : X_\omega^{\mathrm{an}} \to X_\omega$ be the specification map. By Lemma 6.1.19, $j_\omega(F_\omega)$ is Zariski dense in X_ω and hence F_ω is dense in X_ω^{an} with respect to the Berkovich topology (see [9, Corollary 3.4.5]). The proposition is thus proved. □

We assume that K is equipped with the trivial absolute value $|\cdot|_0$. Let F be a finitely generated field over K such that the transcendence degree of F over K is one. Let C_F be a regular projective curve over K such that the function field of C_F is F, that is, C_F is the unique regular model of F over K. It is well-known that, for any absolute value $|\cdot|$ of F over K (i.e. the restriction of $|\cdot|$ to K is trivial), there are a closed point ξ of C_F and $q \in \mathbb{R}_{\geqslant 0}$ such that $|\varphi| = \exp(-q\,\mathrm{ord}_\xi(\varphi))$ for all $\varphi \in F^\times$ (see [85, §I.6] and [117, Proposition II.(3.3)]). Note that in the case where $q = 0$, the absolute value is trivial. We say that q is the *exponent* of $|\cdot|$. The absolute value given by $\exp(-q\,\mathrm{ord}_\xi(\cdot))$ is denoted by $|\cdot|_{(\xi,q)}$. Let X be a projective scheme over $\operatorname{Spec} K$. The Berkovich space associated with X is denote by X^{an} (see Definition 2.1.2). Let $j : X^{\mathrm{an}} \to X$ be the specification map. Let us consider the following subsets X_0^{an}, $X_{1,\mathbb{Q}}^{\mathrm{an}}$ and $X_{\leqslant 1,\mathbb{Q}}^{\mathrm{an}}$ in X^{an}:

$$\begin{cases} X_0^{\mathrm{an}} := \{x \in X^{\mathrm{an}} \,:\, j(x) \text{ is closed}\}, \\ X_{1,\mathbb{Q}}^{\mathrm{an}} := \left\{ x \in X^{\mathrm{an}} \;\middle|\; \begin{array}{l} \text{the Zariski closure of } \{j(x)\} \text{ has dimension one} \\ \text{and the exponent of the corresponding absolute value} \\ \text{is rational} \end{array} \right\}, \\ X_{\leqslant 1,\mathbb{Q}}^{\mathrm{an}} := X_0^{\mathrm{an}} \cup X_{1,\mathbb{Q}}^{\mathrm{an}}. \end{cases}$$

Lemma 6.1.21 *$X_{\leqslant 1,\mathbb{Q}}^{\mathrm{an}}$ is dense in X^{an} with respect to the Berkovich topology.*

Proof To prove the lemma, we need to show that, for any regular function f over an affine open subset $U = \operatorname{Spec} A$ of X and for any point $x \in U^{\mathrm{an}}$, the value $|f|(x)$ belongs to the closure T of the set $\{|f|(z) \,:\, z \in X_{\leqslant 1,\mathbb{Q}}^{\mathrm{an}} \cap U^{\mathrm{an}}\}$ in $\mathbb{R}$. First let us see the following claim:

Claim 6.1.22 *(a) If f is not a nilpotent element, then $1 \in T$.*
(b) If f has a zero point in U, then $0 \in T$. □

Proof (a) As f is not a nilpotent element, there is a closed point z of U such that $f(z) \neq 0$, so that $1 = |f|(z) \in T$.

(b) In this case one can find a closed point z' with $f(z') = 0$. Therefore, $0 = |f|(z') \in T$. □

Let us go back to the proof of the lemma. Let $X' = \operatorname{Spec} A'$ be the Zariki closure of $\{j(x)\}$ in U, where A' is the quotient domain of A by the prime ideal corresponding to $j(x)$. Let $|\cdot|_x$ be the absolute value on the field of fractions of A' corresponding to x. If $\dim X' = 0$, then $j(x)$ is closed, so that the assertion is obvious. Moreover, if $|f'|_x$ is either 0 or 1, then the assertion is also obvious by the above claim. In particular, if $f' = f|_{X'}$ is algebraic over K, then $|f|(x) = |f'|_x$ is either 0 or 1, and hence the assertion is true. Therefore we may assume that $\dim(X') \geqslant 1$, f' is transcendental over K and $|f'|_x \in \mathbb{R}_{\geqslant 0} \setminus \{0, 1\}$.

Consider the ring $A' \otimes_{K[f']} K(f')$, where $K(f')$ is the fraction field of $K[f']$. This is a localisation of the ring A' with respect to the multiplicatively closed subset $K[f'] \setminus \{0\}$. We pick a closed point $\zeta' \in \operatorname{Spec}(A' \otimes_{K[f']} K(f'))$ and let ζ be the canonical image of ζ' in U. Then the point $\zeta \in U$ has dimension 1 and the canonical image f'' of f' in the residue field $\kappa(\zeta)$ is transcendental over K because f' is an element of the constant field $K(f')$ of the variety $\mathrm{Spec}(A' \otimes_{K[f']} K(f'))$. In particular, the natural homomorphism $K[f'] \to K[f'']$ is an isomorphism, which yields an isomorphism $K(f') \xrightarrow{\sim} K(f'')$. Let $|\cdot|'_x$ be the restriction of $|\cdot|_x$ to $K(f')$, and $|\cdot|''_x$ be the absolute value of $K(f'')$ such that the above isomorphism gives rise to an isometry

$$(K(f'), |\cdot|'_x) \xrightarrow{\sim} (K(f''), |\cdot|''_x).$$

Then $|f''|''_x = |f'|'_x = |f|(x)$. Let $|\cdot|_\zeta$ be an extension of $|\cdot|''_x$ to the residue field $\kappa(\zeta)$ and C_ζ be a regular and projective model of $\kappa(\zeta)$ over K. Then there are a closed point ξ of C_ζ and $q \in \mathbb{R}_{>0}$ such that $|\cdot|_\zeta = |\cdot|_{(\xi,q)}$. Thus the assertion follows if we consider a sequence $\{q_n\}_{n=1}^\infty$ of rational numbers such that $\lim_{n\to\infty} q_n = q$. □

Remark 6.1.23 In the case where the absolute value of K is non-trivial, X_0^{an} is dense in X^{an} (cf. Lemma 6.1.19 and [9, Corollary 3.4.5]). However we need one more layer $X_{1,\mathbb{Q}}^{\mathrm{an}}$ if the absolute value of K is trivial. Moreover, if the dimension of every irreducible component of X is greater than or equal to one, then $X_{1,\mathbb{Q}}^{\mathrm{an}}$ is dense in X^{an} with respect to the Berkovich topology. Indeed, it is sufficient to show that X_0^{an} is contained in the closure of $X_{1,\mathbb{Q}}^{\mathrm{an}}$. Let $x \in X_0^{\mathrm{an}}$ and choose a subvariety C' of X such that $\dim C' = 1$ and $j(x) \in C'$. Let $\mu : C \to C'$ be the normalisation of C' and $\xi \in C$ with $\mu(\xi) = j(x)$. Note that $\lim_{\substack{q\to\infty \\ q\in\mathbb{Q}_{>0}}} |\cdot|_{(\xi,q)}$ gives rise to the trivial valuation of the residue field $\kappa(\xi)$, which means that x belongs to the closure of $X_{1,\mathbb{Q}}^{\mathrm{an}}$.

Remark 6.1.24 If K is countable, then $X_{\leqslant 1,\mathbb{Q}}^{\mathrm{an}}$ is also countable. In fact, the set of all closed points of a projective scheme over K is countable. Therefore, X_0^{an} is countable. Moreover, if we fix an increasing sequence

$$K_1 \subseteq K_2 \subseteq \ldots \subseteq K_n \subseteq K_{n+1} \subseteq \ldots$$

of finite extensions of the field $K(T)$ of rational functions such that $\bigcup_{n\in\mathbb{N},\, n\geqslant 1} K_n$ is the algebraic closure of $K(T)$, then any point $z \in X_1^{\mathrm{an}}$ is represented by a point of X valued in certain K_n equipped with an absolute value over K of rational exponent. Suppose that K_n identifies with the rational function field of the projective curve C_n over K, there are only countably many such absolute values since K is assumed to be countable. Hence $X_{1,\mathbb{Q}}^{\mathrm{an}}$ is also countable.

Remark 6.1.25 We assume that K is uncountable and the absolute value of K is trivial. In this case, we can not expect a dense countable subset of X^{an}. Indeed, let S be a countable subset of $\mathbb{P}_K^{1,\mathrm{an}}$. Let $r : \mathbb{P}_K^{1,\mathrm{an}} \to \mathbb{P}_K^1$ be the specification map. As $r(S)$ is countable and K is uncountable, there is a closed point $\xi \in \mathbb{P}_K^1$ such that $\xi \notin r(S)$. We set

$$I := \{\exp(-q\,\mathrm{ord}_\xi(\cdot)) \,:\, q \in]0,\infty[\}.$$

Then I is an open subset of $\mathbb{P}_K^{1,\mathrm{an}}$ and $r(I) = \{\xi\}$, so that $I \cap S = \emptyset$, which shows that S is not dense in $\mathbb{P}_K^{1,\mathrm{an}}$.

Let $S = (K, (\Omega, \mathcal{A}, \nu), \phi)$ be an adelic curve. Denote by Ω_0 the set of $\omega \in \Omega$ such that the absolute value $|\cdot|_\omega$ on K is trivial. Let $\mathcal{A}_0$ be the restriction of the

σ-algebra to Ω_0. Let X be a projective scheme over Spec K. We equip K with the trivial absolute value and denote by X^{an} be the Berkovich space associated with X (see Definition 2.1.2). Suppose given an invertible $\mathcal{O}_X$-module L equipped with a metric family $\varphi = \{\varphi_\omega\}_{\omega\in\Omega}$. For any point $x \in X^{\mathrm{an}}$, the metric φ induces, for any $\omega \in \Omega$, a norm $|\cdot|_{\varphi_\omega}(x)$ on the one-dimensional vector space $L \otimes_{\mathcal{O}_X} \widehat{\kappa}(x)$.

Proposition 6.1.26 *Let $S = (K, (\Omega, \mathcal{A}, \nu), \phi)$ be an adelic curve. We assume that, either the restriction of $\mathcal{A}$ to Ω_0 is discrete, or the field K is countable. Let X be a projective scheme over* Spec K, *L be an invertible $\mathcal{O}_X$-module and φ be a metric family on L. Suppose that, for any $x \in X^{\mathrm{an}}_{\leqslant 1,\mathbb{Q}}$ and any non-zero element ℓ in $L \otimes_{\mathcal{O}_X} \widehat{\kappa}(x)$, the function $(\omega \in \Omega_0) \mapsto |\ell|_{\varphi_\omega}(x)$ is $\mathcal{A}_0$-measurable. Then for any $s \in H^0(X, L)$, the function*

$$(\omega \in \Omega_0) \longmapsto \|s\|_{\varphi_\omega}$$

is measurable on $(\Omega_0, \mathcal{A}_0)$.

Proof It suffices to treat the case where K is countable. By Lemma 6.1.21, we can write $\|s\|_{\varphi_\omega}$ as

$$\|s\|_{\varphi_\omega} = \sup_{z\in X^{\mathrm{an}}_{\leqslant 1,\mathbb{Q}}} |s|_{\varphi_\omega}(z).$$

By the assumption of the proposition, each function $(\omega \in \Omega_0) \mapsto |s|_{\varphi_\omega}(z)$ is $\mathcal{A}_0$-measurable. Since $X^{\mathrm{an}}_{\leqslant 1,\mathbb{Q}}$ is a countable set (see Remark 6.1.24), we deduce that the function $(\omega \in \Omega_0) \mapsto \|s\|_{\varphi_\omega}$ is also $\mathcal{A}_0$-measurable. The proposition is thus proved. □

6.1.4 Measurable metric families

Definition 6.1.27 Let $S = (K, (\Omega, \mathcal{A}, \nu), \phi)$ be an adelic curve, X be a projective scheme over Spec K and L be an invertible $\mathcal{O}_X$-module. We say that a metric family $\varphi = \{\varphi_\omega\}_{\omega\in\Omega}$ on L is *measurable* if the following conditions are satisfied:

(a) for any closed point P in X, the norm family $P^*(\varphi)$ on $P^*(L)$ is measurable,
(b) for any point $x \in X^{\mathrm{an}}_{\leqslant 1,\mathbb{Q}}$ (where we consider the trivial absolute value on K in the construction of the Berkovich space X^{an}) and any element ℓ in $L \otimes_{\mathcal{O}_X} \widehat{\kappa}(x)$, the function $(\omega \in \Omega_0) \mapsto |\ell|_{\varphi_\omega}(x)$ is $\mathcal{A}_0$-measurable, where Ω_0 denotes the set of $\omega \in \Omega$ such that the absolute value $|\cdot|_\omega$ on K is trivial, and $\mathcal{A}_0$ is the restriction of the σ-algebra to Ω_0.

Proposition 6.1.28 *Let $S = (K, (\Omega, \mathcal{A}, \nu), \phi)$ be an adelic curve and X be a projective scheme over* Spec K.

(1) *Let L be an invertible $\mathcal{O}_X$-module equipped with a measurable metric family φ, then the dual metric family $-\varphi$ on $L^\vee$ is measurable.*
(2) *Let L_1 and L_2 be two invertible $\mathcal{O}_X$-modules. If φ_1 and φ_2 are measurable metric families on L_1 and L_2 respectively, then the metric family $\varphi_1 + \varphi_2$ on $L_1 \otimes L_2$ is measurable.*

(3) *Let L be an invertible $\mathcal{O}_X$-module equipped with a metric family φ. Suppose that there exists an integer $n \geqslant 1$ such that $n\varphi$ is measurable, then the metric family φ is also measurable.*

(4) *Let L be an invertible $\mathcal{O}_X$-module equipped with a measurable metric family φ, and $f : Y \to X$ be a projective morphism of K-schemes. Then $f^*(\varphi)$ is measurable.*

Proof (1) Let P be a closed point of X. One has $P^*(-\varphi) = P^*(\varphi)^\vee$. Since φ is measurable, the norm family $P^*(\varphi)$ is measurable. By Proposition 4.1.22 (2), we obtain that $P^*(-\varphi)$ is also measurable.

Let x be a point of $X^{\mathrm{an}}_{\leqslant 1,\mathbb{Q}}$, ℓ be a non-zero element of $L \otimes_{\mathcal{O}_X} \widehat{\kappa}(x)$ and $\ell^\vee$ be the dual element of ℓ in $L^\vee \otimes_{\mathcal{O}_X} \widehat{\kappa}(x)$. Then for any $\omega \in \Omega_0$ one has

$$|\ell|_{\varphi_\omega}(x) \cdot |\ell^\vee|_{-\varphi_\omega}(x) = 1.$$

Since the function $(\omega \in \Omega_0) \mapsto |\ell|_{\varphi_\omega}(x)$ is $\mathcal{A}_0$-measurable, also is the function $(\omega \in \Omega_0) \mapsto |\ell^\vee|_{-\varphi_\omega}(x)$. Therefore the metric family $-\varphi$ is measurable.

(2) Let P be a closed point of X. One has $P^*(\varphi_1+\varphi_2) = P^*(\varphi_1)\otimes P^*(\varphi_2)$. Since φ_1 and φ_2 are both measurable, the norm families $P^*(\varphi_1)$ and $P^*(\varphi_2)$ are measurable. By Proposition 4.1.22 (3), the tensor norm family $P^*(\varphi_1) \otimes P^*(\varphi_2)$ is also measurable.

Let x be a point of $X^{\mathrm{an}}_{\leqslant 1,\mathbb{Q}}$, ℓ_1 and ℓ_2 be non-zero elements of $L_1 \otimes_{\mathcal{O}_X} \widehat{\kappa}(x)$ and $L_2 \otimes_{\mathcal{O}_X} \widehat{\kappa}(x)$, respectively. For any $\omega \in \Omega_0$ one has

$$|\ell_1|_{\varphi_{1,\omega}}(x) \cdot |\ell_2|_{\varphi_{2,\omega}}(x) = |\ell_1 \otimes \ell_2|_{(\varphi_1+\varphi_2)_\omega}(x).$$

Since the metric families φ_1 and φ_2 are measurable, the functions $(\omega \in \Omega_0) \mapsto |\ell_1|_{\varphi_{1,\omega}}$ and $(\omega \in \Omega_0) \mapsto |\ell_2|_{\varphi_{2,\omega}}$ are both $\mathcal{A}_0$-measurable. Hence the function $(\omega \in \Omega_0) \mapsto |\ell_1 \otimes \ell_2|_{(\varphi_1+\varphi_2)_\omega}(x)$ is $\mathcal{A}_0$-measurable.

(3) Let P be a closed point of X. One has $P^*(n\varphi) = P^*(\varphi)^{\otimes n}$. Since $n\varphi$ is measurable, the norm family $P^*(n\varphi)$ is measurable. By Proposition 4.1.22 (4), we obtain that the norm family $P^*(\varphi)$ is also measurable.

Let x be a point of $X^{\mathrm{an}}_{\leqslant 1,\mathbb{Q}}$ and ℓ be a non-zero element of $L \otimes_{\mathcal{O}_X} \widehat{\kappa}(x)$. For any $\omega \in \Omega_0$ one has $|\ell^{\otimes n}|_{n\varphi_\omega}(x) = |\ell|_{\varphi_\omega}(x)^n$ and hence $|\ell|_{\varphi_\omega}(x) = |\ell|_{n\varphi_\omega}(x)^{1/n}$. Since the metric family $n\varphi$ is measurable, we obtain that the function $(\omega \in \Omega_0) \mapsto |\ell^{\otimes n}|_{n\varphi_\omega}(x)$ is $\mathcal{A}_0$-measurable. Hence the function $(\omega \in \Omega_0) \mapsto |\ell|_{\varphi_\omega}(x)$ is also $\mathcal{A}_0$-measurable. Therefore the metric family φ is measurable.

(4) This is obvious by the definition of the measurability of φ. □

The following proposition shows that the measurability of metric family is a property stable by pointwise limit.

Proposition 6.1.29 *Let $S = (K, (\Omega, \mathcal{A}, \nu), \phi)$ be an adelic curve, X be a projective scheme over* Spec K *and L be an invertible $\mathcal{O}_X$-module. Let φ and $\{\varphi_n\}_{n\in\mathbb{N}}$ be metric families on L such that, for any $\omega \in \Omega$ and any $x \in X^{\mathrm{an}}_\omega$ (where X^{an}_ω is the Berkovich space associated with $X_\omega := X \times_{\operatorname{Spec} K}$* Spec K_ω*), one has*

$$\lim_{n\to+\infty} d(|\cdot|_{\varphi_{n,\omega}}(x), |\cdot|_{\varphi_\omega}(x)) = 0.$$

Assume that the metric families φ_n, $n \in \mathbb{N}$, are measurable. Then the limite metric family φ is also measurable.

Proof Let P be a closed point of X and s be a non-zero element in $P^*(L)$. By the assumption of the proposition, for any $\omega' \in \Omega_{K(P)}$ over $\omega \in \Omega$, the norm $\|\cdot\|_{\omega'}$ indexed by ω' in the norm family $P^*(\varphi)$ is given by $|\cdot|_{\varphi_\omega}(P_{\omega'})$, where $P_{\omega'} \in X_\omega^{\mathrm{an}}$ consists of the algebraic point of X_ω determined by P and $|\cdot|_{\omega'}$. As well, for any $n \in \mathbb{N}$, the norm $\|\cdot\|_{n,\omega'}$ indexed by ω' in $P^*(\varphi_n)$ is given by $|\cdot|_{\varphi_{n,\omega}}(P_{\omega'})$. Therefore, by the limit assumption of the proposition, the sequence of functions

$$(\omega' \in \Omega_{K(P)}) \longmapsto \|s\|_{n,\omega'}, \quad n \in \mathbb{N}$$

converges pointwisely to $(\omega' \in \Omega_{K(P)}) \mapsto \|s\|_{\omega'}$, which implies that the latter function is $\mathcal{A}_{K(P)}$-measurable by the measurability assumption of the proposition.

Similarly, for any point $x \in X^{\mathrm{an}}_{\leqslant 1,\mathbb{Q}}$ (where we consider the trivial absolute value on K in the construction of X^{an}), and any element $\ell \in L \otimes_{\mathcal{O}_X} \widehat{\kappa}(x)$, the sequence of functions

$$(\omega \in \Omega_0) \longmapsto |\ell|_{\varphi_{n,\omega}}(x), \quad n \in \mathbb{N}$$

converges pointwisely to $(\omega \in \Omega_0) \mapsto |\ell|_{\varphi_\omega}(x)$. Since each function in the sequence is $\mathcal{A}_0$-measurable, also is the limit function. The proposition is thus proved. □

Proposition 6.1.30 *Let $S = (K, (\Omega, \mathcal{A}, \nu), \phi)$ be an adelic curve, $f : X \to \operatorname{Spec} K$ be a projective K-scheme and L be an invertible $\mathcal{O}_X$-module. We assume that, either the σ-algebra $\mathcal{A}$ is discrete, or K admits a countable subfield which is dense in every K_ω, $\omega \in \Omega$. Let E be a finite-dimensional vector space over K, equipped with a measurable norm family $\xi = \{\|\cdot\|_\omega\}_{\omega \in \Omega}$. Suppose given a surjective homomorphism $\beta : f^*(E) \to L$ and let φ be the quotient metric family on L induced by (E, ξ) and β. Then the metric family φ is measurable.*

Proof It suffices to treat the case where K admits a countable subfield which is dense in all K_ω.

By Proposition 4.1.24 (1.b), the double dual norm family $\xi^{\vee\vee}$ is measurable. By Remark 6.1.4, we can replace ξ by $\xi^{\vee\vee}$ without changing the corresponding quotient metric family. Hence we may assume without loss of generality that the norm $\|\cdot\|_\omega$ is ultrametric for any $\omega \in \Omega \setminus \Omega_\infty$.

Let P be a closed point of X. Then the norm family $P^*(\varphi)$ identifies with the quotient norm family of $\xi \otimes K(P)$ induced by the surjective homomorphism

$$P^*(\beta) : P^*(f^*(E)) \cong E \otimes_K K(P) \longrightarrow P^*(L).$$

By Proposition 4.1.24 (2.a) and (1.a), the norm family $P^*(\varphi)$ is measurable.

Assume that Ω_0 is not empty. In this case the field K itself is countable. Let x be a point of $X^{\mathrm{an}}_{\leqslant 1,\mathbb{Q}}$ and ℓ be a non-zero element of $L \otimes_{\mathcal{O}_X} \widehat{\kappa}(x)$. By Proposition 1.3.26, for any $\omega \in \Omega_0$ one has

$$|\ell|_{\varphi_\omega}(x) = \inf_{\substack{s \in E,\, \lambda \in \widehat{\kappa}(x)^\times \\ s(x) = \lambda \ell}} |\lambda|_x^{-1} \cdot \|s\|_\omega,$$

where $s(x)$ denotes the image of s by the quotient map

$$\beta_x : E \otimes_K \widehat{\kappa}(x) \longrightarrow L \otimes_{\mathcal{O}_X} \widehat{\kappa}(x).$$

Since the norm family ξ is measurable, the function $(\omega \in \Omega_0) \mapsto \|s\|_\omega$ is $\mathcal{A}_0$-measurable. Moreover, since K is a countable field, the vector space E is a countable set. Hence we obtain that the function $(\omega \in \Omega_0) \mapsto |\ell|_{\varphi_\omega}(x)$ is $\mathcal{A}_0$-measurable. Therefore, the metric family φ is measurable. □

Definition 6.1.31 Let $S = (K, (\Omega, \mathcal{A}, \nu), \phi)$ be an adelic curve, $\pi : X \to \operatorname{Spec} K$ be a projective K-scheme, L be an invertible $\mathcal{O}_X$-module and φ be a metric family on L. We denote by $\pi_*(\varphi)$ the norm family $\{\|\cdot\|_{\varphi_\omega}\}_{\omega\in\Omega}$ on $\pi_*(L)$.

The propositions 6.1.20 and 6.1.26 can be resumed as follows.

Theorem 6.1.32 *Let $S = (K, (\Omega, \mathcal{A}, \nu), \phi)$ be an adelic curve, X be a projective K-scheme and L be an invertible $\mathcal{O}_X$-module. We assume that, either $\Omega_0 \in \mathcal{A}$ and the restriction of $\mathcal{A}$ to Ω_0 is discrete, or the field K is countable. For any measurable metric family φ on L, the norm family $\pi_*(\varphi)$ is measurable.*

6.2 Adelic line bundle and Adelic divisors

In this section, we fix a proper adelic curve $S = (K, (\Omega, \mathcal{A}, \nu), \phi)$.

Definition 6.2.1 Let X be a projective scheme over $\operatorname{Spec} K$. We call *adelic line bundle* on X any invertible $\mathcal{O}_X$-module L equipped with a metric family φ which is dominated and measurable.

6.2.1 Height function

Let (L, φ) be an adelic line bundle on X. For any closed point P of X, the norm family $P^*(\varphi)$ on $P^*(L)$ is measurable and dominated (see Propositions 6.1.12 (6) and 6.1.18 for the dominancy of $P^*(\varphi)$). Therefore $(P^*(L), P^*(\varphi))$ is an adelic line bundle on S (cf. Proposition 4.1.29). We denote by $h_{(L,\varphi)}(P)$ the Arakelov degree of this adelic line bundle, called the *height* of P with respect to the adelic line bundle (L, φ). By abuse of notation, we also denote by $h_{(L,\varphi)}(\cdot)$ the function on the set $X(K^{\mathrm{ac}})$ of algebraic points of X sending any K-morphism $\operatorname{Spec} K^{\mathrm{ac}} \to X$ to the height of the image of the K-morphism.

Proposition 6.2.2 *Let X be a projective K-scheme, (L_1, φ_1) and (L_2, φ_2) be adelic line bundles on X.*

(1) *For any closed point P of X, one has*

$$h_{(L_1\otimes L_2,\varphi_1+\varphi_2)}(P) = h_{(L_1,\varphi_1)}(P) + h_{(L_2,\varphi_2)}(P).$$

(2) *Assume that* L_1 *and* L_2 *are the same invertible* $\mathcal{O}_X$*-module* L*. Then, for any closed point* P *of* X*, one has*

$$|h_{(L,\varphi_1)}(P) - h_{(L,\varphi_2)}(P)| \leqslant \mathrm{dist}(\varphi_1, \varphi_2). \tag{6.5}$$

Proof (1) This follows directly from Proposition 6.1.2 and 4.3.6.

(2) Let P be a closed point of X. By definition, for any $\omega \in \Omega$ and any $x \in M_{K(P),\omega}$ (see §3.3 for the notation of $M_{K(P),\omega}$) one has $d_x(P^*(\varphi_1), P^*(\varphi_2)) \leqslant d_\omega(\varphi_1, \varphi_2)$. Therefore, by taking the integral with respect to x, we obtain the inequality (6.5). □

The following proposition shows that, if the adelic curve S has the Northcott property, then the height function associated with an adelic line bundle with ample underlying invertible sheaf has a finiteness property of Northcott type.

Proposition 6.2.3 *Assume that the adelic curve* S *has the Northcott property (cf. Definition 3.5.2). Let* X *be a projective* K*-scheme and* (L, φ) *be an adelic line bundle on* X *such that* L *is ample. For all positive real numbers* δ *and* C*, the set*

$$\{P \in X(K^{\mathrm{ac}}) \,:\, h_{(L,\varphi)}(P) \leqslant C,\ [K(P):K] \leqslant \delta\} \tag{6.6}$$

is finite.

Proof By Proposition 6.2.2, for any integer $n \geqslant 1$, one has $h_{(nL,n\varphi)} = nh_{(L,\varphi)}$ as functions on $X(K^{\mathrm{ac}})$. Therefore, without loss of generality we may assume that L is very ample. Moreover, by Proposition 6.2.2 (2), to prove the proposition it suffices to check the finiteness of (6.6) for a particular choice of metric family φ. Thus we may assume without loss of generality that there exist a finite dimensional vector space E over K and a surjective homomorphism $\beta : E \otimes_K \mathcal{O}_K \rightarrow L$ inducing a closed immersion $X \rightarrow \mathbb{P}(E)$, together with a basis $\boldsymbol{e} = \{e_i\}_{i=0}^r$ of E over K, such that φ identifies with the quotient metric family induced by $(E, \xi_{\boldsymbol{e}})$ and β (see Example 4.1.5 for the definition of $\boldsymbol{e}$). Let P be a closed point of X. Then $(P^*(L)\otimes_{K(P)}K^{\mathrm{ac}}, P^*(\varphi)_{K^{\mathrm{ac}}})$ is a quotient adelic line bundle of $(E_{K^{\mathrm{ac}}}, \xi_{\boldsymbol{e},K^{\mathrm{ac}}})$ and hence $(L_{K^{\mathrm{ac}}}^\vee, P^*(\varphi)_{K^{\mathrm{ac}}}^\vee)$ identifies with an adelic line subbundle of $(E_{K^{\mathrm{ac}}}^\vee, \xi_{\boldsymbol{e},K^{\mathrm{ac}}}^\vee)$. Suppose that, as a vector subspace of rank 1 of $E_{K^{\mathrm{ac}}}^\vee$, $L_{K^{\mathrm{ac}}}^\vee$ is generated by the vector $a_0e_0^\vee + \cdots + a_re_r^\vee$, where

$$[a_0 : \cdots : a_r] \in \mathbb{P}^r(K^{\mathrm{ac}}),$$

then one has

$$\begin{aligned}\widehat{\deg}(P^*(L), P^*(\varphi)) &= \widehat{\deg}(L_{K^{\mathrm{ac}}}, P^*(\varphi)_{K^{\mathrm{ac}}}) = -\widehat{\deg}(L_{K^{\mathrm{ac}}}^\vee, P^*(\varphi)_{K^{\mathrm{ac}}}^\vee) \\ &= \int_{\Omega_{K^{\mathrm{ac}}}} \ln\big(\max\{|a_0|_\chi, \ldots, |a_r|_\chi\}\big)\, \nu_{K^{\mathrm{ac}}}(\mathrm{d}\chi).\end{aligned}$$

Therefore the finiteness of (6.6) follows from Theorem 3.5.3. □

6.2.2 Essential minimum

In this subsection, we fix an *integral* projective scheme X over Spec K. For any adelic line bundle (L, φ) on X, we define the *essential minimum* of (L, φ) as

$$\widehat{\mu}_{\text{ess}}(L, \varphi) := \sup_{Z \subsetneq X} \inf_{P \in (X \setminus Z)(K^{\text{ac}})} h_{(L,\varphi)}(P),$$

where Z runs over the set of all strict Zariski closed subsets of X, and P runs over the set of closed points of the open subscheme $X \setminus Z$ of X. By Proposition 6.2.2 (1), for any integer $n \geqslant 1$, one has $\widehat{\mu}_{\text{ess}}(L^{\otimes n}, n\varphi) = n\,\widehat{\mu}_{\text{ess}}(L, \varphi)$.

Proposition 6.2.4 *Let (L_1, φ_1) and (L_2, φ_2) be adelic line bundles on X. Then*

$$\widehat{\mu}_{\text{ess}}(L_1 \otimes L_2, \varphi_1 + \varphi_2) \geqslant \widehat{\mu}_{\text{ess}}(L_1, \varphi_1) + \widehat{\mu}_{\text{ess}}(L_2, \varphi_2).$$

Proof Let Z_1 and Z_2 be strict Zariski closed subsets of X. Then $Z = Z_1 \cup Z_2$ is also a strict Zariski closed subset of X. If P is an element of $(X \setminus Z)(K^{\text{ac}})$, one has

$$\begin{aligned} & h_{(L_1 \otimes L_2, \varphi_1 + \varphi_2)}(P) = h_{(L_1,\varphi_1)}(P) + h_{(L_2,\varphi_2)}(P) \\ \geqslant & \inf_{Q_1 \in (X \setminus Z_1)(K^{\text{ac}})} h_{(L_1,\varphi_1)}(Q_1) + \inf_{Q_2 \in (X \setminus Z_2)(K^{\text{ac}})} h_{(L_2,\varphi_2)}(Q_2). \end{aligned}$$

Taking the infimum with respect to $P \in (X \setminus Z)(K^{\text{ac}})$ and then the supremum with respect to (Z_1, Z_2), we obtain that

$$\widehat{\mu}_{\text{ess}}(L_1 \otimes L_2, \varphi_1 + \varphi_2) \geqslant \widehat{\mu}_{\text{ess}}(L_1, \varphi_1) + \widehat{\mu}_{\text{ess}}(L_2, \varphi_2).$$

Proposition 6.2.5 *Let (L, φ) be an adelic line bundle on X.*

(1) *The essential minimum of (L, φ) identifies with the infimum of the set of real numbers C such that $\{P \in X(K^{\text{ac}}) : h_{(L,\varphi)}(P) \leqslant C\}$ is Zariski dense in X.*
(2) *If X' is an integral projective K-scheme and $f : X' \to X$ is a birational projective morphism, then one has $\widehat{\mu}_{\text{ess}}(L, \varphi) = \widehat{\mu}_{\text{ess}}(f^*(L), f^*(\varphi))$.*

Proof (1) Let C be a real number such that the set

$$M_C := \{P \in X(K^{\text{ac}}) : h_{(L,\varphi)}(P) \leqslant C\}$$

is Zariski dense. Then, for any Zariski closed subset Z of X such that $Z \subsetneq X$, the set M_C is not contained in Z, namely $(X \setminus Z)(K^{\text{ac}})$ contains at least one element of M_C. Therefore one has

$$\inf_{P \in (X \setminus Z)(K^{\text{ac}})} h_{(L,\varphi)}(P) \leqslant C.$$

We then obtain that $\widehat{\mu}_{\text{ess}}(L, \varphi)$ is bounded from above by the infimum of the set of real numbers C such that M_C is Zariski dense in X.

Conversely, if C is a real number such that M_C is not Zariski dense in X, then one has

$$\widehat{\mu}_{\mathrm{ess}}(L,\varphi) \geqslant \inf_{P\in(X\setminus\overline{M_C}^{\mathrm{Zar}})(K^{\mathrm{ac}})} h_{(L,\varphi)}(P) \geqslant C.$$

Since C is arbitrary, we obtain that $\widehat{\mu}_{\mathrm{ess}}(L,\varphi)$ is bounded from below by the infimum of the set of real numbers C such that M_C is Zariski dense in X.

(2) Denote by (L',φ') the adelic line bundle $(f^*(L), f^*(\varphi))$, and by Y the exceptional locus of f. Note that the restriction of f to $X'\setminus f^{-1}(Y)$ is an isomorphism between $X'\setminus f^{-1}(Y)$ and $X\setminus Y$. Let Z be a Zariski closed subset of X such that $Z\subsetneq X$. Let $Z' = f^{-1}(Z)$. It is a Zariski closed subset of X' such that $Z'\subsetneq X'$. Therefore,

$$\widehat{\mu}_{\mathrm{ess}}(L',\varphi') \geqslant \inf_{P\in(X'\setminus(Z'\cup f^{-1}(Y)))(K^{\mathrm{ac}})} h_{(L',\varphi')}(P) \geqslant \inf_{Q\in(X\setminus Y)} h_{(L,\varphi)}(Q).$$

Since Y is arbitrary, we obtain that $\widehat{\mu}_{\mathrm{ess}}(L',\varphi') \geqslant \widehat{\mu}_{\mathrm{ess}}(L,\varphi)$.

Let C be a real number such that the set M_C of points $Q\in X(K^{\mathrm{ac}})$ with $h_{(L,\varphi)}(Q)\leqslant C$ is Zariski dense. Then the set $M_C\cap(X\setminus Y)(K^{\mathrm{ac}})$ is also Zariski dense in X. This implies that $f^{-1}(M_C\cap(X\setminus Y)(K^{\mathrm{ac}}))$ is Zariski dense in X'. Note that for any $P\in f^{-1}(M_C\cap(X\setminus Y)(K^{\mathrm{ac}}))$ one has $h_{(L',\varphi')}(P) = h_{(L,\varphi)}(f(P))$. Therefore the set of $P\in X'(K^{\mathrm{ac}})$ with $h_{(L',\varphi')}(P)\leqslant C$ is Zariski dense, which implies that $\widehat{\mu}_{\mathrm{ess}}(L',\varphi')\leqslant C$. Since $C > \widehat{\mu}_{\mathrm{ess}}(L,\varphi)$ is arbitrary, we obtain that $\widehat{\mu}_{\mathrm{ess}}(L',\varphi') \geqslant \widehat{\mu}_{\mathrm{ess}}(L,\varphi)$. □

Proposition 6.2.6 *We assume that, either $\Omega_0\in\mathcal{A}$ and the restriction of $\mathcal{A}$ to Ω_0 is discrete, or the field K is countable. Let $f : X\to \operatorname{Spec} K$ be an integral and geometrically reduced projective K-scheme and (L,φ) be an adelic line bundle on X. If s is a non-zero global section of L, then*

$$\widehat{\mu}_{\mathrm{ess}}(L,\varphi) \geqslant \widehat{\deg}_{f_*(\varphi)}(s).$$

In particular, $\widehat{\mu}_{\mathrm{ess}}(L,\varphi) > -\infty$ once L admits a non-zero global section.

Proof By Theorems 6.1.13 and 6.1.32, the norm family $f_*(\varphi)$ is measurable and dominated, so that $\widehat{\deg}_{f_*(\varphi)}(s)$ is well defined (see Definition 4.3.1). For any closed point P outside of the zero locus of s, one has

$$\begin{aligned}
h_{(L,\varphi)}(P) &= -\int_{\chi\in\Omega_{K^{\mathrm{ac}}}} \ln|s|_{\varphi_{\pi_{K^{\mathrm{ac}}/K}(\chi)}}(\sigma_\chi(P))\,\nu_{K^{\mathrm{ac}}}(\mathrm{d}\chi)\\
&\geqslant -\int_{\chi\in\Omega_{K^{\mathrm{ac}}}} \ln\|s\|_{\varphi_{\pi_{K^{\mathrm{ac}}/K}(\chi)}}\,\nu_{K^{\mathrm{ac}}}(\mathrm{d}\chi)\\
&= -\int_{\omega\in\Omega} \ln\|s\|_{\varphi_\omega}\,\nu(\mathrm{d}\omega) = \widehat{\deg}_{f_*(\varphi)}(s),
\end{aligned}$$

where $\sigma_\chi(P)$ denotes the point of $X^{\mathrm{an}}_{K^{\mathrm{ac}},\chi}$ corresponding to P and the absolute value $|\cdot|_\chi$. This leads to the inequality $\widehat{\mu}_{\mathrm{ess}}(L,\varphi)\geqslant\widehat{\deg}_{f_*(\varphi)}(s)$. □

Proposition 6.2.7 *Let (L,φ) be an adelic line bundle on X. One has $\widehat{\mu}_{\mathrm{ess}}(L,\varphi) < +\infty$.*

Proof If φ and φ' are metric families on L such that (L, φ) and (L, φ') are adelic line bundles on X, then for any $P \in X(K^{\mathrm{ac}})$ one has

$$|h_{(L,\varphi)}(P) - h_{(L,\varphi')}(P)| \leqslant \mathrm{dist}(\varphi, \varphi')$$

(see (6.1) for the definition of $\mathrm{dist}(\varphi, \varphi')$). Therefore, to show the proposition, it suffices to prove the assertion for a particular choice of the metric family φ. This observation allows us to change the metric family whenever necessary in the proof.

Let M be a very ample invertible $\mathcal{O}_X$-module such that $M \otimes L$ is also very ample. Let φ_M be a metric family on M such that (M, φ_M) forms an adelic line bundle on X. By Proposition 6.2.6, one has $\widehat{\mu}_{\mathrm{ess}}(M, \varphi_M) > -\infty$. Moreover, by Proposition 6.2.4 one has $\widehat{\mu}_{\mathrm{ess}}(L \otimes M, \varphi + \varphi_M) \geqslant \widehat{\mu}_{\mathrm{ess}}(L, \varphi) + \widehat{\mu}_{\mathrm{ess}}(M, \varphi_M)$. Therefore, by replacing L by $L \otimes M$ we may assume without loss of generality that L is a very ample invertible $\mathcal{O}_X$-module.

By Noetherian normalisation we obtain that there exist a positive integer n, an integral projective K-scheme X', a birational projective K-morphism $f : X' \to X$, together with a generically finite projective K-morphism $g : X' \to \mathbb{P}^r_K$ (where r is the Krull dimension of X) such that $g^*(\mathcal{O}(1)) \cong f^*(L^{\otimes n})$, where $\mathcal{O}(1)$ denotes the universal invertible sheaf on $\mathbb{P}^r_K$. We can for example construct first a rational morphism from X to $\mathbb{P}^r_K$ corresponding to an injective finite homogeneous homomorphism from the polynomial algebra to the Cox ring of some power of L. This step is guaranteed by the fact that the Cox ring $\bigoplus_{m \in \mathbb{N}} H^0(X, L^{\otimes m})$ is finitely generated, by using Noether normalisation, see [56, §13.1]. Then we can take X' as the blowing-up of X along the locus where the rational morphism is not defined. By Proposition 6.2.5 (2), one has $n\widehat{\mu}_{\mathrm{ess}}(L, \varphi) = n\widehat{\mu}_{\mathrm{ess}}(f^*(L), f^*(\varphi)) = \widehat{\mu}_{\mathrm{ess}}(f^*(L^{\otimes n}), nf^*(\varphi))$. Therefore we can reduce the problem to the case where there exists a generically finite projective K-morphism $g : X \to \mathbb{P}^r_K$ such that $L \cong g^*(\mathcal{O}(1))$.

We identify $\mathbb{P}^r_K$ with $\mathbb{P}(K^{r+1})$ and equip K^{r+1} with the norm family ξ associated with the canonical basis (see Example 4.1.5). Let φ_0 be the quotient metric family on $\mathcal{O}(1)$ induced by (K^{r+1}, ξ) and the canonical surjective homomorphism $K^{r+1} \otimes_K \mathcal{O}_{\mathbb{P}^r_K} \to \mathcal{O}(1)$. As explained above, we may assume without loss of generality that $\varphi = g^*(\varphi_0)$. In particular, for any closed point P of X, one has $h_{(L,\varphi)}(P) = h_{(\mathcal{O}(1),\varphi_0)}(g(P))$. Moreover, similarly as in the proof of Proposition 6.2.3, for any element $[a_0 : \ldots : a_r] \in \mathbb{P}^r_K(K^{\mathrm{ac}})$, one has

$$h_{(\mathcal{O}(1),\varphi_0)}([a_0 : \ldots : a_r]) = \int_{\Omega_{K^{\mathrm{ac}}}} \ln\big(\max\{|a_0|_\chi, \ldots, |a_r|_\chi\}\big)\, \nu_{K^{\mathrm{ac}}}(\mathrm{d}\chi).$$

In particular, if $a_0, \ldots, a_r$ are all roots of the unity, then one has

$$h_{(\mathcal{O}(1),\varphi_0)}([a_0 : \ldots : a_r]) = 0.$$

This implies that the set of closed points in X having non-positive height (with respect to (L, φ)) is Zariski dense. Therefore $\widehat{\mu}_{\mathrm{ess}}(L, \varphi) \leqslant 0$. The proposition is thus proved. □

6.2.3 Adelic divisors

In this subsection, we fix a *geometrically integral* projective scheme over $\operatorname{Spec} K$. If D is a Cartier divisor on X, for any $\omega \in \Omega$, D induces by base change a Cartier divisor on X_ω, which we denote by D_ω.

Let D be a Cartier divisor on X. We call *Green function family* of D any family $g = \{g_\omega\}_{\omega\in\Omega}$ parametrised by Ω such that each g_ω is a Green function of D_ω (cf. Subsection 2.5.1). Note that each Green function g_ω determines a continuous metric on the invertible sheaf $\mathcal{O}_{X_\omega}(D_\omega) \cong \mathcal{O}_X(D)\otimes_{\mathcal{O}_X}\mathcal{O}_{X_\omega}$, which we denote by φ_{g_ω}. Thus the collection $\{\varphi_{g_\omega}\}_{\omega\in\Omega}$ forms a metric family on $\mathcal{O}_X(D)$ which we denote by φ_g and call the *metric family associated with* g. We say that the Green function family g is *dominated* (resp. *measurable*) if the associated metric family φ_g is dominated (resp. measurable). In the case where g is dominated and measurable, we say that the couple (D, g) is an *adelic Cartier divisor* on X. Note that this condition is equivalent to the assertion that $(\mathcal{O}_X(D), \varphi_g)$ is an adelic line bundle on X. In this case we denote by $h_{(D,g)}$ the height function $h_{(\mathcal{O}_X(g),\varphi_g)}$ on $X(K^{\mathrm{ac}})$.

Let D and D' be Cartier divisors on X, $g = \{g_\omega\}_{\omega\in\Omega}$ and $g' = \{g'_\omega\}_{\omega\in\Omega}$ be Green function families of D and D', respectively. We denote by $g + g'$ the Green function family $\{g_\omega + g'_\omega\}_{\omega\in\Omega}$ of $D + D'$. Moreover, we denote by $-g$ the Green function family $\{-g_\omega\}_{\omega\in\Omega}$ of $-D$. Note that, if (D, g) and (D', g') are adelic Cartier divisors then $(D+D', g+g')$ and $(-D, -g)$ are also adelic Cartier divisors. This follows from Propositions 6.1.12 and 6.1.28. Therefore, the set of adelic Cartier divisors forms an abelian group, which we denote by $\widehat{\mathrm{Div}}(X)$.

Remark 6.2.8 In the case where the Cartier divisor D is trivial, a Green function family on D can be considered as a family $\{g_\omega\}_{\omega\in\Omega}$ of continuous real-valued functions, where g_ω is a continuous function on X_ω^{an}. It is dominated if and only if the function $(\omega \in \Omega) \mapsto \sup_{x\in X_\omega^{\mathrm{an}}} |g_\omega|(x)$ is ν-dominated. It is measurable if the following two conditions are satisfied (cf. Definition 6.1.27):

(a) for any closed point P of X, the function $(\omega \in \Omega) \mapsto g_\omega(P)$ is $\mathcal{A}$-measurable,
(b) for any point $x \in X^{\mathrm{an}}_{\leqslant 1,\mathbb{Q}}$ (where we consider the trivial absolute value on K in the construction of X^{an}), the function $(\omega \in \Omega_0) \mapsto g_\omega(x)$ is $\mathcal{A}_0$-measurable, where Ω_0 is the set of $\omega \in \Omega$ such that $|\cdot|_\omega$ is trivial.

The set of all dominated and measurable Green function families on the trivial Cartier divisor forms actually a vector space over $\mathbb{R}$, which we denote by $\widehat{C}^0(X)$.

Definition 6.2.9 Let $\mathbb{K}$ be either $\mathbb{Q}$ or $\mathbb{R}$. We denote by $\widehat{\mathrm{Div}}_{\mathbb{K}}(X)$ the $\mathbb{K}$-vector space $\widehat{\mathrm{Div}}(X) \otimes_{\mathbb{Z}} \mathbb{K}$ modulo the vector subspace generated by elements of the form

$$(0, g_1) \otimes \lambda_1 + \cdots + (0, g_n) \otimes \lambda_n - (0, \lambda_1 g_1 + \cdots + \lambda_n g_n),$$

where $\{g_i\}_{i=1}^n$ is a finite family of elements of $\widehat{C}^0(X)$, and $(\lambda_1, \ldots, \lambda_n) \in \mathbb{K}^n$. In the other words, $\widehat{\mathrm{Div}}_{\mathbb{K}}(X)$ consists of pairs (see §2.5.1 for the notation of $C^0_{\mathrm{gen}}(X_\omega^{\mathrm{an}})$)

$$(D, \{g_\omega\}_{\omega\in\Omega}) \in \mathrm{Div}_{\mathbb{K}}(X) \times \prod_{\omega\in\Omega} C^0_{\mathrm{gen}}(X_\omega^{\mathrm{an}})$$

such that $D = a_1D_1 + \cdots + a_nD_n$ and $g_\omega = a_1g_{1,\omega} + \cdots + a_ng_{n,\omega}$ for some $(D_1, g_1), \ldots, (D_n, g_n) \in \widehat{\mathrm{Div}}(X)$ and $a_1, \ldots, a_n \in \mathbb{K}$. For

$$\lambda_1, \lambda_2 \in \mathbb{K} \quad \text{and} \quad (D_1, g_1), (D_2, g_2) \in \widehat{\mathrm{Div}}_{\mathbb{K}}(X),$$

$\lambda_1(D_1, g_1)+\lambda_2(D_2, g_2)$ is defined as $(\lambda_1D_1+\lambda_2D_2, \lambda_1g_1+\lambda_2g_2)$. Note that $\lambda_1(D_1, g_1)+\lambda_2(D_2, g_2) \in \widehat{\mathrm{Div}}_{\mathbb{K}}(X)$. In this sense, $\widehat{\mathrm{Div}}_{\mathbb{K}}(X)$ forms a vector space over $\mathbb{K}$.

The elements of $\widehat{\mathrm{Div}}_{\mathbb{K}}(X)$ are called *adelic* $\mathbb{K}$*-Cartier divisors* on X. For any element $\overline{D}$ written in the form $\lambda_1\overline{D}_1 + \cdots + \lambda_n\overline{D}_n$ with $(\overline{D}_1, \ldots, \overline{D}_n) \in \widehat{\mathrm{Div}}(X)$ and $(\lambda_1, \ldots, \lambda_n) \in \mathbb{K}^n$, we define a function $h_{\overline{D}} : X(K^{\mathrm{ac}}) \to \mathbb{R}$ such that, for any $P \in X(K^{\mathrm{ac}})$,

$$h_{\overline{D}}(P) := \sum_{i=1}^{n} \lambda_i h_{\overline{D}_i}(P).$$

Note the Proposition 6.2.2 (1) shows that this map is actually well defined.

Remark 6.2.10 Let $\overline{D}$ be an element of $\widehat{\mathrm{Div}}_{\mathbb{K}}(X)$, which is written in the form $\lambda_1(D_1, g_1) + \cdots + \lambda_n(D_n, g_n)$, where $(\lambda_1, \ldots, \lambda_n) \in \mathbb{K}^n$, and for any $i \in \{1, \ldots, n\}$, (D_i, g_i) is an element of $\widehat{\mathrm{Div}}(X)$. Then, for any $\omega \in \Omega$, the element $\lambda_1D_{1,\omega} + \cdots + \lambda_nD_{n,\omega}$ of $\mathrm{Div}_{\mathbb{K}}(X)$ is equal to D_ω, where $D = \lambda_1D_1 + \cdots + \lambda_nD_n \in \mathrm{Div}_{\mathbb{K}}(X)$. Moreover, assume that g_i is written in the form $\{g_{i,\omega}\}_{\omega\in\Omega}$, where $g_{i,\omega}$ is a Green function of $D_{i,\omega}$. Then, for any $\omega \in \Omega$, the element $\lambda_1g_{1,\omega} + \cdots + \lambda_ng_{n,\omega}$ is a Green function of the $\mathbb{K}$-Cartier divisor D_ω, which does not depend on the choice of the decomposition $\overline{D} = \lambda_1(D_1, g_1) + \cdots + \lambda_n(D_n, g_n)$. Thus we can write $\overline{D}$ in the form (D, g), where D is a $\mathbb{K}$-Cartier divisor of X and g is a family of Green functions of the form $\{g_\omega\}_{\omega\in\Omega}$, with g_ω being a Green function of D_ω. Note that the measurability of the Green function families $g_1, \ldots, g_n$ implies the following statements:

(a) for any closed point P of X outside of the support of D, the function $(\omega \in \Omega) \mapsto g_\omega(P)$ is well defined and is $\mathcal{A}$-measurable,
(b) for any point $x \in X^{\mathrm{an}}_{\leqslant 1,\mathbb{Q}}$ outside of the analytification of the support of D, the function $(\omega \in \Omega_0) \mapsto g_\omega(x)$ is well defined and is $\mathcal{A}_0$-measurable.

Moreover, if D belongs to $\mathrm{Div}(X)$, then g is a dominated Green function family of D. This statement results directly from the following proposition.

Example 6.2.11 Let s be a non-zero rational function on X. For any $\omega \in \Omega$, we consider s as a non-zero rational function on X_ω. Note that $-\ln|s|_\omega$ is a Green function of the principal Cartier divisor $\mathrm{div}(s)$. Note that the Green function family $\{-\ln|s|_\omega\}_{\omega\in\Omega}$ is measurable and dominated since the corresponding metric family on $\mathcal{O}_X((s)) \cong \mathcal{O}_X$ is trivial. Thus

$$(s \in K(X)^\times) \longmapsto \widehat{(s)} := ((s), \{-\ln|s|_\omega\}_{\omega\in\Omega})$$

defines a morphism of groups from $(K(X)^\times, \times)$ to $\widehat{\mathrm{Div}}(X)$. The adelic Cartier divisors belonging to the image of this morphism are called *principal adelic Cartier divisors*. Moreover, for $\mathbb{K} \in \{\mathbb{Q}, \mathbb{R}\}$ this morphism induces a $\mathbb{K}$-linear map $\widehat{\mathrm{div}}_{\mathbb{K}} : K(X)^\times \otimes_{\mathbb{Z}}$

$\mathbb{K} \to \widehat{\mathrm{Div}}_{\mathbb{K}}(X)$ sending to $s_1^{\lambda_1} \cdots s_n^{\lambda_n}$ to $\lambda_1 \widehat{(s_1)} + \cdots + \lambda_n \widehat{(s_n)}$. The adelic $\mathbb{K}$-Cartier divisors belonging to the image of this $\mathbb{K}$-linear map are said to be *principal*.

Let (D, g) be an adelic $\mathbb{K}$-Cartier divisor on S. For $s \in H^0_{\mathbb{K}}(X, D)$, $|s|_\omega \exp(-g_\omega)$ extends to a continuous function on X_ω^{an} by Proposition 2.5.8. We denote by

$$\|s\|_{g_\omega} := \sup_{x \in X_\omega^{\mathrm{an}}} \{(|s|_\omega \exp(-g_\omega))(x)\}.$$

Proposition 6.2.12 *We assume that, either $\Omega_0 \in \mathcal{A}$ and the restriction of $\mathcal{A}$ to Ω_0 is discrete, or the field K is countable. Let (D, g) be an adelic $\mathbb{K}$-Cartier divisor on X and $s \in H^0_{\mathbb{K}}(X, D) \setminus \{0\}$. The function on Ω given by*

$$(\omega \in \Omega) \mapsto \ln \|s\|_{g_\omega} = \sup_{x \in X_\omega^{\mathrm{an}}} \{(-g_\omega + \log |s|_\omega)(x)\}$$

is ν-integrable.

Proof Note that $D' := D + (s) \geqslant_{\mathbb{K}} 0$, $g'_\omega := g_\omega - \log |s|_\omega$ is a Green function of D'_ω and $|s|_{g_\omega} = |1|_{g'_\omega}$ on X_ω^{an}, so that we may assume that D is $\mathbb{K}$-effective and $s = 1$.

Let X' be the normalisation of X. Since X and X' have the same function field, X' is also geometrically integral over K. Moreover, let D' (resp. g'_ω) be the pull-back of D by $X' \to X$ (resp. $X'_\omega \to X_\omega$). Then $g' = \{g'_\omega\}_{\omega \in \Omega}$ is a family of Green functions of D' over S. Note that $\|1\|_{g_\omega} = \|1\|_{g'_\omega}$, so that we may further assume that X is normal.

First we consider the case $\mathbb{K} = \mathbb{Q}$. Then there is a positive integer N such that ND is a Cartier divisor. Then $s^N \in H^0(X, ND)$ and $\omega \mapsto \ln \|s^N\|_{Ng_\omega}$ is integrable on Ω by Theorem 6.1.13 and Theorem 6.1.32. Note that $\ln \|s^N\|_{Ng_\omega} = N \ln \|s\|_{g_\omega}$, so that $\omega \mapsto \ln \|s\|_{g_\omega}$ is also integrable on Ω.

Next we consider the case $\mathbb{K} = \mathbb{R}$. By Proposition 2.4.16, there are effective Cartier divisors $D_1, \ldots, D_r$ and $a_1, \ldots, a_r \in \mathbb{R}_{\geqslant 0}$ such that $D = a_1 D_1 + \cdots + a_r D_r$. We choose a family of Green functions $g_i = \{g_{i,\omega}\}_{\omega \in \Omega}$ of D_i over S such that (D_i, g_i) is an adelic Cartier divisor over S for each i and

$$(D, g) = (a_1 D_1 + \cdots + a_r D_r, a_1 g_1 + \cdots + a_r g_r).$$

If we set $\psi_i(\omega) = \ln \|1\|_{g_{i,\omega}}$ and $g'_{i,\omega} := g_{i,\omega} + \psi_i(\omega)$ for $i = 1, \ldots, r$, then ψ_i is integrable on Ω and

$$\|1\|_{g'_{i,\omega}} = \|1\|_{g_{i,\omega}} \exp(-\psi_i(\omega)) = 1,$$

so that $g'_{i,\omega} \geqslant 0$ for all i and ω. Note that if we set $g' = a_1 g'_1 + \cdots + a_n g'_n$, then

$$\ln \|1\|_{g',\omega} = \ln \|1\|_{g,\omega} - (a_1 \psi_1(\omega) + \cdots + a_n \psi_n(\omega)).$$

Therefore, we may assume that $g_{i,\omega} \geqslant 0$ for all i and ω.

For each i, we choose a sequence $\{a_{i,n}\}_{n=1}^\infty$ of non-negative rational numbers such that

$$0 \leqslant a_{i,n} - a_i \leqslant \frac{a_i}{n} \quad \text{and} \quad a_{i,n+1} \leqslant a_{i,n}$$

for all n. We set

$$(D_n, h_n) := (a_{1,n}D_1 + \cdots + a_{r,n}D_r, a_{1,n}g_1 + \cdots + a_{r,n}g_r).$$

Then D_n is effective and

$$-h_{n,\omega} \leqslant -g \leqslant \frac{n}{n+1}(-h_{n,\omega}) \leqslant 0 \quad \text{and} \quad -h_{n,\omega} \leqslant -h_{n+1,\omega}$$

for all n and ω. If we set

$$A(\omega) = \sup_{x \in X_\omega^{\mathrm{an}}} \{-g_\omega(x)\} \text{ and } A_n(\omega) = \sup_{x \in X_\omega^{\mathrm{an}}} \{-h_{n,\omega}(x)\},$$

then

$$A_n(\omega) \leqslant A(\omega) \leqslant \frac{n}{n+1} A_n(\omega) \leqslant 0 \quad \text{and} \quad A_n(\omega) \leqslant A_{n+1}(\omega)$$

for all n and ω. Thus $\lim_{n\to\infty} A_n(\omega) = A(\omega)$ and $A_n(\omega) \leqslant A(\omega) \leqslant 0$. Note that $\omega \mapsto A_n(\omega)$ is integrable for all n. Therefore, by monotone convergence theorem, $A(\omega)$ is integrable. □

Corollary 6.2.13 *We keep the hypothesis of Proposition 6.2.12. Let (D, g) be an adelic $\mathbb{K}$-Cartier divisor on X. Let $s \in K(X)^\times \otimes_{\mathbb{Z}} \mathbb{K}$ such that $D + (s) \geqslant_{\mathbb{K}} 0$. Then the function*

$$(\omega \in \Omega) \longmapsto \ln \|s\|_{g_\omega} = \sup_{x \in X_\omega^{\mathrm{an}}} \{(-g_\omega + \log |s|_\omega)(x)\}$$

is ν-integrable.

Proof If we set $D' = (D) + (s)$ and $g'_\omega = g_\omega - \ln |s|_\omega$, then $(D', g' = \{g'_\omega\}_{\omega\in\Omega})$ is an adelic $\mathbb{K}$-Cartier divisor on X. Thus the assertion follows from Proposition 6.2.12.□

Corollary 6.2.14 *We keep the hypothesis of Proposition 6.2.12. Let $(\mathbf{0}, g)$ be an adelic $\mathbb{K}$-Cartier divisor on X whose underlying $\mathbb{K}$-Cartier divisor is trivial. Assume that g is written in the form $\{g_\omega\}_{\omega\in\Omega}$, where g_ω is considered as a continuous function on X_ω^{an}. Then the function*

$$(\omega \in \Omega) \longmapsto \sup_{x \in X_\omega^{\mathrm{an}}} |g_\omega(x)|$$

is ν-integrable.

For any $\overline{D} \in \widehat{\mathrm{Div}}_{\mathbb{K}}(X)$, we define the *essential minimum* of $\overline{D}$ as

$$\widehat{\mu}_{\mathrm{ess}}(\overline{D}) := \sup_{Z \subsetneq X} \inf_{P \in (X \setminus Z)(K^{\mathrm{ac}})} h_{\overline{D}}(P),$$

where Z runs over the set of all strict Zariski closed subsets of X, and P runs over the set of closed points of the open subscheme $X \setminus Z$ of X. It turns out that the analogue of Proposition 6.2.4 and Proposition 6.2.5 (1) holds for adelic $\mathbb{K}$-Cartier divisors (with essentially the same proof). We resume these statements as follows.

Proposition 6.2.15 *Let $\overline{D}$ be an adelic $\mathbb{K}$-Cartier divisor on X. Then $\widehat{\mu}(\overline{D})$ identifies with the infimum of the set of real numbers C such that $\{P \in X(K^{\mathrm{ac}}) \,:\, h_{\overline{D}}(P) \leqslant C\}$ is Zariski dense in X. Moreover, if $\overline{D}_1$ and $\overline{D}_2$ are adelic $\mathbb{K}$-Cartier divisors on X, then $\widehat{\mu}_{\mathrm{ess}}(\overline{D}_1 + \overline{D}_2) \geqslant \widehat{\mu}_{\mathrm{ess}}(\overline{D}_1) + \widehat{\mu}_{\mathrm{ess}}(\overline{D}_2)$.*

Similarly as in the case of adelic line bundles, the essential minimum of adelic $\mathbb{K}$-Cartier divisors never takes $+\infty$ as its value.

Proposition 6.2.16 *Let $\overline{D}$ be an adelic $\mathbb{K}$-Cartier divisor on X. One has $\widehat{\mu}_{\mathrm{ess}}(\overline{D}) < +\infty$.*

Proof Assume that $\overline{D}$ is written in the form $\overline{D} = \lambda_1\overline{D}_1 + \cdots + \lambda_n\overline{D}_n$, where $D_1, \ldots, D_n$ are very ample Cartier divisors on X and $(\lambda_1, \ldots, \lambda_n) \in \mathbb{K}^n$. By Proposition 6.2.6, for any $i \in \{1, \ldots, n\}$ one has $\widehat{\mu}_{\mathrm{ess}}(\overline{D}_i) > -\infty$. We choose $(\lambda'_1, \ldots, \lambda'_n) \in (\mathbb{K} \cap \mathbb{R}_{>0})^n$ such that $\lambda_i + \lambda'_i \in \mathbb{Z}$ for any $i \in \{1, \ldots, n\}$. Let

$$\overline{E} := \sum_{i=1}^{n} (\lambda_i + \lambda'_i)\overline{D}_i.$$

By Proposition 6.2.7, one has $\widehat{\mu}_{\mathrm{ess}}(\overline{E}) < +\infty$. Moreover, by Proposition 6.2.15, one has

$$\widehat{\mu}_{\mathrm{ess}}(\overline{E}) \geqslant \widehat{\mu}_{\mathrm{ess}}(\overline{D}) + \sum_{i=1}^{n} \lambda'_i\, \widehat{\mu}_{\mathrm{ess}}(\overline{D}_i).$$

Since $\widehat{\mu}_{\mathrm{ess}}(\overline{D}_i) > -\infty$ and $\lambda'_i > 0$ for any $i \in \{1, \ldots, n\}$, we deduce that $\widehat{\mu}_{\mathrm{ess}}(\overline{D}) < +\infty$. □

Definition 6.2.17 Let $\overline{D} = (D, g)$ be an adelic $\mathbb{K}$-Cartier divisor, where the Green function family g is written in the form $\{g_\omega\}_{\omega\in\Omega}$. In the case where $\mathbb{K} = \mathbb{Q}$ or $\mathbb{R}$, we assume that X is normal so that $H^0_{\mathbb{K}}(X, D)$ is a K-vector subspace of $K(X)$. For $\omega \in \Omega$, let X'_ω be the normalization of X_ω and D'_ω (resp. g'_ω) be the pull-back of D by $X'_\omega \to X_\omega$ (resp. the pull-back of g_ω by $X'^{\,\mathrm{an}}_\omega \to X^{\mathrm{an}}_\omega$). By using the natural injective homomorphism $H^0_{\mathbb{K}}(X, D) \otimes_K K_\omega \to H^0_{\mathbb{K}}(X'_\omega, D'_\omega)$ and g'_ω, one has a norm $\|\cdot\|_{g_\omega}$ on $H^0_{\mathbb{K}}(X, D) \otimes_K K_\omega$ (cf. Definition 2.5.9). The norm family $\{\|\cdot\|_{g_\omega}\}_{\omega\in\Omega}$ is denoted by ξ_g.

Theorem 6.2.18 *We assume that, either the σ-algebra $\mathcal{A}$ is discrete, or the field K admits a countable subfield which is dense in every K_ω, $\omega \in \Omega$. Suppose that X is normal. Then the couple $(H^0_{\mathbb{K}}(D), \xi_g)$ is a strongly adelic vector bundle on S.*

Proof The measurability of ξ_g is a consequence of Proposition 6.2.12, which implies the measurability of $\xi_g^\vee$ under the hypothesis of the theorem (see Proposition 4.1.24). Let us consider the dominancy of ξ_g. By using [109, Lemma 5.2.3], $\overline{D}$ is written in the form

$$\lambda_1(D_1, g_1) + \cdots + \lambda_n(D_n, g_n),$$

where (D_i, g_i)'s are elements of $\widehat{\mathrm{Div}}(X)$ such that $D_1, \ldots, D_n$ are effective, and $(\lambda_1, \ldots, \lambda_n) \in \mathbb{K}^n$. Let $(\lambda'_1, \ldots, \lambda'_n)$ be an element of $(\mathbb{K} \cap \mathbb{R}_{>0})^n$ such that $\lambda_i + \lambda'_i \in \mathbb{Z}_{>0}$ for any $i \in \{1, \ldots, n\}$. Let

$$(D', g') := (\lambda_1 + \lambda'_1)(D_1, g_1) + \cdots + (\lambda_n + \lambda'_n)(D_n, g_n),$$

which is viewed as an adelic Cartier divisor on X. Since D_i is effective, we obtain that 1 belongs to $H^0(D_i)$. Moreover, by Proposition 6.2.12, the function $(\omega \in \Omega) \mapsto \ln\|1\|_{g_{i,\omega}}$ is ν-integrable.

Let $\boldsymbol{e} = \{e_i\}_{i=1}^m$ be a basis of $H^0_{\mathbb{K}}(D)$. We complete it into a basis $\boldsymbol{e}' = \{e_i\}_{i=1}^r$ of $H^0_{\mathbb{K}}(D')$. By Theorem 6.1.13, the norm family $\xi_{g'} := \{\|\cdot\|_{g'_\omega}\}_{\omega \in \Omega}$ is strongly dominated, so that, by Corollary 4.1.10, the local distance function $(\omega \in \Omega) \mapsto d_\omega(\xi_{g'}, \xi_{e'})$ is ν-dominated. Further, by Proposition 6.2.12, the function $(\omega \in \Omega) \mapsto \ln\|e_i\|_{g_\omega}$ is ν-integrable for each i.

For $\omega \in \Omega$ and $(a_1, \ldots, a_m) \in K_\omega^m$, one has

$$\begin{aligned}\ln\|a_1e_1 + \cdots + a_me_m\|_{g_\omega} &\leqslant \max_{i \in \{1,\ldots,m\}} \{\ln|a_i| + \ln\|e_i\|_{g_\omega}\} + 1\!\!1_{\Omega_\infty}(\omega)\ln(m) \\ &\leqslant \ln\|a_1e_1 + \cdots + a_me_m\|_{\xi_e} + \max_{i \in \{1,\ldots,m\}} \{\ln\|e_i\|_{g_\omega}\} + 1\!\!1_{\Omega_\infty}(\omega)\ln(m).\end{aligned}$$

Moreover,

$$\begin{aligned}\ln\|a_1e_1 + \cdots + a_me_m\|_{g_\omega} &\geqslant \ln\|a_1e_1 + \cdots + a_me_m\|_{g'_\omega} - \sum_{i=1}^n \lambda'_i \ln\|1\|_{g_{i,\omega}} \\ &\geqslant \ln\|a_1e_1 + \cdots + a_me_m\|_{\xi_e} - d_\omega(\xi_{g'}, \xi_{e'}) - \sum_{i=1}^n \lambda'_i \ln\|1\|_{g_{i,\omega}},\end{aligned}$$

and hence one obtains

$$d_\omega(\xi_g, \xi_e) \leqslant \max\left\{\max_{i \in \{1,\ldots,m\}} \{|\ln\|e_i\|_{g_\omega}|\} + 1\!\!1_{\Omega_\infty}(\omega)\ln(m),\; d_\omega(\xi_{g'}, \xi_{e'}) + \sum_{i=1}^n \lambda'_i |\ln\|1\|_{g_{i,\omega}}|\right\}$$

Therefore the local distance function $(\omega \in \Omega) \mapsto d_\omega(\xi_g, \xi_e)$ is ν-dominated, which implies that the norm family ξ_g is strongly dominated (cf. Corollary 4.1.10). □

6.2.4 The canonical compatifications of Cartier divisors with respect to endomorphisms

In this subsection, we assume that, either the σ-algebra $\mathcal{A}$ is discrete, or there exists a countable subfield K_0 of K which is dense in the completion K_ω of K with respect to any $\omega \in \Omega$.

Let X be a projective and geometrically integral scheme over $\operatorname{Spec} K$ and $f : X \to X$ be a surjective endomorphism of X over K. Let D be a Cartier divisor on X. We assume that there are an integer d and $s \in K(X)^\times$ such that $d > 1$ and $f^*(D) = dD + (s)$. For each $\omega \in \Omega$, by Proposition 2.5.11, there is a unique Green function g_ω of D_ω with $(f_\omega^{\mathrm{an}})^*(g_\omega) = dg_\omega - \log|s|_\omega$ on X_ω.

Proposition 6.2.19 *If we set $g = \{g_\omega\}_{\omega\in\Omega}$, then the pair (D, g) is an adelic Cartier divisor on X.*

Proof Fix a family $g_0 = \{g_{0,\omega}\}_{\omega\in\Omega}$ of Green functions on X such that (D, g_0) is an adelic Cartier divisor on X. For each $\omega \in \Omega$, there is a unique continuous function λ_ω on X_ω such that

$$(f_\omega^{\mathrm{an}})^*(g_{0,\omega}) = dg_{0,\omega} - \log|s|_\omega + \lambda_\omega.$$

If we set

$$\begin{cases} h_{n,\omega} = \displaystyle\sum_{i=0}^{n-1} \frac{1}{d^{i+1}}((f_\omega^{\mathrm{an}})^i)^*(\lambda_\omega), & h_n = \{h_{n,\omega}\}_{\omega\in\Omega}, \\ h_\omega = \displaystyle\sum_{i=0}^{\infty} \frac{1}{d^{i+1}}((f_\omega^{\mathrm{an}})^i)^*(\lambda_\omega), & h = \{h_\omega\}_{\omega\in\Omega}, \end{cases}$$

then $g = g_0 + h$ (cf. Proposition 2.5.11), so that it is sufficient to show that φ_h is dominated and measurable. By Proposition 6.1.12 and Proposition 6.2.12, one can see that the function $\omega \mapsto \|\lambda_\omega\|_{\sup}$ is ν-integrable on Ω. Moreover, by Proposition 2.5.11, $\|h_\omega\|_{\sup} \leqslant \|\lambda_\omega\|_{\sup}/(d-1)$, so that φ_h is dominated by Proposition 6.1.12. On the other hand, by Proposition 6.1.28, φ_{h_n} is measurable, so that φ_h is also measurable by Proposition 6.1.29. □

Definition 6.2.20 (1) An adelic arithmetic $\mathbb{R}$-Cartier divisor $\overline{D} = (D, g)$ is called the *canonical compactification* of D with respect to f if $f^*(\overline{D}) = d\overline{D} + \widehat{(s)}$. Note that $\overline{D}$ is uniquely determined by the equation $f^*(\overline{D}) = d\overline{D} + \widehat{(s)}$ (cf. Proposition 2.5.11).
(2) Let $\overline{D}$ be the canonical compactification of D with respect to f. Then the associated height function $h_{\overline{D}}$ is called the *canonical height function* with respect to f and it is often denoted by $\hat{h}_D$.

Proposition 6.2.21 $\hat{h}_D(f(P)) = d\hat{h}_D(P)$ *for all closed points P of X.*

Proof Indeed, by Proposition 6.2.2,

$$h_{\overline{D}}(f(P)) = h_{f^*(\overline{D})}(P) = h_{d\overline{D}+\widehat{\mathrm{div}}(s)}(P) = h_{d\overline{D}}(P) = dh_{\overline{D}}(P),$$

as required. □

Theorem 6.2.22 *If D is effective and $f^*(D) = dD$, then the canonical compactification $\overline{D}$ of D is also effective.*

Proof By Proposition 6.2.12, we can choose a Green function family $g_0 = \{g_{0,\omega}\}_{\omega\in\Omega}$ of D such that (D, g_0) is effective. Let $\lambda = \{\lambda_\omega\}_{\omega\in\Omega}$ be the collection of continuous functions such that

$$f^*(D, g_0) = d(D, g_0) + (0, \lambda).$$

As before, we set

$$\begin{cases} h_n := \displaystyle\sum_{i=0}^{n-1} \frac{1}{d^{i+1}} (f^i)^*(\lambda) & (n \geqslant 1) \\ g_n = g_0 + h_n & (n \geqslant 1). \end{cases}$$

By Proposition 2.5.11, for each $\omega \in \Omega$, $\{h_{n,\omega}\}_{n=1}^{\infty}$ converges uniformly to a continuous function h_ω, and $\overline{D} = (D, g_0 + h)$, where $h = \{h_\omega\}_{\omega\in\Omega}$. By Proposition 2.5.11 again, one has

$$f^*(D, g_{n-1}) = d(D, g_n)$$

for all $n \in \mathbb{N}_{\geqslant 1}$, so that $(D, g_n) \geqslant 0$ for all $n \in \mathbb{N}$ because $(D, g_0) \geqslant 0$. Therefore, $\overline{D} \geqslant 0$. □

Example 6.2.23 We assume that X is the n-dimensional projective space over K, that is, $X = \mathbb{P}^n_K = \mathrm{Proj}(K[T_0, T_1, \ldots, T_n])$, and that f is a polynomial map, that is,

$$f\left(\mathbb{P}^n_K \setminus \{T_0 = 0\}\right) \subseteq \mathbb{P}^n_K \setminus \{T_0 = 0\}.$$

If we set $D = \{T_0 = 0\}$, then the canonical compactification $\overline{D}$ with respect to f is effective because $f^*(D) = dD$.

Definition 6.2.24 We say that an adelic $\mathbb{R}$-Cartier divisor $\overline{D}$ satisfies the *Dirichlet property* if $\overline{D} + \widehat{(s)}$ is effective for some $s \in K(X)^\times \otimes_{\mathbb{Z}} \mathbb{R}$. The reason of the name "the Dirichlet property" comes from the Dirichlet unit theorem (for details, see [110,46]).

Example 6.2.25 The Dirichlet property is very sensitive on the choice of the dynamic system. For example, we set $K := \mathbb{Q}(\sqrt{-1})$, $X := \mathbb{P}^1_K = \mathrm{Proj}(K[T_0, T_1])$ and $z := T_1/T_0$. Let us consider two endomorphisms f_1 and f_2 on X given by

$$f_1(T_0 : T_1) = (2T_0T_1 : T_1^2 - T_0^2) \quad \text{and} \quad f_2(T_0 : T_1) = (2\sqrt{-1}T_0T_1 : T_1^2 - T_0^2),$$

that is, $f_1(z) = (1/2)(z - 1/z)$ and $f_2(z) = (1/2\sqrt{-1})(z - 1/z)$. If we set $D := \{T_1 - \sqrt{-1}T_0 = 0\}$, then $f_1^*(D) = 2D$ because

$$(T_1^2 - T_0^2) - \sqrt{-1}(2T_0T_1) = (T_1 - \sqrt{-1}T_0)^2.$$

Let g be the canonical Green function of D with respect to f_1. Then, by Theorem 6.2.22, $\overline{D} = (D, g)$ has the Dirichlet property. On the other hand, for an infinite place σ of K, it is well-known that the Julia set of f_2 on X_σ is equal to X_σ itself (cf. [114, Theorem 4.2.18]). Therefore, by [46, Theorem 4.5], for any ample Cartier divisor A, the canonical compactification $\overline{A}$ with respect to f_2 does not satisfy the Dirichlet property.

6.3 Newton-Okounkov bodies and concave transform

6.3.1 Reminder on some facts about convex sets

In this subsection, we recall some basic facts about convex sets in finite-dimensional vector spaces, which will be used in the subsequening subsections.

Proposition 6.3.1 *Let V be a finite-dimensional vector space over $\mathbb{R}$. Suppose that C_1 and C_2 are two convex subsets of V which have the same closure in V, then the interiors C_1° and C_2° are also the same.*

Proof It suffices to prove that, if C is a convex subset of V, then the interior of the closure $\overline{C}$ coincides with the interior C° of C. Let x be an interior point of $\overline{C}$. If x does not belong to C, by Hahn-Banach theorem (see [130, Theorem 3.4]), there exists an affine function $q : V \to \mathbb{R}$ such that $q(x) \leqslant 0$ and that the restriction of q to C is non-negative. As the set $\{y \in V : q(y) \geqslant 0\}$ is closed, it contains $\overline{C}$. Hence the interior of $\overline{C}$ is contained in that of $\{y \in V : q(y) \geqslant 0\}$, which is equal to $\{y \in V : q(y) > 0\}$. This leads to a contradiction since $q(x) \leqslant 0$. Therefore we obtain $\overline{C}^\circ \subseteq C$ and hence $\overline{C}^\circ = C^\circ$. □

Proposition 6.3.2 *Let V be a finite-dimensional vector space over $\mathbb{R}$ and $\{C_i\}_{i\in I}$ be a family of convex subsets of W. Suppose that the family $\{C_i\}_{i\in I}$ is filtered, namely, for any couple (i_1, i_2) of indices in I, there exists $j \in I$ such that $C_{i_1} \cup C_{i_2} \subseteq C_j$. Let C be the union of C_i, $i \in I$. Then the interior of C identifies with the union of C_i°, $i \in I$.*

Proof Since the family $\{C_i\}_{i\in I}$ is filtered, for any couple of points (x, y) in C, there exists an index $i \in I$ such that $\{x, y\} \subseteq C_i$. Therefore C is a convex subset of V. As a consequence, for any point x of the interior C°, there exist points $x_1, \ldots, x_n$ in C such that the point x is contained in the interior of the convex hull of $x_1, \ldots, x_n$. Still by the assumption that the family $(C_i)_{i\in I}$ is filtered, there exists $j \in I$ such that $\{x_1, \ldots, x_n\} \subseteq C_j$. Hence one has $x \in C_j^\circ$.

6.3.2 Graded semigroups

Let V be a finite-dimensional vector space over $\mathbb{R}$. As *graded semigroup* in V we refer to a non-empty subset Γ of $\mathbb{N}_{\geqslant 1} \times V$ which is stable by addition. If Γ is a graded semigroup in V, for any $n \in \mathbb{N}_{\geqslant 1}$ we denote by Γ_n the projection of $\Gamma \cap (\{n\} \times V)$ in V. Let $\mathbb{N}(\Gamma)$ be the set of all $n \in \mathbb{N}_{\geqslant 1}$ such that Γ_n is non-empty. This is a non-empty sub-semigroup of $\mathbb{N}_{\geqslant 1}$. We denote by $\mathbb{Z}(\Gamma)$ the subgroup of $\mathbb{Z}$ generated by $\mathbb{N}(\Gamma)$.

Proposition 6.3.3 *Let Γ be a graded semigroup in V. Then there exist at most finitely many positive elements of $\mathbb{Z}(\Gamma) \setminus \mathbb{N}(\Gamma)$.*

Proof The group $\mathbb{Z}(\Gamma)$ is non-zero since Γ is not empty. Hence there exists a positive integer m such that $\mathbb{Z}(\Gamma) = m\mathbb{Z}$. Assume that m is written in the form

$$m = a_1 n_1 + \cdots + a_\ell n_\ell,$$

where $n_1, \ldots, n_\ell$ are elements of $\mathbb{N}(\Gamma)$ and $a_1, \ldots, a_\ell$ are integers. Since $\mathbb{N}(\Gamma) \subseteq \mathbb{Z}(\Gamma)$, there exists a positive integer N such that $n_1 + \cdots + n_\ell = mN$. Let

$$b = N \cdot \max_{i \in \{1,\ldots,\ell\}} |a_i|.$$

We claim that $mn \in \mathbb{N}(\Gamma)$ for any $n \geqslant Nb$. In fact, we can write such n in the form $n = cN + r$ where $c \in \mathbb{N}_{\geqslant b}$ and $r \in \{0, \ldots, N-1\}$. Thus

$$\begin{aligned} mn = cmN + mr = c(n_1 + \cdots + n_\ell) + r(a_1 n_1 + \cdots + a_\ell n_\ell) \\ = (c + ra_1)n_1 + \cdots + (c + ra_\ell)n_\ell. \end{aligned}$$

Since $c \geqslant b$ and $r < N$, we obtain that $c + ra_i \geqslant 0$ for any $i \in \{1, \ldots, \ell\}$. Hence $mn \in \mathbb{N}(\Gamma)$. □

Definition 6.3.4 Let Γ be a graded semigroup in V. We denote by $\Delta(\Gamma)$ the closure of the set

$$\bigcup_{n \in \mathbb{N},\, n \geqslant 1} \{n^{-1}\alpha \,:\, \alpha \in \Gamma_n\} \subset V.$$

Proposition 6.3.5 *Let Γ be a graded semigroup in V. The set $\Delta(\Gamma)$ is a closed convex subset of V.*

Proof It suffices to prove the convexity of the set $\Delta(\Gamma)$. Observe that, if n and m are two positive integers, α and β are elements of Γ_n and Γ_m, respectively. We show that, for any $\epsilon \in [0, 1] \cap \mathbb{Q}$, one has $\epsilon n^{-1}\alpha + (1-\epsilon)m^{-1}\beta \in \Delta(\Gamma)$. Let $\epsilon = p/q$ be a rational number in $[0, 1]$, where $q \in \mathbb{N}_{\geqslant 1}$. One has

$$\epsilon n^{-1}\alpha + (1-\epsilon)m^{-1}\beta = \frac{p}{qn}\alpha + \frac{q-p}{qm}\beta = (qmn)^{-1}(pm\alpha + (q-p)n\beta).$$

Since $\alpha \in \Gamma_n$ and $\beta \in \Gamma_m$, one has $pm\alpha + (q-p)n\beta \in \Gamma_{qmn}$. Therefore

$$\epsilon n^{-1}\alpha + (1-\epsilon)m^{-1}\beta \in \Delta(\Gamma).$$

Let H be the set

$$\bigcup_{n\in\mathbb{N},\, n\geqslant 1} \{n^{-1}\alpha \,:\, \alpha \in \Gamma_n\}.$$

Let x and y be two points in $\Delta(\Gamma)$, and $\epsilon \in [0,1]$. By definition, there exist two sequences $\{x_n\}_{n\in\mathbb{N}}$ and $\{y_n\}_{n\in\mathbb{N}}$ in H such that

$$\lim_{n\to+\infty} x_n = x, \qquad \lim_{n\to+\infty} y_n = y.$$

Let $\{\epsilon_n\}_{n\in\mathbb{N}}$ be a sequence in $[0,1]\cap\mathbb{Q}$ which converges to ϵ. By what we have shown above, for any $n \in \mathbb{N}$ one has $\epsilon_n x_n + (1-\epsilon_n)y_n \in H$. Moreover, one has

$$\lim_{n\to+\infty} \epsilon_n x_n + (1-\epsilon_n)y_n = \epsilon x + (1-\epsilon)y.$$

Therefore $\epsilon x + (1-\epsilon)y \in \Delta(\Gamma)$. □

Let Γ be a graded semigroup in V. We denote by $\Gamma_{\mathbb{R}}$ the $\mathbb{R}$-vector subspace of $\mathbb{R}\times V$ generated by Γ. For $n \in \mathbb{Z}$, let $A(\Gamma)_n$ be the projection of $\Gamma_{\mathbb{R}} \cap (\{n\}\times V)$ in V. Especially, $A(\Gamma)_1$ is denoted by $A(\Gamma)$. Note that $A(\Gamma)_0$ is a vector subspace of V, which is a translation of the affine subspace $A(\Gamma)$. Since $A(\Gamma)_n$ is the image of an affine subspace of $\mathbb{R}\times V$ by a linear map, it is an affine subspace in V. Note that any element in $A(\Gamma) = A(\Gamma)_1$ can be written in the form

$$\lambda_1\gamma_1 + \cdots + \lambda_\ell\gamma_\ell,$$

where for $i \in \{1,\ldots,\ell\}$, $\gamma_i \in \Gamma_{n_i}$, $n_i \in \mathbb{N}$, $n_i \geqslant 1$, and $(\lambda_1,\ldots,\lambda_\ell)$ is an element of $\mathbb{R}^\ell$ such that $\lambda_1 n_1 + \cdots + \lambda_\ell n_\ell = 1$. We denote by $\Gamma_{\mathbb{Z}}$ the subgroup of $\mathbb{R}\times V$ generated by Γ. For any $n \in \mathbb{Z}$, $n \geqslant 1$, let $\Gamma_{\mathbb{Z},n}$ be the image of $\Gamma_{\mathbb{Z}} \cap (\{n\}\times V)$ in V by the canonical projection. Note that $\Gamma_{\mathbb{Z},n}$ is non-empty if and only if $n \in \mathbb{Z}(\Gamma)$.

Proposition 6.3.6 *Let Γ be a graded semigroup in V. We assume that $\Gamma_{\mathbb{Z}}$ is a discrete subset of $\mathbb{R}\times V$.*

(1) *The set $\Gamma_{\mathbb{Z},0}$ is a lattice in $A(\Gamma)_0$.*
(2) *For any $n, n' \in \mathbb{Z}(\Gamma)$ and any $\gamma_0 \in \Gamma_{\mathbb{Z},n}$, the map from $\Gamma_{\mathbb{Z},n'}$ to $\Gamma_{\mathbb{Z},n+n'}$, sending $\gamma \in \Gamma_{\mathbb{Z},n'}$ to $\gamma + \gamma_0$, is a bijection.*
(3) *For any convex and compact subset K of $A(\Gamma)$ which is contained in the relative interior of $\Delta(\Gamma)$, one has*

$$K \cap \{n^{-1}\gamma \,:\, \gamma \in \Gamma_n\} = K \cap \{n^{-1}\gamma \,:\, \gamma \in \Gamma_{\mathbb{Z},n}\} \tag{6.7}$$

for sufficiently positive $n \in \mathbb{N}(\Gamma)$.

Proof (1) Let n be an element in $\mathbb{Z}(\Gamma)$ and $\gamma_0 \in \Gamma_{\mathbb{Z},n}$. By definition, an element $x \in V$ lies in $A(\Gamma)_0$ if and only if $x + n^{-1}\gamma_0 \in A(\Gamma)$. In other words, $A(\Gamma)_0$ is precisely the vector subspace of V of all vectors γ which can be written in the form

$$\gamma = \lambda_1\gamma_1 + \cdots + \lambda_\ell\gamma_\ell, \tag{6.8}$$

where for any $i \in \{1, \ldots, \ell\}$, $\gamma_i \in \Gamma_{n_i}$ with $n_i \in \mathbb{N}(\Gamma)$, and $(\lambda_1, \ldots, \lambda_\ell)$ is an element in $\mathbb{R}^\ell$ such that $\lambda_1 n_1 + \cdots + \lambda_\ell n_\ell = 0$. Note that the set $\Gamma_{\mathbb{Z},0}$ is characterized by the same condition, except that $(\lambda_1, \ldots, \lambda_\ell)$ is required to be in $\mathbb{Z}^\ell$. Therefore $\Gamma_{\mathbb{Z},0}$ is a subset (and hence a subgroup) of $A(\Gamma)_0$. Moreover, we can also rewrite (6.8) as

$$\gamma = \frac{\lambda_1}{n}(n\gamma_1 - n_1\gamma_0) + \cdots + \frac{\lambda_\ell}{n}(n\gamma_\ell - n_\ell\gamma_0).$$

Since $n\gamma_i - n_i\gamma_0$ belongs to $\Gamma_{\mathbb{Z},0}$ for $i \in \{1, \ldots, \ell\}$, we obtain that $A(\Gamma)_0$ is generated by $\Gamma_{\mathbb{Z},0}$ as a vector space over $\mathbb{R}$. Moreover, since $\Gamma_{\mathbb{Z}}$ is a discrete subspace of $\mathbb{R}\times V$, the set $\Gamma_{\mathbb{Z},0} \subseteq V$ is also discrete. Hence it forms a lattice in $A(\Gamma)_0$.

(2) This comes from the definition of $\Gamma_{\mathbb{Z}}$. In particular, the inverse map is given by $(\gamma' \in \Gamma_{\mathbb{Z},n+n'}) \mapsto \gamma' - \gamma_0$.

(3) Let Θ be the family of all sub-semigroups of Γ which are finitely generated. The family of convex sets $\{\Delta(\Gamma')\}_{\Gamma'\in\Theta}$ is filtered. Let C be the union of all $\Delta(\Gamma')$, $\Gamma' \in \Theta$. By definition, the closure of C coincides with $\Delta(\Gamma)$. Therefore (by Proposition 6.3.1), the interior of $\Delta(\Gamma)$ relatively to $A(\Gamma)$ identifies with that of C, which is equal to $\bigcup_{\Gamma'\in\Theta}\Delta(\Gamma')^\circ$, where $\Delta(\Gamma')^\circ$ denotes the relative interior of $\Delta(\Gamma')$ in $A(\Gamma)$. Since K is a compact subset of $\Delta(\Gamma)^\circ$ and since the family $\{\Delta(\Gamma')^\circ\}_{\Gamma'\in\Theta}$ is filtered, there exists $\Gamma' \in \Theta$ such that $K \subseteq \Delta(\Gamma')^\circ$. Moreover, since $\Gamma_{\mathbb{Z}}$ is a discrete subgroup of $\mathbb{R}\times V$, it is actually finitely generated. Hence by possibly enlarging Γ' we may assume that $\Gamma'_{\mathbb{Z}} = \Gamma_{\mathbb{Z}}$. Therefore, without loss of generality, we may assume that the semigroup Γ is finitely generated.

Let $\{x_i\}_{i=1}^\ell$ be a system of generators of Γ, where $x_i = (n_i, \gamma_i)$. Then $\Delta(\Gamma)$ is just the convex hull of $n_i^{-1}\gamma_i$ $(i \in \{1, \ldots, \ell\})$. The set

$$F = \{\lambda_1 x_1 + \cdots + \lambda_\ell x_\ell \mid (\lambda_1, \ldots, \lambda_\ell) \in [0, 1]^\ell\}$$

is a compact subset of $\mathbb{R}\times V$. Therefore the intersection of F with $\Gamma_{\mathbb{Z}}$ is finite since $\Gamma_{\mathbb{Z}}$ is supposed to be discrete. In particular, there exists $x_0 = (n_0, \gamma_0) \in \Gamma$ such that $x_0 + y \in \Gamma$ for any $y \in F \cap \Gamma_{\mathbb{Z}}$. Let n be an element of $\mathbb{Z}(\Gamma)$, $n \geqslant 1$, and let $\gamma \in \Gamma_{\mathbb{Z},n}$. If $n^{-1}\gamma$ belongs to $\Delta(\Gamma)$, then there exists $(a_1, \ldots, a_\ell) \in \mathbb{R}_+^\ell$ such that $a_1 n_1 + \cdots + a_\ell n_\ell = n$ and that $\gamma = a_1\gamma_1 + \cdots + a_\ell\gamma_\ell$. Let $b_i = \lfloor a_i \rfloor$ and $\lambda_i = a_i - b_i$ for any $i \in \{1, \ldots, \ell\}$. We write $x = (n, \gamma)$ in the form $x = x' + y$ with

$$x' = b_1 x_1 + \cdots + b_\ell x_\ell \in \Gamma, \quad y = \lambda_1 x_1 + \cdots + \lambda_\ell x_\ell \in F.$$

Since $x \in \Gamma_{\mathbb{Z}}$, also is y. Hence $y \in F \cap \Gamma_{\mathbb{Z}}$. Thus $x + x_0 = x' + (y + x_0) \in \Gamma$. In particular, one has

$$\gamma + \gamma_0 \in \Gamma_{n+n_0}.$$

Now we introduce an arbitrary norm $\|\cdot\|$ on V. Since K is a compact subset of the relative interior of $\Delta(\Gamma)$, there exists $\epsilon > 0$ such that, for any $u \in K$, the ball

$$B(u, \epsilon) = \{u' \in W \;:\; \|u - u'\| \leqslant \epsilon\}$$

is contained in $\Delta(\Gamma)$. Moreover, the set K is bounded. Therefore, for sufficiently positive integer $n \in \mathbb{N}(\Gamma)$, if β is an element in $\Gamma_{\mathbb{Z},n} \cap nK$, then one has

$$(n - n_0)^{-1}(\beta - \gamma_0) \in \Delta(\Gamma),$$

which implies that $\beta \in \Gamma_n$ by the above argument. The equality (6.7) is thus proved.□

Definition 6.3.7 Let Γ be a graded semigroup in V such that $\Gamma_{\mathbb{Z}}$ is discrete. Let $A(\Gamma)_0$ be the vector subspace of V which is the translation of the affine subspace $A(\Gamma)$. We equip $A(\Gamma)_0$ with the normalised Lebesgue measure such that the mass of a fundamental domain of the lattice $\Gamma_{\mathbb{Z},0}$ in $A(\Gamma)_0$ is 1. This measure induces by translation a Borel measure on $A(\Gamma)$. We denote by η_Γ the restriction of this Borel measure to the closed convex set $\Delta(\Gamma)$, that is, for any function $f \in C_c(A(\Gamma))$ (namely f is continuous on $A(\Gamma)$ and of compact support), one has

$$\int_{A(\Gamma)} f(x)\,\eta_\Gamma(\mathrm{d}x) = \int_{\Delta(\Gamma)} f(\gamma)\,\mathrm{d}\gamma,$$

where $\mathrm{d}\gamma$ denotes the normalised Lebesgue measure.

The following theorem is the key point of the Newton-Okounkov body approach to the study of graded linear series [118,94,101]. Here we adopte the form presented in the Bourbaki seminar lecture of Boucksom [25].

Theorem 6.3.8 *Let Γ be a graded semigroup in V such that $\Gamma_{\mathbb{Z}}$ is discrete. For any integer $n \in \mathbb{N}(\Gamma)$, we denote by $\eta_{\Gamma,n}$ the Radon measure on $A(\Gamma)$ such that, for any function $f \in C_c(A(\Gamma))$ one has*

$$\int_{A(\Gamma)} f(x)\,\eta_{\Gamma,n}(\mathrm{d}x) = \frac{1}{n^\kappa}\sum_{\gamma\in\Gamma_n} f(n^{-1}\gamma),$$

where κ is the dimension of the affine space $A(\Gamma)$. Then the sequence of measures $\{\eta_{\Gamma,n}\}_{n\in\mathbb{N}(\Gamma)}$ converges vaguely (see §A.3) to the Radon measure η_Γ.

Proof Recall that the vague convergence in the statement of the theorem signifies that the sequence $\{\eta_{\Gamma,n}\}_{n\in\mathbb{N}(\Gamma)}$, viewed as a sequence of positive linear functionals on $C_c(A(\Gamma))$, converges pointwisely to η_Γ. In other words, for any continuous function f on $A(\Gamma)$ of compact support, one has

$$\lim_{n\in\mathbb{N}(\Gamma),\, n\rightarrow+\infty} \frac{1}{n^\kappa}\sum_{\gamma\in\Gamma_n} f(n^{-1}\gamma) = \int_{\Delta(\Gamma)} f(\gamma)\,\mathrm{d}\gamma. \tag{6.9}$$

Note that the direct image preserves the vague convergence. Therefore, it suffices to prove that, for any non-negative continuous function f on $\Delta(\Gamma)$ which is of compact support, the equality (6.9) holds.

For any $n \in \mathbb{N}(\Gamma)$ one has

$$\frac{1}{n^\kappa}\sum_{\gamma\in\Gamma_n} f(n^{-1}\gamma) \leqslant \frac{1}{n^\kappa}\sum_{\gamma\in\Gamma_{\mathbb{Z},n}\cap n\Delta(\Gamma)} f(n^{-1}\gamma).$$

Note that the right hand side of the inequality is the nth Riemann sum of the function f on the convex set $\Delta(\Gamma)$. Therefore one has

$$\lim_{n\in\mathbb{N}(\Gamma),\, n\to+\infty} \frac{1}{n^{\kappa}} \sum_{\gamma\in\Gamma_{\mathbb{Z},n}\cap n\Delta(\Gamma)} f(n^{-1}\gamma) = \int_{\Delta(\Gamma)} f(\gamma)\,\mathrm{d}\gamma,$$

which implies

$$\limsup_{n\in\mathbb{N}(\Gamma),\, n\to+\infty} \frac{1}{n^{\kappa}} \sum_{\gamma\in\Gamma_n} f(n^{-1}\gamma) \leqslant \int_{\Delta(\Gamma)} f(\gamma)\,\mathrm{d}\gamma.$$

Moreover, if g is a continuous function on $\Delta(\Gamma)$ whose support is contained in $\Delta(\Gamma)^{\circ}$ (the relative interior of $\Delta(\Gamma)$ in $A(\Gamma)$) and which is bounded from above by f, by Proposition 6.3.6 (3), for sufficiently positive n one has

$$\sum_{\gamma\in\Gamma_{\mathbb{Z},n}\cap n\Delta(\Gamma)} g(n^{-1}\gamma) = \sum_{\gamma\in\Gamma_n} g(n^{-1}\gamma).$$

Hence one has

$$\liminf_{n\in\mathbb{N}(\Gamma),\, n\to+\infty} \frac{1}{n^{\kappa}} \sum_{\gamma\in\Gamma_n} f(n^{-1}\gamma) \geqslant \lim_{n\in\mathbb{N}(\Gamma),\, n\to+\infty} \frac{1}{n^{\kappa}} \sum_{\gamma\in\Gamma_n} g(n^{-1}\gamma) = \int_{\Delta(\Gamma)^{\circ}} g(\gamma)\,\mathrm{d}\gamma.$$

Since the restriction of the function f to $\Delta(\Gamma)^{\circ}$ can be written as the limite of an increasing sequence of continuous functions with support contained in $\Delta(\Gamma)^{\circ}$, by the monotone convergence theorem, one has

$$\liminf_{n\in\mathbb{N}(\Gamma),\, n\to+\infty} \frac{1}{n^{\kappa}} \sum_{\gamma\in\Gamma_n} f(n^{-1}\gamma) \geqslant \int_{\Delta(\Gamma)^{\circ}} f(\gamma)\,\mathrm{d}\gamma.$$

Finally, since the border of $\Delta(\Gamma)$ has Lebesgue measure 0, we obtain the desired result. □

Definition 6.3.9 Let Γ be a graded semigroup in V. The dimension of the affine space $A(\Gamma)$ is called the *Kodaira dimension* of Γ.

Corollary 6.3.10 *We keep the notation and the hypotheses of Theorem 6.3.8. For any convex subset C of $\Delta(\Gamma)$ one has*

$$\lim_{n\in\mathbb{N}(\Gamma),\, n\to+\infty} \frac{\operatorname{card}(\Gamma_n\cap nC)}{n^{\kappa}} = \eta_{\Gamma}(C), \tag{6.10}$$

where κ is the Kodaira dimension of Γ.

Proof Let C° be the relative interior of C in $A(\Gamma)$. If C° is empty, then one has $\eta_{\Gamma}(C) = 0$. Moreover, for $n \in \mathbb{N}_{\geqslant 1}$, one has $\operatorname{card}(\Gamma_{n,\mathbb{Z}} \cap nC) = o(n^{\kappa})$ since $\Gamma_{n,\mathbb{Z}}$ is a translation of a lattice (see Proposition 6.3.6). Therefore, one has

$$\lim_{n\in\mathbb{N}(\Gamma),\, n\to+\infty} \frac{\operatorname{card}(\Gamma_n \cap nC)}{n^{\kappa}} = 0.$$

In the following, we assume that C° is not empty. Let K be a compact convex subset of C°. We can find a function $f \in C_c(A(\Gamma))$ with $0 \leqslant f \leqslant 1\!\!1_C$, $f|_K \equiv 1$. Then one has

$$\forall\, n \in \mathbb{N}(\Gamma),\quad \frac{\operatorname{card}(\Gamma_n \cap nC)}{n^{\kappa}} \geqslant \int_{A(\Gamma)} f(x)\, \eta_{\Gamma,n}(\mathrm{d}x),$$

which leads to (by Theorem 6.3.8)

$$\liminf_{n\in\mathbb{N}(\Gamma),\, n\to+\infty} \frac{\operatorname{card}(\Gamma_n \cap nC)}{n^{\kappa}} \geqslant \int_{A(\Gamma)} f(x)\, \eta_{\Gamma}(\mathrm{d}x) \geqslant \eta_{\Gamma}(K).$$

Since K is arbitrary, we obtain

$$\liminf_{n\in\mathbb{N}(\Gamma),\, n\to+\infty} \frac{\operatorname{card}(\Gamma_n \cap nC)}{n^{\kappa}} \geqslant \eta_{\Gamma}(C^{\circ}) = \eta_{\Gamma}(C).$$

In particular, if C is not bounded, then

$$\lim_{n\in\mathbb{N}(\Gamma),\, n\to+\infty} \frac{\operatorname{card}(\Gamma_n \cap nC)}{n^{\kappa}} = \eta_{\Gamma}(C) = +\infty.$$

In the following, we assume in addition that the convex set C is bounded. Denote by $\overline{C}$ the closure of the convex set C. It is a conex and compact subset of $A(\Gamma)$. Let K be a compact subset of $A(\Gamma)$ such that the relative interior of K contains $\overline{C}$. For any non-negative function $g \in C_c(A(\Gamma))$ with support contained K and such that $0 \leqslant g \leqslant 1$, $g|_C \equiv 1$, one has

$$\forall\, n \in \mathbb{N}(C),\quad \frac{\operatorname{card}(\Gamma_n \cap nC)}{n^{\kappa}} \leqslant \int_{A(\Gamma)} g(x)\, \eta_{\Gamma,n}(\mathrm{d}x).$$

By Theorem 6.3.8, we obtain

$$\limsup_{n\in\mathbb{N}(\Gamma),\, n\to+\infty} \frac{\operatorname{card}(\Gamma_n \cap nC)}{n^{\kappa}} \leqslant \int_{A(\Gamma)} g(x)\, \eta_{\Gamma}(\mathrm{d}x) \leqslant \eta_{\Gamma}(\mathrm{d}x) \leqslant \eta_{\Gamma}(K).$$

Since K is arbitrary, we obtain

$$\limsup_{n\to+\infty} \frac{\operatorname{card}(\Gamma_n \cap nC)}{n^{\kappa}} \leqslant \int_{A(\Gamma)} g(x)\, \eta_{\Gamma}(\mathrm{d}x) \leqslant \eta_{\Gamma}(C).$$

6.3.3 Concave transform

Let V be a finite-dimensional vector space over $\mathbb{R}$ and Γ be a graded semigroup in V such that $\Gamma_{\mathbb{Z}}$ is discrete. We suppose given a map $\delta : \mathbb{N}_{\geqslant 1} \to \mathbb{R}$ such that $\delta(n)/n$ tends to 0 when $n \to +\infty$.

Definition 6.3.11 Let $g : \Gamma \to \mathbb{R}$ be a function. We say that the function g is *strongly δ-superadditive* if for any $\ell \in \mathbb{N}_{\geqslant 2}$ and for all elements $(n_1, \gamma_1), \ldots, (n_\ell, \gamma_\ell)$ in Γ, one has

$$g(n_1 + \cdots + n_\ell, \gamma_1 + \cdots + \gamma_\ell) \geqslant \sum_{i=1}^{\ell} (g(n_i, \gamma_i) - \delta(n_i)). \tag{6.11}$$

The purpose of this subsection is to prove the following result.

Theorem 6.3.12 *Let Γ be a graded semigroup in V. We assume that $\Gamma_{\mathbb{Z}}$ is discrete and that $\Delta(\Gamma)$ is compact. Suppose given a function g on Γ which is strongly δ-superadditive for certain function $\delta : \mathbb{N}_{\geqslant 1} \to \mathbb{R}$ such that*

$$\lim_{n \to +\infty} \frac{\delta(n)}{n} = 0.$$

For any $n \in \mathbb{N}(\Gamma)$, let ν_n be the Borel probability measure on $\mathbb{R}$ given by

$$\forall\, f \in C_c(\mathbb{R}), \quad \int_{\mathbb{R}} f(t)\, \nu_n(\mathrm{d}t) = \frac{1}{\mathrm{card}(\Gamma_n)} \sum_{\gamma \in \Gamma_n} f\left(\tfrac{1}{n} g(n, \gamma)\right).$$

The the sequence of measures $\{\nu_n\}_{n \in \mathbb{N}(\Gamma)}$ converges vaguely to a Borel measure ν_Γ on $\mathbb{R}$. Moreover, ν_Γ is either the zero measure or a probability measure, and in the latter case the sequence $\{\nu_n\}_{n \in \mathbb{N}(\Gamma)}$ actually converges weakly to ν_Γ (see Theorem A.3.2) and there exists a concave function $G_\Gamma : \Delta(\Gamma)^\circ \to \mathbb{R}$ such that ν_Γ identifies with the direct image of

$$\frac{1}{\eta_\Gamma(\Delta(\Gamma))} \eta_\Gamma$$

by the map G_Γ.

Proof We introduce an auxiliary function $\widetilde{g}$ on Γ taking values in $\mathbb{R} \cup \{+\infty\}$ as follows:

$$\forall\, u \in \Gamma, \quad \widetilde{g}(u) = \limsup_{n \to +\infty} \frac{g(nu)}{n}. \tag{6.12}$$

Note that the sequence defining $\widetilde{g}(u)$ is bounded from below and hence the sup limit does not take the value $-\infty$. The proof of the theorem is decomposed into the following steps.

Step 1: The sup limit in the formula (6.12) *is actually a limit.* This follows from the following generalisation of Fekete's lemma (the case where $\delta(n) = 0$ for all n): let $\{a_n\}_{n \geqslant 1}$ be a sequence in $\mathbb{R}$ such that, for any $\ell \in \mathbb{N}_{\geqslant 2}$ and for all $n_1, \ldots, n_\ell$ in $\mathbb{N}_{\geqslant 1}$ one has

$$a_{n_1+\cdots+n_\ell} \geqslant \sum_{i=1}^{\ell}(a_{n_i} - \delta(n_i)),$$

then the sequence $\{a_n/n\}_{n\geqslant 1}$ converges in $\mathbb{R}\cup\{+\infty\}$. In fact, if p is an integer, $p \geqslant 1$ and if $m \in \mathbb{N}$, $r \in \{1, \ldots, p\}$ one has

$$a_{mp+r} \geqslant ma_p + a_r - m\delta(p) - \delta(r),$$

and hence

$$\frac{a_{mp+r}}{mp+r} \geqslant \frac{m}{mp+r}a_p + \frac{a_r}{mp+r} - \frac{m\delta(p)+\delta(r)}{mp+r}.$$

Therefore

$$\liminf_{n\to+\infty} \frac{a_n}{n} \geqslant \frac{a_p}{p} - \frac{\delta(p)}{p}.$$

In particular, $\liminf_{n\to+\infty} a_n/n \geqslant a_1 - \delta(1) > -\infty$. Moreover, this inequality also implies that

$$\liminf_{n\to+\infty} \frac{a_n}{n} \geqslant \limsup_{p\to+\infty}\left(\frac{a_p}{p} - \frac{\delta(p)}{p}\right) = \limsup_{p\to+\infty}\frac{a_p}{p},$$

which leads to the convergence of the sequence $\{a_n/n\}_{n\geqslant 1}$.

Step 2: Some properties of the function $\widetilde{g}$. Let $u_1 = (n_1, \gamma_1)$ and $u_2 = (n_2, \gamma_2)$ be two elements of Γ. For any $n \in \mathbb{N}_{\geqslant 1}$ one has

$$g(n(u_1+u_2)) \geqslant g(nu_1) + g(nu_2) - \delta(nn_1) - \delta(nn_2)$$

and hence

$$\frac{g(n(u_1+u_2))}{n} \geqslant \frac{g(nu_1)}{n} + \frac{g(nu_2)}{n} - \frac{\delta(nn_1)+\delta(nn_2)}{n}.$$

By taking the limit when $n \to +\infty$, we obtain $\widetilde{g}(u_1+u_2) \geqslant \widetilde{g}(u_1) + \widetilde{g}(u_2)$. In other words, the function $\widetilde{g}$ is superadditive.

Let (n, γ) be an element of Γ. Note that for any $N \in \mathbb{N}_{\geqslant 1}$ one has

$$\frac{g(Nn, N\gamma)}{N} \geqslant g(n,\gamma) - \delta(n).$$

By taking the limit when $N \to +\infty$, we obtain

$$\widetilde{g}(n,\gamma) \geqslant g(n,\gamma) - \delta(n). \tag{6.13}$$

Step 3: Construction of the function G_Γ. For any $t \in \mathbb{R}$, let Γ^t be the set of all $(n,\gamma) \in \Gamma$ such that $\widetilde{g}(n,\gamma) \geqslant nt$. It is actually a sub-semigroup of Γ since $\widetilde{g}$ is super-additive. Note that $\{\Gamma^t\}_{t\in\mathbb{R}}$ is a decreasing family of sub-semigroups of Γ and hence $\{\Delta(\Gamma^t)\}_{t\in\mathbb{R}}$ is a decreasing family of closed convex subsets of $\Delta(\Gamma)$. We define the function $G_\Gamma : \Delta(\Gamma) \to \mathbb{R}\cup\{+\infty\}$ as follows:

$$\forall\, x \in \Delta(\Gamma), \quad G_\Gamma(x) = \sup\{t \in \mathbb{R} \,:\, x \in \Delta(\Gamma^t)\}.$$

By definition, if t is a real number, then $G_\Gamma(x) \geqslant t$ if and only if $x \in \bigcap_{s<t} \Delta(\Gamma^s)$. We claim that the function G_Γ is concave. In fact, since the function $\widetilde{g}$ is super-additive, we obtain that, if s and t are two real numbers and if $\epsilon \in [0,1] \cap \mathbb{Q}$, for $u \in \Gamma^s$ and $v \in \Gamma^t$ one has

$$N(\epsilon u + (1-\epsilon)v) \in \Gamma^{\epsilon s+(1-\epsilon)t},$$

where N is an element in $\mathbb{N}_{\geqslant 1}$ such that $N\epsilon \in \mathbb{N}$. Therefore one has

$$\epsilon\Delta(\Gamma^s) + (1-\epsilon)\Delta(\Gamma^t) \subseteq \Delta(\Gamma^{\epsilon s+(1-\epsilon)t}).$$

In general, if we choose a sequence $\{\epsilon_n\}$ of rational numbers such that $\lim_{n\to\infty} \epsilon_n = \epsilon$ and $\epsilon_n s + (1-\epsilon_n)t \geqslant \epsilon s + (1-\epsilon)t$ for all n, then

$$\epsilon_n\Delta(\Gamma^s) + (1-\epsilon_n)\Delta(\Gamma^t) \subseteq \Delta(\Gamma^{\epsilon_n s+(1-\epsilon_n)t}) \subseteq \Delta(\Gamma^{\epsilon s+(1-\epsilon)t}),$$

and hence $\epsilon\Delta(\Gamma^s) + (1-\epsilon)\Delta(\Gamma^t) \subseteq \Delta(\Gamma^{\epsilon s+(1-\epsilon)t})$. Combining with the definition of the function G_Γ, we obtain the concavity of G_Γ. In particular, the restriction of the function G_Γ to $\Delta(\Gamma)^\circ$ is either finite or identically $+\infty$, and it is a continuous function on $\Delta(\Gamma)^\circ$ when it is finite.

Step 4. Abundance of $\Gamma^t_{\mathbb{Z}}$. Let t be an element of $\mathbb{R}$ such that $t < \sup_{x\in\Delta(\Gamma)} G_\Gamma(x)$. We will prove that $\Gamma^t_{\mathbb{Z}} = \Gamma_{\mathbb{Z}}$ (and hence $A(\Gamma^t) = A(\Gamma)$). Note that $\Gamma_{\mathbb{Z}}$ is finitely generated because $\Gamma_{\mathbb{Z}}$ is discrete. Let $u_i = (n_i, \gamma_i)$, $i \in \{1, \ldots, \ell\}$ be a family of elements in Γ which forms a system of generators in $\Gamma_{\mathbb{Z}}$. Since $t < \sup_{x\in\Delta(\Gamma)} G_\Gamma(x)$, there exists $\epsilon > 0$ such that $\Gamma^{t+\epsilon}$ is not empty. Let $u_0 = (n_0, \gamma_0)$ be an element in $\Gamma^{t+\epsilon}$. By definition, one has $\widetilde{g}(u_0) \geqslant n_0(t+\epsilon)$. Therefore, for sufficiently positive integer p, one has

$$\forall\, i \in \{1, \ldots, \ell\}, \quad \widetilde{g}(pu_0 + u_i) \geqslant p\widetilde{g}(u_0) + \widetilde{g}(u_i) \geqslant (pn_0 + n_i)t,$$

namely $pu_0 + u_i \in \Gamma^t$ for any $i \in \{1, \ldots, \ell\}$, which leads to $\Gamma^t_{\mathbb{Z}} = \Gamma_{\mathbb{Z}}$.

Step 5: Lower bound of the function g. We fix a (closed) fundamental domain F of the lattice $\Gamma_{\mathbb{Z},0}$ (see Proposition 6.3.6 (1)). For $n \in \mathbb{N}(\Gamma)$, we call an *n-cell* in $A(\Gamma)_n$ any closed convex subset of $A(\Gamma)_n$ of the form $\gamma_0 + F$, where γ_0 is an element in $\Gamma_{\mathbb{Z},n}$. We say that a compact subset K of $A(\Gamma)$ is *n-tileable* if it can be written as a union of n-cells in $A(\Gamma)_n$. Note that, if K is n-tileable, then, for any integer $p \geqslant 1$, the set pK is pn-tileable since pF can be written as the union of p^κ 0-celles.

Let t be a real number such that $t < \sup_{x\in\Delta(\Gamma)} G_\Gamma(x)$, and ϵ be a positive real number. Let $m \geqslant 1$ be the generator of the group $\mathbb{Z}(\Gamma)$. Suppose given a compact subset K of $\Delta(\Gamma^t)^\circ$. We assume that there exists an integer $n \in \mathbb{N}(\Gamma)$ such that nK is n-tileable.

By Proposition 6.3.6 (3), there exists an integer $n_0 \in \mathbb{N}_{\geqslant 1}$ which verifies the following conditions (in the condition (2) we also use the result of Step 4 to identify $\Gamma_{mn_0,\mathbb{Z}}$ with $\Gamma^t_{mn_0,\mathbb{Z}}$):

(1) mn_0K is mn_0-tileable;
(2) $mnK \cap \Gamma_{mn,\mathbb{Z}} \subseteq \Gamma^t_{mn}$ for any $n \in \mathbb{N}$, $n \geqslant n_0$;
(3) for any integer $q \geqslant mn_0$, $\delta(q)/q < \epsilon/3$.

For simplifying the notation, in the following we denote by Θ the set $mn_0K \cap \Gamma_{mn_0,\mathbb{Z}}$. Note that the condition (2) implies that $\widetilde{g}(mn_0, \gamma) \geqslant t$ for any $\gamma \in \Theta$. Therefore by the definition of the function $\widetilde{g}$ and the finiteness of the set Θ, we obtain that there exists an integer N_0 divisible by n_0 such that

$$\frac{1}{mN_0} g(mN_0, (N_0/n_0)\gamma) \geqslant t - \frac{\epsilon}{3} \tag{6.14}$$

for any $\gamma \in \Theta$.

Let N be an integer, $N \geqslant n_0$. Let α be an element in $mNK \cap \Gamma_{mN}$ and $x = (n_0/N)\alpha$. Since mn_0K is mn_0-tileable, there exists an mn_0-cell C such that x belongs to C. We write C as $\gamma_0 + F$ with $\gamma_0 \in \Gamma_{mn_0}$. Let $\{e_1, \ldots, e_\kappa\}$ be the basis of $\Gamma_{\mathbb{Z}}$ defining the fundamental domain F. Then the point x can be written in a unique way as

$$x = \gamma_0 + \sum_{i=1}^{\kappa} \lambda_i e_i,$$

where

$$\forall\, i \in \{1, \ldots, \kappa\}, \quad \lambda_i \in [0, 1].$$

Moreover, since $N(x - \gamma_0) = n_0\alpha - N\gamma_0 \in \Gamma_{0,\mathbb{Z}}$, we obtain that $N\lambda_i \in \mathbb{Z}$ for any $i \in \{1, \ldots, \kappa\}$. Without loss of generality, we may assume that $\lambda_1 \geqslant \ldots \geqslant \lambda_\kappa$. Then we can rewrite x as

$$x = \sum_{i=0}^{\kappa} (\lambda_i - \lambda_{i+1})\gamma_i,$$

where by convention $\lambda_0 = 1$, $\lambda_{\kappa+1} = 0$, and for $i \in \{1, \ldots, \kappa\}$, $\gamma_i = \gamma_0 + e_1 + \cdots + e_i$. Note that $\gamma_0, \ldots \gamma_\kappa$ are vertices of the mn_0-cell C, hence belong to Θ. For any $i \in \{0, \ldots, \kappa\}$, let b_i be the integral part of

$$\frac{N}{n_0}(\lambda_i - \lambda_{i+1}).$$

One has

$$N - (\kappa + 1)n_0 + 1 \leqslant n_0 \sum_{i=0}^{\kappa} b_i \leqslant N.$$

Therefore, we can write α as

$$\alpha = \sum_{i=0}^{\kappa} b_i \gamma_i + \beta',$$

where $\beta \in \Gamma_{mr',\mathbb{Z}} \cap mr'K$, with

$$r' = N - n_0 \sum_{i=0}^{\kappa} b_i \in \{0, \ldots, (\kappa + 1)n_0 - 1\}.$$

Note that we have assumed that $N \geqslant n_0$. Therefore, if $r' \leqslant n_0 - 1$, then there exists at least an indice b_i which is > 0. In this case, we replace β' by $\beta' + \gamma_i$ and b_i by $b_i - 1$. Thus we obtain the existence of a decomposition of α into the form

$$\alpha = \sum_{i=0}^{\kappa} a_i \gamma_i + \beta$$

with $a_i \in \mathbb{N}$ for $i \in \{0, \ldots, \kappa\}$, and $\beta \in \Gamma_{mr,\mathbb{Z}} \cap mrK$ with

$$r \in \{n_0, \ldots, (\kappa + 1)n_0 - 1\}.$$

The advantage of the new decomposition is that β actually belongs to Γ_{mr} (see the condition (2) above). Finally, we write each a_i in the form $a_i = p_i N_0/n_0 + r_i$ with $p_i \in \mathbb{N}$ and $r_i \in \{0, \ldots, N_0/n_0 - 1\}$. Then we can decompose α as

$$\alpha = \sum_{i=0}^{\kappa} p_i (N_0/n_0)\gamma_i + \omega,$$

where

$$\omega = \beta + \sum_{i=0}^{\kappa} r_i \gamma_i.$$

The element ω belongs to certain $\Gamma_{ms} \cap msK$ with

$$s \in \{n_0, \ldots, (\kappa + 1)N_0 - 1\}.$$

Hence by (6.11) and (6.14) one obtains

$$\begin{aligned}
&\frac{g(mN, \alpha)}{mN} \\
&\geqslant \frac{1}{mN}\bigg(\sum_{i=0}^{\kappa} p_i g(mN_0, (N_0/n_0)\gamma_i) + g(ms, \omega) - \delta(mN_0)\sum_{i=0}^{\kappa} p_i - \delta(ms)\bigg) \qquad (6.15)\\
&\geqslant \frac{N_0 P}{N}(t - \epsilon/3) + \frac{g(ms, \omega)}{mN} - \frac{P}{mN}\delta(mN_0) - \frac{\delta(ms)}{N},
\end{aligned}$$

where

$$P = p_0 + \cdots + p_\kappa = \frac{N - s}{N_0}.$$

Therefore we obtain

$$\liminf_{N \to +\infty} \inf_{\alpha \in mNK \cap \Gamma_{mN}} \frac{g(mN, \alpha)}{mN} \geqslant t - \frac{2\epsilon}{3},$$

where we have used the condition (3) above to obtain

$$\frac{P\delta(mN_0)}{mN} = \frac{N - s}{N} \cdot \frac{\delta(mN_0)}{mN_0} \leqslant \frac{N - s}{N} \cdot \frac{\epsilon}{3}.$$

Therefore, there exists an integer N' depending on t, ϵ and K such that $g(mN,\alpha) \geqslant mN(t-\epsilon)$ for any $N \geqslant N'$ and any $\alpha \in \Gamma_{mN} \cap mNK$.

Step 6: Convergence of measures. We now proceed with the proof of the convergence of the measures. We first consider the case where G_Γ is identically $+\infty$ on the interior of $\Delta(\Gamma)$. Let f be a non-negative continuous function with compact support on $\mathbb{R}$ and $t_0 \in \mathbb{R}$ be a real number which is larger than the supremum of the support of the function f. Let K be a compact subset of $\Delta(\Gamma)^\circ$. By the results in Step 5, we obtain that, there exists $n_0 \in \mathbb{N}$ such that, for any $n \in \mathbb{N}(\Gamma)$, $n \geqslant n_0$ and any $\alpha \in \Gamma_n \cap nK$, one has $g(N,\alpha) \geqslant nt_0$. Hence

$$\int_{\mathbb{R}} f(t)\,\nu_n(\mathrm{d}t) \leqslant \Big(\frac{\mathrm{card}(\Gamma_n \setminus nK)}{\mathrm{card}(\Gamma_n)}\Big)M = \Big(1-\frac{\mathrm{card}(\Gamma_n \cap nK)}{\mathrm{card}(\Gamma_n)}\Big)M,$$

where $M = \sup_{t\in\mathbb{R}} f(t)$. By Corollary 6.3.10, one has

$$\lim_{n\in\mathbb{N}(\Gamma),\, n\to+\infty} \frac{\mathrm{card}(\Gamma_n \cap nK)}{\mathrm{card}(\Gamma_n)} = \frac{\eta_\Gamma(K)}{\eta_\Gamma(\Delta(\Gamma))}.$$

Since K is arbitrary, we obtain

$$\lim_{n\in\mathbb{N}(\Gamma),\, n\to+\infty} \int_{\mathbb{R}} f(t)\,\nu_n(\mathrm{d}t) = 0.$$

In the following, we assume that the function G_Γ is finite. In this case, the direct image ν_Γ of $\eta_\Gamma(\Delta(\Gamma))^{-1}\eta_\Gamma$ by G_Γ is a Borel probability measure on $\mathbb{R}$. We denote by F its probability distribution function, namely

$$\forall\, t \in \mathbb{R},\ F(t) = \nu_\Gamma(]-\infty,t]) = 1 - \frac{\eta_\Gamma(\Delta(\Gamma^t))}{\eta_\Gamma(\Delta(\Gamma))}.$$

By Corollary 6.3.10, one has

$$F(t) = 1 - \lim_{n\in\mathbb{N}(\Gamma),\, n\to+\infty} \frac{\mathrm{card}(\Gamma_n \cap n\Delta(\Gamma^t))}{\mathrm{card}(\Gamma_n)}. \tag{6.16}$$

The function F is continuous on $\mathbb{R}$, except possibly at the point $\sup_{x\in\Delta(\Gamma)} G_\Gamma(x)$ (the discontinuity of the function F happens precisely when the function G_Γ is constant on $\Delta(\Gamma)^\circ$). For any $n \in \mathbb{N}(\Gamma)$, let F_n be the probability distribution function of ν_n.

If (n,γ) is an element of Γ, then one has

$$G(n^{-1}\gamma) \geqslant \frac{\widetilde{g}(n,\gamma)}{n} \geqslant \frac{g(n,\gamma)}{n} - \frac{\delta(n)}{n},$$

where the second inequality comes from (6.13). Therefore we obtain

$$\forall\, t \in \mathbb{R},\ \{(n,\gamma)\in\Gamma\ :\ G(n^{-1}\gamma) > t - \delta(n)/n\} \supseteq \{(n,\gamma)\in\Gamma\ :\ g(n,\gamma)/n > t\},$$

which implies (by (6.16))

$$\forall \epsilon > 0, \quad 1 - \liminf_{n \in \mathbb{N}(\Gamma),\, n \to +\infty} F_n(t) \leqslant 1 - F(t - \epsilon). \tag{6.17}$$

Conversely, for any $t \in \mathbb{R}$, any $\epsilon > 0$ and any compact subset K of $\Delta(\Gamma^{t+\epsilon})^\circ$ (the relative interior of $\Delta(\Gamma^{t+\epsilon})$ with respect to $A(\Gamma)$), by the result obtained in Step 5, we obtain that, there exists $N_0 \in \mathbb{N}$ such that, for any $n \in \mathbb{N}(\Gamma)$, $n \geqslant N_0$, one has

$$\forall \gamma \in \Gamma_n \cap nK, \quad g(n, \gamma) \geqslant nt.$$

Therefore, we get

$$\forall \epsilon > 0, \quad 1 - \limsup_{n \in \mathbb{N}(\Gamma),\, n \to +\infty} F_n(t) \geqslant 1 - F(t + \epsilon). \tag{6.18}$$

The estimates (6.17) and (6.18) lead to the convergence of $\{F_n(t)\}_{n \in \mathbb{N}}$ to $F(t)$ if $t \in \mathbb{R}$ is a point of continuity of the function F, which implies the weak convergence of the sequence $\{\nu_n\}_{n \in \mathbb{N}(\Gamma)}$ to ν_Γ (see [121, §I.4] for more details about weak convergence of Borel probability measures on $\mathbb{R}$). □

Definition 6.3.13 Let Γ be a graded semigroup in V such that $\Gamma_{\mathbb{Z}}$ is discrete, and $g : \Gamma \to \mathbb{R}$ and $\delta : \mathbb{N}_{\geqslant 1} \to \mathbb{R}$ be functions. We say that the function g is δ-*superadditive* if for all elements (n_1, γ_1) and (n_2, γ_2) in Γ, one has

$$g(n_1 + n_2, \gamma_1 + \gamma_2) \geqslant g(n_1, \gamma_1) + g(n_2, \gamma_2) - \delta(n_1) - \delta(n_2). \tag{6.19}$$

Lemma 6.3.14 *Let $\delta : \mathbb{N}_{\geqslant 1} \to \mathbb{R}_{\geqslant 0}$ be an increasing function such that*

$$\sum_{a \in \mathbb{N}} \frac{\delta(2^a)}{2^a} < +\infty.$$

Then one has

$$\lim_{n \to +\infty} \frac{\delta(n)}{n} = 0 \tag{6.20}$$

and

$$\lim_{a \to +\infty} \frac{1}{2^a} \sum_{i=0}^{a} \delta(2^i) = 0. \tag{6.21}$$

Proof For $n \in \mathbb{N}_{\geqslant 1}$, let $a(n) = \lfloor \log_2 n \rfloor$. One has $2^{a(n)} \leqslant n < 2^{a(n)+1}$. Hence

$$\frac{\delta(n)}{n} \leqslant \frac{\delta(2^{a(n)+1})}{2^{a(n)}}.$$

By the hypothesis of the lemma, one has

$$\lim_{n \to +\infty} \frac{\delta(2^{a(n)+1})}{2^{a(n)+1}} = 0,$$

which implies (6.20).

For any $a \in \mathbb{N}$, let

$$S_a := \sum_{i\in\mathbb{N},\, i\geqslant a} \frac{\delta(2^i)}{2^i}.$$

By Abel's summation formula, one has

$$\sum_{i=0}^{a} \delta(2^i) = \sum_{i=0}^{a} (S_i - S_{i+1})2^i = S_0 - S_{a+1}2^a + \sum_{i=1}^{a} S_i 2^{i-1}.$$

As the sequence $\{S_a\}_{a\in\mathbb{N}}$ converges to 0, one has

$$\lim_{a\to+\infty} \frac{1}{2^a}\sum_{i=1}^{a} S_i 2^{i-1} = 0,$$

which implies the relation (6.21). □

Proposition 6.3.15 *Let $\delta : \mathbb{N}_{\geqslant 1} \to \mathbb{R}_{\geqslant 0}$ be an increasing function such that*

$$\sum_{a\in\mathbb{N}_{\geqslant 1}} \frac{\delta(2^a)}{2^a} < +\infty. \tag{6.22}$$

Let $\{b_n\}_{n\in\mathbb{N}}$ be a sequence of real numbers. We assume that there exists an integer $n_0 > 0$ such that, for any couple (n, m) of integers which are $\geqslant n_0$, one has

$$b_{n+m} \geqslant b_n + b_m - \delta(n) - \delta(m). \tag{6.23}$$

Then the sequence $\{b_n/n\}_{n\in\mathbb{N}_{\geqslant 1}}$ converges in $\mathbb{R}\cup\{+\infty\}$.

Proof We first treat the case where $n_0 = 1$. For any $n \in \mathbb{N}_{\geqslant 1}$ one has

$$b_{2n} \geqslant 2b_n - 2\delta(n),$$

and hence by induction we obtain that

$$b_{2^a n} \geqslant 2^a\bigg(b_n - \sum_{i=0}^{a-1}\frac{\delta(2^i n)}{2^i}\bigg). \tag{6.24}$$

In particular, one has

$$\frac{b_{2^a}}{2^a} \geqslant b_1 - \sum_{i=0}^{a-1}\frac{\delta(2^i)}{2^i},$$

which implies that

$$\limsup_{n\to+\infty} \frac{b_n}{n} > -\infty.$$

For any $a \in \mathbb{N}$, let

$$S_a = \sum_{i\in\mathbb{N},\, i\geqslant a} \frac{\delta(2^i)}{2^i}.$$

By the hypothesis (6.22), we have

$$\lim_{a\to+\infty} S_a = 0. \tag{6.25}$$

Let $n \in \mathbb{N}_{\geqslant 1}$ and let $a(n)$ be the unique natural number such that $2^{a(n)} \leqslant n < 2^{a(n)+1}$, namely $a(n) = \lfloor \log_2 n \rfloor$. Let p be an element in $\mathbb{N}_{\geqslant 1}$, which is written in 2-adic basis as

$$p = \sum_{i=0}^{\kappa} \epsilon_i 2^i$$

with $\epsilon_i \in \{0, 1\}$ for $i \in \{0, \dots, \kappa\}$ and $\epsilon_\kappa = 1$. For any $r \in \{0, \dots, n-1\}$, by (6.23) one has

$$b_{np+r} \geqslant b_{np} + b_r - \delta(np) - \delta(r) \geqslant b_{np} + b_r - 2\delta(np). \tag{6.26}$$

Moreover, by induction on κ one has

$$\begin{aligned}
b_{np} &\geqslant \sum_{i=0}^{\kappa} \epsilon_i b_{2^i n} - 2\sum_{i=1}^{\kappa} \epsilon_i \delta(2^i n) \\
&\geqslant \sum_{i=0}^{\kappa} \epsilon_i 2^i b_n - \sum_{i=1}^{\kappa} \epsilon_i \bigg(2^i \sum_{j=0}^{i-1} \frac{\delta(2^j n)}{2^j} + 2\delta(2^i n) \bigg) \\
&\geqslant p b_n - 2p \sum_{j=0}^{\kappa} \frac{\delta(2^j n)}{2^j}.
\end{aligned}$$

Since $n \geqslant 2^{a(n)}$ we deduce that

$$b_{np} \geqslant p b_n - p 2^{\beta+1} S_{a(n)} \geqslant p b_n - pn S_{a(n)} \tag{6.27}$$

Combining (6.26) and (6.27), we obtain

$$\frac{b_{np+r}}{np+r} \geqslant \frac{p b_n + b_r}{np+r} - \frac{np}{np+r} S_{a(n)} - 2\frac{\delta(np)}{np+r}.$$

Taking the infimum limit when $p \to +\infty$, by (6.20) we obtain

$$\liminf_{m\to+\infty} \frac{b_m}{m} \geqslant \frac{b_n}{n} - S_{a(n)},$$

which implies, by (6.25), that

$$\liminf_{m\to+\infty} \frac{b_m}{m} \geqslant \limsup_{n\to+\infty} \frac{b_n}{n}.$$

Therefore the sequence $\{b_n/n\}_{n\in\mathbb{N}_{\geqslant 1}}$ converges in $\mathbb{R} \cup \{+\infty\}$.

For the general case, we apply the obtained result to the sequence $\{b_{n_0 k}\}_{k\in\mathbb{N}_{\geqslant 1}}$ and obtain the convergence of the sequence $\{b_{n_0 k}/k\}_{k\in\mathbb{N}_{\geqslant 1}}$. Moreover, if ℓ is an element in $\{n_0, \dots, 2n_0 - 1\}$, then for any $k \in \mathbb{N}_{\geqslant 1}$ one has

$$b_{n_0(k+2)} - b_{2n_0-\ell} + \delta(n_0k+\ell) + \delta(2n_0-\ell) \geqslant b_{n_0k+\ell} \geqslant b_{n_0k} + b_\ell - \delta(n_0k) - \delta(\ell).$$

Dividing this formula by $n_0k+\ell$ and taking the limit when $k \longrightarrow +\infty$, we obtain

$$\lim_{k\to+\infty} \frac{b_{n_0k+\ell}}{n_0k+\ell} = \lim_{k\to+\infty} \frac{b_{n_0k}}{n_0k}.$$

Since ℓ is arbitrary, we obtain the statement announced in the proposition. □

Theorem 6.3.16 *Let Γ be a graded semigroup in V. We assume that $\Gamma_{\mathbb{Z}}$ is discrete and that $\Delta(\Gamma)$ is compact. Suppose given a function g on Γ which is δ-superadditive for certain increasing function $\delta : \mathbb{N}_{\geqslant 1} \to \mathbb{R}$ such that*

$$\sum_{a\in\mathbb{N}} \frac{\delta(2^a)}{2^a} < +\infty. \tag{6.28}$$

For any $n \in \mathbb{N}(\Gamma)$, let ν_n be the Borel probability measure on $\mathbb{R}$ such that

$$\int_{\mathbb{R}} f(t)\,\nu_n(\mathrm{d}t) = \frac{1}{\#\Gamma_n} \sum_{\gamma\in\Gamma_n} f(n^{-1}g(n,\gamma)).$$

The the sequence of measures $\{\nu_n\}_{n\in\mathbb{N}(\Gamma)}$ converges vaguely to a Borel measure ν_Γ on $\mathbb{R}$. Moreover, ν_Γ is either the zero measure or a probability measure, and in the latter case the sequence $\{\nu_n\}_{n\in\mathbb{N}(\Gamma)}$ actually converges weakly to ν_Γ and there exists a concave function $G_\Gamma : \Delta(\Gamma)^\circ \to \mathbb{R}$, called concave transform *of g, such that ν_Γ identifies with the direct image of*

$$\frac{1}{\eta_\Gamma(\Delta(\Gamma))}\eta_\Gamma$$

by the map G_Γ.

Proof The proof is very similar to that of Theorem 6.3.12. We will sketch it in emphasising the difference. Let $u = (\ell, \gamma)$ be an element in Γ, where $\ell \geqslant 1$. Since the function g is δ-superadditive, for any pair $(n,m) \in \mathbb{N}_{\geqslant 1}$ one has

$$g((n+m)u) \geqslant g(nu) + g(mu) - \delta(n\ell) - \delta(m\ell).$$

Moreover, if we let b be an integer such that $\ell \leqslant 2^b$, then, by the increasing property of the function δ, one has

$$\sum_{a\in\mathbb{N}} \frac{\delta(2^a\ell)}{2^a} \leqslant 2^b \sum_{a\in\mathbb{N}} \frac{\delta(2^{a+b})}{2^{a+b}} < +\infty.$$

By Proposition 6.3.15, for any $u \in \Gamma$, the sequence $\{g(nu)/n\}_{n\in\mathbb{N}_{\geqslant 1}}$ converges in $\mathbb{R}\cup\{+\infty\}$. We denote by $\widetilde{g}(u)$ the limite of the sequence. Moreover, the convergence of the series $\sum_{a\in\mathbb{N}} \delta(2^a)/2^a$ implies that

$$\lim_{a\to+\infty}\frac{\delta(2^a)}{2^a}=0.$$

Still by the hypothesis that the function $\delta(\cdot)$ is increasing, we deduce that

$$\lim_{n\to+\infty}\frac{\delta(n)}{n}=0.$$

Therefore, by the same argument as in the Step 2 of the proof of Theorem 6.3.12, we obtain that the function $\widetilde{g}$ is superadditive, namely, for any pair u_1, u_2 of elements in Γ, one has

$$\widetilde{g}(u_1+u_2)\geqslant\widetilde{g}(u_1)+\widetilde{g}(u_2).$$

Moreover, for any $(n,\gamma)\in\Gamma$ and any $a\in\mathbb{N}_{\geqslant 1}$ one has

$$g(2^a n, 2^a\gamma)\geqslant 2^a g(n,\gamma)-\sum_{i=0}^{a-1}2^{a-i}\delta(2^i n). \tag{6.29}$$

Let $b(n)=\lfloor\log_2 n\rfloor+1$. One has $2^{b(n)-1}\leqslant n<2^{b(n)}$. Let

$$R(n)=2^{b(n)}\sum_{i=b(n)}^{+\infty}\frac{\delta(2^i)}{2^i}.$$

Note that one has

$$\lim_{n\to+\infty}\frac{R(n)}{n}=0$$

by the hypothesis (6.28). By the increasing property of the function δ one has

$$\sum_{i=0}^{a-1}2^{a-i}\delta(2^i n)\leqslant 2^{a+b(n)}\sum_{i=0}^{a-1}\frac{\delta(2^{i+b(n)})}{2^{i+b(n)}}\leqslant 2^a R(n).$$

Therefore the inequality (6.29) leads to

$$\widetilde{g}(n,\gamma)\geqslant g(n,\gamma)-R(n).$$

We then proceed as in the Steps 3-6 of the proof of Theorem 6.3.12, except that in the counterpart of the minoration (6.15) we need more elaborated estimate as in (6.27). □

Remark 6.3.17 We keep the notation of the proof of Theorems 6.3.12 and 6.3.16. By virtue of [26, Lemma 1.6] (see also [43]), we obtain that, for any real number t such that

$$t<\lim_{n\in\mathbb{N}(\Gamma),\,n\to+\infty}\max_{\gamma\in\Gamma_n}\frac{\widetilde{g}(n,\gamma)}{n}=\lim_{n\in\mathbb{N}(\Gamma),\,n\to+\infty}\max_{\gamma\in\Gamma_n}\frac{g(n,\gamma)}{n},$$

the set $\{x\in\Delta(\Gamma)^\circ\,:\,G(x)\geqslant t\}$ has a positive measure with respect to η_Γ (and hence is not empty). In particular, we obtain

$$\sup_{x\in\Delta(\Gamma)^\circ} G(x) = \lim_{n\in\mathbb{N}(\Gamma),\, n\to+\infty} \max_{\gamma\in\Gamma_n} \frac{g(n,\gamma)}{n} \tag{6.30}$$

6.3.4 Applications to the study of graded algebras

Let $d \geqslant 1$ be an integer. We call *monomial order* on $\mathbb{Z}^d$ any total order $\leqslant$ on $\mathbb{Z}^d$ such that $0 \leqslant \alpha$ for any $\alpha \in \mathbb{N}^d$ and that $\alpha \leqslant \alpha'$ implies $\alpha + \beta \leqslant \alpha' + \beta$ for all α, α' and β in $\mathbb{Z}^d$. For example, the lexicographic order on $\mathbb{Z}^d$ is a monomial order.

Given a monomial order $\leqslant$ on $\mathbb{Z}^d$, we construct a $\mathbb{Z}^d$-valuation

$$v : k[\![T_1,\ldots,T_d]\!] \longrightarrow \mathbb{Z}^d \cup \{\infty\}$$

as follows. For any $\alpha = (a_1,\ldots,a_d) \in \mathbb{N}^d$ we denote by T^α the monomial $T_1^{a_1}\cdots T_d^{a_d}$. For any formal series F written as

$$F(T_1,\ldots,T_d) = \lambda_\alpha T^\alpha + \sum_{\alpha<\beta} \lambda_\beta T^\beta, \quad \lambda_\alpha \neq 0,$$

we let $v(F) := \alpha$. If $F = 0$ is the zero formal series, let $v(0) = \infty$. It is easy to check that the map v satisfies the following axioms of valuation : for any $(F,G) \in k[\![T_1,\ldots,T_d]\!]^2$, one has $v(FG) = v(F) + v(G)$ and $v(F+G) \geqslant \min(v(F), v(G))$, and the equality $v(F+G) = \min(v(F), v(G))$ holds when $v(F) \neq v(G)$. In particular, if we denote by R the fraction field of $k[\![T_1,\ldots,T_d]\!]$, then the map $v : k[\![T_1,\ldots,T_d]\!] \to \mathbb{Z}^d \cup \{\infty\}$ extends to a map $v : R \to \mathbb{Z}^d \cup \{\infty\}$ such that, for any $(F,G) \in k[\![T_1,\ldots,T_d]\!]$ with $G \neq 0$, one has $v(F/G) = v(F) - v(G)$. The valuation map $v : R \to \mathbb{Z}^d \cup \{\infty\}$ allows to define a $\mathbb{Z}^d$-filtration $\mathcal{G}$ of R as follows

$$\forall\, \alpha \in \mathbb{Z}^d, \quad \mathcal{G}_{\geqslant\alpha}(R) := \{f \in R \,:\, v(f) \geqslant \alpha\}.$$

Note that for $(f,g) \in R^2$ one has $v(fg) = v(f) + v(g)$ and $v(f+g) \geqslant \min(v(f), v(g))$. Therefore, for $(\alpha,\beta) \in \mathbb{Z}^d \times \mathbb{Z}^d$ one has

$$\mathcal{G}_{\geqslant\alpha}(R) \cdot \mathcal{G}_{\geqslant\beta}(R) \subseteq \mathcal{G}_{\geqslant\alpha+\beta}(R). \tag{6.31}$$

For any $\alpha \in \mathbb{Z}^d$, we let

$$\mathcal{G}_{>\alpha}(R) := \{f \in R \,:\, v(f) > \alpha\} \text{ and } \mathrm{gr}_\alpha(R) := \mathcal{G}_{\geqslant\alpha}(R)/\mathcal{G}_{>\alpha}(R).$$

The relation (6.31) shows that the k-algebra structure on R induces by passing to graduation a k-algebra structure on

$$\mathrm{gr}(R) := \bigoplus_{\alpha\in\mathbb{Z}} \mathrm{gr}_\alpha(R)$$

so that $\mathrm{gr}(R)$ is isomorphic to the group algebra $k[\mathbb{Z}^d]$.

Let $V_\bullet = \bigoplus_{n\in\mathbb{N}} V_n$ be a graded sub-k-algebra of the polynomial ring $R[T]$ (viewed as a graded k-algebra with the grading by the degree on T). The filtration $\mathcal{G}$ on R induces an $\mathbb{Z}^d$-filtration on each homogeneous component V_n. The direct sum of subquotients of V_n form an $\mathbb{N}\times\mathbb{Z}^d$-graded sub-$k$-algebra

$$\mathrm{gr}(V_\bullet) = \bigoplus_{(n,\alpha)\in\mathbb{N}\times\mathbb{Z}^d} \mathrm{gr}_{(n,\alpha)}(V_\bullet)$$

of $\mathrm{gr}(R)[T] \cong k[\mathbb{N}\times\mathbb{Z}^d]$. In particular, $\mathrm{gr}(V_\bullet)$ is an integral ring, and each homogeneous component $\mathrm{gr}_{(n,\alpha)}(V_\bullet)$ is either zero or a k-vector space of dimension 1. In particular, the set

$$\Gamma(V_\bullet) := \{(n,\alpha)\in\mathbb{N}_{\geqslant 1}\times\mathbb{Z}^d \;:\; \mathrm{gr}_{(n,\alpha)}(V_\bullet)\neq\{0\}\}$$

is a sub-semigroup of $\mathbb{N}\times\mathbb{Z}^d$, called the *Newton-Okounkov semigroup* of V. The algebra $\mathrm{gr}(V_\bullet)$ is canonically isomorphic to the semigroup k-algebra associated with $\Gamma(V_\bullet)$. Denote by $\Delta(V_\bullet)$ the closure of the subset

$$\{n^{-1}\alpha \;:\; (n,\alpha)\in\Gamma(V_\bullet)\}$$

of $\mathbb{R}^d$, called the *Newton-Okounkov body* of $V_\bullet$. Let $A(V_\bullet)$ be the affine subspace of $\mathbb{R}^d$ the canonical projection of $\Gamma(V_\bullet)\cap(\{1\}\times\mathbb{R}^d)$ in $\mathbb{R}^d$. By Proposition 6.3.5, $\Delta(V_\bullet)$ is a closed convex subset of $A(V_\bullet)$. Moreover, the relative interior of $\Delta(V_\bullet)$ in $A(V_\bullet)$ is not empty. The dimension of the affine space $A(V_\bullet)$ is called the *Kodaira dimension* of the graded linear series $V_\bullet$.

Proposition 6.3.18 *Let $V_\bullet = \bigoplus_{n\in\mathbb{N}} V_n$ be a graded sub-k-algebra of the polynomial ring $R[T]$. One has*

$$\lim_{n\in\mathbb{N}(V_\bullet),\, n\to+\infty} \frac{\dim_k(V_n)}{n^\kappa} = \mathrm{vol}(\Delta(V_\bullet)),$$

where $\mathbb{N}(V_\bullet)$ is the set of $n\in\mathbb{N}$ such that $V_n\neq\{0\}$, κ is the Kodaira dimension of $V_\bullet$, and $\mathrm{vol}(\cdot)$ is the Lebesgue measure which is normalised with respect to the semi-group $\Gamma(V_\bullet)$ as in Definition 6.3.7.

Proof It is a direct consequence of Corollary 6.3.10. □

Definition 6.3.19 Let $V_\bullet$ be a graded sub-k-algebra of $R[T]$ such that V_n is of finite dimension over k for any $n\in\mathbb{N}$.

(a) We say that $V_\bullet$ is *of subfinite type* if it is contained in a graded sub-k-algebra of $R[T]$ which is of finite type (over k).
(b) We call $\mathbb{R}$*-filtration on* $V_\bullet$ any collection $\mathcal{F}_\bullet = \{\mathcal{F}_n\}_{n\in\mathbb{N}}$, where $\mathcal{F}_n$ is an $\mathbb{R}$-filtration on V_n.
(c) Let $\delta : \mathbb{N}_{\geqslant 1} \to \mathbb{R}_{\geqslant 0}$ be a function. We say that an $\mathbb{R}$-filtration $\mathcal{F}_\bullet$ on $V_\bullet$ is *strongly δ-superadditive* if for any $\ell\in\mathbb{N}_{\geqslant 1}$ and all $(n_1,\ldots,n_\ell)\in\mathbb{N}_{\geqslant 1}^\ell$ and $(t_1,\ldots,t_\ell)\in\mathbb{R}^\ell$, one has

$$\mathcal{F}_{n_1}^{t_1}(V_{n_1})\cdots\mathcal{F}_{n_\ell}^{t_\ell}(V_{n_\ell})\subseteq\mathcal{F}_{n_1+\cdots+n_\ell}^{t_1+\cdots+t_\ell-\delta(n_1)-\cdots-\delta(n_\ell)}V_{n_1+\cdots+n_\ell}.$$

We say that the $\mathbb{R}$-filtration $\mathcal{F}_\bullet$ is *δ-superadditive* if the above relation holds in the particular case where $\ell=2$, namely, for any $(n_1,n_2)\in\mathbb{N}_{\geqslant 1}^2$ and any $(t_1,t_2)\in\mathbb{R}^2$

$$\mathcal{F}_{n_1}^{t_1}(V_{n_1})\mathcal{F}_{n_2}^{t_2}(V_{n_2})\subseteq\mathcal{F}_{n_1+n_2}^{t_1+t_2-\delta(n_1)-\delta(n_2)}V_{n_1+n_2}.$$

In the following theorem, we fix a graded sub-k-algebra $V_\bullet$ of subfinite type of $R[T]$, which is equipped with an $\mathbb{R}$-filtration $\mathcal{F}_\bullet$. We suppose in addition that $\mathbb{N}(V_\bullet):=\{n\in\mathbb{N}\,:\,n\geqslant 1,\ V_n\neq\{0\}\}$ is not empty. For each $n\in\mathbb{N}(V_\bullet)$, let $\mathbb{P}_n$ be the Borel probability measure on $\mathbb{R}$ such that, for any positive Borel function f on $\mathbb{R}$, one has

$$\int_{\mathbb{R}}f(t)\,\mathbb{P}_n(\mathrm{d}t)=\frac{1}{\dim_k(V_n)}\sum_{i=1}^{\dim_k(V_n)}f\Big(\tfrac{1}{n}\widehat{\mu}_i(V_n,\|\cdot\|_{\mathcal{F}_n})\Big),$$

where $\|\cdot\|_{\mathcal{F}_n}$ is the norm on V_n associated with the $\mathbb{R}$-filtration $\mathcal{F}_n$ (see Remark 1.1.40).

Theorem 6.3.20 *Let $\delta:\mathbb{N}_{\geqslant 1}\rightarrow\mathbb{R}_{\geqslant 0}$ be an increasing function. We suppose that, either $\mathcal{F}_\bullet$ is strongly δ-superadditive and* $\lim_{n\rightarrow+\infty}\delta(n)/n=0$, *or $\mathcal{F}_\bullet$ is δ-superadditive and $\sum_{a\in\mathbb{N}}\delta(2^a)/2^a<+\infty$. Then the sequence of measures $\{\mathbb{P}_n\}_{n\in\mathbb{N}(V_\bullet)}$ converges vaguely to a limite Borel measure $\mathbb{P}_{\mathcal{F}_\bullet}$ on $\mathbb{R}$, which is the direct image of the uniform probability measure on $\Delta(V_\bullet)^\circ$ by a concave function $G_{\mathcal{F}_\bullet}:\Delta(V_\bullet)^\circ\rightarrow\mathbb{R}\cup\{+\infty\}$, called the* concave transform *of $\mathcal{F}_\bullet$. Moreover, $\mathbb{P}_{\mathcal{F}_\bullet}$ is either the zero measure or a probability measure, and, in the case where it is a probability measure, $\{\mathbb{P}_n\}_{n\in\mathbb{N}(V_\bullet)}$ also converges weakly to $\mathbb{P}_{\mathcal{F}_\bullet}$.*

Proof Let $\Gamma(V_\bullet)$ be the Newton-Okounkov semigroup of $V_\bullet$. Since $V_\bullet$ is contained in an $\mathbb{N}$-graded sub-algebra of finite type of $R[T]$, the group $\Gamma(V_\bullet)_{\mathbb{Z}}$ is discrete and the Newton-Okounkov body $\Delta(V_\bullet)$ is compact. For any $\gamma=(n,\alpha)\in\Gamma(V_\bullet)$, let $\|\cdot\|_\gamma$ be the subquotient norm on $\mathrm{gr}_\gamma(V_\bullet)$ induced by $\|\cdot\|_{\mathcal{F}_n}$ and let $g_{\mathcal{F}_\bullet}(\gamma)$ be the Arakelov degree of $(\mathrm{gr}_\gamma(V_\bullet),\|\cdot\|_\gamma)$. Since the $\mathbb{R}$-filtration $\mathcal{F}_\bullet$ is strongly δ-superadditive (resp. δ-superadditive), the function $g_{\mathcal{F}_\bullet}$ on $\Gamma(V_\bullet)$ is strongly δ-superadditive (resp. δ-superadditive). Moreover, by Proposition 5.1.2 (7), the sequence of successive slopes of $(V_n,\|\cdot\|_{\mathcal{F}_n})$ identifies with the sorted sequence of $\{g_{\mathcal{F}_\bullet}(n,\alpha)\}_{\alpha\in\Gamma(V_\bullet)_n}$. Therefore the Borel probability measure $\mathbb{P}_n$ verifies

$$\int_{\mathbb{R}}f(t)\,\mathbb{P}_n(\mathrm{d}t)=\frac{1}{\dim_k(V_n)}\sum_{\alpha\in\Gamma(V_\bullet)_n}f\Big(\tfrac{1}{n}g_{\mathcal{F}_\bullet}(n,\alpha)\Big).$$

Therefore the assertion follows from Theorem 6.3.12 (resp. Theorem 6.3.16). □

Remark 6.3.21 We keep the notation and the hypothesis of Theorem 6.3.20.

(1) By (6.30) we obtain that

$$\sup_{x\in\Delta(V_\bullet)^\circ} G_{\mathcal{F}_\bullet}(x) = \lim_{n\in\mathbb{N}(V_\bullet),\, n\to+\infty} \frac{1}{n}\widehat{\mu}_1(V_n, \|\cdot\|_{\mathcal{F}_n}). \tag{6.32}$$

(2) Let m be an integer, we denote by $V_\bullet^{(m)}$ the graded sub-k-algebra of $R[T]$ such that $V_n^{(m)} = V_{nm}$ for any $n \in \mathbb{N}$. Then one has

$$\Gamma(V_\bullet^{(m)}) = \{(n,\alpha) \in \mathbb{N}_{\geqslant 1} \times \mathbb{Z}^d \,:\, \mathrm{gr}_{(nm,\alpha)}(V_\bullet) \neq \{0\}\}.$$

Therefore one has $\Delta(V_\bullet^{(m)}) = m\Delta(V_\bullet)$. Denote by $\mathcal{F}_\bullet^{(m)}$ the family of filtrations $\{\mathcal{F}_{mn}\}_{n\in\mathbb{N}}$ on $V_\bullet^{(m)}$, then one has

$$\forall\, x \in \Delta(V_\bullet)^\circ, \quad G_{\mathcal{F}_\bullet^{(m)}}(mx) = mG_{\mathcal{F}_\bullet}(x).$$

In particular, $\mathbb{P}_{\mathcal{F}_\bullet^{(m)}}$ identifies with the direct image of $\mathbb{P}_{\mathcal{F}_\bullet}$ by the dilatation map $(t \in \mathbb{R}) \mapsto mt$.

Remark 6.3.22 Let $U_\bullet$, $V_\bullet$ and $W_\bullet$ be graded sub-k-algebras of subfinite type of $R[T]$. Suppose that, for any $n \in \mathbb{N}$, one has

$$U_n + V_n := \{x + y \,:\, x \in U_n,\ y \in V_n\} \subseteq W_n.$$

Then, for all $n \in \mathbb{N}$ and $(\alpha,\beta) \in \mathbb{N}^d$ such that $(n,\alpha) \in \Gamma(U_\bullet)$ and $(n,\beta) \in \Gamma(V_\bullet)$, one has $(n, \alpha+\beta) \in \Gamma(W_\bullet)$. Therefore, $\Delta(U_\bullet) + \Delta(V_\bullet) \subseteq \Delta(W_\bullet)$.

Assume that the graded sub-k-algebras $U_\bullet$, $V_\bullet$ and $W_\bullet$ are equipped with $\mathbb{R}$-filtrations $\mathcal{F}_\bullet^U$, $\mathcal{F}_\bullet^V$ and $\mathcal{F}_\bullet^W$ respectively. Let $\delta : \mathbb{N}_{\geqslant 1} \to \mathbb{R}_{\geqslant 0}$ be a map. We suppose that, either $\mathcal{F}_\bullet^U$, $\mathcal{F}_\bullet^V$ and $\mathcal{F}_\bullet^W$ are strongly δ-superadditive and $\lim_{n\to+\infty}\delta(n)/n = 0$, or $\mathcal{F}_\bullet^U$, $\mathcal{F}_\bullet^V$ and $\mathcal{F}_\bullet^W$ are δ-superadditive and $\sum_{a\in\mathbb{N}}\delta(2^a)/2^a < +\infty$. Let $\epsilon : \mathbb{N}_{\geqslant 1} \to \mathbb{R}_{\geqslant 0}$ be a map such that $\lim_{n\to+\infty}\epsilon(n)/n = 0$. Suppose that, for any $n \in \mathbb{N}_{\geqslant 1}$ and any $(t_1,t_2) \in \mathbb{R}^2$, one has

$$\mathcal{F}_n^{U,t_1}(U_n)\cdot\mathcal{F}_n^{V,t_2}(V_n) \subseteq \mathcal{F}_n^{W,t_1+t_2-\epsilon(n)}(W_n).$$

Then, for all $(n,m) \in \mathbb{N}_{\geqslant 1}^2$ and $(\alpha,\beta) \in \mathbb{N}^d$ such that $(n,\alpha) \in \Gamma(U_\bullet)$ and $(n,\beta) \in \Gamma(V_\bullet)$, one has

$$g_{\mathcal{F}_\bullet^U}(mn, m\alpha) + g_{\mathcal{F}_\bullet^V}(mn, m\beta) \leqslant g_{\mathcal{F}_\bullet^W}(mn, m\alpha + m\beta) + \epsilon(mn).$$

Dividing the two sides of the inequality by m, by passing to limit when $m \to +\infty$, we obtain that

$$\widetilde{g}_{\mathcal{F}_\bullet^U}(n,\alpha) + \widetilde{g}_{\mathcal{F}_\bullet^V}(n,\beta) \leqslant \widetilde{g}_{\mathcal{F}_\bullet^W}(n,\alpha+\beta).$$

Therefore, for $(x,y) \in \Delta(U_\bullet)\times\Delta(V_\bullet)$, one has

$$G_{\mathcal{F}_\bullet^U}(x) + G_{\mathcal{F}_\bullet^V}(y) \leqslant G_{\mathcal{F}_\bullet^W}(x+y).$$

6.3.5 Applications to the study of the volume function

Let $S = (K, (\Omega, \mathcal{A}, \nu), \phi)$ be a proper adelic curve such that, either the σ-algebra $\mathcal{A}$ is discrete, or there exists a countable subfield of K which is dense in all K_ω, $\omega \in \Omega$.

Definition 6.3.23 Let C_0 be a non-negative real number. We say that the adelic curve S *satisfies the tensorial minimal slope property of level* $\geqslant C_0$ if, for any couple $(\overline{E}, \overline{F})$ of adelic vector bundles on S, the following inequality holds

$$\widehat{\mu}_{\min}(\overline{E} \otimes_{\varepsilon,\pi} \overline{F}) \geqslant \widehat{\mu}_{\min}(\overline{E}) + \widehat{\mu}_{\min}(\overline{F}) - C_0 \ln(\dim_K(E) \cdot \dim_K(F)). \qquad (6.33)$$

Recall that we have proved in Corollary 5.6.2 that, if the field K is of characteristic 0, then the adelic curve S satisfies the tensorial minimal slope property of level $\geqslant \frac{3}{2}\nu(\Omega_\infty)$.

We let $R = \mathrm{Frac}(K[\![T_1, \ldots, T_d]\!])$ be the fraction field of the K-algebra of formal series of d variables $T_1, \ldots, T_d$ and we equip $\mathbb{Z}^d$ with a monomial order $\leqslant$ and R with the corresponding $\mathbb{Z}^d$-filtration as explained in Subsection 6.3.4.

Definition 6.3.24 We call *graded K-algebra of adelic vector bundles with respect to R* any family $\overline{E}_\bullet = \{(E_n, \xi_n)\}_{n\in\mathbb{N}}$ of adelic vector bundles on S such that the following conditions are satisfied:

(a) $E_\bullet = \bigoplus_{n\in\mathbb{N}} E_n T^n$ forms a graded sub-K-algebra of subfinite type of the polynomial ring $R[T]$;
(b) for any $n \in \mathbb{N}$, the norm family ξ_n is ultrametric on $\Omega \setminus \Omega_\infty$;
(c) assume that ξ_n is of the form $\{\|\cdot\|_{n,\omega}\}_{\omega\in\Omega}$, then, for all $\omega \in \Omega$, $(n_1, n_2) \in \mathbb{N}^2_{\geqslant 1}$, and $(s_1, s_2) \in E_{n_1,K_\omega} \times E_{n_2,K_\omega}$, one has $\|s_1 \cdot s_2\|_{n_1+n_2,\omega} \leqslant \|s_1\|_{n_1,\omega} \cdot \|s_2\|_{n_2,\omega}$.

Proposition 6.3.25 *Assume that the adelic curve S satisfies the tensorial minimal slope property of level $\geqslant C$, where C is a non-negative constant. Let $\overline{E}_\bullet = \{(E_n, \xi_n)\}_{n\in\mathbb{N}}$ be a graded K-algebra of adelic vector bundles with respect to R. For any $n \in \mathbb{N}$, we equip E_n with the Harder-Narasimhan $\mathbb{R}$-filtration $\mathcal{F}_n$. Then the collection $\mathcal{F}_\bullet = \{\mathcal{F}_n\}_{n\in\mathbb{N}}$ forms an $\mathbb{R}$-filtration on $E_\bullet$ which is δ-superadditive, where δ denotes the function $\mathbb{N}_{\geqslant 1} \to \mathbb{R}_{\geqslant 0}$ sending $n \in \mathbb{N}_{\geqslant 1}$ to $C \ln(\dim_K(E_n))$.*

Proof Let n_1 and n_2 be elements of $\mathbb{N}_{\geqslant 1}$. By the condition (c) in Definition 6.3.24, for any $\omega \in \Omega$ and $s_1^{(1)} \otimes s_2^{(1)} + \cdots + s_1^{(N)} \otimes s_2^{(N)} \in E_{n_1,K_\omega} \otimes_{K_\omega} E_{n_2,K_\omega}$, one has

$$\|s_1^{(1)} s_2^{(1)} + \cdots + s_1^{(N)} s_2^{(N)}\|_{n_1+n_2,\omega} \leqslant \begin{cases} \max_{i\in\{1,\ldots,N\}} \|s_1^{(i)}\|_{n_1,\omega} \cdot \|s_2^{(i)}\|_{n_2,\omega}, & \omega \in \Omega \setminus \Omega_\infty, \\ \sum_{i=1}^N \|s_1^{(i)}\|_{n_1,\omega} \cdot \|s_2^{(i)}\|_{n_2,\omega}, & \omega \in \Omega_\infty. \end{cases}$$

Therefore, the canonical K_ω-linear map $E_{n_1,K_\omega} \otimes_{K_\omega} E_{n_2,K_\omega} \to E_{n_1+n_2,K_\omega}$ is of operator norm $\leqslant 1$. Let F_1 and F_2 be non-zero vector subspace of E_{n_1} and E_{n_2}, respectively, and let G be the image of $F_1 \otimes_K F_2$ by the canonical K-linear map $E_{n_1} \otimes_K E_{n_2} \to E_{n_1+n_2}$. By Proposition 4.3.31 (2), one has

$$\begin{aligned}&\widehat{\mu}_{\min}(\overline{G}) \geqslant \widehat{\mu}_{\min}(\overline{F}_1 \otimes_{\varepsilon,\pi} \overline{F}_2)\\ &\geqslant \widehat{\mu}_{\min}(\overline{F}_1) + \widehat{\mu}_{\min}(F_2) - C(\ln(\dim_K(F_1))) - C(\ln(\dim_K(F_2)))\\ &\geqslant \widehat{\mu}_{\min}(\overline{F}_1) + \widehat{\mu}_{\min}(F_2) - C(\ln(\dim_K(E_{n_1}))) - C(\ln(\dim_K(E_{n_2}))),\end{aligned}$$

where the second inequality comes from (7.1). By Proposition 4.3.46, we obtain that, for any $(t_1, t_2) \in \mathbb{R}^2$, one has

$$\mathcal{F}_{n_1}^{t_1}(E_{n_1}) \cdot \mathcal{F}_{n_2}^{t_2}(E_{n_2}) \subseteq \mathcal{F}_{n_1+n_2}^{t_1+t_2-\delta(n_1)-\delta(n_2)}(E_{n_1+n_2}).$$

The proposition is thus proved. □

Corollary 6.3.26 *Let $\overline{E}_\bullet = \{(E_n, \xi_n)\}_{n\in\mathbb{N}}$ be a graded K-algebra of adelic vector bundles with respect to R. We assume that $\mathbb{N}(E_\bullet)$ does not reduce to $\{0\}$ and we denote by $q \in \mathbb{N}$ a generator of the group $\mathbb{Z}(E_\bullet)$. Suppose in addition that*

$$\sum_{a\in\mathbb{N},\, 2^a q\in\mathbb{N}(E_\bullet)} \frac{\ln(\dim_K(E_{2^a q}))}{2^a} < +\infty. \tag{6.34}$$

For each $n \in \mathbb{N}(V_\bullet)$, let $\mathbb{P}_n$ be the Borel probability measure on $\mathbb{R}$ such that, for any positive Borel function f on $\mathbb{R}$, one has

$$\int_{\mathbb{R}} f(t)\,\mathbb{P}_n(\mathrm{d}t) = \frac{1}{\dim_K(E_n)} \sum_{i=1}^{\dim_K(E_n)} f(\tfrac{1}{n}\widehat{\mu}_i(\overline{E}_n)).$$

Then the sequence of measures $\{\mathbb{P}_n\}_{n\in\mathbb{N}(E_\bullet)}$ converges vaguely to a limite Borel measure $\mathbb{P}_{\overline{E}_\bullet}$, which is the direct image of the uniform distribution on $\Delta(E_\bullet)$ by a concave function $G_{\overline{E}_\bullet} : \Delta(\overline{E}_\bullet) \to \mathbb{R} \cup \{+\infty\}$. Moreover, the limite measure is either zero or a Borel probability measure, and in the latter case the sequence $\{\mathbb{P}_n\}_{n\in\mathbb{N}(E_\bullet)}$ also converges weakly to $\mathbb{P}_{\overline{E}_\bullet}$.

Proof This is a direct consequence of Proposition 6.3.25 and Theorem 6.3.20. □

Remark 6.3.27 We keep the notation and the conditions of Corollary 6.3.26. We suppose that $\{\frac{1}{n}\widehat{\mu}_1(\overline{E}_n)\}_{n\in\mathbb{N}(V_\bullet),\, n\geqslant 1}$ is bounded from above. Then the limit measure $\mathbb{P}_{\overline{E}_\bullet}$ is a probability measure. The weak convergence of $\{\mathbb{P}_n\}_{n\in\mathbb{N}(E_\bullet)}$ to $\mathbb{P}_{\overline{E}_\bullet}$ implies that

$$\begin{aligned}\int_{\mathbb{R}} \max\{t, 0\}\,\mathbb{P}_{\overline{E}_\bullet}(\mathrm{d}t) &= \lim_{n\in\mathbb{N}(E_\bullet),\, n\to+\infty} \frac{1}{n\dim_K(E_n)} \sum_{i=1}^{\dim_K(E_n)} \max\{\widehat{\mu}_i(\overline{E}_n), 0\}\\ &= \lim_{n\in\mathbb{N}(E_\bullet),\, n\to+\infty} \frac{\widehat{\deg}_+(\overline{E}_n)}{n\dim_K(E_n)}.\end{aligned} \tag{6.35}$$

If in addition the sequence $\{\frac{1}{n}\widehat{\mu}_{\min}(\overline{E}_n)\}_{n\in\mathbb{N}(V_\bullet),\, n\geqslant 1}$ is bounded from below, then one has

$$\int_{\mathbb{R}} t\,\mathbb{P}_{\overline{E}_\bullet}(\mathrm{d}t) = \lim_{n\in\mathbb{N}(E_\bullet),\,n\to+\infty} \frac{1}{n\dim_K(E_n)} \sum_{i=1}^{\dim_K(E_n)} \widehat{\mu}_i(\overline{E}_n) \tag{6.36}$$
$$= \lim_{n\in\mathbb{N}(E_\bullet),\,n\to+\infty} \frac{\widehat{\mu}(\overline{E}_n)}{n}.$$

Proposition 6.3.28 *Let*

$$\begin{cases} \overline{U}_\bullet = \{(U_n, \{\|\cdot\|_{U_n,\omega}\}_{\omega\in\Omega})\}_{n\in\mathbb{N}}, \\ \overline{V}_\bullet = \{(V_n, \{\|\cdot\|_{V_n,\omega}\}_{\omega\in\Omega})\}_{n\in\mathbb{N}}, \\ \overline{W}_\bullet = \{(W_n, \{\|\cdot\|_{W_n,\omega}\}_{\omega\in\Omega})\}_{n\in\mathbb{N}} \end{cases}$$

be graded K-algebras of adelic vector bundles with respect to R. We assume that

$$\sum_{a\in\mathbb{N},\ 2^a q\in\mathbb{N}(W_\bullet)} \frac{\ln(\dim_K(W_{2^a q}))}{2^a} < +\infty, \tag{6.37}$$

where $q \in \mathbb{N}$ is a generator of the group $\mathbb{Z}(W_\bullet)$. Suppose that, for any $n \in \mathbb{N}$ one has $U_n \cdot V_n \subseteq W_n$, and

$$\forall\,\omega\in\Omega,\ \forall\,(s,s')\in U_{n,K_\omega}\times V_{n,K_\omega},\quad \|ss'\|_{W_n,\omega} \leqslant \|s\|_{U_n,\omega}\cdot\|s'\|_{V_n,\omega}. \tag{6.38}$$

Then, for any $(x,y) \in \Delta(U_\bullet)\times\Delta(V_\bullet)$, one has

$$G_{\overline{W}_\bullet}(x+y) \geqslant G_{\overline{U}_\bullet}(x) + G_{\overline{V}_\bullet}(y). \tag{6.39}$$

Proof Denote by $\delta : \mathbb{N}_{\geqslant 1} \to \mathbb{R}_{\geqslant 0}$ the function sending $n \in \mathbb{N}_{\geqslant 1}$ to

$$C\ln(\dim_K(U_n)) + C\ln(\dim_K(V_n))$$

Let $n \in \mathbb{N}$, $n \geqslant 1$. Suppose that E is a non-zero vector subspace of U_n and F is a non-zero vector subspace of V_n. Since the adelic curve S satisfies the tensorial minimal slope superadditivity of level $\geqslant C$, one has

$$\widehat{\mu}_{\min}(\overline{E}\otimes_{\varepsilon,\pi}\overline{F}) \geqslant \widehat{\mu}_{\min}(E) + \widehat{\mu}_{\min}(F) - \delta(n).$$

Moreover, by (6.38) the canonical K-linear map $E\otimes F \to W_n$ has height $\leqslant 0$ if we consider the adelic vector bundles $\overline{E}\otimes_{\varepsilon,\pi}\overline{F}$ and $(W_n, \{\|\cdot\|_{W_n,\omega}\}_{\omega\in\Omega}$. Therefore, if we denote by $\mathcal{F}_n^U$, $\mathcal{F}_n^V$ and $\mathcal{F}_n^W$ the Harder-Narasimhan $\mathbb{R}$-filtrations of $\overline{U}_n$, $\overline{V}_n$ and $\overline{W}_n$ respectively, then, for any $(t,t')\in\mathbb{R}^2$,

$$\mathcal{F}_n^{U,t}(U_n)\cdot\mathcal{F}_n^{V,t'}(V_n) \subseteq \mathcal{F}_n^{W,t+t'-\delta(n)}.$$

By Remark 6.3.22, we obtain the inequality (6.39). □

Remark 6.3.29 Let $V_\bullet$ be a graded k-algebra. We say that $V_\bullet$ is *of subfinite type* if it is contained in a graded k-algebra of finite type. It is not true that any integral graded k-algebra of subfinite type can be identifies as a graded sub-k-algebra of the ring of polynomials (of one variable) with coefficients in the fraction field of the formal series ring (with finitely many variables) over k since the latter condition implies that $V_\bullet$ admits a valuation of one-dimensional leaves and in particular $V_\bullet$ is geometrically integral. We refer to [45, Remark 5.3] for more details. Moreover, the combination of the methods in [43, §4] and [45, §5] allows to obtain a generalisation of Corollary 6.3.26 and Proposition 6.3.28 to the case of graded algebras of adelic vector bundles whose underlying graded k-algebras are integral domains of subfinite type over k. Note that the $\mathbb{R}$-filtration by slopes of a graded algebra of adelic vector bundles is not necessarily superadditive and we need an argument similar to the Step 2 in the proof of Theorem 6.3.12 in order to replace the $\mathbb{R}$-filtration by slopes by a superadditive $\mathbb{R}$-filtration while keeping the asymptotic behaviour of the distribution of average on the jump points of the $\mathbb{R}$-filtrations. The approach can serve to remove the hypothesis that the scheme X admits a regular rational points in Theorem 6.4.6 and Theorem 6.4.7.

Definition 6.3.30 Let C_1 be a non-negative real number. We say that the adelic curve S satisfies *the strong tensorial minimal slope property of level* $\geqslant C_1$ if, for any integer $n\in\mathbb{N}_{\geqslant 2}$ and any family $\{\overline{E}_i\}_{i=1}^n$ of non-zero adelic vector bundles on S, the following inequality holds

$$\widehat{\mu}(\overline{E}_1\otimes_{\varepsilon,\pi}\cdots\otimes_{\varepsilon,\pi}\overline{E}_n)\geqslant\sum_{i=1}^n\big(\widehat{\mu}_{\min}(\overline{E}_i)-C_1\ln(\dim_K(E_i))\big).$$

Note that Corollary 5.6.2 shows that the adelic curve S satisfies the strong tensorial minimal slope property of level $\geqslant\frac{3}{2}\nu(\Omega_\infty)$, provided that the field K is of characteristic 0.

Remark 6.3.31 Let C be a non-negative real number. We suppose that the adelic curve S satisfies the strong minimal slope property of level $\geqslant C$. Let $\overline{E}_\bullet=\{(E_n,\xi_n)\}_{n\in\mathbb{N}}$ be a graded K-algebra of adelic vector bundles with respect to R. For any $n\in\mathbb{N}$, we equip E_n with the Harder-Narasimhan $\mathbb{R}$-filtration $\mathcal{F}_n$. Then the collection $\mathcal{F}_\bullet=\{\mathcal{F}_n\}_{n\in\mathbb{N}}$ forms an $\mathbb{R}$-filtration on $E_\bullet$ which is strongly δ-superadditive, where δ denotes the function $\mathbb{N}_{\geqslant 1}\rightarrow\mathbb{R}_{\geqslant 0}$ sending $n\in\mathbb{N}_{\geqslant 1}$ to $C\ln(\dim_K(E_n))$. Therefore, by Theorem 6.3.20, if the condition

$$\lim_{n\rightarrow+\infty}\frac{\ln(\dim_K(E_n))}{n}=0$$

is satisfied, then the sequence of Borel probability measures $\{\mathbb{P}_n\}_{n\in\mathbb{N}(E_\bullet)}$, defined by

$$\int_{\mathbb{R}}f(t)\,\mathbb{P}_n(\mathrm{d}t)=\frac{1}{\dim_K(E_n)}\sum_{i=1}^{\dim_K(E_n)}\delta_{\frac{1}{n}\widehat{\mu}_i(\overline{E}_n)},$$

converges vaguely to a Borel measure $\mathbb{P}_{\overline{E}_\bullet}$, which is either the zero measure or a Borel probability measure. In the latter case, the sequence $\{\mathbb{P}_n\}_{n\in\mathbb{N}(E_\bullet)}$ converges weakly to $\mathbb{P}_{\overline{E}_\bullet}$. Similarly, the assertion of Proposition 6.3.28 holds under the condition

$$\lim_{n\to+\infty}\frac{\ln(\dim_K(W_n))}{n}=0.$$

Remark 6.3.32 In the number field setting, Yuan [152, 153] has proposed another method to associate to each adelic line bundle a convex body which computes the arithmetic volume of the adelic line bundle. His method relies on multiplicity estimates of arithmetic global sections with respect to a flag of subvarieties of the fibre of the arithmetic variety over a finite place, which is similar to [101]. Note that in the general setting of adelic curves the set of "arithmetic global sections" is not necessarily finite and the classic formula relating the volume function and the asymptotic behaviour of "arithmetic global sections" does not hold in general.

6.4 Asymptotic invariants of graded linear series

In this section, we fix an integral projective K-scheme X and denote by $\pi : X \to \operatorname{Spec} K$ the structural morphism. Let d be the Krull dimension of the scheme X.

6.4.1 Asymptotic maximal slope

Let (L,φ) be an adelic line bundle on an integral projective K-scheme $\pi : X \to \operatorname{Spec} K$. For any $n \in \mathbb{N}_{\geqslant 1}$, the metric family $n\varphi$ on $L^{\otimes n}$ induces a norm family $\{\|\cdot\|_{n\varphi_\omega}\}_{\omega\in\Omega}$ on the linear series $\pi_*(L^{\otimes n}) = H^0(X, L^{\otimes n})$ which we denote by $\pi_*(n\varphi)$. By Theorems 6.1.13 and 6.1.32, the pair $(\pi_*(L^{\otimes n}), \pi_*(n\varphi))$ forms an adelic vector bundle on S, which we denote by $\pi_*(L^{\otimes n}, n\varphi)$. Note that $\pi_*(n\varphi)$ is ultrametric on $\Omega\setminus\Omega_\infty$. We can then compute diverse arithmetic invariants of these adelic vector bundles. The asymptotic behaviour of these arithmetic invariants describes the positivity of the adelic line bundle (L,φ).

Let (L,φ) be an adelic line bundle on X. We define

$$\nu_1^{\mathrm{asy}}(L,\varphi) := \limsup_{n\to+\infty}\frac{\nu_1(\pi_*(L^{\otimes n}, n\varphi))}{n},$$

called the *asymptotic first minimum* of (L,φ). Similarly, we define

$$\widehat{\mu}_{\max}^{\mathrm{asy}}(L,\varphi) := \limsup_{n\to+\infty}\frac{\widehat{\mu}_{\max}(\pi_*(L^{\otimes n}, n\varphi))}{n}, \tag{6.40}$$

called the *asymptotic maximal slope* of (L,φ). Note that all adelic vector bundles $\pi_*(L^{\otimes n}, n\varphi)$ are ultrametric on $\Omega\setminus\Omega_\infty$. Therefore, by Remark 4.3.48 and the fact that

$$\lim_{n\to+\infty}\frac{\ln(\dim_K(H^0(X, L^{\otimes n})))}{n}=0,$$

we obtain that

$$\widehat{\mu}_{\max}^{\mathrm{asy}}(L,\varphi)=\limsup_{n\to+\infty}\frac{\widehat{\mu}_1(\pi_*(L^{\otimes n},n\varphi))}{n}. \tag{6.41}$$

Let $\mathbb{K}$ be either $\mathbb{Z}$ or $\mathbb{Q}$ or $\mathbb{R}$. From now on we assume that $\mathbb{K}=\mathbb{Z}$ or X is normal. Let (D,g) be an adelic $\mathbb{K}$-Cartier divisor on X. Note that if X is not normal and $\mathbb{K}$ is either $\mathbb{Q}$ or $\mathbb{R}$, then $H^0_{\mathbb{K}}(X,D)$ is not necessarily a vector space over K (cf. Example 2.4.15). Similarly we can define $\nu_1^{\mathrm{asy}}(D,g)$ and $\widehat{\mu}_{\max}^{\mathrm{asy}}(D,g)$ as follows:

$$\begin{cases}\nu_1^{\mathrm{asy}}(D,g):=\limsup\limits_{n\to+\infty}\dfrac{\nu_1(H^0_{\mathbb{K}}(X,nD),\xi_{ng})}{n},\\ \widehat{\mu}_{\max}^{\mathrm{asy}}(D,g):=\limsup\limits_{n\to+\infty}\dfrac{\widehat{\mu}_{\max}(H^0_{\mathbb{K}}(X,nD),\xi_{ng})}{n}.\end{cases}$$

In the case where $\mathbb{K}=\mathbb{Z}$,

$$\nu_1^{\mathrm{asy}}(D,g)=\nu_1^{\mathrm{asy}}(\mathcal{O}_X(D),\varphi_g)\quad\text{and}\quad\widehat{\mu}_{\max}^{\mathrm{asy}}(D,g)=\widehat{\mu}_{\max}^{\mathrm{asy}}(\mathcal{O}_X(D),\varphi_g),$$

where φ_g is the metric family of L defined by g.

Proposition 6.4.1 *Let (D,g) be an adelic $\mathbb{K}$-Cartier divisor on S. Then one has $\nu_1^{\mathrm{asy}}(D,g)\leqslant\widehat{\mu}_{\max}^{\mathrm{asy}}(D,g)$. The equality holds when S satisfies the Minkowski property of certain level (see Definition 4.3.73).*

Proof By (4.83), for any $n\in\mathbb{N}_{n\geqslant 1}$ one has

$$\nu_1(H^0_{\mathbb{K}}(X,nD),\xi_{ng})\leqslant\widehat{\mu}_{\max}(H^0_{\mathbb{K}}(X,nD),\xi_{ng}).$$

Therefore $\nu_1^{\mathrm{asy}}(D,g)\leqslant\widehat{\mu}_{\max}^{\mathrm{asy}}(D,g)$.

If the adelic curve S satisfies the Minkowski property of level $\geqslant C$, where $C\geqslant 0$, then

$$\nu_1(H^0_{\mathbb{K}}(X,nD),\xi_{ng})\geqslant\widehat{\mu}_{\max}(H^0_{\mathbb{K}}(X,nD),\xi_{ng})-C\ln(\dim_K(H^0_{\mathbb{K}}(X,nD))).$$

Since X is a projective scheme, one has

$$\dim_K(H^0_{\mathbb{K}}(X,nD))=O(n^{\dim(X)}).$$

Hence

$$\lim_{n\to+\infty}\frac{\ln(\dim_K(H^0_{\mathbb{K}}(X,nD)))}{n}=0.$$

Therefore $\nu_1^{\mathrm{asy}}(D,g)\geqslant\widehat{\mu}_{\max}^{\mathrm{asy}}(D,g)$. □

Let (D,g) be an adelic $\mathbb{R}$-Cartier divisor on X. Let $\mathbb{K}$ be either $\mathbb{Q}$ or $\mathbb{R}$. We set

$$\begin{cases}\Gamma^{\times}_{\mathbb{K}}(D)=\{s\in K(X)^{\times}\otimes_{\mathbb{Z}}\mathbb{K}\;:\;D+(s)\geqslant_{\mathbb{K}}0\},\\ \nu_{1,\mathbb{K}}^{\mathrm{asy}}(D,g)=\begin{cases}\sup\left\{\widehat{\deg}_{\xi_g}(s)\;:\;s\in\Gamma^{\times}_{\mathbb{K}}(D)\right\} & \text{if }\Gamma^{\times}_{\mathbb{K}}(D)\neq\emptyset,\\ -\infty & \text{if }\Gamma^{\times}_{\mathbb{K}}(D)=\emptyset,\end{cases}\end{cases}$$

where (cf. Corollary 6.2.13)

$$\widehat{\deg}_{\xi_g}(s) = -\int_{\Omega} \ln \|s\|_{g_\omega}\, \nu(d\omega).$$

Note that $\nu_1^{\mathrm{asy}}(D,g) = \nu_{1,\mathbb{Q}}^{\mathrm{asy}}(D,g) \leqslant \nu_{1,\mathbb{R}}^{\mathrm{asy}}(D,g)$.

Proposition 6.4.2 *We assume that X is normal. Let (D,g) and (D',g') be adelic $\mathbb{R}$-Cartier divisors on X. Then one has the following:*

(1) $\nu_{1,\mathbb{K}}^{\mathrm{asy}}((D,g)+(D',g')) \geqslant \nu_{1,\mathbb{K}}^{\mathrm{asy}}(D,g) + \nu_{1,\mathbb{K}}^{\mathrm{asy}}(D',g')$.
(2) $\nu_{1,\mathbb{K}}^{\mathrm{asy}}(a(D,g)) = a\nu_{1,\mathbb{K}}^{\mathrm{asy}}(D,g)$ *for all* $a \in \mathbb{K}_{\geqslant 0}$.

Proof (1) Clearly we may assume that $\Gamma_{\mathbb{K}}^{\times}(D) \neq \emptyset$ and $\Gamma_{\mathbb{K}}^{\times}(D') \neq \emptyset$. If $s \in \Gamma_{\mathbb{K}}^{\times}(D)$ and $s' \in \Gamma_{\mathbb{K}}^{\times}(D')$, then $\|ss'\|_{g+g'} \leqslant \|s\|_g \|s'\|_{g'}$, so that

$$-\log\|s\|_g - \log\|s'\|_{g'} \leqslant -\log\|ss'\|_{g+g'} \leqslant \nu_{1,\mathbb{K}}^{\mathrm{asy}}(D+D', g+g').$$

Therefore one has (1).

(2) Clearly we may assume that $a > 0$ and $\Gamma_{\mathbb{K}}^{\times}(D) \neq \emptyset$. Then one has a bijective correspondence $(s \in \Gamma_{\mathbb{K}}^{\times}(D)) \mapsto (s^a \in \Gamma_{\mathbb{K}}^{\times}(aD))$. Moreover, $\widehat{\deg}_{\xi_{ag}}(s^a) = a\widehat{\deg}_{\xi_g}(s)$ for $s \in \Gamma_{\mathbb{K}}^{\times}(D)$. Thus the assertion follows. □

Theorem 6.4.3 *We assume that X is normal. Let (D,g) be an adelic $\mathbb{R}$-Cartier divisor on X. If $\Gamma_{\mathbb{Q}}^{\times}(D) \neq \emptyset$, then $\nu_{1,\mathbb{Q}}^{\mathrm{asy}}(D,g) = \nu_{1,\mathbb{R}}^{\mathrm{asy}}(D,g)$. In particular, if D is big, then $\nu_1^{\mathrm{asy}}(a(D,g)) = a\nu_1^{\mathrm{asy}}(D,g)$ for all $a \in \mathbb{R}_{\geqslant 0}$.*

Proof By our assumption, we can find $\psi \in \Gamma_{\mathbb{Q}}^{\times}(D)$. Then the map

$$\alpha_\psi : \Gamma_{\mathbb{K}}^{\times}(D) \to \Gamma_{\mathbb{K}}^{\times}(D + (\psi))$$

given by $s \mapsto s\psi^{-1}$ is bijective and, for $s \in \Gamma_{\mathbb{K}}^{\times}(D)$, $\|s\|_g = \|\alpha_\psi(s)\|_{g-\log|\psi|}$, so that

$$\nu_{1,\mathbb{K}}^{\mathrm{asy}}(D,g) = \nu_{1,\mathbb{K}}^{\mathrm{asy}}(D+(\psi), g - \log|\psi|).$$

Therefore we may assume that D is effective. Moreover, for an integrable function φ on Ω,

$$\nu_{1,\mathbb{K}}^{\mathrm{asy}}(D, g+\varphi) = \nu_{1,\mathbb{K}}^{\mathrm{asy}}(D,g) + \int_{\Omega} \varphi\, \nu(d\omega),$$

so that we may further assume that

$$\int_{\Omega} -\log\|1\|_{g_\omega}\, \nu(d\omega) \geqslant 0.$$

For $s' \in \Gamma_{\mathbb{R}}^{\times}(D)$, we choose $s_1, \ldots, s_r \in K(X)^{\times} \otimes_{\mathbb{Z}} \mathbb{Q}$ and $a_1, \ldots, a_r \in \mathbb{R}$ such that $s' = s_1^{a_1} \cdots s_r^{a_r}$ and $a_1, \ldots, a_r$ are linearly independent over $\mathbb{Q}$. We set $\|x\|_0 = |x_1| + \cdots + |x_r|$ and $s^x = s_1^{x_1} \cdots s_r^{x_r}$ for $x = (x_1, \ldots, x_r) \in \mathbb{R}^r$, so that if we denote $(a_1, \ldots, a_r)$ by α, then $s' = s^\alpha$. By Proposition 2.4.18, for any a positive rational number ε, there is a positive number δ such that if $\|\alpha' - \alpha\|_0 \leqslant \delta$ for $\alpha' \in \mathbb{R}^r$, then $(1+\varepsilon)D + (s^{\alpha'})$ is effective. We choose a basis $\{\omega_1, \dots, \omega_r\}$ of $\mathbb{Q}^r$ such that $\|\omega_j - a\|_0 \leqslant \delta$ for all $j \in \{1, \dots, r\}$, so that $(1+\varepsilon)D + (s^{\omega_j}) \geqslant 0$ for all j. Here we set $\alpha = \lambda_1\omega_1 + \cdots + \lambda_r\omega_r$. Further, if we define a norm $\|\cdot\|_\omega$ by $\|x_1\omega_1 + \cdots + x_r\omega_r\|_\omega = |x_1| + \cdots + |x_r|$ for $x_1, \dots, x_r \in \mathbb{R}$, then there is a positive constant C such that $C\|\cdot\|_0 \leqslant \|\cdot\|_\omega$. Note that for any $t > 0$, there is $\alpha' \in \mathbb{Q}^r$ such that if we set $\alpha' = \lambda'_1\omega_1 + \cdots + \lambda'_r\omega_r$, then $\lambda'_j \geqslant \lambda_j$ $(\forall\, j)$ and $\|\alpha' - \alpha\|_\omega \leqslant t$. Indeed, for each j, one can find $\lambda'_j \in \mathbb{Q}$ such that $0 \leqslant \lambda'_j - \lambda_j \leqslant t/r$, and hence $\|\alpha' - \alpha\|_\omega \leqslant t$. Therefore, we can also choose a sequence $\{\alpha_n\}_{n=1}^\infty$ of $\mathbb{Q}^r$ with the following properties:

(i) $\alpha_n \in \alpha + \mathbb{R}_{\geqslant 0}\omega_1 + \cdots + \mathbb{R}_{\geqslant 0}\omega_r$ for all $n \geqslant 1$.
(ii) $\|\alpha_n - \alpha\|_\omega \leqslant \min\{\varepsilon/((1+\varepsilon)n), C\delta\}$ for all $n \geqslant 1$.

Since $\|\alpha_n - \alpha\|_0 \leqslant \delta$ by (ii), $(1+\varepsilon)D + (s^{\alpha_n}) \geqslant 0$ for all $n \geqslant 1$. Moreover, if we set $\alpha_n - \alpha = \sum_{j=1}^r \lambda_j^{(n)}\omega_j$, then $\lambda_j^{(n)} \geqslant 0$ $(\forall\, j)$ and $\sum_{j=1}^r \lambda_j^{(n)} \leqslant \varepsilon/((1+\varepsilon)n)$. Therefore, if we denote $\varepsilon - (1+\varepsilon)\sum_{j=1}^r \lambda_j^{(n)}$ by κ_n, then $\kappa_n \geqslant 0$ and, for each $\omega \in \Omega$,

$$
\begin{aligned}
|s^{\alpha_n}|_{(1+\varepsilon)g_\omega} &= |s^\alpha|_{g_\omega} |s^{\omega_1\lambda_1^{(n)}} \cdots s^{\omega_r\lambda_r^{(n)}}|_{\varepsilon g_\omega} \\
&= |s'|_{g_\omega} |s^{\omega_1}|_{(1+\varepsilon)g_\omega}^{\lambda_1^{(n)}} \cdots |s^{\omega_r}|_{(1+\varepsilon)g_\omega}^{\lambda_r^{(n)}} |1|_{\kappa_n g_\omega},
\end{aligned}
$$

so that $\|s^{\alpha_n}\|_{(1+\varepsilon)g_\omega} \leqslant \|s'\|_{g_\omega} \|s^{\omega_1}\|_{(1+\varepsilon)g_\omega}^{\lambda_1^{(n)}} \cdots \|s_{\omega_r}\|_{(1+\varepsilon)g_\omega}^{\lambda_r^{(n)}} \|1\|_{g_\omega}^{\kappa_n}$. Therefore, since

$$
\int_\Omega -\log\|s^{\alpha_n}\|_{(1+\varepsilon)g_\omega}\nu(\mathrm{d}\omega) \leqslant \nu_{1,\mathbb{Q}}^{\mathrm{asy}}((1+\varepsilon)(D,g))
$$

and $\kappa_n \int_\Omega -\log\|1\|_{g_\omega}\nu(\mathrm{d}\omega) \geqslant 0$, one has

$$
\int_\Omega -\log\|s'\|_{g_\omega}\nu(d\omega) + \sum_{j=1}^r \lambda_j^{(n)} \int_\Omega -\log\|s^{\omega_j}\|_{(1+\varepsilon)g_\omega}\nu(d\omega) \leqslant \nu_{1,\mathbb{Q}}^{\mathrm{asy}}((1+\varepsilon)(D,g)),
$$

so that taking $n \to \infty$, we obtain

$$
\int_\Omega -\log\|s'\|_{g_\omega}\nu(d\omega) \leqslant \nu_{1,\mathbb{Q}}^{\mathrm{asy}}((1+\varepsilon)(D,g)),
$$

and hence, as ε is a rational number, by Proposition 6.4.2, (2), one can see

$$
\nu_{1,\mathbb{R}}^{\mathrm{asy}}(D,g) \leqslant \nu_{1,\mathbb{Q}}^{\mathrm{asy}}((1+\varepsilon)(D,g)) = (1+\varepsilon)\nu_{1,\mathbb{Q}}^{\mathrm{asy}}(D,g),
$$

which implies $\nu_{1,\mathbb{R}}^{\mathrm{asy}}(D,g) \leqslant \nu_{1,\mathbb{Q}}^{\mathrm{asy}}(D,g)$, as required. □

Proposition 6.4.4 *Let (D, g) be an adelic $\mathbb{K}$-Cartier divisor on X. We assume that either $\mathbb{K} = \mathbb{Z}$ or X is normal. Then one has*

$$\widehat{\mu}_{\mathrm{ess}}(D, g) \geqslant \widehat{\mu}_{\max}(H^0_{\mathbb{K}}(X, D), \xi_g).$$

In particular,

$$\widehat{\mu}_{\mathrm{ess}}(D, g) \geqslant \widehat{\mu}^{\mathrm{asy}}_{\max}(D, g). \tag{6.42}$$

Proof The second inequality is a consequence of the first inequality because $\widehat{\mu}_{\mathrm{ess}}(nD, ng) = n\widehat{\mu}_{\mathrm{ess}}(D, g)$.

Let U be a non-empty Zariski open set of X given by

$$\{x \in X \,:\, X \to \operatorname{Spec} K \text{ is smooth at } x \text{ and } x \notin \operatorname{Supp}_{\mathbb{K}}(D)\}.$$

Note that $\exp(-g_\omega)$ is a positive continuous function on U^{an}_ω for each $\omega \in \Omega$ and that, for $s \in H^0_{\mathbb{K}}(X, D)$ and $x \in U$, one has $s \in \mathcal{O}_{X,x}$. Let t be a real number such that $\widehat{\mu}_{\mathrm{ess}}(D, g) < t$. Then there is an infinite subset Λ of $U(K^{\mathrm{ac}})$ such that Λ is Zariski dense in U and $h_{(D,g)}(P) \leqslant t$ for all $P \in \Lambda$.

Let F be a non-zero vector subspace of $H^0_{\mathbb{K}}(X, D)$. Then there exist $P_1, \ldots, P_{\dim F} \in \Lambda$ such that the evaluation map

$$f : F \otimes_K K^{\mathrm{ac}} \longrightarrow \bigoplus_{i=1}^{\dim F} \kappa(P_i)$$

is a bijection. For $\chi \in \Omega_{K^{\mathrm{an}}}$, let $P_{i,\chi}$ be the unique extension of $P_i \in X_{K^{\mathrm{an}}}$ to $(X_{K^{\mathrm{an}}})^{\mathrm{an}}_\chi$. Let $\|\cdot\|_{P_{i,\chi}}$ be a norm of $\kappa(P_i)_\chi$ given by $\|1\|_{P_{i,\chi}} = \exp(-g_{\pi(\chi)}(\mu(P_{i,\chi})))$, where π is the canonical map $\Omega_{K^{\mathrm{an}}} \to \Omega$ and $\mu : (X_{K^{\mathrm{an}}})^{\mathrm{an}}_\chi \to X^{\mathrm{an}}_{\pi(\chi)}$ is also the canonical morphism as analytic spaces. We set $\xi_i := \{\|\cdot\|_{P_{i,\chi}}\}_{\chi \in \Omega_{K^{\mathrm{an}}}}$. We equip $\bigoplus_{i=1}^{\dim F} \kappa(P_i)$ with the ψ_0-direct sum $\xi = \{\|\cdot\|_\chi\}_{\chi \in \Omega_{K^{\mathrm{an}}}}$ of $\xi_1, \ldots, \xi_{\dim F}$, where ψ_0 denotes the function from $[0, 1]$ to $[0, 1]$ sending $x \in [0, 1]$ to $\max(x, 1 - x)$ (see Subsections 1.1.10 and 4.1.1). Note that, if we denote by $\{e_i\}_{i=1}^{\dim F}$ a basis of $\bigoplus_{i=1}^{\dim F} \kappa(P_i)$ such that $e_i \in \kappa(P_i)$, then this basis is orthogonal with respect to $\|\cdot\|_\chi$ for any $\chi \in \Omega_{K^{\mathrm{an}}}$. By Proposition 1.2.23, this basis is also a Hadamard basis with respect to $\|\cdot\|_\chi$ for any $\chi \in \Omega_{K^{\mathrm{an}}}$. In particular, one has

$$\widehat{\deg}\Big(\bigoplus_{i=1}^{\dim_K(F)} (\kappa(P_i), \xi_i)\Big) = \sum_{i=1}^{\dim_K(F)} h_{(D,g)}(P_i) \leqslant \dim_K(F)t.$$

Moreover, for any $\chi \in \Omega_{K^{\mathrm{ac}}}$ the operator norm of f_χ is $\leqslant 1$. Therefore, by Proposition 4.3.18, one has

$$\widehat{\mu}(\overline{F}) = \widehat{\mu}(\overline{F} \otimes K^{\mathrm{an}}) \leqslant \frac{1}{\dim_K(F)} \widehat{\deg}\Big(\bigoplus_{i=1}^{\dim_K(F)} (\kappa(P_i), \xi_i)\Big) \leqslant t.$$

Since F is arbitrary, we obtain $\widehat{\mu}_{\max}(H^0_{\mathbb{K}}(X, D), \xi_g) \leqslant t$. Therefore (6.42) follows because t is an arbitrary real number with $t > \widehat{\mu}_{\mathrm{ess}}(L, \varphi)$. □

6.4.2 Arithmetic volume function

We assume that there exists $C \geqslant 0$ such that the adelic curve S verifies the tensorial minimal slope property.

Definition 6.4.5 Let (L, φ) be an adelic line bundle on X. We define the *arithmetic volume* of (L, φ) as

$$\widehat{\mathrm{vol}}(L, \varphi) := \limsup_{n \to +\infty} \frac{\widehat{\deg}_+(\pi_*(L^{\otimes n}, n\varphi))}{n^{d+1}/(d+1)!}.$$

We say that (L, φ) is *big* if $\widehat{\mathrm{vol}}(L, \varphi) > 0$.

Assume that the K-scheme X admits a regular rational point of X. Then the local ring $\mathcal{O}_{X,P}$ is a regular local ring. By Cohen's structure theorem of complete regular local rings [56, Proposition 10.16], the formal completion of $\mathcal{O}_{X,P}$ is isomorphic to the algebra of formal series $K[\![T_1, \ldots, T_d]\!]$, where d is the Krull dimension of X. If L is an invertible $\mathcal{O}_X$-module, by choosing a local generator of the $\mathcal{O}_{X,P}$-module L_P, we can identify the graded linear series $\bigoplus_{n \in \mathbb{N}} H^0(X, L^{\otimes n})$ as a graded sub-K-algebra (of subfinite type) of $K[\![T_1, \ldots, T_d]\!][T]$. We denote by $\Delta(L)$ the Newton-Okounknov body of this graded algebra (see §6.3.4 for the construction of $\Delta(L)$). For any $n \in \mathbb{N}$, let $r_n := \dim_K(H^0(X, L^{\otimes n})) > 0$. By Proposition 6.3.18 one has

$$\int_{\Delta(L)} 1 \,\mathrm{d}x = \lim_{r_n > 0,\, n \to +\infty} \frac{r_n}{n^\kappa},$$

where κ is the Kodaira-Iitaka dimension of the graded linear series $\bigoplus_{n \in \mathbb{N}} H^0(X, L^{\otimes n})$ (which is also called the *Kodaira-Iitaka dimension* of L). In particular, if L is a big line bundle, namely

$$\mathrm{vol}(L) := \limsup_{n \to +\infty} \frac{r_n}{n^d/d!} > 0,$$

or equivalently, $\kappa = d$, one has

$$\mathrm{vol}(L) = d! \int_{\Delta(L)} 1 \,\mathrm{d}x.$$

If (L, φ) is an adelic line bundle of X, then the family $\{(\pi_*(L^{\otimes n}, n\varphi))\}_{n \in \mathbb{N}}$ forms a graded K-algebra of adelic vector bundles with respect to $\mathrm{Frac}(K[\![T_1, \ldots, T_d]\!])$ (see Definition 6.3.24). For any $n \in \mathbb{N}_{\geqslant 1}$ such that $r_n := \dim_K(H^0(X, L^{\otimes n})) > 0$, let $\mathbb{P}_{(L,\varphi),n}$ be the Borel probability measure on $\mathbb{R}$ such that, for any positive Borel function f on $\mathbb{R}$, one has

$$\int_{\mathbb{R}} f(t)\,\mathbb{P}_{(L,\varphi),n}(\mathrm{d}t) = \frac{1}{r_n}\sum_{i=1}^{r_n} f(\tfrac{1}{n}\widehat{\mu}_i(\pi_*(L^{\otimes n}, n\varphi))).$$

Theorem 6.4.6 *Assume that the scheme X admits a regular rational point. Let (L,φ) be an adelic line bundle on X. For any $n \in \mathbb{N}$, let $r_n = \dim_K(H^0(X, L^{\otimes n}))$. Assume that there exists $n \in \mathbb{N}_{\geqslant 1}$ such that $r_n > 0$. Then the sequence of measures $\{\mathbb{P}_{(L,\varphi),n}\}_{n\in\mathbb{N},\, r_n>0}$ converges weakly to a Borel probability measure $\mathbb{P}_{(L,\varphi)}$, which is the direct image of a concave real-valued function $G_{(L,\varphi)}$ on $\Delta(L)^{\circ}$. In particular, if (L,φ) is big, then the invertible $\mathcal{O}_X$-module is big. Moreover, in the case where L is big, the sequence*

$$\frac{\widehat{\deg}_+(\pi_*(L^{\otimes n}, n\varphi))}{n^{d+1}/(d+1)!}, \quad n \in \mathbb{N},\ r_n > 0 \tag{6.43}$$

converges to $\widehat{\mathrm{vol}}(L,\varphi)$, which is also equal to

$$(d+1)\mathrm{vol}(L)\int_{[0,+\infty[} t\,\mathbb{P}_{(L,\varphi)}(\mathrm{d}t) = (d+1)\int_{\Delta(L)^{\circ}} \max(G_{(L,\varphi)}(x), 0)\,\mathrm{d}x. \tag{6.44}$$

Proof We deduce from Corollary 6.3.26 that the sequence $\{\mathbb{P}_{(L,\varphi),n}\}_{n\in\mathbb{N},\, r_n>0}$ converges vaguely to a Borel measure $\mathbb{P}_{(L,\varphi)}$ on $\mathbb{R}$, which is the direct image of the uniform probability measure on $\Delta(L)^{\circ}$ by a concave function $G_{(L,\varphi)} : \Delta(L)^{\circ} \to \mathbb{R}\cup\{+\infty\}$. Moreover, by Proposition 6.4.4 and 6.2.7, we obtain that the supports of the Borel probability measures $\mathbb{P}_{(L,\varphi),n}$ are uniformly bounded from above. The function $G_{(L,\varphi)}$ is then bounded from above and hence limit measure $\mathbb{P}_{(L,\varphi)}$ is a Borel probability measure and the sequence $\{\mathbb{P}_n\}_{n\in\mathbb{N},\, r_n>0}$ converges weakly to $\mathbb{P}_{(L,\varphi)}$. In particular, the sequence $\{\frac{1}{nr_n}\widehat{\deg}_+(L^{\otimes n}, n\varphi)\}_{n\in\mathbb{N},\, r_n>0}$ converges to

$$\int_{[0,+\infty[} t\,\mathbb{P}_{(L,\varphi)}(\mathrm{d}t) = \frac{1}{\mathrm{vol}(\Delta(L))}\int_{\Delta(L)^{\circ}} \max(G_{(L,\varphi)}(x), 0)\,\mathrm{d}x$$

since

$$\int_{[0,+\infty[} t\,\mathbb{P}_n(\mathrm{d}t) = \int_{[0,\widehat{\mu}_{\max}^{\mathrm{asy}}(L,\varphi)[} t\,\mathbb{P}_n(\mathrm{d}t) = \frac{1}{nr_n}\sum_{i=1}^{r_n}\max\{\widehat{\mu}_i(\pi_*(L^{\otimes n}, n\varphi)), 0\},$$

and by (4.71) and (4.67),

$$\left|\widehat{\deg}_+(\pi_*(L^{\otimes n}, n\varphi)) - \sum_{i=1}^{r_n}\max\{\widehat{\mu}_i(\pi_*(L^{\otimes n}, n\varphi)), 0\}\right| \leqslant \frac{1}{2}\ln(r_n)\nu(\Omega_\infty).$$

In particular, in the case where $\widehat{\mathrm{vol}}(L,\varphi) > 0$, one has

$$\limsup_{n\to+\infty}\frac{r_n}{n^d} > 0,$$

namely the invertible $\mathcal{O}_X$-module L is big. Moreover, in the case where L is big, one has

$$\lim_{n\to+\infty}\frac{r_n}{n^d/d!}=\mathrm{vol}(L)>0.$$

Therefore the sequence (6.43) converges (to $\widehat{\mathrm{vol}}(L,\varphi)$ by definition), which is equal to (6.44). □

Theorem 6.4.7 *Assume that the scheme X admits a regular rational point. Let (L_1,φ_1) and (L_2,φ_2) be big adelic line bundles on X. Then the following inequality of Brunn-Minkowski type holds*

$$\widehat{\mathrm{vol}}(L_1\otimes L_2,\varphi_1+\varphi_2)^{1/(d+1)}\geqslant\widehat{\mathrm{vol}}(L_1,\varphi_1)^{1/(d+1)}+\widehat{\mathrm{vol}}(L_2,\varphi_2)^{1/(d+1)}.\qquad(6.45)$$

Proof For any adelic line bundle (L,φ) on X such that L is big, we denote by $\widehat{\Delta}(L,\varphi)$ the closure of the convex set $\{(x,t)\in\Delta(L)^\circ\times\mathbb{R}\,:\,0\leqslant t\leqslant G_{(L,\varphi)}(x)\}$. Then Theorem 6.4.6 implies that

$$\widehat{\mathrm{vol}}(L,\varphi)=(d+1)\int_{\widehat{\Delta}(L,\varphi)}1\,\mathrm{d}(x,t)$$

By Proposition 6.3.28, one has

$$\widehat{\Delta}(L_1,\varphi_1)+\widehat{\Delta}(L_2,\varphi_2)\leqslant\widehat{\Delta}(L_1\otimes L_2,\varphi_1+\varphi_2).$$

Therefore the relation (6.45) follows from the classic Brunn-Minkowski inequality.□

6.4.3 Volume of adelic $\mathbb{R}$-Cartier divisors

We assume that X is normal and geometrically integral and admits a regular rational point P. We identify the formal completion of $\mathcal{O}_{X,P}$ with $K[\![T_1,\ldots,T_d]\!]$, which allow us to embed the rational function field $K(X)$ into the fraction field $R=\mathrm{Frac}(K[\![T_1,\ldots,T_d]\!])$. We also suppose that there exists $C\geqslant 0$ such that the adelic curve S verifies the tensorial minimal slope property of level $\geqslant C$. In the following, the symbol $\mathbb{K}$ denotes $\mathbb{Z}$, $\mathbb{Q}$ or $\mathbb{R}$.

Let D be a $\mathbb{K}$-Cartier divisor. We identify $\bigoplus_{n\in\mathbb{N}}H^0(nD)$ with a graded sub-K-algebra of subfinite type of $R[T]$. We denote by $\Delta(D)$ the Newton-Okounkov body of this graded algebra (see §6.3.4 for its construction).

Definition 6.4.8 Let (D,g) be an adelic $\mathbb{K}$-Cartier divisor. We define the *arithmetic volume* of (D,g) as

$$\widehat{\mathrm{vol}}(D,g):=\limsup_{n\to+\infty}\frac{\widehat{\deg}_+(H^0_{\mathbb{K}}(nD),\xi_{ng})}{n^{d+1}/(d+1)!}$$

(for the definition of the norm family ξ_{ng}, see Definition 6.2.17). We say that (D, g) is *big* if $\widehat{\mathrm{vol}}(D, g) > 0$. Note that for any $s \in K(X)^{\times}$ one has (see Remark 2.5.10)

$$\widehat{\mathrm{vol}}((D, g) + \widehat{(s)}) = \widehat{\mathrm{vol}}(D, g). \tag{6.46}$$

Moreover, (D, g) is said to be *arithmetically* $\mathbb{K}$*-effective*, which is denoted by $(D, g) \geqslant_{\mathbb{K}} (0, 0)$, if D is $\mathbb{K}$-effective and $g_\omega \geqslant 0$ for all $\omega \in \Omega$. For adelic $\mathbb{K}$-Cartier divisors (D_1, g_1) and (D_2, g_2) on X,

$$(D_1, g_1) \geqslant_{\mathbb{K}} (D_2, g_2) \quad \overset{\text{def}}{\iff} \quad (D_1, g_1) - (D_2, g_1) \geqslant_{\mathbb{K}} (0, 0).$$

Note that if $(D_1, g_1) \geqslant_{\mathbb{K}} (D_2, g_2)$, then $\widehat{\deg}_+(H^0_{\mathbb{K}}(D_1), \xi_{g_1}) \geqslant \widehat{\deg}_+(H^0_{\mathbb{K}}(D_2), \xi_{g_2})$. In particular, $\widehat{\mathrm{vol}}(D_1, g_1) \geqslant \widehat{\mathrm{vol}}(D_2, g_2)$. By using Prposition 6.2.12, if $D \geqslant_{\mathbb{K}} 0$, then there is a family of D-Green functions g over S such that $(D, g) \geqslant_{\mathbb{K}} (0, 0)$.

Let (D, g) be an adelic $\mathbb{K}$-Cartier divisor. The family $\{(H^0_{\mathbb{K}}(nD), \xi_{ng})\}_{n \in \mathbb{N}}$ forms a graded K-algebra of adelic vector bundles with respect to

$$R = \mathrm{Frac}(K[\![T_1, \ldots, T_d]\!]).$$

For any $n \in \mathbb{N}_{\geqslant 1}$ such that $r_n := \dim_K(H^0(nD)) > 0$, we let $\mathbb{P}_{(D,g),n}$ be the Borel probability measure on $\mathbb{R}$ such that, for any positive Borel function on $\mathbb{R}$, one has

$$\int_{\mathbb{R}} f(t)\, \mathbb{P}_{(D,g),n}(\mathrm{d}t) = \frac{1}{r_n} \sum_{i=1}^{r_n} f(\tfrac{1}{n}\widehat{\mu}_i(H^0_{\mathbb{K}}(nD), \xi_{ng})).$$

Theorem 6.4.9 *Let (D, g) be an adelic $\mathbb{K}$-Cartier divisor. For any $n \in \mathbb{N}$, let $r_n = \dim_K(H^0_{\mathbb{K}}(nD))$. Assume that there exists $n \in \mathbb{N}_{n \geqslant 1}$ such that $r_n > 0$. Then the sequence of measures $\{\mathbb{P}_{(D,g),n}\}_{n \in \mathbb{N}, r_n > 0}$ converges weakly to a Borel probability measure $\mathbb{P}_{(D,g)}$, which is the direct image of a concave real-valued function $G_{(D,g)}$ on $\Delta(D)^\circ$. In particular, if (D, g) is big, then the $\mathbb{K}$-Cartier divisor D is big. Moreover, in the case where D is big, the sequence*

$$\frac{\widehat{\deg}_+(H^0_{\mathbb{K}}(nD), \xi_{ng})}{n^{d+1}/(d+1)!}, \quad n \in \mathbb{N},\ r_n > 0$$

converges to $\widehat{\mathrm{vol}}(D, g)$, which is also equal to

$$(d+1)\mathrm{vol}(L) \int_{[0,+\infty[} t\, \mathbb{P}_{(D,g)}(\mathrm{d}t) = (d+1) \int_{\Delta(D)^\circ} \max(G_{(D,g)}(x), 0)\, \mathrm{d}x. \tag{6.47}$$

Proof We omit the proof since it is quite similar to that of Theorem 6.4.6. □

Corollary 6.4.10 *Let (D, g) and (A, h) be adelic $\mathbb{R}$-Cartier divisors on X. We assume that D is big. Then*

$$\lim_{t\to\infty} \frac{\widehat{\deg}_+(H^0_{\mathbb{R}}(X, tD + A), \xi_{tg+h})}{t^{d+1}/(d+1)!} = \widehat{\mathrm{vol}}(D, g),$$

where t is a positive real number.

Proof Let us begin with the following claim:

Claim 6.4.11 $\widehat{\mathrm{vol}}(aD, ag) = a^{d+1}\widehat{\mathrm{vol}}(D, g)$ *for any positive integer a.* □

Proof By Theorem 6.4.9,

$$\begin{aligned}\widehat{\mathrm{vol}}(aD, ag) &= \lim_{n\to\infty} \frac{\widehat{\deg}_+(H^0_{\mathbb{K}}(naD), \xi_{nag})}{n^{d+1}/(d+1)!}\\ &= a^{d+1} \lim_{n\to\infty} \frac{\widehat{\deg}_+(H^0_{\mathbb{K}}(naD), \xi_{nag})}{(na)^{d+1}/(d+1)!} = a^{d+1}\widehat{\mathrm{vol}}(D, g),\end{aligned}$$

as required. □

Claim 6.4.12 *If D is* $\mathbb{R}$*-effective, then the assertion of the corollary holds.* □

Proof Choose positive integers n_0 and n_1 such that $H^0_{\mathbb{R}}(X, n_0D + A) \neq \{0\}$ and $H^0_{\mathbb{R}}(X, n_1D - A) \neq \{0\}$, so that one can take $s \in H^0_{\mathbb{R}}(X, n_0D + A) \setminus \{0\}$ and $s' \in H^0_{\mathbb{R}}(X, n_1D - A) \setminus \{0\}$. Let us consider the following injective homomorphisms

$$\alpha_t : H^0_{\mathbb{R}}((\lfloor t\rfloor - n_0)D) \to H^0_{\mathbb{R}}(tD + A) \quad \text{and} \quad \beta_t : H^0_{\mathbb{R}}(tD + A) \to H^0_{\mathbb{R}}((\lceil t\rceil + n_1)D)$$

given by $f \mapsto fs$ and $f \mapsto fs'$, respectively. Note that

$$\begin{aligned}\|\alpha_t(f)\|_{tg_\omega+h_\omega} &\leqslant \|f\|_{(\lfloor t\rfloor-n_0)g_\omega}\|s\|_{(t-\lfloor t\rfloor+n_0)g_\omega+h_\omega}\\ &\leqslant \|f\|_{(\lfloor t\rfloor-n_0)g_\omega}\|s\|_{n_0g_\omega+h_\omega}\|1\|_{g_\omega}^{t-\lfloor t\rfloor},\end{aligned}$$

so that, by Proposition 4.3.21, (1) and (2),

$$\begin{aligned}\widehat{\deg}_+(H^0_{\mathbb{R}}((\lfloor t\rfloor - n_0)D), \xi_{(\lfloor t\rfloor-n_0)g}) &\leqslant \widehat{\deg}_+(H^0_{\mathbb{R}}(tD + A), \xi_{tg+h})\\ &+ (\dim_K H^0_{\mathbb{R}}((\lfloor t\rfloor - n_0)D)) \int_\Omega \left(\left|\ln \|s\|_{n_0g_\omega+h_\omega}\right| + \left|\ln \|1\|_{g_\omega}\right|\right) \nu(d\omega). \quad (6.48)\end{aligned}$$

In the same way, one has

$$\begin{aligned}\widehat{\deg}_+(H^0_{\mathbb{R}}(tD + A), \xi_{tg+h}) &\leqslant \widehat{\deg}_+(H^0_{\mathbb{R}}((\lceil t\rceil + n_1)D), \xi_{(\lceil t\rceil+n_1)g})\\ &+ (\dim_K H^0_{\mathbb{R}}(tD + A)) \int_\Omega \left(\left|\ln \|s'\|_{n_1g_\omega-h_\omega}\right| + \left|\ln \|1\|_{g_\omega}\right|\right) \nu(d\omega). \quad (6.49)\end{aligned}$$

Note that

$$\lim_{t\to\infty} \frac{\widehat{\deg}_+(H^0_{\mathbb{R}}((\lfloor t\rfloor - n_0)D), \xi_{(\lfloor t\rfloor-n_0)g})}{t^{d+1}/(d+1)!} = \widehat{\mathrm{vol}}(D, g)$$

and

$$\lim_{t\to\infty} \frac{\dim_K H^0_{\mathbb{R}}((\lfloor t\rfloor - n_0)D)}{t^{d+1}/(d+1)!} = 0,$$

so that, by (6.48), one has

$$\widehat{\mathrm{vol}}(D,g) \leqslant \liminf_{n\to\infty} \frac{\widehat{\deg}_+(H^0_{\mathbb{R}}(tD+A), \xi_{tg+h})}{t^d/(d+1)!}.$$

Similarly, by using (6.49),

$$\limsup_{n\to\infty} \frac{\widehat{\deg}_+(H^0_{\mathbb{R}}(tD+A), \xi_{tg+h})}{t^d/(d+1)!} \leqslant \widehat{\mathrm{vol}}(D,g).$$

Thus the assertion of the claim follows. □

Claim 6.4.13 *If there is $s \in K(X)^\times$ such that $D' := D + (s)$ is $\mathbb{R}$-effective, then the assertion of the corollary holds.* □

Proof We set $(D', g') = (D, g) + \widehat{(s)}$. We choose an arithmetically $\mathbb{R}$-effective adelic Cartier divisor (B, k) on X such that $(B,k) \pm \widehat{(s)}$ are arithmetically $\mathbb{R}$-effective. Then, as $(B,k) \pm (t - \lfloor t\rfloor)\widehat{(s)}$ are arithmetically $\mathbb{R}$-effective, one has

$$\begin{aligned} t(D',g') + (A,h) - (B,k) - \lfloor t\rfloor\widehat{(s)} &\leqslant t(D,g) + (A,h) \\ &\leqslant t(D',g') + (A,h) + (B,k) + \lfloor t\rfloor\widehat{(s)}. \end{aligned}$$

Thus

$$\begin{aligned} \widehat{\deg}_+(H^0_{\mathbb{R}}(tD' + A - B), \xi_{tg'+h-k}) &\leqslant \widehat{\deg}_+(H^0_{\mathbb{R}}(tD+A), \xi_{tg+h}) \\ &\leqslant \widehat{\deg}_+(H^0_{\mathbb{R}}(tD' + A + B), \xi_{tg'+h+k}), \end{aligned}$$

so that, by using Claim 6.4.12,

$$\begin{aligned} \widehat{\mathrm{vol}}(D',g') &= \lim_{t\to\infty} \frac{\widehat{\deg}_+(H^0_{\mathbb{R}}(tD' + A - B), \xi_{tg'+h-k})}{t^{d+1}/(d+1)!} \\ &\leqslant \liminf_{t\to\infty} \frac{\widehat{\deg}_+(H^0_{\mathbb{R}}(tD+A), \xi_{tg+h})}{t^{d+1}/(d+1)!} \leqslant \limsup_{t\to\infty} \frac{\widehat{\deg}_+(H^0_{\mathbb{R}}(tD+A), \xi_{tg+h})}{t^{d+1}/(d+1)!} \\ &\leqslant \lim_{t\to\infty} \frac{\widehat{\deg}_+(H^0_{\mathbb{R}}(tD' + A + B), \xi_{tg'+h+k})}{t^{d+1}/(d+1)!} = \widehat{\mathrm{vol}}(D',g'). \end{aligned}$$

Therefore one has the claim because $\widehat{\mathrm{vol}}(D,g) = \widehat{\mathrm{vol}}(D',g')$ □

In general, there are a positive integer a and $f \in K(X)^\times$ such that $aD + (f)$ is $\mathbb{R}$-effective, so that, by using Claim 6.4.11 and Claim 6.4.13, one has

$$\begin{aligned}\widehat{\mathrm{vol}}(D,g) &= \frac{1}{a^{d+1}}\widehat{\mathrm{vol}}(aD,ag) = \frac{1}{a^{d+1}}\lim_{t\to\infty}\frac{\widehat{\deg}_+(H^0_{\mathbb{R}}(taD+A),\xi_{tag+h})}{t^{d+1}/(d+1)!}\\ &= \lim_{t\to\infty}\frac{\widehat{\deg}_+(H^0_{\mathbb{R}}(taD+A),\xi_{tag+h})}{(ta)^{d+1}/(d+1)!}\\ &= \lim_{t\to\infty}\frac{\widehat{\deg}_+(H^0_{\mathbb{R}}(tD+A),\xi_{tg+h})}{t^{d+1}/(d+1)!},\end{aligned}$$

as required. □

Corollary 6.4.14 *Let (D,g) be an adelic $\mathbb{R}$-Cartier divisor on X. Then, for any $a \in \mathbb{R}_{\geqslant 0}$, $\widehat{\mathrm{vol}}(aD,ag) = a^{d+1}\widehat{\mathrm{vol}}(D,g)$.*

Proof Clearly we may assume that $a > 0$. If D is not big, then aD is also not big, so that $\widehat{\mathrm{vol}}(D,g) = 0$ and $\widehat{\mathrm{vol}}(aD,ag) = 0$. Thus the assertion follows in this case. If D is big, then by Corollary 6.4.10,

$$\begin{aligned}\widehat{\mathrm{vol}}(aD,ag) &= \lim_{t\to\infty}\frac{\widehat{\deg}_+(H^0_{\mathbb{K}}(taD),\xi_{tag})}{t^{d+1}/(d+1)!}\\ &= a^{d+1}\lim_{t\to\infty}\frac{\widehat{\deg}_+(H^0_{\mathbb{K}}(taD),\xi_{tag})}{(ta)^{d+1}/(d+1)!} = a^{d+1}\widehat{\mathrm{vol}}(D,g),\end{aligned}$$

as required. □

Remark 6.4.15 Let (D_1,g_1) and (D_2,g_2) be adelic $\mathbb{K}$-Cartier divisors. Proposition 6.3.28 shows that, if $(x,y) \in \Delta(D_1)^\circ \times \Delta(D_2)^\circ$, one has

$$G_{(D_1+D_2,g_1+g_2)}(x+y) \geqslant G_{(D_1,g_1)}(x) + G_{(D_2,g_2)}(y).$$

Similarly to Theorem 6.4.7, an analogue of Brunn-Minkowski inequality holds for adelic $\mathbb{K}$-Cartier divisors.

Theorem 6.4.16 *Let (D_1,g_1) and (D_2,g_2) be big adelic $\mathbb{K}$-Cartier divisors on X. Then the following inequality holds*

$$\widehat{\mathrm{vol}}(D_1+D_2,g_1+g_2)^{1/(d+1)} \geqslant \widehat{\mathrm{vol}}(D_1,g_1)^{1/(d+1)} + \widehat{\mathrm{vol}}(D_2,g_2)^{1/(d+1)}. \tag{6.50}$$

Proof The proof of (6.50) is similar to that of (6.45), which relies on the inequality

$$\forall\,(x,y) \in \Delta(D_1)\times\Delta(D_2),\quad G_{(D_1+D_2,g_1+g_2)}(x+y) \geqslant G_{(D_1,g_1)}(x) + G_{(D_2,g_2)}(y).$$

Let us consider a criterion for the bigness of adelic $\mathbb{K}$-Cartier divisors.

Lemma 6.4.17 *Let (D,g) be an adelic $\mathbb{K}$-Cartier divisor on X such that D is big. Then $\sup_{x\in\Delta(D)^\circ} G_{(D,g)}(x)$ is equal to $\widehat{\mu}^{\mathrm{asy}}_{\max}(D,g)$.*

Proof By (6.32), we obtain that the maximal value of $G_{(D,g)}$ is equal to

$$\lim_{n\to+\infty}\frac{1}{n}\widehat{\mu}_1(H^0_{\mathbb{K}}(nD),\xi_{ng}).$$

Note that all norm families ξ_{ng} are ultrametric on $\Omega\setminus\Omega_\infty$.

By Remark 4.3.48 and the relation

$$\lim_{n\to+\infty}\frac{1}{n}\ln \mathrm{rk}_K(H^0_{\mathbb{K}}(nD))=0,$$

we obtain the equality $\sup_{x\in\Delta(D)^\circ} G_{(D,g)}(x)=\widehat{\mu}^{\mathrm{asy}}_{\max}(D,g)$. □

Proposition 6.4.18 *Let (D,g) be an adelic $\mathbb{K}$-Cartier divisor on X. Then the following are equivalent:*

(1) (D,g) is big.
(2) D is big and $\widehat{\mu}^{\mathrm{asy}}_{\max}(D,g)>0$.

Proof First of all, note that $\sup_{x\in\Delta(D)^\circ} G_{(D,g)}(x)=\widehat{\mu}^{\mathrm{asy}}_{\max}(D,g)$ by Lemma 6.4.17. Moreover, by Theorem 6.4.9,

$$\widehat{\mathrm{vol}}(D,g)=(d+1)\int_{\Delta(D)^\circ}\max(G_{(D,g)}(x),0)\,\mathrm{d}x, \tag{6.51}$$

(1) $\Longrightarrow$ (2): By the above facts, one has

$$\widehat{\mathrm{vol}}(D,g)\leqslant(d+1)\mathrm{vol}(D)\max(\widehat{\mu}^{\mathrm{asy}}_{\max}(D,g),0). \tag{6.52}$$

Therefore, the assertion follows.

(2) $\Longrightarrow$ (1): First of all, as D is big, $\Delta(D)^\circ\neq\emptyset$. Moreover, since $\widehat{\mu}^{\mathrm{asy}}_{\max}(D,g)>0$ and $G_{(D,g)}$ is continuous on $\Delta(D)^\circ$, one can find a non-empty open set U of $\Delta(D)^\circ$ such that $G_{(D,g)}>0$ on U, so that the assertion follows from (6.51). □

Definition 6.4.19 An adelic $\mathbb{K}$-Cartier divisor (D,g) is *strongly big* if D is big and $\nu^{\mathrm{asy}}_1(D,g)>0$, that is, D is big and there are a positive integer a and $s\in H^0_{\mathbb{K}}(aD)\setminus\{0\}$ such that $\widehat{\deg}_{\xi_{ag}}(s)>0$. Note that strong bigness implies bigness by Proposition 6.4.1 and Proposition 6.4.18. Moreover if S satisfies the Minkowski property of certain level, then strong bigness is equivalent to bigness by Proposition 6.4.1 and Proposition 6.4.18.

Proposition 6.4.20 *Let (D,g) be an adelic $\mathbb{K}$-Cartier divisor on X such that D is big. Then there is an integrable function φ on Ω such that $(D,g+\varphi)$ is strongly big.*

Proof Since D is big, there are a positive integer a and $f\in K(X)^\times$ such that $aD+(f)$ is effective. By Proposition 6.2.12, a function given by $\omega\mapsto\ln\|f\|_{ag_\omega}$ is integrable. Thus if we set

$$\varphi(\omega) := \begin{cases} \dfrac{1}{a}(\ln \|f\|_{ag_\omega} + \ln 2) & \text{if } \omega \in \Omega_\infty, \\ \dfrac{1}{a} \ln \|f\|_{ag_\omega} & \text{if } \omega \in \Omega \setminus \Omega_\infty, \end{cases}$$

then $(\omega \in \Omega) \mapsto \varphi(\omega)$ is integrable. Let F_n be a vector subspace of $H^0_{\mathbb{K}}(naD)$ generated by f^n. Then

$$\begin{aligned} \widehat{\deg}_{\xi_{na(g+\varphi)}}(F_n) &= -\int_\Omega \ln \|f^n\|_{na(g_\omega+\varphi(\omega))}\, \nu(d\omega) \\ &= -n \int_\Omega (\ln \|f\|_{ag_\omega} - a\varphi(\omega))\, \nu(d\omega) = n \int_{\Omega_\infty} (\ln 2)\, \nu(d\omega), \end{aligned}$$

so that

$$\nu_1(H^0(naD), \xi_{na(g+\varphi)}) \geqslant \widehat{\deg}_{\xi_{na(g+\varphi)}}(F_n) = n \int_{\Omega_\infty} (\ln 2)\, \nu(d\omega),$$

which shows that $\nu_1^{\mathrm{asy}}(D, g + \varphi) > 0$, so that $(D, g + \varphi)$ is strongly big. □

Definition 6.4.21 Let (D, g) and (D', g') be adelic $\mathbb{K}$-Cartier divisors on X. We define $(D', g') \lesssim (D, g)$ to be

$$(D', g') \lesssim (D, g) \quad \overset{\text{def}}{\Longleftrightarrow} \quad \text{either } (D', g') = (D, g) \text{ or } (D, g) - (D', g') \text{ is big.}$$

Proposition 6.4.22 (1) *The relation $\lesssim$ forms a partial order on the group of adelic $\mathbb{K}$-Cartier divisors on X.*

(2) *For adelic $\mathbb{K}$-Cartier divisors $(D, g), (D', g'), (E, h)$ and (E', h') on X, if $(D', g') \lesssim (D, g)$ and $(E', h') \lesssim (E, h)$, then $(D', g') + (E', h') \lesssim (D, g) + (E, h)$ and $a(D', g') \lesssim a(D, g)$ for $a \in \mathbb{K}_{\geqslant 0}$.*

(3) *For adelic $\mathbb{K}$-Cartier divisors (D, g) and (D', g') on X, if $(D', g') \lesssim (D, g)$, then $\widehat{\mathrm{vol}}(D', g') \leqslant \widehat{\mathrm{vol}}(D, g)$.*

Proof (1) We assume that $(D', g') \lesssim (D, g)$ and $(D, g) \lesssim (D', g')$. If $(D', g') \neq (D, g)$, then $(D, g) - (D', g')$ and $(D', g') - (D, g)$ are big, so that

$$(0, 0) = ((D, g) - (D', g')) + ((D', g') - (D, g))$$

is also big by Theorem 6.4.16, which is a contradiction. Next let us see that if $(D_1, g_1) \lesssim (D_2, g_2)$ and $(D_2, g_2) \lesssim (D_3, g_3)$, then $(D_1, g_1) \lesssim (D_3, g_3)$. Indeed, this is a consequence of Theorem 6.4.16 because

$$(D_3, g_3) - (D_1, g_1) = ((D_3, g_3) - (D_2, g_2)) + ((D_2, g_2) - (D_1, g_1)).$$

(2) follows from Theorem 6.4.16 and Corollary 6.4.14 because

$$\begin{cases} ((D, g) + (E, h)) - ((D', g') + (E', h')) = ((D, g) - (D', g')) + ((E, h) - (E', h')), \\ a(D, g) - a(D', g') = a\,((D, g) - (D', g')). \end{cases}$$

(3) We may assume that $(D,g)-(D',g')$ is big. If (D',g') is big, then the assertion follows from Theorem 6.4.16 because $(D,g)=((D,g)-(D',g'))+(D',g')$. Otherwise, the assertion is obvious because $\widehat{\mathrm{vol}}(D',g')=0$. □

Proposition 6.4.23 *Let (D,g) be a big adelic $\mathbb{K}$-Cartier divisor on X and (A,h) be an adelic $\mathbb{R}$-Cartier divisor on X. Then there is a positive integer n_0 such that $n(D,g)+(A,h)$ is big for all $n\in\mathbb{Z}_{\geqslant n_0}$.*

Proof It is sufficient to find a positive integer n_0 such that $n_0(D,g)+(A,h)$ is big because $n(D,g)+(A,h)=n_0(D,g)+(A,h)+(n-n_0)(D,g)$.

As D is big, one can find a positive integer m such that $mD+A$ is big, so that, by Prioposition 6.4.20, $(mD+A, mg+h+\phi)$ is big for some non-negative integrable function ϕ on Ω. Let a be a positive integer such that

$$\widehat{\mathrm{vol}}(D,g)>\frac{(d+1)\mathrm{vol}(D)}{a}\int_\Omega \phi\,\nu(\mathrm{d}\omega).$$

Since

$$\widehat{\mathrm{vol}}(D,g-\phi/a)\geqslant\widehat{\mathrm{vol}}(D,g)-\frac{(d+1)\mathrm{vol}(D)}{a}\int_\Omega\phi\,\nu(\mathrm{d}\omega)>0$$

by using Proposition 4.3.21, (2), one obtains that $(D,g-\phi/a)$ is big. Thus the assertion follows because

$$(m+a)(D,g)+(A,h)=(mD+A,mg+h+\phi)+a(D,g-\phi/a).$$

Theorem 6.4.24 *Let $(D,g),(D_1,g_1),\ldots,(D_n,g_n)$ be adelic $\mathbb{R}$-Cartier divisors on X. Then*

$$\lim_{\varepsilon_1\to0,\ldots,\varepsilon_n\to0}\widehat{\mathrm{vol}}((D,g)+\varepsilon_1(D_1,g_1)+\cdots+\varepsilon_n(D_n,g_n))=\widehat{\mathrm{vol}}(D,g).$$

Proof Let us begin with the following Claim 6.4.25, Claim 6.4.26, Claim 6.4.27 and Claim 6.4.28:

Claim 6.4.25 *Let (E,h) be an adelic $\mathbb{R}$-Cartier divisor on X. Let $(0,f)$ be an adelic Cartier divisor on X. Then $\lim_{\varepsilon\to0}\widehat{\mathrm{vol}}(E,h+\varepsilon f)=\widehat{\mathrm{vol}}(E,h)$.* □

Proof We set $\varphi_1(\omega)=\sup_{x\in X_\omega}\{f_\omega(x)\}$ and $\varphi_2(\omega)=\sup_{x\in X_\omega}\{-f_\omega(x)\}$. Then, by Proposition 6.2.12, $\varphi_1(\omega)$ and $\varphi_2(\omega)$ are integrable on Ω, so that $\varphi(\omega)=\max\{|\varphi_1(\omega)|,|\varphi_2(\omega)|\}$ is also integrable on Ω and $|f_\omega(x)|\leqslant\varphi(\omega)$ for all $x\in X_\omega$ and $\omega\in\Omega$. Therefore,

$$h_\omega-|\varepsilon|\varphi(\omega)\leqslant h_\omega+\varepsilon f_\omega\leqslant h_\omega+|\varepsilon|\varphi(\omega),$$

so that, by Proposition 4.3.21, (1),

$$\begin{aligned}\widehat{\deg}_+(H^0_{\mathbb{K}}(nE),\mathrm{e}^{n|\varepsilon|\varphi}\xi_{nh})&\leqslant\widehat{\deg}_+(H^0_{\mathbb{K}}(nE),\xi_{n(h+\varepsilon f)})\\&\leqslant\widehat{\deg}_+(H^0_{\mathbb{K}}(nE),\mathrm{e}^{-n|\varepsilon|\varphi}\xi_{nh}).\end{aligned}$$

Moreover, by Proposition 4.3.21, (2),

$$\begin{aligned}\widehat{\deg}_+(H^0_{\mathbb{K}}(nE), \xi_{nh}) &\leqslant \widehat{\deg}_+(H^0_{\mathbb{K}}(nE), \mathrm{e}^{-n|\varepsilon|\varphi}\xi_{nh}) \\ &\leqslant \widehat{\deg}_+(H^0_{\mathbb{K}}(nE), \xi_{nh}) + n|\varepsilon| \dim_K(H^0_{\mathbb{K}}(nE)) \int_\Omega \varphi\, \nu(\mathrm{d}\omega),\end{aligned}$$

and

$$\begin{aligned}\widehat{\deg}_+(H^0_{\mathbb{K}}(nE), \mathrm{e}^{n|\varepsilon|\varphi}\xi_{nh}) &\leqslant \widehat{\deg}_+(H^0_{\mathbb{K}}(nE), \xi_{nh}) \\ &\leqslant \widehat{\deg}_+(H^0_{\mathbb{K}}(nE), \mathrm{e}^{n|\varepsilon|\varphi}\xi_{nh}) + n|\varepsilon| \dim_K(H^0_{\mathbb{K}}(nE)) \int_\Omega \varphi\, \nu(\mathrm{d}\omega).\end{aligned}$$

Therefore the assertion of the claim follows. □

Claim 6.4.26 *Let (B, f) be an adelic $\mathbb{R}$-Cartier divisor on X such that $(B, f)\pm(D_i, g_i)$ is big for every $i = 1, \ldots, n$. Then*

$$\begin{aligned}\widehat{\mathrm{vol}}((D, g) - (|\varepsilon_1| + \cdots + |\varepsilon_n|)(B, f)) &\\ \leqslant \widehat{\mathrm{vol}}((D, g) + \varepsilon_1(D_1, g_1) + \cdots + \varepsilon_n(D_n, g_n)) &\\ \leqslant \widehat{\mathrm{vol}}((D, g) + (|\varepsilon_1| + \cdots + |\varepsilon_n|)(B, f)).&\end{aligned}$$

□

Proof Since

$$\begin{cases}|\varepsilon_i|(B, f) - \varepsilon_i(D_i, g_i) = |\varepsilon_i|((B, f) \pm (D_i, g_i)),\\ \varepsilon_i(D_i, g_i) + |\varepsilon_i|(B, f) = |\varepsilon_i|((B, f) \pm (D_i, g_i)),\end{cases}$$

one has $-|\varepsilon_i|(B, f) \lesssim \varepsilon_i(D_i, g_i) \lesssim |\varepsilon_i|(B, f)$ by Proposition 6.4.22, (2), so that, by using Proposition 6.4.22, (2) again,

$$\begin{aligned}(D, g) - (|\varepsilon_1| + \cdots + |\varepsilon_n|)(B, f) &\\ \lesssim (D, g) + \varepsilon_1(D_1, g_1) + \cdots + \varepsilon_n(D_n, g_n) &\\ \lesssim (D, g) + (|\varepsilon_1| + \cdots + |\varepsilon_n|)(B, f).&\end{aligned}$$

Therefore, by Proposition 6.4.22, (3), one obtains the claim. □

Claim 6.4.27 *Let (H, g_H) be an adelic $\mathbb{R}$-Cartier divisor on X. Then there is an integrable function ψ on S such that $(H, g_H - \psi)$ is not big.* □

Proof Proposition 6.2.16 and Proposition 6.4.4, one obtains $\widehat{\mu}^{\mathrm{asy}}_{\max}(H, g_H) < \infty$, so that one can find an integrable function ψ on S such that

$$\widehat{\mu}^{\mathrm{asy}}_{\max}(H, g_H) < \int_\Omega \psi\, \nu(\mathrm{d}\omega).$$

We choose a positive integer n_0 such that

$$\widehat{\mu}_{\max}(H^0_{\mathbb{R}}(X, nH), \xi_{ng_H}) \leqslant \int_\Omega n\psi\, \nu(\mathrm{d}\omega)$$

for all $n \geqslant n_0$. Thus, as $\xi_{n(g_H - \psi)} = \exp(n\psi)\,\xi_{ng_H}$, by Lemma 4.3.36, (1),

$$\widehat{\mu}_{\max}(H^0_{\mathbb{R}}(X, nH), \xi_{n(g_H - \psi)}) = \widehat{\mu}_{\max}(H^0_{\mathbb{R}}(X, nH), \xi_{ng_H}) - \int_\Omega n\psi\, \nu(\mathrm{d}\omega) \leqslant 0,$$

so that the assertion follows from Lemma 4.3.36, (2).

Claim 6.4.28 *Let (H, g_H) be an adelic $\mathbb{R}$-Cartier divisor on X and φ be an integrable function on Ω. Then*

$$\widehat{\mathrm{vol}}(H, g_H + \varphi) \leqslant \widehat{\mathrm{vol}}(H, g_H) + (d+1)\mathrm{vol}(H) \int_\Omega |\varphi(\omega)|\, \nu(\mathrm{d}\omega).$$

□

Proof As $\xi_{n(g_H+\varphi)} = \exp(-n\varphi)\xi_{ng_H}$, by using Proposition 4.3.21, (2),

$$\begin{aligned}\widehat{\deg}_+(H^0_{\mathbb{R}}(X, nH), \xi_{n(g_H+\varphi)}) \leqslant\ & \widehat{\deg}_+(H^0_{\mathbb{R}}(X, nH), \xi_{ng_H}) \\ & + n(\dim_K H^0_{\mathbb{R}}(nH)) \int_\Omega |\varphi(\omega)|\, \nu(\mathrm{d}\omega),\end{aligned}$$

so that the assertion follows. □

First we assume that D is big. By Proposition 6.4.20, we can choose a D-Green functions family g' such that (D, g') is a big adelic $\mathbb{K}$-Cartier divisor. Then, by Proposition 6.4.23, one can choose a positive integer a such that $a(D, g') \pm (D_i, g_i)$ is big for every $i = 1, \ldots, n$. Then, by Claim 6.4.26,

$$\begin{aligned}&\widehat{\mathrm{vol}}((D, g) - a(|\varepsilon_1| + \cdots + |\varepsilon_n|)(D, g')) \\ &\qquad \leqslant \widehat{\mathrm{vol}}((D, g) + \varepsilon_1(D_1, g_1) + \cdots + \varepsilon_n(D_n, g_n)) \\ &\qquad\qquad \leqslant \widehat{\mathrm{vol}}((D, g) + a(|\varepsilon_1| + \cdots + |\varepsilon_n|)(D, g'))\end{aligned}$$

If we set $f = g' - g$ and $\varepsilon = |\varepsilon_1| + \cdots + |\varepsilon_n|$, then

$$\begin{cases}(D, g) - a(|\varepsilon_1| + \cdots + |\varepsilon_n|)(D, g') = (1 - a\varepsilon)\left((D, g) + \left(0, \frac{a\varepsilon}{1-a\varepsilon} f\right)\right), \\ (D, g) + a(|\varepsilon_1| + \cdots + |\varepsilon_n|)(D, g') = (1 + a\varepsilon)\left((D, g) + \left(0, \frac{a\varepsilon}{1+a\varepsilon} f\right)\right).\end{cases}$$

Therefore, by Claim 6.4.25,

$$\lim_{\varepsilon_1\to 0,\ldots,\varepsilon_n\to 0} \widehat{\mathrm{vol}}((D,g) - a(|\varepsilon_1| + \cdots + |\varepsilon_n|)(D,g'))$$
$$= \lim_{\varepsilon\to 0}(1-a\varepsilon)^{d+1}\widehat{\mathrm{vol}}\left((D,g) + \left(0, \frac{a\varepsilon}{1-a\varepsilon}f\right)\right) = \widehat{\mathrm{vol}}(D,g).$$

In the same way,

$$\lim_{\varepsilon_1\to 0,\ldots,\varepsilon_n\to 0} \widehat{\mathrm{vol}}((D,g) + a(|\varepsilon_1| + \cdots + |\varepsilon_n|)(D,g')) = \widehat{\mathrm{vol}}(D,g).$$

One has the theorem in the case where D is big.

Next we assume that D is not big. Let (A,h) be a big adelic Cartier divisor on X such that $D+A$ is big and $(A,h) \pm (D_i, g_i)$ are big for every $i = 1,\ldots,n$. Then, by Claim 6.4.26, if we set $\varepsilon = |\varepsilon_1| + \cdots + |\varepsilon_n|$, then

$$0 \leqslant \widehat{\mathrm{vol}}((D,g) + \varepsilon_1(D_1,g_1) + \cdots + \varepsilon_n(D_n,g_n)) \leqslant \widehat{\mathrm{vol}}((D,g) + \varepsilon(A,h)),$$

and hence one need show that $\lim_{\varepsilon\downarrow 0}\widehat{\mathrm{vol}}((D,g)+\varepsilon(A,h)) = 0$. By Claim 6.4.27, one can choose a non-negative integrable function φ on Ω such that $(D,g)+(A,h)-(0,\varphi)$ is not big. Then, as $(D,g) - (0,\varphi) + \varepsilon(A,h) \lesssim (D,g) + (A,h) - (0,\varphi) + \varepsilon(A,h)$, one has

$$\widehat{\mathrm{vol}}((D,g) - (0,\varphi) + \varepsilon(A,h)) \leqslant \widehat{\mathrm{vol}}((D,g) + (A,h) - (0,\varphi) + \varepsilon(A,h)).$$

Since $D + A$ is big, by the previous case,

$$\lim_{\varepsilon\downarrow 0}\widehat{\mathrm{vol}}((D,g) + (A,h) - (0,\varphi) + \varepsilon(A,h)) = \widehat{\mathrm{vol}}((D,g) + (A,h) - (0,\varphi)) = 0,$$

and hence

$$\lim_{\varepsilon\downarrow 0}\widehat{\mathrm{vol}}((D,g) - (0,\varphi) + \varepsilon(A,h)) = 0. \tag{6.53}$$

On the other hand, by Claim 6.4.28,

$$\widehat{\mathrm{vol}}((D,g) + \varepsilon(A,h)) \leqslant \widehat{\mathrm{vol}}((D,g) - (0,\varphi) + \varepsilon(A,h))$$
$$+ (d+1)\mathrm{vol}(D+\varepsilon A)\int_\Omega \varphi(\omega)\nu(d\omega).$$

As D is not big, one obtains

$$\lim_{\varepsilon\downarrow 0}\mathrm{vol}(D+\varepsilon A) = \mathrm{vol}(D) = 0,$$

and hence, by (6.53), one has $\lim_{\varepsilon\downarrow 0}\widehat{\mathrm{vol}}((D,g) + \varepsilon(A,h)) = 0$, as required.

Corollary 6.4.29 *Let H be a finite-dimensional vector subspace of $\widehat{\mathrm{Div}}_{\mathbb{R}}(X)$. Then the set $\{(D,g) \in H \mid (D,g)$ is big$\}$ is an open cone in H.*

Proof The openness of it is a consequence of Theorem 6.4.24. One can check that it is a cone by Theorem 6.4.16 and Corollary 6.4.14. □

Corollary 6.4.30 *The volume function* $\widehat{\mathrm{vol}} : \widehat{\mathrm{Div}}_{\mathbb{R}}(X) \to \mathbb{R}$ *factors through* $\widehat{\mathrm{Div}}_{\mathbb{R}}(X)$ *modulo the vector subspace over* $\mathbb{R}$ *generated by principal Cartier divisors, that is,*

$$\widehat{\mathrm{vol}}((D, g) + a_1\widehat{(f_1)} + \cdots + a_r\widehat{(f_r)}) = \widehat{\mathrm{vol}}(D, g)$$

for any $r \in \mathbb{Z}_{\geqslant 1}$, $(D, g) \in \widehat{\mathrm{Div}}_{\mathbb{R}}(X)$, $f_1, \ldots, f_r \in K(X)^{\times}$ *and* $a_1, \ldots, a_r \in \mathbb{R}$.

Proof If $a_1, \ldots, a_r \in \mathbb{Z}$, then the assertion is obvious. Next we assume that $a_1, \ldots, a_r \in \mathbb{Q}$. We choose a positive integer N such that $Na_1, \ldots, Na_r \in \mathbb{Z}$. Then

$$\begin{aligned} N^{d+1}\widehat{\mathrm{vol}}(D, g) &= \widehat{\mathrm{vol}}(ND, Ng) \\ &= \widehat{\mathrm{vol}}((ND, Ng) + (Na_1)\widehat{(f_1)} + \cdots + (Na_r)\widehat{(f_r)}) \\ &= N^{d+1}\widehat{\mathrm{vol}}((D, g) + a_1\widehat{(f_1)} + \cdots + a_r\widehat{(f_r)}), \end{aligned}$$

as required. In general, take sequences $\{a_{1,n}\}_{n=1}^{\infty}, \ldots, \{a_{r,n}\}_{n=1}^{\infty}$ of rational numbers such that $a_1 = \lim_{n\to\infty} a_{1,n}, \ldots, a_r = \lim_{n\to\infty} a_{r,n}$. Then, by Theorem 6.4.24,

$$\begin{aligned} &\widehat{\mathrm{vol}}((D, g) + a_1\widehat{(f_1)} + \cdots + a_r\widehat{(f_r)}) \\ &\qquad = \lim_{n\to\infty} \widehat{\mathrm{vol}}((D, g) + a_{1,n}\widehat{(f_1)} + \cdots + a_{r,n}\widehat{(f_r)}) = \widehat{\mathrm{vol}}(D, g), \end{aligned}$$

so that the assertion follows. □

Chapter 7
Nakai-Moishezon's criterion

In this chapter, we fix a proper adelic curve $S = (K, (\Omega, \mathcal{A}, \nu), \phi)$. We assume that, either the σ-algebra $\mathcal{A}$ is discrete, or the field K admits a countable subfield which is dense in each K_ω, where $\omega \in \Omega$. We let Ω_0 be the set of all $\omega \in \Omega$ such that $|\cdot|_\omega$ is the trivial absolute value. Note that, if Ω_0 is not empty, then the above hypothesis implies that, either the σ-algebra $\mathcal{A}$ is discrete, or the field K is countable.

7.1 Graded algebra of adelic vector bundles

Let C be a non-negative real number. In this section, we assume that the adelic curve S satisfies the *tensorial minimal slope property* of level $\geqslant C_0$. Namely, for any pair $(\overline{E}, \overline{F})$ of non-zero adelic vector bundles on S, one has

$$\widehat{\mu}_{\min}(\overline{E} \otimes_{\varepsilon,\pi} \overline{F}) \geqslant \widehat{\mu}_{\min}(\overline{E}) + \widehat{\mu}_{\min}(\overline{F}) - C(\ln(\dim_K(E) \cdot \dim_K(F))). \tag{7.1}$$

Note that we have shown in Chapter 5 that, if the field K is of characteristic 0, then the adelic curve S satisfies the tensorial minimal slope property of level $\geqslant \frac{3}{2}\nu(\Omega_\infty)$.

Definition 7.1.1 Let $R_\bullet = \bigoplus_{n\in\mathbb{N}} R_n$ be a graded K-algebra. We assume that, for any $n \in \mathbb{N}$, R_n is of finite dimension over K. For any $n \in \mathbb{N}$, let $\xi_n = \{\|\cdot\|_{n,\omega}\}_{\omega\in\Omega}$ be a norm family on R_n. We say that $\overline{R}_\bullet = \{(R_n, \xi_n)\}_{n\in\mathbb{N}}$ is a *normed graded algebra on* S if, for any $\omega \in \Omega$, $\overline{R}_{\bullet,\omega} = \{(R_{n,\omega}, \|\cdot\|_{n,\omega})\}_{n\in\mathbb{N}}$ forms a normed graded algebra over K_ω, where $R_{n,\omega} := R_n \otimes_K K_\omega$ (cf. Subsection 1.1.14). Moreover, if (R_n, ξ_n) forms an adelic vector bundle on S for all $n \in \mathbb{N}$, then $\overline{R}_\bullet$ is called a *graded algebra of adelic vector bundles on S*. Furthermore, we say that $\overline{R}_\bullet$ is *of finite type* if the underlying graded K-algebra $R_\bullet$ is of finite type over K.

Proposition 7.1.2 *Let $\overline{R}_\bullet = \{(R_n, \xi_n)\}$ be a graded algebra of adelic vector bundles on S such that $\overline{R}_0$ is the trivial adelic line bundle, namely $R_0 = K$ and for any $\omega \in \Omega$, one has $\|1\|_\omega = 1$. Suppose in addition that $R_\bullet$ is generated as R_0-algebra by R_1. Then the sequence $\{\widehat{\mu}_{\min}(\overline{R}_n)/n\}_{n\in\mathbb{N}}$ converges to an element of $\mathbb{R} \cup \{+\infty\}$.*

H. Chen, A. Moriwaki, *Arakelov Geometry over Adelic Curves*, Lecture Notes in Mathematics 2258, https://doi.org/10.1007/978-981-15-1728-0_7

Proof Let (n, m) be a couple of positive integers. Since $R_\bullet$ is generated as K-algebra by R_1, the canonical K-linear map $f_{n,m} : R_n \otimes_K R_m \longrightarrow R_{n+m}$ is surjective. Moreover, if we equip $R_n \otimes_K R_m$ with the ε,π-tensor product norm family $\xi_n \otimes_{\varepsilon,\pi} \xi_m$, then, by the submultiplicativity condition, the homomorphism $f_{n,m}$ has height $\leqslant 1$. By Proposition 4.3.31, one has

$$\widehat{\mu}_{\min}(\overline{R}_n \otimes_{\varepsilon,\pi} \overline{R}_m) \leqslant \widehat{\mu}_{\min}(\overline{R}_{n+m}).$$

Moreover, by the assumption of tensorial minimal slope property,

$$\widehat{\mu}_{\min}(\overline{R}_n \otimes_{\varepsilon,\pi} \overline{R}_m) \geqslant \widehat{\mu}_{\min}(\overline{R}_n) + \widehat{\mu}_{\min}(\overline{R}_m) - C_0(\ln(\dim_K(R_n)) + \ln(\dim_K(R_m))).$$

Note that $R_\bullet$ is a quotient K-algebra of $K[R_1]$. Hence $\dim_K(R_n) = O(n^{\dim_K(R_1)-1})$. By [40, Proposition 1.3.5][1], the sequence $\{\widehat{\mu}_{\min}(\overline{R}_n)/n\}_{n\in\mathbb{N}}$ converges to an element in $\mathbb{R} \cup \{+\infty\}$. □

Definition 7.1.3 Let $\overline{R}_\bullet$ be a normed graded algebra on S. Let $M_\bullet = \bigoplus_{n\in\mathbb{Z}} M_n$ be a $\mathbb{Z}$-graded K-linear space and h be a positive integer. We say that $M_\bullet$ is an *h-graded $R_\bullet$-module* if $M_\bullet$ is equipped with a structure of $R_\bullet$-module such that

$$\forall\,(n, m) \in \mathbb{N} \times \mathbb{Z}, \quad \forall (a, x) \in R_n \times M_m, \quad ax \in M_{nh+m}.$$

Let $M_\bullet$ be an h-graded $R_\bullet$-module. Assume that each homogeneous component M_n is of finite dimension over K and is equipped with a norm family $\xi'_n = \{\|\cdot\|'_{n,\omega}\}_{\omega\in\Omega}$. We say that $\overline{M}_\bullet = \{\overline{M}_n\}_{n\in\mathbb{Z}}$ is a *normed h-graded $\overline{R}_\bullet$-module* if, for any $\omega \in \Omega$, $\overline{M}_{\bullet,\omega} = \{(M_{n,\omega}, \|\cdot\|'_{n,\omega})\}_{n\in\mathbb{Z}}$ forms a normed h-graded $\overline{R}_{\bullet,\omega}$-module, where $M_{n,\omega} := M_n \otimes_K K_\omega$ (cf. Subsetion 1.1.14). We say that an h-graded $\overline{R}_\bullet$-module $\overline{M}_\bullet$ is *of finite type* if the underlying h-graded $R_\bullet$-module $M_\bullet$ is of finite type. Moreover, if (M_n, ξ'_n) forms an adelic vector bundle on S for all $n \in \mathbb{Z}$, then $\overline{M}_\bullet$ is called a *h-graded $\overline{R}_\bullet$-module of adelic vector bundles on S.*

Proposition 7.1.4 *Let C_0 be a non-negative constant. We assume that the adelic curve S satisfies the tensorial minimal slope property of level $\geqslant C_0$. Let $\overline{R}_\bullet$ be a graded algebra of adelic vector bundles which is of finite type, and $\overline{M}_\bullet$ be an h-graded $\overline{R}_\bullet$-module of adelic vector bundles on S such that $\overline{M}_\bullet$ is of finite type, where h is a positive integer. Then one has*

$$\liminf_{n\to+\infty} \frac{\widehat{\mu}_{\min}(\overline{M}_n)}{n} \geqslant \frac{1}{h} \liminf_{n\to+\infty} \frac{\widehat{\mu}_{\min}(\overline{R}_n)}{n} > -\infty. \tag{7.2}$$

[1] In the statement of [40, Proposition 1.3.5], we suppose given a *positive* sequence $\{b_n\}_{n\in\mathbb{N},\, n\geqslant 1}$ satisfying the *weak subadditivity* condition $b_{n+m} \leqslant b_n + b_m + f(n) + f(m)$, where $f : \mathbb{N}_{\geqslant 1} \to \mathbb{R}_+$ is a non-decreasing function such that $\sum_{\alpha\geqslant 0} f(2^\alpha)/2^\alpha < +\infty$. Then the sequence $\{b_n/n\}_{n\in\mathbb{N},\, n\geqslant 1}$ converges in $\mathbb{R}_+$. However the same proof applies to a general (not necessarily positive) sequence satisfying the same weak subadditivity condition and leads to the convergence of the sequence $\{b_n/n\}_{n\in\mathbb{N},\, n\geqslant 1}$ in $\mathbb{R} \cup \{-\infty\}$.

Proof If we replace $\overline{R}_0$ by the trivial adelic line bundle, we obtain a new graded algebra of adelic vector bundles (denoted by $\overline{R}'_{\bullet}$) and $\overline{M}_{\bullet}$ is naturally equipped with a structure of h-graded module over this graded algebra of adelic vector bundles. Moreover, $R_{\bullet}$ is a finite $R'_{\bullet}$-algebra since R_0 is supposed to be of finite dimension over K. In particular, $M_{\bullet}$ is a module of finite type over $R'_{\bullet}$. If $\{a_i\}_{i\in I}$ is a basis of R_0 over K which contains $1 \in R_0$ and if $\{b_j\}_{j\in J}$ is a finite family of homogeneous elements of positive degree in $R_{\bullet}$, which generates $R_{\bullet}$ as R_0-algebra, then $R'_{\bullet}$ is generated as K-algebra by $\{a_ib_j\}_{(i,j)\in I\times J}$. This shows that $R'_{\bullet}$ is a K-algebra of finite type. Therefore (by replacing $\overline{R}_{\bullet}$ by $\overline{R}'_{\bullet}$) we may assume without loss of generality that $\overline{R}_0$ is the trivial adelic line bundle.

We first prove the proposition in the particular case where $R_{\bullet}$ is generated as K-algebra by R_1. Let A be the infimum limit of the sequence $\{\widehat{\mu}_{\min}(\overline{R}_n)/n\}_{n\in\mathbb{N},\,n\geqslant 1}$. By [74, Lemma 2.1.6], there exist integers b_1 and $m > 0$ such that, for any integer b with $b \geqslant b_1$ and any integer $\ell \geqslant 1$ the canonical K-linear map $R_{\ell m} \otimes_K M_b \to M_{b+\ell mh}$ is surjective. Hence by Proposition 4.3.31, one has

$$\widehat{\mu}_{\min}(\overline{M}_{b+\ell mh}) \geqslant \widehat{\mu}_{\min}(\overline{R}_{\ell m} \otimes_{\varepsilon,\pi} \overline{M}_b),$$

which leads to

$$\widehat{\mu}_{\min}(\overline{M}_{b+\ell mh}) \geqslant \widehat{\mu}_{\min}(\overline{R}_{\ell m}) + \widehat{\mu}_{\min}(\overline{M}_b) - C_0 \ln(\dim_K(R_{\ell m}) \cdot \dim_K(M_b)).$$

Dividing the two sides of the inequality by ℓmh and then letting ℓ tend to the infinity, we obtain

$$\liminf_{\ell\to+\infty} \frac{\widehat{\mu}_{\min}(\overline{M}_{b+\ell mh})}{\ell mh} \geqslant \frac{1}{h}A,$$

where we have used the fact that

$$\lim_{\ell\to+\infty} \frac{\ln(\dim_K(R_{\ell m}))}{\ell} = 0.$$

Since $b \geqslant b_1$ is arbitrary, we obtain

$$\liminf_{n\to+\infty} \frac{\widehat{\mu}_{\min}(\overline{M}_n)}{n} \geqslant \frac{1}{h}A.$$

We now consider the general case. By [74, Lemma 2.1.6], there exists a positive integer u such that $R_{\bullet}^{(u)} := \bigoplus_{n\in\mathbb{N}} R_{un}$ is generated as K-algebra by $R_1^{(u)} = R_u$. Moreover, $R_{\bullet}$ is a u-graded $R^{(u)}$-module of finite type and hence a finite $R_{\bullet}^{(u)}$-algebra. Therefore $M_{\bullet}$ is an hu-graded $R_{\bullet}^{(u)}$-algebra of finite type. Let B be the infimum limit of the sequence $\{\widehat{\mu}_{\min}(\overline{R}_{nu})/n\}_{n\in\mathbb{N},\,n\geqslant 1}$. By applying the particular case of the proposition established above, we obtain

$$\liminf_{n\to+\infty} \frac{\widehat{\mu}_{\min}(\overline{R}_n)}{n} \geqslant \frac{B}{u} \quad\text{and}\quad \liminf_{n\to+\infty} \frac{\widehat{\mu}_{\min}(\overline{M}_n)}{n} \geqslant \frac{B}{hu}.$$

Note that the first inequality actually implies that

$$\liminf_{n\to+\infty}\frac{\widehat{\mu}_{\min}(\overline{R}_n)}{n}=\frac{B}{u}$$

since $\{\widehat{\mu}_{\min}(\overline{R}_{nu})/n\}_{n\in\mathbb{N},\,n\geqslant 1}$ converges to B. The inequality (7.2) is thus proved.

Finally, if the adelic curve S satisfies the tensorial minimal slope property, by Proposition 7.1.2 the sequence $\{\widehat{\mu}_{\min}(\overline{R}_{nu})/n\}_{n\in\mathbb{N},\,n\geqslant 1}$ converges to an element of $\mathbb{R}\cup\{+\infty\}$. Hence the last statement of the proposition is true. □

Remark 7.1.5 Let $\overline{R}_\bullet$ be a graded algebra of adelic vector bundles, $I_\bullet$ be a homogeneous ideal of $R_\bullet$ and $R'_\bullet$ be the quotient algebra $R_\bullet/I_\bullet$. If we equip R'_n with the quotient norm family of that of R_n, then $\overline{R}'_\bullet$ is a graded algebra of adelic vector bundles, denoted by $\overline{R_\bullet/I_\bullet}$ (cf. Proposition 1.1.71 (1)).

More generally, let $\overline{M}_\bullet$ be an h-graded $\overline{R}_\bullet$-module and $Q_\bullet$ is a graded quotient $R_\bullet$-module of $M_\bullet$. If we equip each Q_n with the quotient norm family of that of $\overline{M}_n$, then $\overline{Q}_\bullet$ becomes an h-graded $\overline{R}_\bullet$-module (cf. Proposition 1.1.71 (2)).

Let $\overline{R}_\bullet$ be a graded algebra of adelic vector bundles, $\overline{M}_\bullet$ be an h-graded module, where $h\in\mathbb{N}$, $h\geqslant 1$. Let $I_\bullet$ be a homogeneous ideal of $R_\bullet$. Assume that $M_\bullet$ is annihilated by $I_\bullet$, then $\overline{M}_\bullet$ is naturally equipped with a structure of h-graded $\overline{R_\bullet/I_\bullet}$-module (cf. Proposition 1.1.71 (2)).

Proposition 7.1.6 *We suppose that the adelic curve S satisfies the tensorial minimal slope property. Let $\overline{R}_\bullet$ be a graded algebra of adelic vector bundles, $I_\bullet$, $J_\bullet$ and $M_\bullet$ be homogeneous ideals of $R_\bullet$ such that $J_\bullet\subseteq M_\bullet$ and $I_\bullet\cdot M_\bullet\subseteq J_\bullet$. Let $R'_\bullet=R_\bullet/I_\bullet$ and $Q_\bullet=M_\bullet/J_\bullet$. For each $n\in\mathbb{N}$, we equip R'_n and Q_n with the quotient norm families of that of $\overline{R}_n$ and $\overline{M}_n$ respectively. Then one has*

$$\liminf_{n\to+\infty}\frac{\widehat{\mu}_{\min}(\overline{Q}_n)}{n}\geqslant\liminf_{n\to+\infty}\frac{\widehat{\mu}_{\min}(\overline{R}'_n)}{n}.$$

Proof By the above remark, $\overline{Q}_\bullet$ is equipped with a structure of graded $\overline{R}'_\bullet$-module. Hence the statement follows from Proposition 7.1.4. □

7.2 Fundamental estimations

In this section, we prove some lower bounds of asymptotic minimal slope. Let $\overline{R}_\bullet=\{(R_n,\xi_n)\}_{n\in\mathbb{N}}$ be a graded algebra of adelic vector bundles which is of finite type. We assume that $R_\bullet$ is an integral ring. Let $X=\mathrm{Proj}(R_\bullet)$ be the projective spectrum of $R_\bullet$. If Y is an integral closed subscheme of X and $P_\bullet\subseteq R_\bullet$ is the defining homogeneous prime ideal of Y, we denote by $R_{Y,\bullet}$ the quotient graded ring $R_\bullet/P_\bullet$. Note that each $R_{Y,n}$ is naturally equipped with the quotient norm family $\xi_{Y,n}$ of ξ_n so that $\overline{R}_{Y,\bullet}$ becomes a graded algebra of adelic vector bundles (cf. Proposition 1.1.71, Proposition 4.1.19 and Proposition 4.1.24).

Theorem 7.2.1 *We assume that the adelic curve S satisfies the tensorial minimal slope property of level $\geqslant C_0$, where $C_0 \geqslant 0$. Let $\mathfrak{S}_X$ be the set of all integral closed subschemes of X. To each $Y \in \mathfrak{S}_X$ we assigne a real number υ_Y, a positive integer n_Y and a non-zero element s_Y in R_{Y,n_Y} such that $\widehat{\deg}_{\xi_{Y,n_Y}}(s_Y) \geqslant n_Y \upsilon_Y$. Then there exists a finite subset $\mathfrak{S}$ of $\mathfrak{S}_X$ such that*

$$\liminf_{n\to+\infty} \frac{\widehat{\mu}_{\min}(\overline{R}_n)}{n} \geqslant \min\{\upsilon(Y) \,:\, Y \in \mathfrak{S}\}. \tag{7.3}$$

In particular, one has

$$\liminf_{n\to+\infty} \frac{\widehat{\mu}_{\min}(R_n, \xi_n)}{n} \geqslant \inf_{Y\in\mathfrak{S}_X} \limsup_{n\to+\infty} \frac{\nu_1(R_{Y,n}, \xi_{Y,n})}{n} \tag{7.4}$$

Proof **Step 1**: For any positive integer h, we set

$$R_n^{(h)} := R_{hn} \quad \text{and} \quad R^{(h)} = \bigoplus_{n\in\mathbb{N}} R_n^{(h)}.$$

If we assign υ_Y^h, hn_Y and s_Y^h to each $Y \in \mathfrak{S}_X$, then $s_Y^h \in R_{Y,hn_Y} \setminus \{0\}$ and

$$\widehat{\deg}_{\xi_{Y,hn_Y}}(s_Y^h) \geqslant h \cdot \widehat{\deg}_{\xi_{Y,n_Y}}(s_Y) \geqslant h\upsilon_Y,$$

so that the above assignment satisfies the condition of the theorem for $R^{(h)}$. Moreover, R is a finitely generated h-graded $R^{(h)}$-module (cf. [111, Lemma 5.44]). By using Proposition 7.1.6, we can see that if the theorem holds for $R^{(h)}$, then it holds for R. Therefore, by [31, Chapitre III, §1, Proposition 3], we may assume that R is generated by R_1 over R_0 and $n_X = 1$. Let $\mathscr{O}_X(1)$ be the tautological invertible sheaf of X arising from R_1.

We prove the theorem by induction on $d = \dim X$.

Step 2: In the case where $d = 0$, $X = \operatorname{Spec}(F)$ for some finite extension field F over K, so that $R_n \subseteq H^0(X, \mathscr{O}_X(n)) \cong F$. Therefore, $\dim_K(R_n) \leqslant [F : K]$ for all $n \in \mathbb{N}$. Let us consider the following sequence of homomorphisms:

$$R_0 \xrightarrow{s_X\cdot} R_1 \xrightarrow{s_X\cdot} R_2 \xrightarrow{s_X\cdot} R_3 \xrightarrow{s_X\cdot} \cdots \xrightarrow{s_X\cdot} R_{n-1} \xrightarrow{s_X\cdot} R_n \xrightarrow{s_X\cdot} \cdots,$$

Note that each homomorphism is injective and $\dim_K(R_n)$ is bounded, so that we can find a positive integer N such that $R_n \xrightarrow{s_X\cdot} R_{n+1}$ is an isomorphism for all $n \in \mathbb{N}_{\geqslant N}$. Therefore, by Proposition 4.3.31,

$$\widehat{\mu}_{\min}(\overline{R}_n) \geqslant \widehat{\mu}_{\min}(\overline{R}_N) + (n-N)\,\widehat{\deg}_{\xi_1}(s_X) \geqslant \widehat{\mu}_{\min}(\overline{R}_N) + (n-N)\upsilon_X,$$

which leads to

$$\liminf_{n\to+\infty} \frac{\widehat{\mu}_{\min}(\overline{R}_n)}{n} \geqslant \upsilon_X.$$

Step 3: We assume $d > 0$. Let $I_\bullet$ be the homogeneous ideal generated by s_X, that is, $I_\bullet = R_\bullet s_X$. By using the same ideas as in [85, Chapter I, Proposition 7.4], we can find a sequence

$$I_\bullet = I_{0,\bullet} \subsetneq I_{1,\bullet} \subsetneq \cdots \subsetneq I_{r,\bullet} = R_\bullet$$

of homogeneous ideals of $R_\bullet$ and non-zero homogeneous prime ideals $P_{1,\bullet}, \ldots, P_{r,\bullet}$ of $R_\bullet$ such that $P_{i,\bullet} \cdot I_{i,\bullet} \subseteq I_{i-1,\bullet}$ for $i \in \{1, \ldots, r\}$.

Step 4: Consider the following sequence:

$$\begin{array}{cccccccccc}
R_0 & \xrightarrow{\cdot s_X} & I_{0,1} & \hookrightarrow \cdots \hookrightarrow & I_{i,1} & \hookrightarrow \cdots \hookrightarrow & I_{r,1} = R_1 \\
\vdots & \vdots & \vdots & & \vdots & \vdots & \vdots \\
 & \xrightarrow{\cdot s_X} & I_{0,j} & \hookrightarrow \cdots \hookrightarrow & I_{i,j} & \hookrightarrow \cdots \hookrightarrow & I_{r,j} = R_j \\
 & \xrightarrow{\cdot s_X} & I_{0,j+1} & \hookrightarrow \cdots \hookrightarrow & I_{i,j+1} & \hookrightarrow \cdots \hookrightarrow & I_{r,j+1} = R_{j+1} \\
\vdots & \vdots & \vdots & & \vdots & \vdots & \vdots \\
 & \xrightarrow{\cdot s_X} & I_{0,n} & \hookrightarrow \cdots \hookrightarrow & I_{i,n} & \hookrightarrow \cdots \hookrightarrow & I_{r,n} = R_n
\end{array}$$

By using Proposition 4.3.33, one has

$$\widehat{\mu}_{\min}(\overline{R}_n) \geqslant \min\Big\{ \min_{\substack{i\in\{1,\ldots,r\}\\ j\in\{1,\ldots,n\}}} \widehat{\mu}_{\min}(\overline{I_{i,j}/I_{i-1,j}}) + (n-j)\upsilon_X, \widehat{\mu}_{\min}(\overline{R}_0) + n\upsilon_X \Big\}. \quad (7.5)$$

For any $i \in \{1, \ldots, r\}$, let Y_i be the integral closed subscheme defined by P_i. By Proposition 7.1.6, one has

$$\liminf_{m\to+\infty} \frac{\widehat{\mu}_{\min}(\overline{I_{i,m}/I_{i-1,m}})}{m} \geqslant \liminf_{m\to+\infty} \frac{\widehat{\mu}_{\min}(\overline{R}_{Y_i,m})}{m}$$

Moreover, by the induction hypothesis, there is a finite subset $\mathfrak{S}_i$ of $\mathfrak{S}_{Y_i}$ such that

$$\liminf_{m\to+\infty} \frac{\widehat{\mu}_{\min}(\overline{R}_{Y_i,m})}{m} \geqslant \min\{\upsilon_Z \,:\, Z \in \mathfrak{S}_i\}.$$

Therefore the estimate (7.5) leads to

$$\liminf_{n\to+\infty} \frac{\widehat{\mu}_{\min}(\overline{R}_n)}{n} \geqslant \min\Big\{\upsilon_Z \,:\, Z \in \{X\} \cup \bigcup_{i=1}^{r} \mathfrak{S}_i\Big\}.$$

The inequality (7.3) is thus proved.

Step 5: We show how to deduce (7.4) from (7.3). Let δ be an arbitrary positive number. For any $Y \in \mathfrak{S}_X$, there exist a positive integer n_Y and a non-zero element s_Y in R_{Y,n_Y} such that

$$\frac{\widehat{\deg}_{\xi_{Y,n_Y}}(s_Y)}{n_Y} \geqslant \limsup_{n\to+\infty} \frac{\nu_1(R_{Y,n}, \xi_{Y,n})}{n} - \delta.$$

Hence the inequality (7.3) leads to

$$\liminf_{n\to+\infty}\frac{\widehat{\mu}_{\min}(R_n,\xi_n)}{n}\geqslant\inf_{Y\in\mathfrak{S}_X}\limsup_{n\to+\infty}\frac{\nu_1(R_{Y,n},\xi_{Y,n})}{n}-\delta.$$

Since $\delta>0$ is arbitrary, the inequality (7.4) holds. □

Remark 7.2.2 Consider the following variante of the above theorem. Assume that $R_\bullet$ is generated as R_0-algebra by R_1. By using Proposition 4.3.13, we obtain that, for integers n and m such that $1\leqslant m\leqslant n$, one has

$$\begin{aligned}\widehat{\deg}(\overline{R}_n)&\geqslant\sum_{j=1}^{n}\sum_{i=1}^{r}\widehat{\deg}(\overline{I_{i,j}/I_{i-1,j}})+\widehat{\deg}(\overline{R}_0)+\upsilon_X\sum_{k=0}^{n-1}\dim_K(R_k)\\&\geqslant\sum_{j=1}^{m}\sum_{i=1}^{r}\widehat{\deg}(\overline{I_{i,j}/I_{i-1,j}})\\&\qquad+\min_{i\in\{1,\ldots,r\}}\inf_{\ell\in\mathbb{N}_{\geqslant m}}\frac{\widehat{\mu}_{\min}(\overline{I_{i,\ell}/I_{i-1,\ell}})}{\ell}\sum_{j=m+1}^{n}j\dim_K(R_j/R_{j-1})\\&\qquad+\widehat{\deg}(\overline{R}_0)+\upsilon_X\sum_{k=0}^{n-1}\dim_K(R_k).\end{aligned}$$

Dividing the two sides by $n\dim_K(R_n)$ and letting n tend to the infinity, we obtain

$$\liminf_{n\to+\infty}\frac{\widehat{\mu}(\overline{R}_n)}{n}\geqslant\frac{d}{d+1}\min_{i\in\{1,\ldots,r\}}\inf_{\ell\in\mathbb{N}_{\geqslant m}}\frac{\widehat{\mu}_{\min}(\overline{I_{i,\ell}/I_{i-1,\ell}})}{\ell}+\frac{1}{d+1}\upsilon_X,$$

where we have used the geometric Hilbert-Samuel theorem asserting that $\mathrm{rk}(R_n)=\deg(X)n^d+O(n^{d-1})$, with d being the Krull dimension of the scheme X, which leads to

$$\lim_{n\to+\infty}\frac{1}{n\dim_K(R_n)}\sum_{j=0}^{n-1}\dim_K(R_j)=\frac{1}{d+1}.$$

Since m is arbitrary, we obtain

$$\liminf_{n\to+\infty}\frac{\widehat{\mu}(\overline{R}_n)}{n}\geqslant\frac{d}{d+1}\min_{i\in\{1,\ldots,r\}}\min_{Z\in\mathfrak{S}_i}\upsilon_Z+\frac{1}{d+1}\upsilon_X.\tag{7.6}$$

Note that in the general case where $R_\bullet$ is not necessarily generated by R_1, the same argument leads to

$$\limsup_{n\to+\infty}\frac{\widehat{\mu}(\overline{R}_n)}{n}\geqslant\frac{d}{d+1}\min_{i\in\{1,\ldots,r\}}\min_{Z\in\mathfrak{S}_i}\upsilon_Z+\frac{1}{d+1}\upsilon_X.\tag{7.7}$$

In the case where the adelic curve S satisfies the Minkowski property and the norm families ξ_n are ultrametric on $\Omega\setminus\Omega_\infty$, by the same argument as in Step 5 of the proof of Theorem 7.2.1, one obtains

$$\limsup_{n\to+\infty}\frac{\widehat{\mu}(\overline{R}_n)}{n}\geqslant\frac{1}{d+1}\limsup_{n\to+\infty}\frac{\widehat{\mu}_{\max}(\overline{R}_n)}{n}+\frac{d}{d+1}\inf_{\substack{Y\in\mathfrak{S}_X\\ Y\subsetneq X}}\limsup_{n\to+\infty}\frac{\widehat{\mu}_{\max}(\overline{R}_{Y,n})}{n},\tag{7.8}$$

and, when $R_\bullet$ is generated as an R_0-algebra by R_1, one obtains

$$\liminf_{n\to+\infty}\frac{\widehat{\mu}(\overline{R}_n)}{n}\geqslant\frac{1}{d+1}\limsup_{n\to+\infty}\frac{\widehat{\mu}_{\max}(\overline{R}_n)}{n}+\frac{d}{d+1}\inf_{\substack{Y\in\mathfrak{S}_X\\ Y\subsetneq X}}\limsup_{n\to+\infty}\frac{\widehat{\mu}_{\max}(\overline{R}_{Y,n})}{n}.\tag{7.9}$$

Corollary 7.2.3 *We keep the notation and hypothesis of Theorem 7.2.1, and assume in addition that the adelic curve S satisfies the Minkowski property and all norm families ξ_n are ultrametric on $\Omega\setminus\Omega_\infty$. Let $\mathfrak{S}_{X,0}$ be the set of all closed points of X. One has*

$$\liminf_{n\to+\infty}\frac{\widehat{\mu}_{\min}(R_n,\xi_n)}{n}\geqslant\inf_{z\in\mathfrak{S}_{X,0}}\limsup_{n\to+\infty}\frac{\widehat{\mu}_{\max}(R_{z,n},\xi_{z,n})}{n}.\tag{7.10}$$

Proof We denote by λ the term on the right hand side of the inequality (7.10) and we will prove the inequality by induction on the Krull dimension d of X. By the Minkowski property and Remark 7.2.2, we obtain that, for any $z\in\mathfrak{S}_{X,0}$, one has

$$\limsup_{n\to+\infty}\frac{\nu_1(R_{z,n},\xi_{z,n})}{n}\geqslant\lambda.$$

Therefore, by Theorem 7.2.1 we obtain that the inequality (7.10) holds when $d=0$. In the following, we assume that $d>0$ and that the statement is true for integral schemes of dimension $<d$. In particular, the induction hypothesis leads to

$$\limsup_{n\to+\infty}\frac{\widehat{\mu}_{\max}(R_{Y,n},\xi_{Y,n})}{n}\geqslant\lambda$$

for any integral closed subscheme $Y\subsetneq X$. Therefore, by Theorem 7.2.1 we obtain that the inequality (7.10) holds once

$$\liminf_{n\to+\infty}\frac{\widehat{\mu}_{\min}(R_n,\xi_n)}{n}<\limsup_{n\to+\infty}\frac{\widehat{\mu}_{\max}(R_n,\xi_n)}{n}.$$

It remains the case where the equality

$$\liminf_{n\to+\infty}\frac{\widehat{\mu}_{\min}(R_n,\xi_n)}{n}=\limsup_{n\to+\infty}\frac{\widehat{\mu}_{\max}(R_n,\xi_n)}{n}=\limsup_{n\to+\infty}\frac{\widehat{\nu}_1(R_n,\xi_n)}{n}.$$

In this case, these infimum and supremum limits are actually limits, and are both equal to

$$\lim_{n\to+\infty}\frac{\widehat{\mu}(R_n,\xi_n)}{n}.$$

Since $d>0$, by (7.8) we still get the inequality (7.10). □

Under the strong tensorial minimal slope property (see Definition 6.3.30), Theorem 7.2.1 admits the following analogue.

Theorem 7.2.4 *We assume that the adelic curve S satisfies the* strong *tensorial minimal slope property of level $\geqslant C_1$, where $C_1 \in \mathbb{R}_{\geqslant 0}$. Let $\mathfrak{S}_X$ be the set of all integral closed subschemes of X. Then one has*

$$\liminf_{n\to+\infty} \frac{\widehat{\mu}_{\min}(\overline{R}_n)}{n} \geqslant \inf_{Y\in\mathfrak{S}_X} \limsup_{m\to+\infty} \frac{\widehat{\mu}_1(\overline{R}_{Y,m})}{m}.$$

Proof We reason by induction on the dimension d of the scheme X.

First we treat the case where $d = 0$. Let m be an integer, $m \geqslant 1$. Let E be a vector subspace of R_m such that $\widehat{\mu}_{\min}(\overline{E}) = \widehat{\mu}_1(\overline{R}_m)$. There exists an integer $N \in \mathbb{N}_{\geqslant 1}$ such that, for any $p \in \mathbb{N}_{\geqslant 1}$, the canonical K-linear map

$$R_{mN} \otimes E^{\otimes p} \longrightarrow R_{m(N+p)}$$

is surjective. Therefore, by Proposition 4.3.31 and the strong tensorial minimal slope property (by an argument similar to the **Step 2** of the proof of Theorem 7.2.1), one has

$$\begin{aligned}\widehat{\mu}_{\min}(\overline{R}_{m(N+p)}) &\geqslant \widehat{\mu}_{\min}(\overline{R}_{mN}) - C_1\ln(\dim_K(R_{mN})) + p\big(\widehat{\mu}_1(\overline{R}_m) - C_1\ln(\dim_K(E))\big)\\ &\geqslant \widehat{\mu}_{\min}(\overline{R}_{mN}) - C_1\ln(\dim_K(R_{mN})) + p\big(\widehat{\mu}_1(\overline{R}_m) - C_1\ln(\dim_K(R_m))\big).\end{aligned}$$

Dividing the two sides by $m(N+p)$ and letting p tend to $+\infty$, by Proposition 7.1.6 we obtain

$$\liminf_{n\to+\infty} \frac{\widehat{\mu}_{\min}(\overline{R}_n)}{n} \geqslant \frac{\widehat{\mu}_1(\overline{R}_m)}{m} - C_1\frac{\ln(\dim_K(R_m))}{m}.$$

Note that

$$\lim_{m\to+\infty} \frac{\ln(\dim_K(R_m))}{m} = 0.$$

Therefore, by taking the limsup when $m \to +\infty$, we obtain

$$\liminf_{n\to+\infty} \frac{\widehat{\mu}_{\min}(\overline{R}_n)}{n} \geqslant \limsup_{m\to+\infty} \frac{\widehat{\mu}_1(\overline{R}_m)}{m}.$$

We now assume that $d \geqslant 1$. Let m be an integer such that $m \geqslant 1$. Let E be a vector subspace of R_m such that $\widehat{\mu}_{\min}(\overline{E}) = \widehat{\mu}_1(\overline{R}_m)$. Let $I_\bullet$ be the homogeneous ideal of $R_\bullet^{(m)} = \bigoplus_{n\in\mathbb{N}} R_{mn}$ generated by E. That is, for any $n \in \mathbb{N}$, I_n is the image of the canonical homomorphism

$$R_{(n-1)m} \otimes E \longrightarrow R_{nm}.$$

As in the the proof of Theorem 7.2.1, we let

$$I_\bullet = I_{0,\bullet} \subsetneq I_{1,\bullet} \subsetneq \ldots \subsetneq I_{r,\bullet} = R_\bullet^{(m)}$$

be a sequence of homogeneous ideals of $R_{\bullet}^{(m)}$ and $P_{1,\bullet},\ldots,P_{r,\bullet}$ be non-zero homogeneous prime ideals of $R^{(m)}$ such that $P_{i,\bullet}\cdot I_{i,\bullet}\subset I_{i-1,\bullet}$ for $i\in\{1,\ldots,r\}$. Let p be an integer in $\mathbb{N}_{\geqslant 1}$. We denote by F_p the image of the canonical K-linear map

$$R_0\otimes E^{\otimes p}\longrightarrow R_{mp}.$$

Consider the following sequence:

$$\begin{array}{ccccccc}
F_p = & I_{0,1}E^{p-1} & \hookrightarrow\cdots\hookrightarrow & I_{i,1}E^{p-1} & \hookrightarrow\cdots\hookrightarrow & I_{r,1}E^{p-1} \\
\vdots & \vdots & \vdots & \vdots & \vdots & \vdots \\
= & I_{0,j}E^{p-j} & \hookrightarrow\cdots\hookrightarrow & I_{i,j}E^{p-j} & \hookrightarrow\cdots\hookrightarrow & I_{r,j}E^{p-j} \\
= & I_{0,j+1}E^{p-j-1} & \hookrightarrow\cdots\hookrightarrow & I_{i,j+1}E^{p-j-1} & \hookrightarrow\cdots\hookrightarrow & I_{r,j+1}E^{p-j-1} \\
\vdots & \vdots & \vdots & \vdots & \vdots & \vdots \\
= & I_{0,p} & \hookrightarrow\cdots\hookrightarrow & I_{i,p} & \hookrightarrow\cdots\hookrightarrow & I_{r,p}=R_{mp}
\end{array}$$

By Proposition 4.3.33 we obtain that

$$\widehat{\mu}_{\min}(\overline{R}_{mp})\geqslant\min\Big\{\widehat{\mu}_{\min}(\overline{F}_p),\min_{\substack{i\in\{1,\ldots,r\}\\ j\in\{1,\ldots,p\}}}\widehat{\mu}_{\min}(\overline{I_{i,j}E^{p-j}/I_{i-1,j}E^{p-j}})\Big\}.\tag{7.11}$$

By Proposition 4.3.31 and the strong tensorial minimal slope property, one has

$$\begin{aligned}\widehat{\mu}_{\min}(\overline{F}_p)&\geqslant\widehat{\mu}_{\min}(\overline{R}_0)-C_1\ln(\dim_K(R_0))+p\big(\widehat{\mu}_{\min}(\overline{E})-C_1\ln(\dim_K(E))\big)\\&\geqslant\widehat{\mu}_{\min}(\overline{R}_0)-C_1\ln(\dim_K(R_0))+p\big(\widehat{\mu}_1(\overline{R}_m)-C_1\ln(\dim_K(R_m))\big).\end{aligned}$$

Similarly, for any $(i,j)\in\{1,\ldots,r\}\times\{1,\ldots,n\}$, one has

$$\begin{aligned}\widehat{\mu}_{\min}(\overline{I_{i,j}E^{p-j}/I_{i-1,j}E^{p-j}})\geqslant\widehat{\mu}_{\min}(\overline{I_{i,j}/I_{i-1,j}})-C_1\ln(\dim_K(I_{i,j}/I_{i-1,j}))\\+(p-j)\big(\widehat{\mu}_{\min}(\overline{E})-C_1\ln(\dim_K(E))\big).\end{aligned}$$

For any $i\in\{1,\ldots,r\}$, let Y_i be the integral closed subscheme defined by P_i. By Proposition 7.1.6, one has

$$\liminf_{j\to+\infty}\frac{\widehat{\mu}_{\min}(\overline{I_{i,j}/I_{i-1,j}})}{j}\geqslant\liminf_{j\to+\infty}\frac{\widehat{\mu}_{\min}(\overline{R}_{Y_i,j})}{j}\geqslant\min_{Z\in\mathfrak{S}_{Y_i}}\limsup_{k\to+\infty}\frac{\widehat{\mu}_1(\overline{R}_{Z,k})}{k},$$

where the second inequality comes from the induction hypothesis. Therefore, if we denote by υ the value

$$\inf_{Z\in\mathfrak{S}_X}\limsup_{k\to+\infty}\frac{\widehat{\mu}_1(\overline{R}_{Z,k})}{k},$$

then the inequality (7.11) leads to

$$\liminf_{n\to+\infty}\frac{\widehat{\mu}_{\min}(\overline{R}_n)}{n}=\liminf_{p\to+\infty}\frac{\widehat{\mu}_{\min}(\overline{R}_{mp})}{mp}\geqslant\min\Big\{\frac{\widehat{\mu}_1(\overline{R}_m)}{m}-C_1\frac{\ln(\dim_K(R_m))}{m},\upsilon\Big\},$$

where the equality comes from Proposition 7.1.6. By taking the limsup when $m \to +\infty$, we obtain

$$\liminf_{n\to+\infty} \frac{\widehat{\mu}_{\min}(\overline{R}_n)}{n} \geqslant \upsilon,$$

as desired. □

7.3 A consequence of the extension property of semipositive metrics

The purpose of this section is to prove the following theorem as a consequence of the extension property of semipositive metrics (cf. Theorem 2.3.32 and Theorem 2.3.36).

Theorem 7.3.1 *Let X be a geometrically reduced projective K-scheme, L be a semiample invertible $\mathcal{O}_X$-module, and $\varphi = \{\varphi_\omega\}_{\omega\in\Omega}$ be a metric family of L. Let Y be a closed subscheme of X, and $\varphi|_Y = \{\varphi|_{Y,\omega}\}_{\omega\in\Omega}$ be the restriction of φ to Y. For each $n \in \mathbb{N}$, let $\xi_n := \{\|\cdot\|_{n\varphi_\omega}\}_{\omega\in\Omega}$, $\xi_n|_Y := \{\|\cdot\|_{n\varphi|_{Y,\omega}}\}_{\omega\in\Omega}$, $R_{Y,n}$ be the image of $H^0(X, L^{\otimes n}) \to H^0(Y, L|_Y^{\otimes n})$ and $\xi_{Y,n} = \{\|\cdot\|_{Y,n,\omega}\}_{\omega\in\Omega}$ be the quotient norm family on $R_{Y,n}$ induced by $H^0(X, L^{\otimes n}) \to R_{Y,n}$ and ξ_n. If φ is dominated and measurable and φ_ω is semipositive for all $\omega \in \Omega$, then we have the following:*

(1) The norm families ξ_n and $\xi_{Y,n}$ are dominated and measurable for all $n \geqslant 0$.
(2) For any $n \in \mathbb{N}$ and $s \in R_{Y,n}$, the function $(\omega \in \Omega) \mapsto \ln\|s\|_\omega$ is measurable and upper dominated.
(3) For $s \in R_{Y,1} \setminus \{0\}$, one has (see Remark 4.3.3 for notation)

$$\lim_{n\to\infty} \frac{\widehat{\deg}_{\xi_{Y,n}}(s^{\otimes n})}{n} = \widehat{\deg}_{\xi_1|_Y}(s). \tag{7.12}$$

Proof (1) First, by Theorem 6.1.13 and Theorem 6.1.32, ξ_n is dominated and measurable for $n \geqslant 0$. Moreover, by Theorem 6.1.13 and Theorem 6.1.32 together with Proposition 6.1.12 and Proposition 6.1.28, $\xi_n|_Y$ is dominated and measurable for $n \geqslant 0$. Finally, by virtue of Proposition 4.1.19 and Proposition 4.1.24, $\xi_{Y,n}$ is dominated and measurable for $n \geqslant 0$.

(2) follows from Remark 6.1.17 (for the upper dominancy), propositions 6.1.20 and 6.1.26 (for the measurability).

Before starting the proof of (3), we need to prepare several facts. We set $\xi_{Y,n} = \{\|\cdot\|_{Y,n,\omega}\}_{\omega\in\Omega}$. We claim the following:

Claim 7.3.2 (a) *For all $\omega \in \Omega$, $n \geqslant 0$ and $s \in R_{Y,n,\omega}$,*

$$\|s\|_{n\varphi|_{Y,\omega}} \leqslant \|s\|_{Y,n,\omega}.$$

(b) *For all $\omega \in \Omega$, $n \geqslant 1$ and $s \in R_{Y,1,\omega} \setminus \{0\}$,*

$$\ln \|s\|_{\varphi|_{Y,\omega}} \leqslant \frac{\ln \|s^{\otimes n}\|_{Y,n,\omega}}{n} \leqslant \ln \|s\|_{Y,1,\omega}.$$

(c) *For all* $\omega \in \Omega$ *and* $s \in R_{Y,1,\omega} \setminus \{0\}$,

$$\lim_{n\to\infty} \frac{\ln \|s^{\otimes n}\|_{Y,n,\omega}}{n} = \ln \|s\|_{\varphi|_{Y,\omega}}.$$

□

Proof (a) Note that, for all $l \in H^0(X, L^{\otimes n})$ with $l|_Y = s$, one has $\|s\|_{n\varphi|_{Y,\omega}} \leqslant \|l\|_{n\varphi_\omega}$, so that the assertion follows.

(b) By Proposition 1.1.71, $\|s^{\otimes n}\|_{Y,n,\omega} \leqslant \left(\|s\|_{Y,1,\omega}\right)^n$. Moreover, by (a),

$$\left(\|s\|_{\varphi|_{Y,\omega}}\right)^n = \|s^{\otimes n}\|_{n\varphi|_{Y,\omega}} \leqslant \|s^{\otimes n}\|_{Y,n,\omega},$$

so that one has (b).

(c) For a positive number ϵ, by Theorem 2.3.32 and Theorem 2.3.36, there is a positive integer n_0 such that, for all $n \geqslant n_0$, we can find $l \in H^0(X_\omega, L_\omega^{\otimes n})$ such that $l|_{Y_\omega} = s^{\otimes n}$ and $\|l\|_{n\varphi_\omega} \leqslant e^{n\epsilon}\left(\|s\|_{\varphi|_{Y,\omega}}\right)^n$, and hence

$$\ln \|s^{\otimes n}\|_{Y,n,\omega} \leqslant \ln \|l\|_{n\varphi_\omega} \leqslant n\epsilon + n \ln \|s\|_{\varphi|_{Y,\omega}}.$$

Therefore, by (b),

$$0 \leqslant \frac{\ln \|s^{\otimes n}\|_{Y,n,\omega}}{n} - \ln \|s\|_{\varphi|_{Y,\omega}} \leqslant \epsilon$$

for all $n \geqslant n_0$, as required. □

(3) By (1) and (2), the function $(\omega \in \Omega) \mapsto \ln \|s\|_{\varphi|_{Y,\omega}}$ is measurable and upper dominated. By the reverse Fatou lemma, (c) leads to

$$\limsup_{n\to+\infty} \frac{1}{n} \int_\Omega \ln\|s^{\otimes n}\|_{Y,n,\omega}\, \nu(\mathrm{d}\omega) \leqslant \int_\Omega \ln \|s\|_{\varphi|_{Y,\omega}}\, \nu(\mathrm{d}\omega),$$

which is equivalent to

$$\liminf_{n\to\infty} \frac{\widehat{\deg}_{\xi_{Y,n}}(s^{\otimes n})}{n} \geqslant \widehat{\deg}_{\xi_1|_Y}(s).$$

In particular, the equality (7.12) holds when $\widehat{\deg}_{\xi_1|_Y}(s) = +\infty$. In the case where $\widehat{\deg}_{\xi_1|_Y}(s)$ is finite, the function

$$(\omega \in \Omega) \mapsto \left|\ln \|s\|_{\varphi|_{Y,\omega}}\right|$$

is integrable. By (1), the function

$$(\omega \in \Omega) \mapsto \left|\ln \|s\|_{Y,1,\omega}\right|$$

is also integrable. Moreover, by (b), one has

$$\left|\frac{\ln\|s^{\otimes n}\|_{Y,n,\omega}}{n}\right| \leqslant \max\left\{\left|\ln\|s\|_{\varphi|_{Y,\omega}}\right|, \left|\ln\|s\|_{Y,1,\omega}\right|\right\},$$

and hence, by Lebesgue's dominated convergence theorem together with (c),

$$\lim_{n\to\infty}\frac{1}{n}\int_{\Omega}\ln\|s^{\otimes n}\|_{Y,n,\omega}\,\nu(\mathrm{d}\omega) = \int_{\Omega}\ln\|s\|_{\varphi|_{Y,\omega}}\,\nu(\mathrm{d}\omega),$$

which shows (7.12). □

Remark 7.3.3 We keep the notation and hypotheses of Theorem 7.3.1. Let z be a closed point of X. For any $n \in \mathbb{N}$, let $R_{z,n}$ be the image of the restriction map $H^0(X, L^{\otimes n}) \to z^*(L^{\otimes n})$ and $\xi_{z,n}$ be the quotient norm family of ξ_n on $R_{z,n}$. Let $\kappa(z)$ be the residue field of z. For any $\omega \in \Omega$ and any $x \in M_{\kappa(z),\omega}$ (namely $|\cdot|_x$ is an absolute value of $\kappa(z)$ extending $|\cdot|_\omega$, see §3.3), if s is a non-zero element of $R_{Y,1}$, one has

$$\ln|s|_{\varphi_\omega}(z_x) \leqslant \frac{\ln\|s^{\otimes n}\|_{z,n,\omega}}{n},$$

where z_x denotes the point of X_ω^{an} given by the couple $(z, |\cdot|_x)$. Moreover, the semi-positivity of the metric φ_ω leads to

$$\lim_{n\to+\infty}\ln|s|_{\varphi_\omega}(z_x) = \lim_{n\to+\infty}\frac{\ln\|s^{\otimes n}\|_{z,n,\omega}}{n}.$$

In particular, one has

$$\begin{aligned}h_{(L,\varphi)}(z) &= -\int_{\Omega}\int_{M_{\kappa(x),\omega}}\ln|s|_{\varphi_\omega}(z_x)\,\mathbb{P}_{\kappa(x),\omega}(\mathrm{d}x)\nu(\mathrm{d}\omega)\\ &= -\lim_{n\to+\infty}\frac{1}{n}\int_{\Omega}\ln\|s^{\otimes n}\|_{z,n,\omega} = \lim_{n\to+\infty}\frac{1}{n}\widehat{\deg}_{\xi_{z,n}}(s^{\otimes n})\\ &\leqslant \limsup_{n\to+\infty}\frac{\nu_1(R_{z,n},\xi_{z,n})}{n}.\end{aligned}$$

7.4 Nakai-Moishezon's criterion in a general settings

In this section, let us consider the following Nakai-Moishezon's criterion in a general settings:

Theorem 7.4.1 *Let X be an integral and geometrically reduced projective K-scheme, L be an invertible $\mathcal{O}_X$-module and $\varphi = \{\varphi_\omega\}_{\omega\in\Omega}$ be a metric family of L. For any $n \in \mathbb{N}$, let $\xi_n := \{\|\cdot\|_{n\varphi_\omega}\}_{\omega\in\Omega}$ and let $\xi_n|_Y := \{\|\cdot\|_{n\varphi|_{Y,\omega}}\}_{\omega\in\Omega}$ for any integral closed subscheme Y of X. We assume the following:*

(1) (Dominancy and measurability) *The metric family φ is dominated and measurable.*
(2) (Semipositivity) *L is semiample and φ_ω is semipositive for all $\omega \in \Omega$.*
(3) (Bigness) *For every integral closed subscheme Y of X, $L|_Y$ is big, and there are a positive number n_Y and $s_Y \in H^0(Y, L^{\otimes n_Y}|_Y)\setminus\{0\}$ such that $\widehat{\deg}_{\xi_{n_Y}|_Y}(s_Y) > 0$.*

Then one has

$$\liminf_{n\to\infty} \frac{\widehat{\mu}_{\min}\left(H^0(X, L^{\otimes n}), \xi_n\right)}{n} > 0. \tag{7.13}$$

Moreover, if the adelic curve S satisfies the strong Minkowski property, then

$$\liminf_{n\to\infty} \frac{\nu_{\min}\left(H^0(X, L^{\otimes n}), \xi_n\right)}{n} > 0, \tag{7.14}$$

so that, there are a positive integer n and a basis $\{e_i\}_{i=1}^N$ of $H^0(X, L^{\otimes n})$ such that $\widehat{\deg}_{\xi_n}(e_i) > 0$ for $i = 1, \ldots, N$.

Proof First of all, for each integral closed subscheme Y of X, as $L|_Y$ is nef and big, one has $(L|_Y^{\dim Y}) > 0$, so that by the classic Nakai-Moishezon criterion, L is ample.

For any subvariety Y of X, we set

$$\begin{cases} R_{Y,n} := \text{the image of the natural homomorphism } H^0(X, L^{\otimes n}) \longrightarrow H^0(Y, L|_Y^{\otimes n}), \\ R_{Y,n,\omega} := R_{Y,n} \otimes_K K_\omega \quad (\omega \in \Omega), \\ \|\cdot\|_{Y,n,\omega} := \text{the quotient norm of } \|\cdot\|_{n\varphi_\omega} \text{ on } R_{Y,n,\omega} \quad (\omega \in \Omega), \\ \xi_{Y,n} := \{\|\cdot\|_{Y,n,\omega}\}_{\omega\in\Omega}. \end{cases}$$

Claim 7.4.2 *There are a positive number n'_Y and $s'_Y \in R_{Y,n'_Y} \setminus \{0\}$ such that $\widehat{\deg}_{\xi_{Y,n'_Y}}(s'_Y) > 0$.* □

Proof Fix a positive integer n_0 such that the natural homomorphism

$$H^0(X, L^{\otimes n}) \to H^0(Y, L|_Y^{\otimes n})$$

is surjective for all $n \geqslant n_0$, that is, $R_{Y,n} = H^0(Y, L|_Y^{\otimes n})$ for all $n \geqslant n_0$, so that $s_Y^{\otimes n_0} \in R_{Y,n_0n_Y} \setminus \{0\}$. By Theorem 7.3.1 (3), one has

$$\lim_{n\to\infty} \frac{\widehat{\deg}_{\xi_{Y,nn_0n_Y}}(s_Y^{\otimes nn_0})}{n} = \widehat{\deg}_{\xi_{n_0n_Y}|_Y}(s_Y^{\otimes n_0}) = n_0\widehat{\deg}_{\xi_{n_Y}|_Y}(s_Y) > 0,$$

so that there is a positive integer n_1 such that $\widehat{\deg}_{\xi_{Y,n_1n_0n_Y}}(s_Y^{\otimes n_1n_0}) > 0$. Therefore, if we set $n'_Y := n_1n_0n_Y$ and $s'_Y := s_Y^{\otimes n_1n_0}$, one has the claim. □

The assertion (7.13) follows from the above claim together with Theorem 7.2.1. Further, if S satisfies the strong Minkowski property, then there is a constant C depending only on S such that

$$\nu_{\min}(H^0(X, L^{\otimes n}), \xi_n) + C\ln(\dim_K H^0(X, L^{\otimes n})) \geqslant \widehat{\mu}_{\min}(H^0(X, L^{\otimes n}), \xi_n),$$

and hence the assertion (7.14) follows. □

Remark 7.4.3 In the case where $\Omega = \Omega_0$ and $\#(\Omega_0) = 1$, S satisfies the strong Minkowski property of level $\geqslant 0$, that is, if E is a finite-dimensional vector space over K and $\|\cdot\|$ is an ultrametric norm of E over $(K, |\cdot|)$, then $\nu_{\min}(E, \|\cdot\|) = \widehat{\mu}_{\min}(E, \|\cdot\|)$, which can be checked as follows:

In general, one has $\nu_{\min}(E, \|\cdot\|) \leqslant \widehat{\mu}_{\min}(E, \|\cdot\|)$ by Proposition 4.3.77, so that it is sufficient to show that $\nu_{\min}(E, \|\cdot\|) \geqslant \widehat{\mu}_{\min}(E, \|\cdot\|)$. Let $\{e_i\}_{i=1}^r$ be an orthogonal basis of E with respect to $\|\cdot\|$ (cf. Proposition 1.2.30). Clearly we may assume that $\|e_r\| = \max\{\|e_1\|, \ldots, \|e_r\|\}$. Let $Q := E/(Ke_1 + \cdots + Ke_{r-1})$ and $\|\cdot\|_Q$ be the quotient norm of $\|\cdot\|$ on Q. Then $\|\pi(e_r)\|_Q = \|e_r\|$, where $\pi : E \to Q$ is the canonical homomorphism. Thus $-\log\|e_r\| \geqslant \widehat{\mu}_{\min}(E, \|\cdot\|)$, and hence $\widehat{\deg}(e_i) \geqslant \widehat{\mu}_{\min}(E, \|\cdot\|)$ for all i. Therefore, one has $\nu_{\min}(E, \|\cdot\|) \geqslant \widehat{\mu}_{\min}(E, \|\cdot\|)$.

Lemma 7.4.4 *We assume that the adelic curve S satisfies the Minkowski property and the tensorial minimal slope property. Let X be an integral and geometrically reduced projective K-scheme, L be an ample invertible $\mathcal{O}_X$-module, $\varphi = \{\varphi_\omega\}_{\omega\in\Omega}$ be a dominated and measurable metric family on L such that φ_ω is semipositive for any $\omega \in \Omega$. Then one has*

$$\liminf_{n\to+\infty} \frac{\widehat{\mu}_{\min}(H^0(X, L^{\otimes n}), \xi_n)}{n} \geqslant \inf_{z\in X(K^{\mathrm{ac}})} h_{(L,\varphi)}(z), \tag{7.15}$$

where for any $n \in \mathbb{N}$, $\xi_n = \{\|\cdot\|_{n\varphi_\omega}\}_{\omega\in\Omega}$.

Proof Denote by $R_\bullet$ the graded sectional algebra $\bigoplus_{n\in\mathbb{N}} H^0(X, L^{\otimes n})$. For any closed point z of X and any $n \in \mathbb{N}$, let $R_{z,n}$ be the image of the restriction map $H^0(X, L^{\otimes n}) \to z^*(L^{\otimes n})$ and $\xi_{z,n}$ be the quotient norm family of ξ_n on $R_{z,n}$. By Corollary 7.2.3, one has

$$\liminf_{n\to+\infty} \frac{\widehat{\mu}_{\min}(R_n, \xi_n)}{n} \geqslant \inf_{z\in\mathfrak{S}_{X,0}} \limsup_{n\to+\infty} \frac{\widehat{\mu}_{\max}(R_{z,n}, \xi_{z,n})}{n},$$

where $\mathfrak{S}_{X,0}$ denotes the set of all closed points of X. Moreover, by Remark 7.3.3, one has

$$\limsup_{n\to+\infty} \frac{\widehat{\mu}_{\max}(R_{z,n}, \xi_{z,n})}{n} \geqslant \limsup_{n\to+\infty} \frac{\nu_1(R_{z,n}, \xi_{z,n})}{n} \geqslant h_{(L,\varphi)}(z)$$

for any $z \in \mathfrak{S}_{X,0}$. Therefore the inequality (7.15) holds. □

Definition 7.4.5 Let X be a geometrically integral scheme over $\operatorname{Spec} K$, L be an ample invertible $\mathcal{O}_X$-module, $\varphi = \{\varphi_\omega\}_{\omega\in\Omega}$ be a dominated and measurable metric family on L. We denote by $\nu_{\mathrm{abs}}(L, \varphi)$ the infimum of the height function $h_{(L,\varphi)}$, called the *absolute minimum* of (L, φ).

Theorem 7.4.6 *We assume that the adelic curve S satisfies the* strong *Minkowski property and the tensorial minimal slope property. Let X be an integral and geometrically reduce projective K-scheme, L be an ample invertible $\mathcal{O}_X$-module,*

$\varphi = \{\varphi_\omega\}_{\omega\in\Omega}$ be a dominated and measurable metric family on L such that φ_ω is semipositive for any $\omega \in \Omega$. Then the following inequality holds.

$$\widehat{\mu}^{\mathrm{asy}}_{\min}(L,\varphi) := \liminf_{n\to+\infty} \frac{\widehat{\mu}_{\min}(H^0(X,L^{\otimes n}),\xi_n)}{n} = \nu_{\mathrm{abs}}(L,\varphi), \tag{7.16}$$

where for any $n \in \mathbb{N}$, $\xi_n = \{\|\cdot\|_{n\varphi_\omega}\}_{\omega\in\Omega}$, and d is the Krull dimension of X.

Proof For any $n \in \mathbb{N}$, let $E_n := H^0(X, L^{\otimes n})$. Since the adelic curve S satisfies the strong Minkowski property, one has

$$\begin{aligned}\widehat{\mu}^{\mathrm{asy}}_{\min}(L,\varphi) &= \liminf_{n\to+\infty} \frac{\nu_{\min}(E_n,\xi_n)}{n}\\ &\leqslant \liminf_{n\to+\infty} \frac{\nu^{\mathrm{a}}_{\min}(E_n,\xi_n)}{n} \leqslant \liminf_{n\to+\infty} \frac{\widehat{\mu}_{\min}(E_{n,K^{\mathrm{ac}}},\xi_{n,K^{\mathrm{ac}}})}{n},\end{aligned}$$

where the second inequality comes from Proposition 4.3.71 and the last inequality comes from Corollary 4.3.78. Let P be an algebraic point of X. For sufficiently positive integer n, the invertible $\mathcal{O}_X$-module $L^{\otimes n}$ is very ample hence defines a closed embedding $X \to \mathbb{P}(E_n)$. Let $\mathcal{O}_{E_n}(1)$ be the universal invertible sheaf on $\mathbb{P}(E_n)$. Then, viewed as a quotient vector space of dimension 1 of $E_{n,K^{\mathrm{ac}}}$, the Arakelov degree of $P^*(\mathcal{O}_{E_n}(1))$ (equipped with the quotient norm family) is bounded from above by $nh_{(L,\varphi)}(P)$ and bounded from below by $\widehat{\mu}_{\min}(E_{n,K^{\mathrm{ac}}},\xi_{n,K^{\mathrm{ac}}})$. Therefore we obtain $\widehat{\mu}^{\mathrm{asy}}_{\min}(L,\varphi) \leqslant h_{(L,\varphi)}(P)$. Since $P \in X(K^{\mathrm{ac}})$ is arbitrary, this leads to the inequality

$$\widehat{\mu}^{\mathrm{asy}}_{\min}(L,\varphi) \leqslant \nu_{\mathrm{abs}}(L,\varphi).$$

Moreover, by Lemma 7.4.4 the converse inequality also holds. Therefore the equality (7.16) is proved.

7.5 Nakai-Moishezon's criterion over a number field

Throughout this section, we fix a number field K and the standard adelic curve $S = (K, (\Omega, \mathcal{A}, \nu), \phi)$ of K as in Subsection 3.2.2. Denote by Ω_{fin} the set $\Omega \setminus \Omega_\infty$ of finite places of K, and by $\mathfrak{o}_K$ the ring of algebraic integers in K. Note that S satisfies the strong Minkowski property (see [44, Theorem 1.1]). Moreover, for $\omega \in \Omega_{\mathrm{fin}}$, the valuation ring of the completion K_ω of K with respect to ω is denoted by $\mathfrak{o}_\omega$.

7.5.1 Invariants λ and σ for a graded algebra of adelic vector bundles

Let $\overline{R}_{\bullet} = \{(R_n, \xi_n)\}_{n \in \mathbb{Z}_{\geqslant 0}}$ be a graded algebra of adelic vector bundles on S such that (R_n, ξ_n) is dominated and coherent for all $n \geqslant 0$. For the definition of the invariants λ and σ, see Subsection 4.4.3.

Definition 7.5.1 We say that $\overline{R}_{\bullet}$ is *asymptotically pure* if

$$\limsup_{n\to\infty} \frac{\sigma(R_n, \xi_n)}{n} = 0.$$

As a consequence of Proposition 4.4.10, we have the following:

Proposition 7.5.2 *One has the following inequalities:*

$$\begin{aligned}[K:\mathbb{Q}] \liminf_{n\to+\infty} \frac{\lambda(R_n, \xi_n)}{n} &\leqslant \liminf_{n\to+\infty} \frac{\nu_{\min}(R_n, \xi_n)}{n} \\ &\leqslant [K:\mathbb{Q}] \liminf_{n\to+\infty} \frac{\lambda(R_n, \xi_n)}{n} + \limsup_{n\to+\infty} \frac{\sigma(R_n, \xi_n)}{n}.\end{aligned}$$

In particular, if $\overline{R}_{\bullet}$ is asymptotically pure, then

$$[K:\mathbb{Q}] \liminf_{n\to+\infty} \frac{\lambda(R_n, \xi_n)}{n} = \liminf_{n\to+\infty} \frac{\nu_{\min}(R_n, \xi_n)}{n}.$$

Let $\overline{M}_{\bullet} = \{(M_n, \xi_{M_n})\}_{n\in\mathbb{Z}}$ be an h-graded $\overline{R}_{\bullet}$-module such that (M_n, ξ_n) is dominated and coherent for all $n \in \mathbb{Z}$.

Proposition 7.5.3 (1) *If $R_{\bullet}$ is generated by R_1 over K, then* $\displaystyle\lim_{n\to\infty} \frac{\lambda(R_n, \xi_n)}{n}$ *exists in* $\mathbb{R} \cup \{\infty\}$.
(2) *If $R_{\bullet} = \bigoplus_{n=0}^{\infty} R_n$ is of finite type over K and $M_{\bullet} = \bigoplus_{n\in\mathbb{Z}} M_n$ is finitely generated over $R_{\bullet}$, then*

$$\frac{1}{h} \liminf_{n\to\infty} \frac{\lambda(R_n, \xi_n)}{n} \leqslant \liminf_{n\to\infty} \frac{\lambda(M_n, \xi_{M_n})}{n}.$$

Proof We set $\mathscr{R}_n := (R_n, \xi_n)^{\mathrm{fin}}_{\leqslant 1}$ for $n \geqslant 0$, and $\mathscr{M}_n := (M_n, \xi_{M_n})^{\mathrm{fin}}_{\leqslant 1}$ for $n \in \mathbb{Z}$.

(1) For $\epsilon > 0$, we choose bases $e_1, \ldots, e_r$ and $e'_1, \ldots, e'_{r'}$ of R_n and R_m over K, respectively, such that

$$\begin{cases} e_1, \ldots, e_r \in \mathscr{R}_n, & \max\{\|e_i\|_{\infty,n}\} \leqslant e^{-\lambda(R_n,\xi_n)+\epsilon}, \\ e'_1, \ldots, e'_{r'} \in \mathscr{R}_m, & \max\{\|e'_j\|_{\infty,n}\} \leqslant e^{-\lambda(R_m,\xi_m)+\epsilon}. \end{cases}$$

Then $e_i e'_j \in \mathscr{R}_{n+m}$ and $\max\{\|e_i e'_j\|_{\infty,n+m}\} \leqslant e^{-\lambda(R_n,\xi_n)-\lambda(R_n,\xi_n)+2\epsilon}$. Note that $\{e_i e'_j\}$ forms generators of R_{n+m} over K because $R_n \otimes R_m \to R_{n+m}$ is surjective, so that $e^{-\lambda(R_{n+m},\xi_{n+m})} \leqslant e^{-\lambda(R_n,\xi_n)-\lambda(R_n,\xi_n)+2\epsilon}$. Therefore, one has

$$\lambda(R_{n+m}, \xi_{n+m}) \geqslant \lambda(R_n, \xi_n) + \lambda(R_n, \xi_n)$$

for all n, m. Thus the assertion follows from Fekete's lemma.

(2) It can be proved in the similar way as in Proposition 7.1.4. First we assume that $R_\bullet$ is generated by R_1 over K. Then there exist integers b_1 and $m > 0$ such that, for any integer b with $b \geqslant b_1$ and any integer $\ell \geqslant 1$ the canonical K-linear map $R_{\ell m} \otimes_K M_b \longrightarrow M_{b+\ell mh}$ is surjective. For $\epsilon > 0$, we choose a basis $e_1, \ldots, e_r$ of $R_{\ell m}$ and a basis $m_1, \ldots, m_{r'}$ of M_b such that $e_1, \ldots, e_r \in \mathscr{R}_{\ell m}$, $m_1, \ldots, m_{r'} \in \mathscr{M}_b$, $\max\{\|e_i\|_{\infty,\ell m}\} \leqslant e^{-\lambda(R_{\ell m}, \xi_{\ell m})+\epsilon}$ and $\max\{\|m_j\|_{\infty,M_b}\} \leqslant e^{-\lambda(M_b, \xi_{M_b})+\epsilon}$. Note that $e_i m_j \in \mathscr{M}_{b+\ell mh}$ and

$$\|e_i m_j\|_{\infty, M_{b+\ell mh}} \leqslant e^{-\lambda(R_{\ell m}, \xi_{\ell m}) - \lambda(M_b, \xi_{M_b}) + 2\epsilon}.$$

Moreover we can find a basis of $M_{b+\ell mh}$ among $\{e_i m_j\}_{1 \leqslant i \leqslant r, 1 \leqslant j \leqslant r'}$, so that

$$e^{-\lambda(M_{b+\ell mh}, \xi_{M_{b+\ell mh}})} \leqslant e^{-\lambda(R_{\ell m}, \xi_{\ell m}) - \lambda(M_b, \xi_{M_b}) + 2\epsilon},$$

and hence one has

$$\lambda(M_{b+\ell mh}, \xi_{M_{b+\ell mh}}) \geqslant \lambda(R_{\ell m}, \xi_{\ell m}) + \lambda(M_b, \xi_{M_b}).$$

Therefore,

$$\liminf_{l \to \infty} \frac{\lambda(M_{b+\ell mh}, \xi_{M_{b+\ell mh}})}{\ell mh} \geqslant \frac{1}{h} \liminf_{l \to \infty} \frac{\lambda(R_{\ell m}, \xi_{\ell m})}{\ell m} \geqslant \frac{1}{h} \liminf_{n \to \infty} \frac{\lambda(R_n, \xi_n)}{n},$$

which implies

$$\liminf_{n \to \infty} \frac{\lambda(M_n, \xi_{M_n})}{n} \geqslant \frac{1}{h} \liminf_{n \to \infty} \frac{\lambda(R_n, \xi_n)}{n}$$

because $b \geqslant b_1$ is arbitrary.

In general, we can find a positive integer u such that $R_\bullet^{(u)} := \bigoplus_{n=0}^{\infty} R_{un}$ is generated by $R_1^{(u)} = R_u$ over K. Note that $R_\bullet$ is a finitely generated $R_\bullet^{(u)}$-module. Therefore, by the previous observation, one has

$$\begin{cases} \liminf_{n \to \infty} \dfrac{\lambda(R_n, \xi_n)}{n} \geqslant \dfrac{1}{u} \liminf_{n \to \infty} \dfrac{\lambda(R_{un}, \xi_{un})}{n}, \\ \liminf_{n \to \infty} \dfrac{\lambda(M_n, \xi_{M_n})}{n} \geqslant \dfrac{1}{hu} \liminf_{n \to \infty} \dfrac{\lambda(R_{un}, \xi_{un})}{n}. \end{cases}$$

Moreover, as

$$\liminf_{n \to \infty} \frac{\lambda(R_{un}, \xi_{un})}{un} \geqslant \liminf_{m \to \infty} \frac{\lambda(R_m, \xi_m)}{m},$$

one obtains

$$\liminf_{n \to \infty} \frac{\lambda(R_n, \xi_n)}{n} = \frac{1}{u} \liminf_{n \to \infty} \frac{\lambda(R_{un}, \xi_{un})}{n}.$$

Thus the assertion follows. □

7.5.2 Dominancy and coherency of generically pure metric

Let X be a geometrically integral projective variety over K and L be an invertible sheaf on X. For $\omega \in \Omega$, let $X_\omega := X \times_{\mathrm{Spec}\, K} K_\omega$ and $L_\omega := L \otimes_{O_X} O_{X_\omega}$. Let φ_ω be a continuous metric of L_ω on X_ω^{an} for each $\omega \in \Omega$, and $\varphi := \{\varphi_\omega\}_{\omega\in\Omega}$.

Let us begin with the definition of the generic purity of the metric family φ.

Definition 7.5.4 We say that φ is *generically pure* if there are a non-empty Zariski open set U of $\mathrm{Spec}(\mathfrak{o}_K)$, a projective integral scheme $\mathscr{X}$ over U and an invertible $O_{\mathscr{X}}$-module $\mathscr{L}$ such that $\mathscr{X} \times_U \mathrm{Spec}(K) = X$, $\mathscr{L}|_X = L$ and, for each $\omega \in U\cap\Omega_{\mathrm{fin}}$, φ_ω coincides with the metric arising from $\mathscr{X}_\omega$ and $\mathscr{L}_\omega$, where $\mathscr{X}_\omega = \mathscr{X} \times_U \mathrm{Spec}(\mathfrak{o}_\omega)$ and $\mathscr{L}_\omega$ is the pull-back of $\mathscr{L}$ to $\mathscr{X}_\omega$.

Proposition 7.5.5 *(1) If L is generated by global sections and φ is generically pure, there exist a non-empty Zariski open set U of* $\mathrm{Spec}(\mathfrak{o}_K)$ *and a basis $\boldsymbol{e} = \{e_i\}_{i=1}^r$ of $H^0(X, L)$ such that $\varphi_\omega = \varphi_{\boldsymbol{e},\omega}$ for all $\omega \in U \cap \Omega_{\mathrm{fin}}$.*
(2) If L is semiample and φ is generically pure, then φ is dominated and $\big(H^0(X, L^{\otimes n}), \{\|\cdot\|_{n\varphi_\omega}\}_{\omega\in\Omega}\big)$ is coherent for all $n \geqslant 0$.

Proof (1): We use the notation in Definition 7.5.4. Shrinking U if necessarily, we may assume that $H^0(\mathscr{X}, \mathscr{L})$ is a free $\mathfrak{o}_U$-module and $H^0(\mathscr{X}, \mathscr{L}) \otimes_{\mathfrak{o}_U} O_{\mathscr{X}} \to \mathscr{L}$ is surjective. Let $\boldsymbol{e} = \{e_i\}_{i=1}^r$ be a free basis of $H^0(\mathscr{X}, \mathscr{L})$ over $\mathfrak{o}_U$. Then $\{e_i\}_{i=1}^r$ yields a free basis of $H^0(\mathscr{X}_\omega, \mathscr{L}_\omega)$ over $\mathfrak{o}_\omega$ for any $\omega \in U\cap\Omega_{\mathrm{fin}}$. Let $\|\cdot\|_{H^0(\mathscr{X}_\omega,\mathscr{L}_\omega)}$ be the norm of $H^0(X_\omega, L_\omega^{\otimes r})$ arising from the lattice $H^0(\mathscr{X}_\omega, \mathscr{L}_\omega)$. Then, by Proposition 1.2.21, $\|\cdot\|_{H^0(\mathscr{X}_\omega,\mathscr{L}_\omega)} = \|\cdot\|_{\boldsymbol{e},\omega}$ for any $\omega \in U \cap \Omega_{\mathrm{fin}}$. Moreover, $H^0(\mathscr{X}_\omega, \mathscr{L}_\omega) \otimes_{\mathfrak{o}_\omega} O_{\mathscr{X}_\omega} \to \mathscr{L}_\omega$ is surjective, so that, by Proposition 2.3.12, one has $\varphi_\omega = \varphi_{\boldsymbol{e},\omega}$, as required.

(2) We choose a positive integer m such that $L^{\otimes m}$ is generated by global sections and $\alpha_n : H^0(X, L^{\otimes m})^{\otimes n} \to H^0(X, L^{\otimes nm})$ is surjective for all $n \geqslant 1$. Then, by (1), there are a non-empty Zariski open set U of $\mathrm{Spec}(\mathfrak{o}_K)$ and a basis $\boldsymbol{e} = \{e_i\}_{i=1}^r$ of $H^0(X, L^{\otimes m})$ such that $m\varphi_\omega = \varphi_{\boldsymbol{e},\omega}$ for all $\omega \in U \cap \Omega_{\mathrm{fin}}$. In particular, $m\varphi$ is dominated, so that φ is also dominated by Proposition 6.1.12.

For $\omega \in U\cap\Omega_{\mathrm{fin}}$, let $\|\cdot\|_{\boldsymbol{e},\omega}^{\otimes n}$ be the ε-tensor products of $\|\cdot\|_{\boldsymbol{e},\omega}$ on $H^0(X_\omega, L_\omega^{\otimes m})^{\otimes n}$. Note that, by Proposition 1.2.19 together with (1.17) in Remark 1.1.56,

$$\Bigg\| \sum_{\substack{(i_1,\ldots,i_r)\in\mathbb{Z}_{\geqslant 0}^r,\\ i_1+\cdots+i_r=n}} a_{i_1,\ldots,i_r} e_1^{\otimes i_1} \otimes \cdots \otimes e_r^{\otimes i_r} \Bigg\|_{\boldsymbol{e},\omega}^{\otimes n} = \max\{|a_{i_1,\ldots,i_r}|_\omega\}$$

for all $a_{i_1,\ldots,i_N} \in K_\omega$. Moreover, by Remark 2.2.19, $n\varphi_{\boldsymbol{e},\omega}$ coincides with the quotient metric induced by the surjective homomorphism $H^0(X_\omega, L_\omega^{\otimes m})^{\otimes n} \otimes_{K_\omega} O_{X_\omega} \to L_\omega^{\otimes nm}$ and $\|\cdot\|_{\boldsymbol{e},\omega}^{\otimes n}$.

Fix $s \in H^0(X, L^{\otimes n})$ ($n \geqslant 1$). Then $s^{\otimes m} \in H^0(X, L^{\otimes mn})$. As α_n is surjective, one can choose

$$f = \sum_{\substack{(i_1,\ldots,i_r)\in\mathbb{Z}^r_{\geqslant 0},\\ i_1+\cdots+i_r=n}} f_{i_1,\ldots,i_r} e_1^{\otimes i_1} \otimes \cdots \otimes e_r^{\otimes i_r} \in H^0(X, L^{\otimes m})^{\otimes n} \quad (f_{i_1,\ldots,i_r} \in K)$$

such that $\alpha_n(f) = s^{\otimes m}$. Then, by Proposition 2.2.23,

$$(\|s\|_{n\varphi_\omega})^m = \|s^{\otimes m}\|_{nm\varphi_\omega} = \|s^{\otimes m}\|_{n\varphi_{e,\omega}} \leqslant \|f\|^{\otimes n}_{e,\omega} = \max\{|f_{i_1,\ldots,i_r}|_\omega\},$$

for all $\omega \in U \cap \Omega_{\text{fin}}$, so that so that $\|s\|_{n\varphi_\omega} \leqslant 1$ for all $\omega \in \Omega$ except finitely many ω because $\Omega \setminus (U \cap \Omega_{\text{fin}})$ is finite and $|f_{i_1,\ldots,i_r}|_\omega = 1$ for all $i_1, \ldots, i_r$ and $\omega \in \Omega_{\text{fin}}$ except finitely many ω. □

7.5.3 Fine metric family

Let X be a geometrically integral projective variety over K and L be an invertible sheaf on X. For $\omega \in \Omega$, let $X_\omega := X \times_{\operatorname{Spec} K} K_\omega$ and $L_\omega := L \otimes_{O_X} O_{X_\omega}$. Let φ_ω be a continuous metric of L_ω on X_ω^{an} for each $\omega \in \Omega$, and $\varphi := \{\varphi_\omega\}_{\omega\in\Omega}$.

Definition 7.5.6 We say that φ is *very fine* if φ is dominated and there are a generically pure continuous metric family $\varphi' = \{\varphi'_\omega\}_{\omega\in\Omega}$ of L and a non-empty Zariski open set U of $\operatorname{Spec}(\mathfrak{o}_K)$ such that $|\cdot|_{\varphi_\omega} \leqslant |\cdot|_{\varphi'_\omega}$ for all $\omega \in U \cap \Omega_{\text{fin}}$. Further, φ is said to be *fine* if $r\varphi$ is very fine for some positive integer r.

Proposition 7.5.7 *Let L and M be invertible O_X-module, and φ and ψ be continuous metric families of L and M, respectively.*

(1) *If φ and ψ are very fine, then $\varphi + \psi$ is very fine.*
(2) *If φ and ψ are fine, then $\varphi + \psi$ is fine.*
(3) *If $a\varphi$ is fine for some positive integer a, then φ is fine.*
(4) *If φ is fine, then $\big(H^0(X, L), \{\|\cdot\|_{\varphi_\omega}\}_{\omega\in\Omega}\big)$ is coherent.*

Proof (1) is obvious.

(2) We choose positive integers r and r' such that $r\varphi$ and $r'\psi$ are very fine. Then, by (1), $rr'\varphi$ and $rr'\psi$ are very fine, so that $rr'(\varphi + \psi)$ is very fine, as required.

(3) Since $a\varphi$ is fine, there is a positive integer r such that $ra\varphi$ is very fine, so that φ is fine.

(4) Let r be a positive integer such that $r\varphi$ is very fine. Then there are a generically pure continuous metric family $\varphi' = \{\varphi'_\omega\}_{\omega\in\Omega}$ of $L^{\otimes r}$ and a non-empty Zariski open set U of $\operatorname{Spec}(\mathfrak{o}_K)$ such that $|\cdot|_{r\varphi_\omega} \leqslant |\cdot|_{\varphi'_\omega}$ for all $\omega \in U \cap \Omega_{\text{fin}}$, so that, for $s \in H^0(X, L) \setminus \{0\}$ and $\omega \in U \cap \Omega_{\text{fin}}$, $\|s^{\otimes r}\|_{r\varphi_\omega} \leqslant \|s^{\otimes r}\|_{\varphi'_\omega}$. By Proposition 7.5.5, $\|s^{\otimes r}\|_{\varphi'_\omega} \leqslant 1$ except finitely many $\omega \in \Omega$. Therefore, the same assertion holds for $\|s^{\otimes r}\|_{r\varphi_\omega}$. Note that $\|s^{\otimes r}\|_{r\varphi_\omega} = \|s\|^r_{\varphi_\omega}$, and hence $\|s\|_{\varphi_\omega} \leqslant 1$ except finitely many $\omega \in \Omega$. □

For $n \geqslant 0$, we set $R_n := H^0(X, L^{\otimes n})$ and $\xi_n := \{\|\cdot\|_{n\varphi_\omega}\}_{\omega\in\Omega}$. Note that $\overline{R}_\bullet = \{(R_n, \xi_n)\}_{n=0}^\infty$ forms a graded algebra of adelic vector bundles over S (cf.

Definition 7.1.1). For $n \geqslant 0$ and $\omega \in \Omega$, we denote $R_n \otimes_K K_\omega = H^0(X_\omega, L_\omega^{\otimes n})$ by $R_{n,\omega}$. Moreover, for $n \geqslant 0$ and $\omega \in \Omega_{\text{fin}}$, we set $\mathscr{R}_{n,\omega} := \{x \in R_{n,\omega} \,:\, \|x\|_{n,\omega} \leqslant 1\}$. Note that $\mathscr{R}_{n,\omega}$ is a locally free $\mathfrak{o}_\omega$-module and $\mathscr{R}_{n,\omega} \otimes_{\mathfrak{o}_\omega} K_\omega = R_{n,\omega}$ (cf. Proposition 1.1.25 and Proposition 1.1.30). Further we set

$$\mathscr{R}_n := \{x \in R_n \,:\, \|x\|_{n,\omega} \leqslant 1 \text{ for all } \omega \in \Omega_{\text{fin}}\}.$$

If (R_n, ξ_n) is dominated and coherent, then, by Proposition 4.4.2 and Proposition 4.4.6, $\mathscr{R}_n$ is finitely generated over $\mathfrak{o}_K$, $\mathscr{R}_n \otimes_{\mathfrak{o}_K} K = R_n$, $\mathscr{R}_n \otimes_{\mathfrak{o}_K} K_\omega = R_{n,\omega}$ and $\mathscr{R}_n \otimes_{\mathfrak{o}_K} \mathfrak{o}_\omega = \mathscr{R}_{n,\omega}$ for all $\omega \in \Omega_{\text{fin}}$.

Proposition 7.5.8 *We assume that L is ample and φ is dominated. Then the following are equivalent:*

(1) *The metric family φ is fine.*
(2) *(R_n, ξ_n) is coherent for all $n \geqslant 0$.*

Proof (1) $\Longrightarrow$ (2): This is a consequence of Proposition 7.5.7.

(2) $\Longrightarrow$ (1): First note that (R_n, ξ_n) is dominated for $n \geqslant 0$ by Proposition 6.1.12 and Theorem 6.1.13. Moreover, by our assumption, (R_n, ξ_n) is coherent for every $n \geqslant 0$.

Let r be a positive integer such that $L^{\otimes r}$ is very ample. Let $\mathscr{X}$ be the Zariski closure of X in $\mathbb{P}(\mathscr{R}_r)$ and $\mathscr{L} = \mathcal{O}_{\mathbb{P}(\mathscr{R}_r)}(1)\big|_{\mathscr{X}}$. Then $\mathscr{L}|_X = L^{\otimes r}$. Moreover, since $\mathscr{R}_r \otimes_{\mathfrak{o}_K} \mathcal{O}_{\mathbb{P}(\mathscr{R}_r)} \to \mathcal{O}_{\mathbb{P}(\mathscr{R}_r)}(1)$ is surjective, $\mathscr{R}_r \otimes_{\mathfrak{o}_K} \mathcal{O}_{\mathscr{X}} \to \mathscr{L}$ is also surjective. For each $\omega \in \Omega_{\text{fin}}$, let ψ_ω be the metric of $L_\omega^{\otimes r}$ arising from $\mathscr{X}_\omega$ and $\mathscr{L}_\omega$, where $\mathscr{X}_\omega = \mathscr{X} \times_{\operatorname{Spec}(\mathfrak{o}_K)} \operatorname{Spec}(\mathfrak{o}_\omega)$ and $\mathscr{L}_\omega$ is the pull-back of $\mathscr{L}$ to $\mathscr{X}_\omega$. Let $\varphi' = \{\varphi'_\omega\}_{\omega \in \Omega}$ be the metric family of $L^{\otimes r}$ given by

$$\varphi'_\omega := \begin{cases} \psi_\omega & \text{if } \omega \in \Omega_{\text{fin}}, \\ r\varphi_\omega & \text{otherwise.} \end{cases}$$

Here let us see

$$\forall\, \omega \in \Omega,\ \forall\, x \in X_\omega^{\text{an}}, \quad |\cdot|_{r\varphi_\omega}(x) \leqslant |\cdot|_{\varphi'_\omega}(x). \tag{7.17}$$

Clearly we may assume that $\omega \in \Omega_{\text{fin}}$. Note that

$$\mathscr{R}_{r,\omega} = \{s \in H^0(\mathscr{X}_\omega, \mathscr{L}_\omega) \,:\, \|s\|_{r\varphi_\omega} \leqslant 1\}$$

and $\mathscr{R}_{r,\omega} \otimes \mathcal{O}_{\mathscr{X}_\omega} \to \mathscr{L}_\omega$ is surjective, so that, by Proposition 2.3.12, (7.17) follows. Therefore, $r\varphi$ is very fine, and hence φ is fine. □

Finally we consider the following theorem:

Theorem 7.5.9 *If φ is very fine, then $\overline{\boldsymbol{R}}_{\bullet}$ is asymptotically pure.*

Proof By our assumption, φ is dominated and there are a generically pure continuous metric family $\varphi' = \{\varphi'_\omega\}_{\omega \in \Omega}$ of L and a non-empty Zariski open set U of $\operatorname{Spec}(\mathfrak{o}_K)$ such that $|\cdot|_{\varphi_\omega} \leqslant |\cdot|_{\varphi'_\omega}$ for all $\omega \in U \cap \Omega_{\text{fin}}$.

First note that (R_n, ξ_n) is dominated for $n \geqslant 0$ by Proposition 6.1.12 and Theorem 6.1.13. Moreover, by Proposition 7.5.7 or Proposition 7.5.8, (R_n, ξ_n) is coherent for every $n \geqslant 0$.

By the generic purity of φ', there are a non-empty Zariski open set U' of $\mathrm{Spec}(\mathfrak{o}_K)$, a projective integral scheme $\mathscr{X}$ over U' and an invertible $\mathcal{O}_{\mathscr{X}}$-module $\mathscr{L}$ such that $\mathscr{X} \times_{U'} \mathrm{Spec}(K) = X$, $\mathscr{L}|_X = L$ and, for each $\omega \in U' \cap \Omega_{\mathrm{fin}}$, φ'_ω coincides with the metric arising from $\mathscr{X}_\omega$ and $\mathscr{L}_\omega$, where $\mathscr{X}_\omega = \mathscr{X} \times_U \mathrm{Spec}(\mathfrak{o}_\omega)$ and $\mathscr{L}_\omega$ is the pull-back of $\mathscr{L}$ to $\mathscr{X}_\omega$. Replacing U and U' by $U \cap U'$, we may assume that $U = U'$. Moreover, as X is geometrically integral over K, by virtue of [76, Théorème 9.7.7], shrinking U if necessarily, we may also assume that, for any $\omega \in U \cap \Omega_{\mathrm{fin}}$, the fiber of $\mathscr{X} \to U$ over ω is geometrically integral over the residue field at ω. Then, by Proposition 2.3.16,

$$\begin{cases} \{x \in R_n \,:\, \|x\|_{n\varphi'_\omega} \leqslant 1\} = H^0(\mathscr{X}_\omega, \mathscr{L}_\omega^{\otimes n}) = H^0(\mathscr{X}, \mathscr{L}^{\otimes n}) \otimes_{\mathfrak{o}_U} \mathfrak{o}_\omega, \\ \|\cdot\|_{n\varphi'_\omega} = \|\cdot\|_{H^0(\mathscr{X}_\omega, \mathscr{L}_\omega^{\otimes n})} \end{cases} \tag{7.18}$$

for all $n \geqslant 1$ and $\omega \in U \cap \Omega_{\mathrm{fin}}$.

Claim 7.5.10 (a) $|\cdot|_{n\varphi_\omega}(x) \leqslant |\cdot|_{n\varphi'_\omega}(x)$ *for all* $\omega \in U \cap \Omega_{\mathrm{fin}}$, $x \in X_\omega^{\mathrm{an}}$ *and* $n \geqslant 1$.
(b) $\|\cdot\|_{n\varphi_\omega} \leqslant \|\cdot\|_{\mathscr{R}_{n,\omega}} \leqslant \min\left\{|\varpi_\omega|_\omega^{-1}\|\cdot\|_{n\varphi_\omega}, \|\cdot\|_{n\varphi'_\omega}\right\}$ *for all* $\omega \in U \cap \Omega_{\mathrm{fin}}$ *and* $n \geqslant 1$, *where* ϖ_ω *is a uniformizing parameter of* $\mathfrak{o}_\omega$. □

Proof (a) is obvious.
(b) First of all, by Proposition 1.1.30,

$$\|\cdot\|_{n\varphi_\omega} \leqslant \|\cdot\|_{\mathscr{R}_{n,\omega}} \leqslant |\varpi_\omega|_\omega^{-1}\|\cdot\|_{n\varphi_\omega}.$$

By (a), one has $\|\cdot\|_{n\varphi_\omega} \leqslant \|\cdot\|_{n\varphi'_\omega}$, so that, by (7.18), one obtains

$$\mathscr{R}_{n,\omega} \supseteq H^0(\mathscr{X}_\omega, \mathscr{L}_\omega^{\otimes n}).$$

Therefore, by (7.18) again, (b) follows. □

Claim 7.5.11 *If we set* $A_\omega = d_\omega(\varphi_\omega, \varphi'_\omega)$ *for* $\omega \in \Omega$, *then one has the following:*

(a) $\displaystyle\sup_{x \in R_n \setminus \{0\}} \ln \frac{\|s\|_{\mathscr{R}_{n,\omega}}}{\|s\|_{n\varphi_\omega}} \leqslant A_\omega n$ *for all* $\omega \in U \cap \Omega_{\mathrm{fin}}$ *and* $n \geqslant 1$.

(b) $\displaystyle\int_\Omega A_\omega \nu(d\omega) = \sum_{\omega \in \Omega} A_\omega \nu(\{\omega\}) < \infty.$ □

Proof (a) By using the inequality (2.5) in Subsection 2.2.2 together with Claim 7.5.10,

$$\begin{aligned} \sup_{x \in R_n \setminus \{0\}} \ln \frac{\|s\|_{\mathscr{R}_{n,\omega}}}{\|s\|_{n\varphi_\omega}} &\leqslant \sup_{x \in R_n \setminus \{0\}} \ln \frac{\|s\|_{n\varphi'_\omega}}{\|s\|_{n\varphi_\omega}} \\ &= d_\omega\left(\|\cdot\|_{n\varphi'_\omega}, \|\cdot\|_{n\varphi'_\omega}\right) \\ &\leqslant d_\omega(n\varphi_\omega, n\varphi'_\omega) = nA_\omega. \end{aligned}$$

(b) Note that φ' is dominated by Proposition 7.5.5. Moreover, φ is dominated by our assumption, so that, by Proposition 6.1.12, the function $\omega \mapsto d_\omega(\varphi_\omega, \varphi'_\omega) = A_\omega$ is ν-dominated. Thus one obtains (b). □

Fix a positive number ϵ. Then, by Claim 7.5.11, there is a non-empty Zariski open set U_ϵ of U such that

$$\sum_{\omega \in U_\epsilon \cap \Omega_{\text{fin}}} A_\omega \nu(\{\omega\}) \leqslant \epsilon.$$

Thus, if we set $B = \sum_{\omega \in \Omega_{\text{fin}} \setminus U_\epsilon} (-\ln |\varpi_\omega|_\omega) \nu(\{\omega\})$, then, by Claim 7.5.10,

$$\begin{aligned}\varpi(R_n, \xi_n) &= \sum_{\omega \in \Omega_{\text{fin}}} \sup_{x \in R_{n,\omega} \setminus \{0\}} \ln\left(\frac{\|x\|_{\mathscr{R}_{n,\omega}}}{\|x\|_{n\varphi_\omega}}\right) \nu(\{\omega\}) \\ &\leqslant \sum_{\omega \in U_\epsilon \cap \Omega_{\text{fin}}} \sup_{x \in R_{n,\omega} \setminus \{0\}} \ln\left(\frac{\|x\|_{\mathscr{R}_{n,\omega}}}{\|x\|_{n\varphi_\omega}}\right) \nu(\{\omega\}) + B \\ &\leqslant n \sum_{\omega \in U_\epsilon \cap \Omega_{\text{fin}}} A_\omega \nu(\{\omega\}) + B \leqslant n\epsilon + B\end{aligned}$$

for $n \geqslant 1$, and hence one has

$$\limsup_{n \to \infty} \frac{\varpi(R_n, \xi_n)}{n} \leqslant \epsilon,$$

so that the assertion of the theorem follows. □

7.5.4 A generalization of Nakai-Moishezon's criterion

Let X be a geometrically integral projective variety over K and L be an invertible sheaf on X. For $\omega \in \Omega$, let $X_\omega := X \times_{\operatorname{Spec} K} K_\omega$ and $L_\omega := L \otimes_{\mathcal{O}_X} \mathcal{O}_{X_\omega}$. Let φ_ω be a continuous metric of L_ω on X_ω^{an} for each $\omega \in \Omega$, and $\varphi := \{\varphi_\omega\}_{\omega \in \Omega}$. For $n \geqslant 0$ and any subvariety Y of X, let $\xi_n := \{\|\cdot\|_{n\varphi_\omega}\}_{\omega \in \Omega}$ and $\xi_n|_Y := \{\|\cdot\|_{n\varphi_\omega|_{Y_\omega}}\}_{\omega \in \Omega}$. In this subsection, let us consider the following Nakai-Moishezon's criterion over a number field, which gives a generalisation of Nakai-Moishezon's criterion due to Shouwu Zhang.

Theorem 7.5.12 *We assume the following:*

(1) (Fineness) *The metric family φ is fine.*
(2) (Semipositivity) *L is semiample and φ_ω is semipositive for every $\omega \in \Omega$.*
(3) (Bigness) *For every subvariety Y of X, $L|_Y$ is big, and there are a positive number n_Y and $s_Y \in H^0(Y, L^{\otimes n_Y}|_Y) \setminus \{0\}$ such that $\widehat{\deg}_{\xi_{n_Y}|_Y}(s_Y) > 0$.*

Then one has

$$\liminf_{n\to\infty} \frac{\lambda\left(H^0(X, L^{\otimes n}), \xi_n\right)}{n} > 0.$$

Proof We set $R_n := H^0(X, L^{\otimes n})$ for $n \geqslant 0$. By Proposition 7.5.3, for a positive number h,

$$\frac{1}{h}\liminf_{n\to\infty} \frac{\lambda(R_n, \xi_n)}{n} \leqslant \liminf_{n\to\infty} \frac{\lambda(R_n, \xi_n)}{n},$$

so that, replacing L, φ, n_Y and s_Y by $L^{\otimes h}$, $h\varphi$, hn_Y and $s_Y^{\otimes h}$ for a sufficiently large integer h, we may assume that φ is very fine. Moreover, by Remark 4.1.25, we can see that φ is measurable. Therefore, by Theorem 7.4.1, one has

$$\liminf_{n\to\infty} \frac{\nu_{\min}(R_n, \xi_n)}{n} > 0.$$

By Proposition 6.1.12, Theorem 6.1.13 and Proposition 7.5.7, (R_n, ξ_n) is dominated and coherent for all $n \geqslant 0$. Therefore, by Proposition 7.5.2 and Theorem 7.5.9, one can see the assertion of the theorem. □

Remark 7.5.13 In [48], Nakai-Moishezon's criterion was proved under the following condition (7.19) instead of (Fineness) in Theroem 7.5.12:

$$\begin{cases} \text{The adelic vector bundle } \left(H^0(X, L^{\otimes n}), \xi_n\right) \text{ over } S \\ \text{is dominaited and coherent for every } n \geqslant 0. \end{cases} \tag{7.19}$$

Appendix A
Reminders on measure theory

A.1 Monotone class theorems

We recall here a monotone class theorem in the functional form and several related results, and we refer to [52, §I.2] and [151, §2.2] for reference. For convenience of readers, we include the proof here. We fix in this section a non-empty set Ω. If $\mathcal{H}$ is a family of real-valued functions on Ω, we denote by $\sigma(\mathcal{H})$ the σ-algebra on Ω generated by $\mathcal{H}$. It is the smallest σ-algebra on Ω with respect to which all functions in $\mathcal{H}$ are measurable.

Definition A.1.1 Let $\mathcal{H}$ be a family of non-negative and bounded functions on Ω. We say that $\mathcal{H}$ is a *λ-family* if it verifies the following conditions:

(i) the constant function 1 belongs to $\mathcal{H}$;
(ii) if f and g are two functions in $\mathcal{H}$, a and b are non-negative numbers, then $af + bg \in \mathcal{H}$;
(iii) if f and g are two functions in $\mathcal{H}$ such that $f \leqslant g$, then $g - f \in \mathcal{H}$;
(iv) if $\{f_n\}_{n\in\mathbb{N}}$ is an increasing and uniformly bounded sequence of functions in $\mathcal{H}$, then the limit of the sequence $\{f_n\}_{n\in\mathbb{N}}$ belongs to $\mathcal{H}$.

Lemma A.1.2 *Let $\mathcal{H}$ be a λ-family of non-negative and bounded functions on Ω. If for any couple (f, g) of functions in $\mathcal{H}$, one has $\min(f, g) \in \mathcal{H}$, then any non-negative, bounded and $\sigma(\mathcal{H})$-measurable function on Ω belongs to $\mathcal{H}$. In particular, the σ-algebra $\sigma(\mathcal{H})$ is equal to the set of all $A \subseteq \Omega$ such that $\mathbb{1}_A \in \mathcal{H}$.*

Proof Let $\mathcal{F}$ be the set of all $A \subseteq \Omega$ such that $\mathbb{1}_A \in \mathcal{H}$. Since $\mathcal{H}$ is a λ-family, we obtain that $\mathcal{F}$ is a λ-system[1] and at the same time a π-system (namely for all $A \in \mathcal{F}$ and $B \in \mathcal{F}$ one has $A \cap B \in \mathcal{F}$) since the family $\mathcal{H}$ is supposed to be stable by the operator $(f, g) \mapsto \min(f, g)$. Therefore $\mathcal{F}$ is actually a σ-algebra.

If (f, g) is a couple of functions in $\mathcal{H}$, one has

[1] Namely $\mathcal{F}$ satisfies the following conditions: (i) $\Omega \in \mathcal{F}$, (ii) if $A \in \mathcal{F}$, $B \in \mathcal{F}$ and $A \subseteq B$, then $B \setminus A \in \mathcal{F}$, (iii) if $\{A_n\}_{n\in\mathbb{N}}$ is an increasing sequence of elements of $\mathcal{F}$, then the union $\bigcup_{n\in\mathbb{N}} A_n$ belongs to $\mathcal{F}$.

H. Chen, A. Moriwaki, *Arakelov Geometry over Adelic Curves*, Lecture Notes in Mathematics 2258, https://doi.org/10.1007/978-981-15-1728-0

$$\max(f, g) = f + g - \min(f, g) \in \mathcal{H}.$$

In particular, if $f \in \mathcal{H}$ and $a \in \mathbb{R}_+$, then

$$\max(f - a, 0) = \max(f, a) - a \in \mathcal{H}.$$

This property actually implies that, for any $f \in \mathcal{H}$ and any integer $n \geqslant 1$, one has $f^n \in \mathcal{H}$. In fact, the function $x \mapsto x^n$ is convex on $\mathbb{R}_+$, which can be written as the supremum of a countable family of functions of the form

$$x \longmapsto \max(na^{n-1}x - (n-1)a^n, 0)$$

with $a \in \mathbb{Q}_+ := \mathbb{Q} \cap \mathbb{R}_+$. Therefore by the condition (iv) in Definition A.1.1 one obtains

$$f^n = \sup_{a \in \mathbb{Q}_+} \max(na^{n-1}f - (n-1)a^n, 0) \in \mathcal{H}.$$

If f is an element of $\mathcal{H}$ and t is a real number, $t > 0$, one has $\min(t^{-1}f, 1) \in \mathcal{H}$. Moreover, the sequence $\{1 - \min(t^{-1}f, 1)^n\}_{n \in \mathbb{N}, n \geqslant 1}$ is increasing and converges to $\mathbb{1}_{\{f<t\}}$, which implies that $\mathbb{1}_{\{f<t\}} \in \mathcal{H}$ and hence $\{f < t\} \in \mathcal{F}$. Therefore every function in $\mathcal{H}$ is $\mathcal{F}$-measurable, and thus $\sigma(\mathcal{H}) \subseteq \mathcal{F}$.

It remains to prove that any non-negative bounded $\mathcal{F}$-measurable function belongs to $\mathcal{H}$. Let f be such a function. For any integer $n \geqslant 1$, let

$$f_n = \sum_{k=0}^{n2^n-1} \frac{k}{2^n} \mathbb{1}_{\{k/2^n \leqslant f < (k+1)/2^n\}} + n\mathbb{1}_{\{f \geqslant n\}}.$$

This is a function in $\mathcal{H}$. Moreover, the sequence $\{f_n\}_{n \in \mathbb{N}, n \geqslant 1}$ is increasing and converges to f. Therefore $f \in \mathcal{H}$. □

Theorem A.1.3 *Let $\mathcal{H}$ be a λ-family of non-negative and bounded functions on Ω and C be a subset of $\mathcal{H}$. Assume that for any couple (f, g) of functions in C, the product function fg belongs to C. Then any non-negative and bounded $\sigma(C)$-measurable function belongs to $\mathcal{H}$.*

Proof By replacing $\mathcal{H}$ by the intersection of all λ-families containing C we may assume that $\mathcal{H}$ is the smallest λ-family which contains C.

We first prove that $\mathcal{H}$ is stable by multiplication. Let $\mathcal{H}_1$ be the set of all non-negative and bounded functions f on Ω such that $fg \in \mathcal{H}$ for any $g \in C$. This is a λ-family containing C. Hence one has $\mathcal{H}_1 \supseteq \mathcal{H}$. Let $\mathcal{H}_2$ be the set of all non-negative and bounded functions f on Ω such that $fg \in \mathcal{H}$ for any $g \in \mathcal{H}$. This is also a λ-family. Moreover, since $\mathcal{H}_1 \supseteq \mathcal{H}$ one obtains $\mathcal{H}_2 \supseteq C$ and hence $\mathcal{H}_2 \supseteq \mathcal{H}$, which implies that $\mathcal{H}$ is stable by multiplication.

Let f and g be two functions in $\mathcal{H}$. We will prove that $|f - g| \in \mathcal{H}$. By dilating the function $|f - g|$ by a positive constant, we may assume that $|f - g|$ is bounded from above by 1. One has

$$(f - g)^2 = f^2 + g^2 - 2fg \in \mathcal{H}.$$

Let $\{f_n\}_{n\in\mathbb{N}}$ be the sequence of functions on Ω defined by the following recursive formula

$$f_0 = 0, \quad f_{n+1} = f_n + \frac{1}{2}((f-g)^2 - f_n^2).$$

By induction on n, we can show that $f_n \in \mathcal{H}$ and $f_n \leqslant |f-g|$. In fact, these properties are trivially satisfied by f_0. If $f_n \in \mathcal{H}$ and $f_n \leqslant |f-g|$, then one has $f_{n+1} \in \mathcal{H}$. Moreover, by the relation $|f-g| \leqslant 1$ one obtains $f_{n+1} \leqslant |f-g|$ since the function $t \mapsto t - \frac{1}{2}t^2$ is increasing on the interval $[0,1]$. The properties $f_n \in \mathcal{H}$ and $f_n \leqslant |f-g|$ show that the sequence $\{f_n\}_{n\in\mathbb{N}}$ is increasing and converges to $|f-g|$. Hence $|f-g| \in \mathcal{H}$, which implies that

$$\min(f,g) = \frac{1}{2}(f+g-|f-g|) \in \mathcal{H}.$$

By Lemma A.1.2, any non-negative, bounded and $\sigma(\mathcal{H})$-measurable function belongs to $\mathcal{H}$. The theorem is thus proved. □

A.2 Measurable selection theorem

In this section, we recall a measurable selection theorem due to Kuratowski and Ryll-Nardzewski [100]. See [119, Chapter 5] for more details.

Theorem A.2.1 *Let Y be a complete separable metric space and $\mathscr{P}(Y)$ be the set of subsets of Y. Let $(\Omega, \mathcal{A})$ be a measurable space and $F : \Omega \to \mathscr{P}(Y)$ be a map. We assume that*

(1) for any $\omega \in \Omega$, the set $F(\omega)$ is a non-empty closed subset of Y,
(2) for any open subset U of Y, the set $\{\omega \in \Omega : F(\omega) \cap U \neq \varnothing\}$ belongs to $\mathcal{A}$.

Then there exists a measurable map $f : \Omega \to Y$ such that $f(\omega) \in F(\omega)$ for any $\omega \in \Omega$.

A.3 Vague convergence and weak convergence of measures

Let X be a locally compact Hausdorff space. Recall that a *Radon measure* is by definition a Borel measure ν on X which satisfies the following conditions:

(1) ν is *tight*, that is, for any Borel subset B of X, $\nu(B)$ is equal to the supremum of $\nu(K)$, where K runs over the set of compact subsets of B;
(2) ν is *outer regular*, that is, for any Borel subset B of X, $\nu(B)$ is equal to the infimum of $\nu(U)$, where U runs over the set of open subsets of X containing B;
(3) ν is *locally finite*, that is, for any $x \in X$ there exists a neighbourhood U of x such that $\nu(U) < +\infty$.

We denote by $\mathscr{M}(X)$ be the set of Radon measures on X. Let $C_c(X)$ be the vector space of continuous real-valued functions of compact support on X. We say that an $\mathbb{R}$-linear map $\varphi : C_c(X) \to \mathbb{R}$ is a *positive linear functional* if $\varphi(f) \geqslant 0$ for any non-negative function f in $C_c(X)$. Recall the Riesz's representation theorem as follows. See [83, §56] for a proof.

Theorem A.3.1 *Let X be a locally compact Hausdorff space. The map sending $\nu \in \mathscr{M}(X)$ to the positive linear functional*

$$f \in C_c(X) \longrightarrow \int_X f \,\mathrm{d}\nu$$

defines a bijection between the set $\mathscr{M}(X)$ and the set of all positive linear functionals on $C_c(X)$.

The *vague topology* on $\mathscr{M}(X)$ is an example of weak-* topology if we identify $\mathscr{M}(X)$ with a subset of the dual space of $C_c(X)$. More precisely, we say that a sequence $\{\nu_n\}_{n\in\mathbb{N}}$ of Radon measures *converges vaguely* if for any function $f \in C_c(X)$, the sequence of integrals $\{\int_X f \,\mathrm{d}\nu_n\}_{n\in\mathbb{N}}$ converges in $\mathbb{R}$. Note that the limit of the above sequence defines a positive linear functional on $C_c(X)$ when f varies, which corresponds to a Radon measure, called the *vague limit* of $\{\nu_n\}_{n\in\mathbb{N}}$.

If $\{\nu_n\}_{n\in\mathbb{N}}$ is a sequence of Radon probability measure which converges vaguely, the limite measure may have a total mass < 1. In probability theory, the notion of weak convergence is also largely used. Let $\mathscr{M}_1(X)$ be the subset of $\mathscr{M}(X)$ of probability measures. Let $C_b(X)$ be the vector space of bounded continuous functions. We say that a sequence $\{\nu_n\}_{n\in\mathbb{N}}$ of measures in $\mathscr{M}_1(X)$ (they are therefore *probability* measures) *converges weakly* if for any bounded continuous function f on X, the sequence of integrals $\{\int_X f \,\mathrm{d}\nu_n\}_{n\in\mathbb{N}}$ converges in $\mathbb{R}$. Clearly, if the sequence $\{\nu_n\}_{n\in\mathbb{N}}$ converges weakly, then it also converges vaguely, and its vague limit is also called its *weak limit*. Note that, in the weak convergence case, the limit measure should be a probability measure. The following criterion provides a criterion of weak convergence for vaguely convergence sequence of Radon probability measures. We refer the readers to [98, Theorem 13.16] for the proof and for more details[2].

Theorem A.3.2 *Let X be a locally compact metrisable space and $\{\nu_n\}_{n\in\mathbb{N}}$ be a sequence of Radon probability measures on X, which converges vaguely to a limite measure ν. Assume the limite measure ν is a probability measure. Then the sequence $\{\nu_n\}_{n\in\mathbb{N}}$ converge weakly to ν.*

---

[2] In [98, Theorem 13.16], it is assumed that the topological space is a locally compact Polish space. This condition is satisfied notably when X is a locally compact Hausdorff space with countable base, see [145]. However, it actually suffices that the topological space is locally compact and metrisable (see Lemma 13.10 of [98] which is used in the proof of Theorem 13.16 of *loc. cit.*).

A.4 Upper and lower integral

Let $(\Omega, \mathcal{A}, \nu)$ be a measure space. We denote by $\mathscr{L}^1(\Omega, \mathcal{A}, \nu)$ the vector space of all real-valued ν-integrable functions on $(\Omega, \mathcal{A})$. We say that a subset A of Ω is ν-*negligible* if there exists a set $B \in \mathcal{A}$ such that $\nu(B) = 0$ and that $A \subseteq B$. We say that two functions h_1 and h_2 on Ω are ν-*indistinguishable* if $\{h_1 \neq h_2\}$ is a ν-negligible set. Any function on Ω which is ν-indistinguishable with the zero function is said to be ν-*negligible*. In other words, a function f on Ω is ν-negligible if and only if $\{f \neq 0\}$ is a ν-negligible set. If a formula depending on a variable $\omega \in \Omega$ is satisfied outside of a ν-negligible set, we say that it holds ν-almost everywhere (written in abbriviation as ν-a.e.).

Definition A.4.1 We construct two non-necessarily linear functional $\overline{I}_\nu(\cdot)$ and $\underline{I}_\nu(\cdot)$ as follows. For any function $h : \Omega \to \mathbb{R}$, let

$$\overline{\int_\Omega} h(\omega)\nu(d\omega) := \inf_{\substack{f \in \mathscr{L}^1(\Omega,\mathcal{A},\nu) \\ f \geqslant h\ \nu\text{-a.e.}}} \int_\Omega f(\omega)\nu(d\omega),$$

$$\underline{\int_\Omega} h(\omega)\nu(d\omega) := \sup_{\substack{g \in \mathscr{L}^1(\Omega,\mathcal{A},\nu) \\ g \leqslant h\ \nu\text{-a.e.}}} \int_\Omega g(\omega)\nu(d\omega).$$

If h is not ν-almost everywhere bounded from above by any integrable function, then $\overline{\int_\Omega} h(\omega)\nu(d\omega)$ is defined as $+\infty$ by convention. Similarly, if h is not ν-almost everywhere bounded from below by any integrable function, then $\underline{\int_\Omega} h(\omega)\nu(d\omega)$ is defined as $-\infty$ by convention. The values $\overline{\int_\Omega} h(\omega)\nu(d\omega)$ and $\underline{\int_\Omega} h(\omega)\nu(d\omega)$ are called *upper integral* and *lower integral* of the function h, respectively. From now on, for simplicity,

$$\overline{\int_\Omega} h(\omega)\nu(d\omega), \quad \underline{\int_\Omega} h(\omega)\nu(d\omega) \quad \text{and} \quad \int_\Omega f(\omega)\nu(d\omega)$$

are denoted by $\overline{I}_\nu(h)$, $\underline{I}_\nu(h)$ and $I_\nu(f)$, respectively, for any function h on Ω and any integrable function f on Ω.

The following properties are straightforward from the definition of upper and lower integrals.

Proposition A.4.2 (1) *For any function* $h : \Omega \to \mathbb{R}$

$$\underline{I}_\nu(h) \leqslant \overline{I}_\nu(h). \tag{A.1}$$

(2) *If* h_1 *and* h_2 *are two real-valued functions on* Ω *such that* $h_1 \leqslant h_2$*, then*

$$\underline{I}_\nu(h_1) \leqslant \underline{I}_\nu(h_2), \quad \overline{I}_\nu(h_1) \leqslant \overline{I}_\nu(h_2). \tag{A.2}$$

Proposition A.4.3 *Let h be a real-valued function on Ω. Then h is ν-indistinguishable with a ν-integrable function if and only if $\overline{I}_\nu(h) = \underline{I}_\nu(h) \in \mathbb{R}$.*

Proof If h is indistinguishable with a ν-integrable function $\widetilde{h}$, then $\overline{I}_\nu(h)$ and $\underline{I}_\nu(h)$ are both equal to the integral of $\widetilde{h}$ with respect to the measure ν, which is a real number.

Conversely, assume that $\overline{I}_\nu(h) = \underline{I}_\nu(h)$, then we can find two sequences $\{f_n\}_{n\in\mathbb{N}}$ and $\{g_n\}_{n\in\mathbb{N}}$ of functions in $\mathscr{L}^1(\Omega, \mathcal{A}, \nu)$ such that $g_n \leqslant h \leqslant f_n$ ν-almost everywhere. and that

$$\lim_{n\to+\infty} I_\nu(f_n) = \overline{I}_\nu(h) = \underline{I}_\nu(h) = \lim_{n\to+\infty} I_\nu(g_n).$$

Without loss of generality, we may assume that the sequence $\{f_n\}_{n\in\mathbb{N}}$ is decreasing and $\{g_n\}_{n\in\mathbb{N}}$ is increasing (otherwise we replace f_n by $\widetilde{f}_n = \min\{f_1, \ldots, f_n\}$ and g_n by $\widetilde{g}_n = \max\{g_1, \ldots, g_n\}$). Let $f = \inf_{n\in\mathbb{N}} f_n$ and $g = \sup_{n\in\mathbb{N}} g_n$. By Lebesgue's dominated convergence theorem, we obtain that f and g are both ν-integrable, and

$$I_\nu(f) = \overline{I}_\nu(h) = \underline{I}_\nu(h) = I_\nu(g).$$

Moreover, one has $g \leqslant h \leqslant f$ ν-almost everywhere, which implies that $f = g = h$ ν-almost everywhere. □

In general the operators $\overline{I}_\nu(\cdot)$ and $\underline{I}_\nu(\cdot)$ are not linear operators. However, they satisfies some convexity property.

Proposition A.4.4 *Let h_1 and h_2 be two real-valued functions on Ω.*

(1) *Assume that $\{\overline{I}_\nu(h_1), \overline{I}_\nu(h_2)\} \neq \{+\infty, -\infty\}$. Then one has*

$$\overline{I}_\nu(h_1 + h_2) \leqslant \overline{I}_\nu(h_1) + \overline{I}_\nu(h_2) \tag{A.3}$$

(2) *Assume that $\{\underline{I}_\nu(h_1), \underline{I}_\nu(h_2)\} \neq \{+\infty, -\infty\}$. Then one has*

$$\underline{I}_\nu(h_1 + h_2) \geqslant \underline{I}_\nu(h_1) + \underline{I}_\nu(h_2). \tag{A.4}$$

Proof (1) We first treat the case where neither of $\overline{I}_\nu(h_1)$ and $\overline{I}_\nu(h_2)$ is $+\infty$. If f_1 and f_2 are two ν-integrable functions on Ω such that $f_1 \geqslant h_1$ and $f_2 \geqslant h_2$ ν-almost everywhere, then the sum $f_1 + f_2$ is ν-integrable, and $f_1 + f_2 \geqslant h_1 + h_2$. Therefore $I_\nu(f_1) + I_\nu(f_2) = I_\nu(f_1 + f_2) \geqslant \overline{I}_\nu(h_1 + h_2)$. Since f_1 and f_2 are arbitrary, we obtain $\overline{I}_\nu(h_1) + \overline{I}_\nu(h_2) \geqslant \overline{I}_\nu(h_1 + h_2)$.

If at least one of the upper integrals $\overline{I}_\nu(h_1)$ and $\overline{I}_\nu(h_2)$ is $+\infty$, then by the hypothesis $\{\overline{I}_\nu(h_1), \overline{I}_\nu(h_2)\} \neq \{+\infty, -\infty\}$ one has $\overline{I}_\nu(h_1) + \overline{I}_\nu(h_2) = +\infty$. Hence the inequality (A.3) is trivial.

The proof of the statement (2) is very similar to that of (1). We omit the details.□

Proposition A.4.5 *Let h be a real-valued function on Ω and φ be a ν-integrable function. Then one has*

$$\overline{I}_\nu(h + \varphi) = \overline{I}_\nu(h) + I_\nu(\varphi), \quad \underline{I}_\nu(h + \varphi) = \underline{I}_\nu(h) + I_\nu(\varphi).$$

Proof Since φ is ν-integrable, one has

$$\overline{I}_\nu(\varphi) = I_\nu(\varphi) = \underline{I}_\nu(\varphi) \in \mathbb{R}.$$

By Proposition A.4.4, one has

$$\overline{I}_\nu(h+\varphi) \leqslant \overline{I}_\nu(h) + I_\nu(\varphi).$$

Moreover, if we apply this inequality to $h+\varphi$ and $-\varphi$, we obtain

$$\overline{I}_\nu(h) \leqslant \overline{I}_\nu(h+\varphi) - I_\nu(\varphi).$$

Therefore the first equality is true. The proof of the second equality is quite similar, we omit the details. □

Proposition A.4.6 *Let h be a real-valued function on Ω. If a is a non-negative number, then one has*

$$\overline{I}_\nu(ah) = a\overline{I}_\nu(h), \quad \underline{I}_\nu(ah) = a\underline{I}_\nu(h).$$

Proof The assertions are trivial when $a = 0$. In the following, we assume that $a > 0$. If f is a ν-integrable function such that $h \leqslant f$ ν-almost everywhere, then af is a ν-integrable function such that $ah \leqslant af$ ν-almost everywhere. Therefore, we obtain that $\overline{I}_\nu(ah) \leqslant a\overline{I}(h)$. If we apply this inequality to a^{-1} and ah we get $\overline{I}_\nu(h) \leqslant a^{-1}\overline{I}_\nu(ah)$. Hence the first equality is true. The proof of the second equality is very similar, we omit the details. □

Proposition A.4.7 *Let h be a real-valued function on Ω. One has*

$$\overline{I}_\nu(-h) = -\underline{I}_\nu(h), \quad \underline{I}_\nu(-h) = -\overline{I}_\nu(h).$$

Proof If f is a ν-integrable function such that $-h \leqslant f$ ν-almost everywhere, then one has $-f \leqslant h$ ν-almost everywhere. Since f is arbitrary, we obtain $-\overline{I}_\nu(-h) \leqslant \underline{I}_\nu(h)$. Similarly, if g is a ν-integrable function such that $g \leqslant -h$ ν-almost everywhere, then one has $h \leqslant -g$ ν-almost everywhere. Since g is arbitrary, we obtain $-\underline{I}_\nu(-h) \geqslant \overline{I}_\nu(h)$. Finally, if we apply the obtained inequality to $-h$, we obtain $-\overline{I}_\nu(h) \leqslant \underline{I}_\nu(-h)$ and $-\underline{I}_\nu(h) \geqslant \overline{I}_\nu(-h)$. Therefore the equalities hold. □

Proposition A.4.8 *Let h_1 and h_2 be two real-valued functions on Ω, and let $h = h_1 + h_2$. Assume that $\{\overline{I}_\nu(h_1), \underline{I}_\nu(h_2)\} \neq \{+\infty, -\infty\}$. Then one has*

$$\underline{I}_\nu(h) \leqslant \overline{I}_\nu(h_1) + \underline{I}_\nu(h_2) \leqslant \overline{I}_\nu(h).$$

Proof By the equality $h = h_1 + h_2$ we obtain $h_1 = (-h_2) + h$. Thus Proposition A.4.4 leads to

$$\overline{I}_\nu(h_1) \leqslant \overline{I}_\nu(-h_2) + \overline{I}_\nu(h) = -\underline{I}_\nu(h_2) + \overline{I}_\nu(h),$$

where the equality comes from Proposition A.4.7. Hence we obtain the inequality

$$\overline{I}_\nu(h) \geqslant \overline{I}_\nu(h_1) + \underline{I}_\nu(h_2).$$

We then apply this inequality to $-h$, $-h_2$ and $-h_1$ to get the other equality. □

Definition A.4.9 Let $h : \Omega \to \mathbb{R}$ be a real valued function on Ω. We say that h is *ν-dominated* if there exists a ν-integrable function f such that $\{\omega \in \Omega : |h(\omega)| \leqslant f(\omega)\}$ is a ν-negligible set (in other words, $|h| \leqslant f$ ν-almost everywhere). Note that this condition is equivalent to

$$\overline{I}_\nu(h) < +\infty \text{ and } \underline{I}_\nu(h) > -\infty.$$

We denote by $\mathscr{D}^1(\Omega, \mathcal{A}, \nu)$ the vector space of ν-dominated functions on Ω. Clearly one has $\mathscr{D}^1(\Omega, \mathcal{A}, \nu) \supseteq \mathscr{L}^1(\Omega, \mathcal{A}, \nu)$, and $\mathscr{D}^1(\Omega, \mathcal{A}, \nu)$ is invariant by the operator $f \mapsto |f|$ of taking the absolute value. Moreover, if f and g are real-valued functions on Ω such that $|f| \leqslant |g|$ ν-almost everywhere and that g is ν-dominated, then the function f is also ν-dominated.

Proposition A.4.10 *Let $\|\cdot\|_{\mathscr{D}^1_\nu}$ be the function on $\mathscr{D}^1(\Omega, \mathcal{A}, \nu)$ sending any ν-dominated function f to $\overline{I}_\nu(|f|)$. Then $\|\cdot\|_{\mathscr{D}^1_\nu}$ is a seminorm. Moreover, a function $f \in \mathscr{D}^1(\Omega, \mathcal{A}, \nu)$ satisfies $\|f\|_{\mathscr{D}^1_\nu} = 0$ if and only if it is ν-negligible.*

Proof Let f be a ν-dominated function and a be a real number. One has $|af| = |a| \cdot |f|$. By Proposition A.4.6 we obtain that

$$\|af\|_{\mathscr{D}^1_\nu} = \overline{I}_\nu(|af|) = \overline{I}_\nu(|a| \cdot |f|) = |a| \cdot \overline{I}_\nu(|f|) = |a| \cdot \|f\|_{\mathscr{D}^1_\nu}.$$

Moreover, if f and g are two ν-dominated functions, then by Proposition A.4.4, one has

$$\|f+g\|_{\mathscr{D}^1_\nu} = \overline{I}_\nu(|f+g|) \leqslant \overline{I}_\nu(|f|+|g|) \leqslant \overline{I}_\nu(|f|) + \overline{I}_\nu(|g|) = \|f\|_{\mathscr{D}^1_\nu} + \|g\|_{\mathscr{D}^1_\nu},$$

where the first inequality comes from (A.2), and the second inequality comes from (A.4.4). Therefore $\|\cdot\|_{\mathscr{D}_\nu}$ is a seminorm on $\mathscr{D}^1_\nu(\Omega, \mathcal{A}, \nu)$.

Let f be a ν-negligible function. Then one has $|f| = 0$ ν-almost everywhere. Hence one has $\|f\|_{\mathscr{D}^1_\nu} = \overline{I}_\nu(|f|) = 0$. Conversely, if f is a ν-dominated function such that $\overline{I}_\nu(|f|) = 0$, then one has $\underline{I}_\nu(|f|) = \overline{I}_\nu(|f|) = 0$. By Proposition A.4.3, $|f|$ is ν-indistinguishable with a ν-integrable function g of integral 0. Moreover, since $|f|$ is non-negative, we obtain that the set $\{g < 0\}$ is ν-negligible. Therefore g vanishes ν-almost everywhere. Thus f is ν-negligible. □

Proposition A.4.11 *Let $\{f_n\}_{n\in\mathbb{N}}$ be an increasing sequence of non-negative functions on Ω and f be the limit of $\{f_n\}_{n\in\mathbb{N}}$. Then one has*

$$\lim_{n\to+\infty} \overline{I}_\nu(f_n) = \overline{I}_\nu(f).$$

Proof Clearly one has $\overline{I}_\nu(f_n) \leqslant \overline{I}_\nu(f)$ for any $n \in \mathbb{N}$. Hence

$$\lim_{n\to+\infty} \overline{I}_\nu(f_n) \leqslant \overline{I}_\nu(f).$$

If one of the functions f_n is not dominated, then neither is f. Hence one has

$$\lim_{n\to+\infty} \overline{I}_\nu(f_n) = +\infty = \overline{I}_\nu(f).$$

In the following, we assume that all the functions f_n are dominated. Let $\epsilon > 0$. For any $n \in \mathbb{N}$, let g_n be an integrable function on Ω such that $f_n \leqslant g_n$ and that $\overline{I}_\nu(f_n) \geqslant I_\nu(g_n) - \epsilon$. Note that $\widetilde{g}_n := \inf_{m\geqslant n} g_m$ is also an integrable function on Ω such that $f_n \leqslant \widetilde{g}_n$ and $\overline{I}_\nu(f_n) \geqslant I_\nu(\widetilde{g}_n) - \varepsilon$. Therefore, by replacing g_n by $\widetilde{g}_n$, we may assume without loss of generality that the sequence $\{g_n\}_{n\in\mathbb{N}}$ is increasing. Let $g = \sup_{n\in\mathbb{N}} g_n$. By the monotone convergence theorem one has

$$\overline{I}_\nu(g) = \lim_{n\to+\infty} I_\nu(g_n) \leqslant \lim_{n\to+\infty} \overline{I}_\nu(f_n) + \epsilon.$$

Moreover, since $g \geqslant f$, one has $\overline{I}_\nu(f) \leqslant \overline{I}_\nu(g)$. Therefore the proposition is proved.□

Corollary A.4.12 *Let $\{f_n\}_{n\in\mathbb{N}}$ be a sequence of non-negative functions on Ω, and f be the sum of the series $\sum_{n\in\mathbb{N}} f_n$. Then one has*

$$\overline{I}_\nu(f) \leqslant \sum_{n\in\mathbb{N}} \overline{I}_\nu(f_n).$$

Proof For any $n \in \mathbb{N}$, let $g_n = \sum_{k=0}^{n} f_k$. The sequence $\{g_n\}_{n\in\mathbb{N}}$ is increasing, and converges to f. Therefore, by Proposition A.4.11, one has

$$\overline{I}_\nu(f) = \lim_{n\to+\infty} \overline{I}_\nu(g_n).$$

Moreover, by Proposition A.4.4, for any $n \in \mathbb{N}$ one has

$$\overline{I}_\nu(g_n) \leqslant \sum_{k=0}^{n} \overline{I}_\nu(f_k).$$

Hence we obtain

$$\overline{I}_\nu(f) \leqslant \sum_{n\in\mathbb{N}} \overline{I}_\nu(f_n).$$

Proposition A.4.13 *Let $\{f_n\}_{n\in\mathbb{N}}$ be a sequence of non-negative functions on Ω and $f = \liminf_{n\to+\infty} f_n$. Then one has*

$$\overline{I}_\nu(f) \leqslant \liminf_{n\to+\infty} \overline{I}_\nu(f_n). \tag{A.5}$$

Proof For any $n \in \mathbb{N}$, let $g_n = \inf_{m\geqslant n} f_m$. Then the sequence $\{g_n\}_{n\in\mathbb{N}}$ is increasing and converges to f. By Proposition A.4.11, one has

$$\overline{I}_\nu(f) = \lim_{n\to+\infty} \overline{I}_\nu(g_n) \leqslant \liminf_{n\to+\infty} \overline{I}_\nu(f_n),$$

where the inequality comes from the fact that $g_n \leqslant f_n$ for any $n \in \mathbb{N}$. The proposition is thus proved. □

Proposition A.4.14 *Let $D^1(\Omega, \mathcal{A}, \nu)$ be the quotient space of $\mathscr{D}^1(\Omega, \mathcal{A}, \nu)$ by the vector subspace of ν-negligible functions. Then the seminorm $\|\cdot\|_{\mathscr{D}^1_\nu}$ on $\mathscr{D}^1(\Omega, \mathcal{A}, \nu)$ induces a norm $\|\cdot\|_{D^1_\nu}$ on $D^1(\Omega, \mathcal{A}, \nu)$, and the vector space $D^1(\Omega, \mathcal{A}, \nu)$ is complete with respect to this norm.*

Proof The first assertion is a direct consequence of Proposition A.4.10. In the following, we prove the second assertion.

Let $\{f_n\}_{n\in\mathbb{N}}$ be a Cauchy sequence in $\mathscr{D}^1(\Omega, \mathcal{A}, \nu)$. For any $\epsilon > 0$ and any $m, n \in \mathbb{N}$, one has $|f_n - f_m| \geqslant \epsilon \mathbb{1}_{\{|f_n - f_m| > \epsilon\}}$, which implies that

$$\|f_n - f_m\|_{\mathscr{D}^1_\nu} = \overline{I}_\nu(|f_n - f_m|) \geqslant \epsilon \overline{I}_\nu(\mathbb{1}_{\{|f_n - f_m| > \epsilon\}}).$$

Since $\{f_n\}_{n\in\mathbb{N}}$ is a Cauchy sequence, one has

$$\lim_{N\to+\infty} \sup_{\substack{(n,m)\in\mathbb{N}^2\\ n\geqslant N,\, m\geqslant N}} \|f_n - f_m\| = 0.$$

Therefore we can construct a subsequence $\{f_{n_k}\}_{k\geqslant 1}$ of $\{f_n\}_{n\in\mathbb{N}}$ such that

$$\forall\, k \in \mathbb{N}_{\geqslant 1}, \quad \overline{I}_\nu(\mathbb{1}_{\{|f_{n_k} - f_{n_{k+1}}| > 2^{-k}\}}) < 2^{-k}.$$

For any $m \in \mathbb{N}_{\geqslant 1}$, let $A_m = \bigcup_{k\geqslant m}\{|f_{n_k} - f_{n_{k+1}}| > 2^{-k}\}$. Then the set

$$B := \{\omega \in \Omega \,:\, \{f_{n_k}(\omega)\}_{k\geqslant 1} \text{ does not converge}\}$$

is contained in $\bigcap_{m\geqslant 1} A_m$. Moreover, for any $m \in \mathbb{N}_{\geqslant 1}$, by Corollary A.4.12 one has

$$\overline{I}_\nu(\mathbb{1}_{A_m}) \leqslant \sum_{k\geqslant m} \overline{I}_\nu(\mathbb{1}_{\{|f_{n_k} - f_{n_{k+1}}| > 2^{-k}\}}) \leqslant 2^{-m+1}.$$

Therefore we obtain $\overline{I}_\nu(\mathbb{1}_B) = 0$, which implies that B is a ν-negligible set. Thus we obtain that the sequence $\{f_{n_k}\}_{k\geqslant 1}$ converges ν-almost everywhere to some function f on Ω. Note that by Proposition A.4.13 one has

$$\overline{I}_\nu(|f|) \leqslant \liminf_{k\to+\infty} \overline{I}_\nu(|f_{n_k}|).$$

Therefore f is a dominated function. Finally, still by Proposition A.4.13, for any $n \in \mathbb{N}$ one has

$$\overline{I}_\nu(|f_n - f|) \leqslant \liminf_{k\to+\infty} \overline{I}_\nu(|f_n - f_{n_k}|).$$

Hence one has

$$\lim_{n\to+\infty} \overline{I}_\nu(|f_n - f|) = 0.$$

The proposition is thus proved. □

A.5 L^1 space

Let $\mathscr{L}^1(\Omega, \mathcal{A}, \nu)$ be the vector space of all real-valued ν-integrable functions on the measurable space $(\Omega, \mathcal{A})$. This vector space is equipped with the seminorm $\|\cdot\|_{\mathscr{L}^1_\nu}$ which sends a function $f \in \mathscr{L}^1(\Omega, \mathcal{A}, \nu)$ to

$$\|f\|_{\mathscr{L}^1_\nu} := \int_\Omega |f(\omega)|\,\nu(\mathrm{d}\omega).$$

Note that the set of all functions $f \in \mathscr{L}^1(\Omega, \mathcal{A}, \nu)$ such that $\|f\|_{\mathscr{L}^1_\nu} = 0$ forms a vector subspace of $\mathscr{L}^1(\Omega, \mathcal{A}, \nu)$. Such functions are said to be ν-*negligible*. The quotient space of $\mathscr{L}^1(\Omega, \mathcal{A}, \nu)$ by the vector subspace of ν-negligible functions is denoted by $L^1(\Omega, \mathcal{A}, \nu)$. The seminorm $\|\cdot\|_{\mathscr{L}^1_\nu}$ induces by quotient a norm on $L^1(\Omega, \mathcal{A}, \nu)$, which we denote by $\|\cdot\|_{L^1_\nu}$. Note that the vector space $L^1(\Omega, \mathcal{A}, \nu)$ is complete with respect to this norm, and the integration with respect to ν induces a continuous linear form on $L^1(\Omega, \mathcal{A}, \nu)$, which we denote by

$$(\zeta \in L^1(\Omega, \mathcal{A}, \nu)) \longmapsto \int_\Omega \zeta(\omega)\,\nu(\mathrm{d}\omega)$$

by abuse of notation.

References

1. Ahmed Abbes. *Éléments de géométrie rigide. Volume I*, volume 286 of *Progress in Mathematics*. Birkhäuser/Springer Basel AG, Basel, 2010. Construction et étude géométrique des espaces rigides, With a preface by Michel Raynaud.
2. Yves André. Slope filtrations. *Confluentes Mathematici*, 1(1):1–85, 2009.
3. Yves André. On nef and semistable Hermitian lattices, and their behaviour under tensor product. *The Tohoku Mathematical Journal. Second Series*, 63(4):629–649, 2011.
4. Suren Yurievich Arakelov. An intersection theory for divisors on an arithmetic surface. *Izvestiya Akademii Nauk SSSR. Seriya Matematicheskaya*, 38:1179–1192, 1974.
5. Suren Yurievich Arakelov. Theory of intersections on the arithmetic surface. In *Proceedings of the International Congress of Mathematicians (Vancouver, B.C., 1974), Vol. 1*, pages 405–408. Canad. Math. Congress, Montreal, Que., 1975.
6. Michael Francis Atiyah, Raoul Harry Bott, and Vijay Kumar Patodi. On the heat equation and the index theorem. *Inventiones Mathematicae*, 19:279–330, 1973.
7. Michael Francis Atiyah and Ian Grant Macdonald. *Introduction to commutative algebra*. Addison-Wesley Publishing Co., Reading, Mass.-London-Don Mills, Ont., 1969.
8. Itaï Ben Yaakov and Ehud Hrushovski. Towards a model theory of global fields. Lecture notes of Ehud Hrushovski at Institut Henri Poincaré, 2016.
9. Vladimir G. Berkovich. *Spectral theory and analytic geometry over non-Archimedean fields*, volume 33 of *Mathematical Surveys and Monographs*. American Mathematical Society, Providence, RI, 1990.
10. Zbigniew Błocki and Sławomir Kołodziej. On regularization of plurisubharmonic functions on manifolds. *Proceedings of the American Mathematical Society*, 135(7):2089–2093, 2007.
11. Enrico Bombieri. The Mordell conjecture revisited. *Annali della Scuola Normale Superiore di Pisa. Classe di Scienze. Serie IV*, 17(4):615–640, 1990.
12. Enrico Bombieri and Walter Gubler. *Heights in Diophantine geometry*, volume 4 of *New Mathematical Monographs*. Cambridge University Press, Cambridge, 2006.
13. Enrico Bombieri and Jeffrey D. Vaaler. On Siegel's lemma. *Inventiones Mathematicae*, 73(1):11–32, 1983.
14. Frank F. Bonsall and John Duncan. *Numerical ranges. II*. Cambridge University Press, New York-London, 1973. London Mathematical Society Lecture Notes Series, No. 10.
15. Thomas Borek. Successive minima and slopes of Hermitian vector bundles over number fields. *Journal of Number Theory*, 113(2):380–388, 2005.
16. Jean-Benoît Bost. Périodes et isogénies des variétés abéliennes sur les corps de nombres (d'après D. Masser et G. Wüstholz). *Astérisque*, (237):Exp. No. 795, 4, 115–161, 1996. Séminaire Bourbaki, Vol. 1994/1995.
17. Jean-Benoît Bost. Hermitian vector bundle and stability. Lecture at Oberwolfach conference "Algebraische Zahlentheorie", July 24th, 1997.

H. Chen, A. Moriwaki, *Arakelov Geometry over Adelic Curves*, Lecture Notes in Mathematics 2258, https://doi.org/10.1007/978-981-15-1728-0

18. Jean-Benoît Bost. Algebraic leaves of algebraic foliations over number fields. *Publications Mathématiques. Institut de Hautes Études Scientifiques*, (93):161–221, 2001.
19. Jean-Benoît Bost. Germs of analytic varieties in algebraic varieties: canonical metrics and arithmetic algebraization theorems. In *Geometric aspects of Dwork theory. Vol. I, II*, pages 371–418. Walter de Gruyter GmbH & Co. KG, Berlin, 2004.
20. Jean-Benoît Bost. Evaluation maps, slopes, and algebraicity criteria. In *Proceedings of the International Congress of Mathematicians, Vol. II (Madrid, 2006)*, pages 537–562, Switzerland, 2007. European Mathematical Society.
21. Jean-Benoît Bost and Antoine Chambert-Loir. Analytic curves in algebraic varieties over number fields. In *Algebra, arithmetic, and geometry: in honor of Yu. I. Manin. Vol. I*, volume 269 of *Progr. Math.*, pages 69–124. Birkhäuser Boston, Inc., Boston, MA, 2009.
22. Jean-Benoît Bost and Huayi Chen. Concerning the semistability of tensor products in Arakelov geometry. *Journal de Mathématiques Pures et Appliquées. Neuvième Série*, 99(4):436–488, 2013.
23. Jean-Benoît Bost, Henri Gillet, and Christophe Soulé. Heights of projective varieties. *Journal of the American Mathematical Society*, 7(4):903–1027, October 1994.
24. Jean-Benoît Bost and Klaus Künnemann. Hermitian vector bundles and extension groups on arithmetic schemes. I. Geometry of numbers. *Advances in Mathematics*, 223(3):987–1106, 2010.
25. Sébastien Boucksom. Corps d'Okounkov (d'après Okounkov, Lazarsfeld-Mustaţă et Kaveh-Khovanskii). *Astérisque*, (361):Exp. No. 1059, vii, 1–41, 2014.
26. Sébastien Boucksom and Huayi Chen. Okounkov bodies of filtered linear series. *Compositio Mathematica*, 147(4):1205–1229, 2011.
27. Nicolas Bourbaki. *Éléments de mathématique. Algèbre commutative. Chapitre 5 à 6*. Hermann, Paris, 1964.
28. Nicolas Bourbaki. *Éléments de mathématique. Algèbre. Chapitres 1 à 3*. Hermann, Paris, 1970.
29. Nicolas Bourbaki. *Éléments de mathématique. Algèbre. Chapitres 4 à 7*. Masson, Paris, 1981.
30. Nicolas Bourbaki. *Espaces vectoriels topologiques. Chapitres 1 à 5*. Masson, Paris, 1981.
31. Nicolas Bourbaki. *Éléments de mathématique. Algèbre commutative. Chapitre 1 à 4*. Masson, Paris, 1985.
32. Nicolas Bourbaki. *Elements of Mathematics. General topology. Chapiters 5 – 10*. Springer-Verlag, Berlin, 1989.
33. José Ignacio Burgos Gil, Atsushi Moriwaki, Patrice Philippon, and Martín Sombra. Arithmetic positivity on toric varieties. *Journal of Algebraic Geometry*, 25(2):201–272, 2016.
34. José Ignacio Burgos Gil, Patrice Philippon, and Martín Sombra. Height of varieties over finitely generated fields. *Kyoto Journal of Mathematics*, 56(1):13–32, 2016.
35. Gregory S. Call and Joseph H. Silverman. Canonical heights on varieties with morphisms. *Compositio Mathematica*, 89(2):163–205, 1993.
36. Antoine Chambert-Loir and Antoine Ducros. Formes différentielles réelles et courants sur les espaces de berkovich. arXiv:1204.6277.
37. Huayi Chen. Positive degree and arithmetic bigness, 2008. arXiv:0803.2583.
38. Huayi Chen. Maximal slope of tensor product of Hermitian vector bundles. *Journal of algebraic geometry*, 18(3):575–603, 2009.
39. Huayi Chen. Arithmetic Fujita approximation. *Annales Scientifiques de l'École Normale Supérieure. Quatrième Série*, 43(4):555–578, 2010.
40. Huayi Chen. Convergence des polygones de Harder-Narasimhan. *Mémoires de la Société Mathématique de France*, 120:1–120, 2010.
41. Huayi Chen. Harder-Narasimhan categories. *Journal of Pure and Applied Algebra*, 214(2):187–200, 2010.
42. Huayi Chen. Majorations explicites de fonctions de hilbert-samuel géométrique et arithmétique. *Mathematische Zeitschrift*, 279(1):99–137, 2015.
43. Huayi Chen. Newton-Okounkov bodies: an approach of function field arithmetic. *Journal de Théorie des Nombres de Bordeaux*, 30(3):829–845, 2018.

44. Huayi Chen. Sur la comparaison entre les minima et les pentes. In *Publications mathématiques de Besançon. Algèbre et théorie des nombres, 2015*, volume 2018 of *Publ. Math. Besançon Algèbre Théorie Nr.*, pages 5–23. Presses Univ. Franche-Comté, Besançon, 2018.
45. Huayi Chen and Hideaki Ikoma. On subfiniteness of graded linear series. *European Journal of Mathematics*, 2019. doi.org/10.1007/s40879-019-00349-0.
46. Huayi Chen and Atsushi Moriwaki. Algebraic dynamical systems and Dirichlet's unit theorem on arithmetic varieties. *International Mathematics Research Notices. IMRN*, (24):13669–13716, 2015.
47. Huayi Chen and Atsushi Moriwaki. Sufficient conditions for the dirichlet property. arXiv:1704.01410, 2017.
48. Huayi Chen and Atsushi Moriwaki. Extension property of semipositive invertible sheaves over a non-archimedean field. *Annali della Scuola Normale Superiore di Pisa. Classe di Scienze. Serie V*, 18(1):241–282, 2018.
49. Claude Chevalley. *Introduction to the Theory of Algebraic Functions of One Variable*. Mathematical Surveys, No. VI. American Mathematical Society, New York, N. Y., 1951.
50. Irvin Sol Cohen. On non-Archimedean normed spaces. *Nederl. Akad. Wetensch., Proc.*, 51:693–698 = Indagationes Math. 10, 244–249, 1948.
51. Pietro Corvaja and Umberto Zannier. Arithmetic on infinite extensions of function fields. *Boll. Un. Mat. Ital. B (7)*, 11(4):1021–1038, 1997.
52. Claude Dellacherie and Paul-André Meyer. *Probabilités et potentiel*. Hermann, Paris, 1975. Chapitres I à IV, Édition entièrement refondue, Publications de l'Institut de Mathématique de l'Université de Strasbourg, No. XV, Actualités Scientifiques et Industrielles, No. 1372.
53. Michel Demazure and Pierre Gabriel. *Groupes algébriques. Tome I: Géométrie algébrique, généralités, groupes commutatifs*. Masson & Cie, Éditeur, Paris; North-Holland Publishing Co., Amsterdam, 1970. Avec un appendice ıt Corps de classes local par Michiel Hazewinkel.
54. Nikolai Durov. New approach to Arakelov geometry. arxiv:0704.2030, 2007.
55. Harold M. Edwards. *Galois theory*, volume 101 of *Graduate Texts in Mathematics*. Springer-Verlag, New York, 1984.
56. David Eisenbud. *Commutative algebra*, volume 150 of *Graduate Texts in Mathematics*. Springer-Verlag, New York, 1995. With a view toward algebraic geometry.
57. Gerd Faltings. Endlichkeitssätze für abelsche Varietäten über Zahlkörpern. *Inventiones Mathematicae*, 73(3):349–366, 1983.
58. Gerd Faltings. Calculus on arithmetic surfaces. *Annals of Mathematics. Second Series*, 119(2):387–424, 1984.
59. Gerd Faltings. Erratum: "Finiteness theorems for abelian varieties over number fields". *Inventiones Mathematicae*, 75(2):381, 1984.
60. Gerd Faltings. Diophantine approximation on abelian varieties. *Annals of Mathematics. Second Series*, 133(3):549–576, 1991.
61. Gerd Faltings and Gisbert Wüstholz. Diophantine approximations on projective spaces. *Inventiones Mathematicae*, 116(1-3):109–138, 1994.
62. Éric Gaudron. Pentes de fibrés vectoriels adéliques sur un corps globale. *Rendiconti del Seminario Matematico della Università di Padova*, 119:21–95, 2008.
63. Éric Gaudron. Minima and slopes of rigid adelic spaces. In Gaël Rémond and Emmanuel Peyre, editors, *Arakelov geometry and Diophantine applications*. 2017. math.univ-bpclermont.fr/ gaudron/art18.pdf.
64. Éric Gaudron and Gaël Rémond. Minima, pentes et algèbre tensorielle. *Israel Journal of Mathematics*, 195(2):565–591, 2013.
65. Éric Gaudron and Gaël Rémond. Corps de Siegel. *Journal für die Reine und Angewandte Mathematik*, 726:187–247, 2017.
66. Éric Gaudron and Gaël Rémond. Espaces adéliques quadratiques. *Mathematical Proceedings of the Cambridge Philosophical Society*, 162(2):211–247, 2017.
67. Henri Gillet, Damian Rössler, and Christophe Soulé. An arithmetic Riemann-Roch theorem in higher degrees. *Université de Grenoble. Annales de l'Institut Fourier*, 58(6):2169–2189, 2008.

68. Henri Gillet and Christophe Soulé. Intersection sur les variétés d'Arakelov. *Comptes Rendus des Séances de l'Académie des Sciences. Série I. Mathématique*, 299(12):563–566, 1984.
69. Henri Gillet and Christophe Soulé. An arithmetic Riemann-Roch theorem. *Inventiones Mathematicae*, 110(3):473–543, 1992.
70. Sarah Glaz. *Commutative coherent rings*, volume 1371 of *Lecture Notes in Mathematics*. Springer-Verlag, Berlin, 1989.
71. Daniel R. Grayson. Reduction theory using semistability. *Commentarii Mathematici Helvetici*, 59(4):600–634, 1984.
72. Alexander Grothendieck. Produits tensoriels topologiques et espaces nucléaires. *Memoirs of the American Mathematical Society*, No. 16:140, 1955.
73. Alexander Grothendieck. Résumé de la théorie métrique des produits tensoriels topologiques. *Resenhas do Instituto de Matemática e Estatística da Universidade de São Paulo*, 2(4):401–480, 1996. Reprint of Bol. Soc. Mat. São Paulo **8** (1953), 1–79.
74. Alexander Grothendieck and Jean Dieudonné. Éléments de géométrie algébrique. II. Étude globale élémentaire de quelques classes de morphismes. *Institut des Hautes Études Scientifiques. Publications Mathématiques*, (8):222, 1961.
75. Alexander Grothendieck and Jean Dieudonné. Éléments de géométrie algébrique. IV. Étude locale des schémas et des morphismes de schémas. II. *Institut des Hautes Études Scientifiques. Publications Mathématiques*, (24):231, 1965.
76. Alexander Grothendieck and Jean Dieudonné. Éléments de géométrie algébrique. IV. Étude locale des schémas et des morphismes de schémas. III. *Institut des Hautes Études Scientifiques. Publications Mathématiques*, (28):255, 1966.
77. Alexander Grothendieck and Jean Dieudonné. Éléments de géométrie algébrique. IV. Étude locale des schémas et des morphismes de schémas IV. *Institut des Hautes Études Scientifiques. Publications Mathématiques*, (32):361, 1967.
78. Alexander Grothendieck and Jean Dieudonné. *Eléments de géométrie algébrique. I*, volume 166 of *Grundlehren der Mathematischen Wissenschaften*. Springer-Verlag, Berlin, 1971.
79. Walter Gubler. Heights of subvarieties over M-fields. In *Arithmetic geometry (Cortona, 1994)*, Sympos. Math., XXXVII, pages 190–227. Cambridge Univ. Press, Cambridge, 1997.
80. Walter Gubler. The Bogomolov conjecture for totally degenerate abelian varieties. *Inventiones Mathematicae*, 169(2):377–400, 2007.
81. Walter Gubler. A guide to tropicalizations. In *Algebraic and combinatorial aspects of tropical geometry*, volume 589 of *Contemp. Math.*, pages 125–189. Amer. Math. Soc., Providence, RI, 2013.
82. Walter Gubler and Klaus Künnemann. Positivity properties of metrics and delta-forms. *Journal für die Reine und Angewandte Mathematik. [Crelle's Journal]*, 752:141–177, 2019.
83. Paul R. Halmos. *Measure Theory*. Springer-Verlag, New York, 1974. Graduate Texts in Mathematics, No. 18.
84. Günter. Harder and Mudumbai S. Narasimhan. On the cohomology groups of moduli spaces of vector bundles on curves. *Mathematische Annalen*, 212:215–248, 1974/1975.
85. Robin Hartshorne. *Algebraic geometry*. Springer-Verlag, New York, 1977. Graduate Texts in Mathematics, No. 52.
86. Martin Henk. Löwner-John ellipsoids. *Documenta Mathematica*, (Extra volume: Optimization stories):95–106, 2012.
87. Paul Hriljac. Heights and Arakelov's intersection theory. *American Journal of Mathematics*, 107(1):23–38, 1985.
88. Ehud Hrushovski. A logic for global fields. Lecture at Séminaire d'Arithmétique et de Géométrie Algébrique, May 31st, 2016.
89. Aubrey William Ingleton. The Hahn-Banach theorem for non-Archimedean valued fields. *Mathematical Proceedings of the Cambridge Philosophical Society*, 48:41–45, 1952.
90. Johan Jensen. Sur un nouvel et important théorème de la théorie des fonctions. *Acta Mathematica*, 22(1):359–364, 1899.
91. Fritz John. Extremum problems with inequalities as subsidiary conditions. In *Studies and Essays Presented to R. Courant on his 60th Birthday, January 8, 1948*, pages 187–204. Interscience Publishers, Inc., New York, N. Y., 1948.

92. Shizuo Kakutani. Some characterizations of Euclidean space. *Japanese Journal of Mathematics*, 16:93–97, 1939.
93. Olav Kallenberg. *Foundations of modern probability*. Probability and its Applications (New York). Springer-Verlag, New York, 1997.
94. Kiumars Kaveh and Askold Khovanskii. Algebraic equations and convex bodies. In *Perspectives in analysis, geometry, and topology*, volume 296 of *Progr. Math.*, pages 263–282. Birkhäuser/Springer, New York, 2012.
95. Shu Kawaguchi and Joseph H. Silverman. Nonarchimedean Green functions and dynamics on projective space. *Mathematische Zeitschrift*, 262(1):173–197, 2009.
96. George R. Kempf. Instability in invariant theory. *Annals of Mathematics*, (108):299–316, 1978.
97. Steven L. Kleiman. Misconceptions about K_x. *L'Enseignement Mathématique. Revue Internationale. IIe Série*, 25(3-4):203–206 (1980), 1979.
98. Achim Klenke. *Probability theory*. Universitext. Springer, London, second edition, 2014. A comprehensive course.
99. Maciej Klimek. *Pluripotential theory*, volume 6 of *London Mathematical Society Monographs. New Series*. The Clarendon Press, Oxford University Press, New York, 1991. Oxford Science Publications.
100. Kazimierz Kuratowski and Czesław Ryll-Nardzewski. A general theorem on selectors. *Bulletin de l'Académie Polonaise des Sciences. Série des Sciences Mathématiques, Astronomiques et Physiques*, 13:397–403, 1965.
101. Robert Lazarsfeld and Mircea Mustaţă. Convex bodies associated to linear series. *Annales Scientifiques de l'École Normale Supérieure. Quatrième Série*, 42(5):783–835, 2009.
102. Masaki Maruyama. The theorem of Grauert-Mülich-Spindler. *Mathematische Annalen*, 255(3):317–333, 1981.
103. Richard Clive Mason. *Diophantine equations over function fields*, volume 96 of *London Mathematical Society Lecture Note Series*. Cambridge University Press, Cambridge, 1984.
104. Hideyuki Matsumura. *Commutative algebra*, volume 56 of *Mathematics Lecture Note Series*. Benjamin/Cummings Publishing Co., Inc., Reading, Mass., second edition, 1980.
105. R. B. McFeat. Geometry of numbers in adele spaces. *Dissertationes Math. Rozprawy Mat.*, 88:49, 1971.
106. Atsushi Moriwaki. Arithmetic height functions over finitely generated fields. *Inventiones Mathematicae*, 140(1):101–142, 2000.
107. Atsushi Moriwaki. The canonical arithmetic height of subvarieties of an abelian variety over a finitely generated field. *Journal für die Reine und Angewandte Mathematik*, 530:33–54, 2001.
108. Atsushi Moriwaki. Diophantine geometry viewed from Arakelov geometry [translation of Sūgaku **54** (2002), no. 2, 113–129]. *Sugaku Expositions*, 17(2):219–234, 2004. Sugaku Expositions.
109. Atsushi Moriwaki. Zariski decompositions on arithmetic surfaces. *Publications of the Research Institute for Mathematical Sciences*, 48(4):799–898, 2012.
110. Atsushi Moriwaki. Toward Dirichlet's unit theorem on arithmetic varieties. *Kyoto Journal of Mathematics*, 53(1):197–259, 2013.
111. Atsushi Moriwaki. Arakelov geometry. 244:x+285, 2014. Translated from the 2008 Japanese original.
112. Atsushi Moriwaki. Semiample invertible sheaves with semipositive continuous hermitian metrics. *Algebra & Number Theory*, 9(2):503–509, 2015.
113. Atsushi Moriwaki. Adelic divisors on arithmetic varieties. *Memoirs of the American Mathematical Society*, 242(1144):v+122, 2016.
114. Shunsuke Morosawa, Yasuichiro Nishimura, Masahiko Taniguchi, and Tetsuo Ueda. *Holomorphic dynamics*, volume 66 of *Cambridge Studies in Advanced Mathematics*. Cambridge University Press, Cambridge, 2000. Translated from the 1995 Japanese original and revised by the authors.

115. David Mumford, John Fogarty, and Frances Clare Kirwan. *Geometric invariant theory*, volume 34 of *Ergebnisse der Mathematik und ihrer Grenzgebiete (2) [Results in Mathematics and Related Areas (2)]*. Springer-Verlag, Berlin, third edition, 1994.
116. André Néron. Quasi-fonctions et hauteurs sur les variétés abéliennes. *Annals of Mathematics. Second Series*, 82:249–331, 1965.
117. Jürgen Neukirch. *Algebraic number theory*, volume 322 of *Grundlehren der Mathematischen Wissenschaften*. Springer-Verlag, Berlin, 1999. Translated from the 1992 German original and with a note by Norbert Schappacher, With a foreword by G. Harder.
118. Andrei Okounkov. Brunn-Minkowski inequality for multiplicities. *Inventiones Mathematicae*, 125(3):405–411, 1996.
119. Thiruvenkatachari Parthasarathy. *Selection theorems and their applications*. Lecture Notes in Mathematics, Vol. 263. Springer-Verlag, Berlin-New York, 1972.
120. Maria Cristina Perez-Garcia and Wilhelmus H. Schikhof. *Locally convex spaces over non-Archimedean valued fields*, volume 119 of *Cambridge Studies in Advanced Mathematics*. Cambridge University Press, Cambridge, 2010.
121. Valentin Vladimirovich Petrov. *Sums of independent random variables*. Springer-Verlag, New York-Heidelberg, 1975. Translated from the Russian by A. A. Brown, Ergebnisse der Mathematik und ihrer Grenzgebiete, Band 82.
122. Patrice Philippon. Critères pour l'indépendance algébrique. *Institut des Hautes Études Scientifiques. Publications Mathématiques*, (64):5–52, 1986.
123. Patrice Philippon. Sur des hauteurs alternatives. II. *Université de Grenoble. Annales de l'Institut Fourier*, 44(4):1043–1065, 1994.
124. Sundararaman Ramanan and Annamalai Ramanathan. Some remarks on the instability flag. *The Tohoku Mathematical Journal. Second Series*, 36(2):269–291, 1984.
125. Hugues Randriambololona. Harder-Narasimhan theory for linear codes (with an appendix on Riemann-Roch theory). *Journal of Pure and Applied Algebra*, 223(7):2997–3030, 2019.
126. Michel Raynaud. Fibrés vectoriels instables—applications aux surfaces (d'après Bogomolov). In *Algebraic surfaces (Orsay, 1976–78)*, volume 868 of *Lecture Notes in Math.*, pages 293–314. Springer, Berlin, 1981.
127. Guy Rousseau. Immeubles sphériques et théorie des invariants. *Comptes Rendus Mathématique. Académie des Sciences. Paris*, 286(5):A247–A250, 1978.
128. Damien Roy and Jeffrey Lin Thunder. An absolute Siegel's lemma. *Journal für die Reine und Angewandte Mathematik*, 476:1–26, 1996.
129. Damien Roy and Jeffrey Lin Thunder. Addendum and erratum to: "An absolute Siegel's lemma". *Journal für die Reine und Angewandte Mathematik*, 508:47–51, 1999.
130. Walter Rudin. *Functional analysis*. McGraw-Hill Book Co., New York-Düsseldorf-Johannesburg, 1973. McGraw-Hill Series in Higher Mathematics.
131. Raymond A. Ryan. *Introduction to tensor products of Banach spaces*. Springer Monographs in Mathematics. Springer-Verlag London, Ltd., London, 2002.
132. John J. Saccoman. On the extension of linear operators. *International Journal of Mathematics and Mathematical Sciences*, 28(10):621–623, 2001.
133. Kichi-Suke Saito, Mikio Kato, and Yasuji Takahashi. Absolute norms on $\mathbf{C}^n$. *Journal of Mathematical Analysis and Applications*, 252(2):879–905, 2000.
134. Wilhelmus H. Schikhof. *Ultrametric calculus*, volume 4 of *Cambridge Studies in Advanced Mathematics*. Cambridge University Press, Cambridge, 2006. An introduction to p-adic analysis, Reprint of the 1984 original.
135. Stephen S. Shatz. The decomposition and specialization of algebraic families of vector bundles. *Compositio Mathematica*, 35(2):163–187, 1977.
136. Christophe Soulé. Géométrie d'Arakelov et théorie des nombres transcendants. *Astérisque*, (198-200):355–371 (1992), 1991. Journées Arithmétiques, 1989 (Luminy, 1989).
137. Christophe Soulé. Hermitian vector bundles on arithmetic varieties. In *Algebraic geometry—Santa Cruz 1995*, volume 62 of *Proc. Sympos. Pure Math.*, pages 383–419. Amer. Math. Soc., Providence, RI, 1997.
138. Ulrich Stuhler. Eine Bemerkung zur Reduktionstheorie quadratischen Formen. *Archiv der Mathematik*, 27:604–610, 1976.

139. Lucien Szpiro. Degrés, intersections, hauteurs. *Astérisque*, (127):11–28, 1985. Seminar on arithmetic bundles: the Mordell conjecture (Paris, 1983/84).
140. Lucien Szpiro, Emmanuel Ullmo, and Shouwu Zhang. équirépartition des petits points. *Inventiones Mathematicae*, 127(2):337–347, 1997.
141. Fumio Takemoto. Stable vector bundles on algebraic surfaces. *Nagoya Mathematical Journal*, 47:29–48, 1972.
142. Jeffrey Lin Thunder. An adelic Minkowski-Hlawka theorem and an application to Siegel's lemma. *Journal für die Reine und Angewandte Mathematik*, 475:167–185, 1996.
143. Burt Totaro. Tensor products in p-adic Hodge theory. *Duke Mathematical Journal*, 83(1):79–104, 1996.
144. Emmanuel Ullmo. Positivité et discrétion des points algébriques des courbes. *Annals of Mathematics. Second Series*, 147(1):167–179, 1998.
145. Herbert E. Vaughan. On locally compact metrisable spaces. *Bulletin of the American Mathematical Society*, 43(8):532–535, 1937.
146. Paul Vojta. Siegel's theorem in the compact case. *Annals of Mathematics. Second Series*, 133(3):509–548, 1991.
147. André Weil. Sur la théorie du corps de classes. *Journal of the Mathematical Society of Japan*, 3:1–35, 1951.
148. Hermann Weyl. *The classical groups*. Princeton Landmarks in Mathematics. Princeton University Press, Princeton, NJ, 1997. Their invariants and representations, Fifteenth printing, Princeton Paperbacks.
149. Kazuhiko Yamaki. Strict supports of canonical measures and applications to the geometric Bogomolov conjecture. *Compositio Mathematica*, 152(5):997–1040, 2016.
150. Kazuhiko Yamaki. Non-density of small points on divisors on abelian varieties and the Bogomolov conjecture. *Journal of the American Mathematical Society*, 30(4):1133–1163, 2017.
151. Jia An Yan. *Lectures in measure theory (in Chinese)*. Lecture Notes of Chinese Academy of Sciences. Science Press, Beijing, 2004. Second edition.
152. Xinyi Yuan. On volumes of arithmetic line bundles. *Compositio Mathematica*, 145(6):1447–1464, 2009.
153. Xinyi Yuan. Volumes of arithmetic Okounkov bodies. *Mathematische Zeitschrift*, 280(3-4):1075–1084, 2015.
154. Oscar Zariski. A new proof of Hilbert's Nullstellensatz. *Bull. Amer. Math. Soc.*, 53:362–368, 1947.
155. Shouwu Zhang. Positive line bundles on arithmetic varieties. *Journal of the American Mathematical Society*, 8(1):187–221, January 1995.
156. Shouwu Zhang. Small points and adelic metrics. *Journal of Algebraic Geometry*, 4(2):281–300, 1995.
157. Shouwu Zhang. Equidistribution of small points on abelian varieties. *Annals of Mathematics. Second Series*, 147(1):159–165, 1998.

Index

H. Chen, A. Moriwaki, *Arakelov Geometry over Adelic Curves*, Lecture Notes in Mathematics 2258, https://doi.org/10.1007/978-981-15-1728-0

LECTURE NOTES IN MATHEMATICS Springer

Editorial Policy

1. Lecture Notes aim to report new developments in all areas of mathematics and their applications – quickly, informally and at a high level. Mathematical texts analysing new developments in modelling and numerical simulation are welcome.

   Manuscripts should be reasonably self-contained and rounded off. Thus they may, and often will, present not only results of the author but also related work by other people. They may be based on specialised lecture courses. Furthermore, the manuscripts should provide sufficient motivation, examples and applications. This clearly distinguishes Lecture Notes from journal articles or technical reports which normally are very concise. Articles intended for a journal but too long to be accepted by most journals, usually do not have this "lecture notes" character. For similar reasons it is unusual for doctoral theses to be accepted for the Lecture Notes series, though habilitation theses may be appropriate.

2. Besides monographs, multi-author manuscripts resulting from SUMMER SCHOOLS or similar INTENSIVE COURSES are welcome, provided their objective was held to present an active mathematical topic to an audience at the beginning or intermediate graduate level (a list of participants should be provided).

   The resulting manuscript should not be just a collection of course notes, but should require advance planning and coordination among the main lecturers. The subject matter should dictate the structure of the book. This structure should be motivated and explained in a scientific introduction, and the notation, references, index and formulation of results should be, if possible, unified by the editors. Each contribution should have an abstract and an introduction referring to the other contributions. In other words, more preparatory work must go into a multi-authored volume than simply assembling a disparate collection of papers, communicated at the event.

3. Manuscripts should be submitted either online at www.editorialmanager.com/lnm to Springer's mathematics editorial in Heidelberg, or electronically to one of the series editors. Authors should be aware that incomplete or insufficiently close-to-final manuscripts almost always result in longer refereeing times and nevertheless unclear referees' recommendations, making further refereeing of a final draft necessary. The strict minimum amount of material that will be considered should include a detailed outline describing the planned contents of each chapter, a bibliography and several sample chapters. Parallel submission of a manuscript to another publisher while under consideration for LNM is not acceptable and can lead to rejection.

4. In general, **monographs** will be sent out to at least 2 external referees for evaluation.

   A final decision to publish can be made only on the basis of the complete manuscript, however a refereeing process leading to a preliminary decision can be based on a pre-final or incomplete manuscript.

   Volume Editors of **multi-author works** are expected to arrange for the refereeing, to the usual scientific standards, of the individual contributions. If the resulting reports can be forwarded to the LNM Editorial Board, this is very helpful. If no reports are forwarded or if other questions remain unclear in respect of homogeneity etc, the series editors may wish to consult external referees for an overall evaluation of the volume.

5. Manuscripts should in general be submitted in English. Final manuscripts should contain at least 100 pages of mathematical text and should always include

   - a table of contents;
   - an informative introduction, with adequate motivation and perhaps some historical remarks: it should be accessible to a reader not intimately familiar with the topic treated;
   - a subject index: as a rule this is genuinely helpful for the reader.
   - For evaluation purposes, manuscripts should be submitted as pdf files.

6. Careful preparation of the manuscripts will help keep production time short besides ensuring satisfactory appearance of the finished book in print and online. After acceptance of the manuscript authors will be asked to prepare the final LaTeX source files (see LaTeX templates online: https://www.springer.com/gb/authors-editors/book-authors-editors/manuscriptpreparation/5636) plus the corresponding pdf- or zipped ps-file. The LaTeX source files are essential for producing the full-text online version of the book, see http://link.springer.com/bookseries/304 for the existing online volumes of LNM). The technical production of a Lecture Notes volume takes approximately 12 weeks. Additional instructions, if necessary, are available on request from lnm@springer.com.

7. Authors receive a total of 30 free copies of their volume and free access to their book on SpringerLink, but no royalties. They are entitled to a discount of 33.3 % on the price of Springer books purchased for their personal use, if ordering directly from Springer.

8. Commitment to publish is made by a *Publishing Agreement*; contributing authors of multiauthor books are requested to sign a *Consent to Publish form*. Springer-Verlag registers the copyright for each volume. Authors are free to reuse material contained in their LNM volumes in later publications: a brief written (or e-mail) request for formal permission is sufficient.

Addresses:

Professor Jean-Michel Morel, CMLA, École Normale Supérieure de Cachan, France
E-mail: moreljeanmichel@gmail.com

Professor Bernard Teissier, Equipe Géométrie et Dynamique,
Institut de Mathématiques de Jussieu – Paris Rive Gauche, Paris, France
E-mail: bernard.teissier@imj-prg.fr

Springer: Ute McCrory, Mathematics, Heidelberg, Germany,
E-mail: lnm@springer.com

Printed in the United States
By Bookmasters